Rough Set Methods and Applications

Adam Mrozek, 1949 - 1999

Lech Polkowski
Shusaku Tsumoto
Tsau Y. Lin
Editors

Rough Set Methods and Applications

New Developments in Knowledge Discovery in Information Systems

With 100 Figures
and 133 Tables

Physica-Verlag
A Springer-Verlag Company

Prof. Lech Polkowski
Warsaw University of Technology
Department of Mathematics and Information Sciences
Pl. Politechniki 1
00650 Warsaw
Poland

Prof. Shusaku Tsumoto
Shimane Medical University
Department of Medical Informatics
89-1 Enya-cho
Izumo City, Shimane 693-8501
Japan

Prof. Tsau Y. Lin
San Jose State University
Department of Mathematics and Computer Science
San Jose, CA 95192
USA

E-mail: polkow@pjwstk.waw.pl
tsumoto@computer.org
tylin@mathcs.sjsu.edu

ISSN 1434-9922
ISBN 3-7908-1328-1 Physica-Verlag Heidelberg New York

Cataloging-in-Publication Data applied for
Die Deutsche Bibliothek – CIP-Einheitsaufnahme
Rough set methods and applications: new developments in knowlegde discovery in information systems; with 133 tables; [Adam Mrozek, 1949–1999; this volume is dedicated to the memory of Professor Adam Mrózek] / Lech Polkowski ... (ed.). – Heidelberg; New York: Physica-Verl., 2000
  (Studies in fuzziness and soft computing; Vol. 56)
  ISBN 3-7908-1328-1

This work is subject to copyright. All rights are reserved, whether the whole or part of the material is concerned, specifically the rights of translation, reprinting, reuse of illustrations, recitation, broadcasting, reproduction on microfilm or in any other way, and storage in data banks. Duplication of this publication or parts thereof is permitted only under the provisions of the German Copyright Law of September 9, 1965, in its current version, and permission for use must always be obtained from Physica-Verlag. Violations are liable for prosecution under the German Copyright Law.

Physica-Verlag Heidelberg New York
a member of BertelsmannSpringer Science+Business Media GmbH

© Physica-Verlag Heidelberg 2000
Printed in Germany

The use of general descriptive names, registered names, trademarks, etc. in this publication does not imply, even in the absence of a specific statement, that such names are exempt from the relevant protective laws and regulations and therefore free for general use.

Hardcover Design: Erich Kirchner, Heidelberg

SPIN 10777447   88/2202-5 4 3 2 1 0 – Printed on acid-free paper

This volume is dedicated to the memory of
Professor Adam Mrózek

Foreword

The collection of papers presented in this volume concerns new progress in rough set methodology and its applications. The Authors of the papers share with us their recent results on various aspects of rough sets and their applications. The papers are not only restricted to "classical" rough set theory but contain many new ideas and extensions of the theory and show new horizons and perspectives for this challenging and fast growing domain.

The twelve papers included in the book comprise the ongoing work on various aspects of rough set theory. The applications of rough set methodology to knowledge discovery, spatial reasoning, data reduction methods, concurrent systems analysis and synthesis, data analysis and data mining in incomplete information systems, classification problems and conflict analysis are thoroughly presented and discussed. The contributed papers may result in giving better insight, tools and algorithms for rough set based knowledge discovery and data analysis.

Furthermore, a list of about four hundred papers contains various new results published in recent years, which may be useful for all interested in the state of the art and recent developments in this area. The rough set concept introduced almost twenty years ago proved to be useful not only for engineering, AI, data mining, knowledge discovery, decision support, logic and others – but it also addresses several important questions raised for a long time in philosophy of sciences, concerning similarity, concurrency and indiscernibility studied already by Leibniz. Thus the book might be also attractive to those who are interested in philosophy of science.

Congratulations are due to the Contributors and the Editors for making this volume possible.

Zdzisław Pawlak, Warszawa, May 2000

Contents

Foreword
    Z. Pawlak   VII

PART 1. INTRODUCTION

Introducing the Book   3
    L. Polkowski, S. Tsumoto and T.Y. Lin

Chapter 1. *A Rough Set Perspective on Knowledge Discovery in Information Systems: An Essay on the Topic of the Book*   9
    L. Polkowski, S. Tsumoto and T.Y. Lin

PART 2. METHODS AND APPLICATIONS: REDUCTS, SIMILARITY, MEREOLOGY

Chapter 2. *Rough Set Algorithms in Classification Problem*   49
    J.G. Bazan, Hung Son Nguyen, Sinh Hoa Nguyen,
    P. Synak and J. Wróblewski

Chapter 3. *Rough Mereology in Information Systems. A Case Study: Qualitative Spatial Reasoning*   89
    L. Polkowski and A. Skowron

Chapter 4. *Knowledge Discovery by Application of Rough Set Models*   137
    J. Stepaniuk

Chapter 5. *Various Approaches to Reasoning with Frequency Based Decision Reducts: A Survey*   235
    D. Ślęzak

PART 3. METHODS AND APPLICATIONS: REGULAR PATTERN EXTRACTION, CONCURRENCY

Chapter 6. *Regularity Analysis and its Applications in Data Mining*   289
    Sinh Hoa Nguyen

Chapter 7. *Rough Set Methods for the Synthesis and Analysis of Concurrent Processes* 379
    Z. Suraj

PART 4. METHODS AND APPLICATIONS: ALGEBRAIC AND STATISTICAL ASPECTS, CONFLICTS, INCOMPLETENESS

Chapter 8. *Conflict Analysis* 491
    R. Deja

Chapter 9. *Logical and Algebraic Techniques for Rough Set Data Analysis* 521
    I. Düntsch and G. Gediga

Chapter 10. *Statistical Techniques for Rough Set Data Analysis* 545
    G. Gediga and I. Düntsch

Chapter 11. *Data Mining in Incomplete Information Systems from Rough Set Perspective* 567
    M. Kryszkiewicz and H. Rybiński

PART 5. AFTERWORD

Chapter 12. *Rough Sets and Rough Logic: A KDD Perspective* 583
    Z. Pawlak, L. Polkowski and A. Skowron

APPENDIX: SELECTED BIBLIOFGRAPHY ON ROUGH SETS

Bibliography 649

PART 1:

INTRODUCTION

Introducing the Book

In a couple of years that elapsed since the monograph edited by T. Y. Lin and N. Cercone: *Rough Sets and Data Mining. Analysis of Imprecise Data* (1997) and two monographs edited by L. Polkowski and A. Skowron: *Rough Sets in Knowledge Discovery 1, 2* (1998) were presented to the research community (cf. [A]2, [A]3, [A]4 in Bibliography, Appendix 1), rough set researchers have made further progress in inventing methods aimed, among others, at knowledge discovery and data mining, spatial reasoning, data reduction, concurrent system analysis and synthesis, conflict analysis as well as in theoretical analysis of information and decision systems.

The eleven chapters of this monograph now presented to the research community give a detailed report about many of these developments.

Classically, the rough set theory, as proposed by Zdzisław Pawlak, assumed that approximations to concepts were based on indiscernibility relations induced from data. Partitions of the universe of objects induced by these relations led to exact concepts by means of which approximations to general concepts were constructed, reducts were defined in terms of these relations, and finally, decision rules and decision algorithms were induced via reducts.

As rough set methods enter more and more complex application contexts and a fortiori data tables analyzed by these methods become also more complex, rough set theory is undergoing an evolution characterized by a plethora of new ideas and algorithmic tools.

These new ideas and tools are discussed briefly in a survey given in Chapter 1, to which the reader is referred in order to have a bird's eye view on the current research interests of rough set community presented in consecutive chapters of this monograph.

Among these new ideas and related tools, we may mention similarity and tolerance based notions of a reduct and of a decision rule, fusing together rough set and mereological paradigms for approximate reasoning, analysis of complex feature synthesis in distributed systems, approximations to reducts based on various information measures, local as well as dynamic reducts, decision rules based on generalized patterns (templates), quantization of attributes (e.g. discretization, value grouping etc.), hybrid algorithms for reduct and rule induction (e.g. based on evolutionary algorithms combined with rough set–theoretic ones), concurrency analysis (e.g. the notion of a rough Petri net).

We now introduce briefly the content of the monograph inviting the reader to make acquaintance with its chapters.

The only chapter in PART 1. INTRODUCTION, Chapter 1: *A rough set perspective on knowledge discovery in information systems: An essay on the topic of the book* by L. POLKOWSKI, S. TSUMOTO, and T. Y. LIN, to which the reader is invited to make the first encounter with the topical subjects of the monograph, brings forth the concise discussion of basic ideas of both classical rough set theory as well as of its currently investigated extensions. Together with the bibliography items collected in Appendix 1, Chapter 1 provides a comprehensive account of main ideas of rough set theory.

PART 2. METHODS AND APPLICATIONS : REDUCTS, SIMILARITY, MEREOLOGY consists of four chapters devoted to various aspects of reduct and generalized reduct generation (cf. Chapter 1, Sect. 3, Sect. 7), tolerance or similarity approach to rough set approximations (cf. Chapter 1, Sect. 6) as well as mereological approach in which rough set approximations are constructed by means of the mereological class operator from objects related one to another by means of the rough mereological predicate of a *part in a degree* (cf. Chapter 1, Sect. 6).

In Chapter 2: *Rough set algorithms in classification problem*, J. BAZAN, NGUYEN HUNG SON, NGUYEN SINH HOA, P. SYNAK , and J. WRÓBLEWSKI present a balanced selection of rough set algorithms for inducing decision rules (cf. Chapter 1, Sect. 5). The Chapter begins (cf. Sect. 2) with a discussion of classical rough set approach to reduct generation, based on boolean reasoning via discernibility functions (cf. Chapter 1, Sect. 3), and to generation of decision rules (cf. Sect. 3). Then (Sect. 4) attribute quantization techniques are discussed with *discretization* as the main example, and some heuristics are described returning semi–minimal sets of cuts. Local reducts (cf. Chapter 1, Sect. 7. 2) are discussed in Sect. 5 and genetic and hybrid algorithms for reduct induction are presented in Sect. 6. Dynamic reducts (cf. Chapter 1, Sect. 7. 3-4) and decision rules generated from them are discussed in Sects. 7–13. Finally, approximate rules (Sect. 14) and mechanisms for negotiations among rules (Sect. 15) are presented.

Chapter 3: *Rough mereology in information systems. A case study: Qualitative spatial reasoning* by L. POLKOWSKI and A. SKOWRON presents a discussion of rough mereology in information systems (cf. Chapter 1, Sect. 6. 2). Set in the classical spirit of ontology–mereology formalization, it begins with an outline of ontology in information systems modeled on ontological theory of S. Leśniewski (Sect. 3) and proceeds with basics of mereology also in the spirit of S. Leśniewski (Sect. 4), with a discussion of particular models in information systems. Rough mereology is presented in Sect. 5 as an extension of mereology and then in the following sections an approach to qualitative spatial reasoning based on rough mereology is proposed. In particular, Sect. 7 is concerned with mereo–topology i.e. quasi–Čech topologies that arise in rough mereological universe. Then, the notion of connection is introduced in rough mereological context (Sect. 8). Finally, mereo–geometry is discussed in Sect. 9 and the chapter concludes with some examples (Sect. 10).

The topic of Chapter 4: *Knowledge discovery by application of rough set models* by J. STEPANIUK is knowledge discovery in generalized rough set systems of which *generalized approximation spaces* (cf. Chapter 1, Sect. 6. 1) are the main example, discussed in Sect. 2. Counterparts of reducts in generalized approximation spaces are discussed in Sect. 3, and a parallel discussion of decision rules is presented in Sect. 4. Examples of applications of the proposed methods are presented in Sect. 5 with medical data. In Sect. 6 problems related to relational learning are discussed and the remainder of the Chapter is devoted to knowledge discovery in distributed systems with information granulation and rough set approximations in distributed systems presented in Sect. 7.

D. ŚLĘZAK in Chapter 5: *Various approaches to reasoning with frequency based decision reducts: A survey* discusses generalized notions of a reduct based on various information measures (cf. Chapter 1, Sect. 7) along with complexity issues. He considers reducts based on generalized decision functions as well as reducts based on approximate discernibility (Sect. 2) then proceeds with a discussion of reducts based on conditional frequencies associated with the decision (Sect. 3). Some more general notions of approximate reducts related to distance measures as well as to normalized decision functions are discussed in Sect. 4 and in Sect. 5 notions of reducts based on some measures of information (e.g. conditional entropy) are presented.

PART 3. METHODS AND APPLICATIONS: REGULAR PATTERN EXTRACTION, CONCURRENCY consists of two chapters dedicated to, respectively, regularity analysis in data tables by rough set methods and analysis of concurrent processes by rough set methods.

In Chapter 6: *Regularity analysis and its applications in Data Mining*, NGUYEN SINH HOA presents an exhaustive study of computational and algorithmic problems related to regular pattern extraction from data tables by means of *templates* (cf. Chapter 1, Sect. 8). The notion of a template is introduced in Sect. 2 and Sect. 3 brings forth methods for template extraction by sequential search. Data segmentation i.e. a generalization of discretization technique (cf. Chapter 1, Sect. 9) to symbolic values is discussed in Sect. 4. *M(aximal) D(iscernibility) by G(rouping)*, a classification method extending decision tree techniques to symbolic attributes is presented in Sect. 5. Schemes for applications of introduced methods to classification tasks are discussed in Sect. 6. Relational patterns defined by tolerance relations (cf. Chapter 1, Sect. 6) are introduced and discussed in Sect. 7 and Sect. 8 is devoted to methods of searching for (optimal) tolerance relations with applications to data clustering and classification in Sect. 9. The Chapter culminates in Sect. 10, where results of experiments testing introduced methods with 10 real–world domains (taken from UC Irvine repository) and artificial domains (Monk's problems) are reported.

Z. SURAJ in Chapter 7: *Rough set methods for synthesis and analysis of concurrent processes* presents a study on rough set based modeling of concurrency. Taking *marked Petri nets* (cf. Chapter 1, Sect. 10) as a model for concurrency (Sect. 2), he presents in Sect. 3 methods for synthesis of a concurrent system specified by an information system as well as by a dynamic information system. The converse problem of discovering a concurrent data model in a given data table is discussed with some solutions in Sect. 4. Applications of the proposed synthesis methods to re–engineering of cooperative information systems are presented in Sect. 5 and Sect. 6 is devoted to real–time decision making where an algorithm constructing a Petri–net represented program for object identification from a given data table is presented. Applications of presented methods to control design are proposed in Sect. 7 and in Sect. 8 software implementations of proposed methods are described.

PART 4. METHODS AND APPLICATIONS: ALGEBRAIC AND STATISTICAL ASPECTS, CONFLICTS, INCOMPLETENESS contains four chapters providing an additional insight into relevant aspects of rough set theory and its applications. They highlight important issues of logical and algebraic aspects of rough sets, statistical techniques in rough set data analysis, missing value treatment and applications of rough set methods to conflict analysis.

The last topic is discussed as first in Chapter 8: *Conflict analysis* by R. DEJA. Conflict analysis (cf. Appendix 1, [B]218, [B]228) is presented here as a problem of finding consensus in a distributed system of (information systems of) local agents and a rough set environment for conflict analysis is constructed. To this end, the author discusses a model of a distributed information system leading to notions of a *local state*, a *situation* and a *constraint* (Sect. 3) and then introduces the notion of a *conflict* (Sect. 4). Consensus problems are discussed in Sect. 5 with some examples in Sect. 6. The idea of coalitions is outlined in Sect. 7 and some strategies based on boolean reasoning are proposed in Sect. 8.

Chapter 9: *Logical and algebraic techniques for rough set data analysis* by I. DUENTSCH and G. GEDIGA recapitulates basic logical, relational and algebraic structures induced in a rough set environment. Logical and algebraic aspects of rough sets have been subject of much research (cf. Appendix 1, [A]2 and the list [B]) and this Chapter provides a valuable resume of this topic. Authors discuss algebraic aspects of approximation spaces (Sect. 4), logics for approximation spaces (Sect. 5), and various logics for information systems (Sect. 6).

In Chapter 10: *Statistical techniques for rough set data analysis*, G. GEDIGA and I. DUENTSCH discuss statistical aspects of rough set based data analysis (cf. Appendix 1, [B]217, [B]406). They begin with basic statistics associated with rough set approximations (Sect. 3) and significance testing of approximation quality based on randomized information systems (Sect. 4). In Sect. 5, the system **SORES** based on *normalized entropy of rough prediction* is presented and Sect. 6 brings forth a discussion of probabilistic decision rules with relevant algorithms.

The problem of missing values along with resulting incompleteness of data (cf. Appendix 1, [B]90, [B]97) has also been given much attention by rough set community. In Chapter 11: *Data Mining in incomplete information systems from rough set perspective*, M. KRYSZKIEWICZ and H. RYBIŃSKI present algorithms which induce certain, possible and generalized decision rules (cf. Chapter 1, Sect. 5) in incomplete information systems. After introductory discussion of information systems and decision rules (Sects. 2, 3), they describe in Sect. 4 three algorithms: *IAprioriCertain*, *IAprioriPossible*, and *IAprioriGeneralized* for inducing respective types of (optimal) decision rules.

PART 5. AFTERWORD contains a final Chapter 12: Rough Sets and Rough Logic: A KDD Perspective by Z. PAWLAK, L. POLKOWSKI, and A. SKOWRON. Intended as a summary of the topics discussed in the consecutive Chapters, it provides the reader with a discussion of rough set approach to Knowledge Discovery in Data set in a general perspective on tasks and problems of this area of research. An important part of this discussion is devoted to problems of logical tools which Knowledge Discovery in Data may employ with particular emphasis on inductive reasoning and the place of rough and related logics in the repertoire of logical tools of KDD.

The monograph ends with Appendix 1 containing bibliography of recent works devoted to rough sets: 7 monographs and Conference Proceedings edited since 1997 and 410 research papers. Along with the bibliography inserted in [A]2, this bibliography provides the reader with an access to about 1500 works on rough sets.

This monograph reflects the research effort of rough set community in the last two years and it would not be possible without the work and cooperation of very many of its members to which the editors of this volume direct their words of gratitude.

We would like to thank Professor Zdzisław Pawlak for his support and encouragement.

Our thanks go to Professor Andrzej Skowron for his initiative to prepare this monograph and his support of this undertaking.

We would like to express thanks to all authors of papers presented in the monograph as well as to all our colleagues who provided us with bibliographical information. Due to their effort and cooperation, this monograph is making its appearance.

Our special thanks go to Professor Janusz Kacprzyk, the Series Editor, for his kind acceptance of our initiative and his constant support. We would like to thank Dr Martina Bihn and the editing staff of Physica-Verlag for their help and cooperation.

The editing of this book has been supported by a Research Grant No. 8T 11C 02417 from the State Committee for Scientific Research (KBN) of the Republic of Poland.

Lech Polkowski, Shusaku Tsumoto, Tsau Young Lin

Warsaw, Izumo City, San Jose   *May 2000*

Chapter 1

A Rough Set Perspective on Knowledge Discovery in Information Systems: An Essay on the Topic of the Book

Lech Polkowski[1], Shusaku Tsumoto[2], and Tsau Young Lin[3]

[1] Polish–Japanese Institute of Information Technology
Koszykowa 86 02008 Warsaw, Poland;
Department of Mathematics and Information Sciences, Warsaw University of Technology, Pl. Politechniki 1, Warsaw, Poland

[2] Department of Medicine Informatics
Shimane Medical University, School of Medicine
89-1 Enya–cho, Izumo City, Shimane 693–8501, Japan

[3] Department of Mathematics and Computer Science
San Jose State University, 129S 10th St., San Jose CA 95192 USA
emails: polkow@pjwstk.waw.pl; tsumoto@computer.org;
tylin@mathcs.sjsu.edu

1 Introduction

The underlying ideas of rough set theory – proposed by Zdzisław Pawlak in the early 1980's – have been developed into a manifold theory for the purpose of data and knowledge analysis, due to a systematic growth of interest in this theory in the scientific community, witnessed by – among others – a bibliography of over four hundred research papers on rough sets and their applications produced in the last couple of years and appendiced to this collection of expository papers (in the sequel, we refer to items in this bibliography (see Appendix 1) with indices like $[X]xyz$ where $X = A$ or $X = B$ and xyz is the number of the item in the list X).

The primary goal of rough set theory has been outlined as a classificatory analysis of data (cf. [B]19, [B]20, [B]21, [B]58, [B]67, [B]79, [B]82, [B]83, [B]85, [B]108, [B]186, [B]188, [B]193, [B]200, [B] 201, [B]202, [B]204, [B]212, [B]213, [B]220, [B]222, [B]261, [B]262, [B]277, [B]287, [B]288, [B]338, [B]339, [B]370, [B]376, [B]377, [B]378, [B]407, [B]408): given a data table, rough set algorithms induce a set of relevant concepts providing a classification of data. In performing this task, rough sets have underwent significant changes, motivated both by theoretical considerations as well as by practical accumulated experience, from their "classical" period to the present state of the art.

Many of those changes could be aptly summarized by marking them with the label "from indiscernibility to similarity".

In order to explain this label, we should begin with an introductory perusal of basic notions and ideas of rough set theory and, this done, we could outline new ideas and methods, which are reflected in a detailed way in many chapters of the presented monograph.

Classical rough set theory has been formulated in terms of indiscernibility relations and derived notions: of a reduct, a decision rule, a decision algorithm (a rough set *classifier*). Indiscernibility relations are induced from data table and they hold for a given pair of objects whenever values of attributes from a given attribute set agree on these objects. Equivalence classes of these relations generate fields of *exact concepts*.

Indiscernibility relations allow for information reduction by selecting sets of attributes preserving classification; minimal sets with this property are called *reducts*. Restricting ourselves to a reduct, we do not loose any knowledge but reduce the amount of information to be processed.

Given a reduct, we may induce *decision rules* (cf. op.cit., op. cit.) expressed in the logical form abstracted from *attribute–value* descriptors provided by the data table.

This scheme has proved its value in many applications (cf. [A]2, [A]3); nevertheless, the raising complexity and size of data table analyzed currently call for new methods and techniques for inducing classifiers and the search is on in the rough set community for new tools in data analysis. These tools have primarily to do with useful feature extraction i.e. with a search for concepts which would model the decision satisfactorily well in a given context.

In search for useful features, classical notions undergo revisions; although these revisions go in many directions, one may select a few main lines of development:

1. Indiscernibility relations are relaxed to *tolerance* or *similarity* relations (cf. [B]67, [B]108, [B]196, [B]197, [B]198, [B]199, [B]200, [B]202, [B]203, [B]275, [B]276, [B]278, [B]280, [B]282, [B]291, [B]292, [B]332, [B]341, [B]344, [B]345, [B]346, [B]347, [B]348, [B]349, [B]372) induced from data tables and usually depending on parameters subject to tuning–up in the optimization of a classifier process. A systematic analysis of similarity–based rough set methods has led to the emergence of some paradigms extending classical rough set theory: we would like to mention here *rough mereology* based on similarity–generating predicate of being a *part in a degree* and *generalized approximation spaces* based on data–induced tolerance relations and rough membership functions. In similarity relation–based paradigms, notions of a reduct, decision rule etc. are introduced analogously to the classical case, with natural modifications, leading to classifiers often better predisposed to model the decision. In this Volume, the reader will find a detailed discussion of generalized approximation spaces in the Chapter by STEPANIUK and the Chapter by POLKOWSKI and SKOWRON will bring forth a discussion of rough mereology in information systems with a case study in which rough mereology is applied to spatial reasoning in information systems. Methods for extracting similarity relations from data tables are discussed also in the

Chapter by NGUYEN SINH HOA.

2. Reducts are defined classically as sets of attributes preserving indiscernibility; this idea may be generalized: given a measure of information related to decision (e.g. a *generalized decision, approximate discernibility measure, conditional frequency, entropy* etc.), one may search for (minimal) sets of attributes preserving or approximately (i.e. up to a given threshold) preserving this measure (cf. [B]19, [B]196, [B]204, [B]299, [B]361, [B]362, [B]363, [B]364, [B]365, [B]367, [B]368, [B]371). In this way a variety of reducts may be induced from data tables, usually shorter then classical reducts hence leading to shorter decision rules better predisposed to the task of classifying new objects. The reader will find an illuminating discussion of generalized reducts in the Chapter by ŚLĘZAK.

3. The idea of a reduct may be approached from another angle: first, it may be localized; given an object x, one may search for (minimal) sets of attributes preserving the local information about x, called *local reducts*. Local reducts generate decision rules which are less specific hence they match more new objects then rules generated from classically defined reducts (cf. [B]19, [B]20, [B]203, [B]204, [B]393, [B]394, [B]395, [B]396). Next, in search for robust, noise-resistant classifiers, the idea of a *dynamic reduct* has been found promising; a dynamic reduct is generated as a reduct common to all tables in a given family of sub-tables of the original data table: as each sub-table may be regarded as a perturbation of the original table, reducts common to all sub-tables may be regarded as stable with respect to given perturbations hence noise-resistant in a higher degree than ordinary reducts (cf. [B]19, [B]20, [B]21, [B]203, [B]204, [B]366). Local reducts and dynamic reducts are discussed in detail from algorithmic and applicational points of view in the Chapter by BAZAN et AL.

4. In reduct computation, hybrid heuristics are applied: dominant idea here is to apply evolutionary algorithms combined with rough set strategies (cf. [B]393, [B]394, [B]395, [B]396). Hybrid algorithms lead to sub-optimal reducts but require a short computing time. Evolutionary strategies and hybrid algorithms for reduct generation are discussed in the Chapter by BAZAN et AL.

5. A more abstract view on classifier induction may refrain from the notion of a reduct in favor of the notion of a *template* as a representation of a regular concept providing the meaning for the predecessor of a decision rule: in inducing classifiers, one actually looks for *patterns* approximating well a given set of concepts (e.g. decision classes). These patterns may be expressed in various data-related languages like

   (a) *Descriptor language* whose expressions are disjunction-and-conjunction-closed formulae generated from descriptors of the form (a, v) (i.e. (attribute, value of attribute)).

   (b) *Generalized descriptor language* with more general expressions generated from *generalized descriptors* of the form $a \in W_a$ where W_a is a set of values of a.

(c) *Relational descriptor language* whose descriptors are induced by a relation, say R, usually a similarity relation, in the form of R–*pre*–*classes*: $[x,R]=\{y : xRy\}$.

A pattern expressed as a conjunction of descriptors of the above form is called a *template*. Methods for finding relevant templates are discussed in e.g. [B]197, [B]198, [B]199, [B]200, [B]201, [B]202, [B]203, [B]393.

The Chapter by NGUYEN SINH HOA brings an exhaustive discussion of computational, algorithmic and applicational problems related to template and (optimal) similarity relation extraction from data.

6. Analysis of data encoded by complex information systems, e.g. with continuous values of attributes or a very large number of attribute values, by rough set methods, requires a pre–procesing stage. In this stage, techniques familiar in AI like *discretization* or *value grouping* are applied to data table (cf. [B]184, [B]185, [B]186, [B]187, [B]188, [B]189, [B]190, [B]194, [B]197, [B]198, [B]199, [B]200, [B]202, [B]203, [B]204). Chapters by BAZAN et AL. and by NGUYEN SINH HOA discuss in detail the discretization problems in rough set theory from computational, algorithmic and applicational points of view.

7. One may search for complex features encoded in data table, related e.g. to concurrency, parallelism, or chaotic behavior or, to the contrary, one may try to represent via an information system a complex system encoded in a different language. A good example of this problem is modeling concurrency via information systems. In this task, two problems emerge

   (a) The problem of synthesis of a concurrent system specified by an information system.

   (b) The problem of synthesis of models of concurrency from experimental data.

   (cf. [B]264, [B]265, [B]266, [B]267, [B]268, [B]334, [B]353, [B]354, [B]355, [B]356).

   A solution to these problems would have an impact on rough set applications in the area of re–engineering of distributed information systems, real–time decision making, and discrete control design. The analysis of concurrent processes by rough set methods is presented in a detailed way in the Chapter by SURAJ.

The above list is by no means exhaustive, but it reflects main areas of interest covered by expository chapters composing this Volume. We now elaborate in more detail on the research aspects outlined above with the intention of introducing the reader to the main topic of this monograph.

2 Information/decision systems

Data for rough set analysis are represented in the form of an *attribute–value table*, each row of which represents an *object* (e.g. a signal, a case, a patient etc. etc.). Columns of the table represent *attributes* (properties, features) characterizing objects; values of attributes are acquired either by measurement or via human expertise. Such a table is called an *information system*.

2.1 Information systems

Information systems may be given a formal definition: any information system A can be regarded as a pair (U, A), where U is a non–empty finite set of objects called the *universe* of A and A is a non–empty finite set of attributes; any attribute a is a map $a : U \to V_a$. The set V_a is called the *value set* of a.

Example 1. A simple information system is presented in Table 1.

	a_1	a_2
x_1	1	3
x_2	1	0
x_3	3	1
x_4	3	1
x_5	4	2
x_6	1	2
x_7	4	2

Table 1. An information system A1.

In this setting, it is the task of rough set methodology to induce from data descriptions of concepts of interest to the user of the system.

2.2 Foundational issues: indiscernibility, concept approximations

The basic assumption of rough set theory is of two–fold nature (cf. [B]212, [B]216, [B]224, [B]226, [B]234, [B]241, [B]277):

1. a given information system presents by itself all information/knowledge available about the world it does represent.
2. objects bear with themselves information about them and any two objects with the same information should be regarded as identical (be *indiscernible*).

For an object $x \in U$, the information about x borne by the information system A with respect to a set $B \subseteq A$ of attributes may be defined as *the B–information set*

$$Inf_B(x) = \{(a, a(x)) : a \in B\}$$

of x.

In consequence of the assumption 2, the relation IND_B of the *B–indiscernibility* is defined as follows

$$(x, y) \in IND_B \iff Inf_B(x) = Inf_B(y).$$

Equivalence classes $[x]_B$ of the relation IND_B represent therefore elementary (atomic) portions of knowledge represented by the subsystem $A_B=(U, B)$, any B.

From these classes, greater aggregates may be formed via the set–theoretic operation of the union: for a concept (set of objects), $X \subseteq U$, one says that X is *B-exact* if and only if

$$X = \bigcup_{i=1}^{k} [x_i]_B$$

for some $x_1, x_2, ..., x_k \in U$ i.e. when X is the union of some B–indiscernibility classes.

Clearly, the family of B–exact sets is closed with respect to the set–theoretic operations of the union, the intersection and the complement; in the language of the algebraic set theory, we would say that B–exact sets form a *field of sets*.

We may notice that any B–exact set is defined in terms of attribute–value pairs of the form (a, v) where $a \in B$ is an attribute and $v \in V_a$ is its value; formally, one may define a *conditional logic* L_A (cf. [B]109, [B]110) whose elementary formulae are of the form (a, v) and the set of formulae is the least extension of the set of elementary formulae with respect to the propositional connectives $\vee, \wedge, \neg$.

Semantics of L_A is defined by structural induction:

1. the meaning $[(a, v)]$ of (a, v) is: $[(a, v)] = \{x \in U : a(x) = v\}$.
2. $[(a, v) \vee (b, w)] = [(a, v)] \cup [(b, w)]$.
3. $[(a, v) \wedge (b, w)] = [(a, v)] \cap [(b, w)]$.
4. $[\neg(a, v)] = U - [(a, v)]$.

Then, clearly,

$$X = [\vee_{i=1}^{k} \wedge_{a \in B} (a, a(x_i))]$$

is the rendering of the B–exact concept X in terms of descriptors induced from the data table.

This approach is not sufficient in one major respect: we may be interested in concepts which are not exact.

Therefore, the need arises for a method by which one could describe non–exact concepts; such a description induced from data may be approximate only.

Rough set theory solves this problem by proposing to approximate a non–exact concept X with two exact concepts, called, respectively, the *lower-* and the *upper-*approximation.

For any attribute set B, one thus defines the B–*lower* and B–*upper approximations of* X, denoted $\underline{B}X$ and $\overline{B}X$, respectively, as follows

$$\underline{B}X = \{x \in U : [x]_B \subseteq X\}$$

and

$$\overline{B}X = \{x : [x]_B \cap X \neq \emptyset\}.$$

We would agree that objects in $\underline{B}X$ could be with certainty classified as elements of X on the basis of knowledge in B, while objects in $\overline{B}X$ can only be classified possibly as elements of X on the basis of knowledge in B. In this way the two approximations play an essential role in semantics of modal operators associated with an information system (cf. the Chapter by DUENTSCH and GEDIGA).

The set
$$BN_B(X) = \overline{B}X - \underline{B}X$$
is called the *B-boundary region of X*, and it consists of objects which neither are certainly members of X nor they are certainly members of $U - X$. Thus, the presence of a non–empty boundary region does indicate that the concept in question is non–exact (cf. [B]222, [B]226). A concept X which is non–B–exact is said to be *B-rough*.

Example 2. Let us look at Table 1. The non-empty subsets of the attribute set A are $\{a_1\}$, $\{a_2\}$ and $\{a_1, a_2\}$.

The indiscernibility relation $IND_{\{.\}}$ defines three partitions of the universe

1. $IND_{\{a_1\}} = \{\{x_1, x_2, x_6\}, \{x_3, x_4\}, \{x_5, x_7\}\}$.
2. $IND_{\{a_2\}} = \{\{x_1\}, \{x_2\}, \{x_3, x_4\}, \{x_5, x_6, x_7\}\}$.
3. $IND_{\{a_1, a_2\}} = \{\{x_1\}, \{x_2\}, \{x_3, x_4\}, \{x_5, x_7\}, \{x_6\}\}$.

It follows that e.g. the concept $X = \{x_1, x_2, x_3\}$ is exact for neither of the three attribute sets while e.g. the concept $Y = \{x_1, x_2\}$ is both $\{a_2\}$– and A–exact.

We have for instance

1. $\underline{a_1}X = \emptyset$.
2. $\overline{a_1}X = \{x_1, x_2, x_3, x_4, x_6\}$.
3. $\underline{a_2}X = \{x_1, x_2\}$.

The above described concept formation may be regarded from Machine Learning point of view as a case of *unsupervised learning*: given a data table, concepts are formed by the system itself on the basis of descriptors extracted from the data. In the most important applications, however, one has to deal with the *supervised case* when objects in the universe are pre–classified by an expert (by means of a distinguished attribute not in A, called the *decision*) and the task of a rough set based algorithm is to induce concepts which would approximate as closely as possible the given pre–classification by an expert's decision.

2.3 Decision systems

To discuss this problem in formal terms, we introduce the notion of a *decision system* understood as a triple $\mathsf{A_d} = (U, A, d)$ where (U, A) is an information system as introduced above, and d is a distinguished attribute called the *decision*; as any other attribute, d is a map $d: U \to V_d$ of the universe U into the value set V_d.

	a_1	a_2	d
x_1	1	3	0
x_2	1	0	0
x_3	3	1	1
x_4	3	1	1
x_5	4	2	1
x_6	1	2	0
x_7	4	2	1

Table 2. A decision system $A2_d$.

Example 3. Look at a decision system in Table 2.

The decision attribute d induces the partition of the universe U into equivalence classes of the relation IND_d of *d–indiscernibility*. Assuming, without any loss of generality, that $V_d = \{1, 2, ..., k(d)\}$, we obtain the partition $\{X_1, X_2, ..., X_{k(d)}\}$ of U into *decision classes*. These classes express the classification of objects done by an expert e.g. a medical practitioner, on the basis of their experience, often intuitive and/or based on long experience. This classification should now be modeled by a rough set algorithm in terms of the attribute–value formulae induced from *conditional* attributes in the attribute set A.

There are few possibilities imposing themselves.

1. Each of decision classes $X_1, X_2, ..., X_{k(d)}$ may be approximated by its lower, resp. upper, approximations over a set $B \subseteq A$ of attributes. The resulting approximations $\underline{B}X_i, \overline{B}X_i$ provide a (local) description of decision classes.
2. Global approximation of decision classes may be produced e.g. by looking at B–indiscernibility classes providing a refinement of the partition by IND_d i.e. we are interested in those B–classes $[x]_B$ which satisfy the condition $[x]_B \subseteq X_i$ for some X_i. This idea leads to the notion of the *positive region of the decision* to which we proceed now.

Given a decision system A_d, and a set $B \subseteq A$ of attributes, we let

$$POS_B(d) = \{x \in U : \exists i \in \{1, 2, .., k(d)\}.[x]_B \subseteq X_i\}.$$

The set $POS_B(d)$ is called the *B–positive region of d.*

Example 4. Let us look at Table 2. Non–empty subsets of the set A are $\{a_1\}$, $\{a_2\}$ and $\{a_1, a_2\}$. There are two decision classes: $X_1 = \{x : d(x) = 0\} = \{x_1, x_2, x_6\}$ and $X_2 = \{x : d(x) = 1\} = \{x_3, x_4, x_5, x_7\}$. We can see that

1. $POS_{\{a_1\}}(d) = U$.
2. $POS_{\{a_2\}}(d) = \{x_1, x_2, x_3, x_4\}$.
3. $POS_A(d) = U$.

We may distinguish between the two cases: $POS_B(d) = U$ in which we say that the decision system A_d is *B–determininistic* (there is an exact description of d in terms of IND_B) and $POS_B(d) \neq U$ in which we say that A_d is *B–non-deterministic* (we may only obtain an approximate description of d in terms of IND_B).

In our example above, we have found that our decision system has been both $\{a_1\}$–deterministic and A–deterministic. This fact poses a problem of *information reduction*: some attributes may be redundant in the classification process and their removal reduces the amount of information encoded in data while preserving knowledge. This remark leads to the fundamental notion of a reduct (cf. [B]108, [B]109, [B]212, [B]214, [B]222, [B]239, [B]241, [B]254). Many aspects of the notion of a reduct and its ramifications are discussed in detail in Chapters by BAZAN et. AL., ŚLĘZAK, NGUYEN SINH HOA, STEPANIUK.

3 Reducts: classical notions

We embark here upon the reduction task consisting in selecting relevant sets of (conditional) attributes that suffice to preserve the indiscernibility relation IND_A and, consequently, lead to the same concept approximations. The remaining attributes are then *redundant*: their removal does not affect (e.g. worsen) the classification. This process goes a two–fold way: for information systems and for decision systems. We begin with the case of an information system.

3.1 Reducts in information systems

Given an information system $A = (U, A)$, we say that a set $B \subseteq A$ of attributes is a *reduct of* A when

1. $IND_B = IND_A$.
2. B is a minimal set of attributes with the property 1.

In other words, a reduct is a minimal set of attributes from A that preserves the partitioning of the universe by IND_A, and hence the original classification. Finding a minimal reduct is computationally an NP-hard problem as shown by Skowron and Rauszer (cf. [SR1] in [B]222). One can also show that the number of reducts of an information system with k attributes can be equal to

$$\binom{m}{\lfloor k/2 \rfloor}$$

Fortunately, there exist good heuristics that e.g. compute sufficiently many reducts in an admissible time, unless the number of attributes is very high (see the Chapter by BAZAN et AL.).

3.2 Reducts via boolean reasoning: discernibility approach

It was shown by Skowron and Rauszer (op. cit.) that the problem of finding reducts of a given information system may be solved as a case in *boolean reasoning*; the idea of boolean reasoning, going back to George Boole, is to represent a problem with a boolean function and to interpret its *prime implicants* (an *implicant* of a boolean function f is any conjunction of literals (variables or their negations) such that for each valuation v of variables, if the values of these literals are true under v then the value of the function f under v is also true; a *prime implicant* is a minimal implicant) as solutions to the problem. This approach to the reduct problem is implemented as follows.

Let A be an information system with n objects and k attributes. The *discernibility matrix* of A is a symmetric $n \times n$ matrix M_A with entries c_{ij} consisting of sets of attributes discerning between objects x_i and x_j where $i, j = 1, ..., n$:

$$c_{ij} = \{a \in A : a(x_i) \neq a(x_j)\}.$$

Once the discernibility matrix M_A is found, the *discernibility function* f_A for the information system A, a Boolean function of k boolean variables $a_1^*, ..., a_k^*$ (corresponding to attributes $a_1, ..., a_k \in A$), is defined as below, where $c_{ij}^* = \{a^* : a \in c_{ij}\}$

$$f_A(a_1^*, ..., a_k^*) : \bigwedge \left\{ \bigvee c_{ij}^* : 1 \leq j, c_{ij} \neq \emptyset \right\}.$$

Then, as shown by Skowron and Rauszer (op.cit.), there is a one–to–one correspondence between the set $PRI - IMP(f_A)$ of prime implicants of the discernibility function f_A and the set $RED(A)$ of reducts of the information system A; specifically, a conjunction $\wedge_{j=1}^m a_{i_j}^*$ is in $PRI - IMP(f_A)$ if and only if the set $\{a_{i_1}, a_{i_2}, ..., a_{i_m}\}$ is in $RED(A)$.

Example 5. Look at the decision system shown in Table 3, below. We compute reducts in its conditional part A

	a_1	a_2	a_3	a_4	Decision
x_1	1	1	1	2	yes
x_2	1	0	1	0	no
x_3	2	0	1	1	no
x_4	3	2	1	0	yes
x_5	3	1	1	0	no
x_6	3	2	1	2	yes
x_7	1	2	0	1	yes
x_8	2	0	0	2	no

Table 3. A decision table $A3_d$

We may check that the discernibility function f_A is (d, e, f, r are boolean variables corresponding to a_1, a_2, a_3, a_4, respectively, and the $\wedge$ sign is omitted):

$$\begin{aligned}
f_A(d,e,f,r) : &(e \vee r)(d \vee e \vee r)(d \vee e \vee r)(d \vee r)(d \vee e)(e \vee f \vee r)(d \vee e \vee f) \\
&(d \vee r)(d \vee e)(d \vee e)(d \vee e \vee r)(e \vee f \vee r)(d \vee f \vee r) \\
&(d \vee e \vee r)(d \vee e \vee r)(d \vee e \vee r)(d \vee e \vee f)(f \vee r) \\
&(e)(r)(d \vee f \vee r)(d \vee e \vee f \vee r) \\
&(e \vee r)(d \vee e \vee f \vee r)(d \vee e \vee f \vee r) \\
&(d \vee f \vee r)(d \vee e \vee f) \\
&(d \vee e \vee r)
\end{aligned}$$

The function f_A can be brought to an equivalent simplified form after boolean absorption laws are applied:

$$f_A(d,e,f,r) : er$$

It follows that the set $\{a_2, a_4\}$ of attributes is the reduct of the information system A.

3.3 Reducts in decision systems

The problem of reduction of the conditional attribute set in decision systems may be approached on similar lines. We resort here to the notion of the positive region of a decision, defined above.

Given a decision system A_d, a subset $B \subseteq A$ of the attribute set A is called a *relative reduct of* A_d if

1. $POS_B(d) = POS_A(d)$.
2. B is a minimal set of attributes with respect to the property 1.

Relative reducts may be found along similar lines to those for reducts i.e. by an appropriate application of Boolean reasoning (cf. Skowron and Rauszer (op. cit.)). It suffices to modify the discernibility matrix:

$M^d_{A_d} = (c^d_{ij})$ where $c^d_{ij} = \emptyset$ in case $d(x_i) = d(x_j)$ and $c^d_{ij} = c_{ij} - \{d\}$, otherwise. The matrix $M^d_{A_d}$ is the *relative discernibility matrix of* A_d.

From the relative discernibility matrix, the *relative discernibility function* $f^d_{A_d}$ is constructed in the same way as the discernibility function was constructed from the discernibility matrix. Again, as shown by Skowron and Rauszer (op. cit.), the set $PRI-IMP(f^d_{A_d})$ of *prime implicants* of $f^d_{A_d}$ and the set $REL-RED(A_d)$ of relative reducts of A_d are in one–to–one correspondence described above.

Example 6. Relative reducts for the decision table $A3_d$ are: $REL - RED(A_d) = \{\{a_1, a_2\}, \{a_2, a_4\}\}$.

4 Numerical measures of goodness of approximation

In deterministic systems, decision is described by exact concepts induced from conditional attributes; in case of non-deterministic systems, there is need to measure the degree of accuracy of approximations in order to discern among them and to select better ones (with respect to a given measure).

4.1 Measures of roughness

A rough set $X \subseteq U$ can be characterized numerically by the following coefficient

$$\alpha_B(X) = \frac{|\underline{B}(X)|}{|\overline{B}(X)|}$$

called the *accuracy of approximation* (cf. [B]222), where $|X|$ denotes the cardinality of $X \neq \emptyset$. Obviously $0 \leq \alpha_B(X) \leq 1$ and $\alpha_B(X) = 1$ means that X is B-exact while $\alpha_B(X) < 1$ indicates that X is B-rough. The coefficient α_B may be regarded as an unbiased estimate of the probability that an object chosen at random is certainly in X given it is possibly in X and it is a coarse measure of exactness/rougness of the set X.

4.2 Measures of goodness of approximation by the positive region

A similar idea may be applied to measure the quality of the approximation to decision d provided by a positive region of the form $POS_B(d)$: the ratio $|POS_B(d)|/|U|$ may be taken as the measure of closeness of the partition defined by the decision d and its approximation defined by attributes from B.

4.3 Rough membership functions

While two measures introduced above may be regarded as global estimates of approximation quality, we may have (and feel the need for) a more precise, local measure indicating the roughness degree of a set X at particular objects. This purpose is served by the *rough membership function* (cf. [B]258). In agreement with the rough set ideology, the rough membership function is information–dependent and a fortiori it is constant on indiscernibility classes.

Given a set $B \subseteq A$ of attributes, the rough membership function $\mu_X^B : U \to [0,1]$ is defined as follows

$$\mu_X^B(x) = \frac{|[x]_B \cap X|}{|[x]_B|}.$$

The rough membership function can be interpreted as an unbiased estimate of $\Pr(x \in X \mid x, B)$, the conditional probability that the object x belongs to the set X, given x is defined uniquely in terms of B (cf. [B]216, [B]217).

Clearly, a set X is B-exact if and only if $\mu_X^B = \chi_X$ i.e. the rough membership function of X is identical with its *characteristic function* χ_X taking values $0, 1$ only ($\chi_X(x) = 1$ when $x \in X$ and 0, otherwise). Thus, rough set theory agrees with the classical set theory when only exact sets are concerned.

5 Dependencies and decision rules

The numerical coefficients defined above, characterizing rough sets, may be applied also to measure degrees of dependencies among sets of attributes (cf. [B] 222). Here, we may extend the idea of a positive region, replacing the decision d with any set D of attributes. The best case of dependency between, say, a set C and a set D of attributes, is the one when values of $Inf_D(x)$ depend functionally on values $Inf_C(x)$ i.e. there exists a function F such that $Inf_D(x) = F(Inf_C(x))$ for each $x \in U$. To express this dependence in terms already familiar, we extend the notion of a positive region.

5.1 The notion of dependency

Given $C, D \subseteq A$, we let

$$POS_C(D) = \{x \in U : [x]_C \subseteq [x]_D\}.$$

Then, we say that D *depends on* C in *degree* $k(C, D)$, denoted symbolically as $C \Longrightarrow_{k(C,D)} D$, where

$$k(C, D) = \frac{|POS_C(D)|}{|U|}.$$

Clearly, D depends functionally on C if and only if $k(C, D) = 1$ and we say in this case that D *depends* on C; when $k(C, D) < 1$, we say that D *depends partially* (in *degree* $k(C, D)$) on C.

5.2 Decision rules

It follows already from our discussion that dependencies between the decision and various sets of conditional attributes are of particular importance. However, in practical applications, we would be interested in local dependencies among information sets of objects defined by – respectively – a set of conditional attributes and the decision. These dependencies are called *decision rules* (cf. [B]17, [B]18, [B]19, [B]20, [B]21, [B]82, [B]86, [B]99, [B]109, [B]183, [B]186, [B]196, [B]198, [B]200, [B]201, [B]202, [B]203, [B]204, [B]214, [B]217, [B]220, [B]222, [B]224, [B]287, [B]298, [B]338, [B] 339, [B]376, [B]377, [B]378, [B]379, [B]407, [B]408). In the process of inducing decision rules, reducts play an important role as they generate *minimal length* decision rules.

Example 7. For Table A3$_d$, we may see that *Decision* depends on the relative reduct $\{a_1, a_2\}$: $(1, 1) \Longrightarrow yes$, $(1, 0) \Longrightarrow no$, $(2, 0) \Longrightarrow no$, $(3, 2) \Longrightarrow yes$, $(3, 1) \Longrightarrow no$, $(1, 2) \Longrightarrow yes$.

We shall make these notions precise.
Let $A_d = (U, A, d)$ be a decision system.
We employ the logic L_{A_d} whose syntax and semantics are already defined in Sect. 2.1.

A *decision rule* for A is an expression of the form $\alpha \implies (d, v)$, where α is a formula of L_{A_d} employing only elementary sub–formulae of the form (a, v) with $a \in B$ for some $B \subseteq A$ and the meaning $[\alpha] \neq \emptyset$. Formulae α and (d, v) are referred to as the *predecessor* and the *successor* of the decision rule $\alpha \implies (d, v)$.

A decision rule $\alpha \implies (d, v)$ is *true* in A_d if and only if $[\alpha] \subseteq [(d, v)]$.

Example 8. In the preceding example, we have shown some dependencies between *Decision* and the set $B = \{a_1, a_2\}$ of attributes. These dependencies may be easily converted into decision rules in the adopted above form:

1. $(a_1, 1) \wedge (a_2, 1) \implies (Decision, yes)$.
2. $(a_1, 1) \wedge (a_2, 0) \implies (decision, no)$.
3. $(a_1, 2) \wedge (a_2, 0) \implies (Decision, no)$.
4. $(a_1, 3) \wedge (a_2, 2) \implies (Decision, yes)$.
5. $(a_1, 3) \wedge (a_2, 1) \implies (Decision, no)$.
6. $(a_1, 1) \wedge (a_2, 2) \implies (Decision, yes)$.

All these rules are true.

Several numerical factors can be associated with a decision rule. For example, the *support* of a decision rule is the number of objects that match the predecessor of the rule.

The *accuracy* of a rule is the quotient $\frac{|[\alpha] \cap [(d,v)]|}{|[\alpha]|}$ while the *coverage* of the rule is defined as the quotient $\frac{|[\alpha] \cap [(d,v)]|}{|[(d,v)]|}$. These numbers characterize the relation between the predecessor and the successor of a rule.

Such numerical factors characterizing a rule are subject to a delicate trade–off and their adequate tuning presents one of substantial problems in rule induction. The reader will find a discussion of this topic in the Chapter by BAZAN et AL.

The main application of decision rules is in classification of new objects, not included in the training decision system. Thus, decision rules adequate to this task should be robust i.e. stable with respect to "sufficiently small" perturbations of information sets of objects and noise–resistant i.e. they should preserve classification in the presence of noise.

It is therefore one of the main research tasks of rough set community to explore various ideas leading to induction of rules well predisposed to the task of classification of new yet unseen objects. These tasks encompass among others

1. Searching for adequate sets of attributes (reducts or approximations to them) which would determine predecessors of decision rules.
2. Searching for adequate techniques for discovery of proper patterns (meanings of rule predecessors) which would ensure the high classification rate by decision rules.
3. Searching for decomposition methods in a two–fold sense; first, in order to decompose (preprocess) complex attributes e.g. to discretize continuous attributes, and next, to decompose complex (large) decision systems into smaller ones, from which it would be computationally feasible to extract local decision rules and then to aggregate them (fuse) into global ones pertaining to the original table. Clearly, this point has close relations to 1 and 2.

We now present an outline of techniques and ideas developed in order to fulfill the program described in 1–3 above. The main purpose of this currently going research can be described as a search for decision–related features in the conditional part of the decision system which would induce classifiers of a satisfactory quality.

6 In search of features: similarity based techniques

Rough set community has found quite recently that one possible and promising way of obtaining good classifiers is to relax the indiscernibility relations to tolerance or similarity relations (cf. [B]21, [B]62, [B]67, [B]69, [B]197, [B]198, [B]199, [B]200, [B]202, [B]277, [B]280, [B]283, [B]332, [B]341, [B]348, [B]349, [B]351, [B]372, [B]367, [B]369).

We will call in the sequel a *tolerance* relation any relation τ which is

1. reflexive i.e. $\tau(x,x)$ for any $x \in U$.
2. symmetric i.e. $\tau(x,y) \Longrightarrow \tau(y,x)$ for any pair $x,y \in U$.

A *similarity* relation will be any relation τ which is merely reflexive.

We will outline a few ways by which tolerance or similarity approach enters rough set methods.

6.1 General approximation spaces

An abstract rendering of classical rough set approach has been proposed in the form of an *approximation space* i.e. a universe U endowed with operators l, u on the power set $P(U)$ of U whose properties reflect the properties of lower and upper approximations. A modification of this approach is defined as a *generalized approximation space*. A detailed analysis of generalized approximation spaces with applications to rule induction will be found in the Chapter by STEPANIUK.

A *generalized approximation space* is a system $GAS = (U, I, \nu)$, where

1. U is a universe (i.e. a non–empty set of objects).
2. $I : U \rightarrow P(U)$ is an *uncertainty function*.
3. $\nu : P(U) \times P(U) \rightarrow [0,1]$ is a *rough inclusion function*.

The uncertainty function assigns to an object x the set $I(x)$ of objects similar in an assumed sense to x.

Practically, such functions may be constructed e.g. from metrics imposed on value sets of attributes:

$$y \in I(x) \Leftrightarrow max_a\{dist_a(a(x), a(y))\} \leq \varepsilon$$

where $dist_a$ is a fixed metric on V_a and ε is a fixed threshold.

Values $I(x)$ of the uncertainty function define usually a covering of U (as it is usually true that $x \in I(x)$ for any x: uncertainty functions defined in rough set theory on the lines sketched above induce usually tolerance relations).

The rough inclusion function ν defines the degree of inclusion between two subsets of U.

An example of a rough inclusion is a *generalized rough membership function*

$$\nu(X,Y) = \begin{cases} \frac{card(X \cap Y)}{card(X)} & \text{if } X \neq \emptyset \\ 1 & \text{if } X = \emptyset \end{cases}.$$

The lower and the upper approximations of subsets of U in the generalized approximation space may be defined as follows.

For an approximation space $GAS = (U, I, \nu)$ and any subset $X \subseteq U$, lower and upper approximations are defined by

1. $\underline{GAS}X = \{x \in U : \nu(I(x), X) = 1\}$.

2. $\overline{GAS}X = \{x \in U : \nu(I(x), X) > 0\}$.

From both theoretical as well as practical points of view, it is desirable to allow I, ν to depend on some parameters, tuned in the process of optimalization of the decision classifier.

The analogy between classical rough set approximations and those in generalized approximation spaces may be carried out further e.g. the notion of a positive region may be rendered here as follows.

Given a classification $\{X_1, .., X_r\}$ of objects (i.e. $X_1, .., X_r \subseteq U$, $\bigcup_{i=1}^{r} X_i = U$ and $X_i \cap X_j = \emptyset$ for $i \neq j$, where $i, j = 1, \ldots, r$), the *positive region* of the classification $\{X_1, \ldots, X_r\}$ with respect to the approximation space GAS is defined by

$$POS(GAS, \{X_1, \ldots, X_r\}) = \bigcup_{i=1}^{r} \underline{GAS}X_i.$$

The quality of approximation of the classification $\{X_1, \ldots, X_r\}$ and other related numerical factors may be then defined on the lines of classical rough set theory, with appropriate modifications.

The notions of a reduct and a relative reduct are preserved in generalized approximation spaces, their definitions adequately modified. For instance, the notion of a reduct may be defined in a generalized approximation space as follows.

A subset $B \subseteq A$ is called a reduct of A if and only if

1. For every $x \in U$, we have $I_B(x) = I_A(x)$.
2. For every proper subset $C \subset B$ condition 1 is not satisfied.

where I_B is a partial uncertainty function obtained by restricting ourselves to attributes in B.

Similarly, one extends the definition of a relative reduct and a fortiori decision rules are induced by classical techniques adapted to this case. Algorithms based on boolean reasoning presented above may be extended to the new case.

One may observe that due to the fact that in this new context we may operate with parameterized tolerance relations, we obtain a greater flexibility in modeling the classification by the decision via conditional attributes; in consequence, we have prospects for better classification algorithms. The reader will find a detailed discussion of this topic in the Chapter by STEPANIUK.

6.2 Rough mereology

In generalized approximation spaces, the search is for tolerance relations, induced from value sets of attributes as ingredients of uncertainty functions and such that decision rules defined from these relations have sufficient quality.

Such approach has obvious advantages of being flexible and practical.

One may look however at this situation with a theoretician's eye and notice that from an abstract point of view this approach actually is about establishing for any two objects (or, concepts) a degree in which one of them is the "part" of the other; the notion of a "part in degree" being non–symmetrical would lead in general to similarity (and non–tolerance) relations. Clearly, one may adapt to this new situation all algorithms for reduct and rule generation presented above.

In addition, a judiciously defined notion of " a part in degree" would lead to a new paradigm having close relations to theories for approximate reasoning based on the primitive notion of a "part" i.e. mereological and meronymical theories; also this paradigm would have affinities with fuzzy set theory based on the notion of being an element on degree. Rough mereology fulfills these expectations.

Rough mereology has been proposed as a paradigm for approximate reasoning in complex information systems (cf. [B]272, [B]280, [B]282, [B]292, [B]296). Its primitive notion is that of a "rough inclusion" functor which gives for any two entities of discourse the degree in which one of them is a part of the other. Rough mereology may be regarded as an extension of rough set theory as it proposes to argue in terms of similarity relations induced from a rough inclusion instead of reasoning in terms of indiscernibility relations; it also proposes an extension of mereology as it replaces the mereological primitive functor of being a part with a more general functor of being a part in a degree. Rough mereology has deep relations to fuzzy set theory as it proposes to study the properties of partial containment which is also the fundamental subject of study for fuzzy set theory. Rough mereology may be regarded as an independent first order theory but it may be also formalized in the traditional mereological scheme proposed first by Stanisław Leśniewski where mereology is regarded and formalized within ontology i.e. theory of names (concepts). We regard this approach as particularly suited for rough set theory which is also primarily concerned with concept approximation in information systems.

We formalize here the functor of "part in degree" as a predicate μ with $X\mu_r Y$ meaning that X is a part of Y in degree r. We recall that a mereological theory of sets may be based on the primitive notion of an *element el* satisfying the following

1. $XelX$ (reflexivity).
2. $XelY \wedge YelZ \Longrightarrow XelZ$ (transitivity).
3. $XelY \wedge YelX \Longrightarrow X = Y$ (weak symmetricity).

The functor *pt* of *part* is defined as follows

$$XptY \iff XelY \land non(X = Y).$$

Rough mereology is constructed in such way as to extend mereology. The following is the list of basic postulates of rough mereology.

(RM1) $X\mu_1 Y \iff XelY$.

This means that being a part in degree 1 is equivalent to being an element; in this way a connection is established between rough mereology and mereology: the latter is a sub-theory of the former.

(RM2) $X\mu_1 Y \implies \forall Z.(Z\mu_r X \implies Z\mu_r Y)$.

Meaning the monotonicity property: any object Z is a part of Y in degree not smaller than that of being a part in X whenever X is an element of Y.

(RM3) $X = Y \land X\mu_r Z \implies Y\mu_r Z$.

This means that the identity of individuals is a congruence with respect to μ.

(RM4) $X\mu_r Y \land s \leq r \implies X\mu_s Y$.

Establishes the meaning " a part in degree at least r".

We may observe that $X\mu_r Y$ may be regarded as the statement that X is similar to Y in degree r i.e. as a parameterized similarity relation. We call the functor μ a *rough inclusion*.

One may ask about procedures for defining rough inclusions in information systems. The following simple procedure defines a rough inclusion in an information system (U, A) in case we would like to start with rows of the information table.

1. Consider a partition $P = \{A_1, A_2, ..., A_k\}$ of A.
2. Select a convex family of coefficients: $W = \{w_1, w_2, ..., w_k\}$ (i.e. $w_i \geq 0$, $\sum_{i=1}^{k} w_i = 1$).
3. Define $IND(A_i)(x, x') = \{a \in A_i : a(x) = a(x')\}$.
4. Let $r = \sum_{i=1}^{k} w_i \frac{card(IND(A_i)(x,x'))}{card(A_i)}$.
5. Let $x\mu_r x'$.

The rough inclusion defined according to this simple recipe is symmetric: $x\mu_r x' \iff x'\mu_r x$.

However, taking an additional care, one may find a bit more intricate recipes leading to similarities not being tolerances.

6.3 Rough mereology in complex information systems: in search of features in distributed systems

Rough mereology is well-suited to the task of concept approximation in complex information systems e.g. distributed systems. We outline here its workings in the case of a distributed/multi-agent system.

Distributed systems of agents We assume that a pair (Inv, Ag) is given where Inv is an *inventory of elementary objects* and Ag is a set of intelligent computing units called shortly *agents*. We consider an agent $ag \in Ag$. The agent ag is endowed with tools for reasoning and communicating about objects in its scope; these tools are defined by components of the agent label.

The *label of the agent ag* is the tuple

$$lab(ag) = (\mathsf{A}(ag), \mu(ag), L(ag), Link(ag), O(ag), St(ag)), Unc - rel(ag),$$

$$Unc - rule(ag), Dec - rule(ag))$$

where

1. $\mathsf{A}(ag)$ is an information system of the agent ag.
2. $\mu(ag)$ is a functor of part in a degree at ag.
3. $L(ag)$ is a set of unary predicates (properties of objects) in a predicate calculus interpreted in the set $U(ag)$ (e.g. in the desciptor logic $L_{A(ag)}$).
4. $St(ag) = \{st(ag)_1, ..., st(ag)_n\} \subset U(ag)$ is the set of *standard objects* at ag.
5. $Link(ag)$ is a collection of strings of the form $ag_1 ag_2 ... ag_k ag$ which are elementary teams of agents; we denote by the symbol $Link$ the union of the family $\{Link(ag) : ag \in Ag\}$.
6. $O(ag)$ is the set of *operations* at ag; any $o \in O(ag)$ is a mapping of the Cartesian product $U(ag_1) \times U(ag_2) \times ... \times U(ag_k)$ into the universe $U(ag)$ where $ag_1 ag_2 ... ag_k ag \in Link(ag)$.
7. $Unc - rel(ag)$ is the set of parameterized *uncertainty relations* $\rho_i = \rho_i(o_i(ag), st(ag_1)_i, st(ag_2)_i, ..., st(ag_k)_i, st(ag))$ where $ag_1 ag_2 ... ag_k ag \in Link(ag)$, $o_i(ag) \in O(ag)$ such that

$$\rho_i((x_1, \varepsilon_1), (x_2, \varepsilon_2), ., (x_k, \varepsilon_k), (x, \varepsilon))$$

holds for $x_1 \in U(ag_1), x_2 \in U(ag_2), .., x_k \in U(ag_k)$ and $\varepsilon, \varepsilon_1, \varepsilon_2, .., \varepsilon_k \in [0, 1]$ iff $x_j \mu(ag_j)_{\varepsilon_j} st(ag_j)$ for $j = 1, 2, .., k$ and $x \mu(ag)_\varepsilon st(ag)$ where

$$o_i(st(ag_1), st(ag_2), .., st(ag_k)) = st(ag) \text{ and } o_i(x_1, x_2, .., x_k) = x.$$

Uncertainty relations express the agents knowledge about relationships among uncertainty coefficients of the agent ag and uncertainty coefficients of its children.

8. $Unc - rule(ag)$ is the set of *uncertainty rules* f_j where $f_j : [0, 1]^k \longrightarrow [0, 1]$ is a function which has the property that
if $x_1 \in U(ag_1), x_2 \in U(ag_2), .., x_k \in U(ag_k)$ satisfy the condition

$$x_i \mu(ag_i)_{\varepsilon(ag_i)} st(ag_i)$$

then $o_j(x_1, x_2, ..., x_k) \mu(ag)_{f_j(\varepsilon(ag_1), \varepsilon(ag_2), .., \varepsilon(ag_k))} st(ag)$ where all parameters are as above.

9. $Dec - rule(ag)$ is a set of *decomposition rules* $dec - rule_i$ and

$$(\Phi(ag_1), \Phi(ag_2), .., \Phi(ag_k), \Phi(ag)) \in dec - rule_i$$

where $\Phi(ag_1) \in L(ag_1), \Phi(ag_2) \in L(ag_2), .., \Phi(ag_k) \in L(ag_k), \Phi(ag) \in L(ag)$ and $ag_1 ag_2..ag_k ag \in Link(ag))$ if there exists a collection of standards $st(ag_1), st(ag_2), ..., st(ag_k), st(ag)$ with the properties that $o_j(st(ag_1), st(ag_2), ..., st(ag_k)) = st(ag), st(ag_i)$ satisfies $\Phi(ag_i)$ for $i = 1, 2, .., k$ and $st(ag)$ satisfies $\Phi(ag)$. Decomposition rules are decomposition schemes in the sense that they describe the standard $st(ag)$ and the standards $st(ag_1), ..., st(ag_k)$ from which the standard $st(ag)$ is assembled under o_i in terms of predicates which these standards satisfy.

Approximate synthesis of features in distributed information systems

The process of synthesis of a complex feature (e.g. a signal, an action) by the above defined scheme of agents consists of the two communication stages viz. the top - down communication/negotiation process and the bottom - up communication/assembling process. We outline the two stages here in the language of approximate formulae.

We assume for simplicity that the relation $ag' \leq ag$, which holds for agents $ag', ag \in Ag$ iff there exists a string $ag_1 ag_2 ..ag_k ag \in Link(ag)$ with $ag' = ag_i$ for some $i \leq k$, orders the set Ag into a tree. We also assume that $O(ag) = \{o(ag)\}$ for $ag \in Ag$ i.e. each agent has a unique assembling operation.

We define a logic $L(Ag)$ (cf. [B]109) in which we can express global properties of the synthesis process.

Elementary formulae of $L(Ag)$ are of the form $< st(ag), \Phi(ag), \varepsilon(ag) >$ where $st(ag) \in St(ag), \Phi(ag) \in L(ag), \varepsilon(ag) \in [0, 1]$ for any $ag \in Ag$. Formulae of $L(ag)$ form the smallest extension of the set of elementary formulae closed under propositional connectives $\vee, \wedge, \neg$ and under the modal operators $\Box, \Diamond$.

The meaning of a formula $\Phi(ag)$ is defined classically as the set $[\Phi(ag)] = \{u \in U(ag) : u$ has the property $\Phi(ag)\}$; we denote satisfaction by $u \models \Phi(ag)$. For $x \in U(ag)$, we say that x *satisfies* $< st(ag), \Phi(ag), \varepsilon(ag) >$, in symbols:

$$x \models < st(ag), \Phi(ag), \varepsilon(ag) >,$$

if

1. $st(ag) \models \Phi(ag)$.
2. $x\mu(ag)_{\varepsilon(ag)} st(ag)$.

We extend satisfaction over formulae by recursion as usual.

By a *selection* over Ag we mean a function sel which assigns to each agent ag an object $sel(ag) \in U(ag)$. For two selections sel, sel' we say that sel *induces* sel', in symbols $sel \to_{Ag} sel'$ when $sel(ag) = sel'(ag)$ for any $ag \in Leaf(Ag)$ and $sel'(ag) = o(ag)(sel'(ag_1), sel'(ag_2), ..., sel'(ag_k))$ for any $ag_1 ag_2 ... ag_k ag \in Link$.

We extend the satisfiability predicate $\models$ to selections: for an elementary formula $< st(ag), \Phi(ag), \varepsilon(ag) >$, we let $sel \models < st(ag), \Phi(ag), \varepsilon(ag) >$ iff $sel(ag) \models < st(ag), \Phi(ag), \varepsilon(ag) >$.

We now let $sel \models \diamond < st(ag), \Phi(ag), \varepsilon(ag) >$ when there exists a selection sel' satisfying the conditions: $sel \to_{Ag} sel'; sel' \models < st(ag), \Phi(ag), \varepsilon(ag) >$.

In terms of $L(Ag)$ it is possible to express the problem of synthesis of an approximate solution to the problem posed to Ag. We denote by $head(Ag)$ the root of the tree $(Ag, \leq)$ and by $Leaf(Ag)$ the set of leaf-agents in Ag. In the process of top – down communication, a requirement Ψ received by the scheme from an external source (which may be called a *customer*) is decomposed into approximate specifications of the form $< st(ag), \Phi(ag), \varepsilon(ag) >$ for any agent ag of the scheme. The decomposition process is initiated at the agent $head(Ag)$ and propagated down the tree. We now are able to formulate the synthesis problem.

Synthesis problem. *Given a formula*

$$\alpha : < st(head(Ag)), \Phi(head(Ag)), \varepsilon(head(Ag)) >$$

find a selection sel over the tree $(Ag, \leq)$ with the property $sel \models \alpha$.

A solution to the synthesis problem with a given formula α is found by negotiations among the agents based on uncertainty rules and their successful result can be expressed by a top – down recursion in the tree $(Ag, \leq)$ as follows: given a local team $ag_1 ag_2 ... ag_k ag$ with the formula $< st(ag), \Phi(ag), \varepsilon(ag) >$ already chosen, it is sufficient that each agent ag_i choose a standard $st(ag_i) \in U(ag_i)$, a formula $\Phi(ag_i) \in L(ag_i)$ and a coefficient $\varepsilon(ag_i) \in [0, 1]$ such that

3. $(\Phi(ag_1), \Phi(ag_2),... \Phi(ag_k), \Phi(ag)) \in Dec - rule(ag)$ with standards $st(ag)$, $st(ag_1),..., st(ag_k)$.
4. $f(\varepsilon(ag_1),.., \varepsilon(ag_k)) \geq \varepsilon(ag)$ where f satisfies $unc - rule(ag)$ with $st(ag)$, $st(ag_1), ..., st(ag_k)$ and $\varepsilon(ag_1), ..., \varepsilon(ag_k), \varepsilon(ag)$.

For a formula α, we call an α - *scheme* an assignment of a formula $\alpha(ag) : < st(ag), \Phi(ag), \varepsilon(ag) >$ to each $ag \in Ag$ in such manner that 3, 4 above are satisfied and $\alpha(head(Ag))$ is $< st(head(Ag)), \Phi(head(Ag)), \varepsilon(head(Ag)) >$. We denote this scheme with the symbol

$$sch(< st(head(Ag)), \Phi(head(Ag)), \varepsilon(head(Ag)) >).$$

We say that a selection sel is *compatible* with a scheme $sch(< st(head(Ag)), \Phi(head(Ag)), \varepsilon(head(Ag)) >)$ in case $sel(ag)\mu(ag)_{\varepsilon(ag)} st(ag)$ for each leaf agent $ag \in Ag$.

The goal of negotiations can be summarized now as follows.

Sufficiency condition. *Given a formula*

$$< st(head(Ag)), \Phi(head(Ag)), \varepsilon(head(Ag)) >$$

if *a selection sel is compatible with a scheme*

$$sch(< st(head(Ag)), \Phi(head(Ag)), \varepsilon(head(Ag)) >)$$

then

$$sel \models \Diamond < st(head(Ag)), \Phi(head(Ag)), \varepsilon(head(Ag)) > .$$

Rough mereological approach to complex information systems has other merits: the dominant one is the presence of the *class operator* of mereology (cf. e.g. [B]292). The class operator allows for representation of a collection of objects as an individual object and – depending on the context – it may be applied as a clustering, granule–forming, or neighborhood–forming operator.

Due to class operator, rough mereology has been applied to problems of granulation of knowledge and computing with words (cf. [B]272, [B]273, [B]274, [B]275, [B]280, [B]282, [B]283, [B]284, [B]285, [B]291).

Rough mereology is presented in more detail in this Volume in the Chapter by POLKOWSKI and SKOWRON where new applications to Spatial Reasoning are discussed.

7 In search of features: generalized reducts and reduct approximations

We have outlined above some approaches to decision rule induction based on application of similarity relations instead of equivalence relations. Under these approaches, reducts undergo changes: we may expect that new reducts will be subsets of classically defined reducts thus leading to shorter decision rules with better classification qualities.

Such motivations are behind the general problem of relevant feature selection to which the rough set community has devoted a considerable attentions. We now present some more general approaches to reducts leading to new notions of reducts and a fortiori to new mechanisms of decision rule generation and to new classifiers based on them.

7.1 Frequency based reducts

In classical deterministic decision systems, reducts are minimal sets of attributes which preserve information about decision. This observation may be a starting point for various ramifications of the notion of a reduct: given a measure of information related to decision, one may search for various sets of attributes preserving or approximately preserving this information with respect to the assumed measure.

Some of these measures may be based on frequency characteristics of decision.

The reader will find a detailed discussion of frequency based measures and reducts in the Chapter by ŚLĘZAK; here, we give a few introductory examples.

In a general setting, one has to accept the fact that values of an attribute a may be complex i.e. they may be intervals, or subsets of the value set V_a; accordingly, the notion of an information set becomes more complex:

$$Inf_B(u) = \prod_{a \in B} \{\langle a, a(u) \rangle\}$$

i.e. elements of the information sets are threads (i.e. elements) in the Cartesian product of values of $a \in B$ on u.

Elements of the set

$$V_B^U = \{Inf_B(u) : u \in U\}$$

of all supported information patterns on B correspond to equivalence classes of indiscernibility relation

$$IND(B) = \{(u_1, u_2) \in U \times U : Inf_B(u_1) = Inf_B(u_2)\}.$$

We denote by w_B an element of Inf_B.

Generalized decision reducts Representing knowledge via decision rules requires attaching to each information pattern over considered conditional attributes a local decision information. As an example, let us consider the notion of a generalized decision function $\partial_{d/B} : V_B \longrightarrow 2^{V_d}$, which, for any subset $B \subseteq A$, labels information patterns with possible decision values

$$\partial_{d/B}(w_B) = \{d(u) : u \in Inf_B^{-1}(w_B)\}.$$

For each $w_B \in V_B^U$, the subset $\partial_{d/B}(w_B) \subseteq V_d$ contains exactly these decision values which occur in the indiscernibility class of w_B. Thus the implication

$$(B, w_B) \longrightarrow \bigvee_{v_d \in \partial_{d/B}(w_B)} (d, v_d)$$

is in some sense optimal with respect to A_d. Namely, disjunction in its right side is the minimal one which keeps consistency of the rule with A_d: any implication with (B, w_B) as predecessor and the disjunction of a smaller number of atomic elements than $|\partial_{d/B}(w_B)|$ as the successor is not true in A_d.

The generalized decision function leads to a new notion of a decision reduct for inconsistent data.

For a decision table $A_d = (U, A, d)$, we say that a subset $B \subseteq A$ ∂-*defines* d if for any $w_A \in V_A^U$ the equality holds

$$\partial_{d/B}\left(w_A^{\downarrow B}\right) = \partial_{d/A}(w_A)$$

where $w_A^{\downarrow B}$ is the thread w_A restricted to B.

A subset $B \subseteq A$ which ∂-defines d is called a ∂-*decision reduct* if none of its proper subsets has this property.

Approximate discernibility reducts Classical reducts were obtained by means of discernibility matrices; they discerned among all possible pairs of objects. Usually, exactly discerning reducts turn out to be much longer than ∂-decision ones, because they have to discern among many more object pairs.

As for consistent decision tables both types are equivalent to the notion of an exact decision reduct, their common extension which could lead to yet shorter (and possibly, better) decision rules might be based on a notion(s) of *approximate discernibility*.

To this end, one may introduce discernibility measures which assign to each $B \subseteq A$ a numerical value of a degree in which B discerns pairs of objects with different decision values.

For instance, let us consider discernibility measure

$$N_\mathbf{A}(B) = \frac{|\{(u,u') \in U \times U : (d(u) \neq d(u')) \wedge (Inf_B(u) \neq Inf_B(u'))\}|}{|\{(u,u') \in U \times U : (d(u) \neq d(u')) \wedge (Inf_A(u) \neq Inf_A(u'))\}|}$$

labeling subsets $B \subseteq A$ with the ratio of object pairs discerned by B, which are:

1. pairs necessary to be discerned because their elements belong to different decision classes.
2. pairs possible to be discerned by the whole of A.

It is easy to notice that a given $B \subseteq A$ exactly discerns d if and only if equality $N_\mathbf{A}(B) = 1$ holds.

Employing this discernibility measure, one may define the notion of an ε-*discerning decision reduct*.

Given a decision table $\mathsf{A}=(U, A, d)$ and an approximation threshold $\varepsilon \in [0,1)$, we say that a subset $B \subseteq A$ ε-*discerns d* if and only if the inequality

$$N_\mathbf{A}(B) \geq 1 - \varepsilon$$

holds.

A subset $B \subseteq A$ which ε-discerns d is called a ε-*discerning decision reduct* if it is a minimal set of attributes with this property.

Approximate reducts keep therefore a judicious balance between the (reduced) number of conditions and the loss of determinism with respect to the initial table.

By adaptive tuning of ε, one is likely to obtain interesting results via appropriate heuristics. The reader is referred to [B]296 for a study of a possible design of such heuristics aimed at extraction of approximate association rules from data.

Approximate reducts based on conditional frequency The idea of finding reducts based on a global discernibility measure may be refined yet: one may search for reducts preserving a given frequency characteristic of decision. To have a glimpse into this venue of research, let us focus on conditional representation based on rough membership functions $\mu_{d/B} : V_d \times V_B^U \to [0, 1]$, defined, for any fixed $B \subseteq A$, by the formula

$$\mu_{d/B}(v_d/w_B) = \frac{|Inf_B^{-1}(w_B) \cap d^{-1}(v_d)|}{|Inf_B^{-1}(w_B)|}$$

where the number of objects with the pattern value w_B on B and the decision value v_d is divided by the number of objects with the pattern value w_B on B.

The value of $\mu_{d/B}(v_d/w_B)$ is, actually, the conditional frequency of occurrence of the given $v_d \in V_d$ under the condition $w_B \in V_B^U$ on $B \subseteq A$.

A rational approach is to choose, for any given $w_B \in V_B^U$, the decision value with the highest conditional frequency of occurrence, i.e., such $v_d \in V_d$ that

$$\mu_{d/B}(v_d/w_B) = \max_{k=1,..,|V_d|} \mu_{d/B}(v_k/w_B)$$

which assigns to each pattern over $B \subseteq A$ a decision value which minimizes the risk of wrong classification.

We then define the *majority decision function* $m_{d/B} : V_B^U \to 2^{V_d}$

$$m_{d/B}(w_B) = \left\{ v_d \in V_d : \mu_{d/B}(v_d/w_B) = \max_k \mu_{d/B}(v_k/w_B) \right\}.$$

Now, one may define a new notion of a reduct preserving the information stored in the majority decision function.

Given a decision table A=(U, A, d), we say that a subset $B \subseteq A$ *m–defines d* if

$$m_{d/B}\left(w_A^{\downarrow B}\right) = m_{d/A}(w_A)$$

for any $w_A \in V_A^U$.

A subset B which m–defines d is called an *m–decision reduct* if no proper subset of it m–defines d.

This line of research is continued and in the search for optimal classifiers various measures of information are employed e.g. *entropy*, *distance measures* etc. As with classical reducts, problems of finding a reduct of any of these types is NP–complete, and the respective problem of finding an (optimal) reduct is NP–hard (cf.the Chapter by ŚLĘZAK for a discussion of complexity issues). The reader will find in [B]296 a discussion of relevant heuristics.

7.2 Local reducts

Another important type of reducts are local reducts (cf. [B]393, [B]394). A *local reduct* $r(x_i) \subseteq A$ (or a *reduct relative to decision and object* $x_i \in U$) where x_i is called a *base object*, is a subset of A such that

1. $\forall\, x_j \in U.(\ d(x_i) \neq d(x_j) \implies \exists\, a_k \in r(x_i).\ a_k(x_i) \neq a_k(x_j)).$
2. $r(x_i)$ is minimal with respect to inclusion.

Local reducts may be better suited to the task of classification as they reflect the local information structure. In the Chapter by BAZAN et AL. a detailed discussion of local reducts and algorithms generating coverings of the universe with local reducts may be found.

7.3 Dynamic reducts

Methods based on calculation of reducts allow to compute, for a given decision table A_d, descriptions of all decision classes in the form of decision rules (see previous sections). In search for robust decision algorithms, the idea of dynamic reducts has turned out to be a happy one: as suggested by experiments, rules calculated by means of dynamic reducts are better predisposed to classify unseen cases, because these reducts are the most stable reducts in a process of random sampling of the original decision table.

For a decision system $A_d=(U,A,d)$, we call any system $B=(U',A,d)$ such that $U' \subseteq U$ a *sub–table* of A. By the symbol $P(A_d)$, we denote the set of all sub–tables of A_d. For $F \subseteq P(A_d)$, by $DR(A,F)$ we denote the set

$$RED(A_d) \cap \bigcap_{B \in F} RED(B).$$

Any element of $DR(A,F)$ is called an F-*dynamic reduct* of A.

From the definition of a dynamic reduct it follows that a relative reduct of A is dynamic if it is also a reduct of all sub–tables from a given family F. This notion can be sometimes too much restrictive and one may relax it to the notion of an (F,ε)-*dynamic reduct*, where $\varepsilon \in [0,1]$. The set $DR_\varepsilon(A,F)$, of all (F,ε)-dynamic reducts is defined by
$DR_\varepsilon(A,F) = \{C \in RED(A_d) :$

$$\frac{card(\{B \in F : C \in RED(B)\})}{card(F)} \geq 1 - \varepsilon\}$$

For $C \in RED(A_d)$, the number:

$$\frac{card(\{B \in F : C \in RED(B)\})}{card(F)}$$

is called the *stability coefficient* of the reduct C (relative to F).

7.4 Generalized dynamic reducts

From the definition of a dynamic reduct it follows that a relative reduct of any table from a given family F of sub-tables of A can be dynamic if it is also a reduct of the table A. This can be sometimes not convenient because we are interested in useful sets of attributes which are not necessarily reducts of the table A. Therefore we have to generalize the notion of a dynamic reduct.

Let $A_d = (U, A, d)$ be a decision table and $F \subseteq P(A_d)$. By the symbol $GDR(A_d, F)$, we denote the set

$$\bigcap_{B \in F} RED(B).$$

Elements of $GDR(A_d, F)$ are called F-*generalized dynamic reducts* of A.

From the above definition it follows that any subset of A is a generalized dynamic reduct if it is a reduct of all sub-tables from a given family F. As with dynamic reducts, one defines a more general notion of an (F, ε)-*generalized dynamic reduct*, where $\varepsilon \in [0, 1]$. The set $GDR_\varepsilon(A_d, F)$ of all (F, ε)-generalized dynamic reducts is defined by
$GDR_\varepsilon(A_d, F) = \{C \subseteq A :$

$$\frac{card(\{B \in F : C \in RED(B)\})}{card(F)} \geq 1 - \varepsilon\}.$$

The number
$$\frac{card(\{B \in F : C \in RED(B)\})}{card(F)}$$

is called the *stability coefficient* of the generalized dynamic reduct C (relative to F).

Dynamic reducts are determined on the basis of a family of sub-tables, each of which may be regarded as a perturbed original table; thus one may expect that these reducts are less susceptible to noise. This expectation is borne out by experiments. In the Chapter by BAZAN ET AL., the reader will find a detailed discussion of algorithmic and applicational issues related to dynamic reducts.

7.5 Genetic and hybrid algorithms in reduct computation

Recently, much attention is paid to hybrid algorithms for reduct generation, employing rough set algorithms in combination with other techniques e.g. evolutionary programming.

To exploit advantages of both genetic and heuristic algorithms, one can use a hybridization strategy. The general scheme of a *hybrid algorithm* is as follows:

1. Find a strategy (a heuristic algorithm) which gives an approximate result.
2. Modify (parameterize) the strategy using a control sequence, so that the result depends on this sequence (recipe).
3. Encode the control sequence as a chromosome.

4. Use a genetic algorithm to produce control sequences. Proceed with the heuristic algorithm controlled by the sequence. Evaluate an object generated by the algorithm and use its quality measure as the fitness of the control sequence.
5. A result of evolution is the best control sequence, i.e. the sequence producing the best object. Send this object to the output of the hybrid algorithm.

Hybrid algorithms proved to be useful and efficient in many areas, including NP-hard problems of combinatorics. The short reduct finding problem also can be solved efficiently by this class of algorithms.

Finding reducts using Genetic Algorithms

An order-based genetic algorithm is one of the most widely used components of various hybrid systems. In this type of genetic algorithm a chromosome is an n-element permutation σ, represented by a sequence of numbers: $\sigma(1)\ \sigma(2)\ \sigma(3)\ \ldots\ \sigma(n)$. Mutation of an order-based individual means one random transposition of its genes (a transposition of random pair of genes). There are various methods of recombination (*crossing–over*) considered in literature like **PMX** (Partially Matched Crossover), **CX** (Cycle Crossover) and **OX** (Order Crossover). After crossing–over, fitness function of every individual is calculated. In the case of a hybrid algorithm a heuristics is launched under control of an individual; a fitness value depends on the result of heuristics. A new population is generated using the "roulette wheel" algorithm: the fitness value of every individual is normalized and treated as a probability distribution on the population;one chooses randomly M new individuals using this distribution. Then all steps are repeated.

In the hybrid algorithm a simple, deterministic method may be used for reduct generation:

Algorithm 8. *Finding a reduct basing on permutation.*
Input:
    1. decision table $\mathbf{A} = (U, \{a_1, ..., a_n\} \cup \{d\})$
    2. permutation τ generated by genetic algorithm
Output: *a reduct R generated basing on permutation τ*
Method:
    $R = \{a_1, ..., a_n\}$
    $(b_1 \ldots b_n) = \tau(a_1 \ldots a_n)$
    for $i = 1$ **to** n **do**
    begin
        $R = R - b_i$
        if not $Reduct(R, \mathbf{A})$ **then** $R = R \cup b_i$
    end
    return R
□

The result of the algorithm will always be a reduct. Every reduct can be found using this algorithm, the result depends on the order of attributes. The genetic algorithm is used to generate the proper order. To calculate the function of fitness for a given permutation (order of attributes) one run of the deterministic algorithm is performed and the length of the found reduct is calculated. In the selection phase of the genetic algorithm fitness function

$$F(\tau) = n - L_\tau + 1,$$

where L_τ is the length of the reduct, may be used. Yet another approach is to select a reduct due to the number of rules it generates rather than to its length. Every reduct generates an indiscernibility relation on the universe and in most cases it identifies some pairs of objects. When a reduct generates a smaller number of rules it means that the rules are more general and they should better recognize new objects.

The hybrid algorithm described above can be used to find reducts generating the minimal number of rules. The only thing to change is the definition of the fitness function:

$$F(\tau) = m - R_\tau + \frac{n - L_\tau + 1}{n},$$

where R_τ denotes the number of rules generated by the reduct. Now, the primary criterion of optimization is the number of rules, while the secondary is the reduct length. The results of experiments show, that the classification system based on the reducts optimized due to the number of rules performs better (or not worse) than the short reduct based one. Moreover, due to the rule set reduction, it occupies less memory and classifies new objects faster.

A discussion of this topic will be found in the Chapter by BAZAN et AL.

8 In search of features: template techniques

Feature extraction may be also regarded abstractly as concerned with developing methods of searching for *regular patterns* (regularities) hidden in data sets.

There are two two fundamental kinds of regularities: regularities defined by *templates* and regularities defined in a *relational language*. The process of extracting regularities should find templates of high quality. In the general case, we distinguish between numerical (orderable) value sets and symbolic (nominal) value sets. A unified treatment of both types of attributes may be provided by means of (generalized) descriptors.

8.1 Templates

Let A$= (U, A)$ be an information system. Any clause of the form $D = (a \in W_a)$ is called the *descriptor* and the value set $W_a \subseteq V_a$ is called the *range* of D. In case a is a numeric attribute, one usually restricts the range of descriptors for a to real intervals, that means $W_a = [v_1, v_2] \subseteq V_a$. In case of symbolic attributes, W_a can be any non-empty subset $W_a \subseteq V_a$. The *volume* of a given descriptor $D = (a \in W_a)$ is defined by

$$Volume(D) = |W_a \cap a(U)|.$$

By $Prob(D)$, where $D = (a \in W_a)$, we denote the *hitting probability* of the set W_a i.e.

$$Prob(D) = \frac{|W_a|}{|V_a|}.$$

For a a numeric attribute,

$$Prob(D) = \frac{v_2 - v_1}{\max(V_a) - \min(V_a)}$$

where $\max(V_a)$, $\min(V_a)$ denote the maximum and the minimum value in the value set V_a, respectively.

A formula $T : \bigwedge_{a \in B} (a \in W_a)$ is called *a template* of A. A template T is *simple*, if any descriptor of T has range of one element. Templates with descriptors consisting of more than one element are called *generalized*. For any $X \subseteq U$, the set $\{x \in X : \forall_{a \in B}\, a(x) \in W_a\}$ of objects from X satisfying T is denoted by $[T]_X$. One defines: $support_X(T) = |[T]_X|$.

Any template $T = D_1 \wedge ... \wedge D_k$, where $D_i = (a_i \in W_i)$, is characterized by the following parameters:

1. $length(T)$, which is the number of descriptors occurring in T.
2. $support(T) = support_U(T)$, which is the number of objects in U satisfying T.
3. $applength(T)$, which is $\sum_{1 \leq i \leq k} \frac{1}{Volume(D_i)}$ called the *approximated length* of the generalized template T.

Functions *applength* and *length* coincide for simple templates.

The complexity and algorithmic problems related to search for good templates with applications are given a throughout examination in the Chapter by NGUYEN SINH HOA to which the reader is invited for details.

In some applications we are interested in templates "well matching" some chosen decision classes. Such templates are called *decision templates*. The template associated with one decision class is called a *decision rule*. The precision of any decision template (e.g. $T \Longrightarrow (d \in V)$) is estimated by its *confidence ratio*. The *confidence* of a decision rule $T \Longrightarrow (d, i)$ (associated with the decision class X_i) is defined as

$$confidence_{X_i}(T) = \frac{support_{X_i}(T)}{support(T)}$$

8.2 Template goodness measures

Template quality should be evaluated taking into account the context of an individual application. In general, templates with high quality have the following properties:

1. They are supported by a *large* number of objects.

2. They are of high *specificity*, i.e. they should be described by a sufficiently *large* number of descriptors and the set of values W_a in any descriptor ($a \in W_a$) should be *small*.

Moreover, decision templates (decision rules) used for classification tasks should be of high *predictive accuracy*. Decision templates (decision rules) of high quality should have the following properties

3. They should be supported by a *large* number of objects.
4. They should have high predictive accuracy.

For a given template T, its support is defined by the function $support(T)$ and its length by the function $applength(T)$. Any quality measure is a combination of these functions. One may use one of the following exemplary functions

$$quality(T) = support(T) + applength(T);$$

$$quality(T) = support(T) \cdot applength(T).$$

In case of decision templates, one should take into consideration the confidence ratios. Let T be a decision template associated with the decision class X_i. The quality of T can be measured by one of the following functions

$$quality(T) = confidence(T);$$

$$quality(T) = support_{C_i}(T) \cdot applength(T);$$

$$quality(T) = support_{C_i}(T) \cdot confidence(T).$$

In search for good template–based classifiers one may applied various strategies; they are discussed in illuminating detail in the Chapter by NGUYEN SINH HOA. Here we comment in passing on the two basic methodologies.

One can start with the *empty descriptor set* and extend it by adding the most relevant descriptors from the original set. One can also begin with the original *full set* of descriptors and remove irrelevant descriptors from it. The former strategy is the *sequential forward* strategy and the latter is the *sequential backward* strategy. Every strategy can be classified as *deterministic* or *random* depending on a method used for descriptor selection. A deterministic strategy always chooses the descriptor with the *optimal fitness*, whereas the random selector chooses templates according to some probability distribution.

Sequential forward generation This strategy begins with the *empty* template T. The template T is extended by adding one descriptor at a time that well fits the existing template. Let T_i be a temporary template obtained in the i-th step of construction. A new descriptor is selected according to the function $fitness_{T_i}$. For the temporary template T_i, the fitness of any descriptor D measured relatively to T_i reflects its potential ability to create a new template $T_{i+1} = T_i \wedge D$ of high quality. In the *Deterministic Sequential Forward Generation* (DSFG), the template T_i is extended by a descriptor with the maximum value of fitness. In general DSFG detects a template of high quality. The drawback of this method is that it can be stuck in a local extreme (the best template at the moment). To avoid this situation, a strategy *Random Sequential Forward Generation* (RSFG) is applied. In RSFG, descriptors are chosen randomly according to a probability distribution defined by fitness function.

Let P be a set of descriptors, which can be used to extend T_i. Let p_0 be a descriptor in P. In the simplest case, the distribution function $Prob$ can be defined by $Prob(p_0) = \frac{fitness_{T_i}(p_0))}{\sum_{p \in P} fitness_{T_i}(p)}$

Sequential backward generation The sequential backward method uses top–down strategy rather than bottom–up strategy used for DSFG. Starting from a *full template* T_{full}, (which is often defined by an information vector $Inf_A(x)$ for some object x), the algorithm finds irrelevant descriptors and removes one descriptor at a time. After a removal, the quality of a new template is estimated. The descriptor p is *irrelevant* if the template T without this descriptor is of better quality. Attributes are selected according to *fitness function*, similarly as in DSFG.

Assume the full template T_{full} is of the form $T_{full} = \bigwedge_{1 \leq i \leq k}(a_i \in W_i)$, where k is a number of attributes in a data table and W_i are fixed subsets of V_{a_i}.

Analogously to the forward strategy, a descriptor to be removed can also be chosen randomly. For $fitness_{T_i}(p_0)$ denoting the fitness of a descriptor p_0 according to the template T_i, the distribution function for irrelevant descriptor elimination can be defined as follows: $Prob(p_0) = 1 - \frac{fitness_{T_i}(p_0))}{\sum_{p \in T_i} fitness_{T_i}(p)}$.

For any algorithm based on sequential schemes, the following three parameters are fixed:

1. estimation of the descriptor fitness.
2. estimation of the template quality.
3. a method of searching for the best descriptor $(a \in W_a)$ of a given attribute a.

8.3 Searching for optimal descriptors

On can treat a descriptor $(a \in W_a)$ as a descriptor of "good quality" if the template $T \wedge (a \in W_a)$ has large support, although the set W_a is small.

Hence the fitness function is defined using the following two parameters

1. the number of objects supporting the template $T \wedge (a \in W_a)$.
2. the cardinality of the set W_a.

The fitness function should be proportional to the first parameter and inversely proportional to the second parameter. It can be defined by the following formula:

$$fitness_T(a \in W_a) = support_{[T]_U}(a \in W_a) \cdot Prob^{-1}(W_a)$$

The descriptor $a \in W_a$ is *optimal* if $fitness(a \in W_a)$ is maximal.

A searching heuristics for an optimal descriptor may be based on a fixed similarity relation.

Let $R_a \subseteq V_a \times V_a$ be a given similarity relation. The range of W_a may be computed by taking the similarity pre–class $[v, R_a]$ of the properly chosen *generator v*.

Searching heuristics for optimal similarity relations are investigated in detail in the Chapter by NGUYEN SINH HOA.

9 In search of features in complex attribute systems: preprocessing via discretization

Complex attributes may have continuous values or the number of values may be large so they require value grouping etc. In such cases a pre–processing step is necessary which would return a data table with a sufficiently small number of attribute values. We focus here on discretization of continuous attributes as an example of pre–processing stage (cf. [B]184, [B]185, [B]186, [B]187, [B]188, [B]189, [B]190, [B]194, [B]197, [B]198, [B]199, [B]200, [B]202, [B]203, [B]204). In Chapters by BAZAN et AL. and by NGUYEN SINH HOA, the reader will find an exhaustive discussion of algorithmic and computational aspects of discretization as well as examples of applications.

Discretization is a step that is not specific to the rough set approach but the majority of rule or tree induction algorithms require it in order to perform well. The search for appropriate cut–off points can be reduced to finding some minimal Boolean expressions cf. [B]191.

The reported results (op.cit., op.cit.) are showing that the discretization problems and symbolic value partition problems are of high computational complexity (i.e. NP-complete or NP-hard) which clearly justifies the importance of designing efficient heuristics.

To implement discretization, one can use the idea of *cuts*. These are pairs (a, c) where $c \in V_a$.

9.1 Value partition via cuts

Any cut defines a new conditional attribute with binary values. For example the attribute corresponding to the cut $(a, 1.2)$ takes 0 on x if $a(x) < 1.2$, otherwise it takes 1 on x.

Similarly, any set P of cuts defines a new conditional attribute a_P for any a. One should consider a partition of the value set of a by cuts from P and assign new names to elements of this partition. Let us take the set of cuts: $P = \{(a, 0.9), (a, 1.5), (b, 0.75), (b, 1.5)\}$. This set of cuts is gluing together all values of a less then 0.9, all values in the interval $[0.9, 1.5)$ and all values in the interval $[1.5, 4)$. Similarly for b.

The problem of finding a (minimal) set of cuts may be solved by means of boolean reasoning.

Let $A_d = (U, A, d)$ be a decision system where $U = \{x_1, x_2, \ldots, x_n\}$, $A = \{a_1, \ldots, a_k\}$, and $d : U \to \{1, \ldots, r\}$. We assume that $V_a = [l_a, r_a) \subset \Re$ is an interval of real numbers for any $a \in A$ and A_d is a deterministic decision system. Any pair (a, c) where $a \in A$ and $c \in \Re$ will be called a *cut on* V_a. Let P_a be a partition of V_a (for $a \in A$) into sub–intervals i.e. $P_a = \{[c_0^a, c_1^a), [c_1^a, c_2^a), \ldots, [c_{k_a}^a, c_{k_a+1}^a)\}$ for some integer k_a, where $l_a = c_0^a < c_1^a < c_2^a < \ldots < c_{k_a}^a < c_{k_a+1}^a = r_a$ and $V_a = [c_0^a, c_1^a) \cup [c_1^a, c_2^a) \cup \ldots \cup [c_{k_a}^a, c_{k_a+1}^a)$. Hence any partition P_a is uniquely defined and often identified as the set of cuts: $\{(a, c_1^a), (a, c_2^a), \ldots, (a, c_{k_a}^a)\} \subset A \times \Re$. Any set of cuts $P = \bigcup_{a \in A} P_a$ defines from A_d a new decision system $A_d^P = (U, A^P, d)$ called the P-*discretization of* A_d, where $A^P = \{a^P : a \in A\}$ and $a^P(x) = i \Leftrightarrow a(x) \in [c_i^a, c_{i+1}^a)$ for $x \in U$ and $i \in \{0, \ldots, k_a\}$.

Two sets of cuts P', P are *equivalent* if $A^P = A^{P'}$.

We say that the set of cuts P is A_d–*consistent* if $\partial_{A_d} = \partial_{A^P}$, where ∂_{A_d} and $\partial_{A_d^P}$ are generalized decisions of A_d and A_d^P, respectively.

The A_d-consistent set of cuts P^{irr} is A_d–*irreducible* if P is not A_d–consistent for any $P \subset P^{irr}$.

The A_d-consistent set of cuts P^{opt} is A_d–*optimal* if $card(P^{opt}) \leq card(P)$ for any A_d–consistent set of cuts P.

One can show that the decision problem of checking if for a given decision system A_d and an integer k there exists an irreducible set of cuts P such that $card(P) < k$ is NP-complete.

The problem of searching for an optimal set of cuts P in a given decision system A_d is NP-hard.

9.2 Heuristics

One can construct efficient heuristics returning semi-minimal sets of cuts. A simple heuristics may be based on the Johnson strategy. Using this strategy one can look for a cut discerning the maximal number of object pairs, then one can eliminate all already discerned object pairs and repeat the procedure until all object pairs to be discerned are discerned. It is interesting to note that this can be realized by computing the minimal relative reduct of the corresponding decision system. heuristics of this type are discussed in the Chapter by BAZAN et AL. and in the Chapter by NHUYEN SINH HOA.

The *M(aximal)D(iscernibility)* heuristics is based on searching for a cut with maximal number of object pairs discerned by this cut. The idea is analogous to the Johnson approximation algorithm and can be briefly formulated as follows.

First, a new decision system is constructed from the decision system $A_d = (U, A, d)$; for each $a \in A$, we denote by C_a the set

$$\{(a, \frac{c_1^a + c_2^a}{2}), ..., (a, \frac{c_{n_a-1}^a + c_{n_a}^a}{2})\}$$

of all possible cuts concerning the attribute a where

$$\{a(x) : x \in U\} = \{c_1^a, ..., c_{n_a}^a\}$$

and

$$C_A = \bigcup_{a \in A} C_a$$

denotes the set of all cuts on A.

The decision system $A_d^c = (U^c, A^c, d^c)$ is defined as follows

1. $U^c = \{(x, y) \in U^2 : d(x) \neq d(y)\} \cup \{new\}$ where $new \notin U^2$.
2. $A^c = \{b_{(a,c)} : (a,c) \in C_A\}$ where $b_{(a,c)}(x) = 0$ in case $a(x) < c$ and 1, otherwise.
3. $d^c(new) = 0$ and $d^c(u) = 1$ in case $u \neq new$.

Now, the algorithms follows.

ALGORITHM MD-heuristic (Semi-optimal family of partitions)

Step 1 Construct the table A_d^c from A_d and denote by B the table A_d^c with the row *new* deleted. Let **D** = ∅.
Step 2 Choose a column c_{max} from B with the maximal number of occurrences of 1.
Step 3 Delete from B the column chosen in Step 2 and all rows marked in this column with 1. Insert c_{max} into **D** and remove it from C_A.
Step 4 If B ≠ ∅ then go to Step 2 **else** Stop.

Let us observe that the new features in the considered case of discretization are of the form $a \in V$, where $V \subseteq V_a$ and V_a is the set of values of attribute a.

One can extend the presented approach to the case of symbolic (nominal, qualitative) attributes as well as to the case when in a given decision system nominal and numeric attributes appear. The received heuristics are of very good quality.

The exhaustive discussion of this topic will be found in the Chapter by BAZAN et AL., and in the Chapter by NGUYEN SINH HOA.

In case of real value attributes one can also search for features determined by characteristic functions of half–spaces determined by hyper–planes or even parts of spaces defined by more complex surfaces in multi–dimensional spaces (cf. [B]197). The reported results are showing substantial increase in the quality of classification of unseen objects at the cost of more time–consuming search for a semi–optimal hyperplane.

10 In search of complex features: extracting a concurrent structure from data tables

Problems posed by needs of real-time control design, real–time object identification, or knowledge discovery in complex systems call for the analysis by rough set methods of data tables representing systems with an additional structure e.g the concurrent one. The rough set approach to modeling concurrency has two main aspects.

1. Synthesis of of a model of a concurrent system from data tables representing the behavior of processes in the given concurrent system.
2. Analysis of a model of a concurrent system with the aim of revealing its important features.

These aspects are discussed throughout in the Chapter by SURAJ. In this Chapter, *Petri nets* are adopted as a model of concurrency.

10.1 Petri nets

A Petri net is a quadruple $N = (P, T, \alpha, \beta)$ where P *places* and T *transitions* two sets of *nodes* and α, β are two sets of directed arcs, each arc representing a connection between a place and a transition (in case of α) or a connection between a transition and a place (in case of β). Places represent storage for input/output and transitions represent actions making input into output. The state of a net is represented by type–less marks (*dots*) assigned to places; such an assignment is a *marking* M of the net N and the pair (N, M) is a *marked Petri net*.

Dynamic actions of a net are represented as *firings* of transitions and any sequence of firings determines the evolution of the system. Globally, dynamic behavior of a net may be represented by means of a *reachability relation* R on markings: a marking M' is *reachable* from a marking M if some sequence of firings makes M into M'; then M' belongs in the *reachability set* $R(M)$.

10.2 Synthesis problem

A *synthesis problem* of a concurrent model can be formulated as follows.

Let $A = \{a_1, a_2, ..., a_m\}$ be a non-empty, finite set of *processes*. With every process $a \in A$, we associate a finite set V_a of its *local states*. We assume that behaviour of such a process system is presented by a designer in a form of a table. Each row in the table includes record of states of processes from A, and each record is labeled by an element from the set U of *global states* of the system. The pair (U, A) is an information system A.

Synthesis problem

For a given information system A, construct its concurrent model, i.e., a marked Petri net (N_A, M_A) with the following property: the reachability set $R(M_A)$ defines an extension A* of A by adding to A all new global states corresponding to markings from $R(M_A)$. All (new) global states in A^* are consistent with all rules true in A. Moreover, A^* is the largest extension of A with that property.

Rough set concept of independence/dependence may be applied to processes.

We recall, that a set of attributes C *depends* on a set of attributes B if one can compute the values of attributes from C knowing the values of attributes from B. A set of attributes C depends partially in degree k ($0 \leq k < 1$) on a set of attributes B if the B-positive region of C consists of k % of global states in A (cf. Sect. 5.1).

A set of processes $B \subseteq A$ in a given information system is called *partially independent* if there is no partition of B into sets C and D such that D is dependent on C.

We say that a set $B \subseteq A$ is *totally independent* if there is no partition of B into C and D such that D depends on C in the degree $0 < k \leq 1$.

One may show that a set B is a maximal partially independent set of processes in A if and only if $B \in RED(A)$.

In order to compute partially independent parts of a given information system one can thus execute the procedure for generating reducts.

From the definition of a totally independent set of processes it follows that for an arbitrary totally independent set B there is a reduct $C \in RED(A)$ such that $B \subseteq C$. Hence to find all maximal totally independent sets of processes it is enough to find for every reduct C all maximal independent subsets of C.

On these notions one may base an algorithm for solving the synthesis problem.

A required Petri net may be constructed in two steps. In the first one, all dependencies between local states of processes in the system are extracted from the given set of global states. In the second step, a Petri net corresponding to these dependencies is constructed.

A detailed analysis of concurrent structures in complex information systems by rough set methods for knowledge discovery and of solutions to problems like presented above, will be found in the Chapter by SURAJ.

Conclusions

We have given, in the above text, an introduction to the rough set approach to knowledge discovery in data tables, laying emphasis on the basic notions and ideas and turning the attention to particular Chapters in this monograph in which respective ideas find their detailed exposition. The reader will find in the final Chapter by PAWLAK, POLKOWSKI, and SKOWRON a throughout discussion of the rough set approach to problems of Knowledge Discovery in Data (KDD) along with an introduction to the logical content of KDD and a discussion of the role of rough set based logics in this area of research.

Acknowledgement This Chapter has been prepared with the support of the research grant No 8T 11C 024 17 from the State Committee for Scientific Research (KBN) of the Republic of Poland.

PART 2:

METHODS AND APPLICATIONS: REDUCTS, SIMILARITY, MEREOLOGY

Chapter 2

Rough Set Algorithms in Classification Problem

Jan G. Bazan[1], Hung Son Nguyen[2,3], Sinh Hoa Nguyen[2,3]
Piotr Synak[3], Jakub Wróblewski[3,2]

[1] Institute of Mathematics, Pedagogical University of Rzeszów
Rejtana 16A, 35-310 Rzeszów, Poland
[2] Institute of Mathematics, Warsaw University
Banacha 2, 02-097 Warsaw, Poland
[3] Polish-Japanese Institute of Information Technology
ul. Koszykowa 86, 02-008 Warsaw, Poland
emails: bazan@univ.rzeszow.pl, {son, hoa, jakubw}@mimuw.edu.pl,
synak@pjwstk.waw.pl

Abstract: We we present some algorithms, based on rough set theory, that can be used for the problem of new cases classification. Most of the algorithms were implemented and included in Rosetta system [43]. We present several methods for computation of decision rules based on reducts. We discuss the problem of real value attribute discretization for increasing the performance of algorithms and quality of decision rules. Finally we deal with a problem of resolving conflicts between decision rules classifying a new case to different categories (classes).
Keywords: knowledge discovery, rough sets, classification algorithms, reducts, decision rules, real value attribute discretization

1 Introduction

The term *"classification"* concerns any context in which some decision is taken or a forecast is made on the basis of currently available knowledge or information. A *classification algorithm* is an algorithm which permits us to repeatedly make a forecast on the basis of accumulated knowledge in new situations. We consider here a classification related to construction of a classification algorithm on the basis of current knowledge. Such algorithm is applied then to classify objects previously unseen. Each new object is assigned to a class belonging to a predefined set of classes on the basis of observed values of suitably chosen attributes (features).

Many approaches have been proposed for constructing classification algorithms, among them we would like to point out classical and modern statistical techniques (see e.g. [33], [20]), neural networks (see e.g. [33], [19]), decision trees (see e.g. [13], [48], [58], [60], [33], [49]), decision rules (see e.g. [14], [31], [59], [10], [32], [61], [22], [55], [34], [43]), inductive logic programming (see e.g. [17]).

In this Chapter we present some methods for extracting laws from data, based on rough set approach (see [44]) and Boolean reasoning (see [11]). Most of them were implemented and included in Rosetta system [43]. Results of performed experiments on different kinds of data show that they are very promising (see e.g. [3], [4], [5], [7], [8]).

Standard rough set methods (see [44], [45], [52]) are not always sufficient for extracting laws from data. One of the reasons is that these methods are not taking into account the fact that part of the reduct set (see Section 2) is chaotic i.e. is not stable in randomly chosen samples of a given decision table. We propose a method for selection of feature (attribute) sets relevant for extracting laws from data. These sets of attributes are called dynamic reducts (see [4]). Dynamic reducts are in some sense the most stable reducts of a given decision table, i.e. they are the most frequently appearing reducts in subtables created by random samples of a given decision table.

The most popular method for classification algorithms construction is based on learning rules from examples. The methods based on calculation of all reducts allow to compute, for a given data, the descriptions of concepts by means of decision rules (see [45], [44]). Unfortunately, the decision rules constructed in this way can often be not appropriate to classify unseen cases. We propose a method of the decision rule generation on the basis of dynamic reducts. We suggest that the rules calculated by means of dynamic reducts are better predisposed to classify unseen cases (see [3], [4], [5], [6]).

The Chapter is structured as follows. In Section 2 we present rough set preliminaries. Standard rough set algorithms for synthesis of decision rules from decision tables are described in Section 3.

For purpose of data preprocessing we propose methods for intelligent scaling of real value attributes (see Section 4). Our algorithms are especially useful if the input data table contains continues attributes with many different values. Applying discretization to data decreases further processing time of many methods as well as increases quality of results, e.g. decision rules containing scaled attributes are more general.

Our methods for decision rule generation are based on algorithm for the reduct set computation (see Section 2). Unfortunately, the searching problem for reduct of minimal length (minimal number of attributes) is NP-hard (see [51]). Therefore we often apply approximation algorithms to obtain some knowledge about the reduct set (see [38], [62]). In Section 5 we propose some heuristic for finding local reducts. In Section 6 we present a genetic algorithm for reduct set computation which is very fast and gives its good approximation.

In Sections 7, 8, 9, and 10 we recall the notion of a dynamic reduct and we give some statistical arguments showing that dynamic reducts offer a good tool for extracting laws from decision table.

Some applications of dynamic reducts, e.g. for decision rules generation and dynamic selection of cuts in discretization, are presented in Sections 4, 11 and 12.

In Section 13 we show how the idea of dynamic reducts can be adapted for new methods of dynamic rules computation.

Sometimes the decision rules generated by application of rough set methods can be not acceptable as laws valid for data encoded in a given decision table.

This occurs, e.g. when the number of examples supporting the decision rule (see Section 3) is relatively small. In Section 14 we introduce *approximate rules* to eliminate this drawback. Different methods (e.g. [1], [35], [46], [67]) are now widely used to generate approximate decision rules.

When a set of decision rules has been computed then it is necessary to decide how to resolve conflicts between rule sets classifying tested objects to different decision classes. In Section 15 we present several measures of *the strength of rule set* matched by a given tested object and classifying this object to decision determined by the rules from this set.

In Section 16 we present a general scheme of classification algorithms based on methods and techniques described in previous sections.

2 Rough Set Preliminaries

In this section we recall some basic notions related to information systems and rough sets.

An *information system* is a pair $\mathbf{A} = (U, A)$, where U is a non-empty, finite set called the *universe* and A – a non-empty, finite set of *attributes*, i.e. $a : U \to V_a$ for $a \in A$, where V_a is called the *value set* of a.

Elements of U are called *objects* and interpreted as, e.g. cases, states, processes, patients, observations. Attributes are interpreted as features, variables, characteristic conditions etc.

We also consider a special case of information systems called decision tables. A *decision table* is an information system of the form $\mathbf{A} = (U, A \cup \{d\})$, where $d \notin A$ is a distinguished attribute called *decision*. The elements of A are called *conditions*.

One can interpret the decision attribute as a kind of classifier on the universe of objects given by an expert, a decision-maker, an operator, a physician, etc. In machine learning decision tables are called training sets of examples (see [28]).

The cardinality of the image $d(U) = \{k : d(s) = k \text{ for some } s \in U\}$ is called the *rank of d* and is denoted by $r(d)$.

We assume that the set V_d of values of the decision d is equal to $\{v_d^1, ..., v_d^{r(d)}\}$.

Let us observe that the decision d determines a partition $CLASS_\mathbf{A}(d) = \{X_\mathbf{A}^1, ..., X_\mathbf{A}^{r(d)}\}$ of the universe U, where $X_\mathbf{A}^k = \{x \in U : d(x) = v_d^k\}$ for $1 \leq k \leq r(d)$. $CLASS_\mathbf{A}(d)$ is called the *classification of objects in* $\mathbf{A}$ *determined by the decision d*. The set $X_\mathbf{A}^i$ is called the *i-th decision class of* $\mathbf{A}$. By $X_\mathbf{A}(u)$ we denote the decision class $\{x \in U : d(x) = d(u)\}$, for any $u \in U$.

Let $\mathbf{A} = (U, A)$ be an information system. For every set of attributes $B \subseteq A$, an equivalence relation, denoted by $IND_\mathbf{A}(B)$ and called the B-*indiscernibility relation*, is defined by

$$IND_\mathbf{A}(B) = \{(u, u') \in U^2 : \text{ for every } a \in B, a(u) = a(u')\} \qquad (1)$$

Objects u, u' satisfying the relation $IND_\mathbf{A}(B)$ are indiscernible by attributes from B.

An attribute $a \in B \subseteq A$ is *dispensable* in B if $IND_\mathbf{A}(B) = IND_\mathbf{A}(B \setminus \{a\})$, otherwise a is *indispensable* in B. A set $B \subseteq A$ is *independent* in $\mathbf{A}$ if every attribute from B is indispensable in B, otherwise the set B is *dependent* in $\mathbf{A}$. A set $B \subseteq A$ is called *a reduct* in $\mathbf{A}$ if B is independent in $\mathbf{A}$ and $IND_\mathbf{A}(B) = IND_\mathbf{A}(A)$. The set of all reducts in $\mathbf{A}$ is denoted by $RED(\mathbf{A})$. This is a classical notion of reduct and it is sometimes referred to as *global reduct*.

Let $\mathbf{A} = (U, A)$ be an information system with n objects. By $M(\mathbf{A})$ (see [51]) we denote an $n \times n$ matrix (c_{ij}), called the *discernibility matrix* of $\mathbf{A}$ such that

$$c_{ij} = \{a \in A : a(x_i) \neq a(x_j)\} \text{ for } i,j = 1,\ldots,n. \tag{2}$$

A *discernibility function* $f_\mathbf{A}$ for an information system $\mathbf{A}$ is a boolean function of m boolean variables $\bar{a}_1, \ldots, \bar{a}_m$ corresponding to the attributes $a_1, \ldots, a_m$ respectively, and defined by

$$f_\mathbf{A}(\bar{a}_1, \ldots, \bar{a}_m) = \bigwedge \{\bigvee \bar{c}_{ij} : 1 \leq j < i \leq n, c_{ij} \neq \emptyset\} \tag{3}$$

where $\bar{c}_{ij} = \{\bar{a} : a \in c_{ij}\}$.

It can be shown (see [51]) that the set of all *prime implicants* of $f_\mathbf{A}$ determines the set of all *reducts* of A.

Below we present two deterministic algorithms (see [15]) for computation of the whole reduct set $RED_\mathbf{A}(A)$. Both algorithms compute discernibility matrix of $\mathbf{A}$.

Algorithm 1. *Reduct set computation*
Input:
 Information system $\mathbf{A} = (U, A)$
Output:
 Set $RED_\mathbf{A}(A)$ of all reducts of $\mathbf{A}$
Method:
 Compute indiscernibility matrix $M(\mathbf{A}) = (C_{ij})$
 Reduce M using absorbtion laws
 d - number of non-empty fields $C_1, C_2, .., C_d$ of reduced M
 Build families of sets $R_0, R_1, \ldots, R_d$ in the following way:
begin
  $R_0 = \emptyset$
  for $i = 1$ **to** d
  begin
   $R_i = S_i \cup T_i$, where $S_i = \{R \in R_{i-1} : R \cap C_i \neq \emptyset\}$
    and $T_i = (R \cup \{a\})_{a \in C_i, R \in R_{i-1}: R \cap C_i = \emptyset}$
  end
end
Remove dispensable attributes from each element of family R_d
Remove redundant elements from R_d
 $RED_\mathbf{A}(A) = R_d$
□

It is easy to see that the complexity of this algorithm is exponential. The second algorithm is a modification of Algorithm 1. It allows to stop computation and get a partially computed reduct set.

Algorithm 2. *Reduct set computation with stop possibility*
Input:
   Information system $\mathbf{A} = (U, A)$
Output:
   Set $RED_\mathbf{A}(A)$ of all reducts of $\mathbf{A}$
Method:
   Compute indiscernibility matrix $M(\mathbf{A}) = (C_{ij})$
   Reduce M using absorbtion laws
   d - number of non-empty fields of reduced M
   Build a families of sets $R_0, R_1, ..., R_d$ in the following way:
   begin
      Compute R_1 (see Algorithm 1)
      $i = 1$
      while $i > 0$ **do**
      begin
         if *stop* **then return**
         if $R_i = \emptyset$ **then**
         begin
            $i = i - 1$
            continue
         end
         Remove from family R_i the first element
         Compute R_{i+1} (see Algorithm 1)
         $i = i + 1$
         if $i = d$ **then**
         begin
            Remove from R_d redundant elements
            $RED_\mathbf{A}(A) = RED_\mathbf{A}(A) \cup R_d$
            $i = i - 1$
         end
      end
   end
□

If $\mathbf{A} = (U, A)$ is an information system, $B \subseteq A$ is a set of attributes and $X \subseteq U$ is a set of objects, then the sets: $\{u \in U : [u]_B \subseteq X\}$ and $\{u \in U : [u]_B \cap X \neq \emptyset\}$ are called the *B-lower* and the *B-upper approximation* of X in $\mathbf{A}$, and they are denoted by $\underline{B}X$ and $\overline{B}X$, respectively.

The set $BN_B(X) = \overline{B}X - \underline{B}X$ is called the *B-boundary* of X. When $B = A$ we also write $BN_\mathbf{A}(X)$ instead of $BN_A(X)$.

Sets which are unions of some classes of the indiscernibility relation $IND_\mathbf{A}(B)$ are called *definable* by B (or, *B-definable*, in short). A set X is thus B-definable iff $\overline{B}X = \underline{B}X$. Some subsets (categories) of objects in an information system cannot be exactly expressed by employing available attributes but they can be defined roughly.

The set $\underline{B}X$ is the set of all elements of U which can be classified with certainty as elements of X, having the knowledge about them represented by attributes from B; the set $BN_B(X)$ is the set of elements which one can classify neither to X nor to $-X$ having knowledge about objects represented by B.

If $X_1, \ldots, X_{r(d)}$ are decision classes of **A** then the set $\underline{B}X_1 \cup \ldots \cup \underline{B}X_{r(d)}$ is called *the B-positive region of* **A** and is denoted by $POS_B(d)$.

If $\mathbf{A} = (U, A \cup \{d\})$ is a decision table and $B \subseteq A$, then we define a function $\partial_B : U \to \mathbf{P}(V_d\})$, called *the B-generalized decision of* **A**, by

$$\partial_B(x) = \{v \in V_d : \exists x' \in U \ (x' IND_\mathbf{A}(B)x \text{ and } d(x) = v)\} . \tag{4}$$

The A-generalized decision ∂_A of **A** is called the generalized decision of **A**.

A decision table **A** is called *consistent (deterministic)* if $card(\partial_A(x)) = 1$ for any $x \in U$, otherwise **A** is *inconsistent (non-deterministic)*. It is easy to see that a decision table **A** is consistent iff $POS_A(d) = U$. Moreover, if $\partial_B = \partial_{B'}$ then $POS_B(d) = POS_{B'}(d)$ for any pair of non-empty sets $B, B' \subseteq A$.

A subset B of the set A of attributes of a decision table $\mathbf{A} = (U, A \cup \{d\})$ is *a relative reduct of* **A** iff B is a minimal set with respect to the following property: $\partial_B = \partial_A$. The set of all relative reducts of **A** is denoted by $RED(\mathbf{A}, d)$.

Let $\mathbf{A} = (U, A \cup \{d\})$ be a consistent decision table and let $M(\mathbf{A}) = (c_{ij})$ be its discernibility matrix. We construct a new matrix $M'(\mathbf{A}) = (c'_{ij})$ assuming $c'_{ij} = \emptyset$ if $d(x_i) = d(x_j)$ and $c'_{ij} = c_{ij} - \{d\}$, otherwise. The matrix $M'(\mathbf{A})$ is called *the relative discernibility matrix of* **A**. Now one can construct the *relative discernibility function* $f_{M'(A)}$ of $M'(\mathbf{A})$ in the same way as the discernibility function has been constructed from the discernibility matrix.

It can be shown (see [51]) that the set of all *prime implicants* of $f_{M'(A)}$ determines the set of all *relative reducts* of A.

Another important type of reducts are local reducts. A *local reduct* $r(x_i) \subseteq A$ (or a *reduct relative to decision and object* $x_i \in U$; where x_i is called a *base object*) is a subset of A such that:

a) $\forall\ x_j \in U,\ d(x_i) \neq d(x_j) \implies \exists\ a_k \in r(x_i): a_k(x_i) \neq a_k(x_j)$

b) $r(x_i)$ is minimal with respect to inclusion.

If $\mathbf{A} = (U, A \cup \{d\})$ is a decision table then any system $\mathbf{B} = (U', A \cup \{d\})$ such that $U' \subseteq U$ is called a *subtable* of **A**.

The problem of new cases classification can be described in the following way. Let $\mathbf{W} = (W, A \cup \{d\})$ be a hypothetical *universal decision table* (including known and unknown objects describing an actual considered aspect of reality) and let $\mathbf{A} = (U, A \cup \{d\})$ be a given subtable of the universal decision table. Let $u \in W$ be a so called *tested object*. Our task consists in assigning the value $d(u)$ of the decision d to the tested objects u, knowing only values of condition attributes of u, and relying on a given decision table **A**. In another words we would like to classify object u to the proper decision class $X_k \in CLASS_\mathbf{A}(d)$ (where $k \in 1, \ldots, r(d)$) on the basis of knowledge included in **A**. A solution of this problem is a classification algorithm sufficiently approximating the decision function d.

3 Decision Rules

Let $\mathbf{A} = (U, A \cup \{d\})$ be a decision table and let $V = \bigcup \{V_a : a \in A\} \cup V_d$.

Atomic formulas over $B \subseteq A \cup \{d\}$ and V are expressions of the form $a = v$, called *descriptors* over B and V, where $a \in B$ and $v \in V_a$. The set $\mathbf{F}(B, V)$ of formulas over B and V is the least set containing all atomic formulas over B and V and closed with respect to the classical propositional connectives $\vee$ (disjunction), $\wedge$ (conjunction), and $\neg$ (negation).

Let $\varphi \in \mathbf{F}(B, V)$. Then by $|\varphi|_\mathbf{A}$ we denote the meaning of φ in a decision table $\mathbf{A}$, i.e. the set of all objects in U with property φ, defined inductively by

1. if φ is of the form $a = v$ then $|\varphi|_\mathbf{A} = \{x \in U : a(x) = v\}$;
2. $|\varphi \wedge \varphi'|_\mathbf{A} = |\varphi|_\mathbf{A} \cap |\varphi'|_\mathbf{A}$; $|\varphi \vee \varphi'|_\mathbf{A} = |\varphi|_\mathbf{A} \cup |\varphi'|_\mathbf{A}$; $|\neg \varphi|_\mathbf{A} = U - |\varphi\}_\mathbf{A}$

The set $\mathbf{F}(A, V)$ is called the set of *conditional formulas of* $\mathbf{A}$ and is denoted by $\mathbf{C}(A, V)$.

Any formula of the form $(a_1 = v_1) \wedge ... \wedge (a_l = v_l)$, where $v_i \in V_{a_i}$ (for $i = 1, ..., l$) and $P = \{a_1, ..., a_l\} \subseteq A$, is called a *P-basic* formula of $\mathbf{A}$.

If φ is a *P-basic* formula of $\mathbf{A}$ and $Q \subseteq P$, then by φ/Q we mean the Q-basic formula obtained from the formula φ by removing from φ all its elementary subformulas $(a = v_a)$ such that $a \in P \setminus Q$.

A *decision rule* for $\mathbf{A}$ is any expression of the form $\varphi \Rightarrow d = v$ where $\varphi \in \mathbf{C}(A, V)$, $v \in V_d$ and $|\varphi|_\mathbf{A} \neq \emptyset$. Formulas φ and $d = v$ are referred to as the *predecessor* and the *successor* of the decision rule $\varphi \Rightarrow d = v$ respectively.

If r is a decision rule in $\mathbf{A}$, then by $Pred(r)$ we denote the predecessor of r and by $Succ(r)$ we denote the successor of r.

An object $u \in U$ is *matched* by a decision rule $\varphi \Rightarrow d = v_d^k$ (where $1 \leq k \leq r(d)$) iff $u \in |\varphi|_\mathbf{A}$. If u is matched by $\varphi \Rightarrow d = v_d^k$ then we say that the rule is classifying u to decision class X_k.

The number of objects matched by a decision rule $\varphi \Rightarrow d = v$, denoted by $Match_\mathbf{A}(\varphi \Rightarrow d = v)$ is equal $card(|\varphi|_\mathbf{A})$.

The number $Supp_\mathbf{A}(\varphi \Rightarrow d = v) = card(|\varphi|_\mathbf{A} \cap |d = v|_\mathbf{A})$ is called *the number of objects supporting a decision rule* $\varphi \Rightarrow d = v$.

A decision rule $\varphi \Rightarrow d = v$ for $\mathbf{A}$ is *true* in $\mathbf{A}$, symbolically $\varphi \Rightarrow_\mathbf{A} d = v$, iff $|\varphi|_\mathbf{A} \subseteq |d = v|_\mathbf{A}$. If a decision rule $\varphi \Rightarrow d = v$ is true in $\mathbf{A}$, we say that the decision rule is *consistent* in $\mathbf{A}$, otherwise the decision rule $\varphi \Rightarrow d = v$ is *inconsistent* or *approximate* in $\mathbf{A}$.

If r is a decision rule in $\mathbf{A}$, then the number $\mu_\mathbf{A}(r) = \frac{Supp_\mathbf{A}(r)}{Match_\mathbf{A}(r)}$ is called *the coefficient of consistency* of the rule r. The coefficient $\mu_\mathbf{A}(r)$ can be understood as a degree of consistency of the decision rule r. It is easy to see that a decision rule r for $\mathbf{A}$ is consistent iff $\mu_\mathbf{A}(r) = 1$.

The coefficient of consistency of r can be also treated as a degree of inclusion of $|Pred(r)|_\mathbf{A}$ in $|Succ(r)|_\mathbf{A}$ (see [47]).

If $\varphi \Rightarrow d = v$ is a decision rule for $\mathbf{A}$ and φ is P-basic formula of $\mathbf{A}$ (where $P \subseteq A$), then the decision rule $\varphi \Rightarrow d = v$ is called a *P-basic decision rule for* $\mathbf{A}$, or *basic decision rule* in short.

Let $\mathbf{A} = (U, A \cup \{d\})$ be a consistent decision table with n objects $u_1, ..., u_n$ and with m condition attributes $a_1, ..., a_m$. We present two methods for basic decision rule synthesis for $\mathbf{A}$ (see [52], [53], [45]).

The first method consists of two steps. In the first step we compute the set $RED(\mathbf{A}, d)$ of all relative reducts of decision table $\mathbf{A}$. In the second step, for each reduct $R = \{b_1, ..., b_l\} \in RED(\mathbf{A}, d)$ (where $l \leq m$) and any object $u \in U$ we generate a decision rule in the following way: as the predecessor of the decision rule we take the conjunction $(b_1 = b_1(u)) \wedge ... \wedge (b_l = b_l(u))$ and as the successor of the rule we take the decision attribute d with the value $d(u)$. Hence, the constructed decision rule for the reduct R and the object u is of the form

$$(b_1 = b_1(u)) \wedge ... \wedge (b_l = b_l(u)) \Rightarrow d = d(u).$$

The second method returns basic decision rules with minimal number of descriptors (see [45], [52]).

Let $\varphi \Rightarrow d = v$ be a P-basic decision rule of $\mathbf{A}$ (where $P \subseteq A$) and let $a \in P$. We say that the attribute a is *dispensable* in the rule $\varphi \Rightarrow d = v$ iff $|\varphi \Rightarrow d = v|_\mathbf{A} = U$ implies $|\varphi/(P \setminus \{a\}) \Rightarrow d = v|_\mathbf{A} = U$, otherwise attribute a is *indispensable* in the rule $\varphi \Rightarrow d = v$. If all attributes $a \in P$ are indispensable in the rule $\varphi \Rightarrow d = v$, then $\varphi \Rightarrow d = v$ is called *independent* in $\mathbf{A}$.

The subset of attributes $R \subseteq P$ is called a *reduct* of a P-basic decision rule $\varphi \Rightarrow d = v$, if $\varphi/R \Rightarrow d = v$ is independent in $\mathbf{A}$ and $|\varphi \Rightarrow d = v|_\mathbf{A} = U$ implies $|\varphi/R \Rightarrow d = v|_\mathbf{A} = U$. If R is a reduct of the P-basic decision rule $\varphi \Rightarrow d = v$, then $\varphi/R \Rightarrow d = v$ is said to be *reduced*. If R is a reduct of the A-basic decision rule $\varphi \Rightarrow d = v$, then $\varphi/R \Rightarrow d = v$ is said to be *an optimal basic decision rule of* $\mathbf{A}$ (*a basic decision rule with minimal number of descriptors*). The set of all optimal basic decision rules of $\mathbf{A}$ is denoted by $RUL(\mathbf{A})$.

Let $\mathbf{A} - (U, A \cup \{d\})$ be a consistent decision table and $M'(\mathbf{A}) = (c'_{ij})$ be its relative discernibility matrix. We construct a new matrix $M(\mathbf{A}, k) = (c^k_{ij})$ for any $x_k \in U$ assuming $c^k_{ij} = c'_{ij}$ if $d(x_i) \neq d(x_j) \& (i = k \vee j = k)$ and $c^k_{ij} = \emptyset$, otherwise. The matrix $M(\mathbf{A}, k)$ is called *the k-relative discernibility matrix of* $\mathbf{A}$. Now one can construct the *k-relative discernibility function* $f_{M(\mathbf{A},k)}$ of $M(\mathbf{A}, k)$ in the same way as the discernibility function has been constructed from the discernibility matrix (see Section 2).

It can be shown that the set of all *prime implicants* of functions $f_{M(\mathbf{A},k)}$ (for $k = 1, ..., card(U)$) determines the set of all basic optimal rules of $\mathbf{A}$.

Let us assume now that considered decision tables are inconsistent. One can transform an arbitrary inconsistent decision table $\mathbf{A} = (U, A \cup \{d\})$ into a consistent decision table $\mathbf{A}_\partial = (U, A \cup \{\partial_A\})$ where $\partial_A : U \to \mathbf{P}(V_d)$ is the generalized decision of $\mathbf{A}$ defined in Section 2. It is easy to see that $\mathbf{A}_\partial$ is a consistent decision table and one can apply to $\mathbf{A}_\partial$ the described methods to construct decision rules. Hence one can compute decision rules for any inconsistent decision table.

4 Discretization

Suppose we have a decision table $\mathbf{A} = (U, A \cup \{d\})$ where $card(V_a)$ is large for some $a \in A$. Then there is a very low chance that a new object is recognized by rules generated directly from this table, because the attribute value vector of a new object will not match any of these rules. Therefore for decision tables with real value attributes some discretization strategies are built to obtain a higher quality of classification rules. This problem is intensively studied in e.g. [39] and we consider discretization methods presented in [39], [37], [38] and [41]. These methods are based on rough set techniques and boolean reasoning.

4.1 Basic Notions

Let $\mathbf{A} = (U, A \cup \{d\})$ be a decision table where $U = \{x_1, x_2, \ldots, x_n\}$. We assume $V_a = [l_a, r_a) \subset \mathbf{R}$ for any $a \in A$ where $\mathbf{R}$ is the set of real numbers. In the sequel we assume that $\mathbf{A}$ is a consistent decision table.

Any pair (a, c), where $a \in A$ and $c \in \mathbf{R}$, defines a partition of V_a into *left-hand-side* and *right-hand-side interval*. Formally, any attribute-value pair (a, c) is associated with a new binary attribute $f_{(a,c)}: U \to \{0, 1\}$ such that

$$f_{(a,c)}(u) = \begin{cases} 0 & \text{if } a(u) < c \\ 1 & \text{otherwise} \end{cases} \quad (5)$$

In this sense, the pair (a, c) is called a *binary discriminators* or *cut on V_a*. Usually, discretization of real value attributes is determined by cuts.

Let us fix an attribute $a \in A$. Any set of cuts

$$D_a = \{(a, c_1^a), (a, c_2^a), \ldots, (a, c_{k_a}^a)\}$$

where $k_a \in \mathbf{N}$ and $c_0^a = l_a < c_1^a < c_2^a < \ldots < c_{k_a}^a < r_a = c_{k_a+1}^a$, defines a partition on V_a (for $a \in A$) into sub-intervals i.e.

$$V_a = [c_0, c_1^a) \cup [c_1^a, c_2^a) \cup \ldots \cup [c_{k_a}^a, c_{k_a+1}^a).$$

The set of cuts D_a on a defines a discretization of a, i.e. new discreet attribute $a_{D_a}: U \to \{0, .., k_a\}$ such that

$$a_{D_a}(x) = i \iff a(x) \in [c_i^a, c_{i+1}^a)$$

for any $x \in U$ and $i \in \{0, .., k_a\}$ (see Figure 1).

Analogously, any global discretization is determined by a set of cuts on all real value attributes $\mathbf{D} = \bigcup_{a \in A} D_a$. Any set of cuts

$$\mathbf{D} = \bigcup_{a_i \in A} D_{a_i} = \{(a_1, c_1^1), \ldots, (a_1, c_{k_1}^1)\} \cup \{(a_2, c_1^2), \ldots, (a_2, c_{k_2}^2)\} \cup \ldots$$

transforms the original decision table $\mathbf{A} = (U, A \cup \{d\})$ into new (discreet) decision table $\mathbf{A}|_\mathbf{D} = (U, A_\mathbf{D} \cup \{d\})$, where $A_\mathbf{D} = \{a_{D_a}: a \in A\}$. The table $\mathbf{A}|_\mathbf{D}$ is called the $\mathbf{D}$-*discretized table of* $\mathbf{A}$.

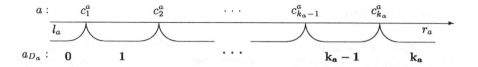

Fig. 1. The discretization of real value attribute $a \in A$ defined by the set of cuts $\{(a, c_1^a), (a, c_2^a), \ldots, (a, c_{k_a}^a)\}$

It is obvious, that discretization process is associated with a loss of information. Usually, the task of discretization is to determine a minimal set of cuts **D** (with respect to inclusion) from a given decision table **A** such that, in spite of losing information, the **D**-discretized table $\mathbf{A}|_\mathbf{D}$ still keeps some useful properties of **A**. In the discretization method based on rough set and Boolean reasoning approach, we are trying to keep the discernibility between objects.

Let $\mathbf{A} = (U, A \cup \{d\})$ be a given decision table and **D** be a set of cuts. We say that

- **D** is *consistent with* **A** (or **A**-*consistent*) if $\partial_\mathbf{A} = \partial_{\mathbf{A}|_\mathbf{D}}$, where ∂_A and $\partial_{\mathbf{A}|_\mathbf{D}}$ are generalized decisions of **A** and $\mathbf{A}|_\mathbf{P}$. In other words, the set of cuts is **A**-consistent iff for any two objects $u, v \in U$:

    if $(u, v$ are discerned by $A)$ **then** $(u, v$ are discerned by $D)$.

- **D** is *irreducible* in **A** if **D** is **A**-consistent and **D**' is not **A**-consistent for any proper subset $\mathbf{D}' \subset \mathbf{D}$.
- **D** is *optimal* in **A** if $card(\mathbf{D}) \leq card(\mathbf{D}')$ for any **A**-consistent set of cuts **D**'.

The problem of searching for optimal set of cuts (Optimal Discretization Problem) has been explored in [39], [37], [38] and [41]. From computational complexity point of view, the Optimal Discretization Problem appears to be hard. We have the following theorem (see [39]):

Theorem 1. *For a given decision table* **A** *and an integer* k.

- *The decision problem for checking if there exists an irreducible set of cuts* **P** *in* **A** *such that* $card(\mathbf{P}) < k$ *(***k$-$minimal partition problem***) is NP-complete.*
- *The problem of searching for an optimal set of cuts* **P** *in* **A** *(***optimal partition problem***) is NP-hard.*

Two sets of cuts **D**', **D** are equivalent (denoted by $\mathbf{D}' \equiv_\mathbf{A} \mathbf{D}$, if and only if $\mathbf{A}|_\mathbf{D} = \mathbf{A}|_{\mathbf{D}'}$. The equivalence relation $\equiv_\mathbf{A}$ has finite number of equivalence classes. In the sequel we do not distinguish between equivalent sets of cuts.

4.2 Maximal Discernibility (MD) Heuristics

We below describe our heuristic for optimal discretization problem.

Let $\mathbf{A} = (U, A \cup \{d\})$ be a decision table. An arbitrary attribute $a \in A$ defines a sequence $v_1^a < v_2^a < \ldots < v_{n_a}^a$, where $\{v_1^a, v_2^a, \ldots, v_{n_a}^a\} = \{a(x) : x \in U\}$ and $n_a \leq n$. Then the set of all possible cuts on a is denoted by

$$\mathbf{C}_a = \left\{ \left(a, \frac{v_1^a + v_2^a}{2}\right), \left(a, \frac{v_2^a + v_3^a}{2}\right), \ldots, \left(a, \frac{v_{n_a-1}^a + v_{n_a}^a}{2}\right) \right\}$$

The set of possible cuts on all attributes is denoted by

$$\mathbf{C}_A = \bigcup_{a \in A} \mathbf{C}_a$$

In [39] we have shown that any irreducible set of cuts of $\mathbf{A}$ is a relative reduct of another decision table $\mathbf{A_1}$ built from $\mathbf{A}$, where $\mathbf{A_1} = (U_1, A_1 \cup \{d_1\})$ is defined as follows:

- $U_1 = \{(x, y) \in U \times U : d(x) \neq d(y)\} \cup \{new\}$, where $new \notin U \times U$.
- $d_1 : U_1 \to \{0, 1\}$ is defined by $d_1(u) = \begin{cases} 0 & \text{if } u = new \\ 1 & \text{otherwise} \end{cases}$
- $A_1 = \{f_{(a,c)} : (a,c) \in \mathbf{C}_A\}$ is a set of all discriminators defined by initial set of cuts from $\mathbf{C}_A$ (see Equation (5)).

The algorithm based on Johnson's strategy described in the previous section is searching for a cut $c \in A_1$ which discerns the largest number of pairs of objects. Then we move the cut c from A_1 to the resulting set of cuts $\mathbf{P}$ and remove from U_1 all pairs of objects discerned by c. Our algorithm is continued until $U_1 = \{new\}$. Now we present the details of our algorithm.

Algorithm 3: *MD-discretization*

Input: *The consistent decision table $\mathbf{A}$.*
Output: *The semi-minimal set of cuts $\mathbf{D}$ consistent with $\mathbf{A}$.*
Method:
    $\mathbf{D} = \emptyset$; $\mathbf{C}_A = $ *initial set of cuts on $\mathbf{A}$;*
    $\mathbf{L} = \{(x, y) \in U \times U : d(x) \neq d(y)\}$;
    while $(\mathbf{L} \neq \emptyset)$ **do**
    begin
        Choose the cut $c_{\max} \in \mathbf{C}_A$ which discerns the largest number of pairs of objects in $\mathbf{L}$. (i.e. the cut corresponding to the column with the largest number of "1")
        Insert c_{max} into $\mathbf{D}$ and remove it from $\mathbf{C}_A$.
        Remove all pairs of objects from $\mathbf{L}$ discerned by c_{max}.
    end

□

A	a	b	d
u_1	1	2	1
u_2	1.2	0.5	0
u_3	1.3	3	0
u_4	1.4	1	1
u_5	1.4	2	0
u_6	1.6	3	1
u_7	1.3	1	1

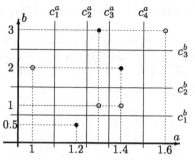

Fig. 2. The decision table with two real value attributes and its geometric interpretation

A_1	$f_{(a,1.1)}$	$f_{(a,1.25)}$	$f_{(a,1.35)}$	$f_{(a,1.5)}$	$f_{(b,0.75)}$	$f_{(b,1.5)}$	$f_{(b,2.5)}$	d_1
(u_1, u_2)	1	0	0	0	1	1	0	1
(u_1, u_3)	1	1	0	0	0	0	1	1
(u_1, u_5)	1	1	1	0	0	0	0	1
(u_4, u_2)	0	1	1	0	1	0	0	1
(u_4, u_3)	0	0	1	0	0	1	1	1
(u_4, u_5)	0	0	0	0	0	1	0	1
(u_6, u_2)	0	1	1	1	1	1	1	1
(u_6, u_3)	0	0	1	1	0	0	0	1
(u_6, u_5)	0	0	0	1	0	0	1	1
(u_7, u_2)	0	1	0	0	1	0	0	1
(u_7, u_3)	0	0	0	0	0	1	1	1
(u_7, u_5)	0	0	1	0	0	1	0	1
new	0	0	0	0	0	0	0	0

Fig. 3. Table A_1 constructed from table A

Example 1. We consider the decision table with two attributes and seven objects (Figure 2).
The set of all cuts consists of four cuts on the attribute a and three cuts on the attribute b.

$$\mathbf{C}_A = \{(a, 1.1), (a, 1.25), (a, 1.35), (a, 1.5)\} \cup \{(b, 0.75), (b, 1.5), (b, 2.5)\}$$

The new decision table $\mathbf{A}_1$ consists of 7 attributes and 13 objects (Figure 3). The MD-discretization algorithm chooses the cut $(b, 1.5)$ first because it discerns 6 pairs of objects, than it chooses the cuts $(a, 1.25)$ and $(a, 1.5)$. The result of the MD-discretization algorithm is the set of cuts $\mathbf{D} = \{(a, 1.25), (a, 1.5), (b, 1.5)\}$.

Let n be a number of objects and k be a number of attributes of decision table $\mathbf{A}$, then $card(A_1) \leq (n-1)k$ and $card(U_1) \leq \frac{n(n-1)}{2}$. It is easy to note that for any cut $c \in A_1$ we need $O(n^2)$ steps to find the number of all pairs of objects discerned by c. Hence the straightforward realization of this algorithm requires $O(kn^2)$ of memory space and $O(kn^3)$ steps to determine one *cut*, so it is not useful in practice. The algorithm presented below determines the best cut in $O(kn)$ steps using $O(kn)$ space only.

At first we show that the number of pairs of objects discerned by a given cut can be computed faster than $O(n^2)$. For the given cut $(a,c) \in \mathbf{C}_A$ on an attribute $a \in A$ and a given subset of objects $X \subseteq U$ we introduce the following notation:

1. $W^X(a,c) =$ number of pairs of objects from X discerned by (a,c).
2. for $j = 1, ..., r$:
   $l_j^X(a,c) = card\{x \in X : [a(x) < c] \wedge [d(x) = j]\}$ and
   $r_j^X(a,c) = card\{x \in X : [a(x) > c] \wedge [d(x) = j]\}$
   are numbers of objects from X belonging to the jth decision class and being on the left-hand-side and right-hand-side of the cut (a,c), respectively.
3. The number of objects on either side of (a,c) is:
   $l^X(a,c) = \sum_{j=1}^{r} l_j^X(a,c) = card\{x \in X : a(x) < c\}$;
   $r^X(a,c) = \sum_{j=1}^{r} r_j^X(a,c) = card\{x \in X : a(x) > c_m^a\}$.

We obtain the following lemma:

Lemma 2. *For any cut $(a,c) \in \mathbf{C}_A$, and $X \subseteq U$:*

$$W^X(a,c) = l^X(a,c) \cdot r^X(a,c) - \sum_{i=1}^{r} l_i^X(a,c) \cdot r_i^X(a,c)$$

This Lemma shows, that the number of pairs of objects discerned by a cut $(a,c) \in \mathbf{C}_A$ can be computed in $O(n)$ time. The next theorem is showing that if (a, c_m) and (a, c_{m+1}) are consecutive cuts on the attribute a, than the value $W^X(a, c_{m+1})$ can be derived from $W^X(a, c_m)$ in $O(1)$ time (because the attribute a is fixed, we will use the notations: $l_j^X(c)$, $r_j^X(c)$, $l^X(c)$, $r^X(c)$ and $W^X(c)$ for simplification, instead of $l_j^X(a,c)$, $r_j^X(a,c)$, $l^X(a,c)$, $r^X(a,c)$ and $W^X(a,c)$).

Theorem 3. *If there is exactly one object $x \in X \subseteq U$ such that $(c_m < a(x) < c_{m+1}$ and let $t = d(x)$ then:*

$$W^X(c_{m+1}) = W^X(c_m) + [r^X(c_m) - l^X(c_m)] - [r_t^X(c_m) - l_t^X(c_m)]$$

Proof: We have:

$$l_j^X(c_{m+1}) = \begin{cases} l_j^X(c_m) & \text{if } j \neq t \\ l_t^X(c_m) + 1 & \text{if } i = t \end{cases} \quad \text{and} \quad r_j^X(c_{m+1}) = \begin{cases} r_j^X(c_m) & \text{if } j \neq t, \\ r_t^X(c_m) - 1 & \text{if } j = t. \end{cases}$$

From those equations it follows that

$$l^X(c_{m+1}) = l^X(c_m) + 1 \text{ and } r^X(c_{m+1}) = r^X(c_m) - 1.$$

From Lemma 2 we have:

$$W^X(c_m) = l^X(c_m) r^X(c_m) - \sum_{i=1}^{r} l_i^X(c_m) r_i^X(c_m);$$

$$W^X(c_{m+1}) = l^X(c_{m+1}) r^X(c_{m+1}) - \sum_{1}^{r} l_i^X(c_{m+1}) r_i^X(c_{m+1})$$

Thus

$$\begin{aligned}
W^X(c_{m+1}) - W^X(c_m) &= l^X(c_{m+1})\,r^X(c_{m+1}) - l^X(c_m)\,r^X(c_m) + \\
&\quad l_t^X(c_m)\,r_t^X(c_m) - l_t^X(c_{m+1})\,r_t^X(c_{m+1}) \\
&= [l^X(c_m) + 1] \cdot [r^X(c_m) - 1] - l^X(c_m)\,r^X(c_m) + \\
&\quad l_t^X(c_m)\,r_t^X(c_m) - [l_t^X(c_m) + 1] \cdot [r_t^X(c_m) - 1] \\
&= [r^X(c_m) - l^X(c_m)] - [r_t^X(c_m) - l_t^X(c_m)].
\end{aligned}$$

□

Let us assume that the number $W^X(a, c_m)$ of pairs of objects from X discerned by the cut (a, c_m) has been determined. Theorem 3 shows that we can compute the value $W^X(a, c_{m+1})$ in time $O(1)$. In the consequence the best cut on any attribute can be determined in time $O(n)$.

We propose two strategies of searching for semi-optimal set of cuts. The first, called *local strategy*, after finding the best cut and dividing the object set into two subsets of objects, repeats this procedure for each object set separately until some stop condition holds. The quality of a cut (i.e. number of objects discerned by cut) in local strategy is computed locally on a subset of objects. In the second strategy, called *global strategy*, quality of cuts are computed on whole set of objects. Usually, the local strategy is easier for realization than the global one, but the set of cuts obtained by the global strategy is smaller.

4.3 Local Strategy

The local strategy can be realized by using *decision tree*. A typical algorithm for decision tree generation for a given decision table $\mathbf{A} = (U, A \cup \{d\})$ is described below:

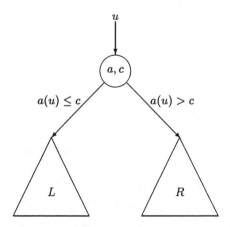

Fig. 4. The decision tree used for local discretization

Algorithm 4: *Local discretization*

Input: *The consistent decision table* **A**.
Output: *The semi-minimal set of cuts* **D** *consistent with* **A**.
Method:
   Initialize the binary tree variable **T** *with the empty tree.*
   Label the root by the set of all objects U and fix the status of the root to be unready.
   while *there is a leaf marked by* unready **do**
   begin
     for *any* unready *leave N of the tree* **T**
     begin
       if *objects labeling N have the same decision value* **then**
       begin
         Replace the object set at N by its common decision
         Change the status of N to ready.
       end
       else
       begin
         Compute the value $W^N(a,c)$ for all cuts from $\mathbf{C}_A$ and search for cut (a^*, c^*) maximizing the function $W^N(.)$, i.e.

$$(a^*, c^*) = \arg\max_{(a,c)} W^N(a, c)$$

         Replace the label of N by (a^*, c^*) and mark it as ready;
         Create two new nodes N_1 and N_2 with status unready *as the left and right subtrees of N, where:*

$$N_1 = \{u \in N : a^*(u) < c^*\} \quad \text{and} \quad N_2 = \{u \in N : a^*(u) \geq c^*\}$$

     end
   end
end
return T

□

4.4 Global Strategy

Now we describe some properties of the **D**-discretized table $\mathbf{A}_\mathbf{D}$, where $\mathbf{D} \subseteq \mathbf{C}_A$ is an arbitrary set of cuts. Assuming $X_1, X_2, ..., X_m$ to be equivalence classes of the discernibility relation $IND(A_\mathbf{D})$ of table $\mathbf{A}|_\mathbf{D}$, one can note that the family **PART(D)** $= \{X_1, X_2, ..., X_m\}$ defines a partition of the set of objects U into m disjoint subsets i.e.

$$U = X_1 \cup ... \cup X_m \text{ and } \forall_{i \neq j} X_i \cap X_j = \emptyset$$

In practice the classes $X_1, .., X_m$ are stored in memory instead of $\mathbf{A}|_{\mathbf{D}}$. Observe that objects from X_i ($i = 1, .., m$) are not discerned by any cut from $\mathbf{D}$. Hence the number of pairs of objects discerned by a cut $c \notin \mathbf{D}$ but not discerned by cuts from $\mathbf{D}$ is equal to

$$W_{\mathbf{D}}(a, c) = W^{X_1}(a, c) + W^{X_2}(a, c) + ... + W^{X_m}(a, c). \tag{6}$$

Let us consider the situation, when we have the set of cuts $\mathbf{D}$ defining the equivalence classes $X_1, .., X_m$ and two consecutive cuts c_j^a, c_{j+1}^a on the attributes a. We can derive the value $W_{\mathbf{D}}(a, c_{j+1}^a)$ from $W_{\mathbf{D}}(a, c_j^a)$ in time $O(1)$ applying the following theorem:

Theorem 4. *Let $\mathbf{D}$ be a given set of cuts. If there is exactly one object $x \in U$ such that $a(x) \in (c_j^a, c_{j+1}^a)$ then:*

$$W_{\mathbf{D}}(a, c_{j+1}^a) - W_{\mathbf{D}}(a, c_j^a) = W^{X_i}(a, c_{j+1}^a) - W^{X_i}(a, c_j^a) =$$
$$= \left(r^{X_i}(c_j^a) - l^{X_i}(c_j^a)\right) - \left(r_t^{X_i}(c_j^a) - l_t^{X_i}(c_j^a)\right)$$

where $t = d(x)$ and $X_i \in \mathbf{PART}(\mathbf{D})$ is the equivalence class containing x.

Proof: This fact follows from Theorem 3 and Equation 6. □

Now we present the details of our algorithm.

Algorithm 5: *Global discretization*

Input: *The consistent decision table $\mathbf{A}$.*
Output: *The semi-minimal set of cuts $\mathbf{P}$ consistent with $\mathbf{A}$.*
Data Structure: $\mathbf{D}$ – *the semi-minimal set of cuts;* $\mathbf{L} = \mathbf{PART}(\mathbf{D})$ – *the partition of U defined by $\mathbf{D}$; $\mathbf{C}_A$ – the set of all possible cuts on $\mathbf{A}$.*
Method:
  1. $\mathbf{D} = \emptyset; \mathbf{L} = \{U\}$; $A_1 = $ *initial set of cuts on $\mathbf{A}$*
  2. *Compute the value $W_{\mathbf{D}}(a, c)$ for all cuts from $\mathbf{C}_A$ and search for cut (a^*, c^*) maximizing the function $W_{\mathbf{D}}(.)$, i.e.*

$$(a^*, c^*) = \arg\max_{(a,c)} W_{\mathbf{D}}(a, c)$$

  3. *Set*

$$\mathbf{D} = \mathbf{D} \cup \{(a^*, c^*)\}; \mathbf{C}_A = \mathbf{C}_A \setminus \{(a^*, c^*)\}$$

  4. **for** $X \in \mathbf{L}$ **do**
      if X consists of objects from one decision class **then** *remove X from $\mathbf{L}$;*
      if (a^*, c^*) *divides the set X into X_1, X_2* **then**
        – *Remove X from $\mathbf{L}$*
        – *Add to $\mathbf{L}$ two sets X_1, X_2*
  5. **if** $\mathbf{L}$ *is empty* **then** *Stop* **else** *Go to 2.*

□

Theorem 5. *Algorithm 5 needs time of order $O(kn(|\mathbf{P}|+\log n))$ and $O(kn)$ memory space for computing of the semi-minimal set of cuts $\mathbf{P}$.*

Proof: Step 1 takes $O(kn \log n)$ steps. From Theorem 4 it follows that Loop 2 and Loop 4 require together $O(kn)$ memory space and $O(kn)$ time. □

Example 2. We illustrate the local and global strategy on the decision table presented in the Figure 5.

In both cases our algorithms begin with choosing the best cut $(a_3, 4.0)$ discerning 20 pairs of objects from $\mathbf{A}$. Theorem 3 assures that this cut can be found in linear time.

In the local strategy, the cut $(a_3, 4.0)$ divides the set of objects into two subsets $U_1 = \{u_1, u_3, u_4, u_5, u_9, u_{10}\}$ and $U_2 = \{u_2, u_6, u_7, u_8\}$. Then the local discretization algorithm chooses best cuts locally on U_1 and U_2. The cuts $(a_1, 3.5)$, $(a_2, 3.5)$ and $(a_2, 5.5)$ are best cuts for U_1 (because they are discerning 6 pairs of objects from U_1); the cuts $(a_1, 5.5)$ and $(a_3, 7.0)$ are the best cuts for U_2 (4 pairs of objects from U_2). Assume that the cuts $(a_2, 3.5)$ and $(a_1, 5.5)$ have been chosen for U_1 and U_2, respectively. Hence U_1 is divided into two subsets $X_1 = \{u_1, u_9, u_{10}\}$ and $X_2 = \{u_3, u_4, u_5\}$, but the set U_2 is divided into two sets $X_3 = \{u_2, u_6\}$ and $X_4 = \{u_7, u_8\}$. One can see that the sets X_1, X_3 and X_4 consists of objects from one decision class, then our algorithm continues the searching process for X_2. The result of our algorithm is the set of cuts $\{(a_3, 4.0), (a_2, 3.5), (a_1, 5.5), (a_1, 3.5), (a_2, 6.5)\}$ (see Figure 6).

In the global strategy, the best cut $(a_3, 4.0)$ defines a partition $\mathbf{L} = \{U_1, U_2\}$ of U, where $U_1 = \{u_1, u_3, u_4, u_5, u_9, u_{10}\}$ and $U_2 = \{u_2, u_6, u_7, u_8\}$. Next, the global discretization algorithm chooses the cut $(a_1, 3.5)$ because it discerns 8 pairs of objects (6 pairs in U_1 and 2 pairs in U_2). The set of two cuts $\{(a_3, 4.0), (a_1, 3.5)\}$ defines a new partition $\mathbf{L} = \{\{u_1, u_3, u_4\}, \{u_5, u_9, u_{10}\}, \{u_2\}, \{u_6, u_7, u_8\}\}$. After removing two sets $\{u_5, u_9, u_{10}\}, \{u_2\}$, which consists of objects from one decision class, the algorithm chooses the cut $(a_2, 3.5)$ (3 pairs). The set of cuts $\{(a_3, 4.0), (a_1, 3.5), (a_2, 3.5)\}$ defines a partition $\mathbf{L} = \{\{u_1\}, \{u_3, u_4\}, \{u_6, u_8\}, \{u_7\}\}$. Finally, the algorithm chooses the cut $(a_2, 6.5)$, which discerns all remaining pairs of objects. As the result we have the set of cuts $\{(a_3, 4.0), (a_1, 3.5), (a_2, 3.5), (a_2, 6.5)\}$.

4.5 Application of Dynamic Reducts to Finding Set of Cuts

Now, we propose another method of searching for an irreducible set of cuts of a given decision table $\mathbf{A}_1 = (U_1, A_1 \cup \{d_1\})$ (see Section 4.2). This method is based on the dynamic reduct notion (see Section 7, 8). We calculate dynamic reducts (or generalized dynamic reducts) for the table $\mathbf{A}_1$ and we choose one with the best stability coefficient. Next, as an irreducible set of cuts we select cuts belonging to the chosen dynamic reduct. Finally, we remove from U_1 and respectively U all objects not belonging to the reduct domain (see Section 9) of the chosen dynamic reduct.

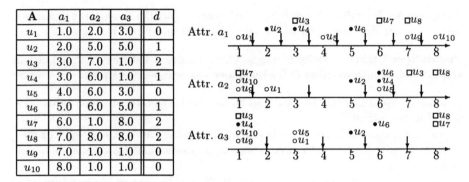

Fig. 5. The exemplary decision table with 10 objects, 3 attributes and 3 decision classes and the illustration of cuts for this table

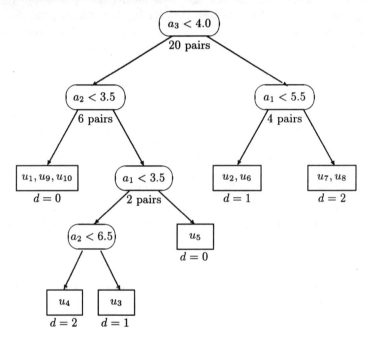

Fig. 6. The construction of decision tree by applying local discretization algorithm

5 Local Reducts Computation

A notion of local reduct (see Section 2 and [65]) seems to be very useful in classification problems. A set of rules generated basing on these reducts is usually less specific and fits more new (unseen) objects than in a classical case.

Let $\mathbf{A} = (U, A \cup \{d\})$ be an information system, where U - set of objects, A - set of attributes, d - decision. A rule generated by a local reduct is concerned with the base object and may not recognize any other object from U. To assure that a set of rules will recognize (at least) all objects from the training set, we have to generate a local reduct for every object. A simple algorithm checking whether a subset is a local superreduct works at a time complexity of $O(mn)$: we have to compare our base object with all other objects and check whether condition a) (see local reduct definition) holds. It takes $O(mn^2)$ time to check this property for all objects (in case we are looking for a local reduct for every object), where n = number of objects, m = number of attributes.

An algorithm presented below realizes the following objective: assuming that the information system $\mathbf{A}$ is consistent, find a family of subsets R_1, R_2,... R_k such that for any object x_i from U at least one R_j is a local reduct (we will say, that R_j covers x_i). We look for a possibly small family R_1,... R_k, i.e. we prefer these subsets which cover possibly many objects. We assume, that these subsets reflect regularities in data and generate more general rules and that fact means better classification of new samples and less memory required to store rules.

Algorithm 6 *Optimal covering with local reducts.*
Input: *decision table* $\mathbf{A} = (U, \{a_1, ..., a_m\} \cup \{d\})$
Output: *a set* $\mathbf{R}$ *of local reducts covering* U
Method:
    $N = \{N_1, \ldots, N_n\} = 0, \ldots, 0$ *-numbers of local reducts found for each object*
    repeat
        $R = \{a_1, \ldots, a_m\}$
        repeat
            $b = \{b_1, \ldots, b_n\} = 0, \ldots, 0$
            $c = \{c_1, \ldots, c_n\} = 0, \ldots, 0$ *-tables of (locally) covered objects*
            $LocRed\,(\mathbf{A}, R, c, N)$ *- check which objects are covered by* R
            for $a_i \in R$ **do**
            begin
                $R = R - a_i$
                $M_i = LocRed\,(\mathbf{A}, R, b, N)$ *- number of objects covered by* R
                $R = R \cup a_i$
            end
            for $i = 0$ **to** n **if** $c_i = 1$ **and** $b_i = 0$ **then**
            begin
                $\mathbf{R} = \mathbf{R} \cup R(u_i)$ *- update set of reducts*
                $N_i = 1$
            end
            $j = \arg\max(M_i)$ *- if there is more than one maximum, select random*
            $R = R - a_j$
        until $R = \emptyset$
    until $\forall_{i \in \{1,\ldots,n\}} N_i = 1$
    return R
$\square$

Lemma. *The algorithm described above generates a covering for all objects in at most* $n = |U|$ *cycles of the outer loop.*

Proof: see [65].

The algorithm is not deterministic, because we use a random selection of maximal M_i. We need to have a method of determining whether a subset is a local superreduct to complete our algorithm. Function *LocRed* checks for which objects a subset R is a local reduct and returns a number of these objects as well as a list of them (in the form of binary table):

Algorithm 7. *Function LocRed(**A**, R, b, N).*
Input:
  1. Decision table $\mathbf{A} = (U, A \cup \{d\})$
  2. Subset $R \subseteq A$
  3. Table of already covered objects N
Output:
  1. Table b of objects for which R is a local superreduct
  2. Number of newly covered objects
Method:
  Sort U by values of attributes in R
  Create partition $U = U_1 \cup \ldots \cup U_k$ into indiscernibility classes,
     i.e. $u_1, u_2 \in U_i \Rightarrow (u_1, u_2) \in IND(R)$
  $newcover = 0$
  for $i = 1$ **to** k **do**
  begin
    $uniform = 1$
    for $j = 2$ **to** $|U_i|$ **do**
    begin
      if $d(u_{i,j}) \neq d(u_{i,j-1})$ **then** $uniform = 0$
         – where $U_i = \{u_{i,1}, \ldots, u_{i,|U_i|}\}$
    end
    if $uniform = 1$ **then**
    begin
      for $j = 1$ **to** $|U_i|$ **do**
      begin
        $t = $ *a number of object $u_{i,j}$ in table U*
        $b_t = 1$
        if $N_t = 0$ **then** $newcover = newcover + 1$
      end
    end
  end
  return $newcover$
□

The partition of U can be done by n operations if U is properly sorted. Since we may use a fast method of sorting, our algorithm has the complexity of $mn \, log(n)$, where $n = $ number of objects, $m = $ number of attributes.

6 Computing Reducts Using Genetic Algorithms

Our methods of decision rules generation from decision tables are based on the reduct set computation. The time cost of the reduct set computation can be too high in case the decision table consists of too many: objects or attributes or different values of attributes. The reason is that in general the size of the reduct set can be exponential with respect to the size of the decision table and the problem of computing a minimal reduct is NP-hard (see [51]). Therefore we are often forced to apply approximation algorithms to obtain some knowledge about the reduct set. One way is to use approximation algorithms that do not give the optimal solutions but require short computing time. Among these algorithms are the following ones: Johnson's algorithm, algorithms based on simulated annealing and Boltzmann machines, algorithms using neural networks and algorithms based on genetic algorithms, which we would like to present in this section.

6.1 Genetic and Hybrid Algorithms

The main idea of genetic algorithms is based on the Darwinian principle of "survival of the fittest" (natural selection). In a case of *classical genetic algorithms* (see [21], [24]) we are given a state space S (finite, but large) and a function: $f : S \rightarrow R_+$. Our goal is to find x_o: $f(x_o) = max\{f(x): x \in S\}$. Elements of set S are *"individuals"*. We treat a value of the function f as ability to survive in the environment (*"fitness"*), and we simulate the process of evolution as follows:

1. We choose the representation scheme: a mapping from a space of "individuals" into "chromosomes" - usually bit strings.
2. We randomly choose the set of chromosomes as an initial population.
3. We calculate "fitness" $F(c)$ of each chromosome c as a value of $f(s(c))$, where $s(c)$ is the individual encoded by c. Then we create a new population, replacing the chromosomes with low fitness by those with higher fitness.
4. We randomly affect the new population by *genetic operators*, e.g. *mutation* (small, random modifications of chromosomes) and *crossing-over* (exchange of "genetic material" between some pairs of chromosomes).
5. We repeat 3-4 with the new population, until a stopping criterion is satisfied.

The result of evolution is the best individual x_{max} which is usually nearly as good as the global optimum x_o.

The scheme presented above is general and domain-independent. On the other hand, in particular problems we often have some approximation algorithms and heuristics producing maybe not optimal, but good results. To exploit advantages of both genetic and heuristic algorithms, one can use a hybridization strategy [64]. The general scheme of *hybrid algorithm* is as follows:

1. Find a strategy (heuristic algorithm) which gives an approximate result.
2. Modify (parameterize) the strategy using a control sequence, so that the result depends on this sequence (recipe).

3. Encode the control sequence to a chromosome.
4. Use a genetic algorithm to produce control sequences. Proceed with the heuristic algorithm controlled by the sequence. Evaluate an object generated by the algorithm and use its quality measure as a fitness of the control sequence.
5. A result of evolution is the best control sequence, i.e. the sequence producing the best object. Send this object to the output of the hybrid algorithm.

Hybrid algorithms proved to be useful and efficient in many areas, including NP-hard problems of combinatorics. As we will see in the next section, short reduct finding problem also can be solved efficiently by this class of algorithms.

6.2 Finding Reducts Using GA

An order-based genetic algorithm is one of the most widely used component of various hybrid systems. Theoretical foundations and practical construction of this algorithm are presented in [64]. In this type of genetic algorithm a chromosome is an n-element permutation σ, represented by a sequence of numbers: $\sigma(1)\ \sigma(2)\ \sigma(3)\ \ldots\ \sigma(n)$. Mutation of an order-based individual means one random transposition of its genes (a transposition of random pair of genes). There are various methods of recombination (*crossing-over*) considered in literature. In [21] such methods as PMX (Partially Matched Crossover), CX (Cycle Crossover) and OX (Order Crossover) are described. Another type of crossing-over operator is presented in [63]. After crossing-over, fitness function of every individual is calculated. In the case of a hybrid algorithm a heuristic part is launched under control of individual; a fitness value depends on the result of heuristic algorithm. Then, new population is generated using "roulette wheel" algorithm: the fitness value of every individual is normalized and treated as probability distribution on population; then we randomly choose M new individuals using this distribution. Then all these steps are repeated.

In the hybrid algorithm [62] a simple, deterministic method was used for reduct generation:

Algorithm 8. *Finding a reduct basing on permutation.*
Input:
    1. decision table $\mathbf{A} = (U, \{a_1, ..., a_n\} \cup \{d\})$
    2. permutation τ generated by genetic algorithm
Output: *a reduct R generated basing on permutation τ*
Method:
    $R = \{a_1, ..., a_n\}$
    $(b_1 \ldots b_n) = \tau(a_1 \ldots a_n)$
    for $i = 1$ **to** n **do**
    begin
        $R = R - b_i$
        if not $Reduct(R, \mathbf{A})$ **then** $R = R \cup b_i$
    end

return R

□

A fast algorithm for determining whether R is a superreduct is presented in [40]:

Algorithm 9. *Function Reduct$(R, \mathbf{A})$.*
Input:
   1. Decision table $\mathbf{A} = (U, A \cup \{d\})$
   2. Subset $R \subseteq A$
   3. Binary tree T_{red} *of reducts found so far, list* L_{subred} *of subreducts found so far.*
Output:
   1. "True" if R is a reduct or superreduct of A, "False" otherwise.
   2. Updated T_{red} *and* L_{subred} *structures.*
Method:
   if $R \in T_{red}$ **then return** *True*
   for $s \in L_{subred}$ **do**
   begin
     if $R \subseteq s$ **return** *False*
   end
   Sort U by values of attributes in R
   for $i = 2$ **to** m **do**
   begin
     if $d(u_i) \neq d(u_{i-1})$ **then**
     begin
        all_equal $= 1$
        for $a_j \in R$ **do**
        begin
            if $a_j(u_i) \neq a_j(u_{i-1})$ **then** *all_equal* $= 0$
        end
        if *all_equal* $= 1$ **then**
        begin
            $Add(L_{subred}, R)$
            return *False*
        end
     end
   end
   $Add(T_{red}, R)$
   return *True*

□

The result of the algorithm will always be a reduct. Every reduct can be found using this algorithm, the result depends on the order of attributes (proof: see [62]). The genetic algorithm is used to generate the proper order. To calculate the function of fitness for a given permutation (order of attributes) we have to perform one run of the deterministic algorithm and calculate the length of the found reduct. In the selection phase of genetic algorithm we used linear scaling [21] and the following fitness function:

$$F(\tau) = n - L_\tau + 1$$

The hybrid algorithm described above performs much slower that the classical genetic algorithm. On the other hand, the reducts obtained by this algorithm are usually shorter. Moreover, the hybrid algorithm generates from 50 to 500 different reducts in comparison with 5 to 50 reducts generated by the classical GA at the same time.

An algorithm described above generates possibly shortest reducts, i.e. the reducts with as few attributes as possible. On the other hand, our goal is not to calculate reducts, but to construct an efficient system for classification or decision making. Another approach is to select a reduct due to the number of rules it generates rather than to its length. Every reduct generates an indiscernibility relation on the universe and in most cases it identifies some pairs of objects. If a reduct generates less rules, it means, that the rules are more general and they should better recognize new objects.

The number of rules can be easily computed due to the improvement of the reduct generation system described in [40]. The hybrid algorithm described above can be used to find reducts generating the minimal number of rules. The only thing we have to change is the definition of the fitness function:

$$F(\tau) = m - R_\tau + \frac{n - L_\tau + 1}{n}$$

where R_τ denotes the number of rules generated by the reduct. Now, the primary criterion of optimization is the number of rules, while the secondary is the reduct length. The results of experiments [64] show, that the classification system based on the reducts optimized due to the number of rules performs better (or not worse) than the short reduct based one. Moreover, due to the rule set reduction, it occupies less memory and classifies new objects faster.

7 Dynamic Reducts

The methods based on calculation of reducts allow to compute, for a given decision table $\mathbf{A}$, the descriptions of all decision classes of $\mathbf{A}$ in the form of decision rules (see previous sections). Unfortunately, decision rules calculated in this way can often be inappropriate to classify unseen cases. We suggest that the rules calculated by means of dynamic reducts are better predisposed to classify unseen cases, because these reducts are the most stable reducts in a process of random sampling of the original decision table.

Let $\mathbf{A} = (U, A \cup \{d\})$ be a decision table. By $\mathbf{P}(\mathbf{A})$ we denote the set of all subtables of $\mathbf{A}$. If $\mathbf{F} \subseteq \mathbf{P}(\mathbf{A})$ then by $DR(\mathbf{A}, \mathbf{F})$ we denote the set

$$RED(\mathbf{A}, d) \cap \bigcap_{\mathbf{B} \in \mathbf{F}} RED(\mathbf{B}, d).$$

Any element of $DR(\mathbf{A}, \mathbf{F})$ is called an **F-*dynamic reduct*** of **A**.

From the definition of a dynamic reduct it follows that a relative reduct of **A** is dynamic if it is also a reduct of all subtables from a given family **F**. This notion can be sometimes too much restrictive so we apply also a generalization of dynamic reducts - $(\mathbf{F}, \varepsilon)$-dynamic reducts, where $\varepsilon \in [0,1]$. The set $DR_\varepsilon(\mathbf{A}, \mathbf{F})$, of all $(\mathbf{F}, \varepsilon)$-dynamic reducts is defined by
$DR_\varepsilon(\mathbf{A}, \mathbf{F}) = \{C \in RED(\mathbf{A}, d) :$

$$\frac{card(\{\mathbf{B} \in \mathbf{F} : C \in RED(\mathbf{B}, d)\})}{card(\mathbf{F})} \geq 1 - \varepsilon\}$$

If $C \in RED(\mathbf{A}, d)$ then the number:

$$\frac{card(\{\mathbf{B} \in \mathbf{F} : C \in RED(\mathbf{B}, d)\})}{card(\mathbf{F})}$$

is called the *stability coefficient* of the reduct C (relative to **F**).

Proposition 6. *Let* $\mathbf{A} = (U, A \cup \{d\})$ *be a decision table.*

1. *If* $\mathbf{F} = \{\mathbf{A}\}$ *then* $DR(\mathbf{A}, \mathbf{F}) = RED(\mathbf{A}, d)$.
2. *If* $\varepsilon_1 \leq \varepsilon_2$ *then* $DR_{\varepsilon_1}(\mathbf{A}, \mathbf{F}) \subseteq DR_{\varepsilon_2}(\mathbf{A}, \mathbf{F})$.
3. $DR(\mathbf{A}, \mathbf{F}) \subseteq DR_\varepsilon(\mathbf{A}, \mathbf{F})$, *for any* $\varepsilon \in [0,1]$.
4. $DR_0(\mathbf{A}, \mathbf{F}) = DR(\mathbf{A}, \mathbf{F})$.

8 Generalized Dynamic Reducts

From the definition of a dynamic reduct it follows that a relative reduct of any table from a given family **F** of subtables of **A** can be dynamic if it is also a reduct of the table **A**. This notion can be sometimes not convenient because we are interested in useful sets of attributes which are not necessarily reducts of the table **A**. Therefore we have to generalize the notion of a dynamic reduct.

Let $\mathbf{A} = (U, A \cup \{d\})$ be a decision table and $\mathbf{F} \subseteq \mathbf{P}(\mathbf{A})$. By $GDR(\mathbf{A}, \mathbf{F})$ we denote the set

$$\bigcap_{\mathbf{B} \in \mathbf{F}} RED(\mathbf{B}, d).$$

Elements of $GDR(\mathbf{A}, \mathbf{F})$ are called **F-*generalized dynamic reducts*** of **A**.

From the above definition it follows that any subset of **A** is a generalized dynamic reduct if it is also a reduct of all subtables from a given family **F**. Analogously to dynamic reducts we define a more general notion of generalized dynamic reducts - $(\mathbf{F}, \varepsilon)$-*generalized dynamic reducts*, where $\varepsilon \in [0,1]$. The set $GDR_\varepsilon(\mathbf{A}, \mathbf{F})$ of all $(\mathbf{F}, \varepsilon)$-generalized dynamic reducts is defined by
$GDR_\varepsilon(\mathbf{A}, \mathbf{F}) = \{C \subseteq A :$

$$\frac{card(\{\mathbf{B} \in \mathbf{F} : C \in RED(\mathbf{B}, d)\})}{card(\mathbf{F})} \geq 1 - \varepsilon\}.$$

If $C \in RED(\mathbf{B}, d)$ (for any $\mathbf{B} \in \mathbf{F}$) then the number:

$$\frac{card(\{\mathbf{B} \in \mathbf{F} : C \in RED(\mathbf{B}, d)\})}{card(\mathbf{F})}$$

is called the *stability coefficient* of the generalized dynamic reduct C (relative to $\mathbf{F}$).

Proposition 7. *Let* $\mathbf{A} = (U, A \cup \{d\})$ *be a decision table.*

1. $DR(\mathbf{A}, \mathbf{F}) \subseteq GDR(\mathbf{A}, \mathbf{F})$.
2. $DR_\varepsilon(\mathbf{A}, \mathbf{F}) \subseteq GDR_\varepsilon(\mathbf{A}, \mathbf{F})$ *for any* $\varepsilon \in [0, 1]$.
3. *If* $\mathbf{A} \in \mathbf{F}$ *then* $DR(\mathbf{A}, \mathbf{F}) = GDR(\mathbf{A}, \mathbf{F})$.

9 Reduct Domain

Let $\mathbf{A} = (U, A \cup \{d\})$ be a decision table, $\mathbf{F} \subseteq \mathbf{P}(\mathbf{A})$, and let $GDR_\varepsilon(\mathbf{A}, \mathbf{F})$ (for any $\varepsilon \in [0, 1]$) be the set of $(\mathbf{F}, \varepsilon)$-generalized dynamic reducts.
For any $R \in GDR_\varepsilon(\mathbf{A}, \mathbf{F})$ we define the *reduct domain* (denoted by $RD(\mathbf{A}, \mathbf{F}, R)$) as the set:

$$\bigcup \{U_\mathbf{B} : R \in RED(\mathbf{B}, d) \text{ and } \mathbf{B} = (U_\mathbf{B}, A \cup \{d\}) \in \mathbf{F}\} \cap POS_R(d).$$

The domain of the reduct R is the set of all objects belonging to decision tables $\mathbf{B} \in \mathbf{F}$ satisfying $R \in RED(\mathbf{B}, d)$ on which the decision can be uniquely determined by attributes from R.

The notion of a reduct domain is very important for decision rules generation (see Section 12).

10 Statistical Inference about Dynamic Reducts

Statistics is frequently defined as the science of collecting and studying numerical data in which deductions are made on the assumption that the relationships between a sufficient sample of numerical data are characteristic of those between all such data (see [12], [18], [25]). The data in most statistical problems relate to a sample drawn from some parent population or universe (as it is called in rough set theory). When a sample is used to make inferences about the population, we generally assume that the sample is random. This usually means (when the population is finite) that any individual in the population has an equal chance of being included in the sample. It is desirable that sampling should be as nearly random as possible, although this is often difficult to be achieved in practice, because sometimes we do not have equal access to all elements of the universe. This problem can be described in language of rough set theory. If we want to calculate the set of decision rules based on a given decision table, we would like to construct rules which are proper not only for objects from this decision table but also for still unknown examples of objects. Dynamic reducts (defined in the previous sections) are calculated with respect to a family **F** of subtables created by random samples of a given decision table. The family **F** is, of course, only a subfamily of all subtables of the hypothetical universal decision table (including known and unknown objects describing a currently considered aspect of reality). The family of all subtables of the universal decision table is denoted by **G**. We are interesting in reducts which most frequently appear in the family **G**, because we expect that the decision rules generated from these reducts are better predisposed to designate a value of the decision for objects from the universal decision table **W**. In other words, we are interested in the probability of the event that a given **F**-generalized dynamic reduct R is a reduct for any subtable from **G**. This probability is denoted by $P_\mathbf{G}(R)$ and defined as the quotient $\frac{card(\mathbf{G}_R)}{card(\mathbf{G})}$, where

$$\mathbf{G}_R = \{\mathbf{B} \in \mathbf{G} : R \in RED(\mathbf{B}, d)\}. \tag{7}$$

We would like to select an **F**-dynamic reduct R for which the probability $P_\mathbf{G}(R)$ is not less than the probability $P_\mathbf{G}$ for other dynamic reducts. However we cannot calculate $P_\mathbf{G}$ for any **F**-generalized dynamic reduct R because we do not know the whole family **G**. In this section we show that the so called stability coefficient of any $(\mathbf{F}, \varepsilon)$-generalized dynamic reduct R is a proper measure of the probability $P_\mathbf{G}(R)$. Unfortunately, we usually do not have an access to all subtables from the family **G**, therefore we construct the family **F** based only on subtables of a given decision table **A**. We have to assume that the decision table **A** is a representative sample from the universal decision table **W**.

Theorem 8. *Stability coefficients as maximum likelihood estimator (see [6])*
Let us assume that

- $\mathbf{W} = (W, A \cup \{d\})$ *is a universal decision table,*
- $\mathbf{A} = (U, A \cup \{d\})$ *is a given decision table ($U \subseteq W$),*
- $\mathbf{G} = \mathbf{P}(\mathbf{W})$ *is a family, called the parent population,*
- $\mathbf{F} \subseteq \mathbf{P}(\mathbf{A})$,
- R *is an* $(\mathbf{F}, \varepsilon)$-*generalized dynamic reduct for some* $\varepsilon \in [0, 1]$.

Then we have: the stability coefficient of the $(\mathbf{F}, \varepsilon)$-*generalized dynamic reduct R of the decision table* **A** *is the maximum likelihood estimator of the probability* $P_\mathbf{G}(R)$ *(see [25]).*

The maximum likelihood estimator of the probability $P_\mathbf{G}(R)$ is denoted by $MLE(P_\mathbf{G}(R))$.
Proof: Let us first introduce the simple binomial distribution $X_\mathbf{G}^R(\mathbf{B}) : \mathbf{G} \to \{0, 1\}$ (for the family **G** and the reduct R) defined for any $\mathbf{B} \in \mathbf{G}$:

$$X_\mathbf{G}^R(\mathbf{B}) = \begin{cases} 1 \text{ for } R \in RED(\mathbf{B}, d) (\text{success}) \\ 0 \text{ for } R \notin RED(\mathbf{B}, d) (\text{defeat}) \end{cases} \tag{8}$$

Let $\mathbf{G}^1 = \{\mathbf{B} \in \mathbf{G} : X_\mathbf{G}^R(\mathbf{B}) = 1\}$ and $\mathbf{G}^0 = \{\mathbf{B} \in \mathbf{G} : X_\mathbf{G}^R(\mathbf{B}) = 0\}$. Now it is easy to observe that the probability P of the success in our binomial distribution is: $P[X_\mathbf{G}^R(\mathbf{B}) = 1] = P_\mathbf{G}(R)$ and the probability of the defeat in our binomial distribution is: $P[X_\mathbf{G}^R(\mathbf{B}) = 0] = 1 - P_\mathbf{G}(R)$.

From [25] we know that the probability of a success in our distribution $X_\mathbf{G}^R$ may be estimated by taking a sample of subtables from the family $\mathbf{G}$ (for example $\mathbf{F}$) and next using the method of maximum likelihood estimator. The maximum likelihood estimator of success probability in our distribution is the arithmetic mean of values $X_\mathbf{G}^R$ for all subtables from the sample (see for instance [25]). Hence:

$$MLE(P_\mathbf{G}(R)) = \frac{\sum\limits_{\mathbf{B} \in \mathbf{F}} X_\mathbf{G}^R(\mathbf{B})}{card(\mathbf{F})} =$$

$$= \frac{\sum\limits_{\mathbf{B} \in \mathbf{F} \cap \mathbf{G}^1} X_\mathbf{G}^R(\mathbf{B}) + \sum\limits_{\mathbf{B} \in \mathbf{F} \cap \mathbf{G}^0} X_\mathbf{G}^R(\mathbf{B})}{card(\mathbf{F})} = \frac{card(\mathbf{F} \cap \mathbf{G}^1)}{card(\mathbf{F})}. \quad (9)$$

From our definition of $\mathbf{F}$-generalized dynamic reducts (Section 8) we conclude that the number $\frac{card(\mathbf{F} \cap \mathbf{G}^1)}{card(\mathbf{F})}$ is equal to the stability coefficient of the generalized dynamic reduct R.
This completes the proof.

$\square$

Remark 1. *Dynamic reducts as the tools for classification*
Theorem 8 is showing that dynamic reducts with large stability coefficients are "good" candidates for decision rules generation. They allow to construct rules with better classification quality of unseen objects than reducts with smaller stability coefficients.

Remark 2. *Minimal size of family* $\mathbf{F}$
It is easy to observe (see the proof of Theorem 8) that for any dynamic reduct R the problem of calculating the stability coefficient of R is equivalent to the problem of calculating an unknown success probability $P_\mathbf{G}(R)$ in binomial distribution $X_\mathbf{G}^R$. Therefore, the maximum likelihood estimator of the probability $P_\mathbf{G}(R)$ is equal to $\frac{card(\mathbf{F} \cap \mathbf{G}^1)}{card(\mathbf{F})}$. For calculating the necessary size of family $\mathbf{F}$, we need to make some assumption about a confidence coefficient: $1 - \alpha$. The confidence coefficient can be understood as a measure of probability estimation correctness (see for instance [18]). From the Moivre-Laplace theorem (see for instance [18] - Theorem 6.7.1) we know that

$$\frac{MLE(P_\mathbf{G}(R)) - P_\mathbf{G}(R)}{\sqrt{\frac{MLE(P_\mathbf{G}(R)) \cdot (1 - MLE(P_\mathbf{G}(R)))}{card(\mathbf{F})}}} \quad (10)$$

has approximately a standard normal distribution. Hence,

$$P\left[-t_\alpha < \frac{MLE(P_\mathbf{G}(R)) - P_\mathbf{G}(R)}{\sqrt{\frac{MLE(P_\mathbf{G}(R)) \cdot (1 - MLE(P_\mathbf{G}(R)))}{card(\mathbf{F})}}} < t_\alpha\right] = 1 - \alpha, \quad (11)$$

where the number t_α is satisfying the equation:

$$1 - \alpha = \sqrt{\frac{2}{\pi}} \int_{-t_\alpha}^{t_\alpha} \exp(-\frac{t^2}{2}) dt. \qquad (12)$$

From equation (11) we have

$$P\left[MLE(P_\mathbf{G}(R)) - t_\alpha \cdot \sqrt{\frac{MLE(P_\mathbf{G}(R)) \cdot (1 - MLE(P_\mathbf{G}(R)))}{card(\mathbf{F})}} < P_\mathbf{G}(R) <\right.$$

$$\left. MLE(P_\mathbf{G}(R)) + t_\alpha \cdot \sqrt{\frac{MLE(P_\mathbf{G}(R)) \cdot (1 - MLE(P_\mathbf{G}(R)))}{card(\mathbf{F})}}\right] = 1 - \alpha. \qquad (13)$$

Hence, the $100 \cdot (1 - \alpha)\%$ confidence interval for $P_\mathbf{G}(R)$ is given by

$$MLE(P_\mathbf{G}(R)) - t_\alpha \cdot \sqrt{\frac{MLE(P_\mathbf{G}(R)) \cdot (1 - MLE(P_\mathbf{G}(R)))}{card(\mathbf{F})}} < P_\mathbf{G}(R) <$$

$$MLE(P_\mathbf{G}(R)) + t_\alpha \cdot \sqrt{\frac{MLE(P_\mathbf{G}(R)) \cdot (1 - MLE(P_\mathbf{G}(R)))}{card(\mathbf{F})}}. \qquad (14)$$

If $\Delta MLE(P_\mathbf{G}(R))$ is a maximal acceptable estimation error of $MLE(P_\mathbf{G}(R))$, then we conclude from inequality (14) that

$$t_\alpha \cdot \sqrt{\frac{P_\mathbf{G}(R) \cdot (1 - P_\mathbf{G}(R))}{card(\mathbf{F})}} \leq \Delta MLE(P_\mathbf{G}(R)). \qquad (15)$$

Hence

$$card(\mathbf{F}) \geq \frac{t_\alpha^2 \cdot P_\mathbf{G}(R) \cdot (1 - P_\mathbf{G}(R))}{(\Delta MLE(P_\mathbf{G}(R)))^2}. \qquad (16)$$

It is easy to observe that if $P_\mathbf{G}(R) = \frac{1}{2}$ then the product $P_\mathbf{G}(R) \cdot (1 - P_\mathbf{G}(R))$ takes the maximum value of $\frac{1}{4}$. Therefore it is enough to require that the size of $\mathbf{F}$ is no less than $\frac{t_\alpha^2}{4 \cdot (\Delta MLE(P_\mathbf{G}(R)))^2}$.

The value of t_α one can read from the table of standardized normal distribution function (see e.g. [12], [25]).

For example, if we take $1 - \alpha = 0.9$ and $\Delta MLE(P_\mathbf{G}(R)) = 0.05$ than the value t_α is 1.64 and $card(\mathbf{F}) \geq \frac{1.64^2}{4 \cdot 0.05^2} = 268.96$.

11 Techniques for Dynamic Reduct Computation

In this section we present a method for computing generalized dynamic reducts and reduct domains. In our method a random set of subtables **F** from a given data table $\mathbf{A} = (U, A \cup \{d\})$ is taken and reducts for all these tables are calculated. The number of samples $card(\mathbf{F})$ can be selected by taking into account the minimal family size (see previous section). The method of the random choice of sub-table consists of the two steps. In the first step we randomly choose some numbers with the probability:

$$P(k) = \frac{\binom{n}{k}}{\sum_{i=1}^{n} \binom{n}{i}} \qquad (17)$$

where $k = 1, ..., n$ and $n = card(U)$. Next, we randomly choose a sub-table of **A** consisting of k objects.

Thus we receive a family **F** of decision tables and for each $\mathbf{A} \in \mathbf{F}$ we compute reducts. Next step is to compute the reduct domain for every reduct (as the positive region of the table constructed from the sum of all objects from proper subtables). In the following step reducts with the stability coefficients higher than a fixed threshold are extracted. The reducts distinguished in such a way are treated as the true generalized dynamic reducts of the table **A** together with their reduct domains.

We would like to mention another method of dynamic reduct computation for decision tables. In this case we assume that the reduct set $RED(\mathbf{A})$ is already computed. Instead of computing reduct sets for subtables from **F** it is enough to check which reducts from $RED(\mathbf{A})$ are also reducts for subtables from **F**. This can save computing time because time necessary for checking which reducts from $RED(\mathbf{A})$ are F-dynamic reducts of **A** is polynomial with respect to the size of **F** and $RED(\mathbf{A})$. However this method can be used only in case we are able to compute the set of all reducts.

12 Decision Rules Computed from Dynamic Reducts

After a set of dynamic reducts with their reduct domains has been computed it is necessary to decide how to compute the set of decision rules. We discuss methods based on the decision rule set calculation for any chosen dynamic reduct. The rules for each reduct are calculated separately. For example one can calculate decision rules for any chosen dynamic reduct R. Let us consider the two following methods.

In the first method for any object from the reduct domain R we take the value vector of conditional attributes from R and the corresponding decision of the object. Unfortunately, decision rules generated by this method have poor performance, because the number of objects supporting such rules is usually very small.

In the second method we generate decision rules with minimal number of descriptors (see Section 3) for the table consisting of the object set equal to the reduct domain of R, the conditional attribute set equal to R, and assuming the decision attribute to be the same as in the original decision table. The final decision rule set is equal to the union of all these sets.

When a new object is to be classified, it is first matched against all decision rules from the constructed decision rule set. Next, the final decision is predicted by applying some strategy constructing the final decision from all "votes" of decision rules (see Section 15).

13 Dynamic Rules

From [3] and [4] we know that the quality of unseen object classification based on dynamic reducts (see Section 12) is usually better than the quality of classification based on the whole set of attributes. Therefore one can adopt an idea of dynamic reducts as a method for decision rules computation. By analogy with dynamic reducts, we propose the following method for dynamic rules computation. At the beginning, from a given data table a random set of subtables is chosen. Next the optimal decision rule sets for all these tables are calculated. In the following step the rule memory is constructed where all rule sets are stored. Intuitively, any dynamic rule is appearing in all (or almost all) of experimental subtables. The decision rules can be computed from the so called *k-relative discernibility matrix* used to generate decision rules with the minimal number of descriptors (see Section 3).

Let $\mathbf{A} = (U, A \cup \{d\})$ be a decision table and $\mathbf{F} \subseteq \mathbf{P(A)}$. A decision rule $r \in \bigcup_{\mathbf{B} \in \mathbf{F}} RUL(\mathbf{B})$ is called an **F**-*dynamic rule* of **A** iff

$$Supp_{\mathbf{B}}(r) \neq 0 \Rightarrow r \in RUL(\mathbf{B}), \text{ for any } \mathbf{B} \in \mathbf{F}.$$

By $DRUL(\mathbf{A}, \mathbf{F})$ we denote the set of all **F**-dynamic rules of **A**.

From the definition of dynamic rules it follows that any optimal decision rule for $\mathbf{B} \in \mathbf{F}$ is an **F**-dynamic rule of **A** if it is also an optimal rule for all subtables from a given family **F** (having some objects matched by the considered decision rule). This notion can be sometimes too much restrictive, so we apply also a more general notion of a dynamic rule - $(\mathbf{F}, \varepsilon)$-*dynamic rules*, where $\varepsilon \in [0, 1]$. The set $DRUL_\varepsilon(\mathbf{A}, \mathbf{F})$ of all $(\mathbf{F}, \varepsilon)$-dynamic rules is defined as

$$DRUL_\varepsilon(\mathbf{A}, \mathbf{F}) = \{r \in \bigcup_{\mathbf{B} \in \mathbf{F}} RUL(\mathbf{B}) : \frac{card(\{\mathbf{B} \in \mathbf{F} : r \in RUL(\mathbf{B})\})}{card(\{\mathbf{B} \in \mathbf{F} : Supp_{\mathbf{B}}(r) \neq 0\})} \geq 1 - \varepsilon\}.$$

If $r \in RUL(\mathbf{B})$ (for any $\mathbf{B} \in \mathbf{F}$) then the number:

$$\frac{card(\{\mathbf{B} \in \mathbf{F} : r \in RUL(\mathbf{B})\})}{card(\{\mathbf{B} \in \mathbf{F} : Supp_{\mathbf{B}}(r) \neq 0\})}$$

is called the *stability coefficient* of the dynamic rule r (relatively to **F**) and it is denoted $SC_{\mathbf{A}}^{\mathbf{F}}(r)$.

Proposition 9. *Let* $\mathbf{A} = (U, A \cup \{d\})$ *be a decision table.*

1. *If* $\mathbf{F} = \{\mathbf{A}\}$ *then* $DRUL(\mathbf{A}, \mathbf{F}) = RUL(\mathbf{A})$.
2. *If* $\varepsilon_1 \leq \varepsilon_2$ *then* $DRUL_{\varepsilon_1}(\mathbf{A}, \mathbf{F}) \subseteq DRUL_{\varepsilon_2}(\mathbf{A}, \mathbf{F})$.
3. $DRUL(\mathbf{A}, \mathbf{F}) \subseteq DRUL_\varepsilon(\mathbf{A}, \mathbf{F})$, *for any* $\varepsilon \geq 0$.
4. $DRUL_0(\mathbf{A}, \mathbf{F}) = DRUL(\mathbf{A}, \mathbf{F})$.

Our methods for dynamic rules generation from decision tables are based on the reduct set computation. A random set of subtables from a given data table is taken (see Section 11) and the optimal rules for all these tables are calculated (see Section 3). The time cost of the reduct set computation can be very high when the decision table has too many: objects or attributes or different values of attributes (see Section 3). Therefore we often apply some approximate algorithms to obtain knowledge about optimal rule sets (see Section 12, 15, [38], [62]) and next we use the following proposition to compute stability coefficient of calculated decision rules.

Proposition 10. *Let us assume that*

- $\mathbf{A} = (U, A \cup \{d\})$ *is a decision table, where* $card(U) = n$ *and* $card(A) = m$,
- $r = ((a_1 = v_1) \wedge ... \wedge (a_l = v_l) \Rightarrow (d = v_d)) \in DRUL_\varepsilon(\mathbf{A}, \mathbf{P}(\mathbf{A}))$ *for some* $\varepsilon \in [0, 1]$, *where* $a_i \in A$, $v_i \in V_{a_i}$ *for* $i = 1, ..., l$ $(l \leq m)$ *and* $v_d \in V_d$,
- $h_{-1} = card(H_{-1})$, *where* $H_{-1} = \{u \in U : \forall i \in \{1, ..., l\} : a_i(u) = v_i \wedge d(u) \neq v_d\}$,
- $h_0 = card(H_0)$, *where* $H_0 = \{u \in U : \forall i \in \{1, ..., l\} : a_i(u) = v_i \wedge d(u) = v_d\}$,
- $h_i = card(H_i)$, *where* $H_i = \{u \in U : \forall j \in \{1, ..., l\} \setminus \{i\} : a_j(u) = v_j \wedge a_i(u) \neq v_i \wedge d(u) \neq v_d\}$ *for* $i = 1, ..., l$.

Then we have: the stability coefficient of the $(\mathbf{P}(\mathbf{A}), \varepsilon)$-*dynamic rule* r *of the decision table* $\mathbf{A}$ *can be computed using the following equation:*

$$SC_{\mathbf{A}}^{\mathbf{P}(\mathbf{A})}(r) = \frac{1}{2^{h_{-1}}} \text{ for } l = 1, \tag{18}$$

and by

$$SC_{\mathbf{A}}^{\mathbf{P}(\mathbf{A})}(r) = \frac{\prod_{i=1}^{l}(2^{h_i} - 1)}{2^{h_{-1} + \sum_{i=1}^{l} h_i}} \text{ for } l > 1. \tag{19}$$

One can prove the above proposition using some basic facts from combinatorics and rough set theory.

It is easy to construct an algorithm, based on the equation (18) and (19), calculating the stability coefficient $SC_{\mathbf{A}}^{\mathbf{P}(\mathbf{A})}(r)$ for the rule r in time $O(m \cdot n)$ and space $O(C)$, where C is a constant.

14 Approximate Rules

One can use approximate decision rules instead of optimal decision rules to construct the classification algorithm for a decision table **A** (see Section 3). We have implemented a method for computing approximate rules. We begin with algorithm for synthesis of optimal decision rules from a given decision table (see Section 3). Next, we compute approximate rules from already calculated optimal decision rules. Our method is based on the notion of consistency of a decision rule. The original optimal rule is reduced to an approximate rule with the coefficient of consistency exceeding a fixed threshold.

Let $\mathbf{A} = (U, A \cup \{d\})$ be a decision table and $r_0 \in RUL(\mathbf{A})$. The approximate rule (based on rule r_0) is computed using the following algorithm.

Algorithm 10 *Approximate rule synthesis (by descriptor dropping)*
Input:
    1. decision table $\mathbf{A} = (U, A \cup \{d\})$,
    2. decision rule $r_0 \in RUL(\mathbf{A})$,
    3. threshold of consistency μ_0 *(e.g.* $\mu_0 = 0.9$*).*
Output: the approximate rule r_{app} (based on rule r_0).
Method:
    Calculate the coefficient of consistency $\mu_\mathbf{A}(r_0)$
    if $\mu_\mathbf{A}(r_0) < \mu_0$ **then** *STOP (in this case no approximate rule).*
    $\mu_{max} = \mu_\mathbf{A}(r_0)$ *and* $r_{app} = r_0$.
    while $\mu_{max} > \mu_0$ **do**
    begin
        $\mu_{max} = 0$
        for $i = 1$ **to** *the number of descriptors from* $Pred(r_{app})$ **do**
        begin
            $r = r_{app}$.
            Remove i-th descriptor from $Pred(r)$.
            Calculate the coefficient of consistency $\mu_\mathbf{A}(r)$ *and* $\mu = \mu_\mathbf{A}(r)$.
            if $\mu > \mu_{max}$ **then** $\mu_{max} = \mu$ *and* $i_{max} = i$.
        end
        if $\mu_{max} > \mu_0$ *then remove* i_{max} *-th conditional descriptor from* r_{app}.
    end
    return r_{app}.
□

It is easy to see that the time and space complexity of Algorithm 10 are of order $O(l^2 \cdot m \cdot n)$ and $O(C)$, respectively (where l is the number of conditional descriptors in the original optimal decision rule r_0 and C is a constant).

The approximate rules, generated by the above method, can help to extract interesting laws from decision table. By applying approximate rules instead of optimal rules one can slightly decrease the quality of classification of objects from the training set but we expect in return to receive more general rules with the higher quality of classification for new objects (see [9]).

15 Negotiations among Rules

Suppose we have a set of decision rules. In most cases it is necessary to decide how to resolve conflicts between sets of rules classifying tested objects to different decision classes. In this section we present several methods for constructing the measure called *the strength of rule set*. The strength of rule set is a rational number belonging to [0,1] representing the importance of the sets of decision rules relative to the considered tested object.

Let us assume that:

- $\mathbf{W} = (W, A \cup \{d\})$ is a universal decision table,
- $\mathbf{A} = (U, A \cup \{d\})$ is a given decision table ($U \subseteq W$),
- $u_t \in W$ is a tested object,
- $Rul(X_j)$ is a set of all calculated basic decision rules for $\mathbf{A}$, classifying objects to the decision class X_j (where $v_d^j \in V_d$)
- $MRul(X_j, u_t) \subseteq Rul(X_j)$ is a set of all decision rules from $Rul(X_j)$ matching tested objects u_t.

We define several measures for the rule set $MRul(X_j, u_t)$ depending on the number of rules from this set matching tested object, the number of objects supporting decision rules from this set and the stability coefficient of rules.

1. *The simple strength* of a decision rule set is defined by

$$SimpleStrength(X_j, u_t) = \frac{card(MRul(X_j, u_t))}{card(Rul(X_j))}. \qquad (20)$$

2. *The maximal strength* of a decision rule set is defined by

$$MaximalStrength(X_j, u_t) = max_{r \in MRul(X_j, u_t)} \left\{ \frac{Supp_\mathbf{A}(r)}{card(|d = v_d^j|_\mathbf{A})} \right\}. \qquad (21)$$

3. *The basic strength* of a decision rule set is defined by

$$BasicStrength(X_j, u_t) = \frac{\sum_{r \in MRul(X_j, u_t)} Supp_\mathbf{A}(r)}{\sum_{r \in Rul(X_j)} Supp_\mathbf{A}(r)}. \qquad (22)$$

4. *The global strength* of a decision rule set is defined by

$$GlobalStrength(X_j, u_t) = \frac{card\left(\bigcup_{r \in MRul(X_j, u_t)} |Pred(r)|_\mathbf{A} \cap |d = v_d^j|_\mathbf{A} \right)}{card(|d = v_d^j|_\mathbf{A})}. \qquad (23)$$

5. *The stability strength* of a decision rule set is defined by

$$StabilityStrength(X_j, u_t) = max_{r \in MRul(X_j, u_t)} \{SC_\mathbf{A}^{P(\mathbf{A})}(r)\}. \qquad (24)$$

6. *The maximal stability strength* of a decision rule set is defined by

$$MaxStabStrength(X_j, u_t) = max_{r \in MRul(X_j, u_t)} \frac{Supp_\mathbf{A}(r)}{card(|d = v_d^j|_\mathbf{A})} \cdot SC_\mathbf{A}^{\mathbf{P(A)}}(r).$$
(25)

7. *The basic stability strength* of a decision rule set is defined by

$$BasicStabStrength(X_j, u_t) = \frac{\sum\limits_{r \in MRul(X_j, u_t)} Supp_\mathbf{A}(r) \cdot SC_\mathbf{A}^{\mathbf{P(A)}}(r)}{\sum\limits_{r \in Rul(X_j)} Supp_\mathbf{A}(r) \cdot SC_\mathbf{A}^{\mathbf{P(A)}}(r)}.$$
(26)

8. *The global stability strength* of a decision rule set (denoted *GlobStabS*) is defined inductively by

$$\begin{cases} GlobStabS(\{r\}) = SC_\mathbf{A}^{\mathbf{P(A)}}(r), \text{ for } r \in MRul(X_j, u_t) \\ GlobStabS(R \setminus \{r\} \cup \{r\}) = \\ \quad = GlobStabS(\{r\}) + GlobStabS(R \setminus \{r\}) - \\ \quad\quad GlobStabS(\{r\}) \cdot GlobStabS(R \setminus \{r\}), \\ \quad\quad \text{ for } R \neq \emptyset \text{ and } R \subseteq MRul(X_j, u_t). \end{cases}$$
(27)

The maximal strength of a decision rule set is similar to the strength of rule presented in [31] and [57]. The basic strength of a decision rule set is similar to the strength of rule presented in [22] and [23]. The global strength of a decision rule set is similar to the strength of rule presented in [31], [22] and [23]. Measures of strengths of rules defined above can be applied in constructing classification algorithms (see next section).

16 General Scheme of Classification Algorithm

In this section we present general scheme of classification algorithms based on methods and techniques described in previous sections. One can choose options presented in the general scheme (see Figure 7) for the construction of a particular classification algorithm.

17 Summary

In this Chapter we discussed some methods for extracting laws from data. We presented some techniques based on standard rough set methods (see [44], [45], [52]) like reduct set and rule set computation. We also described dynamic techniques e.g. dynamic reducts and dynamic rules computation (see [3], [4], [5], [6]), that give potentially more general decision rules more capable to new cases classification. For the case of larger data tables we proposed a genetic algorithm based method for computation of a approximate reduct set. We also proposed some discretization algorithms of real value attributes that can be used in the

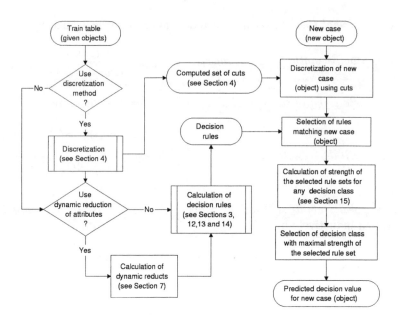

Fig. 7. General scheme of classification algorithm

process of data preprocessing. All presented methods can be successfully used together in order to construct a good classification algorithms.

Acknowledgments: This work was partially supported by the grants of Polish National Committee for Scientific Research (KBN) No. $8T11C01011$ and No. $8T11C02417$ and also by the *ESPRIT* project 20288 CRIT-2.

References

1. Agrawal, R., Mannila, H., Srikant, R., Toivonen, H., Verkamo, I.: Fast Discovery of Association Rules, *Proceedings of the Advances in Knowledge Discovery and Data Mining*. AAAI Press/The MIT Press, CA (1996) 307–328.
2. Almuallim, H., Dietterich, T. G.: Learning with many irrelevant features, *Proceedings of the Ninth National Conference on Artificial Intelligence* (1991) 547–552.
3. Bazan, J., Skowron, A., Synak, P.: Discovery of Decision Rules from Experimental Data, *Proceedings of the Third International Workshop on Rough Sets and Soft Computing*. San Jose, California (1994) 526–533.
4. Bazan, J., Skowron, A., Synak, P.: Dynamic reducts as a tool for extracting laws from decision tables, *Proceedings of the Eighth International Symposium on Methodologies for Intelligent Systems (ISMIS'94), Lecture Notes in Artificial Intelligence 869*. Berlin: Springer-Verlag (1994) 346–355,.
5. Bazan, J., Skowron, A., Synak, P.: Market data analysis: A rough set approach. *ICS Research Report 6/94*, Warsaw University of Technology (1994).

6. Bazan, J.: Dynamic reducts and statistical inference, *Proceedings of Information Processing and Management of Uncertainty on Knowledge Based Systems (IPMU-96)*, July 1-5, Granada, Spain, Universidad de Granada, vol. III, (1996) 1147–1152.
7. Bazan, J., Nguyen, H. S., Nguyen, T. T., Skowron, A., Stepaniuk, J.: Synthesis of Decision Rules for Object Classification, Orowska E. (ed.): *Incomplete Information: Rough Set Analysis*. Heidelberg: Physica-Verlag (1998) 23-57.
8. Bazan, J.: Discovery of Decision Rules by Matching New Objects Against Data Tables. *Proceedings of the First International Conference on Rough Sets and Current Trends in Computing (RSCTC-98)*, Warsaw, June 22-26 (1998), *Lecture Notes in Artificial Intelligence 1424*. Berlin: Springer-Verlag (1998) 521-528.
9. Bazan, J.: A Comparison of Dynamic and non-Dynamic Rough Set Methods for Extracting Laws from Decision Table, Polkowski L., Skowron A. (eds.): *Rough Sets in Knowledge Discovery*. Heidelberg: Physica-Verlag (1998) 321-365.
10. Bloedorn, E., Michalski, R. S.: Data Driven Constructive Induction in AQ17-PRE: A Method and Experiments, *Proceedings of the Third International Conference on Tools for AI*. San Jose, CA (1991)
11. Brown, E. M.: *Boolean reasoning*. Dordrecht: Kluwer (1990).
12. Brownlee, K. A.: *Statistical theory and methodology in science and engineering*. New York: John Wiley&Sons (1965).
13. Cestnik, B., Kononenko, I., Bratko, I.: ASSISTANT 86: A Knowledge Elicitation Tool for Sophisticated Users, *Proceedings of EWSL-87*. Bled, Yugoslavia (1987) 31–47.
14. Clark, P., Niblett, T.: The CN2 Induction Algorithm, *Machine Learning 3*. Kluwer Academic, Boston, MA (1989) 261–284.
15. Cykier, A.: Prime Implicants of Boolean Functions - Applications and Methods of Computations (in Polish), MSc Thesis, University of Warsaw, Warsaw, Poland (1997).
16. Downton, A. C., Tregidgo, R. W. S., Leedham, C. G.: Recognition of handwritten British postal addresses. From Pixels to Features III. Frontiers in Handwriting Recognition North-Holland (1992) 129–144.
17. Dzeroski, S.: *Handling Noise in Inductive Logic Programming*. MS Thesis, Dept. of EE and CS, University of Ljubljana, Slovenia (1991).
18. Fisz, M.: *Probability theory and mathematical statistics*, New York (1961).
19. Fahlman, S. E., Lebiere, C.: The Cascade-Correlation Learning Architecture, in *Advances in Neural Information Processing Systems*, vol. II. Morgan Kaufmann, San Mateo, CA (1990).
20. Friedman, J.: Smart user's guide. *Technical Report 1*. Laboratory of Computational Statistics, Department of Statistics, Stanford University (1984).
21. Goldberg, D. E.: GA in Search, Optimisation, and Machine Learning. Addison-Wesley (1989).
22. Grzymała-Busse, J. W.: LERS - a system for learning from examples based on rough sets. In R. Słowiński, (ed.) *Intelligent Decision Support*, Kluwer Academic Publishers, Dordrecht, Boston, London (1992) 3–18.
23. Grzymała-Busse, J. W.: A new version of the rule induction system LERS. Fundamenta Informaticae **31** (1997) 27–39.
24. Holland, J. H.: Adaptation in Natural and Artificial Systems. The MIT Press, Cambridge (1992).
25. Keeping, E. S.: *Introduction to statistical inference*. Prinston, New Jersey: D.Van Nostrand Company, Inc. (1962).

26. Kira, K., Rendell, L. A.: A practical approach to feature selection, In D.Sleeman (ed.), *Proceedings of the Ninth International Workshop on Machine Learning (ML92)*, Morgan Kaufmann (1992) 249–256.
27. Kittler, J.: Feature selection and extraction, In Young and Fu (ed.), *Handbook of pattern recognition and image processing*. New York: Academic Press (1996).
28. Kodratoff, Y., Michalski, M. (ed.): *Machine Learning vol. III*. Morgan Kaufmann, San Mateo, CA (1990).
29. Michalski, R., Carbonell, J. G. and Mitchel, T. M. (ed): *Machine Learning vol. I*. Tioga/Morgan Kaufmann, Los Altos, CA (1983).
30. Michalski, R., Carbonell, J. G. and Mitchel, T. M. (ed): *Machine Learning vol. II*. Morgan Kaufmann, Los Altos, CA (1986).
31. Michalski, R. S., Mozetic, I., Hong, J. and Lavrac, N.: The Multi-Purpose Incremental Learning System AQ15 and its Testing to Three Medical Domains, *Proceedings of AAAI-86*. Morgan Kaufmann, San Mateo, CA (1986) 1041–1045.
32. Michalski, R., Wnęk, J.: Constructive Induction: An Automated Improvement of Knowledge Representation Spaces for Machine Learning, *Proceedings of a Workshop on Intelligent Information Systems, Practical Aspect of AI II*, Augustów (Poland) (1993) 188–236..
33. Michie, D., Spiegelhalter, D. J., Taylor, C. C.: *Machine learning, neural and statistical classification*. England: Ellis Horwood Limited (1994).
34. Mienko, R., Słowiński, R., Stefanowski, J., Susmaga, R.: RoughFamily - software implementation of rough set based data analysis and rule discovery techniques, Tsumoto S. (ed.), *Proceedings of the Fourth International Workshop on Rough Sets, Fuzzy Sets and Machine Discovery*, Tokyo, November 6-8 (1996), 437–440.
35. Mollestad, T.: A rough set approach to default rules data mining. PhD Thesis, supervisor J. Komorowski, Norwegian Institute of Technology, Trondheim, Norway (1996)
36. Muggelton, S. (ed.): *Inductive logic programming*. Academic Press (1992).
37. Nguyen, H. S., Nguyen, S. H., Skowron, A.: Searching for features defined by hyperplanes, Z.W. Ras, M. Michalewicz (ed.), *Proceedings of Ninth International Symposium on Methodologies for Intelligent Systems (ISMIS-96)*, Zakopane, Poland, June 10-13, (1996). Lecture Notes in Artificial Intelligence vol. 1079, Springer, Berlin (1996) 366–375; see also: *ICS Research Report 16/95*, Warsaw University of Technology.
38. Nguyen, S. H., Nguyen, H. S.: Some efficient algorithms for rough set methods, *Proceedings Information Processing and Management of Uncertainty on Knowledge Based Systems (IPMU-96)*, July 1-5, Granada, Spain, Universidad de Granada, vol. III, (1996) 1451–1456.
39. Nguyen, H. S., Skowron, A.: Quantization of real value attributes, *Proceedings of Second Joint Annual Conf. on Information Sciences*, Wrightsville Beach, North Carolina, September 28 - October 1, USA (1995) 34–37.
40. Nguyen, S. H., Skowron, A., Synak, P., Wróblewski, J.: Knowledge Discovery in Databases: Rough Set Approach. *Proc. of The Seventh International Fuzzy Systems Association World Congress*, vol. II, pp. 204-209, IFSA97, Prague, Czech Republic (1997).
41. Nguyen, H. S.: Discretization of Real Value Attributes: Boolean reasoning Approach. Ph.D. Thesis, Warsaw University, Warsaw, Poland (1997).
42. Nguyen, T., Świniarski, R., Skowron, A., Bazan, J., Thyagarajan, K.: Application of Rough Sets, Neural Networks and Maximum Likelihood for Texture Classification Based on Singular Value Decomposition, *Proceedings of the Third Interna-*

tional Workshop on Rough Sets and Soft Computing. San Jose, California (1994) 332–339.
43. Oehrn, A., Komorowski, J.: ROSETTA - A rough set tool kit for analysis of data. *Proceedings of the Fifth International Workshop on Rough Sets and Soft Computing (RSSC'97) at the Third Joint Conference on Information Sciences (JCIS'97)*, Research Triangle Park, NC, March 2-5 (1997) 403–407.
44. Pawlak, Z.: *Rough sets: Theoretical aspects of reasoning about data*. Dordrecht: Kluwer 1991.
45. Pawlak, Z., Skowron, A.: A rough set approach for decision rules generation, *ICS Research Report 23/93*, Warsaw University of Technology, *Proceedings of the IJCAI'93 Workshop W12: The Management of Uncertainty in AI*, France 1993.
46. Piasta, Z., Lenarcik, A., Tsumoto S.: Machine discovery in databases with probabilistic rough classifiers. In: S. Tsumoto, S. Kobayashi, T. Yokomori, H. Tanaka and A. Nakamura (eds.), *Proceedings of The fourth International Workshop on Rough Sets, Fuzzy Sets, and Machine Discovery (RS96FD)*, November 6-8, The University of Tokyo (1996) 353–359
47. Polkowski, L., Skowron, A.: Synthesis of decision systems from data tables. In: T. Y. Lin and N. Cecerone (eds.), Rough Sets and Data Mining. Analysis for Imprecise Data, Kluwer Academic Publishers, Boston, London, Dordrecht (1997) 259–299.
48. Quinlan, J. R.: Induction of Decision Trees, *Machine Learning 1*. Kluwer Academic, Boston, MA (1986) 81–106.
49. Quinlan, J. R.: *C4.5: Programs for Machine Learning*. San Mateo, California: Morgan Kaufmann (1993).
50. De Raedt, L.: *Interactive Theory Revision. An Inductive Logic Programming*. Academic Press (1992).
51. Skowron, A., Rauszer, C.: The Discernibility Matrices and Functions in Information Systems. In R. Słowiński (ed.), *Intelligent Decision Support. Handbook of Applications and Advances of the Rough Sets Theory*. Dordrecht: Kluwer (1992) 331–362.
52. Skowron, A.: Boolean reasoning for decision rules generation, *Proceedings of the 7-th International Symposium ISMIS'93*, Trondheim, Norway 1993, In Komorowski J. and Ras Z. (ed.), *Lecture Notes in Artificial Intelligence, vol. 689*. Springer-Verlag (1993) 295–305.
53. Skowron, A.: A synthesis of decision rules: Applications of discernibility matrix properties, *Proceedings of the Workshop Intelligent Information Systems, Augustów (Poland)*, 7–11 June, 1993.
54. Słowiński, R. (ed.): Intelligent Decision Support. *Handbook of Applications and Advances of the Rough Sets Theory*. Dordrecht: Kluwer (1992).
55. Słowiński, R., Stefanowski, J.:'RoughDAS' and 'RoughClass' software implementations of the rough set approach, Słowiński R. (ed.): *Intelligent Decision Support. Handbook of Applications and Advances of the Rough Sets Theory*. Dordrecht: Kluwer (1992) 445–456.
56. Thrun, S. B, Bala, J., Bloedorn, E., Bratko, I., Cestnink, B., Cheng, J., DeJong, K. A., Dzeroski, S., Fahlman, S. E., Hamann, R., Kaufman, K., Keller, S., Kononenko, I., Kreuziger, J., Michalski, R., S., Mitchell, T., Pachowicz, P., Vafaie, H., Van de Velde, W., Wenzel, W., Wnęk, J., and Zhang, J.: The MONK's Problems: A Performance Comparison of Different Learning Algorithms. *Technical Report*, Carnegie Mellon University (1991).

57. Tsumoto, S., Tanaka H.: Incremental learning of probabilistic rules from clinical databases. *Proceedings Information Processing and Management of Uncertainty on Knowledge Based Systems (IPMU-96)*, July 1–5, Granada, Spain, Universidad de Granada, vol. II, (1996) 1457–1462.
58. Utgoff, P. E.: Incremental Learning of Decision Trees, *Machine Learning, vol. IV*. Kluwer Academic, Boston, MA (1990) 161–186.
59. Vafaie, L. G., De Jong, K.: Improving the Performance of a Rule Induction System Using Genetic Algorithm, *Proceedings of the First International Workshop on Multistrategy Learning*. Harpers Ferry WV, George Mason University, Center for Artificial Intelligence (1991) 305–315.
60. Van De Velde, W.: IDL, or Taming the Multiplexer, *Proceedings of the 4th European Working Session on Learning*. Pitman, London (1989).
61. Wnęk, J., Michalski, R. S.: Hypothesis-driven Constructive Induction in AQ17: A Method and Experiments, *Proceedings of the IJCAI-91 Workshop on Evaluating and Changing Representation*, K. Morik, F. Bergadano and W. Buntine (ed.). Sydney, Australia (1991) 13–22.
62. Wróblewski, J.: Finding minimal reducts using genetic algorithm (extended version), *Proceedings of Second Joint Annual Conference on Information Sciences*, Wrightsville Beach, North Carolina, 28 September - 1 October, USA, (1995) 186–189; see also: *ICS Research Report 16/95*, Warsaw University of Technology.
63. Wróblewski, J.: Theoretical Foundations of Order-Based Genetic Algorithms. *Fundamenta Informaticae*, vol. 28 (3, 4), pp: 423-430. IOS Press, (1996).
64. Wróblewski, J.: Genetic algorithms in decomposition and classification problem. In: L. Polkowski, A. Skowron (eds.). Rough Sets in Knowledge Discovery. Physica Verlag, (1998).
65. Wróblewski, J.: Covering with reducts – a fast algorithm for rule generation, *Proceedings of RSCTC'98*, Springer-Verlag (LNAI 1424), Berlin Heidelberg (1998) 402 – 407.
66. Ziarko, W., Shan, N.: An incremental learning algorithm for constructing decision rules, *Proceedings of the International Workshop on Rough Sets and Knowledge Discovery*. Banff. (1993) 335–346.
67. Ziarko, W.: Variable Precision Rough Set Model. Journal of Computer and System Sciences **40** (1993) 39–59.

Chapter 3

Rough Mereology in Information Systems. A Case Study: Qualitative Spatial Reasoning

Lech Polkowski[1], Andrzej Skowron[2]

[1] Polish–Japanese Institute of Information Technology
Koszykowa 86, 02-008 Warsaw, Poland
Department of Mathematics and Information Sciences
Warsaw University of Technology
Pl. Politechniki 1, Warsaw, Poland
[2] Institute of Mathematics
Warsaw University
Banacha 2, 02-097 Warsaw, Poland
emails: polkow@pjwstk.waw.pl; skowron@mimuw.edu.pl

Abstract. *Rough Mereology* has been proposed as a paradigm for approximate reasoning in complex information systems [65], [66], [67], [68], [76]. Its primitive notion is that of a *rough inclusion functor* which gives for any two entities of discourse the degree in which one of them is a part of the other. Rough Mereology may be regarded as an extension of Rough Set Theory as it proposes to argue in terms of similarity relations induced from a rough inclusion instead of reasoning in terms of indiscernibility relations (cf. Chapter 1); it also proposes an extension of Mereology as it replaces the mereological primitive functor of being a part with a more general functor of being a part in a degree. Rough Mereology has deep relations to Fuzzy Set Theory as it proposes to study the properties of partial containment which is also the fundamental subject of study for Fuzzy Set Theory.

Rough Mereology is also a generalization of Mereology i.e. a theory of reasoning based on the notion of a *part*. Classical languages of mathematics are of twofold kind: the language of set theory (naive or formal) expressing classes of objects as sets consisting of "elements", "points" etc. suitable for objects perceived as built of "atoms" and the language of part relations suitable for e.g. continuous objects like solids, regions, etc. where two objects are related to each other by saying that one of them is a part of the other. In the sequel, we will rely on Mereology proposed by Stanisław Leśniewski.

In the scheme envisioned by Leśniewski, Mereology is constructed on the basis of Ontology i.e. Theory of Names (Concepts).

Ontology plays here the role of set theory: objects are represented by their names: some objects are perceived as *atomic* and given individual names while other are perceived as (distributive) classes i.e. collections of objects and given general names. Once the taxonomy of names is established, relations like *part* may be introduced (via their names formed in agreement with ontological principles). This course is adopted by contemporary theory of Spatial Reasoning: in application–oriented spatial reasoning systems, ontology appears as typology of concepts and their successive taxonomy cf. e.g. [57] (to quote a small excerpt: *edge* is *frontier, barrier, dam, cliff, shoreline*). This may be interpreted as statement that a general name (concept) i.e. *edge* is a class (a set) of individual concepts (names): *frontier, barrier* etc. Let us remark in passing that ontological usage of *is*, about which below, is opposite to the usage quoted in the last sentence i.e. *is* syntactically acts like the *esti* symbol ∈ in set theoretical notation (thus, e.g. *frontier* is *edge*).

The copula "is" used informally above is formalized in Ontology and given a precise meaning proposing thus an alternative language to set theory in which it is convenient to express properties of objects in particular their mereological structure.

In this Chapter we take this course. We regard this approach as particularly suited for Rough Set Theory which is also primarily concerned with Concept Approximation in Information Systems.

In this Chapter, we give a description of Rough Mereology in Information Systems along the lines outlined above: we give an introduction to Ontology and Mereology according to Leśniewski and we show how one may introduce them in Information Systems on the basis of Rough Set Theory. In this framework, we introduce Rough Mereology and we present some ways for defining rough inclusions. We demonstrate applications of Rough Mereology to approximate reasoning taking as the case subject Qualitative Spatial Reasoning. This topic seems to be especially suitable for rough mereological approach as it relies very essentially on Ontology and Mereology; we address its mereo–topological as well as mereo–geometrical aspects.

Keywords: rough sets, information systems/tables, ontology, mereology, rough mereology, spatial reasoning

1 Introduction

Rough Mereology has been proposed as a tool for reasoning under uncertainty (approximate reasoning) with data collected in information systems as well as a general paradigm within which it would be possible to formally describe schemes for synthesis of approximate solutions to problems posed uncertainly, vaguely or incompletely [65], [66], [67], [68], [76]. It has been shown to constitute a general framework (cf. Chapter 3: A Perspective, in ([69], vol.1) in which it is possible to develop a theory of rough computation with applications to distributed computing, knowledge granulation and computing with words (cf. [99], [100]).

Two guiding paradigms for Rough Mereology were: the Theory of Rough Sets [62] and the Theory of Fuzzy Sets [101]. From Rough Set Theory inherits Rough Mereology the idea of approximation of general concepts with particular i.e. exact concepts: in Rough Set Theory exact concepts are defined by means of attribute–value descriptors and a fortiori they are finite unions of indiscernibility classes with the indiscernibility relations providing partitions of the universe of objects. In recent investigations the need for more relaxed approximations, preferably induced by similarity or tolerance relations has been stressed and experimentally verified to yield better classification results (see e.g. [79], [75], [88], [74], [8], [40]).

Rough Mereology proposes as its primitive notion the notion of a *rough inclusion* which is a parameterized functor μ_r such that for any pair of individual entities X, Y the formula $Y\varepsilon\mu_r(X)$ means that Y is a *part of X in degree r* where $r \in [0, 1]$.

The rough inclusion may be regarded thus as a parameterized family of similarity relations: fixing r, we may define a similarity (i.e. in general, reflexive only) relation sim_r viz. $X sim_r Y$ if and only if $X\varepsilon\mu_r(Y)$ (meaning: X is similar to Y in degree r if X is a part of Y in degree r). One may then define approximations of concepts using the family $\{sim_r : r \in [0,1]\}$ along the lines of Rough Set Theory with modifications described e.g. in [88] (cf. Chapter by STEPANIUK).

As any rough inclusion is concerned with relations among objects expressed by means of degrees of partness of objects, Rough Mereology has clear connections to Fuzzy Set Theory whose basic subject of study are partial containment as well as partial membership cf. [101].

Yet another source of ideas and points of reference for Rough Mereology are Mereological Theories of Sets. We refer here to two mainstream theories of Mereology viz. Mereology due to Stanisław Leśniewski [51], [53], [54], [84], [85], [20], [90], [15], [50] and Mereology based on Connection [14], [48], [96], [17], [18], [58], [4], [6].

Of the two theories, Mereology based on Connection offers a richer variety of mereo–topological functors; yet, as Mereology of Leśniewski is based on the notion of *part*, it offers a formalism of which the formalism of Rough Mereology may be – under a suitable choice of primitive expressions – a direct extension and generalization. This is actually the case: Rough Mereology was proposed [65], [66], [67], [68] to contain Mereology of Leśniewski as the theory of the functor μ_1 and this feature is preserved in the formalization proposed here.

In the scheme envisioned by Stanisław Leśniewski, Mereology was to follow Ontology i.e. the Theory of Names (Concepts) and with this purpose in mind he proposed his Ontology [37], [49], [52], [80]. Ontological theories play an important role in Approximate Reasoning [12], [34], [81] witnessed with particular clearness in Spatial Reasoning [57], [24] where Ontology plays a basic role as it sets spatial concepts and their taxonomy; for the same reason Ontology is an immanent, although frequently implicit, component of Rough Set Theory as the latter discusses concepts and their approximations hence it is vitally interested in taxonomies of concepts and relations among them.

Mereology may be applied in exact schemes of reasoning (similarly to logic on which exact schemes of reasoning are based) related to objects whose structure (in terms of their decomposition into parts) is well–known; however, when our reasoning is applied to situations where our knowledge is uncertain, incomplete etc. we may outline a decomposition scheme only approximatively i.e. we may evaluate only that our object in question is composed of some other objects as parts up to a certain degree. In consequence, other constructs used in spatial reasoning like a neighborhood, an interior, etc. are defined approximatively only, the degree of approximation determined by degrees of being a part in an intricate way.

From this place a twofold way stretches forth; first, we may try to localize Rough Mereology within Ontology in particular within Ontology of Information Systems (i.e. Rough Set Ontology) and second, we may use Rough Mereology to create Ontology for a particular domain of interest. Here we select as such Case Study domain the domain of Spatial Reasoning because of the eminent role played in its development by mereology–based methods.

In this Chapter we take both ways throughout the text and they meet again in the final sections where we discuss Rough Mereo–topology and Rough Mereo–geometry in the context of Spatial Reasoning in Information Systems.

The purpose of mereo–topology is to express approximate topological relations among objects, while mereo–geometry attempts at description of geometric features of sets of objects, like *betweenness*, *orthogonality* etc., again, approximatively. In these approximations, we exploit rough inclusions regarded as weak metrics.

Let us observe here in passing that rough set–theoretic ideas have already been applied in problems of Spatial Reasoning e.g. in problems of multi–resolution spatial reasoning [87], [98], in problems of localization (rough localization) [10], and in the *egg–yolk approach* [19], where a region with uncertain boundary is enclosed in two regions with definite boundaries, one may find ideas very close to rough set–theoretic ideas of approximation. This fact prompts us even more towards a discussion of synthesis of concepts pertaining to Spatial Reasoning by means of Rough Mereology.

We propose to begin in Section 2 with a brief introduction to Rough Set Theory with emphasis on the nature of concepts encountered therein. Then we propose to introduce in respective Sections 3 and 4 Ontology, respectively, Mereology of Leśniewski along with interpretations of these theories in Information Systems.

Rough Mereology is discussed in Section 5 with examples of rough inclusions induced in information systems for various ontologies discussed in Section 3.

Section 6 introduces basic aspects of Qualitative Spatial Reasoning and in Section 7 devoted to Mereo–topology we show that rough inclusions induce quasi-Čech topologies which in certain models become naturally quasi-topologies (i.e. topologies without the null element).

Section 8 outlines how one may define basic primitives of geometry by means of rough inclusions and this Chapter concludes in Section 9 with examples concerning rough mereo–topology as well as rough mereo–geometry.

2 Rough Set Theory

Rough Set Theory has to do with data represented as an *information system*; the information system is formally presented as a pair $A=(U, A)$ where U is a universe of objects (represented as rows in the *information table* $U \times A$) and A is a set of *attributes* (represented as columns in the *information table* $U \times A$).

Each attribute $a \in A$ is formalized as a function $a : U \longrightarrow V_a$ where V_a is the *value set* of a. We may assume that for any a, the values in the a–column in the information table exhaust V_a.

The ontological assumption about the information system (U, A) is that it does represent the whole content of knowledge about the real world to which the data refer. A fortiori, any inference about the world is to be made on the basis of the knowledge represented by the given information system and what bears on the knowledge content is actually not individual objects (rows) but rather their information sets.

For an object $u \in U$, we call an *information set* of u, the set $Inf_A(u) = \{(a, a(u)) : a \in A\}$ (when no confusion arises, we write simply $Inf(u)$ or v when mentioning u and/or A is not necessary).

In consequence of the above assumption, we identify any two objects (rows) whose information sets are identical; formally, we define the *A–indiscernibility relation* IND_A as follows: $(u, u') \in IND_A \iff Inf_A(u) = Inf_A(u')$.

Then, the A-indiscernibility classes $[u]_a$ are represented by information sets $Inf_A(u)$ in the one–one way, and the information system (U, A) may be reduced to the information system (V, A) where $V = \{Inf_a(u) : u \in U\}$ and $a(v) = a(u)$ where $v = Inf_A(u)$ for any $v \in V$ and any $a \in A$.

A basic Ontology of an information system may be based on individuals being objects in U i.e. rows of the data table and general names being *concepts* i.e. sets of individual entities. Among concepts one has to make a distinction: there are concepts which may be described in terms of attributes and their values completely and certainly and there are concepts whose description is by necessity uncertain and incomplete.

To make this distinction more clear, we first observe that notions of an information set as well as of the indiscernibility relation may be taken relative to a set of attributes: given a non–empty set $B \subset A$ of attributes, we define for any $u \in U$ its *B–information set* $Inf_B(u)$ as the set $\{(a, a(u)) : a \in B\}$ and the *B–indiscernibility relation* IND_B is defined accordingly: $(u, u') \in IND_B \iff Inf_B(u) = Inf_B(u')$.

Now, given $B \subseteq A$ ($B \neq \emptyset$), we may define a *B–exact concept* as a concept X such that X may be represented as a union of B–indiscernibility classes i.e. **if** $u \in X \wedge (u, u') \in IND_B$ **then** $u' \in X$.

It may happen (cf. a discussion in Chapter 12, Sect. 1) that given $\mathsf{A} = (U, A)$, we are interested also in some of its extensions of the form $\mathsf{A}^\infty = (U^\infty, A^\infty)$ where $A \subseteq A^\infty$, $B \subseteq B^\infty$ and A^∞, U^∞ are countably infinite (or, potentially infinite). In such case, the following notion makes sense.

An *A–elementary exact concept* will be defined now as a B-exact concept for some $B \subset A$ (we assume that such B is finite and non-empty).

Returning to B, we may now describe all concepts approximately in terms of B–attributes and their values, viz. given a non–empty concept $X \subseteq U$, we define:

the B–lower approximation $B_- X = \{u \in U : [u]_B \subseteq X\}$;
the B–upper approximation $B^+ X = \{u \in U : [u]_B \cap X \neq \emptyset\}$.

For each concept X, one may now describe X approximately – by means of knowledge represented by the set B of attributes – as the pair $(B_- X, B^+ X)$ of two B–exact sets. The general properties of this approximation may be found in [62].

Clearly, a concept X is a B–exact concept if and only if the condition $B_- X = B^+ X$ holds. Otherwise, X is said to be a B–inexact (rough) concept. More generally, we may extend this definition taking into account all classes $[u]_B$, any finite, non–empty $B \subset A$: for a concept X, we let

the A–lower approximation $\mathsf{A}_- X = \{u \in U : \exists B. [u]_B \subseteq X\}$;
the A–upper approximation $\mathsf{A}^+ X = \{u \in U : \forall B. [u]_B \cap X \neq \emptyset\}$.

We will call a concept X an A–*exact concept* in case $\mathsf{A}_- X = \mathsf{A}^+ X$; otherwise, X will be called an A–*inexact concept*. Clearly, any elementary A–exact concept is an A–exact concept.

The taxonomy of concepts outlined above may be also rendered by means of algebraic, topological or logical structures. To this end, let us observe that

Proposition 1. *The set–theoretic operations of the union, the intersection and the complement preserve A–exact sets as well as B–exact sets for any B.*

Let us recall, that given a universe U, a family F of subsets of U closed under the set–theoretic operations of the union, the intersection and the complement is called a *field of sets*; any field of sets is a special case of a *boolean algebra* i.e. a set endowed with the operations of the join $\vee$, the meet $\wedge$ and the complement $'$ having formal properties analogous to those of $\cup, \cap, -$ cf. [72]. An *atom* in a boolean algebra $(U, \vee, \wedge, ')$ is any non–zero u with the property that $\forall v \neq 0. u \wedge v = v \iff u = v$ i.e. u is a minimal non–zero element with respect to inclusion cf. [72].

A boolean algebra is *atomic* if and only if any non–zero element contains an atom. Thus, Proposition 1 may be restated in an algebraic form cf. [13], [61]:

Proposition 2. 1. *For any B, the set of all B–exact sets with the operations of the union, the intersection and the complement is an atomic boolean algebra with atoms being B–indiscernibility classes;*
2. *The set of all A–exact sets with operations of the union, the intersection and the complement is a boolean algebra.*

Similarly, topological paraphrase follows. Recall that a topological space cf. [46] is a set U along with a family O of its subsets closed under any union, finite intersection and containing the empty set. Sets in O are *open sets*; *closed* sets then are complements to open sets. A topology may be induced into a set U also by means of operators called the *interior*, resp. the *closure* op.cit.: the interior of a set X is the union of all elements of O contained in X while the closure of

X is the intersection of all closed sets containing X. A topology is then defined by saying that a set is open (closed) if it is equal to its interior (closure). A set X which is open as well as closed is called a *clopen*. Now, we may restate Propositions 1 and 2 and earlier discussion as follows cf. [78], [97].

Proposition 3. *1. For any B, the family of all B-exact sets induces a topology in the collection of all subsets of the universe U in which any B-exact set is a clopen;*
2. The family of all A-exact sets induces a topology in the collection of all subsets of the universe U^∞ in which any A-exact set is a clopen;
3. For any B, the lower approximation operator B_- is the interior operator in the topological space of all subsets of U; similarly, the upper approximation operator B^+ is the closure operator in that topological space. The same is true for approximations A_-, A_+, respectively.

Finally, one may give a logical frame to our discussion cf. [62], [77]. Let us call a *descriptor* any pair of the form (a, v) where $a \in A$ and $v \in V_a$; descriptors may be regarded as elementary formulae from which formulae of the *descriptor logic* are formed by means of the propositional connectives $\vee$ and $\wedge$ (let us note that the negation is not necessary in the finite case considered here). The meaning $[\alpha]$ of a formula α is defined inductively:

$[(a, v)] = \{u \in U : a(u) = v\}$;
$[\alpha \vee \beta] = [\alpha] \cup [\beta]$;
$[\alpha \wedge \beta] = [\alpha] \cap [\beta]$.

For any pair $B \subseteq A, v = Inf_B(u)$, where $u \in U$, the formula $\alpha_{B,v}$ is the conjunction $\wedge_{a \in B}(a, v_a)$; clearly, $[\alpha_{B,v}] = [u]_B$.

In the sequel, we will call a pair (B, v) as above a *template*. Templates (defined also equivalently as $\alpha_{B,v}$) play an important role in rough set–theoretic methods in Knowledge Discovery and Data Mining cf. a throughout discussion in [75], [74], [8] (cf. Chapter by NGUYEN SINH HOA).

In the above discussion, we actually have introduced several types of concepts (names) like an A-exact set, B-exact set, etc. and we have also some examples of objects (represented as subsets of U) which answer to some of these names. For instance, when $U = \{u_1, u_2\}$, $A = \{a_1, a_2\}$ and $V_a = \{0, 1\}$ for each a, then the set $X = \{u : a_1(u) = 0\}$ is both A-exact and B-exact where $B = \{a_1\}$. In an informal way, we would write these facts down using phrases of the form: "X is B-exact" etc.

Similar situation happens in other contexts where approximate reasoning is carried out within other paradigms; e.g. in developing schemes for Qualitative Spatial Reasoning (see below), one comes at a very early stage at the necessity to introduce concepts, or, names, for spatial entities in question, and to set relations of hierarchy among those names i.e. at the necessity for Ontology. In application-oriented spatial reasoning systems, ontology appears as typology of concepts and their successive taxonomy cf. e.g. [57] (to quote a small excerpt: *edge is frontier, barrier, dam, cliff, shoreline*).

The informal connective "is" may undergo a formalization in which it is given a precise meaning. So we now give a formal scheme of Ontology and next we show that it does agree with our taxonomy presented above. The ontological scheme we choose to apply here is the Leśniewski Ontology cf. [37], [49], [52], [80].

We first give an introduction to formal Ontology and then we propose some interpretations of this formal scheme in Information Systems.

The reader may throughout the formal part of the next Section think of a formula $X \varepsilon Y$ as true if and only if X is a name of an individual element and Y is a name of a set of elements (actually, our specific ontologies are of this type).

3 Ontology: An Introduction to the Leśniewski Ontology

Ontology was intended by Stanisław Leśniewski as a formulation of general principles of *being* [52] cf. also [38], [49], [80], [37].

The only primitive notion of Ontology of Leśniewski is the copula *"is"* denoted by the symbol ε.

We will present now the scheme of Ontology and for the completeness sake, we give a fairly detailed account of it. Our exposition is based on the sources quoted in bibliography, in particular on [80].

All well–formed expressions of Ontology belong in classes called *semantic categories*. Categories are constructed starting with the two *basic semantic categories*: the *semantic category of names* and the *semantic category of propositions*. Either of those categories does contain *constants* as well as *variables* of the given category.

Higher order categories are constructed as functors under the agreement that each functor does belong either in the category of names (a *name–forming* functor) or in the category of propositions (a *proposition–forming* functor); however, a stratification of functors in a usual way is achieved by assigning functors to the same category if both are either name–forming or proposition–forming and if they have the same number of arguments falling into same categories, respectively.

We will use standard symbols for propositional constants as well as quantifier, parentheses etc. symbols. Formulas of Ontology will be constructed as in standard predicate calculus.

We begin with the axiom of Ontology i.e. a formula which introduces the copula ε.

3.1 Axiom of Ontology

The original axiom of Ontology defining the meaning of ε is as follows

The ontological axiom

$X \varepsilon Y \iff \exists Z. Z \varepsilon X \wedge \forall U, W.(U \varepsilon X \wedge W \varepsilon X \implies U \varepsilon W) \wedge \forall Z.(Z \varepsilon X \implies Z \varepsilon Y).$

In this axiom, the defined copula ε happens to occur in both sides of the equivalence: however, the definiendum $X\varepsilon Y$ belongs in the left side only and we may perceive the axiom as the definition of the meaning of $X\varepsilon Y$ via the meaning of terms of "lower level" $Z\varepsilon X$, $Z\varepsilon Y$ etc.

According to this reading of the axiom, we gather that the proposition $X\varepsilon Y$ is true if and only if the conjunction holds of the following three propositions:

1. $\exists Z.Z\varepsilon X$; this proposition asserts the existence of an object (name) Z which is X and so X is not an empty name.
2. $\forall U,W.(U\varepsilon X \wedge W\varepsilon X \implies U\varepsilon W)$; this proposition asserts that any two objects which are X are each other (a fortiori, they will be identified later on): thus means that X is an individual name (or, X is an individual entity, representable as a singleton).
3. $\forall Z.(Z\varepsilon X \implies Z\varepsilon Y)$; this proposition asserts that every object which is X is Y as well (or, X is contained in Y).

The meaning of $X\varepsilon Y$ can be made clear now: X is an individual (name) and this individual is Y (i.e. belongs in Y, responds to the name of Y, etc.).

Remark. Assume that we have a family F of (finite, non–empty) sets. Then we may construct a model for elementary Ontology by expressing the functor ε as the inclusion functor $\subset$. Individual objects are then singletons. The formula $X\varepsilon Y$ may be rendered as $\exists a.X = \{a\} \wedge X \subseteq Y$. Clearly, in this context we may introduce the structure of an atomic boolean algebra of sets (with singletons as atoms) by introducing functors of union, intersection and difference of sets with the provision that these operations are not accessible in cases when they would result in an empty set. We will see below that this is also the general case.

The development of Ontology rests now with consequences to the Ontology Axiom derived according to rules of derivation and rules of category formation. We first recall basic scheme of Elementary Ontology. Our presentation is based in large part on [80].

We apply Ontology in what follows as a framework to discuss Mereology and Rough Mereology: it will aid us in forming names of objects being relations among simpler names or names of individuals being collective classes of such names. The language of Ontology seems to us a suitable vehicle to code approximate reasoning schemes as e.g. we avoid deeper assumptions about nature of sets (like regularity axioms) which are particular to set theory.

3.2 Elementary Ontology

Here, we present the basic ontological scheme for dealing with individuals and their distributive classes.

Rules for functor formation. Rules for new functor formation are as follows:

1. **for proposition–forming functors:** $f(X_1, X_2, ..., X_n) \iff \alpha$
   where α is a propositional expression of ontology; an example is the ontological axiom itself.
2. **for name–forming functors:**

$$X \varepsilon f(X_1, X_2, ..., X_n) \iff$$

$$X \varepsilon X \wedge \alpha(X_1, ..., X_n)$$

   where α is a propositional expression of ontology; examples of such functors are e.g. in this Section (cf. Definition 6).
3. **for nominal constants:** $X \varepsilon C \iff X \varepsilon X \wedge \alpha$
   where α is a propositional expression of ontology.

As basic examples, we introduce two nominal constants:

Definition 4. 1. $X \varepsilon \Lambda \iff X \varepsilon X \wedge non(X \varepsilon X)$; the constant Λ is the *empty name* (here $\alpha : non(X \varepsilon X)$).
2. $X \varepsilon V \iff \exists Y. X \varepsilon Y$; the constant V is the *universal name*. $X \varepsilon V$ reads as "X is an object" (as $\exists Y. X \varepsilon Y$ is equivalent to $X \varepsilon X$, we may take as α any tautology).

Basic facts of elementary ontology. We begin with stating the three basic implications which follow from the Axiom of Ontology.

Proposition 5. *1.* $X \varepsilon Y \implies \exists Z. Z \varepsilon X$;
2. $X \varepsilon Y \implies \forall U, W. (U \varepsilon X \wedge W \varepsilon X \implies U \varepsilon W)$;
3. $X \varepsilon Y \implies \forall Z. (Z \varepsilon X \implies Z \varepsilon Y)$.

We now introduce a name forming functor $\subseteq$ via

Definition 6. $X \varepsilon \subseteq (Y) \iff X \varepsilon X \wedge \forall Z. (Z \varepsilon X \implies Z \varepsilon Y)$

We will read $X \varepsilon \subseteq (Y)$ as "X is *contained* in Y".
The basic properties of the functor $\subseteq$ are summarized below.

Proposition 7.
1. $X \varepsilon X \implies X \varepsilon \subseteq (X)$;
2. $X \varepsilon \subseteq (Y) \wedge Y \varepsilon \subseteq (Z) \implies X \varepsilon \subseteq (Z)$;
3. $X \varepsilon X \implies X \varepsilon \subseteq (V)$.

Proof. The property (1) follows from the tautology $p \implies p$ via instantiation $Z \varepsilon X \implies Z \varepsilon X$ and universal quantification: $\forall Z. Z \varepsilon X \implies Z \varepsilon X$. Similarly, (2) follows by omitting the general quantifier in the premises $X \varepsilon \subseteq (Y), Y \varepsilon \subseteq (Z)$, applying the inference rule

$$\frac{U \varepsilon X \implies U \varepsilon Y, U \varepsilon Y \implies U \varepsilon Z}{U \varepsilon X \implies U \varepsilon Z}$$

and quantifying universally. For (3), we observe that $Z \varepsilon X \implies \exists Y. Z \varepsilon Y \implies Z \varepsilon V$.

We introduce yet another name forming functor via

Definition 8. $X\varepsilon = (Y) \iff X\varepsilon X \wedge Y\varepsilon Y \wedge X\varepsilon Y \wedge Y\varepsilon X$

The functor $=$ is the individual identity functor ("X is *identical* to Z"). With its help, we may write down the individuality condition as: $\forall U, W.(U\varepsilon X \wedge W\varepsilon X \implies U = W)$. This has to be discerned from the set identity functor defined for pairs of sets, not necessarily singletons.

We state the basic properties of the existential statement $X\varepsilon Y$.

Proposition 9.
1. $X\varepsilon Y \wedge Z\varepsilon X \implies Z\varepsilon Y$;
2. $X\varepsilon Y \implies X\varepsilon X$;
3. $X\varepsilon Y \wedge Z\varepsilon X \implies Z = X$;
4. $X\varepsilon X \iff X\varepsilon V$.

Proof. Indeed, (1) follows from the ontological axiom; (2), (3) follow by virtue of definitions. Finally, $X\varepsilon X \implies X\varepsilon V$ is obvious and $X\varepsilon V \implies X\varepsilon X$ patterns (2).

On a higher level of generality, the counterpart of the identity functor $=$ is the scope equality functor $=_E$ defined as follows.

Definition 10. $X =_E Y \iff \forall Z.(Z\varepsilon X \iff Z\varepsilon Y)$

We have the following properties of this notion.

Proposition 11.
1. $X =_E Y \iff X\varepsilon \subseteq (Y) \wedge Y\varepsilon \subseteq (X)$;
2. $X = Y \iff X\varepsilon Y \wedge X =_E Y$;
3. $X\varepsilon V \implies (X = Y \iff X =_E Y)$.

All these properties follow immediately from definitions.

Finally, we give, following [80], a logical (extensional) content to individual identity. Let $\alpha(Z)$ be a propositional expression. We denote by $\alpha(X/Z)$ the expression formed from α by replacing Z with X. Then

Proposition 12. $X = Y \implies (\alpha(X/Z) \iff \alpha(Y/Z))$

Proof. (Słupecki) Define a name-forming functor $f(Y, U, W, ..)$ via

$$Z\varepsilon f(Y, U, W, ..) \iff Z\varepsilon Z \wedge \alpha(Z).$$

Assume that $X = Y$ and $\alpha(X/Z)$; hence $Y\varepsilon X$ and $X\varepsilon\ f(Y, U, W, ...)$ implying $Y\varepsilon f(Y, U, W, ..)$ and $\alpha(Y/Z)$. The conclusion follows by symmetry.

We may now pass to non-elementary ontology.

3.3 Non–Elementary Ontology

In this extension, we may encounter functors of higher order i.e. functors having as arguments also functors. We state additional rules and axioms.

The rule of substitution (cf. [80]). In expressions of ontology, one may substitute for variables of any semantic category either variables or constants of the same semantic category.

Non-elementary ontology deals with functors of higher orders whose arguments are functors as well. In original expositions of ontology, the difference among functors of various orders was stressed by different notation and symbolics e.g usage of brackets of various shapes. Here, we do not pay attention to these distinction, stressing rather the basic principles.

In addition to the usual rule of adding definitions to the system at any stage, non-elementary ontology uses the other rule: the Extensionality rule which guarantees that functors applied to the arguments already found identical give identical objects. Formally, this rule may be formulated as follows.

The Extensionality Rule. Assume that Φ is a proposition–forming functor of nominal variables $\varphi_1, ..., \varphi_k$ and that $\forall x. x\varepsilon\varphi_i \iff x\varepsilon\psi_i$ for $i = 1, 2, .., k$ and nominal variables $\psi_1, ..., \psi_k$; then, $\Phi(\varphi_1, ..., \varphi_k) \iff \Phi(\psi_1, ..., \psi_k)$; in case $\varphi_1, ..., \varphi_k, \psi_1, ..., \psi_k$ are propositional variables the extensionality rule is: $\forall i. \varphi_i \iff \psi_i \implies \Phi(\varphi_1, ..., \varphi_k) \iff \Phi(\psi_1, ..., \psi_k)$. In case Φ is a name forming functor, the consequent of the rule is of the form $\forall x. x\varepsilon\Phi(\varphi_1, ..., \varphi_k) \iff x\varepsilon\Phi(\psi_1, ..., \psi_k)$.

3.4 Basic Theorems of Non-Elementary Ontology

Here also we follow [80] in presenting an outline of non–elementary Ontology. In non–elementary Ontology it is possible to introduce meta–statements.

We begin with a statement that any proposition–forming functor induces a name.

Definition 13. For a proposition–forming functor φ, we define a name– forming functor $prop(\varphi)$ via $X\varepsilon prop(\varphi) \iff X\varepsilon V \wedge \varphi(X)$

Using the functor $prop(\varphi)$, we may prove the *law of identity for extensionality*.

Proposition 14. $X = Y \wedge \varphi(X) \implies \varphi(Y)$

Proof. $X = Y$ hence $X\varepsilon Y$ so $X\varepsilon V$; as $\varphi(X)$ we have $X\varepsilon prop(\varphi)$ and $Y\varepsilon X$ implies $Y\varepsilon prop(\varphi)$ which implies $\varphi(Y)$.

Obviously

Proposition 15. $X = Y \implies \varphi(X) \iff \varphi(Y)$

Conversely, every name induces a proposition–forming functor.

Definition 16. For a name X, let ε_X be a proposition–forming functor defined as $\varepsilon_X(Y) \iff Y\varepsilon X$.

We may prove the converse to the last proposition.

Proposition 17. $X = Y \wedge \forall \varphi.\varphi(X) \iff \varphi(Y) \implies X = Y$.

Proof. Let $X = Y$; as $\forall \varphi.\varphi(X) \iff \varphi(Y)$ we have $\varepsilon_X(X) \iff \varepsilon_X(Y)$ i.e. $X\varepsilon X \iff Y\varepsilon X$ hence $Y\varepsilon X$. Similarly, $Y\varepsilon X$ i.e. $X = Y$.

As a corollary, the *Leibnizian principle of indiscernibility* follows cf.[80].

Corollary 18. $X = Y \iff X\varepsilon V \wedge Y\varepsilon V \wedge \forall \varphi.(\varphi(X) \iff \varphi(Y))$.

To produce the counterparts for non–individual names, we need more name–induced functors.

Definition 19. For a name X, let $\varepsilon^X(Y) \iff X\varepsilon Y$.

Then we have the following proposition:

Proposition 20. $\forall \varphi.(\varphi(X) \iff \varphi(Y)) \implies X =_E Y$.

Proof. By premise, $\varepsilon^Z(X) \iff \varepsilon^Z(Y)$ i.e. $Z\varepsilon X \iff Z\varepsilon Y$ hence $\forall Z.Z\varepsilon X \iff Z\varepsilon Y$ i.e. $X =_E Y$.

To prove the converse, we need an instance of the extensionality rule.

Proposition 21. $\forall Z.(Z\varepsilon X \iff Z\varepsilon Y) \implies \forall \varphi.(\varphi(X) \iff \varphi(Y))$.

Then a corollary follows

Corollary 22. $X =_E Y \iff \forall \varphi.(\varphi(X) \iff \varphi(Y))$.

Counterparts for name–forming functors may be proved similarly.

Definition 23. For a name–forming functor f, we let $\varepsilon(f, X)(Y) \iff X\varepsilon f(Y)$.

We can now prove

Proposition 24. $X =_E Y \implies \forall f.(f(X) =_E f(Y))$.

Proof. As $X =_E Y$, we have by Corollary 22 $\varepsilon(f, Z)(X) \iff \varepsilon(f, Z)(Y)$ i.e. $Z\varepsilon f(X) \iff Z\varepsilon f(Y)$ so $\forall Z.(Z\varepsilon f(X) \iff Z\varepsilon f(Y))$ i.e. $(f(X) =_E f(Y))$.

The converse follows in a similar way via the functor $\varepsilon(Z)$ defined as $\varepsilon(Z)(X) \iff Z\varepsilon X$.

Proposition 25. $\forall f.(f(X) =_E f(Y)) \implies X =_E Y$.

Indeed, by the premises, we have $\varepsilon(Z)(X) \iff \varepsilon(Z)(Y)$ i.e. $X =_E Y$.

Corollary 26. $X =_E Y \iff \forall f.(f(X) =_E f(Y))$.

By applying a reduction technique based on the formal equivalence $(\varphi, X_1)(X_2, .., X_n) \iff \varphi(X_1, X_2, .., X_n)$ and inducting on the arity n, we may prove a general statement

Proposition 27. $\forall_i.(X_i =_E Y_i) \iff \forall f.(f(X_1, X_2, .., X_n) =_E f(Y_1, Y_2, .., Y_n))$.

The above facts correspond to indiscernibility principles of Rough Set Theory when we regard rows in data table as objects and attributes as functors: the rough set principle of extensionality may be formulated in the simplest case as follows: if x, y are B–indiscernible and $\Phi(u)$ is a formula of descriptor logic with occurrences of descriptors of the form (a, v) where $a \in B$ only then $\Phi(x) \iff \Phi(y)$.

Calculus of relations. In ontology one may define the notions of a relation, a function, the equipotency etc.cf. [80] which we briefly recapitulate here.

Definition 28. $\text{rel}(\varphi) \iff (\varphi(X_1, X_2, ..., X_n) \implies X_1 \varepsilon V \wedge X_2 \varepsilon V \wedge .... \wedge X_n \varepsilon V)$

Defines a functor *rel* stating that φ is a relation of arity n.
The notions of the domain and co-domain are introduced easily.

Definition 29. 1. $X \varepsilon \text{dom}(\varphi) \iff X \varepsilon V \wedge \text{rel}(\varphi) \wedge \exists Y.\varphi(X, Y)$.
2. $Y \varepsilon \text{co} - \text{dom}(\varphi) \iff Y \varepsilon V \wedge \text{rel}(\varphi) \wedge \exists X.\varphi(X, Y)$.

The definition of a function is straightforward

Definition 30. $\text{func}(\varphi) \iff \text{rel}(\varphi) \wedge \forall X, Z, U.(\varphi(X, Z) \wedge \varphi(X, U) \implies Z = U)$.

Defines the functor of being a function.
The functor of being an injective function may be defined similarly.

Definition 31. $\text{inj}(\varphi) \iff \text{func}(\varphi) \wedge \forall X, Z, U.(\varphi(X, Z) \wedge \varphi(U, Z) \implies X = U)$.

The functor of equipotency follows.

Definition 32. $\text{equip}(X,Y) \iff \exists \varphi.\text{inj}(\varphi) \wedge X =_E \text{dom}(\varphi) \wedge Y =_E \text{co} - \text{dom}(\varphi)$.

An example of new name–forming functors may be given by means of *fusion operations* cf. [90], [80], [37]:
given names X, Y, we define new names: $X + Y$, $X \cdot Y$, $X - Y$ as follows:

1. $Z\varepsilon X + Y \iff Z\varepsilon X \vee Z\varepsilon Y$;
2. $Z\varepsilon X \cdot Y \iff Z\varepsilon X \wedge Z\varepsilon Y$;
3. $Z\varepsilon X - Y \iff Z\varepsilon X \wedge non(Z\varepsilon Y)$.

One may notice the similarity of these operations with the operations of the union, the intersection and the difference yielding a field of sets or more generally with the operations of the join, the meet and the complement, leading to a boolean algebra. It is actually the case with Ontology and we will present a demonstration of this fact but in a slightly different context viz. we include a discussion of some equivalent axiom schemes for Ontology.

3.5 Equivalent Axiom Schemes

It turns out that one may axiomatize Ontology by means of other notions a fortiori by means of distinct schemes of axioms. The most important are axiomatic characterizations of Ontology by means of ordering functors (relations) as they lead directly to well-known mathematical structures viz. quasi–boolean algebras cf. also [37], [80].

We define a binary proposition–forming functor $ord(X,Y)$ as follows.

Definition 33. $ord(X,Y) \iff \exists Z.Z\varepsilon X \wedge \forall Z.(Z\varepsilon X \implies Z\varepsilon Y)$.

Then we have

Proposition 34.
$X\varepsilon Y \iff ord(X,Y) \wedge \forall U,W.(ord(U,X) \wedge ord(W,X) \implies ord(U,W))$.

For a better visualization of the formulae, we introduce a shortcut notation $X < Y$ for $ord(X,Y)$. This is however only a notational convenience, not a new type of a functor.

Proof. Assume first that $X\varepsilon Y$; then clearly $X < Y$. Now assume that $U < X, W < X$. Observe that $U < X$ implies that U is an individual i.e. $U\varepsilon U$ hence $U\varepsilon X$; similarly, $W\varepsilon X$ and as X is an individual, it follows that $U = W$. From $U\varepsilon V, W\varepsilon V$ it follows by Proposition 11 that $U =_E W$ and finally $U < W$.

Now assume that $ord(X,Y) \wedge \forall U,W.(ord(U,X) \wedge ord(W,X) \implies ord(U,W))$. It suffices to check that $U\varepsilon X \wedge W\varepsilon X \implies U\varepsilon W$. Assume $U\varepsilon X \wedge W\varepsilon X$; then $U < X, W < X$ and thus $U = W$ implying $X\varepsilon X$ hence $X\varepsilon Y$.

Proposition 35. *[Lejewski [49]]* $\exists Z.Z\varepsilon X \iff \exists Z.Z < X$

Proof. Clearly, $\exists Z.Z\varepsilon X$ implies $\exists Z.Z < X$. Conversely, $\exists Z.Z < X$ gives $A\varepsilon X$, for some A hence $\exists B.B\varepsilon A$ and $\forall T.(T\varepsilon A \implies T\varepsilon X)$ so $B\varepsilon X$ and finally $\exists Z.Z\varepsilon X$.

The following proposition establishes a deeper parallelism between "is" and "entails".

Proposition 36. *[Sobociński cf. [49]]*
$$\forall U, W.(U\varepsilon X \wedge W\varepsilon X \implies U\varepsilon W) \iff \forall U, W.(U < X \wedge W < X \implies U < W)$$

Proof. Assume $\forall U, W.(U\varepsilon X \wedge W\varepsilon X \implies U\varepsilon W)$ and $U < X \wedge W < X$. For any P, it suffices to show that $P\varepsilon U \implies P\varepsilon W$. Let $P\varepsilon U$; $Q\varepsilon W$; then, $P\varepsilon X$, $Q\varepsilon X$ and by the premises, $P\varepsilon Q$ hence $P\varepsilon W$.

Conversely, from $\forall U, W.(U < X \wedge W < X \implies U < W)$ and $U\varepsilon X \wedge W\varepsilon X$ it follows that $U < X, W < X$ and so $U < W$; by Proposition , $U\varepsilon W$.

We may now establish

Proposition 37. *[Lejewski [49]]*
$$X < Y \iff \exists Z.Z < X \wedge \forall Z.(Z < X \implies \exists W.(W\varepsilon Z \wedge W\varepsilon Y))$$

Proof. First, let $X < Y$. Then, $\exists Z.Z < X$ hence $\forall T.(T\varepsilon Z \implies T\varepsilon Y)$ and $\exists W.W\varepsilon Z$ which yields $A\varepsilon Z$ hence $A\varepsilon Y$ for some A and finally $\exists W.W\varepsilon Z \wedge W\varepsilon Y$.

Now, let $\exists Z.Z < X \wedge \forall Z.(Z < X \implies \exists W.(W\varepsilon Z \wedge W\varepsilon Y))$; assume $Z\varepsilon X$ so $Z < X$. Then $W\varepsilon Z \wedge W\varepsilon Y$ for some W hence $Z\varepsilon Y$. It follows that $\forall Z.(Z\varepsilon X \implies Z\varepsilon Y)$ so $X < Y$.

The following is an immediate consequence of the above propositions and Definition 33.

Proposition 38. *[Lejewski [49]]*
(ATB) $X < Y \iff \exists Z.Z < X \wedge \forall Z.(Z < X \implies$
$\exists W.(W < Z \wedge W < Y) \wedge \forall P, Q.(P < W \wedge Q < W \implies P < Q))$

Proposition 39. *[Lejewski [49]]* (ATB) *is equivalent to the Axiom of Ontology*

Proof. First, we show that (ATB) implies the Ontology Axiom. It is sufficient to check that: $(Z < X \implies \exists W.W\varepsilon Z \wedge W\varepsilon Y) \iff (Z\varepsilon X \implies Z\varepsilon Y)$. Assume then that $(Z < X \implies \exists W.W\varepsilon Z \wedge W\varepsilon Y)$ and $Z\varepsilon X$. Hence $Z < X$ and so $W\varepsilon Z, W\varepsilon Y$ for some W and thus $Z\varepsilon Y$. Conversely, from $(Z\varepsilon X \implies Z\varepsilon Y)$ and $Z < X$ it follows $\exists W.W\varepsilon Z$ and $\forall W.(W\varepsilon Z \implies W\varepsilon X)$ hence for some W we have $W\varepsilon Z$ hence $W\varepsilon X$ and finally $W\varepsilon Y$ proving $\exists W.W\varepsilon Z \wedge W\varepsilon Y$. That the Ontology Axiom implies (ATB) follows similarly.

We now introduce a constant name AT.

Definition 40. $X\varepsilon AT \iff X\varepsilon X \wedge \forall P, Q.(P < X \wedge Q < X \implies P < Q)$.

The term $X\varepsilon AT$ reads "X is an atom".
We may check that

Proposition 41. *(ATB) is an axiom for atomic boolean algebra without the zero element.*

The proof consists in a straightforward checking of equivalence.
We have therefore

Proposition 42. *[cf. [37]] Theorems of ontology are those which are true in every model for an atomic boolean algebra without the null element.*

3.6 Ontology in Information Systems

Let us observe that in an information system the Leibnizian principle of indiscernibility is observed in e.g. the following form:

$$a(x) = a(y) \text{ for each } a \in A \text{ if and only if } [x]_A = [y]_A.$$

It follows that we have three types of entities when considering an information system A: objects (corresponding to rows in the information table), indiscernibility classes of objects (over various distinct subsets of attributes), collections of entities of two former types.

We propose to discuss here two types of Ontology, related to each other, but distinct in the light of rough mereology (see Section 5.2). We discuss them below.

Ontology of B–indiscernibility. We form first individual entities. Recall that the symbol $[u]_B$ denotes the indiscernibility class of u over B. We call a B–*individual* any B–exact subset of the universe U. Then, we introduce B–*names* as distributive classes (sets) of B–individuals. Then, $X \varepsilon Y$ means that X is a B–individual and Y is a set of such individuals containing X.

Let us observe that some (regular) names may be introduced via templates e.g. given $C \subset B$ and $v = Inf_C(u)$ for some $u \in U$, we may interpret the template (C, v) in a twofold way.

1. We may take the meaning $[C, v]$ of the template (C, v) as denoting the *individual* i.e. the set $\{u \in U : \forall a \in C . a(u) = v_a\}$;
2. We may take (C, v) as a name i.e. the set of all individuals $[B, w]$ such that $w|C = v$.

Thus, $[B, w]\varepsilon(C, v) \iff w|C = v$. In this way we have created names for classes (sets) without the need of specifying these sets. We may propose a name-forming functor + defined as:

$$X\varepsilon(C, v) + (D, w) \iff X\varepsilon(C, v) \vee X\varepsilon(D, w)$$

and then we may write down any name as $+_{i+1}^k (C_i, v_i)$.

Ontology of A–indiscernibility. Our next Ontology will formally be an extension of the former one. We define individual entities as A–exact sets. General names will be defined as sets (collections, lists) of individual entities. Again, we may use templates to express certain (regular) names (recall that $C, D, ..$ are finite subsets of A^∞):

1. We may take the meaning $[C, v]$ of the template (C, v) as denoting the A–*individual* i.e. the set $\{u \in U : \forall a \in C . a(u) = v_a\}$;
2. We may take (C, v) as a name i.e. the set of all individuals $[D, w]$ such that $C \subseteq D \wedge w|C = v$.

We may introduce propositional connectives $\vee, \wedge$ into templates to form the *template logic* corresponding to the boolean structure in Calculus of Names viz.

1. $(C,v)+(D,w)$ is the name containing those individuals which either fall in (C,v) or they fall in (D,w);
2. $(C,v) \cdot (D,w)$ is the name containing those individuals which fall both in (C,v) and (D,w);
3. $(C,v)-(D,w)$ is the name containing those individuals which fall in (C,v) but not in (D,w).

One may see that a general name for A–exact sets – in case all value sets V_a of attributes $a \in A^\infty$ are finite – may be written down in the form $+_{i=1}^{k}(C_i, v_i)$ for an appropriate k and a choice of $C_i \subseteq A, v_i = Inf_{C_i}(u_i), u_i \in U$ where $i = 1, 2, .., k$ (recall our assumption that C_i's are finite). Thus, we have for instance: $X \varepsilon Y$ when $Y =_E +_{i=1}^{k}(C_i, v_i) \wedge \exists i.X = [C_i, v_i]$. Similar considerations are valid for the case of B–exact sets.

We return to these examples in Sections on Mereology and Rough Mereology.

4 The Leśniewski Mereology

Mereology is a theory of collective classes i.e. individual entities representing general names as opposed to Ontology which is a theory of distributive classes i.e. general names. The distinction between a distributive class and its collective class counterpart is like the distinction between a family of sets and its union i.e. a specific set. For instance, we may represent United States through the list of its states (i.e. as a distributive class) or we may mean "United States" as the collective class of its states e.g. their union (i.e. as the individual having spatial location in the northern hemisphere and containing as parts all of its states).

This explains the purpose of Mereology which is to express objects though their parts and define some complex objects in terms of their parts.

Mereology may be based on each of a few notions like those of a *part*, an *element (ingredient)*, a *class* etc. Historically it has been conceived by Stanisław Leśniewski [51], [53], [54] cf. [20], [90], [84], [85], as a theory of the relation *part* and we here follow this line of development. In particular, we present the development of Mereology within Ontology: names of mereological constructs will be formed by means of ontological rules. The meaning of the copula ε explained above, we may now make use of this notation in what follows.

We assume that the copula ε is given and that the Ontology Axiom holds. Under these assumptions, we introduce the notion of the name–forming functor *pt* of *part*. Our presentation is based on [53] in the first place.

4.1 Mereology Axioms

(A1) $X \varepsilon pt(Y) \Longrightarrow X \varepsilon X \wedge Y \varepsilon Y$;
this means that the functor *pt* is defined for **individual entities** only.
(A2) $X \varepsilon pt(Y) \wedge Y \varepsilon pt(Z) \Longrightarrow X \varepsilon pt(Z)$;
meaning that the functor *pt* is transitive i.e. a part of a part is a part.
(A3) $non(X \varepsilon pt(X))$;

which means that the functor pt is non–reflexive (or, equivalently, if $X\varepsilon pt(Y)$ then $non(Y\varepsilon pt(X))$.

On the basis of the notion of part, we define the notion of an *element* (an improper, possibly, part; called originally, an *ingredient*) as a name–forming functor el.

Definition 43. $X\varepsilon el(Y) \iff X\varepsilon pt(Y) \lor X = Y$

It is clearly possible to introduce mereology in terms of the functor el as a partial order functor (i.e. being, consecutively, *reflexive*: $X\varepsilon el(X)$, *transitive*: $X\varepsilon el(Y) \land Y\varepsilon el(Z) \implies X\varepsilon el(Z)$, *weakly symmetric*: $X\varepsilon el(Y) \land Y\varepsilon el(X) \implies X = Y$) cf. Proposition 45.

The remaining axioms of mereology are related to the class functor which converts distributive classes (general names) into individual entities: thus, it may be used to represent "United States" as an individual comprising all US states. The class operator Kl is a principal tool in applications of Rough Mereology to problems of Distributed Systems, Knowledge Granulation, Computing with Words where it does play the role of granulating (clustering) operator allowing for forming granules of knowledge and subsequently instrumental in calculi on them cf. [65], [66], [67], [68], [76].

We may now introduce the notion of a (collective) class via a name–forming functor Kl.

Definition 44. $X\varepsilon Kl(Y) \iff$
$\exists Z.Z\varepsilon Y \land \forall Z.(Z\varepsilon Y \implies Z\varepsilon el(X)) \land \forall Z.(Z\varepsilon el(X) \implies$
$\exists U, W.(U\varepsilon Y \land W\varepsilon el(U) \land W\varepsilon el(Z)))$.

Let us disentangle the meaning of this Definition. First, we may realize that the class operator Kl is intended as the operator converting names (general sets of entities) into individual entities i.e. collective classes; its role may be fully compared to the role of the union of sets operator in the classical set theory. The analogy is indeed not only functional but also formal.

Let us look at the subsequent conjuncts in the defining formula above.

1. $\exists Z.Z\varepsilon Y$;
   this means that Y is a non–empty name (recall that the union of the empty family of sets is the empty set hence prohibited in Ontology).
2. $\forall Z.(Z\varepsilon Y \implies Z\varepsilon el(X))$;
   meaning that any individual listed in Y is an element of $Kl(Y)$ (compare with: any element of the family of sets is a subset of the union of that family).
3. $\forall Z.(Z\varepsilon el(X) \implies \exists U, W.(U\varepsilon Y \land W\varepsilon el(U) \land W\varepsilon el(Z)))$;
   this means that any element of $Kl(Y)$ has an element in common with an individual in Y (similarly, any element in the union of a family of sets is an element in at least one member of this family).

Thus, the class functor pastes together individuals in Y by means of their common elements

The class functor is subject to the following postulates.

(A4) $X\varepsilon Kl(Y) \wedge Z\varepsilon Kl(Y) \implies X\varepsilon Z$;
this means that $Kl(Y)$ is an individual name (entity), for any (non–empty) Y.

(A5) $\exists Z.Z\varepsilon Y \iff \exists Z.Z\varepsilon Kl(Y)$;
meaning that $Kl(Y)$ exists (i.e. is a non–empty individual name) if and only if Y is a non–empty name.

4.2 First Consequences

From axioms above, we start a build-up of mereology. We begin with simple consequences of axioms.

Proposition 45.
1. $X\varepsilon el(Y) \wedge Y\varepsilon el(Z) \implies X\varepsilon el(Z)$;
2. $X\varepsilon el(Y) \wedge Y\varepsilon el(X) \implies X = Y$;
3. $X\varepsilon el(X)$.

Proposition 46. $X\varepsilon X \implies X\varepsilon Kl(elX)$ where $Z\varepsilon elX \iff Z\varepsilon el(X)$.

Proof. Assume $X\varepsilon X$; hence $X\varepsilon el(X)$ and so $\exists Z.Z\varepsilon el(X)$. For each $Z\varepsilon el(X)$ the formula $\exists U, W.\ U\varepsilon el(X) \wedge W\varepsilon el(U) \wedge W\varepsilon el(Z))$ is satisfied with $U = W = Z$ and so $X\varepsilon Kl(elX)$.

Proposition 47. $X\varepsilon el(Y) \iff \exists Z.X\varepsilon Z \wedge Y\varepsilon Kl(Z)$

Proof. $X\varepsilon el(Y)$ implies by Proposition 46 that $Z = Kl(Y)$ satisfies the right hand formula; the converse follows from the class definition.

We define new name-forming functors.

Definition 48.
1. $X\varepsilon partY \iff X\varepsilon pt(Y)$;
2. $X\varepsilon(\varepsilon Y) \iff X\varepsilon Y$.

We have counterparts of Proposition 47.

Proposition 49. $X\varepsilon X \implies X\varepsilon Kl(\varepsilon X)$.

Proof. Assume $X\varepsilon X$; then $\exists Z.Z\varepsilon X, \forall Z.(Z\varepsilon X \implies Z\varepsilon el(X))$,
$\forall Z.(Z\varepsilon el(X) \implies \exists U, W.(U\varepsilon X \wedge W\varepsilon el(U) \wedge W\varepsilon el(Z))$ are satisfied.

Proposition 50. $\exists Z.Z\varepsilon pt(X) \implies X\varepsilon Kl(partX)$.

Proof. Assume $\exists Z.Z\varepsilon pt(X)$; it suffices to check that $T\varepsilon el(X) \implies \exists U,W.W\varepsilon el(U) \wedge W\varepsilon el(T) \wedge U\varepsilon pt(X)$; in case $T = X$, let $U = W = Z$ where $Z\varepsilon pt(X)$ and in case $T\varepsilon pt(X)$ let $U = W = T$.

Proposition 51. $X\varepsilon X \implies X\varepsilon Kl(X)$.

We now define the notion of a set, weaker than that of a class; we may observe that class is the set with the universality property: $\forall Z.(Z\varepsilon Y \implies Z\varepsilon el(Kl(Y)))$.

Definition 52. $X\varepsilon set(Y) \iff \exists Z.Z\varepsilon Y \wedge \forall Z.(Z\varepsilon el(X) \implies \exists U,W.(U\varepsilon el(Z) \wedge U\varepsilon Y \wedge W\varepsilon el(U) \wedge W\varepsilon el(Z)))$.

We now recall following [51], [53] some technical propositions leading to thesis (Proposition 61) equivalent to (A4) and giving an inference rule about the functor el.

Proposition 53. $X\varepsilon set(Y) \wedge \forall Z.(Z\varepsilon Y \implies Z\varepsilon W) \wedge T\varepsilon el(X) \implies \exists P,R.(P\varepsilon el(T) \wedge P\varepsilon el(R) \wedge R\varepsilon W \wedge R\varepsilon el(X))$.

Proof. From $X\varepsilon set(Y)$ and $T\varepsilon el(X)$ it follows that $A\varepsilon el(T)$, $A\varepsilon el(B)$, $B\varepsilon el(X)$, $B\varepsilon Y$ hence $\forall Z.(Z\varepsilon Y \implies Z\varepsilon W)$ implies $B\varepsilon W$ for some A, B so A, B satisfy the consequent.

Proposition 54. $X\varepsilon set(Y) \wedge \forall Z.(Z\varepsilon Y \implies Z\varepsilon W) \implies X\varepsilon set(W)$.

Proposition 55. $X\varepsilon Y \implies X\varepsilon set(Y)$.

Proposition 56. $X\varepsilon Kl(Y) \implies X\varepsilon set(Y)$.

Proofs are obvious.

Proposition 57. $X\varepsilon Kl(set(Y)) \wedge Z\varepsilon el(X) \implies \exists U,W.(U\varepsilon el(Z) \wedge U\varepsilon el(W) \wedge W\varepsilon Y \wedge W\varepsilon el(X))$.

Proof. Assume that $X\varepsilon Kl(set(Y)), Z\varepsilon el(X)$; there exist U,W with the properties $U\varepsilon el(Z), U\varepsilon el(W), W\varepsilon set(Y), W\varepsilon el(X)$ hence there exist P,Q with $P\varepsilon el(U)$, $P\varepsilon el(Q), Q\varepsilon el(W), Q\varepsilon Y$ implying $Q\varepsilon el(X)$. Thus P,Q satisfy the consequent.

Corollary 58. $X\varepsilon Kl(set(Y)) \implies X\varepsilon Kl(Y)$.

Proof. Assume that $X\varepsilon Kl(set(Y))$; then, $Z\varepsilon Y$ implies $Z\varepsilon set(Y)$ by Proposition 55 and finally, by the assumption, $Z\varepsilon el(X)$. Now, for $Z\varepsilon el(X)$, there exist U,W with $U\varepsilon el(Z), U\varepsilon el(W), W\varepsilon set(Y)$, so by the definition of a set (Definition 52), we may assume that $\exists P,Q.P\varepsilon el(U), P\varepsilon el(Q), Q\varepsilon Y$. It follows that $X\varepsilon Kl(Y)$.

Corollary 59. $X\varepsilon Kl(set(Y)) \iff X\varepsilon Kl(Y)$.

Sets are elements of classes.

Proposition 60. $X\varepsilon set(Y) \implies X\varepsilon el(Kl(Y))$.

Proof. $X\varepsilon set(Y)$ implies $\exists U.U\varepsilon Y$ hence $\exists Z.Z\varepsilon Kl(Y)$ by (A5) and $Z\varepsilon Kl(set(Y))$ by Corollary 59 so finally $X\varepsilon el(Z)$.

Proposition 61. $X\varepsilon X \wedge \forall Z.(Z\varepsilon el(X) \implies \exists T, T\varepsilon el(Z) \wedge T\varepsilon el(Y)) \implies X\varepsilon el(Y)$.

Actually, one may prove that this proposition is equivalent to (A4). It may be regarded as an inference rule about the functor *el* and also as an alternative axiom.

Proof. From $X\varepsilon X, \forall Z.(Z\varepsilon el(X) \implies \exists T.T\varepsilon el(Z) \wedge T\varepsilon el(Y))$ it follows that $X \varepsilon\ set(el(Y))$ and by Proposition 60, $X\varepsilon el(Kl(el(Y))$ hence by Proposition 46, $X\varepsilon el(Y)$.

4.3 Subset, Complement, Relations

We define the notions of a subset, a complement and we will look at relations and functions in mereological context. We define first the notion of a subset as a name-forming functor *sub* of an individual variable.

Definition 62. $X\varepsilon sub(Y) \iff X\varepsilon X \wedge Y\varepsilon Y \wedge \forall Z(Z\varepsilon el(X) \implies Z\varepsilon el(Y))$.

Proposition 63. $X\varepsilon sub(Y) \implies X\varepsilon el(Y)$.

Proof. As $X\varepsilon el(X)$ by Proposition 45, it follows that $X\varepsilon sub(Y)$ implies $X\varepsilon el(Y)$.

Proposition 64. $X\varepsilon el(Y) \implies X\varepsilon sub(Y)$.

Corollary 65. $X\varepsilon el(Y) \iff X\varepsilon sub(Y)$.

We now define the notion of being external as a binary proposition-forming functor *ext* of individual variables.

Definition 66. $ext(X,Y) \iff X\varepsilon X \wedge Y\varepsilon Y \wedge non(\exists Z.Z\varepsilon el(X) \wedge Z\varepsilon el(Y))$.

Proposition 67.
1. $X\varepsilon X \implies non(ext(X,X))$;
2. $ext(X,Y) \iff ext(Y,X)$.

The notion of a complement is rendered as a name-forming functor *comp* of two individual variables. We first define a new name Θ as follows: $U\varepsilon\Theta \iff U\varepsilon el(Z) \wedge ext(U,Y)$.

Definition 68. $X\varepsilon comp(Y,Z) \iff Y\varepsilon sub(Z) \wedge X\varepsilon Kl(\Theta)$.

Proposition 69. $X\varepsilon comp(Y,Z) \implies ext(X,Y)$.

Proof. As $X\varepsilon comp(Y,Z)$, if $T\varepsilon el(X)$ then we have U,W with $U\varepsilon el(T), U\varepsilon el(W)$, $W\varepsilon el(Z)$, $ext(W,Y)$ hence $non(U\varepsilon el(Y))$ and thus $non(U\varepsilon el(Y))$.

Corollary 70. $X\varepsilon X \implies non(X\varepsilon comp(X,Z))$.

Proposition 71. $X\varepsilon comp(Y,Z) \implies X\varepsilon el(Z)$

Proof. As $X\varepsilon comp(Y,Z)$ implies $X\varepsilon Kl(\Theta)$ hence if $T\varepsilon el(X)$ then we have U,W with $U\varepsilon el(T), U\varepsilon el(W), W\varepsilon el(Z)$ hence by Proposition 61, $X\varepsilon el(Z)$.

Coming to relations, already defined in Ontology, we will introduce a Proposition–forming functor $\longrightarrow (\varphi, X, Y)$ where X,Y are individual names and φ is a relation.

Definition 72.
$\longrightarrow (\varphi, X, Y)$ iff $X\varepsilon X \wedge Y\varepsilon Y \wedge \forall T.(T\varepsilon el(X) \wedge \varphi(T,U) \implies U\varepsilon el(Y))$.

Proposition 73.
1. $\longrightarrow (\varphi, X, Y) \wedge Z\varepsilon el(X) \implies \longrightarrow (\varphi, Z, Y)$;
2. $\longrightarrow (\varphi, X, Z) \wedge Z\varepsilon el(Y) \implies \longrightarrow (\varphi, X, Y)$;
3. $\longrightarrow (\varphi, X, Y) \implies \longrightarrow (\varphi, X, Kl(el(Y)))$;
4. $\longrightarrow (\varphi, X, Y) \implies \longrightarrow (\varphi, Kl(el(X)), Y)$.

Proof. Obvious.

4.4 Other Axiomatics, Completeness

As with Ontology, Mereology may be axiomatized in terms of notions derived from the *part* functor. We begin with an axiom due to Sobociński [84], which formalizes mereology in terms of the functor *el*.

The Sobociński Axiom

(S) $X\varepsilon el(Y) \iff Y\varepsilon Y \wedge \forall f, Z.[\forall C.(C\varepsilon f(Z) \iff$
$(\forall D.D\varepsilon Z \implies D\varepsilon el(C)) \wedge (\forall D.D\varepsilon el(C) \implies \exists E, F.E\varepsilon Z \wedge F\varepsilon el(D) \wedge F\varepsilon el(E)))$
$\wedge Y\varepsilon el(Y) \wedge Y\varepsilon Z] \implies X\varepsilon el(f(Z))$.

It is not difficult to see that (S) is a theorem of mereology.

Proposition 74. *(S) is a thesis of mereology*

Proof. It is easily seen that $f(Z)$ is $Kl(Z)$ so (S) reads as the thesis:

$$X\varepsilon el(Y) \iff Y\varepsilon Y \wedge \forall Z.(Y\varepsilon Z \implies X\varepsilon el(Kl(Z))$$

which is true in Mereology.

Proposition 75. *(S) implies axioms of mereology*

Proof. Assume (S); then $\exists Z.Z\varepsilon el(Y)$ *(the left side)* implies $Y\varepsilon el(Y)$. Similarly, taking into account that $f(Z)$ denotes $Kl(Z)$, we find that $X\varepsilon Kl(elX)$ and from this we obtain that $X\varepsilon el(Y) \wedge Y\varepsilon el(Z) \implies X\varepsilon el(Z)$. Letting $X\varepsilon pt(Y) \iff X\varepsilon el(Y) \wedge non(X=Y)$, we arrive at (A1) and (A2). The uniqueness of $Kl(Z)$ follows from the subformula $C\varepsilon f(Z) \iff$
$(\forall D.D\varepsilon Z \implies D\varepsilon el(C)) \wedge (\forall D.D\varepsilon el(C) \implies \exists E, F.E\varepsilon Z \wedge F\varepsilon el(D) \wedge F\varepsilon el(E)$
of (S) and similarly, it does imply that $\exists E.E\varepsilon Z \iff \exists C.C\varepsilon Kl(Z)$.

A similar axiom has been proposed by Lejewski [50].
The Lejewski Axiom:
(L) $X\varepsilon el(Y) \iff Y\varepsilon Y \wedge \forall f, Z, C.\{\forall D.[D\varepsilon f(Z) \iff$
$(\forall E.(\exists F.F\varepsilon el(D) \wedge F\varepsilon el(F)) \iff \exists G, H.G\varepsilon Z \wedge H\varepsilon el(E) \wedge H\varepsilon el(G))]$
$\wedge Y\varepsilon el(Y) \wedge Y\varepsilon el(C) \wedge Y\varepsilon Z]\} \implies X\varepsilon el(f(Z))$.

One can show similarly that (L) and (S) are equivalent.

A paraphrase of these axioms has been proposed by Clay [15].
The Clay Axiom:
(Cl) $X\varepsilon el(Y) \iff \{X\varepsilon X \wedge Y\varepsilon Y \wedge Y\varepsilon el(Y) \implies \forall U, W.(Y\varepsilon U \wedge$
$(\forall C.(C\varepsilon W \iff \forall D.(D\varepsilon U \implies D\varepsilon el(C)) \wedge \forall D.(D\varepsilon el(C) \implies$
$\exists E, F.(E\varepsilon U \wedge F\varepsilon el(D) \wedge F\varepsilon el(E)) \implies X\varepsilon el(W)\}$.

We may formally replace the functor el with a new name $\leq$ and we may read (and write down) $X\varepsilon X$ as $X\varepsilon V$; replacing U, W by small letters u, w symbolizing not necessarily individual names and denoting by f the field of $\leq$, we arrive at (Cl) in the form

(M) $X\varepsilon \leq (Y) \iff \{X\varepsilon V \wedge Y\varepsilon V \wedge Y\varepsilon \leq (Y) \implies \forall u, w.(u \subset f \wedge w \subset f \wedge Y\varepsilon u \wedge$
$(\forall C.(C\varepsilon w \iff \forall D.(D\varepsilon u \implies D\varepsilon \leq (C)) \wedge \forall D.(D\varepsilon \leq (C) \implies$
$\exists E, F.(E\varepsilon u \wedge F\varepsilon \leq (D) \wedge F\varepsilon \leq (D)) \implies X\varepsilon \leq (w)\}$.

Substituting for V the universe U, neglecting the copula and taking into account that w must be an individual name as pointed to by (M), one gets following [15],

(CBA) $X \leq Y \iff \{X \in U \wedge Y \in U \wedge Y \leq Y \implies \forall u, w.(u \subset U \wedge w \subset U \wedge Y \in u \wedge$
$(\forall C.(C \in w \iff \forall D.(D \in u \implies \overline{D \leq C}) \wedge \forall D.(D \leq C \implies$
$\exists E, F.(E \in u \wedge F \leq D \wedge F \leq E)) \implies \exists L. w = \{L\} \wedge X \leq L$.

It may be checked that (CBA) is the axiom for a complete boolean algebra without the null element.

Therefore

Proposition 76. *[Tarski [92]]*
 Models of mereology are models of complete boolean algebras without zero.

Mereology is therefore complete with respect to algebraic structures which are models for complete boolean algebras without zero.

4.5 Mereology in Information Systems

As with Ontology, we discern between two basic types of mereological structures. We refer to two kinds of Ontology (cf.Section 3.6).

Mereology of B–indiscernibility. Recall that individual entities in this case are B–exact sets and general names are collections (sets, lists) of B–exact sets. Recalling our usage of templates, we may write down any general name as $+_{i=1}^{k}(B,v_i)$ where $v_i = Inf_B(u_i)$ for some $u_i \in U$ while $[B,v]$ denotes the B–indiscernibility class $[u]_B$ where $v = Inf_B(u)$.

Let us observe that : given a (regular) name $Y = +_{i=1}^{k}(B,v_i)$, we have: $Kl(Y) = \bigcup_{i=1}^{k}[B,v_i]$ i.e. the class of Y is the individual $X = \bigcup_{i=1}^{k}[B,v_i]$; indeed, by the class definition, it suffices to check that any part of Y i.e. any $Z_i = [B,v_i]$ has an element (e.g. Z_i itself) which is in turn an element of an individual (again, Z_i itself) which is in Y (and obviously, $Z_i \varepsilon Y$).

Thus, we may denote any individual either via set theoretic union as

$$\bigcup_{i=1}^{k}[B,v_i]$$

or via the class operator (as $Kl(Y)$ where Y is the name $+_{i=1}^{k}(B,v_i)$).

In particular, we have: $Kl((B,v)) = [B,v]$ for any (B,v).

We define the functor pt in this case: for an individual $Kl(Y)$ as above, we let $X\varepsilon pt(Kl(Y)) \Longleftrightarrow k \geq 2 \wedge \exists i.X = [B,v_i]$.

Thus, we propose that parts of which B–exact sets are formed be simply B–indiscernibility classes; clearly, a B–indiscernibility class has no parts itself–it is an *atom*.

Mereology of A–indiscernibility: all A–indiscernibility classes are atoms. We assume A to be a finite set. We may just repeat the considerations in the preceding paragraph. We define the functor pt in this case: for an individual $Kl(Y)$ with Y: $+i = 1^k(C_i,v_i)$, we let $X\varepsilon pt(Kl(Y))$ if either

1. $k \geq 2 \wedge \exists i, D, w.X = [D,w] \wedge C_i \subseteq D \wedge w|C_i = v_i$;
or
2. $k = 1 \wedge C_1 \subset A \wedge \exists D, w.X = [D,w] \wedge C_1 \subset D \wedge w|C_1 = v_1$.

Thus, we propose that parts of which A–exact sets are formed are simply A–indiscernibility classes. Let us observe that : given a (regular) name $Y = +_{i=1}^{k}(C_i,v_i)$, we have: $Kl(Y) = \bigcup_{i=1}^{k}[C_i,v_i]$ i.e. the class of Y is the individual

$X = \bigcup_{i=1}^{k}[C_i, v_i]$; indeed, by the class definition, it suffices to check that any part of Y i.e. any $Z_i = [C_i, v_i]$ has an element (e.g. Z_i itself) which is in turn an element of an individual (again, Z_i itself) which is in Y (and obviously, $Z_i \varepsilon Y$).

Mereology of A–indiscernibility: the non–atomic case. We assume that A is infinite (and countable). We adopt the definition of a part given above; it follows immediately, that in this case there are no atoms: any individual has an infinite decomposition scheme into parts.

Again, we have here that: given a (regular) name $Y = +_{i=1}^{k}(C_i, v_i)$, we have: $Kl(Y) = \bigcup_{i=1}^{k}[C_i, v_i]$ i.e. the class of Y is the individual $X = \bigcup_{i=1}^{k}[C_i, v_i]$.

The above examples explain the role of the class operator in information systems; we may exploit mereology also in the task of creating Ontology of non–exact concepts (names). Here, we find usage for the mereological functor of a subset.

Ontology of inexact (rough) concepts. We define a relation *rough–approx* by letting *rough–approx*$(X, Y) \iff X \varepsilon X \wedge Y \varepsilon Y \wedge X \varepsilon sub(Y)$. We give the relation *rough–approx* a new name via $(X, Y)\varepsilon\ ROUGH \iff$ *rough–approx*(X, Y).

Then $ROUGH$ is the name which contains all pairs approximating inexact concepts from below as well as from above (possibly, with some surplus: some pairs (X, Y) may not define an inexact concept).

The reader may decide for themselves whether ontological/mereological language adopted here is to be preferred to the standard language of naive set theory; anyway all ideas expressed here in the former language, may be rendered in the latter. However, an advantage of the former language is that we may pass smoothly to rough mereology as set in the discussed up to now framework.

5 Rough Mereology

Rough mereology is an extension of mereology based on the predicate of being a part in a degree; this predicate is rendered here as a family of name–forming functors μ_r parameterized by a real parameter r in the interval $[0, 1]$ with the intent that $X\varepsilon\mu_r(Y)$ reads "X *is a part of Y in degree* r". We begin with the set of axioms and we construct the axiom system as an extension of systems for ontology and mereology.

We assume thus that a functor *el* of an element satisfying the mereology axiom system is given; around this, we develop a system of axioms for rough mereology.

5.1 The Axiom System

The following is the list of basic postulates.

(RM1) $X\varepsilon\mu_1(Y) \iff X\varepsilon el(Y)$;
this means that being a part in degree 1 is equivalent to being an element: this establishes the connection between rough mereology and mereology.

(RM2) $X\varepsilon\mu_1(Y) \Longrightarrow \forall Z.(Z\varepsilon\mu_r(X) \Longrightarrow Z\varepsilon\mu_r(Y))$;
meaning the monotonicity property: any object Z is a part of Y in degree not smaller than that of being a part in x whenever X is an element of Y.
(RM3) $X = Y \wedge X\varepsilon\mu_r(Z) \Longrightarrow Y\varepsilon\mu_r(Z))$;
this means that the identity of individuals is a congruence with respect to μ.
(RM4) $X\varepsilon\mu_r(Y) \wedge s \leq r \Longrightarrow X\varepsilon\mu_s(Y)$;
this does establish the meaning " a part in degree at least r".

It follows that the functor μ_1 coincides with the given functor el establishing a link between rough mereology and mereology while functors μ_r with $r < 1$ diffuse the functor el to a hierarchy of functors expressing being an element (or, part) in various degrees.

We introduce a new name–forming functor μ_r^+ via

Definition 77. $X\varepsilon\mu_r^+(Y) \Longleftrightarrow X\varepsilon\mu_r(Y) \wedge \forall s.(s > r \Longrightarrow non(X\varepsilon\mu_s(Y)))$.

We find a usage for this functor in two new postulates.

(RM5) $ext(X,Y) \Longrightarrow X\varepsilon\mu_0^+(Y)$.

This postulate expresses our intuition that objects which are external to each other should be elements of each other in no positive degree. This assumption however reflects a high degree of certainty of our knowledge and it will lead to models in which connection coincides with overlapping (see below). It will be more realistic to assume that our knowledge is uncertain to the extent that we may not be able to state beyond doubt that two given objects are external to each other, rather we will be pleased with the atatement that they are in such case elements of each other in a bounded degree. Hence, we introduce a weaker form of the postulate (RM5).

(RM5*) $ext(X,Y) \Longrightarrow \exists s < 1.X\varepsilon\mu_s^+(Y)$.

5.2 Rough Mereology in Information Systems

We will refer to Section 4.5, where examples of mereological decompositions into parts were given. We will propose here a few measures of partness based on either individual frequency count or on attribute–value frequency count.

Rough membership functions. Here, we apply the idea of a rough membership function of Pawlak and Skowron [64]; we recall that given a subset (in the set–theoretic sense) X of the universe U of an information system $\mathbf{A}=(U, A)$, one defines the rough membership function μ_X by letting $\mu_X(u) = \frac{card(X \cap [u]_A)}{card([u]_A)}$ for $u \in U$, cf. Chapter 1.

This notion may be extended [65] to a notion of a rough membership function defined on concepts (i.e. subsets of U): given two non–empty concepts X, Y, we let: $\mu_X(Y) = \frac{card(X \cap Y)}{card(Y)}$. Thus, $\mu_X(Y)$ is a measure of the degree in which Y is contained in X.

Rough Mereology of B–exact sets: the row frequency count. We apply the rough membership function in its generalized form. Given two B–exact sets, X, Y, we let $r = \mu_X(Y)$ and accordingly $Y \varepsilon \mu_r(X)$. The measure μ is thus based on the frequency count of rows of the information table in respectively, X and Y. From probabilistic point of view, it may be regarded as an unbiased estimate of the conditional probability $Pr(X|Y)$ cf. [63].

Rough Mereology of B–exact sets: the B–class frequency count. Here, we propose yet another measure based on the rough membership function; we apply it counting this time the number of B–indiscernibility classes in respectively $X \cap Y$ and Y. Accordingly, for $X = Kl(+_{i=1}^k [B, v_i])$ and $Y = Kl(+_{j=1}^m [B, w_j])$, we let:
$$r = \frac{card(\{[B, v_i] : i \leq k\} \cap \{[B, w_j] : j \leq m\})}{m}$$
and accordingly, $Y \varepsilon \mu_r(X)$.

Rough Mereology of A–exact sets: the case when all A–indiscernibility classes are atoms. We may apply here either of the two frequency count measures defined earlier: in the first case, for two A–exact sets $X = Kl(+_{i=1}^k [B_i, v_i])$ and $Y = Kl(+_{j=1}^m [C_j, w_j])$, we let:
$$r = \frac{card(X \cap Y)}{card(Y)}$$
and accordingly, $Y \varepsilon \mu_r(X)$.

In the second case, we let:
$$r = \frac{card(\{[B_i, v_i] : i \leq k\} \cap \{[C_j, w_j] : j \leq m\})}{m}$$
and accordingly, $Y \varepsilon \mu_r(X)$.

We now propose a method for extending a measure defined for elements of two concepts to a measure on these two concepts; this idea will be applied in the following section.

Extending rough inclusions. Assume that we are given two individuals X, Y being classes of (finite) names: $X = Kl(X')$, $Y = Kl(Y')$ and that we have defined values of μ for pairs T, Z of individuals where $T \varepsilon X'$, $Z \varepsilon Y'$.

We extend μ to a measure μ^* on X, Y by letting:
$r = min_{Z \in Y'} \{max_{T \in X'} max\{s : Z \varepsilon \mu_s(T)\}\}$ and $Y \varepsilon \mu_r^*(X)$.
It may be proved straightforwardly that

Proposition 78. *The measure μ^* satisfies (RM1)–(RM4).*

Rough Mereology of A–exact sets: the case when atoms are all A–indiscernibility classes. We apply the idea of the last section to this case. First, we define μ on elementary individuals: given $X = [B, v]$ and $Y = [C, w]$, we define the set $IND(X, Y) = \{a \in A : a \in B \cap C \wedge v_a = w_a\}$ and then we let: $r = \frac{card(IND(X,Y))}{card(B)}$ and finally $Y \varepsilon \mu_r(X)$. Thus, the degree of partial containment of

T in X is determined by the frequency count of identical elementary descriptors in templates (B,v) and (C,w).

Now, given individual entities

$$X = Kl(+_{i=1}^{k}[B_i, v_i])$$

and

$$Y = Kl(+_{j=1}^{m}[C_j, w_j])$$

we let

$$r = min_{[C_j,w_j]}\{max_{[B_i,v_i]}max\{s : [C_j, w_j]\varepsilon\mu_s([B_i, v_i])\}\}$$

and $X\varepsilon\mu_r^*(X)$.

Example 1. We give a simple example concerning the last method of calculating the measure μ. We begin with an example of an information table.

	a_1	a_2	a_3
u_1	1	0	1
u_2	1	0	0
u_3	1	1	0
u_4	0	1	1
u_5	0	1	0
u_6	1	0	1
u_7	1	1	0

Table 1. *Binary1*: An example of an information table

Consider $B = \{a_1, a_2\}$, $C = \{a_2, a_3\}$, $v = <1, 0>$, $w = <0, 1>$; for $X = [B, v], Y = [C, w]$, we have $IND(X, Y) = \{a_2\}$ and accordingly, $Y\varepsilon\mu_{0.5}^*(X)$.

We now have at our disposal some recipes for introducing rough inclusions in information systems. The choice may depend on the context; let us observe that we may also have also some parameterized formulae, subject to optimization in a given context.

Rough Mereology in Information systems: measures induced from rows. The following procedure may define a rough inclusion in an information system (U, A) in case we would like to start with rows of the information table.

1. Consider a partition $P = \{A_1, A_2, ..., A_k\}$ of A.
2. Select a convex family of coefficients: $W = \{w_1, w_2, ..., w_k\}$ i.e. $w_i \geq 0$ for each i and $\sum_{i+1}^{k} w_i = 1$.
3. Define $IND(A_i)(u, u') = \{a \in A_i : a(u) = a(u')\}$.
4. Let $r = \sum_{i=1}^{k} w_i \frac{card(IND(A_i)(u,u'))}{card(A_i)}$.
5. Let $u\varepsilon\mu_r(u')$.

In this way we may relate rows one to another. Observe that this rough inclusion is symmetric: $u\varepsilon\mu_r(u') \iff u'\varepsilon\mu_r(u)$.

Now, we may extend this measure to elementary individuals and then to individuals as indicated in Example. For instance, taking $w_1 = 1$, we have $u_1 \varepsilon \mu_{0.66(6)} u_2$. Applying the extension μ^*, we have for X, Y in Example 1 above: $X = \{u_1, u_2, u_6\}$, $Y = \{u_2, u_6\}$ and $Y \varepsilon \mu_1^*(X)$ (in this model, elementary individuals (i.e. atoms) are rows of the information table hence $Y \varepsilon el(X)$).

5.3 Renormalization: t-Norm Modifiers

We introduce now, following Polkowski and Skowron [65], a modification to our functors μ_r; it is based on an application of residuated implication [41] and a measure of containment defined within the fuzzy set theory (the necessity measure) [36], [7]. Combining the two ideas, we achieve a formula for μ_r which allows for a transitivity rule; this rule will in turn allow to introduce into our universe rough mereological topologies.

We therefore recall the notion of a $t-norm$ T as a function of two arguments $\mathsf{T}: [0,1]^2 \longrightarrow [0,1]$ which satisfies the following requirements:

1. $\mathsf{T}(x,y) = \mathsf{T}(y,x)$;
2. $\mathsf{T}(x, \mathsf{T}(y,z)) = \mathsf{T}(\mathsf{T}(x,y), z)$;
3. $\mathsf{T}(x,1) = x$;
4. $x' \geq x \wedge y' \geq y \implies \mathsf{T}(x', y') \geq \mathsf{T}(x, y)$.

We also invoke a notion of fuzzy containment $\subset_r$ based on necessity cf. [36]; it relies on a many-valued implication Υ i.e. on a function $\Upsilon : [0,1]^2 \longrightarrow [0,1]$ according to the formula:

$$X \subset_r Y \iff \forall Z.(\Upsilon(\mu_X(Z), \mu_Y(Z)) \geq r)$$

where μ_A is the fuzzy membership function [41] of the fuzzy set A.

We replace Υ with a specific implication viz. the residuated implication $\overrightarrow{\mathsf{T}}$ induced by T and defined by the following prescription.

Definition 79. $\overrightarrow{\mathsf{T}}(r,s) \geq t \iff \mathsf{T}(t,r) \leq s$.

We define a functor $\mu_{\mathsf{T},r}$ where $r \in [0,1]$, according to the recipe

Definition 80. $X \varepsilon \mu_{\mathsf{T},r}(Y) \iff X \varepsilon X \wedge \forall Z.(\exists t, w. Z \varepsilon \mu_t(X) \wedge Z \varepsilon \mu_w(Y) \wedge \overrightarrow{\mathsf{T}}(t,w) \geq r)$

It turns out, as first proved in a different context in [65], that $\mu_{\mathsf{T},r}$ satisfies axioms (RM1-RM5*); we include the proof in our case.

Proposition 81. *Functors $\mu_{\mathsf{T},r}$ satisfy (RM1)-(RM5*)*

Proof. For (RM1): assume that $X\varepsilon\mu_{\mathsf{T},1}(Y)$ so for any Z from $Z\varepsilon\mu_u(X), Z\varepsilon\mu_v(Y)$ it follows that $\overrightarrow{\mathsf{T}}(u,v) \geq 1$ hence $\mathsf{T}(1,u) \leq v$ i.e. $u \leq v$; this implies that $\forall Z.(Z\varepsilon\mu_1(X) \Longrightarrow Z\varepsilon\mu_1(Y))$ i.e. $\forall Z.(Z\varepsilon el(X) \Longrightarrow Z\varepsilon el(Y))$ and thus $X\varepsilon el(Y)$. Conversely, $X\varepsilon el(Y)$ implies $\forall Z.(Z\varepsilon\mu_u(X) \wedge Z\varepsilon\mu_v(Y) \Longrightarrow u \leq v$ so $\overrightarrow{\mathsf{T}}(u,v) \geq 1)$ and finally $X\varepsilon\mu_{\mathsf{T},1}(Y)$.

Concerning (RM2), let $X\varepsilon\mu_{\mathsf{T},1}(Y)$ and $Z\varepsilon\mu_{\mathsf{T},u}(X)$ hence for any T from $T\varepsilon\mu_\alpha(Z), T\varepsilon\mu_v(X), T\varepsilon\mu_w(Y)$ it follows that $v \leq w$ hence $\overrightarrow{\mathsf{T}}(\alpha,w) \geq \overrightarrow{\mathsf{T}}(\alpha,v)$ a fortiori $\overrightarrow{\mathsf{T}}(\alpha,v) \geq u$ implies $\overrightarrow{\mathsf{T}}(\alpha,w) \geq u$ i.e. $Z\varepsilon\mu_{\mathsf{T},u}(Y)$.

In case of (RM3), assume that $X = Y$; then for any $T, \alpha : T\varepsilon\mu_\alpha(X) \Longleftrightarrow T\varepsilon\mu_\alpha(Y)$ hence for any T from $T\varepsilon\mu_\gamma(X), T\varepsilon\mu_\delta(Z)$ with $\overrightarrow{\mathsf{T}}(\gamma,\delta) \geq r$ it follows that $T\varepsilon\mu_\gamma(Y), T\varepsilon\mu_\delta(Z)$ with $\overrightarrow{\mathsf{T}}(\gamma,\delta) \geq r$ i.e. $X\varepsilon\mu_{\mathsf{T},r}(T) \Longrightarrow Y\varepsilon\mu_{\mathsf{T},r}(T)$.

(RM4) is obviously satisfied by virtue of definition of $\mu_{\mathsf{T},r}$.

Now, for (RM5): assume that $ext(X,Y)$ hence $X\varepsilon\mu_0^+(Y)$. Let $X\varepsilon\mu_{\mathsf{T},r}(Y)$; then from $T \varepsilon\mu_\gamma(X), T\varepsilon\mu_\delta(Y)$ it follows that $\overrightarrow{\mathsf{T}}(\gamma,\delta) \geq r$ for any T and some γ, δ hence $\mathsf{T}(r,\gamma) \leq \delta$. In particular, for $T = X$, we have $\mathsf{T}(r,1) \leq 0$ i.e. $r \leq 0$.

Consider finally (RM5*): assume $ext(X,Y)$; then $\exists s < 1.X\varepsilon\mu_s^+(Y)$. We have then $X\varepsilon\mu_1(X), X\varepsilon\mu_s^+(Y)$ and $\overrightarrow{\mathsf{T}}(1,s) \geq r$ implies $\mathsf{T}(r,1) \leq s$ i.e. $r \leq s$ so $X\varepsilon\mu_{\mathsf{T},r}(Y)$ implies $r \leq s$ i.e. $X\varepsilon\mu_{\mathsf{T},t}^+(Y)$ with $t \leq s$.

An advantage of this rough inclusion is the fact that it does satisfy a deduction rule of the form

$$(DR)\frac{X\varepsilon\mu_r(Y), Y\varepsilon\mu_s(Z)}{X\varepsilon\mu_u(Z)}$$

where $u = f(r,s)$ depends functionally on r, s. Clearly, the obvious candidate for f is T. Again, we include a short proof for completeness sake.

Proposition 82. *(DR) holds in the form* : $\frac{X\varepsilon\mu_{\mathsf{T},r}(Y), Y\varepsilon\mu_{\mathsf{T},s}(Z)}{X\varepsilon\mu_{\mathsf{T},\mathsf{T}(r,s)}(Z)}$.

Proof. Assume that $X\varepsilon\mu_{\mathsf{T},r}(Y), Y\varepsilon\mu_{\mathsf{T},s}(Z)$; we have then: $T\varepsilon\mu_\alpha(X), T\varepsilon\mu_\beta(Y), T\varepsilon\mu_\delta(Z)$ imply $\overrightarrow{\mathsf{T}}(\alpha,\beta) \geq r, \overrightarrow{\mathsf{T}}(\beta,\delta) \geq s$ i.e. $\mathsf{T}(r,\alpha) \leq \beta, \mathsf{T}(s,\beta) \leq \delta$ and monotonicity of T implies $\mathsf{T}(s,\mathsf{T}(r,\alpha)) \leq \delta$ so $\mathsf{T}(\mathsf{T}(s,r),\alpha)) \leq \delta$ and $\mathsf{T}(\mathsf{T}(r,s),\alpha)) \leq \delta$ which finally yields $\overrightarrow{\mathsf{T}}(\alpha,\delta) \geq \mathsf{T}(r,s)$ a fortiori $X\varepsilon\mu_{\mathsf{T},\mathsf{T}(r,s)}(Z)$.

We will exploit this advantage of $\mu_{\mathsf{T},r}$ in the sequel.

Remark. We may observe that we do not decide the status of
(CRM5) $X\varepsilon\mu_0^+(Y) \Longrightarrow ext(X,Y)$ i.e. the converse to (RM5).

Remark. We may also notice that accepting (RM5) bears inconveniently on $\mu_{\mathsf{T},r}$ as witnessed by the following

Proposition 83. *Under (RM5), for any X, Y* :

1. $X\varepsilon\mu_{\mathsf{T},1}(Y)$ whenever $X\varepsilon el(Y)$;
2. $X\varepsilon\mu_{\mathsf{T},0}^+(Y)$ whenever $non(X\varepsilon el(Y))$.

Proof. Case (1) has been settled already by proving (RM1) for μ_T; in case (2), there exists Z with $Z\varepsilon el(X), ext(Z,Y)$ i.e. $Z\varepsilon\mu_1(X), Z\varepsilon\mu_0^+(Y)$. As $\vec{T}(1,0) \geq r$ implies $T(r,1) = r \leq 0$, it follows that $X\varepsilon\mu_{T,0}^+(Y)$.

The moral of the last proposition is that under (RM5), μ_T becomes a 0-1 measure discerning only between being an element and not being an element. For this reason, we will apply (RM5*) in the sequel as a more realistic approach.

6 Introduction to Qualitative Spatial Reasoning

Qualitative Reasoning aims at studying concepts and calculi on them that arise often at early stages of problem analysis when one is refraining from qualitative or metric details cf. [16]; as such it has close relations to the design cf. [11] as well as planning stages cf. [31] of the model synthesis process. Classical formal approaches to spatial reasoning i.e. to representing spatial entities (points, surfaces, solids etc.) and their features (dimensionality, shape, connectedness degree etc.) rely on Geometry or Topology i.e. on formal theories whose models are spaces (universes) constructed as sets of points; contrary to this approach, qualitative reasoning about space often exploits pieces of space (regions, boundaries, walls, membranes etc.) and argues in terms of relations abstracted from a commonsense perception (like *connected, discrete from, adjacent, intersecting*). In this approach, points appear as ideal objects (e.g. ultrafilters of regions of solids [89]). Qualitative Spatial Reasoning has a wide variety of applications, among them, to mention only a few, representation of knowledge, cognitive maps and navigation tasks in robotics (e.g. [42], [43], [44], [1], [3], [23], [39], [28]), Geographical Information Systems and spatial databases including *Naive Geography* (e.g. [26], [27], [35], [24]), high-level Computer Vision (e.g. [95]), studies in semantics of orientational lexemes and in semantics of movement (e.g. [6], [5]). Spatial Reasoning establishes a link between Computer Science and Cognitive Sciences (e.g. [29]) and it has close and deep relationships with philosophical and logical theories of space and time (e.g. [71], [9], [2]). A more complete perspective on Spatial Reasoning and its variety of themes and techniques may be acquired by visiting one of the following sites: [86], [94], [59].

Any formal approach to Spatial Reasoning, however, would require a formal approach to Ontology as well cf. [34], [81], [12]. In this Chapter we adopt as formal Ontology the ontological theory of Stanisław Leśniewski (cf. [51], [52], [80], [49], [38], [20]. This theory is briefly introduced in Section 3.

For expressing relations among entities, mathematics proposes two basic languages: the language of set theory, based on the opposition element–set, where distributive classes of entities are considered as sets consisting of (discrete) atomic entities, and languages of mereology, for discussing entities continuous in their nature, based on the opposition part–whole. It is thus not surprising that Spatial Reasoning relies to great extent on mereological theories of part cf. [4], [5], [6], [14], [17], [32], [33], [30], [82], [83], [58].

Mereological ideas have been early applied toward axiomatization of geometry of solids cf. [47], [89]. Mereological theories dominant nowadays come from ideas proposed independently by Stanisław Leśniewski and Alfred North Whitehead.

Mereological theory of Leśniewski is based on the notion of a part (proper) and the notion of a (collective) class cf. [51], [53], [20], [84], [54].

Mereological ideas of Whitehead based on the dual to part notion of an extension [96] were formulated in the Calculus of Individuals [48] and given a formulation in terms of the notion of a Connection [14]. Mereology based on connection gave rise to spatial calculi based on topological notions derived therefrom (mereotopology) cf. [18], [16], [22], [25], [5], [6], [17], [32], [33], [30], [82], [58].

In this Chapter, we are adopting mereological theory of Stanisław Leśniewski formalized according to His program within His Ontology. We study in this Chapter possible applications of Rough Mereology to Spatial Reasoning in the frame of Information Systems. We demonstrate that in the framework of Rough Mereology one may define a quasi–Čech topology [21] (a quasi– topology was introduced in the connection model of Mereology [14], [5] under additional assumptions of regularity); see Section 7 for this topic. By a *quasi–topology* we mean a topology without the null element.

Finally, we apply Rough Mereology toward inducing geometrical notions. It is well known [90], [9] that geometry may be introduced via notions of nearness, betweenness etc. In Section 8, we define these notions by means of a rough mereological notion of distance and we show that in this way a geometry may be defined in the rough mereological universe. This geometry is clearly of approximative character, approaching precise notions in a degree due to uncertainty of knowledge encoded in rough inclusions.

As rough mereological constructs (rough inclusions) may be induced from data tables (information systems) as indicated in Section 5, reasoning about spatial entities by means of Rough Mereology may be carried out on the basis of data (e.g. in spatial databases).

7 Mereotopology

We now are concerned with topological structures arising in mereological universe endowed with a rough inclusion.

As mentioned few lines above, topological structures may be defined within the connection framework via the notion of a non–tangential part. Interior entities are formed then by means of some fusion operators cf. e.g. [5], [58]. The functor of connection allows also for some calculi of topological character based directly on regions e.g. $RCC - calculus$ cf. [33]. For a different approach where connection may be derived from the axiomatized notion of a boundary cf. [83].

These topological structures provide an environment in which it is possible to carry out spatial reasoning (cf. op.cit., op.cit.). We now demonstrate that in rough mereological framework one defines in a natural way Čech topologies i.e. topologies which may be deficient in that the closure operator may not have some properties of the topological closure (viz., idempotency and the finite intersection property).

7.1 Mereotopology: Čech Topologies

It has been demonstrated that in mereological setting a quasi–Čech topology may be defined (cf. [14]) which under additional artificial assumptions (op.cit.) may be made into a quasi–topology. Here, we induce a quasi–Čech topology (i.e. topology without the null object) in any rough mereological universe.

We would like to recall that a topology on a given domain U may be introduced by means of a closure operator cl satisfying the *Kuratowski axioms* [46]:

1. (Cl1) $cl\emptyset = \emptyset$;
2. (Cl2) $clclX = clX$;
3. (Cl3) $X \subseteq clX$;
4. (Cl4) $cl(X \cup Y) = clX \cup clY$.

The dual operator int of *interior* is then defined by means of the formula: $intX = U - cl(U - X)$ and it has properties expressed by this duality: $int\emptyset = \emptyset$, $intintX = intX$, $intX \subseteq X$, $int(X \cap Y) = intX \cap intY$.

The Čech topology [21] is a weaker structure as it is required here only that the closure operator satisfy the following:

1. (ČCl1) $cl\emptyset = \emptyset$;
2. (ČCl2) $X \subseteq clX$;
3. (ČCl3) $X \subseteq Y \Longrightarrow clX \subseteq clY$.

so the associated Čech interior operator int should only satisfy the following: $int\emptyset = \emptyset$; $intX \subseteq X$; $X \subseteq Y \Longrightarrow intX \subseteq intY$.

Čech topologies arise naturally in problems related to information systems when one considers coverings induced by similarity relations instead of partitions induced by indiscernibility relations [56].

We now introduce a quasi–Čech topology into a rough mereological universe: we may remember that in our context of Ontology, the empty set (name) may not be used.

To this end, we define the class $Kl_r X$ for any object X and $r < 1$ by the following. First, we introduce a name $(M_r X)$ for the property of being a part in a degree r.

Definition 84. $Z \varepsilon M_r X \Longleftrightarrow Z \varepsilon Z \wedge Z \varepsilon \mu_r(X)$

Now, we define the individual entity $Kl_r X$.

Definition 85. $Z \varepsilon K l_r X \Longleftrightarrow Z \varepsilon Z \wedge Z \varepsilon M_r X$.

Thus $Kl_r X$ is the class of objects having the property $\mu_r(X)$. We now give a direct characterization of $Kl_r X$. With this aim, we introduce a name $B_r X$ defined by means of the condition:

$$Z \varepsilon el(B_r X) \Longleftrightarrow \exists T(Z \varepsilon el(T) \wedge T \varepsilon \mu_r(X)).$$

This definition is correct, as $B_r X = Kl(el(B_r X))$. Then we have

Proposition 86. $Kl_rX = B_rX$

Proof. Assume first that $Z\varepsilon el(Kl_rX)$; then by the class definition (Definition 44), for some U,W, we have $U\varepsilon el(Z)$, $U\varepsilon el(W)$, $W\varepsilon\mu_r(X)$, hence $U\varepsilon el(B_rX)$ and the inference rule (Proposition 61) implies that $Kl_rX\varepsilon el(B_rX)$.

Conversely, $Z\varepsilon el(B_rX)$ implies that for some T we have $Z\varepsilon el(T)$, $T\varepsilon\mu_r(X)$ hence by he class definition $T\varepsilon el(Kl_rX)$ and so $Z\varepsilon el(Kl_rX)$ implying by the inference rule that $B_rX\ \varepsilon\ el(Kl_rX)$. Hence $Kl_rX = B_rX$.

From this the following corollary follows

Corollary 87. For $s \leq r$, $Kl_rX\varepsilon el(Kl_sX)$.

Indeed, $Z\varepsilon el(Kl_rX)$ means that $Z\varepsilon el(T)$, $T\varepsilon\mu_r(X)$ for some T so by (RM4) we have $T\varepsilon\mu_s(X)$ and thus $Z\varepsilon el(Kl_sX)$. The corollary follows.

We mention yet a monotonicity property.

Proposition 88. $X\varepsilon el(Y) \implies Kl_rX\varepsilon el(Kl_rY)$.

Example 2. We recall the Table *Binary*1 from Example 1. With the notation of that Example, we find the class $Kl_{0.5}(X)$ for $X = \{u_1, u_2, u_6\}$; we mark this class in the Table *Binary*2 below with boldface.

	a_1	a_2	a_3
u_1	**1**	**0**	**1**
u_2	**1**	**0**	**0**
u_3	1	1	0
u_4	0	1	1
u_5	0	1	0
u_6	**1**	**0**	**1**
u_7	1	1	0

Table 2. *Binary2*: The class $Kl_{0.5}(X)$

We admit B defined as follows as a base for open sets with which we define the interior operator.

Definition 89. $Z\varepsilon B \iff Z\varepsilon Z \land \exists X, r < 1. Z\varepsilon Kl_rX$.

Following this we define a new functor *int*. Again, we introduce first a new name $I(X)$.

Definition 90. $Z\varepsilon I(X) \iff Z\varepsilon Z \land \exists s < 1(Kl_sZ\varepsilon el(X))$.

Now, we define *int*.

Definition 91. $int(X) = Kl(I(X))$.

Then we have the following properties of *int*.

Proposition 92. *For any X, Y :*

1. $int(X)\varepsilon el(X)$;
2. $X\varepsilon el(Y) \implies int(X)\varepsilon el(int(Y))$;

Proof. For (1): assume that $Z\varepsilon el(int(X))$; there exist U, W with $U\varepsilon el(Z)$, $U\varepsilon el(W)$, $Kl_s W\varepsilon el(X)$ for some $s < 1$; hence, $W\varepsilon el(X)$ and $U\varepsilon el(X)$ so the inference rule (Proposition 61) implies that $int(X)\varepsilon el(X)$.

In case (2), assume that $X\varepsilon el(Y)$ and let $Z\varepsilon el(int(X))$. We have U, W with $U\varepsilon el(Z)$, $U\varepsilon el(W)$, $Kl_s W\varepsilon el(X)$ for an $s < 1$ hence $Kl_s W\varepsilon el(Y)$ and thus $W\varepsilon I(Y)$ hence $W \ \varepsilon \ el(int(Y))$ so a fortiori $U\varepsilon el(int(Y))$ so the inference rule implies that $int(X)\varepsilon el(int(Y))$.

Properties (1)-(2) witness that the quasi–topology introduced by B is a *quasi–Čech topology*. We denote it by the symbol τ_μ.

Proposition 93. *Rough mereotopology τ_μ induced by the rough inclusion μ_r is a quasi–Čech topology.*

Remark. Let us observe that under (RM5) the topology τ_μ is discrete: every individual is an open set (a singleton).

We now study the case of mereo–topology under functors $\mu_{T,r}$; in this case, the quasi– Čech topology τ_μ turns out to be a quasi–topology.

7.2 Mereo–Topology: the Case of μ_T

We begin with an application of deduction rule (DR). We denote by the symbol $Kl_{T,r}X$ the set $Kl_r X$ in case of the rough inclusion μ_T. We assume that $T(r, s) < 1$ when $rs < 1$. We propose a new direct characterization of $Kl_{T,r}X$.

Proposition 94. $Z\varepsilon el(Kl_{T,r}X) \iff Z\varepsilon \mu_{T,r}(X)$.

Proof. By Proposition 89, $Z\varepsilon el(Kl_{T,r}X)$ means that $Z\varepsilon \ el(T)$, $T\varepsilon\mu_{T,r}(X)$ for some T hence $Z\varepsilon\mu_{T,1}(T)$, $T\varepsilon\mu_{T,r}(X)$ imply by (DR) that $Z\varepsilon\mu_{T,T(1,r)}(X)$ i.e. $Z\varepsilon\mu_{T,r}(X)$.

This Proposition means that $Kl_{T,r}X$ may be regarded as "*an open ball of radius r centered at X*".

We assume now, additionally, that the t-norm T has the property that: for every $r < 1$ there exists $s < 1$ such that $T(r, s) \geq r$. With this assumption, we have the following.

Proposition 95. *For $Z\varepsilon el(Kl_{\mathsf{T},r}(X))$,*
if $s_0 = \arg\_\min\{s : \mathsf{T}(r,s) \geq r\}$ then $Kl_{\mathsf{T},s_0}(Z)\varepsilon el(Kl_{\mathsf{T},r}(X))$.

Proof. Let $s \geq s_0$; consider $T\varepsilon el(Kl_{\mathsf{T},s}(Z))$ so $T\varepsilon\mu_{\mathsf{T},s}(Z)$. Then $T\varepsilon\mu_{\mathsf{T},\mathsf{T}(s,r)}(X)$ hence $T\varepsilon\mu_{\mathsf{T},r}(X)$ so $T\varepsilon el(Kl_{\mathsf{T},r}(X))$ implying finally by the inference rule (Proposition 61) that $Kl_{\mathsf{T},s_0}(Z)\varepsilon\ el(Kl_{\mathsf{T},r}(X))$.

We define a functor of two nominal individual variables Ov (of rough mereological overlap) and a functor of two nominal individual variables AND.

Definition 96.
1. $Ov(X,Y) \iff \exists Z.Z\varepsilon el(X) \wedge Z\varepsilon el(Y)$;
2. $Z\varepsilon el(AND(X,Y)) \iff Ov(X,Y) \wedge Z\varepsilon el(X) \wedge Z\varepsilon el(Y)$.

Proposition 97. *The rough mereo–topology τ_{μ_T} has the property:*
$AND(int(X), int(Y)) = int(AND(X,Y))$ *holds whenever*
$AND(int(X), int(Y))$ *is non-empty.*

Proof. The intersection of two open basic classes may be described effectively by means of Proposition 98 : assume that $Z\varepsilon el(Kl_{\mathsf{T},r}(X))$ and $Z\varepsilon el(Kl_{\mathsf{T},s}(Y))$ for some $r, s < 1$. Then for $t_0 = \arg\_\min\{t : \mathsf{T}(r,t) \geq r, \mathsf{T}(s,t) \geq s\}$ and $1 > t \geq t_0$, we have $Kl_{\mathsf{T},t}(Z)\varepsilon el(Kl_{\mathsf{T},r}(X))$ and $Kl_{\mathsf{T},t}(Z)\varepsilon el(Kl_{\mathsf{T},s}(Y))$. The general case follows easily.

Finally, we check that under our assumptions, the operator of interior is idempotent: $intint = int$ which will conclude our verification that the rough mereological topology is a topology.

Proposition 98. $int(int(X)) = int(X)$.

Proof. It suffices to show that $int(X)\varepsilon el(int(int(X)))$ so we consider Z with $Z\varepsilon el(int(X))$. For some U, W, we have $U\varepsilon el(Z)$, $U\varepsilon el(W)$, $Kl_sW\varepsilon el(X)$, some $s < 1$.
   Now, we check that: $Kl_sW = int(Kl_sW)$; we consider then P with $P\ \varepsilon\ el(Kl_sW)$. There exist R, Q with $R\varepsilon el(P)$, $R\varepsilon el(Q)$, $Q\varepsilon\mu_sW$. It follows that $R\varepsilon\mu_sW$ hence $R\varepsilon el(kl_sW)$ so $Kl_tR\varepsilon Kl_sW$ for some $t < 1$ and finally $R\ \varepsilon\ el(int(Kl_sW))$. By the inference rule (Proposition 61), $Kl_sW\varepsilon el(int(Kl_sW))$ and the identity follows.
   Returning to the main proof, as $Kl_sW\varepsilon el(X)$, we have $int(Kl_sW)\varepsilon el(int(X))$ hence $Kl_sW\ \varepsilon el(int(X))$ and thus $Kl_tU\varepsilon el(int(X))$ for some $t < 1$ so, finally, $U\varepsilon el(int(int(X)))$ and the inference rule shows that $int(X)\varepsilon el(int(int(X)))$.

Corollary 99. *The rough mereological topology induced by the rough inclusion $\mu_{\mathsf{T},r}$ is a quasi–topology.*

8 Mereogeometry

Functors μ_r may also be regarded as weak metrics leading to a geometry. From this point of view, we may apply μ in order to define basic notions of mereology–based geometry (mereo–geometry).

In the language of this geometry, we may approximately describe and approach geometry of objects described by data tables; a usage for this geometry may be found e.g. in navigation and control tasks of mobile robotics [1], [3], [23], [39], [43], [44].

It is well–known (cf. [91], [9]) that the geometry of euclidean spaces may be based on some postulates about the basic notions of a point and the ternary equidistance functor. In [91] postulates for euclidean geometry over a real-closed field were given based on the functor of betweenness and the quaternary equidistance functor. Similarly, in [9], a set of postulates aimed at rendering general geometric features of geometry of finite-dimensional spaces over reals has been discussed, the primitive notion there being that of nearness.

Geometrical notions have been applied in e. g. studies of semantics of spatial prepositions [6] and in inferences via cardinal directions cf. e.g [45].

It may not be expected that a geometry induced from a rough mereological context proves to be a euclidean one, however, we demonstrate that we may introduce in the rough mereological context functors of nearness, betweenness and equidistance that satisfy basic postulates about these functors valid in euclidean spaces.

8.1 Rough Mereological Distance, Betweenness

We first introduce a notion of distance in our rough mereological universe by letting

Definition 100. $\kappa_r(X,Y) \iff r = \min\{u,w : X\varepsilon\mu_u^+(Y) \wedge Y\varepsilon\mu_w^+(X)\}$.

We now introduce the notion of betweenness as a functor $T(X,Y)$ of two individual names; the statement $Z\varepsilon T(X,Y)$ reads as "Z is between X and Y".

Definition 101. $Z\varepsilon T(X,Y) \iff Z\varepsilon Z \wedge \forall W. \kappa_r(Z,W) \wedge \kappa_s(X,W) \wedge \kappa_t(Y,W) \implies s \leq r \leq t \vee t \leq r \leq s$.

Thus, $Z\varepsilon T(X,Y)$ holds when the rough mereological distance κ between Z and W is in the non–oriented interval (i.e. between) [distance of X to W, distance of Y to W] for any W.

We check that T satisfies the axioms of Tarski [91] for *betweenness*.

Proposition 102. *The following properties hold:*

1. $Z\varepsilon T(X,X) \implies Z = X$ *(identity)*;
2. $Y\varepsilon T(X,U) \wedge Z\varepsilon T(Y,U) \implies Y\varepsilon T(X,Z)$ *(transitivity)*;

3. $Y \varepsilon T(X,Z) \wedge Y \varepsilon T(X,U) \wedge X \neq Y \implies Z \varepsilon T(X,U) \vee U \varepsilon T(X,Z)$ *(connectivity)*.

Proof. By means of κ, the properties of betweenness in our context are translated into properties of betweenness in the real line which hold by the Tarski theorem [91], Theorem 1.

8.2 Nearness

We may also apply κ to define in our context the functor N of nearness proposed in van Benthem [9].

Definition 103. $Z \varepsilon N(X,Y) \iff Z \varepsilon Z \wedge (\kappa_r(Z,X) \wedge \kappa_s(X,Y) \implies s < r)$.

Here, nearness means that Z is closer to X than to Y (recall that rough mereological distance is defined in an opposite way: the smaller r, the greater distance).

Then the following hold i.e. N does satisfy all axioms for nearness in [9].

Proposition 104.
1. $Z \varepsilon N(X,Y) \wedge Y \varepsilon N(X,W) \implies Z \varepsilon N(X,W)$ *(transitivity)*;
2. $Z \varepsilon N(X,Y) \wedge X \varepsilon N(Y,Z) \implies X \varepsilon N(Z,Y)$ *(triangle inequality)*;
3. $non(Z \varepsilon N(X,Z))$ *(irreflexivity)*;
4. $Z = X \vee Z \varepsilon N(Z,X)$ *(selfishness)*;
5. $Z \varepsilon N(X,Y) \implies Z \varepsilon N(X,W) \vee W \varepsilon N(X,Y)$ *(connectedness)*.

Proof. (4) follows by (RM1); (3) is obvious. In proofs of the remaining properties, we introduce a symbol $\mu(X,Y)$ as that value of r for which $\kappa_r(X,Y)$. Then, for (1), assume that $Z \varepsilon N(X,Y), Y \varepsilon N(X,W)$ i.e. $\mu(Z,X) > \mu(X,Y), \mu(X,Y) > \mu(X,W)$ hence $\mu(Z,X) > \mu(X,W)$ i.e. $Z \varepsilon N(X,W)$. In case (2), $Z \varepsilon N(X,Y)$, $X \varepsilon N(Y,Z)$ mean $\mu(Z,X) > \mu(X,Y)$, $\mu(X,Y) > \mu(Y,Z)$ so $\mu(Z,X) > \mu(Y,Z)$ i.e. $X \varepsilon N(Z,Y)$. Concerning (v), $Z \varepsilon N(X,Y)$ implies that $\mu(Z,X) > \mu(X,Y)$ hence either $\mu(Z,X) > \mu(X,W)$ meaning $Z \varepsilon N(X,W)$ or $\mu(X,W) > \mu(X,Y)$ implying $W \varepsilon N(X,Y)$.

We now may introduce the notion of equidistance in the guise of either a functor $Eq(X,Y)$ or a functor $D(X,Y,Z,W)$ defined as follows.

Definition 105. $Z \varepsilon Eq(X,Y) \iff Z \varepsilon Z \wedge (non(X \varepsilon N(Z,Y)) \wedge non(Y \varepsilon N(Z,X)))$.

It follows that

Proposition 106. $Z \varepsilon Eq(X,Y) \iff Z \varepsilon Z \wedge (\forall r.\kappa_r(X,Z) \iff \kappa_r(Y,Z))$.

We may define a functor of equidistance following Tarski [91].

Definition 107. $D(X,Y,Z,W) \iff (\forall r.\kappa_r(X,Y) \iff \kappa_r(Z,W))$.

These functors do clearly satisfy the following (cf. [9], [91]).

Proposition 108.
1. $Z \varepsilon Eq(X,Y) \wedge X \varepsilon Eq(Y,Z) \implies Y \varepsilon Eq(Z,X)$ *(triangle equality)*;
2. $Z \varepsilon T(X,Y) \wedge W \varepsilon Eq(X,Y) \implies D(Z,W,X,W)$ *(circle property)*;
3. $D(X,Y,Y,X)$ *(reflexivity)*;
4. $D(X,Y,Z,Z) \implies X = Y$ *(identity)*;
5. $D(X,Y,Z,U) \wedge D(X,Y,V,W) \implies D(Z,U,V,W)$ *(transitivity)*.

One may also follow van Benthem's proposal for a betweenness functor defined via the nearness functor as follows

Definition 109. $Z \varepsilon T_B(X,Y) \iff (\forall W. Z \varepsilon W \vee Z \varepsilon N(X,W) \vee Z \varepsilon N(Y,W))$.

One checks in a straightforward way that

Proposition 110. *The functor T_B of betweenness defined according to Definition 130 does satisfy the Tarski axioms.*

8.3 Points

The notion of a point may be introduced in a few ways; e.g. following Tarski [89], one may introduce points as classes of names forming ultrafilters under the ordering induced by the functor of being an element el. Another way, suitable in practical cases, where the universe, or more generally, each ultrafilter F as above is finite i.e. principal (meaning that there exists an object X such that F consists of those Y's for which $X \varepsilon el(Y)$ holds) is to define points as atoms of our universe under the functor of being an element i.e. we define a constant name AT as follows

Definition 111. $X \varepsilon AT \iff X \varepsilon X \wedge non(\exists Y. Y \varepsilon el(X) \wedge non(X \varepsilon = (Y)))$.

We will refer to such points as to *atomic points*. We adopt here this notion of a point.

Clearly, restricting ourselves to atomic points, we preserve all properties of functors of betweenness, nearness and equidistance proved above to be valid in the universe V.

9 Examples

In this section, we will give some examples related to notions and applications thereof presented in the preceding sections.

Our universe will be selected from a quadtree in the Euclidean plane formed by squares $[k + \frac{i}{2^s}, k + \frac{i+1}{2^s}] \times [l + \frac{j}{2^s}, l + \frac{j+1}{2^s}]$ where $k, l \in Z, i, j = 0, 1, ..., 2^s - 1$ and $s = 0, 1, 2, ...$.

The choice of atomic points will depend on the level of granularity of knowledge we assume; we may suppose that our objects are localized in space with a positive degree of uncertainty. We will express this uncertainty assuming that our sensoric system perceives each square X as the square X' whose each side length is that of X plus 2α where $\alpha = 2^{-s}$ for some $s > 1$ (we then express uncertainty as uncertainty of location applying "hazing" of objects cf. [93]). By this assumption, we may restrict ourselves to squares with the side length at least 4α (as smaller squares would be localized with uncertainty too high); in consequence, atomic points will be all squares of the above form having the side length equal to 4α. In our example we let for simplicity $4\alpha = 1$. Our atomic points are therefore squares of the form $[k, k+1] \times [l, l+1]$, $k, l \in \mathbf{Z}$.

We will define functors μ_r by letting

$$X \varepsilon \mu_r(Y) \Leftrightarrow \frac{\lambda(X' \cap Y')}{\lambda(X')} \geq r$$

where X', Y' are enlargements of X, Y defined above and λ is the area (Lebesgue) measure in the two-dimensional plane. We may check straightforwardly that

Proposition 112. *Functors μ_r satisfy (RM1)-(RM4)+(RM5)*.*

Let us remark that this measure is a continuous extension of the measure proposed in [64] for the case of discrete information systems (cf. in this respect [63]).

9.1 Mereogeometry

We will adopt the notion of betweenness T_B based on the nearness functor. Then we find that e.g. the following triples (X, Z, Y) do satisfy the formula $Z \varepsilon T_B(X, Y)$:

1. $([0,1] \times [0,1], [1,2] \times [0,1], [2,3] \times [0,1])$;
2. $([0,1] \times [0,1], [0,1] \times [1,2], [0,1] \times [2,3])$;
3. $([0,1] \times [0,1], [1,2] \times [1,2], [2,3] \times [2,3])$;
4. $([2,3] \times [2,3], [1,2] \times [1,2], [0,1] \times [0,1])$;
5. $([0,1] \times [0,1], [1,2] \times [0,1], [2,3] \times [1,2])$.

Clearly, all translates over the digital space $\mathbf{Z}^2$ of the above triples as well as all their rotations by a multiplicity of $\pi/2$ preserve the functor T_B.

The equidistance functor Eq may be used to define spheres; for instance, admitting as Z the square $[0,1] \times [0,1]$, we have the sphere

$$S(Z; 1/3) = Kl([0,1] \times [1,2], [0,1] \times [-1,0], [1,2] \times [0,1], [-1,0] \times [0,1]\}.$$

A line segment may be defined via the auxiliary notion of a pattern; we introduce this notion as a functor Pt.

We let

$$Pt(X, Y, Z) \Longleftrightarrow Z \varepsilon T_B(X, Y) \vee X \varepsilon T_B(Z, Y) \vee Y \varepsilon T_B(X, Z).$$

We will say that a finite sequence $X_1, X_2, ..., X_n$ of points belong in a line segment whenever $Pt(X_i, X_{i+1}, X_{i+2})$ for $i = 1, , ..., n-2$; formally, we introduce the functor *Line* of finite arity defined via

$$Line(X_1, X_2, ..., X_n) \iff \forall i < n - 1.Pt(X_i, X_{i+1}, X_{i+2})$$

and then we let

$$Line\_seg(X_1, X_2, ..., X_n) \varepsilon Kl(X_1, X_2, ..., X_n : Line(X_1, X_2, ..., X_n)).$$

In particular, classes of sequences:

1. $([0,1] \times [i, i+1])_i$;
2. $([i, i+1] \times [0,1])_i$;
3. $([i, i+1] \times [i, i+1])_i$;
4. $([i, i+1] \times [-i, -i-1])_i$

for $i \in [-n, n]$ where $n = 1, 2, ...$, are line segments.

It is clearly possible to introduce line segments of various types by means of specialized pattern functors.

The notion of orthogonality may be introduced in a well-known way; we introduce a functor *Ortho*: for two line segments A, B, with $Z\varepsilon el(A)$, $Z\varepsilon el(B)$, we let

$$Ortho(A, B) \iff \exists X, Y, U, W. X, Y \varepsilon el(A) \wedge U, W \varepsilon el(B) \wedge non(X\varepsilon = (Y)) \wedge$$

$$non(U\varepsilon = (W)) \wedge U\varepsilon Eq(X, Y) \wedge W \varepsilon Eq(X, Y)$$

(read: A, B are orthogonal). In particular, line segments being classes of sets

1. $([0,1] \times [i, i+1])_i$;
2. $([i, i+1] \times [0,1])_i$;
as well as classes of
3. $([i, i+1] \times [i, i+1])_i$;
4. $([i, i+1] \times [-i, -i-1])_i$.

are orthogonal.

Acknowledgement This work has been prepared under the grant no 8T11C 024 17 from the State Committee for Scientific Research (KBN) of the Republic of Poland. A. Skowron has been supported by grants from KBN as well as from the Wallenberg Foundation.

References

1. AISB-97: Spatial Reasoning in Mobile Robots and Animals, Proceedings AISB-97 Workshop, Manchester Univ., Manchester, 1997.
2. J. Allen, Towards a general theory of action and time, Artificial Intelligence 23(20), 1984, pp. 123–154.
3. R. C. Arkin, Behavior-Based Robotics, MIT Press, Cambridge, MA, 1998.
4. N.Asher, L. Vieu, Toward a geometry of commonsense: a semantics and a complete axiomatization of mereotopology, in: Proceedings IJCAI'95, Morgan Kauffman, San Mateo, CA, 1995, pp. 846-852.
5. N. Asher, M. Aurnague, M. Bras, P. Sablayrolles, L. Vieu, De l'espace-temps dans l'analyse du discours, Rapport interne IRIT/95-08-R, Institut de Recherche en Informatique, Univ. Paul Sabatier, Toulouse, 1995.
6. M. Aurnague and L. Vieu, A theory of space-time for natural language semantics, in: K. Korta and J. M. Larrazábal, eds., Semantics and Pragmatics of Natural Language: Logical and Computational Aspects, ILCLI Series I, Univ. Pais Vasco, San Sebastian, 1995, pp. 69-126.
7. W. Bandler, L. J. Kohout, Fuzzy power sets and fuzzy implication operators, Fuzzy Sets and Systems 4, 1980, 13-30.
8. J.G. Bazan, Hung Son Nguyen, Sinh Hoa Nguyen, P. Synak, J. Wróblewski, Rough set classification algorithms, this volume.
9. J. van Benthem, The Logic of Time, Reidel, Dordrecht, 1983.
10. T. Bittner, On ontology and epistemology of rough location, in: C. Freksa, D. M. Mark (eds.), Spatial Information Theory. Cognitive and Computational Foundations of Geographic Information Science, Lecture Notes in Computer Science 1661, Springer–Verlag, Berlin, 1999, pp. 433–448.
11. G. Booch, Object–Oriented Analysis and Design with Applications, Addison–Wesley Publ., Menlo Park, 1994.
12. R. Casati, B.Smith, A.C. Varzi, Ontological tools for geographic representation, in: N. Guarino (ed.), Formal Ontology in Information Systems, IOS Press, Amsterdam, 1998, pp. 77–85.
13. G. Cattaneo, Abstract approximation spaces for rough theories, in: [69], pp. 59–98.
14. B. L. Clarke, A calculus of individuals based on connection, Notre Dame Journal of Formal Logic 22(2), 1981, pp. 204-218.
15. R. Clay, Relation of Le'sniewski's Mereology to Boolean Algebra, The Journal of Symbolic Logic 39, 1974, pp. 638–648.
16. A. G. Cohn, Calculi for qualitative spatial reasoning, in: J. Calmet, J. A. Campbell, J. Pfalzgraf (eds.), Artificial Intelligence and Symbolic Mathematical Computation, Lecture Notes in Computer Science 1138, Springer–Verlag, Berlin, 1996, pp. 124–143.
17. A. G. Cohn, N. M. Gotts, Representing spatial vagueness: a mereological approach, in: Principles of Knowledge Representation and Reasoning: Proceedings of the 5th International Conference KR'96, Morgan Kaufmann, San Francisco, 1996, pp. 230–241.
18. A. G. Cohn, A. C. Varzi, Connections relations in mereotopology, in: H. Prade (ed.), Proceedings ECAI'98. 13th European Conference on Artificial Intelligence, Wiley and Sons, Chichester, 1998, pp. 150–154.
19. A. G. Cohn, N. M. Gotts, The "egg-yolk" representation of regions with indeterminate boundaries, in: P. Burrough, A. M. Frank (eds.), Proceedings GISDATA

Specialist Meeting on Spatial Objects with Undetermined Boundaries, Fr. Taylor, 1996, pp. 171–187.
20. Collected Works of Stanisław Leśniewski, J. Srzednicki, S. J. Surma, D. Barnett, V. F. Rickey (eds.), Kluwer, Dordrecht, 1992.
21. E. Čech, Topological Spaces, Academia, Praha, 1966.
22. Z. Cui, A. G. Cohn, D. A. Randell, Qualitative and topological relationships, in: Advances in Spatial Databases, Lecture Notes in Computer Science 692, Springer-Verlag, Berlin, 1993, pp. 296–315.
23. M. Dorigo, M. Colombetti, Robot Shaping. An Experiment in Behavior Engineering, MIT Press, Cambridge, MA, 1998.
24. M. J. Egenhofer, R. G. Golledge (eds.), Spatial and Temporal Reasoning in Geographic Information Systems, Oxford U. Press, Oxford, 1997.
25. C. Eschenbach, A predication calculus for qualitative spatial representations, in: C. Freksa, D. M. Mark (eds.), Spatial Information Theory. Cognitive and Computational Foundations of Geographic Information Science, Lecture Notes in Computer Science 1661, Springer–Verlag, Berlin, 1999, pp. 157–172.
26. A. U. Frank, I. Campari (eds.), Spatial Information Theory: A Theoretical Basis for GIS, Lecture Notes in Computer Science 716, Springer–Verlag, Berlin, 1993.
27. A. U. Frank, W. Kuhn (eds.), Spatial Information Theory: A Theoretical Basis for GIS, Lecture Notes in Computer Science, 988, Springer–Verlag, Berlin, 1995.
28. C. Freksa, D. M. Mark (eds.), Spatial Information Theory. Cognitive and Computational Foundations of Geographic Information Science. Proceedings COSIT'99, Lecture Notes in Computer Science 1661, Springer–Verlag, Berlin, 1999.
29. C. Freksa, C. Habel, Repraesentation und Verarbeitung raeumlichen Wissens, Informatik-Fachberichte, Springer–Verlag, Berlin, 1990.
30. A. Galton The mereotopology of discrete space, in: C. Freksa, D. M. Mark (eds.), Spatial Information Theory. Cognitive and Computational Foundations of Geographic Information Science, Lecture Notes in Computer Science 1661, Springer–Verlag, Berlin, 1999, pp. 250–266.
31. J. Glasgow, A formalism for model–based spatial planning, in: A. U. Frank, W. kuhn (eds.), Spatial Information theory - A Theoretical Basis for GIS, Lecture Notes in Computer Science 988, Springer–Verlag, Berlin, 1995, pp. 501–518.
32. N. M. Gotts, A. G. Cohn, A mereological approach to representing spatial vagueness,in: Working papers, the Ninth International Workshop on Qualitative Reasoning, QR'95, 1995.
33. N. M. Gotts, J. M. Gooday, A. G. Cohn, A connection based approach to commonsense topological description and reasoning, The Monist 79(1), 1996, pp. 51–75.
34. N. Guarino, The ontological level, in: R. Casati, B. Smith, G. White (eds.), Philosophy and the Cognitive Sciences, Hoelder-Pichler-Tempsky, Vienna, 1994.
35. S. C. Hirtle, A. U. Frank (eds.), Spatial Information Theory: A Theoretical Basis for GIS, Lecture Notes in Computer Science 1329, Springer–Verlag, Berlin, 1997.
36. M. Inuiguchi, T. Tanino, Level cut conditioning approach to the necessity measure specification, in: Ning Zhong, A. Skowron, S. Ohsuga (eds.), New Directions in Rough Sets, Data Mining and Granular-Soft Computing, Lecture Notes in Artificial Intelligence 1711, Springer–Verlag, Berlin, 1999, pp. 193–202.
37. B. Iwanuś, On Leśniewski's Elementary Ontology, Studia Logica 31, 1973, pp. 73–119.
38. T. Kotarbiński, Elements of the Theory of Knowledge, Formal Logic and Methodology of Science, Polish Sci. Publ., Warsaw, 1966.

39. D. Kortenkamp, R. P. Bonasso, R. Murphy (eds.), Artificial Intelligence and Mobile Robots, AAAI Press/MIT Press, Menlo Park, CA, 1998.
40. K. Krawiec, R. Słowiński, D. Vanderpooten, Learning decision rules from similarity based rough approximations, in: [70], pp.37–54.
41. R. Kruse, J. Gebhardt, F. Klawonn, Foundations of Fuzzy Systems, John Wiley & Sons, Chichester, 1984.
42. B. Kuipers, Qualitative Reasoning: Modeling and Simulation with Incomplete Knowledge, MIT Press, Cambridge MA, 1994.
43. B. J. Kuipers, Y. T. Byun, A qualitative approach to robot exploration and map learning, in: Proceedings of the IEEE Workshop on Spatial Reasoning and Multi-Sensor Fusion, Morgan Kaufmann, San Mateo CA, 1987, pp. 390–404.
44. B. J. Kuipers, T. Levitt, Navigation and mapping in large-scale space, AI Magazine 9(20), 1988, pp. 25–43.
45. L. Kulik, A. Klippel, Reasoning about cardinal directions using grids as qualitative geographic coordinates, in: C. Freksa, D. M. Mark (eds.), Spatial Information Theory. Cognitive and Computational Foundations of Geographic Information Science, Lecture Notes in Computer Science 1661, Springer–Verlag, Berlin, 1999, pp. 205–220.
46. C. Kuratowski, Topology I, II, Academic Press and Polish Scientific Publishers, New York-Warsaw, 1966.
47. T. De Laguna, Point, line, surface as sets of solids, J. Philosophy 19, 1922, pp. 449–461.
48. H. Leonard, N. Goodman, The calculus of individuals and its uses, The Journal of Symbolic Logic 5, 1940, pp. 45–55.
49. Cz. Lejewski, On Leśniewski's Ontology, Ratio I(2), 1958, pp. 150–176.
50. Cz. Lejewski, A contribution to leśniewski's mereology, Yearbook for 1954–55 of the Polish Society of Arts and Sciences Abroad V, 1954–55, pp. 43–50.
51. St. Leśniewski, Grundzüge eines neuen Systems der Grundlagen der Mathematik, Fundamenta Mathematicae 24, 192 , pp. 242–251.
52. St. Leśniewski, Über die Grundlegen der Ontologie, C.R. Soc. Sci. Lettr. Varsovie III, 1930, pp.111–132.
53. St. Leśniewski, On the Foundations of Mathematics, (in Polish), Przegląd Filozoficzny: 30, 1927, pp. 164-206; 31, 1928, pp. 261-291; 32, 1929, pp. 60-101; 33, 1930, pp. 77-105; 34, 1931, pp. 142-170.
54. St. Leśniewski, On the foundations of mathematics, Topoi 2, 1982, pp. 7–52 (an abridged version of the preceding position).
55. G. Link, Algebraic Semantics in Language and Philosophy, CSLI Lecture Notes 74, Stanford Center for the Study of Language and Information, 1998.
56. S. Marcus, Tolerance rough sets,Čech topologies, learning processes, Bull. Polish Acad. Ser. Sci. Tech 42(3), 1994, pp. 471–487.
57. D. M. Mark, M. J. Egenhofer, K. Hornsby, Formal models of commonsense geographic worlds, NCGIA Tech. Report 97-2, Buffalo, 1997.
58. C. Masolo, L. Vieu, Atomicity vs. infinite divisibility of space, in: C. Freksa, D. M. Mark (eds.), Spatial Information Theory. Cognitive and Computational Foundations of Geographic Information Science, Lecture Notes in Computer Science 1661, Springer–Verlag, Berlin, 1999, pp. 235–250.
59. http:/www.cs.albany.edu/‘amit
60. E. Orłowska (ed.), Incomplete Information: Rough Set Analysis, Studies in Fuzziness and Soft Computing (J. Kacprzyk, ed.) vol. 13, Physica–Verlag/Springer–Verlag, Heidelberg, 1998.

61. P. Pagliani, A practical introduction to the modal–relational approach to approximation spaces, in: [69], pp. 209–232.
62. Z. Pawlak, Rough Sets: Theoretical Aspects of Reasoning about Data, Kluwer, Dordrecht, 1992.
63. Z. Pawlak, Decision rules, Bayes' rule and rough sets, in: Ning Zhong, A. Skowron, S. Ohsuga (eds.), New Directions in Rough Sets, Data Mining and Granular-Soft Computing, Lecture Notes in Artificial Intelligence 1711, Springer–Verlag, Berlin, 1999, pp. 1–9.
64. Z. Pawlak and A. Skowron, Rough membership functions, in: R. R. Yager, M. Fedrizzi, J. Kacprzyk (eds.), Advances in the Dempster-Schafer Theory of Evidence, John Wiley and Sons, New York, 1994, pp. 251–271.
65. L. Polkowski and A. Skowron, Rough mereology: a new paradigm for approximate reasoning, International Journal of Approximate Reasoning 15(4), 1997, pp. 333–365.
66. L. Polkowski and A. Skowron, Adaptive decision-making by systems of cooperative intelligent agents organized on rough mereological principles, Intelligent Automation and Soft Computing. An International Journal 2(2), 1996, pp.123–132.
67. L. Polkowski, A. Skowron, Grammar systems for distributed synthesis of approximate solutions extracted from experience, in: Gh. Paun, A. Salomaa (eds.), Grammatical Models of Multi-Agent Systems, Gordon and Breach Sci. Publ./OPA, Amsterdam, 1999, pp. 316–333.
68. L. Polkowski , A. Skowron, Towards an adaptive calculus of granules, in: L. A. Zadeh, J. Kacprzyk (eds.), Computing with Words in Information/Intelligent Systems I, Studies in Fuzziness and Soft Computing (J. Kacprzyk ed.) vol. 33, Physica–Verlag/Springer–Verlag, Heidelberg, 1999, pp. 201–228.
69. L. Polkowski, A. Skowron (eds.), Rough Sets in Knowledge Discovery. Methodology and Applications,in: J. Kacprzyk, ed., Studies in Fuzziness and Soft Computing, vol. 18, Physica–Verlag/Springer–Verlag, Heidelberg, 1998.
70. L. Polkowski, A. Skowron (eds.), Rough Sets in Knowledge Discovery. Applications,Case Studies and Software Systems in: J. Kacprzyk, ed., Studies in Fuzziness and Soft Computing, vol. 19, Physica–Verlag/Springer–Verlag, Heidelberg, 1998.
71. H. Reichenbach, The Philosophy of Space and Time (repr.), Dover, New York, 1957.
72. R. Sikorski, Boolean Algebras, Springer–Verlag, Berlin, 1960.
73. P. Simons, Parts - A Study in Ontology, Clarendon, Oxford, 1987.
74. Sinh Hoa Nguyen, Regularity analysis and its applications in data mining, this volume.
75. Sinh Hoa Nguyen, A. Skowron, P. Synak, Discovery of dat a patterns with applications to decomposition and classification problems, in: [70], pp. 55–97.
76. A. Skowron and L. Polkowski, Rough mereological foundations for design, analysis, synthesis and control in distributed systems, Information Sciences. An Intern. J. 104, 1998, pp. 129–156.
77. A. Skowron, C. Rauszer, The discernibility matrices and functions in information systems, in: R. Słowiński (ed.), Intelligent Decision Systems. Handbook of Applications and Advances of the Rough Sets Theory, Kluwer, Dordrecht, 1992, pp. 331–362.
78. A. Skowron, On topology in information systems, Bull. Polish Acad. Ser. Sci. Math. 36, 1988, pp. 477–479.
79. R. Słowiński, D. Vanderpooten, A generalized definition of rough approximations based on similarity, IEEE Transactions on Data and Knowledge Engineering, in

print.
80. J. Słupecki, S. Leśniewski's Calculus of Names, Studia Logica 3, 1955, pp. 7-72.
81. B. Smith, Logic and formal ontology, in: J. N. Mohanty, W. McKenna (eds.), Husserl's Phenomenology: A Textbook, Lanham: University Press of America, 1989, pp. 29-67.
82. B. Smith, Agglomerations, in: C. Freksa, D. M. Mark (eds.), Spatial Information Theory. Cognitive and Computational Foundations of Geographic Information Science, Lecture Notes in Computer Science 1661, Springer-Verlag, Berlin, 1999, pp. 267-282.
83. B. Smith, Boundaries: an essay in mereotopology, in: L. Hahn (ed.), The Philosophy of Roderick Chisholm, Library of Living Philosophers, La Salle: Open Court, 1997, pp. 534-561.
84. B. Sobociński, Studies in Leśniewski's Mereology, Yearbook for 1954-55 of the Polish Society of Art and Sciences Abroad 5, 1954, pp. 34-43.
85. B. Sobociński, L'analyse de l'antinomie Russellienne par Leśniewski, Methodos, I, 1949, pp. 94-107, 220-228, 308-316; II, 1950, pp. 237-257.
86. http:/agora.leeds.ac.uk/spacenet/spacenet.html
87. J. G. Stell, Granulation for graphs, in: C. Freksa, D. M. Mark (eds.), Spatial Information Theory. Cognitive and Computational Foundations of Geographic Information Science, Lecture Notes in Computer Science 1661, Springer-Verlag, Berlin, 1999, pp. 416-432.
88. J. Stepaniuk, Knowledge discovery by application of rough set models, this volume.
89. A. Tarski, Les fondements de la géométrie des corps, in: Księga Pamiątkowa I Polskiego Zjazdu Matematycznego (Memorial Book of the Ist Polish Mathematical Congress), a supplement to Annales de la Sociéte Polonaise de Mathématique, Cracow, 1929, pp. 29-33.
90. A. Tarski, Appendix E, in: J. H. Woodger, The Axiomatic Method in Biology, Cambridge Univ. Press, Cambridge, 1937.
91. A. Tarski, What is elementary geometry?, in: L. Henkin, P. Suppes, A. Tarski (eds.), The Axiomatic Method with Special Reference to Geometry and Physics, Studies in Logic and Foundations of Mathematics, North-Holland, Amsterdam, 1959, pp. 16-29.
92. A. Tarski, Zur Grundlegung der Boolesche Algebra I, Fundamenta Mathematicae 24, 1935, pp.177-198.
93. T. Topaloglou, First-order theories of approximate space, in Working Notes of the AAAI Workshop on Spatial and Temporal Reasoning, Seattle, 1994, pp. 47-53.
94. http:/www.cs.utexas.edu/users/qr/
95. http:/www.cs.utexas.edu/users/qr/robotics/argus
96. A. N. Whitehead, Process and Reality. An Essay in Cosmology, Macmillan, New York, 1929 (corr. ed. : D. R. Griffin, D. W. Sherbourne (eds.), 1978).
97. A. Wiweger, On topological rough sets, Bull. Polish Acad. Ser. Sci. Math. 37, 1988, pp. 89-93.
98. M. F. Worboys, Imprecision in finite resolution spatial data, Geoinformatica 2, 1998, pp. 257-279.
99. L.A. Zadeh, Fuzzy logic = computing with words, IEEE Trans. on Fuzzy Systems 4, 1996, pp. 103-111.
100. L. A. Zadeh, Toward a theory of fuzzy information granulation and its certainty in human reasoning and fuzzy logic, Fuzzy Sets and Systems 90, 1997, pp. 111-127.
101. L. A. Zadeh, Fuzzy sets, Information and Control 8, 1965, pp. 338-353.

Chapter 4

Knowledge Discovery by Application of Rough Set Models

Jarosław Stepaniuk

Institute of Computer Science
Bialystok University of Technology
Wiejska 45A, 15-351 Bialystok, Poland
e-mail: jstepan@ii.pb.bialystok.pl

Abstract. The amount of electronic data available is growing very fast and this explosive growth in databases has generated a need for new techniques and tools that can intelligently and automatically extract implicit, previously unknown, hidden and potentially useful information and knowledge from these data. These tools and techniques are the subject of the field of Knowledge Discovery in Databases. In this Chapter we discuss selected rough set based solutions to two main knowledge discovery problems, namely the description problem and the classification (prediction) problem.

1 Introduction

Knowledge Discovery in Databases (KDD, for short) has been defined as "the nontrivial extraction of implicit, previously unknown, and potentially useful information from data" [40], [21]. Among others, it uses machine learning, rough sets, statistical and visualization techniques to discover and present knowledge in a form easily comprehensible to humans. Knowledge discovery is a process which helps to make sense of data in a more readable and applicable form. It usually involves at least one of two different goals: description and classification (prediction). Description focuses on finding user-interpretable patterns describing the data. Classification (prediction) involves using some attributes in the data table to predict values (future values) of other attributes.

The theory of rough sets provides a powerful foundation for discovery of important regularities in data and for objects classification. In recent years numerous successful applications of rough set methods for real-life data have been developed (see e.g. [110], [119], [120], [106]).

We list main contributions of this Chapter (for every subject we put references to some previous author's papers related to the discussed topic):

1. Extension of rough set methodology by introducing uncertainty function and rough inclusion ([151], [182]).
2. New methods of Boolean reasoning for reducts and rule computation in tolerance rough set model ([151], [182]).

3. Application of rough set methods to knowledge discovery in real life medical data set ([191], [186]), including new methods of searching for relevant attributes and a hybrid method of rough set and nearest neighbors approaches for prediction.
4. New methods using rough set concepts and relational learning with application to document understanding problem ([189], [187]).
5. Information granules synthesis in knowledge discovery ([155], [156], [157]).

We will now describe in some detail the above mentioned, highly correlated, topics.

Rough set models. Rough set approach has been used in a lot of applications aimed to description of concepts. In most cases only approximate descriptions of concepts can be constructed because of incomplete information about them. Let us consider a typical example for classical rough set approach when concepts are described by positive and negative examples. In such situations it is not always possible describe concepts exactly, since some positive and negative examples of the concepts being described inherently can not be distinguished one from another. Rough set theory was proposed [110] as a new approach to vague concept description from incomplete data. The rough set approach to processing of incomplete data is based on the lower and the upper approximation. The rough set is defined as the pair of two crisp sets corresponding to approximations. If both approximations of a given subset of the universe are exactly the same, then one can say that the subset mentioned above is definable with respect to available information. Otherwise, one can consider it as roughly definable. Suppose we are given a finite non-empty set U of objects, called the universe. Each object of U is characterized by a description constructed, for example from a set of attribute values. In standard rough set approach [110] introduced by Pawlak an equivalence relation (reflexive, symmetric and transitive relation) on the universe of objects is defined from equivalence relations on the attribute values. In particular, this equivalence relation is constructed assuming the existence of the equality relation on attribute values. Two different objects are indiscernible in view of available information, because with these objects the same information can be associated. Thus, information associated with objects from the universe generates an indiscernibility relation in this universe. In the standard rough set model the lower approximation of any subset $X \subseteq U$ is defined as the union of all equivalence classes fully included in X. On the other hand the upper approximation of X is defined as the union of all equivalence classes with a non-empty intersection with X.

In real data sets usually there is some noise, caused for example from imprecise measurements or mistakes made during collecting data. In such situations the notions of "full inclusion" and "non-empty intersection" used in approximations definition are too restrictive. Some extensions in this direction have been proposed by Ziarko in the variable precision rough set model [225].

The indiscernibility relation can be also employed in order to define not only approximations of sets but also approximations of relations [109], [176], [144], [147], [183], [104], [34], [46]. Investigations on relation approximation are well motivated both from theoretical and practical points of view. Let us bring two examples. The equality approximation is fundamental for a generalization of the rough set approach based on a similarity relation approximating the equality relation in the value sets of attributes. Rough set methods in control processes require function approximation.

However, the classical rough set approach is based on the indiscernibility relation defined by means of the equality relations in different sets of attribute values. In many applications instead of these equalities some similarity (tolerance) relations are given only. This observation has stimulated some researchers to generalize the rough set approach to deal with such cases, i.e., to consider similarity (tolerance) classes instead of the equivalence classes as elementary definable sets. There is one more basic notion to be considered, namely the rough inclusion of concepts. This kind of inclusion should be considered instead of the exact set equality because of incomplete information about the concepts. The two notions mentioned above, namely the generalization of equivalence classes to similarity classes (or in more general cases to some neighborhoods) and the equality to rough inclusion have lead to a generalization of classical approximation spaces defined by the universe of objects together with the indiscernibility relation being an equivalence relation. We discuss applications of such approximation spaces for solution of some basic problems related to concept descriptions.

One of the problems we are interested in is the following: given a subset $X \subseteq U$ or a relation $R \subseteq U \times U$, define X or R in terms of the available information. We discuss an approach based on generalized approximation spaces introduced and investigated in [147, 151]. We combine in one model not only some extension of an indiscernibility relation but also some extension of the standard inclusion used in definitions of approximations in the standard rough set model. Our approach allows to unify different cases considered for example in [110, 225].

There are several modifications of the original approximation space definition [110]. The first one concerns the so called uncertainty function. Information about an object, say x is represented for example by its attribute value vector. Let us denote the set of all objects with similar (to attribute value vector of x) value vectors by $I(x)$. In the standard rough set approach [110] all objects with the same value vector create the indiscernibility class. The relation $y \in I(x)$ is in this case an equivalence relation. The second modification of the approximation space definition introduces a generalization of the rough membership function [111]. We assume that to answer a question whether an object x belongs to an object set X we have to answer a question whether $I(x)$ is in some sense included in X.

Boolean reasoning for tolerance approximation spaces. The ability to discern between perceived objects is important for constructing many entities like reducts, decision rules or decision algorithms. In the classical rough set approach the discernibility relation is defined as the complement of the indiscernibility relation. However, this is, in general, not the case for the generalized approximation spaces. The idea of Boolean reasoning is based on construction for a given problem P a corresponding Boolean function f_P with the following property: the solutions for the problem P can be decoded from prime implicants of the Boolean function f_P. Let us mention that to solve real-life problems it is necessary to deal with Boolean functions having large number of variables. A successful methodology based on the discernibility of objects and Boolean reasoning has been developed for computing of many important for applications entities like reducts and their approximations, decision rules, association rules, discretization of real value attributes, symbolic value grouping, searching for new features defined by oblique hyperplanes or higher order surfaces, pattern extraction from data as well as conflict resolution or negotiation (for references see the papers and bibliography in [106], [119], [120]). Most of the problems related to generation of the above mentioned entities are NP-complete or NP-hard. However, it was possible to develop efficient heuristics returning suboptimal solutions of the problems. The results of experiments on many data sets are very promising. They show very good quality of solutions generated by the heuristics in comparison with other methods reported in literature (e.g. with respect to the classification quality of unseen objects). Moreover, they are very efficient from the point of view of time necessary for computing of the solution. It is important to note that the methodology allows to construct heuristics having a very important *approximation property* which can be formulated as follows: expressions generated by heuristics (i.e., implicants) *close* to prime implicants define approximate solutions for the problem. The detailed comparison of rough set classification methods based on combination of Boolean reasoning methodology and discernibility notion with other classification methods one can find in books [119], [120], [106].

In this Chapter we discuss an extension of Boolean reasoning methods for the case of generalized approximation spaces. In particular, we extend methods of data reduction and decision rule generation for this more general case.

Knowledge discovery in medical data. Developed so far rough set methods have shown to be very useful in many real life applications. Rough set based software systems, such as ROSETTA [102], KDD-R [226], PRIMEROSE [199], LERS [50], [51] and Rough Family [165] have been applied to KDD problems. The patterns discovered by the above systems are expressed in attribute-value languages. There are numerous areas of successful applications of rough set software systems (for reviews see books [119], [120], [106]).

We present the results of knowledge discovery in real life diabetes data. We consider three sub-tasks:

– identification of the most relevant condition attributes,
– application of nearest neighbor algorithms for rough set based reduced data,
– discovery of decision rules characterizing the dependency between values of condition attributes and decision attribute.

The nearest neighbor paradigm provides an effective approach to classification. A major advantage of nearest neighbor algorithms is that they are nonparametric, with no assumptions imposed on the data other than the existence of a metric. However, nearest neighbor paradigm is especially susceptible to the presence of irrelevant attributes. We use the rough set approach for selection of the most relevant attributes within the diabetes data set. Next nearest neighbor algorithms are applied with respect to reduced set of attributes.

The presented approach has been applied to knowledge discovery in the medical information system containing children with diabetes mellitus. This is a real life problem coming from the Second Department of Children's Diseases, University Medical School in Bialystok, Poland. The following aspects are evaluated by rough set methods on 107 patients aged 5-22 and suffering from insulin dependent diabetes for 2-13 years: sex, age of disease diagnosis, disease duration, family anamnesis, criteria of the metabolic balance, type of the applied insulin therapy, hypertension, body mass and presence or absence of microalbuminuria.

Rough sets and relational learning. In learning approximations of concepts there is a need to choose a description language. This choice may limit the domains to which a given algorithm can be applied. There are at least two basic types of objects: structured and unstructured. An unstructured object is usually described by attribute-value pairs. For objects having an internal structure first order logic language is often used. In this Chapter we investigate both types of objects. In the former case we use the propositional language with atomic formulas being selectors (i.e. pairs *attribute=value*), in the latter case we consider the first order language.

Attribute-value languages have the expressive power of propositional logic. These languages sometimes do not allow for proper representation of complex structured objects and relations among objects or their components. The background knowledge that can be used in the discovery process is of a restricted form and other relations from the database cannot be used in the discovery process. Using first-order logic (or FOL for short) has some advantages over propositional logic. First order logic provides a uniform and very expressive means of representation. The background knowledge and the examples, as well as the induced patterns, can all be represented as formulas in a first order language. Unlike propositional learning systems, the first order approaches do not require that the relevant data be composed into single relation but, rather can take into account data, which is organized in several database relations with various connections existing among them. First order logic can face problems which cannot be reduced to propositional logics, such as recurrent structures. On the other hand, even if a problem can be reduced to propositional logics, the solutions found in FOL are more readable and simpler than the corresponding ones in propositional logics.

We consider two directions in applications of rough set methods to discovery of interesting patterns expressed in a first order language. The first direction is based on translation of data represented in first-order language to decision table [110] format and next on processing by using rough set methods based on the notion of a reduct. Our approach is based on the iterative checking whether a new attribute adds to the information [189]. The second direction concerns reduction of the size of the data in first-order language and is related to results described in [82], [83], [189]. The discovery process is performed only on well-chosen portions of data which correspond to approximations in the rough set theory. Our approach is based on iteration of approximation operators [189].

Granular computing. One of the rapidly developing areas of soft computing is now granular computing (see e.g. [222], [223]). Several approaches have been proposed toward formalization of the Computing with Words paradigm formulated by Lotfi Zadeh. Information granulation is a very natural concept, and appears (under different names) in many methods related to e.g. data compression, divide and conquer, interval computations, neighborhood systems, and rough sets among others. Notions of a granule [220], [221], [118] and granule similarity (inclusion or closeness) are also very natural in knowledge discovery.

We present a rough set approach for calculus of information granules. This approach is complementary to the calculus of information granules presented in [118]. The presented approach seems to be important for knowledge discovery in distributed environment and for extracting generalized patterns from data. We discuss the basic notions related to information granulation, namely the information granule syntax and semantics as well as the inclusion and closeness (similarity) relations of granules. We discuss some problems of generalized pattern extraction from data assuming knowledge is represented in the form of information granules. We emphasize the importance of information granule application to extract robust patterns from data. We also propose to use complex information granules to extract patterns from data in distributed environment. These patterns can be treated as a generalization of association rules.

The organization of the Chapter is as follows.

In Section 2 we discuss standard and extended rough set models.

In Section 3 we discuss reducts and representatives in standard and tolerance rough set models.

In Section 4 we investigate decision rules generation in standard and tolerance rough set models. We discuss also different quantitative measures associated with rules.

In Section 5 we investigate knowledge discovery in real life diabetes data table.

In Section 6 we apply rough set concepts to relational learning.

In Section 7 we discuss information granules in knowledge discovery.

2 Rough Set Models

The section is organized as follows. In Subsection 2.1 properties of approximations in generalized approximation spaces are discussed. In Subsection 2.2 approximations of relations are investigated. In Subsection 2.3 different constructions of approximation spaces are described.

2.1 Approximations

We present general definition of an approximation space [147], [151], [182] which can be used for example for introducing the tolerance based rough set model and the variable precision rough set model.

For every non-empty set U, let $P(U)$ denote the set of all subsets of U.

Definition 1. A *parameterized approximation space* is a system
$AS_{\#,\$} = (U, I_{\#}, \nu_{\$})$, where

- U is a non-empty set of objects,
- $I_{\#} : U \to P(U)$ is an uncertainty function,
- $\nu_{\$} : P(U) \times P(U) \to [0,1]$ is a rough inclusion function.

The uncertainty function defines for every object x a set of similarly described objects. A constructive definition of uncertainty function can be based on the assumption that some metrics (distances) are given on attribute values. For example, if for some attribute $a \in A$ a metric $\delta_a : V_a \times V_a \to [0, \infty)$ is given, where V_a is the set of all values of attribute a then one can define the following uncertainty function:

$$y \in I_a^{f_a}(x) \text{ if and only if } \delta_a(a(x), a(y)) \leq f_a(a(x), a(y)),$$

where $f_a : V_a \times V_a \to [0, \infty)$ is a given threshold function.

A set $X \subseteq U$ is *definable in* $AS_{\#,\$}$ if and only if it is a union of some values of the uncertainty function.

The rough inclusion function defines the degree of inclusion between two subsets of U [151], [118].

In rough mereology six conditions were formulated for rough inclusions (see for example [116, 140]) for any objects not necessarily being sets. Here, we will usually consider the standard rough inclusion

$$\nu_{SRI}(X, Y) = \begin{cases} \frac{card(X \cap Y)}{card(X)} & \text{if } X \neq \emptyset \\ 1 & \text{if } X = \emptyset \end{cases}.$$

The function $\nu_{l,u}(X, Y) = f_{l,u}(\nu_{SRI}(X, Y))$, where

$$f_{l,u}(t) = \begin{cases} 0 & \text{if } 0 \leq t \leq l \\ \frac{t-l}{u-l} & \text{if } l < t < u \\ 1 & \text{if } t \geq u \end{cases}$$

and $0 \leq l < u \leq 1$ is an example of a rough inclusion for the variable precision rough set model. Let us mention that one can also consider rough inclusion $\xi(X, Y) = \frac{card((U-X) \cup Y)}{card(U)}$ defined in [28].

The lower and the upper approximations of subsets of U are defined as follows.

Definition 2. For an approximation space $AS_{\#,\$} = (U, I_{\#}, \nu_{\$})$ and any subset $X \subseteq U$ the lower and the upper approximations are defined by
$LOW(AS_{\#,\$}, X) = \{x \in U : \nu_{\$}(I_{\#}(x), X) = 1\}$,
$UPP(AS_{\#,\$}, X) = \{x \in U : \nu_{\$}(I_{\#}(x), X) > 0\}$, respectively.

Approximations of concepts (sets) are constructed on the basis of background knowledge. Obviously, concepts are also related to unseen so far objects. Hence it is very useful to define parameterized approximations with parameters tuned in the searching process for approximations of concepts. This idea is crucial for construction of concept approximations using rough set methods. In our notation #, $ are denoting vectors of parameters which can be tuned in the process of concept approximation.

Rough sets can approximately describe sets of patients, events, outcomes, keywords, etc. that may be otherwise difficult to circumscribe.

Example 1. We consider generalized approximation spaces in information retrieval problem [43, 44]. At first, in order to determine an approximation space, we choose the universe U as the set of all keywords. Let DOC be a set of documents, which are described by keywords. Let $key : DOC \longrightarrow P(U)$ be a function mapping documents into sets of keywords. Denote by $c(x_i, x_j)$, where $c : U \times U \longrightarrow \{0, 1, 2, \ldots\}$, the frequency of co-occurrence between two keywords x_i and x_j i.e.

$$c(x_i, x_j) = card(\{doc \in DOC : \{x_i, x_j\} \subseteq key(doc)\}).$$

We define the uncertainty function I_θ depending on a threshold $\theta \in \{0, 1, 2, \ldots\}$ as follows:

$$I_\theta(x_i) = \{x_j \in U : c(x_i, x_j) \geq \theta\} \cup \{x_i\}.$$

One can consider the standard rough inclusion function.

A query is defined as a set of keywords. Different strategies of information retrieval based on the lower and the upper approximations of queries and documents are investigated in [43, 44].

A rough set can be also characterized numerically by the coefficient called the *accuracy of approximation*.

Definition 3. The accuracy of approximation is equal to the number
$$\alpha\left(AS_{\#,\$}, X\right) = \frac{card\left(LOW\left(AS_{\#,\$}, X\right)\right)}{card\left(UPP\left(AS_{\#,\$}, X\right)\right)}.$$

If $\alpha\left(AS_{\#,\$}, X\right) = 1$, then X is *crisp* with respect to $AS_{\#,\$}$ (X is *precise* with respect to $AS_{\#,\$}$), and otherwise, if $\alpha\left(AS_{\#,\$}, X\right) < 1$, then X is *rough* with respect to $AS_{\#,\$}$ (X is *vague* with respect to $AS_{\#,\$}$).

We recall the notions of the positive region and the quality of approximation of classification in the case of generalized approximation spaces.

Definition 4. Let $AS_{\#,\$} = (U, I_\#, \nu_\$)$ be an approximation space. Let $\{X_1, \ldots, X_r\}$ be a classification of objects (i.e. $X_1, \ldots, X_r \subseteq U$, $\bigcup_{i=1}^{r} X_i = U$ and $X_i \cap X_j = \emptyset$ for $i \neq j$, where $i, j = 1, \ldots, r$).

1. The positive region of the classification $\{X_1, \ldots, X_r\}$ with respect to the approximation space $AS_{\#,\$}$ is defined by
$$POS\left(AS_{\#,\$}, \{X_1, \ldots, X_r\}\right) = \bigcup_{i=1}^{r} LOW\left(AS_{\#,\$}, X_i\right).$$

2. The quality of approximation of the classification $\{X_1, \ldots, X_r\}$ in the approximation space $AS_{\#,\$}$ is defined by
$$\gamma\left(AS_{\#,\$}, \{X_1, \ldots, X_r\}\right) = \frac{card\left(POS\left(AS_{\#,\$}, \{X_1,\ldots,X_r\}\right)\right)}{card(U)}.$$

The quality of approximation of the classification coefficient expresses the ratio of the number of all $AS_{\#,\$}$-correctly classified objects to the number of all objects in the data table.

One can also consider other coefficients characterizing classification in the approximation space $AS_{\#,\$}$, for example based on entropy [192].

The presented above definition of approximation space can be treated as a semantic part of the approximation space definition. Usually there is also specified a set of formulas Φ expressing properties of objects. Hence we assume that together with the approximation space $AS_{\#,\$}$ there are given

- a set of formulas Φ over some language,
- semantics $\|\bullet\|$ of formulas from Φ, i.e., a function from Φ into the power set $P(U)$.

Let us consider an example [110]. We define a language L_{IS} used for elementary granule description, where $IS = (U, A)$ is an information system. The syntax of L_{IS} is defined recursively by

1. $(a \in V) \in L_{IS}$, for any $a \in A$ and $V \subseteq V_a$.
2. If $\alpha \in L_{IS}$ then $\neg \alpha \in L_{IS}$.
3. If $\alpha, \beta \in L_{IS}$ then $\alpha \wedge \beta \in L_{IS}$.
4. If $\alpha, \beta \in L_{IS}$ then $\alpha \vee \beta \in L_{IS}$.

The semantics of formulas from L_{IS} with respect to an information system IS is defined recursively by

1. $\|a \in V\|_{IS} = \{x \in U : a(x) \in V\}$.
2. $\|\neg \alpha\|_{IS} = U - \|\alpha\|_{IS}$.
3. $\|\alpha \wedge \beta\|_{IS} = \|\alpha\|_{IS} \cap \|\beta\|_{IS}$.
4. $\|\alpha \vee \beta\|_{IS} = \|\alpha\|_{IS} \cup \|\beta\|_{IS}$.

A typical method used by the classical rough set approach [110] for constructive definition of the uncertainty function is the following: for any object $x \in U$ there is given information $Inf_A(x)$ (information vector, attribute value vector of x) which can be interpreted as a conjunction $EF_B(x)$ of selectors $a = a(x)$ for $a \in A$ and the set $I_\#(x)$ is equal to $\|EF_B(x)\|_{IS} = \left\|\bigwedge_{a \in A} a = a(x)\right\|_{IS}$. One can consider a more general case taking as possible values of $I_\#(x)$ any set $\|\alpha\|_{IS}$ containing x. Next from the family of such sets the resulting neighborhood $I_\#(x)$ can be selected or constructed. One can also use another approach by considering more general approximation spaces in which $I_\#(x)$ is a family of subsets of U [18], [79].

Now, we list properties of approximations in generalized approximation spaces. Next, we present definitions and algorithms for checking rough definability, internal undefinability etc.

Let $AS = (U, I, \nu)$ be an approximation space. For two sets $X, Y \subseteq U$ the equality with respect to the rough inclusion ν is defined in the following way: $X =_\nu Y$ if and only if $\nu(X,Y) = 1 = \nu(Y,X)$.

Proposition 5. *Assuming that for every $x \in U$ we have $x \in I(x)$ and that ν_{SRI} is the standard rough inclusion one can show the following properties of approximations:*

1. $\nu_{SRI}(LOW(AS,X), X) = 1$ and $\nu_{SRI}(X, UPP(AS,X)) = 1$.
2. $LOW(AS, \emptyset) =_{\nu_{SRI}} UPP(AS, \emptyset) =_{\nu_{SRI}} \emptyset$.
3. $LOW(AS, U) =_{\nu_{SRI}} UPP(AS, U) =_{\nu_{SRI}} U$.
4. $UPP(AS, X \cup Y) =_{\nu_{SRI}} UPP(AS, X) \cup UPP(AS, Y)$.
5. $\nu_{SRI}(UPP(AS, X \cap Y), UPP(AS, X) \cap UPP(AS, Y)) = 1$.
6. $LOW(AS, X \cap Y) =_{\nu_{SRI}} LOW(AS, X) \cap LOW(AS, Y)$.
7. $\nu_{SRI}(LOW(AS, X) \cup LOW(AS, Y), LOW(AS, X \cup Y)) = 1$.
8. $\nu_{SRI}(X, Y) = 1$ implies $\nu_{SRI}(LOW(AS, X), LOW(AS, Y)) = 1$.
9. $\nu_{SRI}(X, Y) = 1$ implies $\nu_{SRI}(UPP(AS, X), UPP(AS, Y)) = 1$.
10. $LOW(AS, U - X) =_{\nu_{SRI}} U - UPP(AS, X)$.
11. $UPP(AS, U - X) =_{\nu_{SRI}} U - LOW(AS, X)$.
12. $\nu_{SRI}(LOW(AS, LOW(AS, X)), LOW(AS, X)) = 1$.
13. $\nu_{SRI}(LOW(AS, X), UPP(AS, LOW(AS, X))) = 1$.
14. $\nu_{SRI}(LOW(AS, UPP(AS, X)), UPP(AS, X)) = 1$.
15. $\nu_{SRI}(UPP(AS, X), UPP(AS, UPP(AS, X))) = 1$.

By analogy with the standard rough set theory, we define the following four types of sets:

1. X is *roughly AS-definable* if and only if $LOW(AS, X) \neq_{\nu_{SRI}} \emptyset$ and $UPP(AS, X) \neq_{\nu_{SRI}} U$.
2. X is *internally AS-undefinable* if and only if $LOW(AS, X) =_{\nu_{SRI}} \emptyset$ and $UPP(AS, X) \neq_{\nu_{SRI}} U$.
3. X is *externally AS-undefinable* if and only if $LOW(AS, X) \neq_{\nu_{SRI}} \emptyset$ and $UPP(AS, X) =_{\nu_{SRI}} U$.
4. X is *totally AS-undefinable* if and only if $LOW(AS, X) =_{\nu_{SRI}} \emptyset$ and $UPP(AS, X) =_{\nu_{SRI}} U$.

The intuitive meaning of this classification is the following.

If X is roughly AS-definable, then with the help of AS we are able to decide for some elements of U whether they belong to X or $U - X$.

If X is internally AS-undefinable, then we are able to decide whether some elements of U belong to $U - X$, but we are unable to decide for any element of U, whether it belongs to X, using AS.

If X is externally AS-undefinable, then we are able to decide for some elements of U whether they belong to X, but we are unable to decide, for any element of U whether it belongs to $U - X$, using AS.

If X is totally AS-undefinable, then we are unable to decide for any element of U whether it belongs to X or $U - X$, using AS.

The algorithms for checking corresponding properties of sets have $O\left(n^2\right)$ time complexity, where $n = card\left(U\right)$. Let us also note that using two properties of approximations:

$$LOW\left(AS, U - X\right) =_{\nu_{SRI}} U - UPP\left(AS, X\right),$$

$$UPP\left(AS, U - X\right) =_{\nu_{SRI}} U - LOW\left(AS, X\right)$$

one can obtain internal AS-undefinability of X if and only if $U - X$ is externally AS-undefinable. Having that property, we can utilize an algorithm that check internal undefinability of X to examine if $U - X$ is externally undefinable.

2.2 Rough Relations

One can distinguish several directions in research on relation approximations. Below we list some examples of them. In [109], [176] properties of the rough relations are presented. The relationships of rough relations and modal logics have been investigated by many authors (see e.g. [208], [144]). We refer to [144], where the upper approximation of the input-output relation $R(P)$ of a given program P with respect to indiscernibility relation IND is treated as the composition $IND \circ R(P) \circ IND$ and where a special symbol for the lower approximation of $R(P)$ is introduced. Properties of relation approximations in generalized approximation spaces are presented in [147], [183]. The relationships of rough sets with algebras of relations are investigated for example in [104], [34]. Relationships between rough relations and a problem of objects ranking are presented for example in [46], where it is shown that the classical rough set approximations based on indiscernibility relation do not take into account the ordinal properties of the considered criteria. This drawback is removed by considering rough approximations of the preference relations by graded dominance relations [46].

In this subsection we discuss approximations of relations with respect to different rough inclusions. For simplicity of the presentation we consider only binary relations.

Let $AS = (U, I, \nu)$ be an approximation space, where $U \subseteq U_1 \times U_2$ and U, U_1, U_2 are non-empty sets.

By $\pi_i(R)$ we denote the projection of the relation $R \subseteq U$ onto the $i-th$ axis i.e. for example for $i = 1$

$$\pi_1(R) = \{x_1 \in U_1 : \exists_{x_2 \in U_2} (x_1, x_2) \in R\}.$$

Definition 6. For any relations $S, R \subseteq U$ the rough inclusion functions ν_{π_1} and ν_{π_2} based on the cardinality of the projections are defined as follows:

$$\nu_{\pi_i}(S, R) = \begin{cases} \frac{card(\pi_i(S \cap R))}{card(\pi_i(S))} & \text{if } S \neq \emptyset \\ 1 & \text{if } S = \emptyset \end{cases}, \text{ where } i = 1, 2.$$

We describe the intuitive meaning of the approximations in approximation spaces $AS_\$ = (U, I, \nu_\$)$, where $\$ \in \{SRI, \pi_1, \pi_2\}$. The standard lower approximation $LOW(AS_{SRI}, R)$ of a relation $R \subseteq U$ has the following property:

any objects $(x_1, x_2) \in U$ are connected by the lower approximation of R if and only if any objects (y_1, y_2) from $I((x_1, x_2))$ are in the relation R. One can obtain some less restrictive definitions of the lower approximation using the rough inclusions ν_{π_1} and ν_{π_2}. The pair (x_1, x_2) is in the lower approximation $LOW(AS_{\pi_1}, R)$ if and only if for every y_1 there is y_2 such that the pair (y_1, y_2) is from $I((x_1, x_2)) \cap R$. One can obtain similar interpretation for ν_{π_2}. The upper approximation with respect to all introduced rough inclusions is exactly the same, namely, the pair $(x_1, x_2) \in U$ is in the upper approximation $UPP(AS_\$, R)$, where $\$ \in \{SRI, \pi_1, \pi_2\}$ if and only if there is a pair (y_1, y_2) from $I((x_1, x_2)) \cap R$.

Proposition 7. *For the lower and the upper approximations the following conditions are satisfied:*

1. $LOW(AS_{SRI}, R) \subseteq R$.
2. $LOW(AS_{SRI}, R) \subseteq LOW(AS_{\pi_1}, R)$.
3. $LOW(AS_{SRI}, R) \subseteq LOW(AS_{\pi_2}, R)$.
4. $R \subseteq UPP(AS_{SRI}, R) = UPP(AS_{\pi_1}, R) = UPP(AS_{\pi_2}, R)$.

Example 2. We give some example which illustrates that the inclusions from the last proposition can not to be replaced by equalities. Let us also observe that the universe U need not be equal to the Cartesian product of two sets. Let the universe $U = \{(1,2), (2,1), (2,3), (3,2), (1,3), (3,1)\}$ and the binary relation $R = \{(1,2), (2,1), (2,3), (3,2)\}$.

The definition of an uncertainty function $I : U \to P(U)$ and the rough inclusions are described in Table 1.

U	I	ν_{SRI}	ν_{π_1}	ν_{π_2}
$(1,2)$	$\{(1,2), (1,3)\}$	0.5	1	0.5
$(2,1)$	$\{(2,1), (2,3), (3,1)\}$	0.67	0.5	1
$(2,3)$	$\{(2,1), (2,3), (3,1)\}$	0.67	0.5	1
$(3,2)$	$\{(3,2)\}$	1	1	1
$(1,3)$	$\{(1,2), (1,3)\}$	0.5	1	0.5
$(3,1)$	$\{(2,1), (2,3), (3,1)\}$	0.67	0.5	1

Table 1. Uncertainty Function and Rough Inclusions

The lower and the upper approximations of R in the approximation spaces $AS_\$ = (U, I, \nu_\$)$, where $\$ \in \{SRI, \pi_1, \pi_2\}$ are described in Table 2.

Proposition 8. *The time complexity of algorithms for computing approximations of relations is equal to $O\left((card(U))^2\right)$.*

$LOW\,(AS_{SRI}, R)$	$\{(3,2)\}$
$LOW\,(AS_{\pi_1}, R)$	$\{(1,2),(3,2),(1,3)\}$
$LOW\,(AS_{\pi_2}, R)$	$\{(2,1),(2,3),(3,2),(3,1)\}$
$UPP\,(AS_\$, R)$	U

Table 2. Approximations

2.3 Searching for Approximation Spaces

In this subsection we consider problem of searching for approximation space. The searching for proper approximation space is the crucial and the most difficult task related to decision algorithm synthesis based on approximation spaces.

Discretization and approximation spaces. Discretization is considered for real or integer valued attributes. Discretization is based on searching for "cuts" that determine intervals. All values that lie within each interval are then treated as indiscernible. Thus the uncertainty function for an attribute $a \in A$ is defined as shown below.

$y \in I_a(x)$ if and only if values $a(x)$ and $a(y)$ are from the same interval.

Relations obtained on attribute values by using discretization are reflexive, symmetric and transitive.

A simple discretization process consists of the following two steps:

1. Deciding the number of intervals, which is usually done by a user.
2. Determining the width of these intervals, which is usually done by the discretization algorithm.

For example, the equal width discretization algorithm first finds the minimal and maximal values for each attribute. Then it divides this range of the attribute value into a number of user specified, equal width intervals. In equal frequency discretization the algorithm first sorts the values of each attribute in an ascending order, and then divides them into the user specified number of intervals, in such a way that each interval contains the same number of sorted sequential attribute values. These methods are applied to each attribute independently. They make no use of decision class information.

Several more sophisticated algorithms for automatic discretization (with respect to different optimization criteria) exist, for example one rule discretization [54], entropy-based discretization [39], [20], Boolean reasoning discretization [93]. For overviews of different discretization methods and discussion of computational complexity of discretization problems see [94]. We outline here only Boolean reasoning discretization. Let $DT = (U, A \cup \{d\})$ be a decision table. For the sake of simplifying the exposition, we will assume that all condition attributes A are numerical. First, for each attribute $a \in A$ we can sort its value set V_a to obtain the following ordering:

$min_a < \ldots < v_a^i < v_a^{i+1} < \ldots max_a$

Next, we place cuts in the middle of $\left[v_a^i, v_a^{i+1}\right]$, except for the situation when all objects that have these values also have equal generalized decision values. Boolean reasoning discretization is based on combining the cuts found by above procedure with Boolean reasoning procedure for discarding all but a small number of cuts such that the discernibility of objects in DT is preserved. The set of solutions to the problem of finding minimal subsets of cuts that preserve the original discernibility of objects in DT with respect to the decision attribute d, are defined by means of the prime implicants of a suitable Boolean function. Often we are interested in employing as few cuts as possible. Then the set covering heuristic [59] is typically used to arrive at a single solution.

Sometimes the best result is obtained by manual discretization. For example, in Section 5 discretization of numeric attributes in real life diabetes data was done manually according to medical norms.

Metrics and approximation spaces. Other approach to searching for an uncertainty function is based on the assumption that there are given some metrics (distances) on attribute values. Relations obtained on attribute values by using metrics are reflexive. For review of different metrics defined on attribute values see [214]. Here we only present three examples of such metrics.

Let $(U, A \cup \{d\})$ be a decision table. One can use the overlap metric, which defines the distance for an attribute as 0 if the values are equal, or 1 if they are different, regardless of which two values they are. The function *overlap* defines the distance between two values v and v' of an attribute $a \in A$ as:

$$overlap_a(v, v') = \begin{cases} 1 \text{ if } v \neq v' \\ 0 \text{ if } v = v' \end{cases}.$$

The Value Difference Metric (VDM) provides an appropriate distance function for nominal attributes. A version of VDM (without the weighting schemes but with normalization into zero-one interval) defines the distance between two values v and v' of an attribute $a \in A$ as:

$$vdm_a(v, v') = \frac{1}{r(d)} \sum_{i=1}^{r(d)} (\nu_{SRI}(X_v, X_i) - \nu_{SRI}(X_{v'}, X_i))^2,$$

where $r(d)$ is a number of decision classes, ν_{SRI} is the standard rough inclusion, $X_i = \{x \in U : d(x) = i\}$ and $X_v = \{x \in U : a(x) = v\}$.

Using the distance measure VDM, two values are considered to be closer if they have more similar classifications. For example, if an attribute color has three values red, green and blue, and the application is to identify whether or not an object is an apple, then red and green would be considered closer than red and blue because the former two have correlations with decision apple.

If this distance function is used directly on continuous attributes, all values can potentially be unique. Some approaches to the problem of using VDM on continuous attributes are presented in [214].

One can also use some other distance function for real or integer valued attributes, for example

$$diff_a(v, v') = \frac{|v - v'|}{\max_a - \min_a},$$

where $\max_a$ and $\min_a$ are the maximum and minimum values, respectively, for an attribute $a \in A$.

One should specify in searching for optimal uncertainty function at least two elements:

- a class of parameterized uncertainty functions,
- an optimization criterion.

Definition 9. Let $\delta_a : V_a \times V_a \longrightarrow [0, \infty)$ be a given distance function on attribute values, where V_a is the set of all values of attribute $a \in A$. One can define the following uncertainty function

$$y \in I_a^{f_a}(x) \text{ if and only if } \delta_a(a(x), a(y)) \leq f_a(a(x), a(y)),$$

where $f_a : V_a \times V_a \to [0, \infty)$ is a threshold function.

Example 3. We present two examples of a threshold function f_a.

1. For every $x, y \in U$ one can define $f_a(a(x), a(y)) = \varepsilon_a$, where $0 \leq \varepsilon_a \leq 1$ is a given real number.
2. One can define for real or integer valued attribute $a \in A$,

$$f_a(a(x), a(y)) = \varepsilon_a^\alpha * a(x) + \varepsilon_a^\beta * a(y) + \varepsilon_a,$$

where ε_a^α, ε_a^β and ε_a are given real numbers.

Some special examples of uncertainty functions one can also derive from the literature. In [164] strict and weak indiscernibility relations were considered which can define some kind of uncertainty functions. In some cases, it is natural to consider relations defined by so-called ε-indiscernibility [70]. The global uncertainty function for a set of attributes A is usually defined as the intersection i.e. $I_A(x) = \bigcap_{a \in A} I_a(x)$. For some other examples of local and global uncertainty functions see [182].

Different methods of searching for parameters of proper uncertainty functions are discussed for example in [70], [140], [72], [188], [97], [94]. In [188], [72] genetic algorithm was applied for searching for adequate uncertainty functions of the type $I_a^{\varepsilon_a}$, where $\varepsilon_a = f_a(a(x), a(y))$ for every $x, y \in U$. The optimization criterion can be based for example on maximization of the function which combines three quantities. The first quantity can be based on the quality of approximation of classification $\gamma(AS, \{X_1, \ldots, X_r\})$. For specification of the second quantity we first define two relations:

$$R_d = \{(x, y) \in U \times U : d(x) = d(y)\},$$

$$R_{I_A} = \{(x, y) \in U \times U : y \in I_A(x)\}.$$

Using introduced rough inclusions of relations one can measure degree of inclusion of R_d in R_{I_A} in at least three ways:

- $\nu_{SRI}(R_d, R_{I_A})$,

- $\nu_{\pi_1}(R_d, R_{I_A})$,
- $\nu_{\pi_2}(R_d, R_{I_A})$.

The third quantity can be based on the result of the cross-validation test.

In the simplest case one can optimize the combination of three quantities as follows:

$weight_\gamma * \gamma(AS, \{X_1, \ldots, X_r\}) + weight_\nu * \nu_{SRI}(R_d, R_{I_A}) + weight_{test} * test(AS)$,

where

$weight_\gamma, weight_\nu, weight_{test} \geq 0$ and $weight_\gamma + weight_\nu + weight_{test} = 1$.

In the case of $weight_{test} = 0$ first part of the objective function is introduced to prevent shrinking of the positive region of partition. The second part of the objective function is responsible for an increase in the number of connections (see Figure 1). We inherit the notion of connections from simple observation, that if $y \in I_A(x)$ then we can say that there is a connection between x and y. We propose to discern two kinds of connections between objects, namely "good" and "bad":

- x has a good connection with y if and only if $(x,y) \in R_{I_A}$ and $(x,y) \in R_d$,
- x has a bad connection with y if and only if $(x,y) \in R_{I_A}$ and $(x,y) \notin R_d$.

We are interested only in connections between objects with the same decision ("good" connections). So, the objective function tries to find out some kind of balance between enlarging R_{I_A} and preventing the shrinking of the positive region $POS(AS, \{X_1, \ldots, X_r\})$. If we decrease the value of ε_a then for every object $x \in U$ the $I_a^{\varepsilon_a}(x)$ will not change or become larger. So, starting from $\varepsilon_a = 0$ and increasing the value of threshold we can use above property to find all values when $I_a^{\varepsilon_a}(x)$ changes. We can create lists of such thresholds ε_a for each $a \in A$. Next we can check all possible combinations of thresholds to find out the best for our purpose. Of course it can be a long process because, in the worst case, the number of combinations is equal to: $\frac{1}{2} \prod_{a \in A} \left(card(V_a)^2 - card(V_a) \right) + 1$. So, it is visible that we need some heuristics to find, maybe not the best of all, but a close to optimal solution in reasonable time. In [188], [72] we use genetic algorithms for this purpose. One application of this method to handwritten numerals recognition is reported in [63].

3 Data Reduction in Standard and Tolerance Rough Set Models

One of the problems related to practical applications of rough set methods is whether the whole set of attributes is necessary and if not, how to determine the simplified and still sufficient subset of attributes equivalent to the original. The rejected attributes are redundant since their removal cannot worsen the classification. There are usually several such subsets of attributes and those which are minimal with respect to inclusion are called reducts. Finding a minimal reduct (i.e. reduct with a minimal cardinality of attributes among all reducts) is

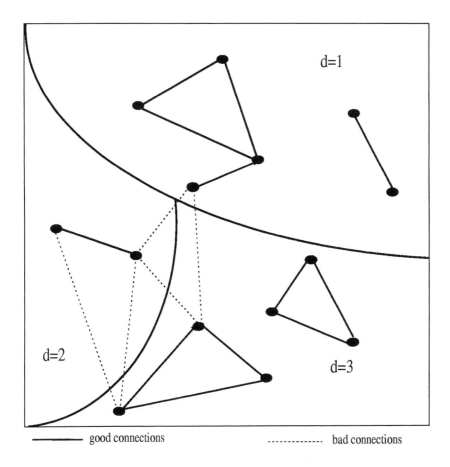

Fig. 1. Three Decision Classes and Connections Between Objects

NP-hard [143]. One can also show that the number of reducts of an information system with m attributes may be as large as

$$\binom{m}{\lfloor m/2 \rfloor}.$$

This means that computation of reducts is a non-trivial task and it can not be solved by a simple increase of computational resources. Significant results in this area have been achieved in [143]. The problem of finding reducts can be transformed to the problem of finding prime implicants of a monotone Boolean function. An implicant of a Boolean function g is any conjunction of literals (variables or their negations) such that, if the values of these literals are true under an arbitrary valuation val of variables, then the value of the function g under val is also true. A prime implicant is a minimal implicant (with respect to the number of literals). Here we are interested in implicants of monotone Boolean functions i.e. functions constructed without the use of negation. The problem of finding prime implicants is known to be NP-hard, but many heuristics are presented for the computation of one prime implicant.

The general scheme of applying Boolean reasoning to a problem RED of reducts computation can be formulated as follows:

1. Encode the problem RED as a Boolean function g_{RED}.
2. Compute the prime implicants of g_{RED}.
3. Solutions to RED are obtained by interpreting the prime implicants of g_{RED}.

In this section we discuss reducts in tolerance rough set model which is a very natural extension of standard rough set model. We consider an approximation spaces AS of the form $AS = (U, I, \nu_{SRI})$ with two conditions for an uncertainty function I:

1. For every $x \in U$, we have $x \in I(x)$ (called reflexivity).
2. For every $x, y \in U$, if $y \in I(x)$, then $x \in I(y)$ (called symmetry).

The section is organized as follows. In Subsection 3.1 the equivalence between reducts and prime implicants of monotonic Boolean functions is investigated. The significance of attributes and the stability of reducts are also considered. In Subsection 3.2 selection of representative objects is investigated.

3.1 Reducts

In this subsection we discuss different definitions of a reduct for a single object and for all objects in a given information system [151]. We also propose similar definitions for a decision table. Those definitions have a property that, like in the standard rough set model (see [110], [143]) and in the variable precision rough set model (see [225], [73]) the set of prime implicants of a corresponding discernibility function is equivalent to the set of reducts.

The computation of all types of reducts is based on generalized discernibility matrix. Discernibility matrix was introduced in [143]. We consider dissimilarity instead of discernibility.

Definition 10. Let $IS = (U, A)$ be an information system. By the generalized discernibility matrix we mean the square matrix $(c_{x,y})_{x,y \in U}$, where

$$c_{x,y} = \{a \in A : y \notin I_a(x)\}.$$

Information systems and reducts. Reduct computation can be translated to computing prime implicants of a Boolean function. The type of reduct controls how the Boolean function is constructed.

In the case of reducts for an information system, the minimal sets of attributes that preserve dissimilarity between objects. Thus the full tolerance relation is considered. Therefore resulting reduct is a minimal set of attributes that enables one to introduce the same tolerance relation on the universe as the whole set of attributes does.

In the case of object-related reducts we consider the dissimilarity relation relative to each object. For each object, there are determined the minimal sets of attributes that preserve dissimilarity of that object from all the others. Thus, we construct a Boolean function by restricting the conjunction to only run over the row corresponding to a particular object x of the discernibility matrix (instead of over all rows). Hence, we obtain the discernibility function related to object x. The set of all prime implicants of this function determines the set of reducts of A related to the object x. These reducts reveal the minimum amount of information needed to preserve dissimilarity of x from all other objects.

Let $IS = (U, A)$ be an information system such that the set $A = \{a_1, \ldots, a_m\}$. We assume that $a_1^*, \ldots, a_m^*$ are Boolean variables corresponding to attributes $a_1, \ldots, a_m$, respectively.

In the following definitions we present more formally notions of both types of reducts.

Definition 11. A subset $B \subseteq A$ is called a reduct of A for an object $x \in U$ if and only if

1. $I_B(x) = I_A(x)$.
2. For every proper subset $C \subset B$ the first condition is not satisfied.

Definition 12. A subset $B \subseteq A$ is called a reduct of A if and only if

1. For every $x \in U$, we have $I_B(x) = I_A(x)$.
2. For every proper subset $C \subset B$ the first condition is not satisfied.

In the following theorems, we present equivalence between reducts and prime implicants of suitable Boolean functions called discernibility functions.

Theorem 13. *For every object $x \in U$ we define the following Boolean function*

$$g_{A,x}(a_1^*, \ldots, a_m^*) = \bigwedge_{y \in U} \bigvee_{a \in c_{x,y}} a^*.$$

The following conditions are equivalent:

1. $\{a_{i_1}, \ldots, a_{i_k}\}$ *is a reduct for the object $x \in U$ in the information system (U, A).*
2. $a_{i_1}^* \wedge \ldots \wedge a_{i_k}^*$ *is a prime implicant of the Boolean function $g_{A,x}$.*

Theorem 14. *We define the following Boolean function*

$$g_A(a_1^*, \ldots, a_m^*) = \bigwedge_{x,y \in U} \bigvee_{a \in c_{x,y}} a^*.$$

The following conditions are equivalent:

1. $\{a_{i_1}, \ldots, a_{i_k}\}$ *is a reduct for the information system (U, A).*
2. $a_{i_1}^* \wedge \ldots \wedge a_{i_k}^*$ *is a prime implicant of the Boolean function g_A.*

Example 4. The flags data table was adopted from the book [131]. The data set is a table listing various features of the flags of different states of the USA, along with the information whether or not the state was a union (U) or confederate (C) state during the civil war. For simplicity of presentation we only consider part of this data set, namely $DT = (U, A \cup \{d\})$, where $U = \{x_1, \ldots x_9\}$ and $A = \{a_1, a_2, a_3, a_4\}$ (see Table 3).

Flag		a_1 Stars	a_2 Hues	a_3 Number	a_4 Word	d Type
Alabama	x_1	0	2	0	0	C
Virginia	x_2	0	5	0	4	C
Tennesee	x_3	3	3	0	0	C
Texas	x_4	1	3	0	0	C
Louisiana	x_5	0	4	0	4	C
Illinois	x_6	0	6	2	6	U
Iowa	x_7	0	5	0	10	U
Ohio	x_8	17	3	0	0	U
New Jersey	x_9	0	5	1	3	U

Table 3. Flags Data Table

For presentation of reducts in the information system (U, A), we assume that the last column ("Type") is not given.

We consider for all attributes $a \in A$, the following uncertainty function: $y \in I_a^{\varepsilon_a}(x)$ if and only if $diff_a(a(x), a(y)) \leq \varepsilon_a$ i.e. $\frac{|a(x) - a(y)|}{\max_a - \min_a} \leq \varepsilon_a$.

We choose the following thresholds:

$\varepsilon_{a_1} = \frac{5}{17}$, $\varepsilon_{a_2} = 0.5$, $\varepsilon_{a_3} = 0.5$ and $\varepsilon_{a_4} = 0.7$.

We consider an approximation space $AS_A = (U, I_A, \nu_{SRI})$, where the global uncertainty function $I_A : U \to P(U)$ is defined by $I_A(x) = \bigcap_{a \in A} I_a^{\varepsilon_a}(x)$.

We construct the discernibility matrix (see Table 4). For example for the object x_3 we obtain the following discernibility function

$$g_{A,x}(a_1^*, a_2^*, a_3^*, a_4^*) = (a_2^* \vee a_3^*) \wedge a_4^* \wedge a_1^*.$$

Object related reducts are presented in Table 5.

For the information system we can not reduce the set of attributes, namely there is only one reduct equal to $\{a_1, a_2, a_3, a_4\}$.

Decision tables and reducts. We present methods of attributes set reduction in a decision table. General scheme of this approach is represented in Figure 2.

If we consider a decision table instead of an information system, this translates to a modification of the discernibility function constructed for the information system. Since we do not need to preserve dissimilarity between objects with the same decision, we can delete those expressions from the discernibility function that preserve dissimilarity between objects within the same decision class. Thus,

	x_1	x_2	x_3	x_4	x_5	x_6	x_7	x_8	x_9
x_1	—	a_2	—	—	—	a_2, a_3	a_2, a_4	a_1	a_2
x_2	a_2	—	—	—	—	a_3	—	a_1	—
x_3	—	—	—	—	—	a_2, a_3	a_4	a_1	—
x_4	—	—	—	—	—	a_2, a_3	a_4	a_1	—
x_5	—	—	—	—	—	a_3	—	a_1	—
x_6	a_2, a_3	a_3	a_2, a_3	a_2, a_3	a_3	—	a_3	a_1, a_2, a_3	—
x_7	a_2, a_4	—	a_4	a_4	—	a_3	—	a_1, a_4	—
x_8	a_1	a_1	a_1	a_1	a_1	a_1, a_2, a_3	a_1, a_4	—	a_1
x_9	a_2	—	—	—	—	—	—	a_1	—

Table 4. Discernibility Matrix for Information System

	Object related reducts
x_1	$\{a_1, a_2\}$
x_2	$\{a_1, a_2, a_3\}$
x_3	$\{a_1, a_2, a_4\}, \{a_1, a_3, a_4\}$
x_4	$\{a_1, a_2, a_4\}, \{a_1, a_3, a_4\}$
x_5	$\{a_1, a_3\}$
x_6	$\{a_3\}$
x_7	$\{a_3, a_4\}$
x_8	$\{a_1\}$
x_9	$\{a_1, a_2\}$

Table 5. Object Related Reducts in Information System

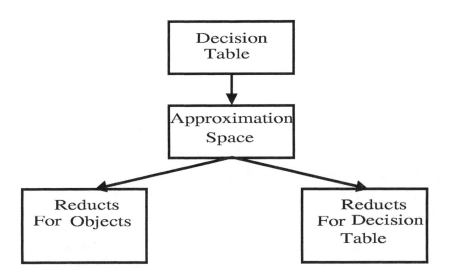

Fig. 2. Two Kinds of Reducts in Decision Table

a resulting reduct is a minimal set of attributes that enables to make the same decisions as the whole set of attributes.

In the case of computing object-related reducts in a decision table, decision rules can be also computed at the same time for efficiency reasons. Let us note that object-related reducts will typically produce shorter rules than reducts calculated for decision table, where length of rules is measured in the number of selectors used in an induced rule.

Let $DT = (U, A \cup \{d\})$ be a decision table. In the following definitions we present more formally all types of reducts.

Definition 15. A subset $B \subseteq A$ is called a relative reduct of A for an object $x \in U$ if and only if

1. $\{y \in I_B(x) : d(y) \neq d(x)\} = \{y \in I_A(x) : d(y) \neq d(x)\}$.
2. For every proper subset $C \subset B$ the first condition is not satisfied.

Definition 16. A subset $B \subseteq A$ is called a relative reduct of A if and only if

1. $POS(AS_B, \{d\}) = POS(AS_A, \{d\})$.
2. For every proper subset $C \subset B$ the first condition is not satisfied.

In the following theorems we demonstrate an equivalence between relative reducts and prime implicants of suitable Boolean functions.

Theorem 17. *For every object $x \in U$ and the Boolean function defined by*

$$g_{A \cup \{d\}, x}(a_1^*, \ldots, a_m^*) = \bigwedge_{y \in U, d(y) \neq d(x)} \bigvee_{a \in c_{x,y}} a^*$$

the following conditions are equivalent:

1. $\{a_{i_1}, \ldots, a_{i_k}\}$ *is a relative reduct for the object $x \in U$ in the decision table DT.*
2. $a_{i_1}^* \wedge \ldots \wedge a_{i_k}^*$ *is a prime implicant of the Boolean function $g_{A \cup \{d\}, x}$.*

Theorem 18. *For the Boolean function defined by*

$$g_{A \cup \{d\}}(a_1^*, \ldots, a_m^*) = \bigwedge_{x, y \in U, d(y) \neq d(x)} \bigvee_{a \in c_{x,y}} a^*$$

the following conditions are equivalent:

1. $\{a_{i_1}, \ldots, a_{i_k}\}$ *is a relative reduct of A.*
2. $a_{i_1}^* \wedge \ldots \wedge a_{i_k}^*$ *is a prime implicant of the Boolean function $g_{A \cup \{d\}}$.*

Example 5. We consider Table 3 with decision attribute "Type". We obtain the relative (with respect to decision) discernibility matrix described in Table 6. For example for object x_7 the discernibility function $g_{A \cup \{d\}, x_7}(a_1^*, a_2^*, a_3^*, a_4^*) = (a_2^* \vee a_4^*) \wedge a_4^* \wedge a_4^* \equiv a_4^*$. In Table 7 we present the set of all relative reducts related to particular objects.

	x_1	x_2	x_3	x_4	x_5	x_6	x_7	x_8	x_9
x_1	–	–	–	–	–	a_2, a_3	a_2, a_4	a_1	a_2
x_2	–	–	–	–	–	a_3	–	a_1	–
x_3	–	–	–	–	–	a_2, a_3	a_4	a_1	–
x_4	–	–	–	–	–	a_2, a_3	a_4	a_1	–
x_5	–	–	–	–	–	a_3	–	a_1	–
x_6	a_2, a_3	a_3	a_2, a_3	a_2, a_3	a_3	–	–	–	–
x_7	a_2, a_4	–	a_4	a_4	–	–	–	–	–
x_8	a_1	a_1	a_1	a_1	a_1	–	–	–	–
x_9	a_2	–	–	–	–	–	–	–	–

Table 6. Relative Discernibility Matrix

	Object related reducts
x_1	$\{a_1, a_2\}$
x_2	$\{a_1, a_3\}$
x_3	$\{a_1, a_2, a_4\}, \{a_1, a_3, a_4\}$
x_4	$\{a_1, a_2, a_4\}, \{a_1, a_3, a_4\}$
x_5	$\{a_1, a_3\}$
x_6	$\{a_3\}$
x_7	$\{a_4\}$
x_8	$\{a_1\}$
x_9	$\{a_2\}$

Table 7. Relative Object Related Reducts

Now we discuss a heuristic [188], [72] which can be applied to the computation of relative reducts without explicit use of discernibility function. We can also use the presented method to simplification of discernibility function.

To find one relative reduct, we build a discernibility matrix. Next, we make reduction of superfluous entries in this matrix. We set an entry to be empty if it is a superset of another non-empty entry. At the end of this process we obtain the set $COMP$ of so called components. From the set of components the described type of reduct can be generated by applying Boolean reasoning. We present heuristics for computing one reduct of the considered type with the possibly minimal number of attributes. These heuristics can produce sets which are supersets of considered reducts but they are much more efficient than the generic procedure.

First, we introduce a notion of a minimal distinction. By a minimal distinction (md, in short) we understand a minimal set of attributes sufficient to discern between two objects. Let us note that the minimal component com consists of minimal distinctions and $card(com)$ is equal or greater than $card(md)$. We say, that md is indispensable if there is a component made of only one md. We include all attributes from the indispensable md to R. Then from $COMP$ we eliminate all these components which have at least one md equal to md in R. It is important that the process of selecting attributes to R will be finished when

the set $COMP$ is empty. We calculate for any md from $COMP$:
$c(md) = w_1 * c_1(md) + w_2 * c_2(md)$, where
$c_1(md) = \left(\frac{card(md \cap R)}{card(md)}\right)^p$ and
$c_2(md) = \left(\frac{card(\{com \in COMP : \exists_{md' \subset com} md' \subset (R \cup md)\})}{card(COMP)}\right)^q$.

For example, we can assume $p = q = 1$.

The first function is a "measure of extending" of R. Since we want to minimize cardinality of R, we are interested in finding md with the largest intersection with actual R. In this way, we always add to R an almost minimal number of new attributes. The second measure is used to examine profit by adding attributes from md to R. We want to include in R the most frequent md in $COMP$ and minimize $COMP$ as much as possible. When $c_2(md) = 1$, then after "adding this md" to R we will obtain a pseudo-reduct i.e. a superset of a reduct.

Significance of attributes and stability of reducts. A problem of relevant attribute selection is one of the importance and have been studied in machine learning and knowledge discovery [65]. There are also several attempts to this problem based on rough sets.

One of the first ideas in rough set based attribute selection [110] was to consider as relevant those attributes which are in the *core* of an information system, i.e. attributes that belong to the intersection of all reducts of that system.

It is also possible to consider as relevant attributes those from some approximate reducts of sufficiently high quality. As it follows from the considerations concerning reduction of attributes, they can not be equally important and some of them can be eliminated from a data table without loose information contained in the table. The idea of attribute reduction can be generalized by introduction of the concept of *attribute significance*. This concept enables evaluation of attributes not only over a two-valued scale, *dispensable – indispensable*, but by associating with an attribute a real number from the $[0,1]$ closed interval. This number expresses the relevance of the attribute in the information table.

Significance of an attribute $a \in B \subseteq A$ in a decision table $DT = (U, A \cup \{d\})$ can be evaluated by measuring the effect of removing of an attribute $a \in B$ from the attribute set B on the positive region defined by the table DT. As shown previously, the number $\gamma(AS_B, \{d\})$ expresses the degree of dependency between attributes B and d. We can ask how the coefficient $\gamma(AS_B, \{d\})$ changes when an attribute a is removed, i.e., what is the difference between $\gamma(AS_B, \{d\})$ and $\gamma(AS_{B-\{a\}}, \{d\})$. We can normalize the difference and define the significance of an attribute a as

$$\sigma_{(AS_B, \{d\})}(a) = \frac{\gamma(AS_B, \{d\}) - \gamma(AS_{B-\{a\}}, \{d\}))}{\gamma(AS_B, \{d\})}.$$

Thus the coefficient $\sigma(a)$ can be understood as the error of classification which occurs when attribute a is dropped.

Example 6. We consider Table 3 with decision attribute "Type". Let $AS_A = (U, I_A, \nu_{SRI})$ be an approximation space defined in Example 4. The uncertainty function I_A is presented in Table 8. We obtain

$$LOW\,(AS_A, \|d = C\|_{DT}) = \{x_1\},$$

$$LOW\,(AS_A, \|d = U\|_{DT}) = \{x_6, x_8\},$$

hence $\gamma(AS_A, \{d\}) = \frac{3}{9}$. For every attribute $a \in A$ the significance of a is presented in Table 9.

	$I_A(\bullet)$
x_1	$\{x_1, x_3, x_4, x_5\}$
x_2	$\{x_2, x_3, x_4, x_5, x_7, x_9\}$
x_3	$\{x_1, x_2, x_3, x_4, x_5, x_9\}$
x_4	$\{x_1, x_2, x_3, x_4, x_5, x_9\}$
x_5	$\{x_1, x_2, x_3, x_4, x_5, x_7, x_9\}$
x_6	$\{x_6, x_9\}$
x_7	$\{x_2, x_5, x_7, x_9\}$
x_8	$\{x_8\}$
x_9	$\{x_2, x_3, x_4, x_5, x_6, x_7, x_9\}$

Table 8. Uncertainty Function

Attribute	$\gamma(AS_{A-\{a\}}, \{d\})$	Significance
a_1	$\frac{1}{9}$	$\frac{2}{3}$
a_2	$\frac{2}{9}$	$\frac{1}{3}$
a_3	$\frac{2}{9}$	$\frac{1}{3}$
a_4	$\frac{3}{9}$	0

Table 9. Significance of Attributes

Another approach to the problem of relevant attributes selection is related to dynamic reducts (see e.g. [7]) i.e. condition attribute sets appearing "sufficiently often" as reducts of samples of the original decision table. The attributes belonging to the "majority" of dynamic reducts are defined as relevant. The value thresholds for "sufficiently often" and "majority" need to be tuned for the given data. Several of the reported experiments show that the set of decision rules based on such attributes is much smaller than the set of all decision rules and the quality of classification of new objects is increasing or at least not significantly decreasing if only rules constructed over such relevant attributes are considered.

We recall the notion of a stability coefficient of a reduct [7]. Let $DT = (U, A \cup \{d\})$ be a decision table. One can say that $DT' = (U', A \cup \{d\})$ is a subtable of DT if and only if $U' \subset U$. Let $B \subseteq A$ be a relative reduct for $(AS_A, \{d\})$. Let F denote a set of subtables of $DT = (U, A \cup \{d\})$. The number

$$\frac{card(\{DT' \in F : B \text{ is a reduct in } DT'\})}{card(F)}$$

is called the stability coefficient of the reduct B.

3.2 Representatives

In some sense dual to the problem of attribute set reduction is the problem of object number reduction (selection). In the standard rough set approach it seems that the first idea is to take one element from every equivalence class defined by a set of attributes. When we consider overlapping classes the above idea should be modified. In this subsection we discuss the problem of proper representative object selection from data tables. We discuss equivalence of the problem of object number reduction to the problem of prime implicants computation for a suitable Boolean function.

The general problem can be described as follows:

Given a set of objects U, the reduction process of U consists in finding a new set $U' \subset U$. The objects which belong to the set U' are chosen for example by using an evaluation criterion. The main advantage of the evaluation criterion approach is that a simple evaluation criterion can be defined which ensures a high level of efficiency. On the other hand, the definition of the evaluation criterion is a difficult problem, because in the new data set some objects are dropped and only a good evaluation criterion preserves the effectiveness of the knowledge acquired during the subsequent learning process.

There are many methods of adequate representative selection (see for example [26], [45], [88], [195], [188], [72]).

In the standard rough set model, representatives can be computed from every indiscernibility class. In this subsection we discuss representative selection based on generalized approximation spaces and Boolean reasoning. This approach was suggested in [140], [195].

We assume that $AS = (U, I_A, \nu_{SRI})$ is an approximation space, where $U = \{x_1, \ldots, x_n\}$ is a set of objects and let $x_1^*, \ldots, x_n^*$ be Boolean variables corresponding to objects $x_1, \ldots, x_n$, respectively.

Definition 19. Let (U, A) be an information system. A subset $U' \subseteq U$ is a minimal set of representatives if and only if the following two conditions are satisfied:

1. For every $x \in U$ there is $y \in U'$ such that $x \in I_A(y)$.
2. For every proper subset $U'' \subset U'$ the first condition is not satisfied.

In the next theorem we obtain a characterization of minimal sets of representatives in information systems.

Theorem 20. *We define the Boolean function*

$$g_{(U,A)}(x_1^*, \ldots, x_n^*) = \bigwedge_{x_i \in U} \bigvee_{x_j \in I_A(x_i)} x_j^*.$$

The following conditions are equivalent:

1. $\{x_{i_1}, \ldots, x_{i_k}\}$ *is a minimal set of representatives.*
2. $x_{i_1}^* \wedge \ldots \wedge x_{i_k}^*$ *is a prime implicant of the Boolean function $g_{(U,A)}$.*

Example 7. We consider Table 3 without decision attribute "Type". From the uncertainty function I_A presented in Table 8, we obtain the Boolean function
$g_{(U,A)}(x_1^*, \ldots, x_9^*) = (x_1^* \vee x_3^* \vee x_4^* \vee x_5^*) \wedge (x_2^* \vee x_3^* \vee x_4^* \vee x_5^* \vee x_7^* \vee x_9^*) \wedge \ldots \wedge (x_2^* \vee x_3^* \vee x_4^* \vee x_5^* \vee x_6^* \vee x_7^* \vee x_9^*)$.
Computing prime implicants of $g_{(U,A)}$, we obtain the following sets of representatives:
$\{x_1, x_2, x_6, x_8\}$, $\{x_1, x_6, x_7, x_8\}$, $\{x_2, x_3, x_6, x_8\}$, $\{x_3, x_6, x_7, x_8\}$,
$\{x_2, x_4, x_6, x_8\}$, $\{x_4, x_6, x_7, x_8\}$, $\{x_5, x_6, x_8\}$, $\{x_1, x_8, x_9\}$, $\{x_3, x_8, x_9\}$,
$\{x_4, x_8, x_9\}$ and $\{x_5, x_8, x_9\}$.

In decision tables, we also consider the decision in computation of minimal sets of representatives.

Definition 21. *Let $(U, A \cup \{d\})$ be a decision table. A subset $U' \subseteq U$ is a relative minimal set of representatives if and only if the following two conditions are satisfied:*

1. *For every $x \in U$ there is $y \in U'$ such that $x \in I_A(y)$ and $d(x) = d(y)$.*
2. *For every proper subset $U'' \subset U'$ the first condition is not satisfied.*

We can formulate a similar theorem for computation of representatives in a decision table as with computation of relative reducts.

Theorem 22. *Let $ST(x_i) = \{x_j \in U : x_j \in I_A(x_i), d(x_i) = d(x_j)\}$.*
We define the Boolean function

$$g_{(U,A \cup \{d\})}(x_1^*, \ldots, x_n^*) = \bigwedge_{x_i \in U} \bigvee_{x_j \in ST(x_i)} x_j^*.$$

The following conditions are equivalent:

1. $\{x_{i_1}, \ldots, x_{i_k}\}$ *is a relative minimal set of representatives.*
2. $x_{i_1}^* \wedge \ldots \wedge x_{i_k}^*$ *is a prime implicant of $g_{(U, A \cup \{d\})}$.*

Below we sketch an algorithm for computation of one set of representatives with minimal or near minimal number of elements.

The main difference between finding out one set of representatives and one relative reduct is in the way in which we calculate and interpret components. In case of the relative set of representatives we do not build the discernibility matrix, but we replace it by a similar table containing for any object x_i all objects similar to x_i and with the same decision:

$$ST\left(x_{i}\right)=\left\{x_{j}\in U:x_{j}\in I_{A}\left(x_{i}\right),d\left(x_{i}\right)=d\left(x_{j}\right)\right\}.$$

After reduction, we obtain components as essential entries in ST. For $COMP$ we can apply the algorithm used to compute a reduct assuming $card(md) = 1$. We add to the constructed relative absorbent set any object which is the most frequent in $COMP$ and then eliminate from $COMP$ all components having this object. This process terminates when $COMP$ is empty. For more details see [188], [72].

4 Selected Classification Algorithms

Any classification algorithm should consists of some classifiers together with a method of conflicts resolving between the classifiers when new objects are classified. In this section we discuss two classes of classification algorithms. Algorithms from the first class are using sets of decision rules as classifiers together with some methods of conflict resolving. The rules are generated from decision tables with tolerance relations using Boolean reasoning approach. They create decision classes descriptions. However, to predict (or classify) new object to a proper decision class it is necessary to fix some methods for conflict resolving between rules recognizing the object and voting for different decisions. We also discuss how such decision rules can be generated using Boolean reasoning. Algorithms of the second kind are based on the nearest neighbor method $(k-NN)$. We show how this method can be combined with some rough set method for relevant attribute selection.

The organization of this section is as follows. In Subsection 4.1 decision rules in standard and tolerance rough set model are discussed. Next we give some overview of quantitative measures for decision rule ranking. The received ranking can be used for rule filtration or for conflict between decision rules resolving when new objects are classified. In Subsection 4.4 we discuss a hybrid method received by combining the $k-NN$ method with some methods for relevant attribute selection.

4.1 Decision Rules in Rough Set Models

In this subsection we show how to use Boolean reasoning approach for decision rule generation from decision tables extended by tolerance relations defined on the attribute value vectors. It is important to note that the aim of Boolean reasoning in considered problems is to preserve some properties described by discernibility relations. In the classical rough set approach the discernibility relation is equal to the complement of the indiscernibility relation. However, in more general cases, (e.g. related to tolerance relation) one can define the discernibility relation in a quite different way (e.g. one can define the discernibility not by $I(x) \cap I(y) = \emptyset$, not by $y \notin I(x)$). We discuss how the Boolean reasoning method works for the classical case of discernibility relation $(y \notin I(x))$. However, one can easily modify it for other cases of discernibility relations. The decision rule generation process for, so called minimal rules, can be reduced to the generation of relative object related reducts. These reducts can be computed as prime implicants of an appropriate Boolean function. Let us mention that the presented methods can be extended for computing of approximate decision rules e.g. association rules [2], [139].

We present some definitions concerning decision rules also called classification rules. A rule

If λ then ξ

is composed of a condition λ and a decision part ξ. The decision part usually describes the predicted class. The condition part λ is a conjunction of selectors, each selector being a condition involving a single attribute. For nominal attributes, this condition is a simple equality test, for example $Sex = f$. For numerical attributes, the condition is typically inclusion in an interval, for example $7 \leq Age < 13$. Decision rules for nominal attributes are represented as statements in the following form:

If $a_1 = v_1$ and ... and $a_k = v_k$ then $d = i$.

Decision rules are generated from reducts. So in order to compute decision rules, reducts have to be computed first. One can use different kinds of reducts (object-related or for data table) and different methods of reducts computation. For example the reducts one can compute by exhaustive calculation. This method finds all reducts by computing prime implicants of a Boolean function.

For example one can use object-related reducts. One can conceptually overlay each reduct over the decision table it was computed from, and read off the attribute values. We give two examples of selectors with respect to two different types of uncertainty functions:

If an uncertainty function I_a for an attribute $a \in A$ is defined by

$y \in I_a(x)$ if and only if $a(x) = a(y)$

(as in the standard rough set model) then for the attribute a from some reduct related to an object $x \in U$ a constructed selector is of the form $a = a(x)$.

If an uncertainty function $I_a^{\varepsilon_a}$ for a numeric attribute $a \in A$ is defined by

$y \in I_a^{\varepsilon_a}(x)$ if and only if $\delta_a(a(x), a(y)) \leq \varepsilon_a$,

where δ_a is a distance function and $\varepsilon_a \in [0, 1]$ is a real number then

for the attribute a from some relative reduct related to an object $x \in U$ a constructed selector should exploit the form of a distance function. For example, for the distance function $diff_a$ a constructed selector is of the form $a \in [a(x) - \varepsilon_a \bullet (max_a - min_a), a(x) + \varepsilon_a \bullet (max_a - min_a)]$.

Example 8. Let us consider data from Table 3. Using relative object related reducts (see Table 7) we obtain rules presented in Table 10.

4.2 Performance and Explanatory Features

Decision rules induced from a data table can be evaluated along at least two dimensions: performance (prediction) and explanatory features (description). By

Objects	Rules
x_1	if $a_1 \in [0,5]$ and $a_2 \in [2,4]$ then $d = C$
x_2	if $a_1 \in [0,5]$ and $a_3 \in [0,1]$ then $d = C$
x_3	if $a_1 \in [0,8]$ and $a_2 \in [2,5]$ and $a_4 \in [0,7]$ then $d = C$
x_3	if $a_1 \in [0,8]$ and $a_3 \in [0,1]$ and $a_4 \in [0,7]$ then $d = C$
x_4	if $a_1 \in [0,6]$ and $a_2 \in [2,5]$ and $a_4 \in [0,7]$ then $d = C$
x_4	if $a_1 \in [0,6]$ and $a_3 \in [0,1]$ and $a_4 \in [0,7]$ then $d = C$
x_5	if $a_1 \in [0,5]$ and $a_3 \in [0,1]$ then $d = C$
x_6	if $a_3 \in [1,2]$ then $d = U$
x_7	if $a_4 \in [3,10]$ then $d = U$
x_8	if $a_1 \in [12,17]$ then $d = U$
x_9	if $a_2 \in [3,6]$ then $d = U$

Table 10. Rules Based on Relative Object Related Reducts

performance is meant assessment of how well the set of rules does in classifying new objects, according to some specified performance criterion. By explanatory features is meant how interpretable the rules are, so that one might gain some insight into how the classification or decision making process is carried out. How these two evaluation dimensions are to weighted is a matter of the intended role of the generated rules. If the set of rules is to operate in a fully automated environment, then performance may be the main feature of interest. Conversely, if the set of rules induction is part of a knowledge discovery process, then the interpretability of the rules will be more important.

Different methods for classification vary in how much they facilitate the knowledge discovery aspect, depending on the type of classifiers they produce. A point that is often held forth in favor of methods that produce rule sets is that the models are directly readable and interpretable.

For example, classification one can perform by the following algorithm. Presented with a given object (information vector) to classify, the algorithm scans through the rule set and determines if each rule fires (is applicable). Rules whose antecedent are not in direct conflict with the contents of the information vector fire. If no rules fire, the most frequent decision in the decision table is taken. This means that the dominating decision in the data set is suggested. If more than one rule fires, these may indicate more than one possible decision class. An election process among the firing rules is then performed in order to resolve conflicts and rank the decisions. First, one can accumulate the votes for each possible decision by each firing rule. Second, the accumulated number of votes for each possible decision defines a certainty coefficient for each decision class. The decision class with the largest certainty coefficient is selected. Ties are resolved by the majority voter algorithm.

4.3 Quantitative Measures Associated with Rules

Let $DT = (U, A \cup \{d\})$ be a given data table. We use the set-theoretical interpretation of rules. It relates a rule to data sets from which the rule is discovered.

Using the cardinalities of sets, we obtain the 2 × 2 contingency table representing the quantitative information about the rule **if** λ **then** ξ (see Table 11).

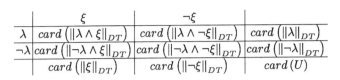

Table 11. Contingency Table Representing the Quantitative Information about the Rule

Using the elements of the contingency table, we may define the support of a decision rule *Rule* of the form **if** λ **then** ξ by
$support_{DT}(Rule) = card(\|\lambda\|_{DT} \cap \|\xi\|_{DT})$.
and its accuracy by
$accuracy_{DT}(Rule) = \frac{card(\|\lambda\|_{DT} \cap \|\xi\|_{DT})}{card(\|\lambda\|_{DT})}$.

This quantity shows the degree to which λ implies ξ. It may be viewed as the conditional probability of a randomly selected object satisfying ξ given that the element satisfies λ. In set-theoretic terms, it is the degree to which $\|\lambda\|_{DT}$ is included in $\|\xi\|_{DT}$ and is equal to $\nu_{SRI}(\|\lambda\|_{DT}, \|\xi\|_{DT})$. Different names were given to this measure, including, the confidence (for mining association rules [2]) and the absolute support [206].

The coverage of *Rule* is defined by
$coverage_{DT}(Rule) = \frac{card(\|\lambda\|_{DT} \cap \|\xi\|_{DT})}{card(\|\xi\|_{DT})}$.

In set-theoretic term, it is the degree to which $\|\xi\|_{DT}$ is included in $\|\lambda\|_{DT}$ and is equal to $\nu_{SRI}(\|\xi\|_{DT}, \|\lambda\|_{DT})$.

In Figure 3 are depicted four decision rules types for ξ defined by condition $d = 1$. Each dashed set represents $\|\lambda\|_{DT}$.

The rule quality for the rules in a rule set *Rule_Set* is determined by a quality function:
$q : Rule\_Set \rightarrow [0, 1]$.

Michalski [85] suggests that high accuracy and coverage are requirements of decision rules. q_M is a weighted sum of the measures of the accuracy and the coverage properties. In the Torgo quality function q_T [196] the accuracy value is judged to be the more important than coverage. The weight is made dependent on accuracy. The Brazdil quality function q_B [10] is a product of accuracy and coverage. The Pearson quality function q_P [12] is based on the theory of contingency tables.

An overview of rule quality formulas is given in [12], [1] and selected formulas are summarized in Table 12.

It is easy to observe that the presented formulas yield values from the interval $[0, 1]$. One can also observe [12] that discussed quality functions are non-

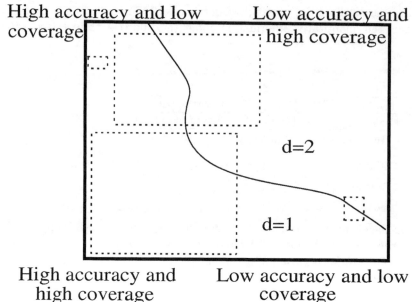

Fig. 3. Accuracy and Coverage

	Formula
q_M	$w \bullet accuracy_{DT}(Rule) + (1-w) \bullet coverage_{DT}(Rule)$
q_T	q_M with $w = \frac{1}{2} + \frac{1}{4} accuracy_{DT}(Rule)$
q_B	$accuracy_{DT}(Rule) \bullet e^{coverage_{DT}(Rule)-1}$
q_P	$\frac{(card(\|\lambda \wedge \xi\|_{DT}) \bullet card(\|\neg\lambda \wedge \neg\xi\|_{DT}) - card(\|\lambda \wedge \neg\xi\|_{DT}) \bullet card(\|\neg\lambda \wedge \xi\|_{DT}))^2}{card(\|\lambda\|_{DT}) \bullet card(\|\xi\|_{DT}) \bullet card(\|\neg\lambda\|_{DT}) \bullet card(\|\neg\xi\|_{DT})}$

Table 12. Selected Rule Quality Formulas

decreasing functions of accuracy and coverage.

Example 9. Let us consider data from Table 3. We compute different coefficients, for the rule

Rule : **if** $a_1 \in [0, 5]$ **and** $a_3 \in [0, 1]$ **then** $d = C$

obtained from the relative object related reduct $\{a_1, a_3\}$ for the objects x_2 and x_5.

For *Rule* we obtain the 2×2 contingency Table 13.

We compute support, accuracy and coverage:

$support_{DT}(Rule) = card(\|a_1 \in [0,5] \wedge a_3 \in [0,1]\|_{DT} \cap \|d = C\|_{DT}) = 5,$

$accuracy_{DT}(Rule) = \frac{card(\|a_1 \in [0,5] \wedge a_3 \in [0,1]\|_{DT} \cap \|d=C\|_{DT})}{card(\|a_1 \in [0,5] \wedge a_3 \in [0,1]\|_{DT})} = \frac{5}{7},$

$coverage_{DT}(Rule) = \frac{card(\|a_1 \in [0,5] \wedge a_3 \in [0,1]\|_{DT} \cap \|d=C\|_{DT})}{card(\|d=C\|_{DT})} = 1.$

	$d=C$	$\neg(d=C)$	
$a_1 \in [0,5] \wedge a_3 \in [0,1]$	5	2	7
$\neg(a_1 \in [0,5] \wedge a_3 \in [0,1])$	0	2	2
	5	4	9

Table 13. Example of Contingency Table Representing the Quantitative Information about *Rule*

We obtain the following qualities of *Rule* :
$q_M(Rule) = w \bullet accuracy_{DT}(Rule) + (1-w) \bullet coverage_{DT}(Rule) = w \bullet \frac{5}{7} + (1-w) \bullet 1 = 1 - \frac{2}{7} \bullet w$, where $0 \leq w \leq 1$ is a parameter,
$q_T(Rule) = \left(\frac{1}{2} + \frac{1}{4} accuracy_{DT}(Rule)\right) \bullet accuracy_{DT}(Rule) + \left(1 - \left(\frac{1}{2} + \frac{1}{4} accuracy_{DT}(Rule)\right)\right) \bullet coverage_{DT}(Rule) = \left(\frac{1}{2} + \frac{5}{28}\right) \bullet \frac{5}{7} + \left(\frac{1}{2} - \frac{5}{28}\right) \bullet 1 = 0.81$,
$q_B(Rule) = accuracy_{DT}(Rule) \bullet e^{coverage_{DT}(Rule)-1} = \frac{5}{7}$,
$q_P(Rule) = \frac{(5 \bullet 2 - 2 \bullet 0)^2}{7 \bullet 5 \bullet 2 \bullet 4} = 0.36$.

4.4 Nearest Neighbor Algorithms

Learning in nearest neighbors algorithm consists of storing the presented data table. When a new object is encountered, a set of similar related objects is retrieved from memory and used to classify the new object.

One advantage of nearest neighbors algorithm is that training is very fast. The second advantage is that the algorithm can learn complex relationships between condition and decision attributes.

One disadvantage of nearest neighbor approach is that the cost of classifying new objects can be high. This is due to the fact that nearly all computation takes place at classification time.

The second disadvantage is based on observation that nearest neighbor approach can be easily fooled by irrelevant attributes. Namely, the distance between objects is calculated based on all attributes of the data table. This lies in contrast to methods based on rough set approach that select only a subset of the attributes when forming a decision algorithm (set of rules). For example, consider applying nearest neighbors approach a problem in which each object is described by twelve attributes, but where only three of these attributes are relevant to determining the classification. In this case, objects that have identical values for the three relevant attributes may nevertheless be distant from one another in twelve dimensional space. As a result, the metric used by nearest neighbors algorithm - depending on all 12 attributes - will be misleading. The distance between neighbors will be dominated by the large number of irrelevant attributes.

One approach to overcoming this problem is to completely eliminate the least relevant attributes from the set of all attributes. The relevant attributes are extracted using rough set approach based, for example, on so called dynamic reducts. The details of the approach are discussed in Section 5. Next $k-NN$ method is applied to the relevant attributes only.

5 Knowledge Discovery in Diabetes Data

Many interesting applications of rough set methods are reported. Let us mention only some of medical applications: treatment of duodenal ulcer by HSV ([112], [41], [159]), analysis of data from peritoneal lavage in acute pancreatitis ([160]), knowledge acquisition in nursing ([14], [51]), medical databases (e.g. headache, meningitis, CVD) analysis ([197], [201], [198], [199], [200]), image analysis for medical applications ([90], [58]), surgical wound infection ([60]), preterm birth prediction ([51]), medical decision–making on board space station Freedom (NASA Johnson Space Center) ([51]), analysis of factors affecting the differential diagnosis between viral and bacterial meningitis ([201]), diagnosing in progressive encephalopathy ([107], [211]), automatic detection of speech disorders ([24]), rough set-based filtration of sound applicable to hearing prostheses ([25]), modeling cardiac patient set residuals ([101]), discovery of attribute dependences in diabetes data ([191], [210], [186]), multistage analysis of therapeutic experience with acute pancreatitis ([161]), analysis of medical data of patients with suspected acute appendicitis ([19]).

In this section the applications of the rough set theory to identify the most relevant attributes and to induce decision rules from a real life medical data set are discussed [191], [210], [186]. The real life medical data set concerns children with diabetes mellitus. Three methods are considered for identification of the most relevant attributes. The first method is based on the notion of reduct and its stability. The second method is based on particular attribute significance measured by relative decrease of positive region after its removal. The third method is inspired by the wrapper approach, where the classification accuracy is used for ranking attributes. The rough set approach additionally offers the set of decision rules. For the rough set based reduced data application of nearest neighbor algorithms is also investigated. The presented methods are general and one can apply all of them to different kinds of data sets.

The structure of the section is as follows. The description of clinical data is presented in Subsection 5.1. The searching for optimal attribute subsets is discussed in Subsection 5.2. In Subsection 5.3 application of nearest neighbors algorithms is investigated. In Subsection 5.4 discovery of decision rules is investigated. In Subsection 5.5 experiments with tolerance thresholds are presented.

5.1 Description of the Clinical Data

There are two main forms of diabetes mellitus:

- type 1(insulin-dependent),
- type 2 (non-insulin-dependent).

Type 1 usually occurs before age 30, although it may strike at any age. The person with this type is usually thin and needs insulin injections to live and dietary modifications to control his or her blood sugar level. Type 2 usually occurs in obese adults over age 40. It's most often treated with diet and exercise (possibly in combination with drugs that lower the blood sugar level), although treatment sometimes includes insulin therapy.

We consider data about children with insulin-dependent diabetes mellitus (type 1). Insulin-dependent diabetes mellitus is a chronic disease of the body's metabolism characterized by an inability to produce enough insulin to process carbohydrates, fat, and protein efficiently. Treatment requires injections of insulin.

Complications may happen when a person has diabetes. Some effects, such as hypoglycemia, can happen any time. Others develop when a person has had diabetes for a long time. These include damage to the retina of the eye (retinopathy), the blood vessels (angiopathy), the nervous system (neuropathy), and the kidneys (nephropathy). The typical form of diabetic nephropathy has large amounts of urine protein, hypertension, and is slowly progressive. It usually doesn't occur until after many years of diabetes, and can be delayed by tight control of the blood sugar. Usually the best lab test for early detection of diabetic nephropathy is measurement of microalbumin in the urine. If there is persistent microalbumin over several repeated tests at different times, the risk of diabetic nephropathy is higher. Normal albumin excretion is less than 20 microgram/min (less than 30 mg/day). Microalbuminuria is 20-200 microgram/min (30-300 mg/day).

Twelve condition attributes, which include the results of physical and laboratory examinations and one decision attribute (microalbuminuria) describe the database used in our experiments. The database is shown at the end of the Chapter in Tables 44 and 45. The data collection so far consists of 107 cases. The collection is growing continuously as more and more cases are analyzed and recorded. Out of twelve condition attributes eight attributes describe the results of physical examinations, one attribute describes insulin therapy type and three attributes describe the results of laboratory examinations. The former eight attributes include sex, the age at which the disease was diagnosed and other diabetological findings. The latter three attributes include the criteria of the metabolic balance, hypercholesterolemia and hypertriglyceridemia. The decision attribute describes the presence or absence of microalbuminuria. All this information is collected during treatment of diabetes mellitus.

Attribute names and types can be found in Table 14.

The range, the mean and the standard deviation of the numerical attributes can be found in Table 15. Additionally attributes with numeric values were discretized. Although several algorithms for automatic discretization exist (for overviews see [94]), in this analysis discretization was done manually according to medical norms. Attributes and their values after discretization are presented in Table 16. Basic data information after discretization is presented in Table 17.

Symbol	Attribute name	Attribute type
a_1	Sex	string
a_2	Age of disease diagnosis (years)	integer
a_3	Disease duration (years)	integer
a_4	Appearance diabetes in the family	string
a_5	Insulin therapy type	string
a_6	Respiratory system infections	string
a_7	Remission	string
a_8	HbA1c	float
a_9	Hypertension	string
a_{10}	Body mass	string
a_{11}	Hypercholesterolemia	string
a_{12}	Hypertriglyceridemia	string
d	Microalbuminuria	string

Table 14. Names and Types of Attributes

Class	Attribute	Range	Mean	Std. dev.
Yes	a_2	$[1, 17]$	9.75	4.01
&	a_3	$[2, 13]$	5.86	2.02
No	a_8	$[4.3, 12.73]$	8.31	1.48
	a_2	$[2, 16]$	10.82	3.02
Yes	a_3	$[2, 13]$	5.93	1.98
	a_8	$[4.3, 11.61]$	8.46	1.56
	a_2	$[1, 17]$	8.57	4.63
No	a_3	$[3, 13]$	5.78	2.08
	a_8	$[5.46, 12.73]$	8.14	1.39

Table 15. Ranges, Means and Standard Deviations of the Numerical Attributes

Attribute	Attribute values
a_1	f, m
a_2	$< 7, [7, 13), [13, 16), \geq 16$
a_3	$< 6, [6, 11), \geq 11$
a_4	yes, no
a_5	KIT, KIT_IIT
a_6	yes, no
a_7	yes, no
a_8	$< 8, [8, 10), \geq 10$
a_9	yes, no
a_{10}	<3, 3-97, >97
a_{11}	yes, no
a_{12}	yes, no
d	yes, no

Table 16. Attributes and Their Values after Discretization

	%	Count
Total number of patients	100	107
Sex		
Male	54.21	58
Female	45.79	49
Age of disease diagnosis (years)		
< 7	22.43	24
[7, 13)	49.53	53
[13, 16)	22.43	24
≥ 16	5.61	6
Disease duration (years)		
< 6	51.40	55
[6, 11)	42.99	46
≥ 11	5.61	6
HbA1c		
< 8	42.99	46
[8, 10)	42.06	45
≥ 10	14.95	16
Microalbuminuria		
yes	52.34	56
no	47.66	51

Table 17. Characterization of Patients Group after Discretization

5.2 Relevance of Attributes

One can measure the importance of attributes with respect to different aspects. One can also consider different strategies searching for the most important subset of attributes. For example one can exhaust all possible subsets of the set of condition attributes and find the optimal ones. In general, its complexity (the number of subsets need to be generated) is $O\left(2^{card(A)}\right)$, where $card(A)$ is a number of attributes. This strategy is very time consuming. Therefore we consider less time consuming strategies.

In this subsection the relevance of attributes is evaluated and compared using three methods.

Reducts application. We compute the accuracy of approximation of decision classes. From Table 18 one can observe that both decision classes are definable by twelve condition attributes.

There are six reducts. Three reducts with nine attributes and three reducts with ten attributes. Reducts are presented in Table 19. Sign " + " means occurrence of the attribute in a reduct. Stability of reducts was verified on subtables. This idea was inspired by the concept of dynamic reducts [7]. Based on experimental verification, reducts for full data table are more stable than other attribute subsets. For example in one experiment we choose 30 subtables starting from 90% to 99% of all objects in data table, thus we consider 300 subtables. Six mentioned above reducts were also reducts at least in 69% from 300 subtables

Decision class	Yes	No
Number of patients	56	51
Cardinality of lower approximation	56	51
Cardinality of upper approximation	56	51
Accuracy of approximation (α)	1.0	1.0

Table 18. Accuracy of Approximation of Decision Classes

and other subsets were reducts in less than 10% of subtables.

Attribute/Reduct	B_1	B_2	B_3	B_4	B_5	B_6
a_1	+	+	+	+	+	+
a_2	+	+	+	+	+	+
a_3	+	+	+	+	+	+
a_4	-	+	+	+	+	+
a_5	+	+	+	+	+	+
a_6	+	-	-	-	+	+
a_7	-	+	+	+	-	-
a_8	+	+	+	+	+	+
a_9	+	-	-	+	-	+
a_{10}	+	-	+	-	+	+
a_{11}	+	+	-	-	+	-
a_{12}	+	+	+	+	+	+
Stability of the reduct	65%	59%	58%	54%	47%	43%
Classification accuracy	63%	77%	71%	76%	68%	70%

Table 19. Reducts, Their Stability and Classification Accuracy

In the Table 19 stability of the reducts based on four experiments is also presented. We consider 300 subtables in every experiment. The sampling strategy is the following: subtables are sampled on 10 equally spaced levels with 30 samples per level. In the following four experiments we consider different sampling levels:
Experiment 1: 60%, 64%, ..., 96% of the original table.
Experiment 2: 70%, 73%, ..., 97% of the original table.
Experiment 3: 80%, 82%, ..., 98% of the original table.
Experiment 4: 90%, 91%, ..., 99% of the original table.

In all experiments we consider subtables with at least 60% of the original table to preserve representability. On the other hand from evaluations presented in [7] we deduce that the number of at least 300 subtables is enough for good estimation of the stability coefficient.

For every reduct one can also compute classification accuracy based on leave-one-out method. The results are presented in the last row of the Table 19.

From the above analysis we infer that the reduct B_2 is a relatively stable subset of attributes with high classification accuracy of generated rules.

Method based on significance of attributes. For the set of all condition attributes the quality of approximation of classification is equal to 1.

In the first step we consider attributes that are in all six reducts. Thus we consider attributes in core. The quality of approximation is equal to 0.76. The ranking of core attributes is presented in Table 20. The idea is to evaluate each individual attribute with the significance measure. This evaluation results in a value attached to an attribute. Attributes are then sorted according to the values. The attribute with the least significance (the smallest contribution to the quality of the approximation) is removed and the process is repeated. One can stop the algorithm when the quality of approximation equals zero.

Attribute removed	Resulting quality of approximation
a_{12}	0.74
a_5	0.55
a_1	0.40
a_3/a_8	0.09
a_2	0

Table 20. Ranking Core Attributes

Method inspired by wrapper approach. We consider method inspired by wrapper approach [65]. The subsets of attributes are evaluated based on the cross-validation result.

We recall that the cross-validation is based on the following procedure. Choose $k > 1$ and partition the available data table $DT = (U, A \cup \{d\})$ into disjoint data subtables $DT_i = (U_i, A \cup \{d\})$ of equal size, where $i = 1, \ldots, k$. For i from 1 to k use DT_i for the test set, and the remaining data for training set. Compute classification accuracy for all k experiments. The average classification accuracy is a result of cross-validation test. Leave-one-out is a special case of cross-validation procedure. In leave-one-out cross-validation the set of $card(U) = n$ objects is repeatedly divided into a training set of size $n - 1$ and test set of size 1, in all possible ways.

In the succeeding steps of the analysis that attribute is removed which removal leads to the best result of the cross-validation test. Let $DT = (U, A \cup \{d\})$ be a data table. The general scheme of the algorithm is as follows:

$B := A$;

Repeat $B := B - \{a\}$, where $a = \arg\max_{a \in B} \{AC(DT_{B-\{a\}})\}$.

Until Stop_Condition;

where $DT_{B-\{a\}} = (U, (B - \{a\}) \cup \{d\})$ and the resulting accuracy coefficient is $AC(DT_{B-\{a\}})$.

The partial results of the analysis are presented in Table 21. The leave-one-out test was used for accuracy estimation. The best result 79.44% was obtained for six attributes. The further removal of attributes thus not led to the increase of classification accuracy.

Attribute	I	II	III	IV	V	VI
a_1	63.55	62.62	64.49	62.62	62.62	60.75
a_2	67.29	64.49	68.22	66.36	71.96	65.42
a_3	69.16	66.36	67.29	66.36	66.36	69.16
a_4	64.49	63.55	71.03	71.03	71.96	69.16
a_5	69.16	69.16	68.22	71.96	72.90	**79.44**
a_6	70.09	**72.90**	-	-	-	-
a_7	69.16	68.22	71.03	72.90	71.03	74.77
a_8	62.62	62.62	64.49	65.42	66.36	62.62
a_9	**71.03**	-	-	-	-	-
a_{10}	68.22	70.09	**76.64**	-	-	-
a_{11}	68.22	70.09	71.03	73.83	**73.83**	-
a_{12}	68.22	70.09	73.83	**74.77**	-	-
γ	1.0	1.0	1.0	0.98	0.95	0.80

Table 21. Classification Accuracy and Quality of Approximation (γ)

Every method allows to analyze data from different angle. Combining the results of the three methods one can find the following three condition attributes as the most relevant: *Age of disease diagnosis, HbA1c* and *Disease duration*. This result is consistent with the general medical knowledge about this disease.

5.3 Rough Set Approach as Preprocessing for Nearest Neighbors Algorithms

In this subsection we discuss the experiments with nearest neighbor algorithms (see for example [213], [87], for more details). The nearest neighbor algorithm retains the entire training data set during learning. This algorithm assumes all objects correspond to points in n-dimensional space. In our experiments the nearest neighbors of an object are defined in terms of the Euclidean distance. More precisely, let $DT = (U, A \cup \{d\})$ be a decision table, for every two objects $x, y \in U$ the Euclidean distance is defined by $E(x,y) = \sqrt{\sum_{a \in A} (a(x) - a(y))^2}$.

Nearest neighbor algorithms are especially susceptible to the inclusion of irrelevant attributes in the data set, and several studies has shown that the classification accuracy degrades as the number of irrelevant attributes is increased (see e.g. [74]). Therefore we use in our experiments relevant subsets of attributes (based on rough set analysis).

Before applying nearest neighbors method to diabetes data, the values of non-numerical attributes were converted into numerical data in the manner described in Table 22.

For number $k \in \{1, \ldots, 10\}$ of nearest neighbors and different attribute subsets (three most important attributes, all attributes and six reducts) we obtain the leave-one-out results presented in Table 23. The best results are obtained for the set $A_3 = \{a_2, a_3, a_8\}$ and are also presented in Figure 4.

	Attribute values
a_1	f - 0, m - 1
a_4	yes - 1, no - 0
a_5	KIT - 1, KIT_IIT - 0
a_6	yes - 1, no - 0
a_7	yes - 1, no - 0
a_9	yes - 1, no - 0
a_{10}	<3 - 0, 3-97 - 1, >97 - 2
a_{11}	yes - 1, no - 0
a_{12}	yes - 1, no - 0
d	yes - 1, no - 0

Table 22. Conversion of Non-Numeric Attributes

k	A_3	A	B_1	B_2	B_3	B_4	B_5	B_6
1	73.83	73.83	71.96	76.64	70.09	70.09	81.31	73.83
2	90.65	85.98	86.92	87.85	87.85	86.92	87.85	86.92
3	75.70	72.90	72.90	75.70	71.96	73.83	73.83	71.96
4	83.18	79.44	79.44	82.24	80.37	80.37	80.37	78.50
5	79.44	72.90	71.96	72.90	73.83	75.70	73.83	73.83
6	85.98	83.18	82.24	84.11	85.05	85.05	83.18	83.18
7	79.44	76.64	76.64	76.64	76.64	80.37	75.70	78.50
8	82.24	82.24	80.37	82.24	84.11	83.18	80.37	82.24
9	78.50	77.57	77.57	74.77	80.37	80.37	77.57	77.57
10	81.31	82.24	82.24	82.24	82.24	82.24	81.31	82.24

Table 23. Nearest Neighbors Method

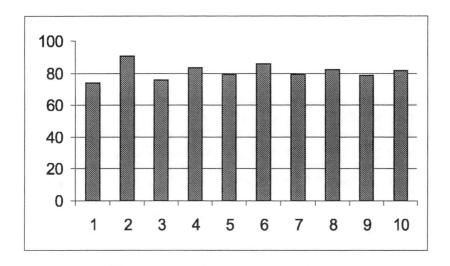

Fig. 4. Nearest Neighbors Method for A_3

We also compare the obtained results with linear discriminant classifier (for subsets of attributes as in nearest neighbors method). The idea of this procedure is to divide object set by a series of lines in two dimensions, planes in three dimensions and, generally hyperplanes in many dimensions (see for example [213], [87], for more details). New objects are classified according to the side of the hyperplane that they fall on.

The method requires numerical attribute vector, therefore we use conversion of the values of non-numerical attributes into numerical data (see Table 22) before applying linear discriminants.

In Table 24 we present the results obtained by linear discriminant classifier (and attribute subsets as for nearest neighbors method) with leave-one-out test.

A_3	A	B_1	B_2	B_3	B_4	B_5	B_6
72.53	67.17	66.37	65.21	58.88	60.66	66.37	66.37

Table 24. Linear Discrimination

The best result is also in the case of the set A_3 with three attributes, but is about 18% worse than for two nearest neighbors method.

5.4 Discovery of Decision Rules

Further analysis of the data table consists in determining the relationships between values of the attributes and presence or absence of microalbuminuria i.e. looking for representation of these relationships in the form of decision rules.

The exhaustive generation of object related reducts and rules from reducts was used to perform the experiment.

In the current experiment the data set has been split three times and all parts of the experiment have been duplicated for each of the three splits. This approach lessens the impact of randomness in the results.

The following procedure was repeated for $i = 1, 2, 3$. The data table $DT = (U, A \cup \{d\})$, where $card(U) = 107$ and $A = \{a_1, \ldots, a_{12}\}$ was split randomly into a testing set $DT_{test_i} = (U_{test_i}, A \cup \{d\})$ and a training set $DT_{train_i} = (U_{train_i}, A \cup \{d\})$ containing approximately 33% ($card(U_{test_i}) = 35$) and 67% ($card(U_{train_i}) = 72$) of the objects in DT, respectively.

We used Michalski's quality function (with weight w of accuracy equal to 0) discussed in Subsection 4.3 for rule filtering. Thus, this experiment corresponds to filtering according to coverage. For summary of obtained results see Table 25.

The difference between classification accuracy obtained for unfiltered rule set (quality threshold equal to zero) and the rules with quality greater or equal than a given threshold is presented in Table 26.

We present examples of rules selected by medical experts from the set of all rules with $q_M \geq 0.1$, where rules are generated from full data table DT.

if $a_2 \in [7, 13)$ **and** $a_3 \in [6, 11)$ **and** $a_{11} = yes$ **then** $d = yes$

Quality	Number of rules			
	DT	DT_{train_1}	DT_{train_2}	DT_{train_3}
$(0, 0.05)$	472	223	291	304
$[0.05, 0.1)$	204	189	262	220
$[0.1, 0.15)$	43	88	73	51
$[0.15, 0.2)$	16	32	19	17
$[0.2, 0.25)$	2	13	2	7
≥ 0.25	1	9	4	1
> 0	738	554	651	600

Table 25. Quality and Number of Rules

Quality threshold	DT_{test_1}	DT_{test_2}	DT_{test_3}
0.00	0%	0%	0%
0.05	3%	3%	3%
0.10	0%	3%	0%
0.15	-3%	0%	-20%
0.20	-11%	-29%	-34%
0.25	-20%	-46%	-57%

Table 26. Difference in Classification Accuracy

$accuracy_{DT} = 1$, $coverage_{DT} = 0.25$
if $a_1 = m$ **and** $a_2 < 7$ **and** $a_4 = KIT$ **then** $d = no$
$accuracy_{DT} = 1$, $coverage_{DT} = 0.20$.
if $a_1 = m$ **and** $a_3 < 6$ **and** $a_4 = KIT$ **and** $a_8 < 8$ **and** $a_9 = no$ **then** $d = no$
$accuracy_{DT} = 1$, $coverage_{DT} = 0.20$.
if $a_2 < 7$ **and** $a_3 \in [6, 11)$ **then** $d = no$
$accuracy_{DT} = 1$, $coverage_{DT} = 0.18$.
if $a_1 = f$ **and** $a_3 \in [6, 11)$ **and** $a_{11} = yes$ **then** $d = yes$
$accuracy_{DT} = 1$, $coverage_{DT} = 0.14$.

5.5 Experiments with Tolerance Thresholds

In this subsection we discuss selected results of numerous experiments performed by the author with uncertainty functions defined by distance measures and threshold functions.

We present results for three data tables $(U, A \cup \{d\})$, $(U, A_3 \cup \{d\})$ and $(U, A - A_3 \cup \{d\})$ with twelve, three $A_3 = \{a_2, a_3, a_8\}$ and nine condition attributes, respectively. We assume that for all non-numerical attributes the uncertainty function is defined in the standard way, i.e., for all $a \in A - A_3$ $I_a(x) = \{y \in U : a(x) = a(y)\}$.

The following thresholds have been found for attributes from A_3:
$\varepsilon_{a_2} = 0.063$, $\varepsilon_{a_3} = 0.182$, $\varepsilon_{a_8} = 0.16$ in the optimization process with respect to classification accuracy.

We compute the accuracy of approximation of decision classes. From Table 27 one can observe that both decision classes are only roughly definable by condition attributes with above thresholds.

Attribute set	Decision class	Yes	No
A	Cardinality of lower approximation	53	46
	Cardinality of upper approximation	61	54
	Accuracy of approximation (α)	0.87	0.85
A_3	Cardinality of lower approximation	11	13
	Cardinality of upper approximation	94	96
	Accuracy of approximation (α)	0.12	0.14
$A - A_3$	Cardinality of lower approximation	42	34
	Cardinality of upper approximation	73	65
	Accuracy of approximation (α)	0.58	0.52

Table 27. Accuracy of Approximation of Decision Classes with Tolerance Thresholds

We compare the results of leave-one-out test for attribute subsets A_3, A and $A - A_3$. The best results are presented in Table 28.

Attribute subset	Classification accuracy
A_3	79.44
A	66.36
$A - A_3$	63.55

Table 28. Leave-One-Out Results

Comparing results from Table 28 we obtain one more argument that attributes a_2, a_3 and a_8 are very relevant for prediction of d.

Discussing obtained decision rules with medical doctors, we observe that sometimes the obtained rules are not very informative for medical experts. More precisely the selectors in obtained rules are not necessarily in exact correspondence with norms used by medical doctors.

This observation can be extended to two different knowledge discovery problems, namely, prediction and description. Experiments with tolerance thresholds and also in some sense with the nearest neighbors method are showing that on the one hand the discretization of numerical attributes based on medical norms can be not optimal with respect to classification accuracy. On the other hand the understanding of decision rules by medical doctors is better when the discretization is based on medical norms, than in case of decision rules with selectors based on tolerance thresholds.

6 Relational Learning

Knowledge discovery is the process of discovering particular patterns over data. In this context data is typically stored in a database. Approaches using First Order Logic (FOL, for short) languages for the description of such patterns offer data mining the opportunity of discovering more complex regularities which may be out of reach for attribute-value languages. Knowledge discovery based on FOL still has other advantages. Complex background knowledge provided by experts can be encoded as first order formulas and be used in the discovery task. The expressiveness of FOL enables discovered patterns to be described in a concise way, which in most cases increases readability of the output. Multiple relations can be naturally handled without explicit (and expensive) joins.

Knowledge discovery and data mining systems have to face several difficulties, in particular related to the huge amount of input data. This problem is especially related to relational learning (or RL for short) systems (see for example [125], [76]) which employ algorithms that are computationally complex. The obvious drawback of RL is efficiency. A relational learning algorithm must consider a much larger set of possible hypothesis (language complexity). Learning time can be reduced by feeding the RL algorithm only a well-chosen portion of the original input data. Such transformation of the input data should throw away unimportant formulas but leave ones that are potentially necessary to obtain proper results.

In this section two approaches to data reduction problem are proposed. Both are based on rough set theory. Rough set techniques serve as data reduction tools to reduce the size of input data fed to more time-expensive (search-intensive) RL techniques. First approach transforms input formulas into decision table form, then uses reducts to select only meaningful data. Second approach introduces a special kind of approximation space. When properly used, iterated lower and upper approximations of target concept have the ability to preferably select facts that are more relevant for concept approximation, at the same time throwing out the non-relevant facts.

6.1 Selected Notions

In this subsection we recall selected definitions of first-order language and we also sketch learning of sets of first-order rules by FOIL system [125, 86].

Before moving on to algorithms for learning sets of rules, let us introduce some basic terminology from relational learning.

Relational learning (also called empirical inductive logic programming) algorithms learn classification rules for a concept. The program typically receives a large collection of positive and negative examples from real-world databases as well as background knowledge in the form of relations. The prototypical example for this research is FOIL [125] and its various successors, but there are several other approaches like LINUS and DINUS [76]. Let p be a target predicate of arity m and $r_1, \ldots, r_l$ be background predicates, where $m, l > 0$ are given natural numbers. We denote the constants by $con_1, \ldots, con_n$, where $n > 0$. A term is either a variable or a constant. An atomic formula is of the form $p(t_1, \ldots, t_m)$ or $r_i(t_1, \ldots)$ where the t's are terms and $i = 1, \ldots, l$. A literal is an atomic formula or its negation. If a literal contains a negation ($\neg$) symbol, we call it a negative literal, otherwise a positive literal. A clause is any disjunction of literals, where all variables are assumed universally quantified. A Horn clause is a clause containing at most one positive literal.

We assume that the expressions are not permitted to contain function symbols (this reduces the complexity of the hypothesis space search).

The learning task for relational learning systems is as follows:

Given:

- a set of positive and negative training examples (expressed by literals without variables) for the target relation,
- background knowledge (or BK for short) expressed by literals without variables and not including the target predicate.
- a language of hypothesis (used for rules searching)
- a notion of satisfaction

Find:

- a set of **if** λ **then** ξ rules, where ξ is an atomic formula of the form $p(var_1^p, \ldots, var_m^p)$ with the target predicate p and λ is a conjunction of literals over background predicates $r_1, \ldots, r_l$, such that the set of rules satisfies positive examples relatively to background knowledge.

Let us note that the learning problems can be also formulated in terms of first-order logic. Consider the background knowledge as a relational structure over the universe of constants. Then for example the concept defined by the **if** part of the rule

if $r(var_1^p, var_1)$ **and** $r(var_1, var_2)$ **and** $r(var_2, var_2^p)$
then $p(var_1^p, var_2^p)$

is equivalent to the relation defined by

$\exists var_1 \exists var_2 \, (r(var_1^p, var_1) \land r(var_1, var_2) \land r(var_2, var_2^p))$.

The class of concepts definable by a non-recursive rules over the background knowledge is equivalent to the class of relations definable over the relational structure corresponding to the background knowledge, by existential formulas such that their quantifier-free part is a conjunction of literals.

We discuss this problem more precisely.

Let k, l and m be given natural numbers and let $r_1, \ldots, r_l$ be predicate symbols. Let Φ_k^m be a set of formulas of the form

$$\exists var_{i_1} \ldots \exists var_{i_j} \varphi \, (var_1^p, \ldots, var_m^p, var_{i_1}, \ldots, var_{i_j})$$

with m free variables $var_1^p, \ldots, var_m^p$ and with at most k existential quantifiers, and φ is a conjunction of literals over $r_1, \ldots, r_l$ with variables

$var_1^p, \ldots, var_m^p, var_{i_1}, \ldots, var_{i_j}$.

For a relational structure $M = (\{con_1, \ldots, con_n\}, r_1^M, \ldots, r_l^M)$, where $r_i^M \subseteq \{con_1, \ldots, con_n\}^{ar(r_i)}$ and $ar(r_i)$ is the arity of the predicate r_i we consider relations definable by disjunctions of formulas from the set Φ_k^m.

Example 10. In this example we sketch how the language used in the standard rough set approach [110] can be translated into the discussed language. Assume that a data table $DT = (U, A \cup \{d\})$ is given. Assume without lost of generality that a set of objects $U = \{con_1, \ldots, con_n\}$ and A is a set of condition attributes and d is a decision attribute. For every attribute-value pair (a, v), where $v \in V_a$ and V_a is a set of values for attribute $a \in A$ one can define an unary predicate symbol $r_{(a,v)}$. One can construct the background knowledge by the following rule:

$r_{(a,v)}(con_i)$ is in the background knowledge if and only if $a(con_i) = v$.

Positive and negative examples can be defined using the following equivalence:

$p_{(d,v)}(con_i)$ is a positive example if and only if $d(con_i) = v$.

The relational structure M based on a given background knowledge is defined by $M = \left(\{con_1, \ldots, con_n\}, \left(r_{(a,v)}^M \right)_{a \in A, v \in V_a} \right)$.

For example a decision rule of the form
if $a_1 = v_1$ **and** ... **and** $a_k = v_k$ **then** $d = v$
one can express by
if $r_{(a_1,v_1)}(var_1^p)$ **and** ... **and** $r_{(a_k,v_k)}(var_1^p)$ **then** $p_{(d,v)}(var_1^p)$.

A variety of algorithms has been proposed for learning first-order rules (see for example [125],[76],[92],[35]). We consider a system called FOIL [125] that employs the following algorithm:

$FOIL(Target\_predicate, BK, X_{target}^+ \cup X_{target}^-)$
$Pos := X_{target}^+$;
$Neg := X_{target}^-$;
$Learned\_rules := \emptyset$;
while $Pos \neq \emptyset$ do
begin
Learn a NewRule;
$NewRule :=$ most general rule possible;
$NewRuleNeg := Neg$;
while $NewRuleNeg \neq \emptyset$ do
begin
Add a new literal to specialize NewRule;
$Candidate\_literals :=$ generate candidates;
$Best\_literal := \arg\max_{L \in Candidate\_literals} Foil\_Gain(L, NewRule)$;
Add $Best\_literal$ to $NewRule$ preconditions;
$NewRuleNeg :=$ subset of $NewRuleNeg$ that satisfies $NewRule$ preconditions;
end;
$Learned\_rules := Learned\_rules \cup \{NewRule\}$;
$Pos := Pos - \{$members of Pos covered by $NewRule\}$;
end;
Return $Learned\_rules$;

To generate candidate specializations of the current rules, FOIL generates a variety of new literals. More precisely, suppose that the current rule being considered is

Rule : if L_1 and ... and L_j then $p(var_1^p, var_2^p, \ldots, var_m^p)$.

FOIL generates candidate specializations of this rule by considering new literal L that fit one of the following forms:

- $r(var_1, \ldots, var_s)$, where $r \in \{r_1, \ldots, r_l\}$ and at least one of the var_i in the created literal must already exist as a variable in the rule.
- The negation of the above form of literal.

To select the most promising literal from the candidates generated at each step, FOIL considers the performance of the rule over the training data. The evaluation function used by FOIL to estimate the utility of adding a new literal is based on the numbers of positive and negative bindings covered before and after adding the new literal. Let us recall that binding is a substitution mapping each variable to a constant. Consider some rule $Rule$, and a candidate literal L that might be added to the body of $Rule$. The value $Foil\_Gain(L, Rule)$ of adding L to $Rule$ is defined as

$$Foil\_Gain(L, Rule) = t \left(\log_2 \frac{p_1}{p_1 + n_1} - \log_2 \frac{p_0}{p_0 + n_0} \right),$$

where
L is the candidate literal to add to rule $Rule$
p_0 = number of positive bindings of $Rule$
n_0 = number of negative bindings of $Rule$
p_1 = number of positive bindings of $Rule + L$
n_1 = number of negative bindings of $Rule + L$
t is the number of positive bindings of $Rule$ also covered by $Rule + L$.

One can observe that $-\log_2 \frac{p_0}{p_0 + n_0}$ is optimal number of bits to indicate the class of a positive binding covered by $Rule$.

6.2 Transforming First–Order Data into Attribute–Value Form

In this subsection we discuss the following approach:

1. The data is transformed from first-order logic into decision table format by the iterative checking whether a new attribute adds any relevant information to the decision table.
2. The reducts and rules from reducts are computed from obtained decision table.

Data represented as a set of formulas can be transformed into attribute–value form. The idea of translation was inspired by LINUS and DINUS systems [75], [76], [35]. We start with a decision table directly derived from the target relations positive and negative examples. Assuming we have m-ary target predicate, the set U of objects in the decision table is a subset of $\{con_1, \ldots, con_n\}^m$. Decision attribute $d_p : U \to \{+, -\}$ is defined by the target predicate with possible values "+" or "−". All positive and negative examples of the target predicate are now put into the decision table. Each example creates a separate row in the table. Then background knowledge is applied to the decision table. We determine all the possible applications of the background predicates to the arguments of the target relation. Each such application introduces a new Boolean attribute.

To analyze the complexity of the obtained data table, let us consider the number of condition attributes. Let A_{r_i} be a set of attributes constructed for every predicate symbol r_i, where $i = 1, \ldots, l$. The number of condition attributes in constructed data table is equal to $\sum_{i=1}^{l} card(A_{r_i})$ resulting from the possible applications of the l background predicates on the variables of the target relation. The cardinality of A_{r_i} depends on the number of arguments of target predicate p (denoted by m) and the arity of r_i. Namely $card(A_{r_i})$ is equal to $m^{ar(r_i)}$, where $ar(r_i)$ is the arity of the predicate r_i. The number of condition attributes in obtained data table is polynomial in the arity m of the target predicate p and the number l of background knowledge predicates, but its size is usually so large that its processing will be not feasible. Therefore one can check interactively if a new attribute is relevant i.e. adds any information to the decision table and next we add to the decision table only relevant attributes.

Three conditions for testing if a new attribute is relevant are proposed [189]:

1. $\gamma(AS_{B \cup \{a\}}, \{X_+, X_-\}) > \gamma(AS_B, \{X_+, X_-\})$,
   where X_+ and X_- denote decision classes corresponding to the target concept. An attribute is added to the decision table if it results in a positive region growth with respect to previously selected attributes.
2. $\nu_{SRI}(X_+ \times X_-, \{(x, y) \in X_+ \times X_- : a(x) \neq a(y)\}) \geq theta$,
   where $theta \in [0, 1]$ is a given real number. An attribute is added to the decision table if it introduces some discernibility between objects belonging to different non-empty classes X_+ and X_-.
3. $\arg\max \{\gamma(AS_{B \cup \{a\}}, \{X_+, X_-\}) - \gamma(AS_B, \{X_+, X_-\})\}$.
   Given several potential attributes, only the attribute with maximal positive region gain is selected to be added to the decision table.

First two conditions can be applied to a single attribute before it is introduced to the decision table. If this attribute does not meet a condition it is not included in the decision table. The third condition is applied when we have several candidate attributes and must select the one that is potentially the best.

The received data table is then analyzed by a rough set based systems (for example ROSETTA, see [102]). First, reducts are computed. Next, decision rules are generated.

Example 11. The daughter problem, adopted from [35], can be used to demonstrate the transformation of relational learning problem into attribute–value form. For simplicity, the names of the persons are 1, 2, ... instead of "Mary", "Ann", ... Suppose that there are the following positive and negative examples of target predicate *daughter*:

$X_{target} =$
$\{daughter(1,2), daughter(3,4), daughter(5,6), daughter(7,6),$
$\neg daughter(4,2), \neg daughter(3,2), \neg daughter(7,5), \neg daughter(2,4)\}.$

Consider the background knowledge about family relations, *parent* and *female* :

$parent(2,1), parent(2,4), parent(4,3), parent(6,5), parent(6,7),$
$female(1), female(2), female(3), female(5), female(7).$

We then transform the data into attribute–value form (decision table). In Table 29 potential attributes are presented.

Symbol	Attribute
a_1	$female(var_1)$
a_2	$female(var_2)$
a_3	$parent(var_1, var_1)$
a_4	$parent(var_1, var_2)$
a_5	$parent(var_2, var_1)$
a_6	$parent(var_2, var_2)$

Table 29. Potential Attributes

Using conditions introduced in this subsection some attributes will not be included in the resulting decision table. For example the second condition with $theta = 0.1$ would not permit the following attributes into the decision table: a_3 ($parent(var_1, var_1)$) and a_6 ($parent(var_2, var_2)$) (see Table 30).

(var_1, var_2)	a_1	a_2	a_4	a_5	$daughter(var_1, var_2)$
$(1,2)$	true	true	false	true	+
$(3,4)$	true	false	false	true	+
$(5,6)$	true	false	false	true	+
$(7,6)$	true	false	false	true	+
$(4,2)$	false	true	false	true	-
$(3,2)$	true	true	false	false	-
$(7,5)$	true	true	false	false	-
$(2,4)$	true	false	true	false	-

Table 30. Resulting Decision Table

Therefore finally $DT_{0.1} = (U, A_{0.1} \cup \{d\})$, where

$$A_{0.1} = \{a_1, a_2, a_4, a_5\}.$$

We have two decision classes: $X_+ = \{(1,2),(3,4),(5,6),(7,6)\}$ and $X_- = \{(4,2),(3,2),(7,5),(2,4)\}$. For the obtained decision table we construct an approximation space $AS_{A_{0.1}} = (U, I_{A_{0.1}}, \nu_{SRI})$ such that the uncertainty function and the rough inclusion are defined in Table 31.

U	$I_{A_{0.1}}$	$\nu_{SRI}(\bullet, X_+)$	$\nu_{SRI}(\bullet, X_-)$
$(1,2)$	$\{(1,2)\}$	1	0
$(3,4)$	$\{(3,4),(5,6),(7,6)\}$	1	0
$(5,6)$	$\{(3,4),(5,6),(7,6)\}$	1	0
$(7,6)$	$\{(3,4),(5,6),(7,6)\}$	1	0
$(4,2)$	$\{(4,2)\}$	0	1
$(3,2)$	$\{(3,2),(7,5)\}$	0	1
$(7,5)$	$\{(3,2),(7,5)\}$	0	1
$(2,4)$	$\{(2,4)\}$	0	1

Table 31. Universe, Uncertainty Function and Rough Inclusion

One can compute the positive region $POS(AS_{A_{0.1}}, \{X_+, X_-\}) = U$. Then we compute reducts. We obtain one reduct for decision table $DT_{0.1} : \{a_1, a_5\} = \{female(var_1), parent(var_2, var_1)\}$. We generate rules for decision attribute $daughter(var_1, var_2)$ based on this reduct obtaining:
if $female(var_1)$ **and** $parent(var_2, var_1)$ **then** $daughter(var_1, var_2)$.

6.3 Selection of Relevant Facts

An approach presented in this subsection consists of the following steps:

1. Selection of potentially relevant facts from background knowledge.
2. Application of relational learning system such as FOIL [125] or PROGOL [92] to selected formulas.

The selection is based on constants occurring in positive and negative examples of a target relation. The set of all constants occurring in a fact x is denoted by $CON(x)$. CON can be treated as a set valued attribute. A set of constants for a set of facts X is defined by $CON(X) = \bigcup_{x \in X} CON(x)$.

Training set reduction begins with determining the set of constants in all positive and negative examples for the target predicate. Such set is denoted as $CON(X_{target})$. We consider a data table $(U, \{CON\} \cup \{d\})$, where U is the set of all facts from background knowledge, $CON : U \to P(\{con_1, \ldots, con_n\})$, where $P(\{con_1, \ldots, con_n\})$ is the set of all subsets of constants and $d : U \to \{0, 1\}$. For every $x \in U$ we assume

$d(x) = 1$ if and only if $CON(x) \subseteq CON(X_{target})$.

The selections can be represented as lower and upper approximations of $X_{d=1} = \{x \in U : d(x) = 1\}$ in the family of approximation spaces $AS_{CON}^{f_{CON}} = \left(U, I_{CON}^{f_{CON}}, \nu_{SRI} \right)$, where

$f_{CON}(CON(x), CON(x')) = w_1 \bullet \frac{card(CON(x))}{card(CON(x) \cup CON(x'))} + w_2 \bullet \frac{card(CON(x'))}{card(CON(x) \cup CON(x'))} + \varepsilon$

and w_1, w_2 and ε are parameters.

Definition 23. Let $AS_{CON}^{fcon} = \left(U, I_{CON}^{fcon}, \nu_{SRI}\right)$ be an approximation space, where

1. U is the set of all formulas from background knowledge.
2. The uncertainty function I_{CON}^{fcon} is defined by
   $x' \in I_{CON}^{fcon}(x)$ if and only if
   $1 - \frac{card(CON(x) \cap CON(x'))}{card(CON(x) \cup CON(x'))} \leq f_{CON}(CON(x), CON(x'))$.

Any uncertainty function contributes to a different approximation space which results in different kinds of approximations that show different properties.

We then define two transformations $LOW : P(U) \to P(U)$ and $UPP : P(U) \to P(U)$ based on the lower and upper approximations in AS_{CON}^{fcon}.

Starting with $X_{d=1}$ one can construct a sequence of approximations by constantly applying one of these transformations first on $X_{d=1}$ and then on the approximation resulting from the previous step.

Thus, the problem of selection is reduced to constantly applying upper (lower) approximation in the same approximation space to the upper (lower) approximation set obtained in the previous step.

The input data reduction problem is then defined as taking into account facts that are included in $LOW\left(AS_{CON}^{fcon}, X_{d=1}\right)$. If this approximation appears to be too restrictive, which results in bad quality of discovered knowledge, we then consider $UPP\left(AS_{CON}^{fcon}, X_{d=1}\right)$. If it also does not meet our expectations, we proceed to consider the following approximations:

$UPP\left(AS_{CON}^{fcon}, UPP\left(AS_{CON}^{fcon}, X_{d=1}\right)\right)$ and so on. We can stop when the approximation is sufficient to learn up to satisfactory definition of the target concept.

Since $X_{target} = X_{target}^+ \cup X_{target}^-$ (the union of positive and negative examples of the target relation) we may also consider separate approximations of sets corresponding to X_{target}^+ and X_{target}^- which are added after the approximation process. This approach results in a more restrictive approximation.

We give an illustrate example of the proposed approach.

Example 12. The experimental data set is related to document understanding and has been an object of previous studies, see for example [37],[82], [189]. The learning task involves identifying the purposes served by components of single-page letters such as that in Figure 5. Predicate data describes thirty single page documents containing 364 components in all. Fifty seven background predicates describe properties of components such as their width and height, and relationships such as horizontal and vertical alignment with other components. Target predicates describe whether a block is one of the five predetermined types: sender, receiver, logo, reference and date.

The experiment was performed by randomly selecting twenty documents for the training set and ten for the test set. Since the problem concerns the classi-

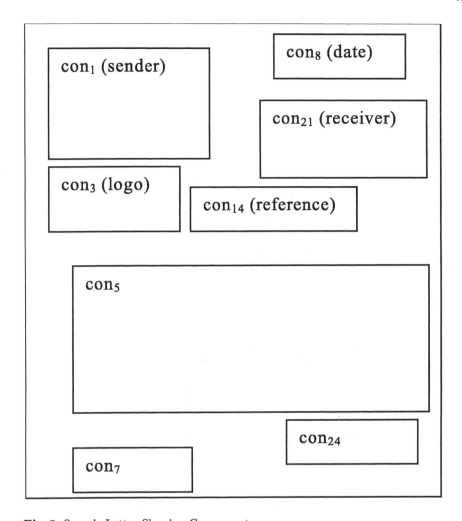

Fig. 5. Sample Letter Showing Components

fication of parts of a document, there are more than twenty training instances (positive and negative) per concept.

We consider an approximation space $AS_{CON}^{fCON} = \left(U, I_{CON}^{fCON}, \nu_{SRI}\right)$ such that $w_1 = w_2 = 0$, thus $f_{CON}\left(CON\left(x\right), CON\left(x'\right)\right) = \varepsilon$, where $\varepsilon \in [0,1]$ is a parameter.

The lower approximation of order one and upper approximations of order one, two and three have been calculated. By applying approximations in different approximation spaces (with respect to ε), several levels of data reduction were obtained. In the training data set approximation spaces were divided into four groups, each displaying different data reduction levels. Overall there were eight data levels, ranging from the empty set to the full training data set. Figure 6 shows the results for different approximation space groups and eight possible reduction levels resulting from four previously mentioned approximations. Bars with different patterns represent the gain in input data resulting from applying the next approximation. Experiments with FOIL system show that any non-empty approximation is sufficient to obtain satisfactory definitions of the target predicates (classification accuracy above 90% on testing data set).

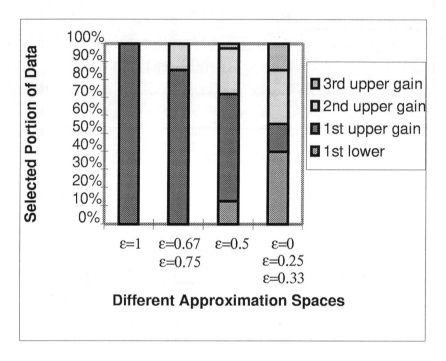

Fig. 6. Four Approximation Space Groups and Eight Approximation Levels

7 Information Granules in Knowledge Discovery

Notions of a granule [220], [221], [118] and granule similarity (inclusion or closeness) are very natural in knowledge discovery. The exact interpretation between granule languages of different information sources (agents) often does not exist. Hence (rough) inclusion and closeness of granules are considered instead of their equality.

For example, the left and right hand sides of association rules [2] describe granules and the *support* and *accuracy* coefficients specify the inclusion degree of granule represented by the formula on the left hand side into the granule represented by the formula on the right hand side of the association rule.

We generalize a simple notion of granules represented by attribute value vectors as well as closeness (inclusion) relation to the case of hierarchical granules representing concepts. We claim that in many application areas related to knowledge discovery and data mining there is a need for algorithmic methods to discover much more complex information granules and relations between them than investigated so far. We discuss examples of information granules and we consider two kinds of basic relations between them, namely (rough) inclusion and closeness. The relations between more complex information granules can be defined by extension of the relations defined on parts of the information granules.

Reasoning in distributed environment requires a construction of interfaces between agents to enable effective learning by agents of concepts definable by different agents. We emphasize the fact concepts definable by some agents can be not definable by the other ones. However, they can be approximated. In this section we suggest one solution based on exchanging views of agents on objects with respect to a given concept. An agent delivering concept is giving positive and negative examples (objects) with respect to a given concept. The agent receiving this information can describe objects using its own attributes. In this way a data table (called a *decision table*) is created and the approximate description of concept can be extracted by the receiving agent. Our solution is based on rough set approach.

An analogous method can be used in case of the *customer-agent* (agent specifying tasks) searching for a top-level cooperating agent (*root-agent*). The customer-agent is presenting examples and counter examples of objects with respect to her/his concept. The concept specified by customer-agent is approximated by agents and an agent returning the best approximation of the customer-agent concept is chosen to be the root agent. The goal of cooperating agents is to produce a concept sufficiently close (or included) to the concept specified by the customer-agent. This concept has to be constructed from some elementary concepts available for agents called *inventory* or *leaf-agents* [117]. This is realized by searching for an agent scheme [117]. The schemes are represented in this section by expressions called *terms*. Our approach can be treated as a method for extracting generalized association rules in distributed environment.

7.1 Syntax and Semantics of Information Granules

In this subsection, we will consider several general kinds of information granules. We present now their syntax and semantics. In the following subsection we discuss the inclusion and closeness relations for granules.

Elementary granules. In an information system $IS = (U, A)$, elementary granules are defined by $EF_B(x)$, where EF_B is a conjunction of selectors of the form $a = a(x)$, $B \subseteq A$ and $x \in U$. For example, the meaning of an elementary granule $a = 1 \wedge b = 1$ is defined by

$$\|a = 1 \wedge b = 1\|_{IS} = \{x \in U : a(x) = 1 \; \& \; b(x) = 1\}.$$

Sequences of granules. Let us assume that S is a sequence of granules and the semantics $\|\bullet\|_{IS}$ in IS of its elements have been defined. We extend $\|\bullet\|_{IS}$ on S by $\|S\|_{IS} = \{\|g\|_{IS}\}_{g \in S}$.

Example 13. Granules defined by rules in information systems are examples of sequences of granules. Let IS be an information system and let (α, β) be a new information granule received from the rule **if** α **then** β where α, β are elementary granules of IS. The semantics $\|(\alpha, \beta)\|_{IS}$ of (α, β) is the pair of sets $(\|\alpha\|_{IS}, \|\beta\|_{IS})$.

Sets of granules. Let us assume that a set G of granules and the semantics $\|\bullet\|_{IS}$ in IS for granules from G have been defined. We extend $\|\bullet\|_{IS}$ on the family of sets $H \subseteq G$ by $\|H\|_{IS} = \{\|g\|_{IS} : g \in H\}$.

Example 14. One can consider granules defined by sets of rules. Assume that there is a set of rules $Rule\_Set = \{(\alpha_i, \beta_i) : i = 1, \ldots, k\}$. The semantics of $Rule\_Set$ is defined by

$$\|Rule\_Set\|_{IS} = \{\|(\alpha_i, \beta_i)\|_{IS} : i = 1, \ldots, k\}.$$

Example 15. One can also consider as set of granules a family of all granules $(\alpha, Rule\_Set(DT_\alpha))$, where α belongs to a given subset of elementary granules.

Example 16. Granules defined by sets of decision rules corresponding to a given evidence are also examples of sequences of granules. Let $DT = (U, A \cup \{d\})$ be a decision table and let α be an elementary granule of $IS = (U, A)$ such that $\|\alpha\|_{IS} \neq \emptyset$. Let $Rule\_Set(DT_\alpha)$ be the set of decision rules (e.g. in minimal form [66]) of the decision table $DT_\alpha = (\|\alpha\|_{IS}, A \cup \{d\})$ being the restriction of DT to objects satisfying α. We obtain a new granule $(\alpha, Rule\_Set(DT_\alpha))$ with the semantics

$$\|(\alpha, Rule\_Set(DT_\alpha))\|_{DT} = (\|\alpha\|_{IS}, \|Rule\_Set(DT_\alpha)\|_{DT}).$$

This granule describes a decision algorithm applied in the situation characterized by α.

Extension of granules defined by tolerance relation. We present examples of granules obtained by application of a tolerance relation.

Example 17. One can consider extension of elementary granules defined by tolerance relation. Let $IS = (U, A)$ be an information system and let τ be a tolerance relation on elementary granules of IS. Any pair (α, τ) is called a τ-*elementary granule*. The semantics $\|(\alpha, \tau)\|_{IS}$ of (α, τ) is the family $\{\|\beta\|_{IS} : (\beta, \alpha) \in \tau\}$.

Example 18. Let us consider granules defined by rules of tolerance information systems. Let $IS = (U, A)$ be an information system and let τ be a tolerance relation on elementary granules of IS. If **if** α **then** β is a rule in IS then the semantics of a new information granule $(\tau : \alpha, \beta)$ is defined by $\|(\tau : \alpha, \beta)\|_{IS} = \|(\alpha, \tau)\|_{IS} \times \|(\beta, \tau)\|_{IS}$.

Example 19. We consider granules defined by sets of decision rules corresponding to a given evidence in tolerance decision tables. Let $DT = (U, A \cup \{d\})$ be a decision table and let τ be a tolerance on elementary granules of $IS = (U, A)$. Now, any granule $(\alpha, Rule\_Set(DT_\alpha))$ can be considered as a representative of information granule cluster

$$(\tau : (\alpha, Rule\_Set(DT_\alpha)))$$

with the semantics

$$\|(\tau : (\alpha, Rule\_Set(DT_\alpha)))\|_{DT} =$$
$$\{\|(\beta, Rule\_Set(DT_\beta))\|_{DT} : (\beta, \alpha) \in \tau\}.$$

Dynamic granules. An elementary granule α of the information system IS is non-empty if $\|\alpha\|_{IS} \neq \emptyset$. A non-empty elementary granule β of IS is an extension of α if $\beta = \alpha \wedge \gamma$, where γ is an elementary granule. Let us consider granules defined by some subsets of

$$\{(\beta, Rule\_Set(DT_\beta)) : \beta \text{ is an extension of } \alpha\}.$$

The semantics of these new granules is defined as in the case of sets of granules. Any set G of granules and a granule α are specifying new granules

$$\{(\beta, Rule\_Set(DT_\beta)) : \beta \text{ is an extension of } \alpha \text{ and } \beta \in G\}$$

important for decision making in dynamically changing environment. A DT-path is any sequence $\pi = ((\alpha_1, R_1), \ldots, (\alpha_k, R_k))$ such that α_i is an elementary non-empty granule of IS, $R_i = Rule\_Set(DT_{\alpha_i})$ for $i = 1, \ldots, k$ and $\alpha_i = \alpha_{i-1} \wedge \gamma_{i-1}$ for some elementary atomic granule γ_{i-1} (e.g. selector $a = v$) with an attribute $a \in A$ not appearing in α_{i-1} for $i = 2, \ldots, k$. A granule α_{i-1} is called a guard of π if R_{i-1} is not sufficiently close to R_i (what we denote by $non(cl_p(R_{i-1}, R_i))$, where p is the closeness degree). By $Guard(\pi)$ we denote the subsequence of $\alpha_1, \ldots, \alpha_k$ consisting all guards of π. In applications it is important to search for a minimal (in cardinality) granule G satisfying the following condition: for any maximal DT-path π of extensions of α all guards β from $Guard(\pi)$ (i.e. all points in which it is sufficient to change the decision algorithm represented by the set of decision rules) are from G. Sets of guards are symbolized in Figure 7. One can also consider dynamic granules with tolerance relation. Let $DT = (U, A \cup \{d\})$ be a decision table and let τ be a tolerance relation on elementary granules of $IS = (U, A)$. Two DT-paths $\pi = ((\alpha_1, R_1), \ldots, (\alpha_k, R_k))$ and $\pi' = ((\beta_1, R'_1), \ldots, (\beta_l, R'_l))$ are τ-*similar* if $(\alpha_{i_s}, \beta_{j_s}) \in \tau$ for $s = 1, \ldots, r$, where $Guard(\pi) = (\alpha_{i_1}, \ldots, \alpha_{i_r})$ and $Guard(\pi') = (\beta_{j_1}, \ldots, \beta_{j_r})$. Let us assume τ has the following property:
if $(\beta, \alpha) \in \tau$ then the granules $Rule\_Set(DT_\alpha)$ and $Rule\_Set(DT_\beta)$ are sufficiently close.

We observe that having such tolerance relation one can search for a set G of guards of the smaller size. To specify the task is enough to change in the above formulated problem the condition for the maximal path to the following one: for

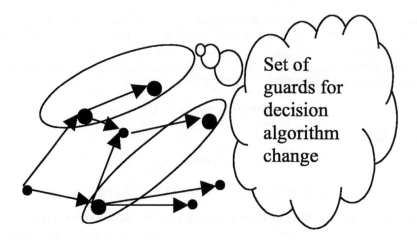

Fig. 7. Two Sets of Guards

any maximal path π of extensions of α there exists a τ-similar path π' to π such that all guards β from $Guard(\pi')$ (i.e. all points where it is necessary to change the decision algorithm represented by the set of decision rules) are from G.

7.2 Granule Inclusion and Closeness

In this subsection we will discuss inclusion and closeness of different information granules introduced in the previous subsection. Let us mention that the choice of inclusion or closeness definition depend very much on area of application and data analyzed. This is the reason that we have decided to introduce a separate subsection with this more subjective part of granule semantics.

The inclusion relation between granules G, G' of degree at least p will be denoted by $\nu_p(G, G')$. Similarly, the closeness relation between granules G, G' of degree at least p will be denoted by $cl_p(G, G')$. By p we denote a vector of parameters (e.g. positive real numbers).

A general scheme for construction of hierarchical granules and their closeness can be described by the following recursive meta-rule: if granules of order $\leq k$ and their closeness have been defined then the closeness $cl_p(G, G')$ (at least in degree p) between granules G, G' of order $k+1$ can be defined by applying an appropriate operator F to closeness values of components of G, G', respectively.

A general scheme of defining more complex granule from simpler ones can be explored using rough mereological approach [117].

Inclusion and closeness of elementary granules. We have introduced the simplest case of granules in information system $IS = (U, A)$. They are defined by $EF_B(x)$, where EF_B is a conjunction of selectors of the form $a = a(x)$, $B \subseteq A$ and $x \in U$. Let $G_{IS} = \{EF_B(x) : B \subseteq A \ \& \ x \in U\}$. In the standard rough set model [110] elementary granules describe indiscernibility classes with respect to some subsets of attributes. In a more general setting see e.g. [151], [182] tolerance (similarity) classes are described.

The crisp inclusion of α in β, where $\alpha, \beta \in \{EF_B(x) : B \subseteq A \ \& \ x \in U\}$ is defined by $\|\alpha\|_{IS} \subseteq \|\beta\|_{IS}$, where $\|\alpha\|_{IS}$ and $\|\beta\|_{IS}$ are sets of objects from IS satisfying α and β, respectively. The non-crisp inclusion, known in KDD, for the case of association rules is defined by means of two thresholds t and t' :
$support_{IS}(\alpha, \beta) = card(\|\alpha \wedge \beta\|_{IS}) \geq t$, and
$accuracy_{IS}(\alpha, \beta) = \frac{support_{IS}(\alpha,\beta)}{card(\|\alpha\|_{IS})} \geq t'$.

Elementary granule inclusion in a given information system IS can be defined using different schemes, e.g., by

$\nu_{t,t'}^{IS}(\alpha, \beta)$ if and only if $support_{IS}(\alpha, \beta) \geq t$ & $accuracy_{IS}(\alpha, \beta) \geq t'$

or

$\nu_t^{IS}(\alpha, \beta)$ if and only if $accuracy_{IS}(\alpha, \beta) \geq t$.

The closeness of granules can be defined by

$cl_{t,t'}^{IS}(\alpha, \beta)$ if and only if $\nu_{t,t'}^{IS}(\alpha, \beta)$ and $\nu_{t,t'}^{IS}(\beta, \alpha)$ hold.

Decision rules as granules. One can define inclusion and closeness of granules corresponding to rules of the form **if** α **then** β using accuracy coefficients (see e.g. [118]).

Having such granules $g = (\alpha, \beta)$, $g' = (\alpha', \beta')$ one can define inclusion and closeness of g and g' by $\nu_{t,t'}(g, g')$ if and only if $\nu_{t,t'}(\alpha, \alpha')$ and $\nu_{t,t'}(\beta, \beta')$.

The closeness can be defined by

$cl_{t,t'}(g, g')$ if and only if $\nu_{t,t'}(g, g')$ and $\nu_{t,t'}(g', g)$.

Another way of defining inclusion of granules corresponding to decision rules is as follows

$\nu_t^{IS}((\alpha, \beta), (\alpha', \beta'))$ if and only if $\nu_{t_1,t_2}(\alpha, \alpha')$ and $\nu_{t_1,t_2}(\beta, \beta')$ and $t = w_1 \bullet t_1 + w_2 \bullet t_2$, where w_1, w_2 are some given weights satisfying $w_1 + w_2 = 1$ and $w_1, w_2 \geq 0$.

Extensions of elementary granules by tolerance relation. For extensions of elementary granules defined by similarity (tolerance) relation, i.e., granules of the form $(\alpha, \tau), (\beta, \tau)$ one can consider the following inclusion measure:
$\nu_{t,t'}^{IS}((\alpha, \tau)(\beta, \tau))$ if and only if

$\nu_{t,t'}^{IS}(\alpha', \beta')$ for any α', β' such that $(\alpha, \alpha') \in \tau$ and $(\beta, \beta') \in \tau$

and the following closeness measure:
$cl_{t,t'}^{IS}((\alpha, \tau)(\beta, \tau))$ if and only if $\nu_{t,t'}^{IS}((\alpha, \tau)(\beta, \tau))$ and $\nu_{t,t'}^{IS}((\beta, \tau)(\alpha, \tau))$.

Sets of rules. It can be important for some applications to define closeness of an elementary granule α and the granule (α, τ). The definition reflecting an intuition that α should be a representation of (α, τ) sufficiently close to this granule is the following one:

$cl_{t,t'}^{IS}(\alpha, (\alpha, \tau))$ if and only if $cl_{t,t'}(\alpha, \beta)$ for any $(\alpha, \beta) \in \tau$.

An important problem related to association rules is that the number of such rules generated even from simple data table can be large. Hence, one should search for methods of aggregating close association rules. We suggest that this can be defined as searching for some close information granules.

Let us consider two finite sets *Rule_Set* and *Rule_Set'* of association rules defined by
$$Rule\_Set = \{(\alpha_i, \beta_i) : i = 1, \ldots, k\},$$
$$Rule\_Set' = \{(\alpha'_i, \beta'_i) : i = 1, \ldots, k'\}.$$
One can treat them as higher order information granules. These new granules *Rule_Set*, *Rule_Set'* can be treated as close in a degree at least t (in IS) if and only if there exists a relation *rel* between sets of rules *Rule_Set* and *Rule_Set'* such that:

1. For any *Rule* from the set *Rule_Set* there is *Rule'* from *Rule_Set'* such that $(Rule, Rule') \in rel$ and *Rule* is close to *Rule'* (in IS) in degree at least t.
2. For any *Rule'* from the set *Rule_Set'* there is *Rule* from *Rule_Set* such that $(Rule, Rule') \in rel$ and *Rule* is close to *Rule'* (in IS) in degree at least t.

Another way of defining closeness of two granules G_1, G_2 represented by sets of rules can be described as follows.

Let us consider again two granules *Rule_Set* and *Rule_Set'* corresponding to two decision algorithms. By $I(\beta'_i)$ we denote the set $\{j : cl_p(\beta'_j, \beta'_i)\}$ for any $i = 1, \ldots, k'$.

Now, we assume $\nu_p(Rule\_Set, Rule\_Set')$ if and only if for any $i \in \{1, \ldots, k'\}$ there exists a set $J \subseteq \{1, \ldots, k\}$ such that

$$cl_p\left(\bigvee_{j \in I(\beta'_i)} \beta'_j, \bigvee_{j \in J} \beta_j\right) \text{ and } cl_p\left(\bigvee_{j \in I(\beta'_i)} \alpha'_j, \bigvee_{j \in J} \alpha_j\right)$$

and for closeness we assume

$cl_p(Rule\_Set, Rule\_Set')$ if and only if
$\nu_p(Rule\_Set, Rule\_Set')$ and $\nu_p(Rule\_Set', Rule\_Set)$.

For example if the granule G_1 consists of rules: **if** α_1 **then** $d = 1$, **if** α_2 **then** $d = 1$, **if** α_3 **then** $d = 1$, **if** β_1 **then** $d = 0$, **if** β_2 **then** $d = 0$ and the granule G_2 consists of rules: **if** γ_1 **then** $d = 1$, **if** γ_2 **then** $d = 0$, then

$cl_p(G_1, G_2)$ if and only if $cl_p(\alpha_1 \vee \alpha_2 \vee \alpha_3, \gamma_1)$ and $cl_p(\beta_1 \vee \beta_2, \gamma_2)$.

One can consider a searching problem for a granule *Rule_Set'* of minimal size such that *Rule_Set* and *Rule_Set'* are close (see e.g. [1]).

Granules defined by sets of granules. The previously discussed methods of inclusion and closeness definition can be easily adopted for the case of granules defined by sets of already defined granules. Let G, H be sets of granules.

The inclusion of G in H can be defined by

$\nu^{IS}_{t,t'}(G, H)$ if and only if for any $g \in G$ there is $h \in H$ for which $\nu^{IS}_{t,t'}(g, h)$
and the closeness by

$cl^{IS}_{t,t'}(G, H)$ if and only if $\nu^{IS}_{t,t'}(G, H)$ and $\nu^{IS}_{t,t'}(H, G)$.

Inclusion for complex granules specified by inclusion of their parts is symbolized in Figure 8.

Let G be a set of granules and let φ be a property of sets of granules from G (e.g. $\varphi(X)$ if and only if X is a tolerance class of a given tolerance $\tau \subseteq G \times G$).

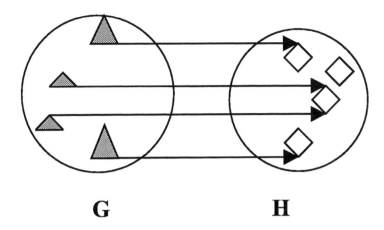

Fig. 8. Two Sets of Granules

Then $P_\varphi(G) = \{X \subseteq G : \varphi(X) \text{ holds}\}$. Closeness of granules $X, Y \in P_\varphi(G)$ can be defined by

$cl_t(X, Y)$ if and only if $cl_t(g, g')$ for any $g \in G$ and $g' \in G'$.

One can also extend other methods for measuring the inclusion and closeness of granules defined by sets of already defined granules.

7.3 Mutual Understanding of Concepts by Agents

An important task for Knowledge Discovery and Data Mining (KDD) [2], [40] in distributed environment is to develop tools for modeling mutual understanding of concepts definable by different agents. Mutual understanding through communication is one of the key issues to enable collaboration among agents [56]. We assume agents specify their knowledge using data tables.

Understanding of concept definable by single agent. Let us consider two agents. There are two data tables $IS_1 = (U, A_1)$ and $IS_2 = (U, \{a\})$ corresponding to agents. We assume that $a : U \to \{0, 1\}$ is a characteristic function of a concept $X = \{x \in U : a(x) = 1\}$.

In this, typical for rough set approach, situation the first agent is specifying the characteristic function of its concept on examples of objects. The second agent is trying to describe the concept using values of its own attributes from A_1 on objects considered by the first agent. In this way a decision table is constructed with condition attributes from A_1 and the decision a. Next, the lower and the upper approximation of the decision class X are computed. The size of the boundary region of X with respect to A_1 can be used a measure of uncertainty in understanding X by the agent with attributes A_1.

Closeness of X to its approximations in the language used by the first agent can be represented by *accuracy of approximation*, i.e., by the coefficient
$$\alpha(AS_{A_1}, X) = \frac{card(LOW(AS_{A_1}, X))}{card(UPP(AS_{A_1}, X))}.$$

The presented above approach can be used for learning by one agent of concepts definable by another agent. Let us consider again two agents. There are two data tables $IS_1 = (U, A_1)$ and $IS_2 = (U, A_2)$ corresponding to agents. We assume that in both data tables there is the same set U of objects and $A_1 = \{a_1^1, \ldots, a_l^1\}$, and $A_2 = \{a_1^2, \ldots, a_k^2\}$ are two sets of attributes, where $l > 0$ and $k > 0$ are given natural numbers. Let us consider concepts definable by attributes from the set A_2. For example suppose that we consider concept defined by formula $(a_1^2 = 1 \wedge a_2^2 = 1) \vee a_3^2 = 1$. This is a concept definable by the second agent. Hence this agent can compute values of the characteristic function of the concept on objects from U and the first agent can find approximations of the concept following the procedure described above.

In this way we define approximations by the first agent of concepts definable by the second one.

Let us mention that the approximation operations are in general not distributive with respect to disjunction or conjunction. Hence one can not expect to construct concept approximations of the good quality from approximation of atomic concepts (e.g. selectors).

The parameterized approximation space of an agent can be used to tune up to satisfactory degree the approximation of concepts definable by another agent. Let us illustrate this idea by example.

Example 20. We consider two agents. The second agent tries to understand a concept defined by the first agent in Table 32. The data table of the second agent is presented in Table 34.

We consider an approximation space $AS^{A_2} = (U, I_{A_2}, \nu_{SRI})$, where $I_{A_2}(x) = \bigcap_{a \in A_2} I_a(x)$ and $y \in I_{Age}^{f_{Age}}(x)$ if and only if

$$diff_{Age}(Age(x), Age(y)) \leq f_{Age}(Age(x), Age(y)),$$

$$y \in I_R^{f_R}(x) \text{ if and only if } R(x) = R(y),$$

where $diff_{Age}(v, v') = \frac{|v-v'|}{\max_{Age} - \min_{Age}}$ and $\max_{Age}$ and $\min_{Age}$ are the maximum and minimum values, respectively, for attribute Age.

Let us consider two cases of constant threshold functions:
$f_{Age}^{0.3}(Age(x), Age(y)) = 0.3$ and $f_{Age}^{0.1}(Age(x), Age(y)) = 0.1$.

In the case of $f_{Age}^{0.3}$ we obtain the following approximations:
$LOW(AS^{A_2}, X_1) = \{x_1, x_5\}$,
$UPP(AS^{A_2}, X_1) = \{x_1, x_2, x_4, x_5, x_6, x_7, x_{10}\}$.
Thus the accuracy of approximation is less than 1.
On the other hand in the case of $f_{Age}^{0.1}$ we obtain the following approximations:

U	d
x_1	1
x_2	1
x_3	0
x_4	1
x_5	1
x_6	0
x_7	1
x_8	0
x_9	0
x_{10}	0

Table 32. Concept Defined by First Agent

$LOW\left(AS^{A_2}, X_1\right) = \{x_1, x_2, x_4, x_5, x_7\}$,
$UPP\left(AS^{A_2}, X_1\right) = \{x_1, x_2, x_4, x_5, x_7\}$.
Thus the accuracy of approximation is equal to 1.

One can consider a more general case when one agent is trying to approximate some concepts using approximations of these concepts delivered by another agent. This can be used when top-down search for synthesis of concepts is performed.

There are two data tables $DT_1 = (U, A_1 \cup \{d\})$ and $DT_2 = (U, A_2 \cup \{d\})$ corresponding to agents. We assume that in both data tables there is the same set of objects U, the same decision d and $A_1 = \{a_1^1, \ldots, a_l^1\}$, and $A_2 = \{a_1^2, \ldots, a_k^2\}$ are two sets of attributes, where $l > 0$ and $k > 0$ are given natural numbers. Let $\{X_1, \ldots, X_r\}$ be a set of $r > 1$ decision classes defined by the decision d. For every agent one can consider a parameterized approximation space $AS_{\#_j, \$_j}^j$, where $j = 1, 2$. The question is how to tune parameters of approximation spaces such that the second agent can well approximate (understand) the approximations of decision classes obtained by the first agent (and vice versa).

The inclusion of approximations can be defined by the average accuracy coefficient:

$$\alpha_{avg}\left(AS_{\#_2,\$_2}^{A_2}, AS_{\#_1,\$_1}^{A_1}\right) =$$

$$\frac{1}{r} \bullet \sum_{i=1}^{r} \alpha\left(AS_{\#_2,\$_2}^{A_2}, LOW\left(AS_{\#_1,\$_1}^{A_1}, X_i\right)\right) +$$

$$\frac{1}{r} \bullet \sum_{i=1}^{r} \alpha\left(AS_{\#_2,\$_2}^{A_2}, UPP\left(AS_{\#_1,\$_1}^{A_1}, X_i\right)\right).$$

Example 21. We consider two agents. The second agent tries to understand a concept approximately defined by the first agent. There are two languages L_{DT_1} and L_{DT_2} corresponding to agents and two data tables $DT_1 = (U, A_1 \cup \{d\})$ and $DT_2 = (U, A_2 \cup \{d\})$. Informally the first agent is identified by a tuple

(L_{DT_1}, DT_1) and the second agent is identified by a tuple (L_{DT_2}, DT_2). We assume that in both data tables there is the same set of objects U.

Let $DT_1 = (U, A_1 \cup \{d\})$ be a hypothetical medical data table (see Table 33) such that $U = \{x_1, \ldots, x_{10}\}$ and the set of attributes $A_1 = \{Sex, Infections\}$.

U	Sex	$Infections$	d
x_1	f	yes	1
x_2	m	yes	1
x_3	f	no	0
x_4	m	yes	1
x_5	m	yes	1
x_6	m	yes	0
x_7	f	yes	1
x_8	f	no	0
x_9	m	no	0
x_{10}	m	yes	0

Table 33. First Agent Data Table

Let $DT_2 = (U, A_2 \cup \{d\})$ be a second hypothetical medical data table (see Table 34), where $A_2 = \{Age, R\}$ and R means remission.

U	Age	R	d	$I_{Age}^{0.1}(\bullet)$	$I_R(\bullet)$
x_1	4	no	1	$\{x_1, x_5\}$	$\{x_1, x_3, x_8, x_9\}$
x_2	6	yes	1	$\{x_2, x_4, x_5, x_7\}$	$\{x_2, x_4, x_5, x_6, x_7, x_{10}\}$
x_3	8	no	0	$\{x_3, x_4, x_{10}\}$	$\{x_1, x_3, x_8, x_9\}$
x_4	7	yes	1	$\{x_2, x_3, x_4, x_7\}$	$\{x_2, x_4, x_5, x_6, x_7, x_{10}\}$
x_5	5	yes	1	$\{x_1, x_2, x_5, x_7\}$	$\{x_2, x_4, x_5, x_6, x_7, x_{10}\}$
x_6	10	yes	0	$\{x_6, x_{10}\}$	$\{x_2, x_4, x_5, x_6, x_7, x_{10}\}$
x_7	6	yes	1	$\{x_2, x_4, x_5, x_7\}$	$\{x_2, x_4, x_5, x_6, x_7, x_{10}\}$
x_8	13	no	0	$\{x_8, x_9\}$	$\{x_1, x_3, x_8, x_9\}$
x_9	14	no	0	$\{x_8, x_9\}$	$\{x_1, x_3, x_8, x_9\}$
x_{10}	9	yes	0	$\{x_3, x_6, x_{10}\}$	$\{x_2, x_4, x_5, x_6, x_7, x_{10}\}$

Table 34. Second Agent Data Table and Uncertainty Functions

There are two decision classes:
$X_1 = \{x \in U : d(x) = 1\} = \{x_1, x_2, x_4, x_5, x_7\}$,
$X_0 = \{x \in U : d(x) = 0\} = \{x_3, x_6, x_8, x_9, x_{10}\}$.

We define the approximation space $AS_{l,u}^{A_1} = (U, I_{A_1}, \nu_{l,u})$ with parameters $0 \leq l < u \leq 1$, where

- $I_{A_1}(x) = \bigcap_{a \in A_1} I_a(x)$ and $I_a(x) = \{y \in U : a(x) = a(y)\}$ for any $x \in U$.

- $\nu_{l,u}(X,Y) = f_{l,u}(\nu_{SRI}(X,Y))$, where

$$f_{l,u}(t) = \begin{cases} 0 & \text{if } t < l \\ \frac{t-l}{u-l} & \text{if } l \leq t \leq u \\ 1 & \text{if } t > u \end{cases}$$

and $\nu_{SRI}(X,Y) = \begin{cases} \frac{card(X \cap Y)}{card(X)} & \text{if } X \neq \emptyset \\ 1 & \text{if } X = \emptyset \end{cases}$ for any $X, Y \subseteq U$.

The lower and the upper approximations for $l = 0$ and $u = 1$ are the following:
$LOW\left(AS_{0,1}^{A_1}, X_1\right) = \{x_1, x_7\}$,
$UPP\left(AS_{0,1}^{A_1}, X_1\right) = \{x_1, x_2, x_4, x_5, x_6, x_7, x_{10}\}$,
$LOW\left(AS_{0,1}^{A_1}, X_0\right) = \{x_3, x_8, x_9\}$,
$UPP\left(AS_{0,1}^{A_1}, X_0\right) = \{x_2, x_3, x_4, x_5, x_6, x_8, x_9, x_{10}\}$.

Using other parameters one can obtain other approximations, for example for $l = 0.4$ and $u = 0.6$ one can obtain:
$LOW\left(AS_{0.4,0.6}^{A_1}, X_1\right) = \{x_1, x_2, x_4, x_5, x_6, x_7, x_{10}\}$,
$UPP\left(AS_{0.4,0.6}^{A_1}, X_1\right) = \{x_1, x_2, x_4, x_5, x_6, x_7, x_{10}\}$,
$LOW\left(AS_{0.4,0.6}^{A_1}, X_0\right) = \{x_3, x_8, x_9\}$,
$UPP\left(AS_{0.4,0.6}^{A_1}, X_0\right) = \{x_3, x_8, x_9\}$.

We consider an approximation space $AS^{A_2} = (U, I_{A_2}, \nu_{SRI})$, where $I_{A_2}(x) = \bigcap_{a \in A_2} I_a(x)$ and $y \in I_{Age}^{f_{Age}}(x)$ if and only if

$$diff_{Age}(Age(x), Age(y)) \leq f_{Age}(Age(x), Age(y)),$$

$$y \in I_R^{f_R}(x) \text{ if and only if } R(x) = R(y),$$

where $diff_{Age}(v, v') = \frac{|v-v'|}{\max_{Age} - \min_{Age}}$ and $\max_{Age}$ and $\min_{Age}$ are the maximum and minimum values, respectively, for attribute Age.

A function $f_{Age}(Age(x), Age(y)) = 0.1$.

We would like to know how well the second agent is understanding the first agent. We compare the results obtained for two pairs of approximation spaces $\left(AS_{0,1}^{A_1}, AS^{A_2}\right)$ and $\left(AS_{0.4,0.6}^{A_1}, AS^{A_2}\right)$.

Therefore one can compute the following approximations for $l = 0$, $u = 1$ and X_1:
$LOW\left(AS^{A_2}, LOW\left(AS_{0,1}^{A_1}, X_1\right)\right) = \{x_1\}$,
$UPP\left(AS^{A_2}, LOW\left(AS_{0,1}^{A_1}, X_1\right)\right) = \{x_1, x_2, x_4, x_5, x_7\}$,
$LOW\left(AS^{A_2}, UPP\left(AS_{0,1}^{A_1}, X_1\right)\right) = \{x_1, x_2, x_4, x_5, x_6, x_7, x_{10}\}$,
$UPP\left(AS^{A_2}, UPP\left(AS_{0,1}^{A_1}, X_1\right)\right) = \{x_1, x_2, x_4, x_5, x_6, x_7, x_{10}\}$,
and similarly for $l = 0.4$, $u = 0.6$ and X_0.

Finally we obtain $\alpha_{avg}\left(AS^{A_2}, AS_{0,1}^{A_1}\right) = \frac{31}{45}$.

On the other hand $\alpha_{avg}\left(AS^{A_2}, AS_{0.4,0.6}^{A_1}\right) = 1$.

Hence better understanding of the first agent concepts by the second agent is received in the case of parameters $l = 0.4$ and $u = 0.6$.

Understanding of concepts definable by team of agents. Assume that a set of agents $Ag = \{ag_1, \ldots, ag_p\}$ is given, where $p > 0$ is a natural number. Let us consider a data table $IS_{ag} = (U, A_{ag})$ for any agent $ag \in Ag$. We assume any agent from Ag is defining a concept X using the above procedure. One can construct a decision table DT with condition attributes being the characteristic functions of the lower and upper approximations of X defined by all agents from Ag and the decision being the characteristic function of X on given examples of objects. The lower and upper approximation of X with respect to condition attributes of DT describe the vagueness in understanding of X by agents from Ag. One can also use other features summarizing the result of voting by different agents. Examples of such features are the majority voting feature, accepting object as belonging to concept if the number of voting agents is greater than a given threshold or the characteristic function of the intersection of the upper approximations $\bigcap_{ag \in Ag} UPP\left(AS_{A_{ag}}, X\right)$ or the intersection of the lower approximations $\bigcap_{ag \in Ag} LOW\left(AS_{A_{ag}}, X\right)$. One can observe that in some cases the above intersections can be undefinable by single agent.

The described problem is analogous to resolving conflict between decision rules voting for decision when they are classifying new objects.

One can extend our approach to the case when e.g. one agent is trying to understand concepts definable by the second agent on the basis of understanding these concepts by the third agent. Common knowledge of a given team of agents about concepts definable by members of this team [38], [216], [130] can also be considered in this framework.

One can also consider the discussed above new features as the characteristic functions of concepts definable in some new approximation spaces constructed from approximation spaces of agents from Ag.

Operations on approximation spaces. The result of cooperation of agents to solve a given problem will be a scheme of agents. One can consider this scheme as a construction of new approximation space. Before we discuss this case let us consider some basic operations on approximation spaces.

The operations are natural in a sense that they correspond to different views on global approximation space represented by groups of agents. These new approximation spaces can be used e.g. for better description of concepts.

For simplicity in all examples we use the standard rough inclusion

$$\nu_{SRI}(X, Y) = \begin{cases} \frac{card(X \cap Y)}{card(X)} & \text{if } X \neq \emptyset \\ 1 & \text{if } X = \emptyset \end{cases} \text{ for any } X, Y \subseteq U.$$

Intersection of approximation spaces. Let $DT = (U, A \cup \{d\})$ be a data table. For every $a \in A$ one can consider an approximation space $AS_a = (U, I_a, \nu_{SRI})$. Using the intersection operator one can construct an approximation space $AS^A = (U, I_A, \nu_{SRI})$, where for every $x \in U$, $I_A(x) = \bigcap_{a \in A} I_a(x)$.

Let us assume that different $a \in A$ are accessible by different agents ag_a. One can see that for any $X \subseteq U$ and any $x \in U$:

- if x is in the lower approximation of X for all approximation spaces corresponding to any agent, then x is in the lower approximation of X with respect to the approximation space represented by all agents,
- if x is in the upper approximation of X with respect to the approximation space represented by all agents, then it is in the upper approximation of X for all approximation spaces corresponding to any agent.

Example 22. In this example we consider the intersection of approximation spaces of two agents. We assume that there are two agents corresponding to two data tables. Let DT_1 and DT_2 be data tables from Tables 33 and 34, respectively. Let $DT_3 = (U, A_3 \cup \{d\})$ be a third hypothetical medical data table (see Table 35), where $A_3 = \{DD, H\}$, where DD and H means disease duration and hypertension, respectively.

U	DD	H	d	$I_{DD}^{f_{DD}}(\bullet)$	$I_H^{f_H}(\bullet)$
x_1	5	no	1	$\{x_1, x_3, x_7, x_8\}$	$\{x_1, x_3, x_6, x_8, x_9\}$
x_2	2	yes	1	$\{x_2, x_3, x_7, x_9\}$	$\{x_2, x_4, x_5, x_7, x_{10}\}$
x_3	3	no	0	$\{x_1, x_2, x_3, x_7, x_8, x_9\}$	$\{x_1, x_3, x_6, x_8, x_9\}$
x_4	8	yes	1	$\{x_4, x_{10}\}$	$\{x_2, x_4, x_5, x_7, x_{10}\}$
x_5	12	yes	1	$\{x_5, x_6\}$	$\{x_2, x_4, x_5, x_7, x_{10}\}$
x_6	12	no	0	$\{x_5, x_6\}$	$\{x_1, x_3, x_6, x_8, x_9\}$
x_7	4	yes	1	$\{x_1, x_2, x_3, x_7, x_8, x_9\}$	$\{x_2, x_4, x_5, x_7, x_{10}\}$
x_8	5	no	0	$\{x_1, x_3, x_7, x_8\}$	$\{x_1, x_3, x_6, x_8, x_9\}$
x_9	2	no	0	$\{x_2, x_3, x_7, x_9\}$	$\{x_1, x_3, x_6, x_8, x_9\}$
x_{10}	8	yes	0	$\{x_4, x_{10}\}$	$\{x_2, x_4, x_5, x_7, x_{10}\}$

Table 35. Third Agent Data Table and Uncertainty Functions

We consider an approximation space $AS^{A_3} = (U, I_{A_3}, \nu_{SRI})$, where $I_{A_3}(x) = \bigcap_{a \in A_3} I_a(x)$ and $y \in I_{DD}^{f_{DD}}(x)$ if and only if

$$diff_{DD}(DD(x), DD(y)) \leq f_{DD}(DD(x), DD(y)),$$

$$y \in I_H^{f_H}(x) \text{ if and only if } H(x) = H(y),$$

where $diff_{DD}(v, v') = \frac{|v - v'|}{\max_{DD} - \min_{DD}}$ and $f_{DD}(DD(x), DD(y)) = 0.2$.
One can construct combined approximation space

$$AS^{A_2, A_3} = (U, I_{A_2 \cup A_3}, \nu_{SRI}),$$

where $I_{A_2 \cup A_3}(x) = I_{A_2}(x) \cap I_{A_3}(x)$ (see Table 36).

Finally we obtain $\alpha_{avg}\left(AS^{A_2, A_3}, AS_{0,1}^{A_1}\right) = 1$. Hence understanding of the first agent concepts by combined agents (second and third) is better than understanding of the first agent concepts only by the second agent.

U	$I_{A_2}(\bullet)$	$I_{A_3}(\bullet)$	$I_{A_2 \cup A_3}(\bullet)$
x_1	$\{x_1\}$	$\{x_1, x_3, x_8\}$	$\{x_1\}$
x_2	$\{x_2, x_4, x_5, x_7\}$	$\{x_2, x_7\}$	$\{x_2, x_7\}$
x_3	$\{x_3\}$	$\{x_1, x_3, x_8, x_9\}$	$\{x_3\}$
x_4	$\{x_2, x_4, x_7\}$	$\{x_4, x_{10}\}$	$\{x_4\}$
x_5	$\{x_2, x_5, x_7\}$	$\{x_5\}$	$\{x_5\}$
x_6	$\{x_6, x_{10}\}$	$\{x_6\}$	$\{x_6\}$
x_7	$\{x_2, x_4, x_5, x_7\}$	$\{x_2, x_7\}$	$\{x_2, x_7\}$
x_8	$\{x_8, x_9\}$	$\{x_1, x_3, x_8\}$	$\{x_8\}$
x_9	$\{x_8, x_9\}$	$\{x_3, x_9\}$	$\{x_9\}$
x_{10}	$\{x_6, x_{10}\}$	$\{x_4, x_{10}\}$	$\{x_{10}\}$

Table 36. Combined Uncertainty Function

Union of approximation spaces. Assume that there are two approximation spaces $AS_1 = (U, I_1, \nu_{SRI})$ and $AS_2 = (U, I_2, \nu_{SRI})$. One can construct new approximation space $AS_3 = (U, I_3, \nu_{SRI})$ such that the uncertainty function I_3 is defined by

$$I_3(x) = \bigcup_{y \in I_1(x)} I_2(y).$$

Cartesian product of approximation spaces with constraints. We consider two data tables $DT_1 = (U_1, A_1 \cup \{d_1\})$ and $DT_2 = (U_2, A_2 \cup \{d_2\})$. Let $\partial_{A_i}(x) = \{d_i(y) : y \in I_{A_i}(x)\}$, where $i = 1, 2$. The data table $DT = (U, A \cup \{d\})$ is called an independent product of data tables DT_1 and DT_2 [138] if and only if

- $U = U_1 \times U_2 - \{(y, z) \in U_1 \times U_2 : \partial_{A_1}(y) \cap \partial_{A_2}(z) = \emptyset\}$,
- For any $(y, z) \in U$ $d((y, z)) = \partial_{A_1}(y) \cap \partial_{A_2}(z)$,
- A is the disjoint union of A_1 and A_2 defined by $(A_1 \times \{1\}) \cup (A_2 \times \{2\})$,
- For $(a, i) \in A$, $i = 1, 2$ and $(x_1, x_2) \in U$ $(a, i)((x_1, x_2)) = a(x_i)$.

One can construct an approximation space for the independent product $DT_1 \odot DT_2$ of DT_1 and DT_2 as follows:
$AS^{DT_1 \odot DT_2} = (U_1 \odot U_2, I_A, \nu_{SRI})$, where

- $U_1 \odot U_2 = U_1 \times U_2 - \{(y, z) \in U_1 \times U_2 : \partial_{A_1}(y) \cap \partial_{A_2}(z) = \emptyset\}$,
- $I_A((y, z)) = (I_{A_1}(y) \times I_{A_2}(z)) \cap U_1 \odot U_2$.

Example 23. We consider two data tables $DT_1 = (U_1, A_1 \cup \{d_1\})$ and $DT_2 = (U_2, A_2 \cup \{d_2\})$, which are described in Tables 37 and 38, respectively. We assume that $U_1 = \{y_1, y_2, y_3, y_4\}$, $U_2 = \{z_1, z_2, z_3, z_4\}$ and $A_1 = \{a_1^1, a_2^1, a_3^1\}$, and $A_2 = \{a_1^2, a_2^2, a_3^2\}$ are two sets of attributes. There are two decision classes in both tables: $X_1^1 = \{y_1, y_3\}$, $X_2^1 = \{y_2, y_4\}$ and $X_1^2 = \{z_4\}$, $X_2^2 = \{z_1, z_2, z_3\}$. The approximation spaces are defined by $AS_1 = (U_1, I_{A_1}, \nu_{SRI})$ and $AS_2 = (U_2, I_{A_2}, \nu_{SRI})$, where the uncertainty function I_{A_i} is defined by

$$y \in I_{A_i}(x) \text{ if and only if for every } a \in A_i \quad a(x) = a(y).$$

U_1	a_1^1	a_2^1	a_3^1	d_1	∂_{A_1}
y_1	yes	yes	no	1	$\{1\}$
y_2	no	yes	no	2	$\{1,2\}$
y_3	no	yes	no	1	$\{1,2\}$
y_4	no	no	yes	2	$\{2\}$

Table 37. Data Table DT_1

U_2	a_1^2	a_2^2	a_3^2	d_2	∂_{A_2}
z_1	yes	yes	yes	2	$\{2\}$
z_2	yes	yes	yes	2	$\{2\}$
z_3	no	no	yes	2	$\{1,2\}$
z_4	no	no	yes	1	$\{1,2\}$

Table 38. Data Table DT_2

Let us recall that the generalized decision $\partial_{A_i}(x) = \{d_i(y) : y \in I_{A_i}(x)\}$. In the approximation space AS_1 we obtain the following approximations:

$$LOW\left(AS_1, X_1^1\right) = \{y_1\}, \quad UPP\left(AS_1, X_1^1\right) = \{y_1, y_2, y_3\},$$
$$LOW\left(AS_1, X_2^1\right) = \{y_4\}, \quad UPP\left(AS_1, X_2^1\right) = \{y_2, y_3, y_4\}.$$

Similarly, in the approximation space AS_2 we obtain the following approximations:

$$LOW\left(AS_2, X_1^2\right) = \emptyset, \quad UPP\left(AS_2, X_1^2\right) = \{z_3, z_4\},$$
$$LOW\left(AS_2, X_2^2\right) = \{z_1, z_2\}, \quad UPP\left(AS_2, X_2^2\right) = \{z_1, z_2, z_3, z_4\}.$$

We construct an independent product $DT_1 \odot DT_2$ of data tables DT_1 and DT_2. In our example the universe of the independent product $DT_1 \odot DT_2$ of DT_1 and DT_2 is equal to (see Table 39)

$$U_1 \odot U_2 = \{(y,z) \in U_1 \times U_2 : \partial_{A_1}(y) \cap \partial_{A_2}(z) \neq \emptyset\}$$
$$= U_1 \times U_2 - \{y_1\} \times \{z_1, z_2\}$$

Let us recall that one can construct an approximation space for the independent product $DT_1 \odot DT_2$ of DT_1 and DT_2 as follows:

$$AS^{DT_1 \odot DT_2} = (U_1 \odot U_2, I_A, \nu_{SRI}),$$

where $I_A((y,z)) = (I_{A_1}(y) \times I_{A_2}(z)) \cap U_1 \odot U_2$.

Let us consider an information granule in DT_1 defined by $(a_3^1, 1) = yes$ and an information granule in DT_2 defined by $(a_2^2, 2) = yes \wedge (a_3^2, 2) = yes$. In $DT_1 \odot DT_2$ one can combine both granules and obtain an information granule $(a_3^1, 1) = yes \wedge (a_2^2, 2) = yes \wedge (a_3^2, 2) = yes$.

$U_1 \odot U_2$	$(a_1^1,1)$	$(a_2^1,1)$	$(a_3^1,1)$	$(a_1^2,2)$	$(a_2^2,2)$	$(a_3^2,2)$	$d_{A1\odot A2}$
(y_1,z_3)	yes	yes	no	no	no	yes	$\{1\}$
(y_1,z_4)	yes	yes	no	no	no	yes	$\{1\}$
(y_2,z_1)	no	yes	no	yes	yes	yes	$\{2\}$
(y_2,z_2)	no	yes	no	yes	yes	yes	$\{2\}$
(y_2,z_3)	no	yes	no	no	no	yes	$\{1,2\}$
(y_2,z_4)	no	yes	no	no	no	yes	$\{1,2\}$
(y_3,z_1)	no	yes	no	yes	yes	yes	$\{2\}$
(y_3,z_2)	no	yes	no	yes	yes	yes	$\{2\}$
(y_3,z_3)	no	yes	no	no	no	yes	$\{1,2\}$
(y_3,z_4)	no	yes	no	no	no	yes	$\{1,2\}$
(y_4,z_1)	no	no	yes	yes	yes	yes	$\{2\}$
(y_4,z_2)	no	no	yes	yes	yes	yes	$\{2\}$
(y_4,z_3)	no	no	yes	no	no	yes	$\{2\}$
(y_4,z_4)	no	no	yes	no	no	yes	$\{2\}$

Table 39. The Product of DT_1 and DT_2

7.4 Rough Sets in Distributed Systems

In this subsection we consider operations on approximation spaces which seem to be important for approximate reasoning in distributed systems. We consider a set of agents Ag. Each agent is equipped with some approximation spaces. Agents are cooperating to solve a problem specified by a special agent called *customer-agent*. The result of cooperation is a scheme of agents. In the simplest case the scheme can be represented by a tree labeled by agents. In this tree leaves are delivering some concepts and any non-leaf agent $ag \in Ag$ is performing an operation $o(ag)$ on approximations of concepts delivered by its children. The root agent returns a concept being the result of computation by the scheme on concepts delivered by leaf agents. It is important to note that different agents use different languages. Hence concepts delivered by one agent ay_1 can be only perceived in an approximate sense by another agent ag, for illustration see Figures 9, 10 and 11.

We assume any non leaf-agent ag is equipped with an operation $o(ag): U_{ag}^{(1)} \times \ldots \times U_{ag}^{(k)} \to U_{ag}^{(0)}$ and has different approximation spaces $AS_{ag}^{(i)} = \left(U_{ag}^{(i)}, I_{ag}^{(i)}, \nu_{SRI}\right)$, where $i = 0, \ldots, k$. We assume that the agent ag is perceiving objects by measuring values of some available attributes. Hence some objects can become indiscernible [110]. This influences the specification of any operation $o(ag)$. We consider a case when arguments and values of operations are represented by attribute value vectors. Hence instead of the operation $o(ag)$ we have its inexact specification $o^*(ag)$ taking as arguments $I_{ag}^{(1)}(x_1), \ldots, I_{ag}^{(k)}(x_k)$ for some $x_1 \in U_{ag}^{(1)}, \ldots, x_k \in U_{ag}^{(k)}$ and returning the value $I_{ag}^{(0)}(o(ag)(x_1, \ldots, x_k))$ if $o(ag)(x_1, ..., x_k)$ is defined, otherwise the empty set. This operation can be extended to the operation $o^*(ag)$ with arguments being definable sets (in approximation spaces attached to arguments) and with values in the family of all

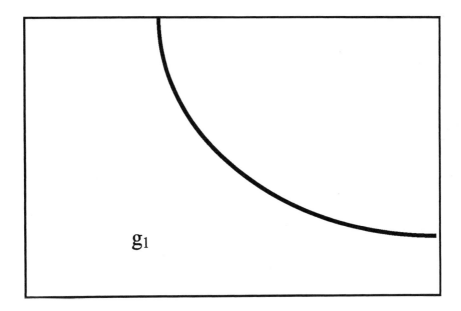

Fig. 9. Concept g_1 – Information Granule of $ag_1 \in Ag$

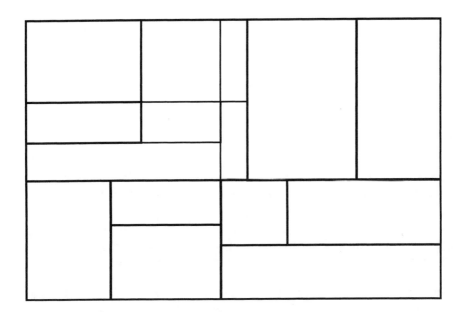

Fig. 10. Communication Interface Defined by Data Table

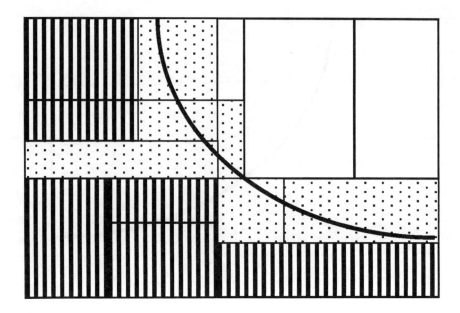

Fig. 11. Lower and Upper Approximation of g_1 by $ag \in Ag$

non-empty subsets of $U_{ag}^{(0)}$. Let $X_1, \ldots, X_k$ be definable sets. We define

$$o^*(ag)(X_1, \ldots, X_k) = \bigcup_{x_1 \in X_1, \ldots, x_k \in X_k} o^*(ag)\left(I_{ag}^{(1)}(x_1), \ldots, I_{ag}^{(k)}(x_k)\right).$$

In the sequel, for simplicity of notation, we write $o(ag)$ instead of $o^*(ag)$.

This idea can be formalized as follows. First we define terms representing schemes of agents.

Let $X_{ag}, Y_{ag}, \ldots$ be agent variables for any leaf-agent $ag \in Ag$. Let $o(ag)$ denote a function of arity k. We have mentioned that it is an operation from Cartesian product of $Def\_Sets(AS_{ag}^{(1)}), \ldots, Def\_Sets(AS_{ag}^{(k)})$ into $P\left(U_{ag}^{(0)}\right)$, where $Def\_Sets(AS_{ag}^{(i)})$ denotes the family of sets definable in $AS_{ag}^{(i)}$. Using the above variables and functors we define terms in a standard way, for example

$$t = o(ag)(X_{ag_1}, X_{ag_2}).$$

Such terms can be treated as description of complex information granules. By a valuation we mean any function val defined on the agent variables with values being definable sets satisfying $val(X_{ag}) \subseteq U_{ag}$ for any leaf-agent $ag \in Ag$. Now we can define the lower and the upper values of any term t under the valuation val with respect to a given approximation space $AS_{ag}^{(i)}$ of an agent ag

1. If t is of the form X_{ag_i} and $val(t) \subseteq U_{ag}^{(i)}$ then

$$val\left(LOW, AS_{ag}^{(i)}\right)(t) = LOW\left(AS_{ag}^{(i)}, val(t)\right)$$

$$val\left(UPP, AS_{ag}^{(i)}\right)(t) = UPP\left(AS_{ag}^{(i)}, val(t)\right)$$

else the lower and the upper values are undefined.

2. If $t = o(ag)(t_1, \ldots, t_k)$, where $t_1, \ldots, t_k$ are terms and $o(ag)$ is an operation of arity k, then

   (a) if for $i = 1, \ldots, k$ $\quad val\left(LOW, AS_{ag}^{(i)}\right)(t_i)$ is defined then
   $$val\left(LOW, AS_{ag}^{(0)}\right)(t) = LOW\left(AS_{ag}^{(0)}, o(ag)\right.$$
   $$\left.\left(val\left(LOW, AS_{ag}^{(1)}\right)(t_1), \ldots, val\left(LOW, AS_{ag}^{(k)}\right)(t_k)\right)\right)$$
   else $val\left(LOW, AS_{ag}^{(0)}\right)(t)$ is undefined,

   (b) if for $i = 1, \ldots, k$ $\quad val\left(UPP, AS_{ag}^{(i)}\right)(t_i)$ is defined then
   $$val\left(UPP, AS_{ag}^{(0)}\right)(t) = UPP\left(AS_{ag}^{(0)}, o(ag)\right.$$
   $$\left.\left(val\left(UPP, AS_{ag}^{(1)}\right)(t_1), \ldots, val\left(UPP, AS_{ag}^{(k)}\right)(t_k)\right)\right)$$
   else $val\left(UPP, AS_{ag}^{(0)}\right)(t)$ is undefined.

Example 24. Let $Ag = \{ag, ag_1, ag_2\}$ be a set of agents and let $cust$ be an agent called customer. We explain how agents from Ag produce a granule defined by a given term and how this granule is related to the concept specified by $cust$. A binary operation $o(ag)$ of ag and an information system

$$IS_{cust} = (\{w_1, w_2, w_3, w_4, w_5, w_6\}, d)$$

are described in Table 43. We assume that objects w_i, where $i = 1, \ldots, 6$ are perceived by ag using a_1^0. Two information systems IS_{ag_1}, IS_{ag_2} presented in Tables 40(a),(b) describe input information granules. Data tables $DT_1 = \left(U_{ag}^{(1)}, A_{ag}^{(1)} \cup \{d_1\}\right)$ and $DT_2 = \left(U_{ag}^{(2)}, A_{ag}^{(2)} \cup \{d_2\}\right)$ described in Table 41 and Table 42 characterize communication interfaces between agents ag_1, ag_2 and ag. A binary operation, input information granules and communication interfaces are illustrated in Figure 12. The first four columns of Table 41 (42) define the information system $IS_{ag}^{(i)}$ and the approximation space $AS_{ag}^{(i)} = \left(U_{ag}^{(i)}, I_{ag}^{(i)}, \nu_{SRI}\right)$, where $i = 1, 2$.

$U_{ag_1}^{(0)}$	d_1
y_1	1
y_2	0
y_3	1
y_4	0

$U_{ag_2}^{(0)}$	d_2
z_1	1
z_2	1
z_3	1
z_4	0

Table 40. (a) Information System IS_{ag_1} (b) Information System IS_{ag_2}

Let $t = o(ag)(X_{ag_1}, X_{ag_2})$ and $val(X_{ag_1}) = \{y_1, y_3\}$. Hence

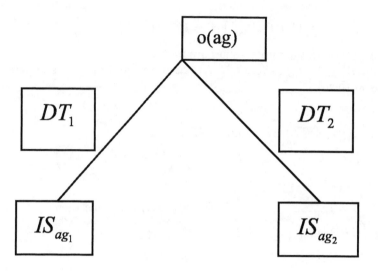

Fig. 12. Operation, Input Granules and Communication Interfaces

$U_{ag}^{(1)}$	a_1^1	a_2^1	a_3^1	d_1	$I_{ag}^{(1)}$
y_1	yes	yes	no	1	$\{y_1\}$
y_2	no	yes	no	0	$\{y_2, y_3\}$
y_3	no	yes	no	1	$\{y_2, y_3\}$
y_4	no	no	yes	0	$\{y_4\}$

Table 41. Data Table DT_1 and Uncertainty Function $I_{ag}^{(1)}$

$U_{ag}^{(2)}$	a_1^2	a_2^2	a_3^2	d_2	$I_{ag}^{(2)}$
z_1	yes	yes	yes	1	$\{z_1, z_2\}$
z_2	yes	yes	yes	1	$\{z_1, z_2\}$
z_3	no	no	yes	1	$\{z_3, z_4\}$
z_4	no	no	yes	0	$\{z_3, z_4\}$

Table 42. Data Table DT_2 and Uncertainty Function $I_{ag}^{(2)}$

a_1^1	a_2^1	a_3^1	a_1^2	a_2^2	a_3^2	a_1^0	$U_{ag}^{(0)}$	d
yes	yes	no	yes	yes	yes	1	w_1	+
yes	yes	no	no	no	yes	2	w_2	+
no	yes	no	yes	yes	yes	3	w_3	+
no	yes	no	no	no	yes	4	w_4	-
no	no	yes	yes	yes	yes	5	w_5	-
no	no	yes	no	no	yes	6	w_6	-

Table 43. Operation $o(ag)$ and Customer Information System

$$val\left(LOW, AS_{ag}^{(1)}\right)(X_{ag_1}) = LOW\left(AS_{ag}^{(1)}, \{y_1, y_3\}\right) = \{y_1\},$$
$$val\left(UPP, AS_{ag}^{(1)}\right)(X_{ag_1}) = UPP\left(AS_{ag}^{(1)}, \{y_1, y_3\}\right) = \{y_1, y_2, y_3\}.$$

Let $val(X_{ag_2}) = \{z_1, z_2, z_3\}$. Hence

$$val\left(LOW, AS_{ag}^{(2)}\right)(X_{ag_2}) = LOW\left(AS_{ag}^{(2)}, \{z_1, z_2, z_3\}\right) = \{z_1, z_2\},$$
$$val\left(UPP, AS_{ag}^{(2)}\right)(X_{ag_2}) = UPP\left(AS_{ag}^{(2)}, \{z_1, z_2, z_3\}\right) = \{z_1, z_2, z_3, z_4\}.$$

We obtain the lower value
$val\left(LOW, AS_{ag}^{(0)}\right)(o(ag)(X_{ag_1}, X_{ag_2})) = LOW\left(AS_{ag}^{(0)}, o(ag)\right.$
$\left(val\left(LOW, AS_{ag}^{(1)}\right)(X_{ag_1}), val\left(LOW, AS_{ag}^{(2)}\right)(X_{ag_2})\right)\right) =$
$LOW\left(AS_{ag}^{(0)}, o(ag)(\{y_1\}, \{z_1, z_2\})\right) = LOW\left(AS_{ag}^{(0)}, o(ag)\right.$
$\left(\left\|a_1^1 = yes \wedge a_2^1 = yes \wedge a_3^1 = no\right\|_{IS_{ag}^{(1)}}, \left\|a_1^2 = yes\right\|_{IS_{ag}^{(2)}}\right)\right) =$
$LOW\left(AS_{ag}^{(0)}, \{w_1\}\right) = \{w_1\}.$

The support of the rule

if t then $d = +$

under the valuation val with respect to the lower approximations is equal to

$$card\left(val\left(LOW, AS_{ag}^{(0)}\right)(t) \cap \|d = +\|_{IS_{ag}^{(0)}}\right) = 1$$

and the accuracy is also equal to 1.
We also obtain the upper value
$val\left(UPP, AS_{ag}^{(0)}\right)(o(ag)(X_{ag_1}, X_{ag_2})) = UPP\left(AS_{ag}^{(0)}, o(ag)\right.$
$\left(val\left(UPP, AS_{ag}^{(1)}\right)(X_{ag_1}), val\left(UPP, AS_{ag}^{(2)}\right)(X_{ag_2})\right)\right) =$
$UPP\left(AS_{ag}^{(0)}, o(ag)(\{y_1, y_2, y_3\}, \{z_1, z_2, z_3, z_4\})\right) =$
$UPP\left(AS_{ag}^{(0)}, o(ag)\left(\left\|a_2^1 = yes\right\|_{IS_{ag}^{(1)}}, \left\|a_1^2 = yes \vee a_2^2 = no\right\|_{IS_{ag}^{(2)}}\right)\right) =$
$UPP\left(AS_{ag}^{(0)}, \{w_1, w_2, w_3, w_4\}\right) = \{w_1, w_2, w_3, w_4\}.$

The support of the rule **if t then $d = +$** under the valuation val with respect to the upper approximations is equal to

$$card\left(val\left(UPP, AS_{ag}^{(0)}\right)(t) \cap \|d = +\|_{IS_{ag}^{(0)}}\right) = 3$$

and the accuracy is equal to 0.75.

Let us observe that the set $val(UPP, AS_{ag}^{(0)})(t) - val(LOW, AS_{ag}^{(0)})(t)$ can be treated as the boundary region of t under val. Moreover, in the process of term construction we have additional parameters to be tuned for obtaining sufficiently high support and accuracy, namely the approximation operations.

A concept X specified by the customer-agent is *sufficiently close to t under a given set Val of valuations* if X is included in the upper approximation of t under any $val \in Val$ and X includes the lower approximation of t under any $val \in Val$ as well as the size of the boundary region of t under Val, i.e.,

$$card\left(\bigcap_{val \in Val} val\left(UPP, AS_{ag}^{(0)}\right)(t) - \bigcup_{val \in Val} val\left(LOW, AS_{ag}^{(0)}\right)(t) \right),$$

is sufficiently small relatively to $\bigcap_{val \in Val} val\left(UPP, AS_{ag}^{(0)}\right)(t)$.

We conclude by formulating some examples of basic algorithmic problems.

- *Synthesis of generalized association rules.* Searching for a scheme (term t) over a given set Ag of agents and for a valuation val such that the rule **if** t **then** α, where α is a concept description specified by customer-agent, has the support at least s and the accuracy at least c under the valuation val.
- *Synthesis of concepts close to the concept specified by the customer-agent.* Searching for a scheme (term t) over a given set Ag of agents and a set Val of valuations such that the concept specified by the customer-agent is sufficiently close to t under Val and the total size of the term t and the set Val is minimal.

7.5 Robustness of Granules

Extracting robust patterns (i.e. patterns with properties not deviating to much under small disturbances of parameters) from data is an important task for many applications, in particular related to KDD.

We restrict our considerations for some initial remarks. Some aspects of this problem have been considered using rough mereological framework [117], [118].

Let us consider granules specified by parameterized formulas $\alpha(p)$, $\beta(p)$ together with parameterized closeness relation $cl_q(\bullet, \bullet)$ between sets of objects representing the semantics of these granules. The parameter $q \in [0, 1]$ is representing the degree of closeness. We assume that values of parameter p are points in the metric space with metric ρ. Hence the intended meaning of $cl_q(\alpha(p), \beta(p))$ is that granules $\alpha(p), \beta(p)$ (parameterized by p) are close at least in the degree q. Let us assume that $\varepsilon > 0$ is a given threshold and let $cl_{q_0}(\alpha(p_0), \beta(p_0))$ holds for some p_0, q_0. We would like to check if there exists $\delta > \delta_0$ (δ_0 is a given threshold) such that for any p

if $\rho(p_0, p) \leq \delta$ then $cl_q(\alpha(p), \beta(p))$ for some q such that $q_0 - \varepsilon \leq q \leq q_0$.

The above condition is specifying that granules $\alpha(p_0), \beta(p_0)$ are not only close at least in degree q_0 but their closeness is also robust with respect to changes of parameter p.

Example 25. Let us assume that the rule
$$\text{if } a \in [1, 1.5) \wedge b \in [-1, 0) \text{ then } d = 1$$
is an association rule in a given decision table $DT = (U, A \cup \{d\})$ with coefficients
$$support_{DT}(a \in [1, 1.5) \wedge b \in [-1, 0), d = 1) = 0.6 * card(U),$$

$accuracy_{DT}(a \in [1, 1.5) \wedge b \in [-1, 0), d = 1) = 0.7$

Let $q_0 = (0.7, 0.6)$, $p_0 = (0, 0)$ and $\varepsilon = 0.1$. We ask if there exists $\delta > \delta_0 = 0.01$ such that for any $p = (p_1, p_2)$ if $\sqrt{p_1^2 + p_2^2} \leq \delta$ then the rule

if $a \in [1 - p_1, 1.5 + p_1) \wedge b \in [-1 - p_2, p_2)$ **then** $d = 1$

is true in a given decision table DT with coefficients

$$support_1 = support_{DT}(a \in [1 - p_1, 1.5 + p_1) \wedge \\ b \in [-1 - p_2, p_2), d = 1),$$

$$accuracy_1 = accuracy_{DT}(a \in [1 - p_1, 1.5 + p_1) \wedge \\ b \in [-1 - p_2, p_2), d = 1),$$

such that $\sqrt{\left(\frac{support_1}{card(U)} - 0.6\right)^2 + (accuracy_1 - 0.7)^2} \leq \varepsilon$.

The robustness of constructed granules can be defined by a notion analogous to the continuity of function in a given point.

Definition 24. Let $f : X_1 \times \ldots \times X_k \to X$ and let us assume closeness relations $cl_p^{(i)}(\bullet, \bullet)$ and $cl_p(\bullet, \bullet)$ in X_i for $i = 1, \ldots, k$ and X are given. We say that f is $(\varepsilon; \varepsilon_1, \ldots, \varepsilon_k)$-*robust* in a given point $(x_1, \ldots, x_k)$ from $X_1 \times \ldots \times X_k$ if and only if for any $(y_1, \ldots, y_k)$ from $X_1 \times \ldots \times X_k$ if $cl_{1-\varepsilon_i}^{(i)}(x_i, y_i)$ for $i = 1, \ldots, k$ then $cl_{1-\varepsilon}(f(x_1, \ldots, x_k)), f(y_1, \ldots, y_k))$.

We can formulate useful property for robustness checking.

Proposition 25. *Let an operation*

$$F(ag) : \prod_{i=1}^{k} Def\_Sets\left(AS_{ag}^{(i)}\right) \to Def\_Sets\left(AS_{ag}^{(0)}\right)$$

be $(\varepsilon; \varepsilon_1, \ldots, \varepsilon_k)$ - *robust in* $(g_1, \ldots, g_k)$ *and let for every agent* ag_i *operation*

$$F(ag_i) : \prod_{j=1}^{l_i} Def\_Sets\left(AS_{ag}^{(j)}\right) \to Def\_Sets\left(AS_{ag}^{(i)}\right)$$

be $(\varepsilon_i; \varepsilon_1^{(i)}, \ldots, \varepsilon_{l_i}^{(i)})$ - *robust in* $(g_1^{(i)}, \ldots, g_{l_i}^{(i)})$ *for* $i = 1, \ldots, k$. *Then the operation*

$$F(ag)(F(ag_1)(\ldots), \ldots, F(ag_k)(\ldots))$$

being the superposition of $F(ag)$ *and* $F(ag_1), \ldots, F(ag_k)$ *is*

$$(\varepsilon_i; \varepsilon_1^{(1)}, \ldots, \varepsilon_{l_1}^{(1)}, \ldots, \varepsilon_1^{(k)}, \ldots, \varepsilon_{l_k}^{(k)}) - robust$$

in

$$\left(g_1^{(1)}, \ldots, g_{l_1}^{(1)}, \ldots, g_1^{(k)}, \ldots, g_{l_k}^{(k)}\right).$$

In Proposition 25 we consider operations with values being definable sets. These operations can be received from $o^*(ag)$ by applying the lower (or upper) approximation operation in $AS_{ag}^{(0)}$ to the values of $o^*(ag)$. Using Proposition 25 one can easily derive the robustness condition for terms describing the construction of information granules.

An important practical problem is to discover rules for decomposition of a given threshold ε into thresholds $\varepsilon_1, \ldots, \varepsilon_k$ for a given operation

$$F(ag) : \prod_{i=1}^{k} Def\_Sets\left(AS_{ag}^{(i)}\right) \to Def\_Sets\left(AS_{ag}^{(0)}\right)$$

in such a way that $F(ag)$ is $(\varepsilon; \varepsilon_1, \ldots, \varepsilon_k)$ - robust in a given point [66].

However, often a more general problem should be solved. Let us consider an agent ag from a set of agents Ag (together with their closeness measures $cl_p^{ag}(\bullet, \bullet)$ for ag from Ag) as well as a given standard (prototype) g from $Def\_Sets\left(AS_{ag}^{(0)}\right)$ together with an uncertainty coefficient $\varepsilon \in [0, 1)$. Let us assume that $In(Ag)$ is a given subset of input (inventory [66]) agents from Ag. A task is to synthesize a term $t(X_{ag_1}, \ldots, X_{ag_k})$, where $X_{ag_1}, \ldots, X_{ag_k}$ are some variables of agents from $In(Ag)$ and a valuation val of variables such that:

- $val(LOW, AS_{ag}^{(0)})(t) \subseteq g \subseteq val(UPP, AS_{ag}^{(0)})(t)$,
- $cl_{1-\varepsilon/2}^{ag}(g, val(LOW, AS_{ag}^{(0)})(t))$ and $cl_{1-\varepsilon/2}^{ag}(g, val(UPP, AS_{ag}^{(0)})(t))$,
- the operations

$$t_{LOW}(val) = val(LOW, AS_{ag}^{(0)})(t),$$

$$t_{UPP}(val) = val(UPP, AS_{ag}^{(0)})(t)$$

defined by $t(X_{ag_1}, \ldots, X_{ag_k})$ are $(\varepsilon/2; \varepsilon_1, \ldots, \varepsilon_k)$-robust in a point

$$(val(X_{ag_1}), \ldots, val(X_{ag_k}))$$

for some $\varepsilon_1, \ldots, \varepsilon_k$ greater than a given threshold $\delta > 0$.

We discuss applications of closeness between granules for deducing that they share close sets of properties. Informally speaking, close granules should share similar properties. Assuming $\mathcal{G}$ is a given set of granules and $R \subseteq \mathcal{G} \times \mathcal{G}$ is a relation describing similarity, indiscernibility or similar functionality of information granules and $\mathcal{P}$ is a given set of properties (unary predicates) defined on information granules we can formulate the relationship between the relation R and the set $\mathcal{P}$ of properties as follows:

$$\forall g, g' \in \mathcal{G} \forall P \in \mathcal{P}[R(g, g') \to (P(g) \to P(g'))].$$

The property is called $(R, \mathcal{P})$-*relationship*. Discovery of this type of properties seems to be important. It consists of discovery of the relation R satisfying the following condition: if for any information granule $g \in \mathcal{G}$ and property $P \in \mathcal{P}$ we have that $P(g)$ holds than $P(g')$ holds for any granule g' from the neighborhood $R(g) = \{g' : (g, g') \in R\}$.

In case of multi-agent environment one should consider different $(R, \mathcal{P})$ - relationships for different agents. One of the fundamental question is how to deduce that close complex information granules have similar properties. Following an idea developed in rough mereological approach we can present now the main steps for such deduction.

The main role in such deduction play, so called *decomposition rules*. They have the following form:

$$o(ag): \frac{P_{ag}(\bullet), \varepsilon, st(ag), cl^{ag}_{1-\varepsilon}(\bullet, \bullet)}{P^{(1)}_{ag_1}(\bullet), \varepsilon_1, st(ag_1), cl^{ag_1}_{1-\varepsilon_1}(\bullet, \bullet); \ldots; P^{(k)}_{ag_k}(\bullet), \varepsilon_k, st(ag_k), cl^{ag_k}_{1-\varepsilon_k}(\bullet, \bullet)}$$

where $o(ag)$ is k-ary operation of ag; $st(ag), st(ag_1), \ldots, st(ag_k)$ are standard information granules (prototypes); $P_{ag}(\bullet), P^{(1)}_{ag_1}(\bullet), \ldots, P^{(k)}_{ag_k}(\bullet)$ are properties; $\varepsilon, \varepsilon_1, \ldots, \varepsilon_k$ are uncertainty coefficients; $cl^{ag}_{1-\varepsilon}, cl^{ag_1}_{1-\varepsilon_1}, \ldots, cl^{ag_k}_{1-\varepsilon_k}$ are closeness relations attached to agents $ag, ag_1, \ldots, ag_k$, respectively.

We assume the decomposition rule is true if the following conditions are satisfied:

- $o(ag)(st(ag_1), \ldots, st(ag_k)) = st(ag)$,
- $P^{(1)}_{ag_1}(st(ag_1)) \wedge \ldots \wedge P^{(k)}_{ag_k}(st(ag_k)) \rightarrow P_{ag}(st(ag))$,
- $cl^{ag_1}_{1-\varepsilon_1}(st(ag_1), g_1) \wedge \ldots \wedge cl^{ag_k}_{1-\varepsilon_k}(st(ag_k), g_k) \rightarrow cl^{ag}_{1-\varepsilon}(st(ag), g)$
  for any $g_1, \ldots, g_k$ from $\mathcal{G}$, where $g = o(ag)(g_1, \ldots, g_k)$,
- $cl^{ag}_{1-\varepsilon}(g, g') \rightarrow (P_{ag}(g) \rightarrow P_{ag}(g'))$ for any $g, g' \in \mathcal{G}$.

The first condition states that the standard granule at ag is equal to the result of operation $o(ag)$ on standard granules $st(ag_1), \ldots, st(ag_k)$. The second condition allows to deduce that the standard granule $st(ag)$ has property P_{ag} if the standard granules $st(ag_1), \ldots, st(ag_k)$ have properties $P^{(1)}_{ag_1}, \ldots, P^{(k)}_{ag_k}$, respectively. The next condition allows to infer that $st(ag)$ is close to $o(ag)(g_1, \ldots, g_k)$ in degree at least $1 - \varepsilon$ if the standard granule $st(ag_i)$ is close to g_i in degree at least $1 - \varepsilon_i$ for any $i = 1, \ldots, k$. The last condition means that if granules g, g' are close in degree at least $1 - \varepsilon$ and g has property P_{ag} then g' has also this property.

Hence we obtain the following basic lemma:

Lemma 26. *Assuming the decomposition rule for $o(ag)$ is true and $P^{(1)}_{ag_1}(g_1)$, $P^{(1)}_{ag_1}(st(ag_1))$, $cl^{ag_1}_{1-\varepsilon_1}(st(ag_1), g_1), \ldots, P^{(k)}_{ag_k}(g_k), P^{(k)}_{ag_k}(st(ag_k))$, $cl^{ag_k}_{1-\varepsilon_k}(st(ag_k), g_k)$ hold we obtain that $P_{ag}(g)$ holds, where $g = o(ag)(g_1, \ldots, g_k)$.*

One of the challenges for knowledge discovery is to discover decomposition rules for operations possessed by agents.

It is easy to observe that by repeated application of decomposition rules one can derive sufficient conditions for synthesis of robust complex information granules.

Conclusions

In the standard rough set model, an equivalence relation is used to define an approximation space. In Section 2 we have discussed a generalization of the approximation space notion to cover some of the extended rough set models, for example the tolerance based rough set model and the variable precision rough set model. Our approach is based simultaneously on extension of indiscernibility class notion and on extension of the inclusion relation between indiscernibility classes and the approximated set.

We point out the role of searching for optimal parameters of approximation space. The space of essential coefficients describing an approximation space is usually relatively huge. Therefore some heuristics are valuable, for example, based on genetic algorithms. In such a way solutions of high quality can be found in a reasonable time.

In Section 3 we have investigated reducts and representatives in standard and tolerance rough set models. Decision rules can be generated using only set of obtained representatives reducing number of generated rules.

In Section 4 we have discuss generation of decision rules from decision tables with tolerance relations. We also discuss different quantitative measures associated with rules.

In Section 5 knowledge discovery in diabetes mellitus data were considered. This data set has been drawn from a real life medical problem. The rough set based analysis showed that the most relevant attributes are the following: age of disease diagnosis, criteria of the metabolic balance and disease duration. The above aspects influence incidence of microalbuminuria in children suffering from diabetes type I. The results of our analysis and the extracted laws are also consistent with general clinical knowledge about diabetes type I and accepted by medical experts. The presented methods go beyond the individual application to diabetes mellitus data analysis and can be applied to mining in different data sets.

In Section 6 two approaches have been presented which aim at overcoming the difficulty met by knowledge discovery systems namely the huge amount of data. First approach, based on the concept of reducts can be applied to a certain class of problems that can be transformed into attribute-value form without the loss of significant data. Second approach uses the parameterized approximation spaces. By employing a new kind of approximation space we are able to select formulas that are more relevant to the problem. If the selection appears to be too restrictive approximation can be used in multiple passes, each of them expanding the set of formulas in a way that includes only the most relevant facts from the ones that were previously thrown out.

In Section 7 we have introduced information granules in knowledge discovery. Our approach can be treated as a step towards understanding of complex information granules and their role in applications.

We have discussed information granule syntax and semantics as well their inclusion and closeness. Several examples of information granules have been presented. We have shown that some higher order patterns, important for knowledge discovery and data mining are expressible by means of complex information granules.

We have also outlined an approach for synthesis and analysis of complex information granules in distributed environment. The approximate understanding by one agent of concepts definable by other agents is an important aspect of our approach for calculus of information granules.

The discussed methods and results have proved to be useful in solving some KDD problems or they create foundations for solving some more complex knowledge discovery problems. Let us mention some of them

- methods for decision rule generation from decision tables with tolerance relations,
- methods for relevant attributes selection defined by propositional and first order formulas,
- methods for concept description in tolerance approximation spaces,
- synthesis of complex information granules in distributed environment.

Acknowledgments

The author would like to thank Professor Andrzej Skowron for his encouragement and advice. The author would like to express his gratitude to members of the Logic Group at Warsaw University, Poland and the Knowledge Systems Group at NTNU, Norway taking part in creation of ROSETTA and to members of the Second Department of Children's Diseases, University Medical School in Bialystok, Poland taking part in construction of the medical information system. This research was supported by the grant No. 8 T11C 023 15 from the State Committee for Scientific Research and Research Program of the European Union - ESPRIT-CRIT 2 No. 20288.

References

1. Ågotnes T., Komorowski J., Loken T.: Taming Large Rule Models in Rough Set Approaches, 3rd European Conference of Principles and Practice of Knowledge Discovery in Databases, September 15-18, 1999, Prague, Czech Republic, Lecture Notes in Artificial Intelligence 1704, 1999, pp. 193-203.
2. Agrawal R., Mannila H., Srikant R., Toivonen H., Verkano A.: Fast Discovery of Association Rules, Fayyad U.M., Piatetsky-Shapiro G., Smyth P., Uthurusamy R. (Eds.): Advances in Knowledge Discovery and Data Mining, The AAAI Press/The MIT Press 1996, pp. 307-328.
3. An A., Chan C., Shan N., Cercone N., Ziarko W.: Applying Knowledge Discovery to Predict Water-Supply Consumption. IEEE Expert 12/4, 1997, pp. 72-78.
4. Bazan J., Nguyen H.S., Nguyen T.T., Skowron A., Stepaniuk J.: Some Logic and Rough Set Applications for Classifying Objects. Institute of Computer Science, Warsaw University of Technology, ICS Research Report, 38/94, 1994.

5. Bazan J., Nguyen H.S., Nguyen T.T., Skowron A., Stepaniuk J.: Application of Modal Logics and Rough Sets for Classifying Objects. In: M. De Glas, Z. Pawlak (Eds.), Proceedings of the Second World Conference on Fundamentals of Artificial Intelligence (WOCFAI'95), Paris, July 3-7, Angkor, Paris, pp. 15-26.
6. Bazan J., Nguyen H.S., Nguyen T.T., Skowron A., Stepaniuk J.: Synthesis of Decision Rules for Object Classification, E. Orlowska (Ed.), Incomplete Information: Rough Set Analysis, Physica-Verlag, Heidelberg, 1998, pp. 23-57.
7. Bazan J.G.: A Comparison of Dynamic and Non-Dynamic Rough Set Methods for Extracting Laws from Decision Tables. L. Polkowski, A. Skowron, (Eds.), Rough Sets in Knowledge Discovery 1. Methodology and Applications. Physica-Verlag, Heidelberg, 1998, pp. 321-365.
8. Bodjanova S.: Approximation of Fuzzy Concepts in Decision Making. Fuzzy Sets and Systems 85, 1997, pp. 23-29.
9. Bonikowski Z., Bryniarski E., Wybraniec-Skardowska U.: Extensions and Intensions in the Rough Set Theory. Information Sciences 107, 1998, pp. 149-167.
10. Brazdil P., Torgo L.: Knowledge Acquisition via Knowledge Integration, Current Trends in Knowledge Acqusition, IOS Press, 1990.
11. Brown F. M.: Boolean Reasoning. Kluwer Academic Publishers, Dordrecht, 1990.
12. Bruha I.: Quality of Decision Rules: Definitions and Classification Schemes for Multiple Rules, G. Nakhaeizadeh, C. C. Taylor (Eds.), Machine Learning and Statistics, The Interface, John Wiley and Sons, 1997, pp. 107-131.
13. Bryniarski E., Wybraniec-Skardowska U.: Generalized Rough Sets in Contextual Spaces, T. Y. Lin, N. Cercone (Eds.), Rough Sets and Data Mining. Analysis of Imprecise Data, Kluwer Academic Publishers, Boston 1997, pp. 339-354.
14. Budihardjo A., Grzymała-Busse J.W., Woolery L., Program LERS-LB 2.5 as a Tool for Knowledge Acquisition in Nursing. In: Proceedings of the Fourth International Conference on Industrial Engineering Applications of Artificial Intelligence Expert Systems, Koloa, Kauai, Hawaii, June 2-5, 1991, pp. 735-740.
15. Cattaneo G.: A Unified Algebraic Approach to Fuzzy Algebras and Rough Approximations. R. Trappl (Ed.), Proceedings of the 13th European Meeting on Cybernetics and Systems Research (CSR'96), April 9-12, 1996, The University of Vienna 1, pp. 352-357.
16. Cattaneo G.: Generalized Rough Sets. Preclusivity Fuzzy-Intuitionistic (BZ) Lattices. Studia Logica 58, 1997, pp. 47-77.
17. Cattaneo G.: Fuzzy Extension of Rough Sets Theory, Proceedings of the International Conference on Rough Sets and Current Trends in Computing, Warsaw, Poland, June 22-26, 1998, Lecture Notes in Artificial Intelligence 1424, pp. 275-282.
18. Cattaneo G.: Abstract Approximation Spaces for Rough Theories, L. Polkowski, A. Skowron (Eds.), Rough Sets in Knowledge Discovery 1. Methodology and Applications, Physica-Verlag, Heidelberg, 1998, pp. 59-98.
19. Carlin U.S., Komorowski J., Ohrn A.: Rough Set Analysis of Patients with Suspected Acute Appendicitis, Proceedings of IPMU'98, Paris, France, July 1998, pp. 1528-1533.
20. Chmielewski M.R., Grzymała-Busse J.W.: Global Discretization of Attributes as Preprocessing for Machine Learning, T.Y. Lin, A.M. Wildberger (Eds.) Soft Computing, Simulation Councils Inc., San Diego, 1995, pp. 294-297.
21. Cios J., Pedrycz W., Świniarski R.W.: Data Mining in Knowledge Discovery, Kluwer Academic Publishers, Dordrecht, 1998.

22. Comer S.: An Algebraic Approach to the Approximation of Information. Fundamenta Informaticae 14, 1991, pp. 492–502.
23. Czyżewski A.: Speaker–Independent Recognition of Digits – Experiments with Neural Networks, Fuzzy Logic Rough Sets. Journal of the Intelligent Automation and Soft Computing 2/2, 1996, pp. 133–146.
24. Czyżewski A., Królikowski R., Skórka P.: Automatic Detection of Speech Disorders. Proceedings of the Fourth European Congress on Intelligent Techniques and Soft Computing, Aachen, Germany, September 2–5, 1996, vol. 1, pp. 183–187.
25. Czyżewski A., Kostek B.: Rough Set-Based Filtration of Sound Applicable to Hearing Prostheses, Tsumoto S., Kobayashi, S., Yokomori, T., Tanaka, H. (Eds.), Proceedings of the Fourth International Workshop on Rough Sets, Fuzzy Sets and Machine Discovery (RSFD'96), Tokyo November 6–8, 1996, pp. 168–175.
26. Dasarathy B. V. (Ed.): Nearest Neighbor Pattern Classification Techniques. IEEE Computer Society Press 1991.
27. Dougherty J., Kohavi R., Sahami M.: Supervised Unsupervised Discretization of Continuous Features. Proceedings of the Twelfth International Conference on Machine Learning, Morgan Kaufmann, San Francisco, CA, 1995, pp. 194–202.
28. Drwal G., Mrózek A.: System RClass - Software Implementation of the Rough Classifier, Proceedings of the Seventh International Workshop on Intelligent Information Systems, Malbork, Poland, June 15–19, 1998, pp. 392–395.
29. Dubois D., Prade H.: Twofold Fuzzy Sets and Rough Sets - Some Issues in Knowledge Representation. Fuzzy Sets and Systems 23, 1987, pp. 3–18.
30. Dubois D., Prade H.: Similarity-Based Approximate Reasoning. J.M. Zurada, R.J. Marks II, and X.C.J. Robinson (Eds.), Proceedings of the IEEE Symposium, Orlando, FL, June 17–July 1st, 1997, IEEE Press, pp. 69–80.
31. Dubois D., Prade H.: Similarity Versus Preference in Fuzzy Set-Based Logics, E. Orłowska (Ed.), Incomplete Information: Rough Set Analysis, Physica Verlag, Heidelberg, 1998, pp. 440–460.
32. Düntsch I.: A Logic for Rough Sets. Theoretical Computer Science 179/1-2, 1997 pp. 427–436.
33. Düntsch I., Gediga G.: Statistical Evaluation of Rough Set Dependency Analysis. International Journal of Human-Computer Studies 46, 1997, pp. 589–604.
34. Düntsch I.: Rough Sets and Algebras of Relations, E. Orlowska (Ed.), Incomplete Information: Rough Set Analysis, Physica-Verlag, Heidelberg, 1998, pp. 95–108.
35. Dzeroski S.: Inductive Logic Programming and Knowledge Discovery in Databases, U. M. Fayyad, G. Piatetsky-Shapiro, P. Smyth, R. Uthurusamy (Eds.), Advances in Knowledge Discovery and Data Mining, The MIT Press, 1996, pp. 117–152.
36. El-Mouadib F.A., Koronacki J., Żytkow J.M.: Taxonomy Formation by Approximate Equivalence Relations, Revisited, 3rd European Conference of Principles and Practice of Knowledge Discovery in Databases, September 1999, Prague, Czech Republic, Lecture Notes in Artificial Intelligence 1704, 1999, pp. 71–79.
37. Esposito F., Malerba D., Semeraro G., Pazzani M.: A Machine Learning Approach to Document Understanding, Proceedings of the Second International Workshop on Multistrategy Learning, West Virginia, 1993, pp. 276–292.
38. Fagin R., Halpern J.Y., Moses Y., Vardi M.: Reasoning about Knowledge, MIT Press, 1996.
39. Fayyad U.M., Irani K.B.: On the Handling of Continuous-Valued Attributes in Decision Tree Generation, Machine Learning 8, 1992, pp. 87–102.
40. Fayyad U.M., Piatetsky-Shapiro G., Smyth P., Uthurusamy R. (Eds.): Advances in Knowledge Discovery and Data Mining, The AAAI Press/The MIT Press 1996.

41. Fibak J., Pawlak Z., Słowiński K., Słowiński R.: Rough Sets Based Decision Algorithm for Treatment of Duodenal Ulcer by HSV. Bulletin of the Polish Academy of Sciences, Biological Sciences, 34/10-12, 1986, pp. 227–246.
42. Fedrizzi M., Kacprzyk J., Nurmi H.: How Different are Social Choice Functions, A Rough Set Approach. Quality & Quantity 30, 1996, pp. 87–99.
43. Funakoshi K., Ho T. B.: Information Retrieval by Rough Tolerance Relation, Tsumoto S., Kobayashi, S., Yokomori, T., Tanaka, H. (Eds.), Proceedings of the Fourth International Workshop on Rough Sets, Fuzzy Sets and Machine Discovery (RSFD'96), Tokyo November, 6–8 1996, pp. 31–35.
44. Funakoshi K., Ho T. B.: A Rough Set Approach to Information Retrieval, L. Polkowski, A. Skowron (Eds.), Rough Sets in Knowledge Discovery 2. Applications, Case Studies and Software Systems, Physica-Verlag, Heidelberg, 1998, pp. 166–177.
45. Gemello R., Mana F.: An Integrated Characterization and Discrimination Scheme to Improve Learning Efficiency in Large Data Sets, Proceedings of the Eleventh International Joint Conference on Artificial Intelligence, Detroit MI, 20–25 August 1989, pp. 719–724.
46. Greco S., Matarazzo B., Słowiński R.: Rough Approximation of a Preference Relation in a Pairwise Comparison Table, L. Polkowski, A. Skowron (Eds.), Rough Sets in Knowledge Discovery 2. Applications, Case Studies and Software Systems, Physica-Verlag, Heidelberg, 1998, pp. 13–36.
47. Greco S., Matarazzo B., Słowiński R.: Fuzzy Similarity Relation as a Basis for Rough Approximations, Proceedings of the International Conference on Rough Sets and Current Trends in Computing, Warsaw, Poland, June 22–26, 1998, Lecture Notes in Artificial Intelligence 1424, pp. 283–289.
48. Greco S., Matarazzo B., Słowiński R.: On Joint Use of Indiscernibility, Similarity and Dominance in Rough Approximation of Decision Classes, 5th International Conference Integrating Technology and Human Decisions: Global Bridges Into The 21st Century, July 4–7, 1999, Athens, Greece.
49. Grzymała-Busse J.W.: Managing Uncertainty in Expert Systems, Kluwer Academic Publishers, Dordrecht, 1991.
50. Grzymała-Busse J.W.: A New Version of the Rule Induction System LERS. Fundamenta Informaticae 31, 1997, pp. 27–39.
51. Grzymała-Busse J.W.: Applications of the Rule Induction System LERS. L. Polkowski, A. Skowron, (Eds.), Rough Sets in Knowledge Discovery 1. Methodology and Applications. Physica–Verlag, Heidelberg, 1998, pp. 366–375.
52. Grzymała-Busse J.W., Goodwin L.K.: Predicting Preterm Birth Risk Using Machine Learning from Data with Missing Values. S. Tsumoto (Ed.), Bulletin of International Rough Set Society 1/2, 1997, pp. 17–21.
53. Grzymała-Busse J.W., Gunn J.D.: Global Temperature Analysis based on the Rule Induction System LERS. In: Proceedings of the Fourth International Workshop on Intelligent Information Systems, Augustów, Poland, June 5–9, 1995, Institute od Computer Science, Polish Academy of Sciences, Warsaw, pp. 148–158.
54. Holte R.C.: Very Simple Classification Rules Perform Well on Most Commonly Used Datasets, Machine Learning 11, 1993, pp. 63–90.
55. Hu X., Cercone N.: Rough Sets Similarity-Based Learning from Databases, Proceedings of the First International Conference on Knowledge Discovery and Data Mining, Montreal, Canada, August 20–21 1995, pp. 162–167.
56. Huhns M.N., Singh M.P.(Eds.): Readings in Agents, Morgan Kaufmann, San Mateo, 1998.

57. Iwiński T.: Algebraic Approach to Rough Sets. Bulletin of the Polish Academy of Sciences Mathematics 35, 1987, pp. 673–683.
58. Jelonek J., Krawiec K., Słowiński R., Szymaś J.: Rough Set Reduction of Features for Picture–Based Reasoning, T.Y. Lin, A.M. Wildberger (Eds.), Soft Computing: Rough Sets, Fuzzy Logic, Neural Networks, Uncertainty Management, Knowledge Discovery, Simulation Councils, Inc., San Diego, 1995, pp. 89–92.
59. Johnson D.S.: Approximation Algorithms for Combinatorial Problems, Journal of Computer and System Sciences, 9, 1974, pp. 256–278.
60. Kandulski M., Marciniec J., Tukałło K.: Surgical Wound Infection – Conductive Factors and Their Mutual Dependencies, R. Slowinski (Ed.), Intelligent Decision Support - Handbook of Applications and Advances of the Rough Sets Theory. Kluwer Academic Publishers, Dordrecht, 1992, pp. 95–110.
61. Katzberg J. D., Ziarko W.: Variable Precision Extension of Rough Sets. Fundamenta Informaticae 27, 1996, pp. 155–168.
62. Kent R.E.: Rough Concept Analysis: A Synthesis of Rough Sets and Formal Concept Analysis. Fundamenta Informaticae 27/2–3, 1996, pp. 169–181.
63. Kim D., Kim C.: A Handwritten Numeral Character Classification Using Tolerant Rough Set, 1998, manuscript.
64. Kodratoff Y., Michalski R.: Machine Learning, An Artificial Intelligence Approach 3, Morgan Kaufmann, 1990.
65. Kohavi R., John G.H.: Wrappers for Feature Subset Selection, Artificial Intelligence Journal, 97, 1997, pp. 273–324.
66. Komorowski J., Pawlak Z., Polkowski L., Skowron A.: Rough Sets: A Tutorial, S.K. Pal, A.Skowron (Eds.), Rough-Fuzzy Hybridization: A New Trend in Decision Making, Springer Verlag, Singapore, 1999, pp. 3–98.
67. Konikowska B.: A logic for Reasoning about Similarity, E. Orłowska (Ed.), Incomplete Information: Rough Set Analysis, Physica-Verlag, Heidelberg, 1998, pp. 462–491.
68. Kostek B., Czyżewski A.: Automatic Classification of Musical Timbres based on Learning Algorithms Applicable to Cochlear Implants. In: Proceedings of IASTED International Conference – Artificial Intelligence, Expert Systems and Neural Networks, August 19–21, 1996, Honolulu, Hawaii, USA, pp. 98–101.
69. Krawiec K., Słowiński R., Vanderpooten D.: Construction of Rough Classifiers Based on Application of a Similarity Relation. In: Tsumoto S., Kobayashi, S., Yokomori, T., Tanaka, H. (Eds.), Proceedings of the Fourth International Workshop on Rough Sets, Fuzzy Sets and Machine Discovery (RSFD'96), Tokyo November 6–8 1996, pp. 23–30.
70. Krawiec K., Słowiński R., Vanderpooten D.: Learning Decision Rules from Similarity Based Rough Approximations, L. Polkowski, A. Skowron (Eds.), Rough Sets in Knowledge Discovery 2. Applications, Case Studies and Software Systems, Physica-Verlag, Heidelberg, 1998, pp. 37–54.
71. Krętowski M., Polkowski L., Skowron A., Stepaniuk J.: Data Reduction Based on Rough Set Theory, Y. Kodratoff, G. Nakhaeizadeh, Ch. Taylor (Eds.), Proceedings of the International Workshop on Statistics, Machine Learning and Knowledge Discovery in Databases, Heraklion April 25–27 1995, pp. 210–215 see also: Institute of Computer Science, Warsaw University of Technology, ICS Research Report 13/95 1995.
72. Krętowski M., Stepaniuk J.: Selection of Objects and Attributes, a Tolerance Rough Set Approach, Proceedings of the Poster Session of Ninth International Symposium on Methodologies for Intelligent Systems, June 10–13, 1996, Zakopane,

Poland, pp. 169–180 see also Institute of Computer Science, Warsaw University of Technology, ICS Research Report 54/95 1995.
73. Kryszkiewicz M.: Maintenance of Reducts in the Variable Precision Rough Set Model, T. Y. Lin, N. Cercone (Eds.), Rough Sets and Data Mining Analysis of Imprecise Data, Kluwer Academic Publishers, Dordrecht 1997, pp. 355–372.
74. Langley P., Iba W.: Average-Case Analysis of a Nearest Neighbor Algorithm, Proceedings of the 13th International Joint Conference on Artificial Intelligence, Morgan Kaufmann, San Mateo, CA, 1993, pp. 889–894.
75. Lavrac N., Dzeroski S., Grobelnik M.: Learning Non-Recursive Definitions of Relations with LINUS, Proceedings of Fifth European Working Session on Learning, 1991, pp. 265–281.
76. Lavrac N., Dzeroski S.: Inductive Logic Programming, Ellis Horwood, Chichester, UK, 1994.
77. Lavrac N., Gamberger D., Turney P.: A Relevancy Filter for Constructive Induction, IEEE Intelligent Systems and Their Applications, 13(2), March/April 1998, pp. 50–56.
78. Lenarcik A., Piasta Z.: Probabilistic Approach to Decision Algorithm Generation in the case of Continuous Condition Attributes. Foundations of Computing and Decision Sciences 18/3–4, 1993, pp. 213–223.
79. Lin T.Y.: Granular Computing on Binary Relations I Data Mining and Neighborhood Systems, L. Polkowski, A. Skowron (Eds.), Rough Sets in Knowledge Discovery 1. Methodology and Applications, Physica–Verlag, Heidelberg, 1998, pp. 107–121.
80. Marcus S.: Tolerance Rough Sets, Cech Topologies, Learning Processes. Bulletin of the Polish Academy of Sciences, Technical Sciences 42/3, 1994, pp. 471–487.
81. Marek W., Pawlak Z.: Rough Sets and Information Systems. Fundamenta Informaticae 17, 1984, pp. 105–115.
82. Martienne E., Quafafou M.: Learning Logical Descriptions for Document Understanding: a Rough Sets-Based Approach, Proceedings of the International Conference on Rough Sets and Current Trends in Computing, Warsaw, Poland, June 22–26, 1998, Lecture Notes in Artificial Intelligence 1424, Springer Verlag, pp. 202–209.
83. Martienne E., Quafafou M.: Vagueness and Data Reduction in Concept Learning, Proceedings of the 13th European Conference on Artificial Intelligence (ECAI-98), Brighton, UK, August 23–28, 1998, pp. 351–355.
84. Michalewicz Z.: Genetic Algorithms + Data Structures = Evolution Programs, Springer-Verlag, Berlin 1996.
85. Michalski R.: A Theory and Methodology of Inductive Learning, R. S. Michalski, J.G. Carbonell, T.M. Mitchell (Eds.), Machine Learning, An Artificial Intelligence Approach, 1983, pp. 83–134.
86. Mitchell T.M.: Machine Learning, McGraw-Hill, New York 1997.
87. Michie D., Spiegelhalter D.J., TaylorC.C., (Eds.): Machine learning, Neural and Statistical Classification. Ellis Horwood, New York, 1994.
88. Michalski R. S., Larson J. B.: Selection of most Representative Training Examples and Incremental Generation of VL1 Hypotheses. Report 867 Department of Computer Science University of Illinois at Urbana-Champaign 1978.
89. Mrózek A.: Information Systems and Control Algorithms. Bulletin of the Polish Academy of Sciences Technical Sciences 33, 1985, pp. 195–212.
90. Mrózek A., Płonka L.: Rough Sets in Image Analysis. Foundations of Computing Decision Sciences 18/3–4, 1993, pp. 259–273.

91. Mrózek A., Płonka L.: Analiza Danych Metodą Zbiorów Przybliżonych. Zastosowania w Ekonomii, Medycynie i Sterowaniu, PLJ, Warszawa, 1999.
92. Muggleton S.: Inverse Entailment and Progol, New Generation Computing, 13, 1995, pp. 245–286.
93. Nguyen H.S., Skowron A.: Quantization of Real Value Attributes, P.P. Wang (Ed.) Second Annual Joint Conference on Information Sciences, September 28–October 1, 1995, North Carolina, USA, pp. 34–37.
94. Nguyen S.H., Nguyen H.S.: Pattern Extraction from Data, Fundamenta Informaticae 34, 1998, pp. 129–144.
95. Nguyen H.S., Nguyen S.H.: Discretization Methods in Data Mining, L. Polkowski, A. Skowron (Eds.): Rough Sets in Knowledge Discovery 1. Methodology and Applications. Physica-Verlag, Heidelberg 1998, pp. 451–482.
96. Nguyen S.H., Skowron A.: Searching for Relational Patterns in Data, Proceedings of the First European Symposium on Principles of Data Mining and Knowledge Discovery (PKDD'97) Trondheim, Norway, June 25–27 Lecture Notes in Artificial Intelligence 1263, 1997, pp. 265–276.
97. Nguyen S. H., Skowron A., Synak P.: Discovery of Data Patterns with Applications to Decomposition and Classification Problems, L. Polkowski, A. Skowron (Eds.), Rough Sets in Knowledge Discovery 2. Applications, Case Studies and Software Systems, Physica-Verlag, Heidelberg, 1998, pp. 55–97.
98. Nieminen J.: Rough Tolerance Equality. Fundamenta Informaticae 11, 1988, pp. 289–296.
99. Nowicki R., Słowiński R., Stefanowski J.: Rough Sets Analysis of Diagnostic Capacity of Vibroacoustic Symptoms. Journal of Computers Mathematics with Applications 24, 1992, pp. 109–123.
100. Novotny M., Pawlak Z.: On Problem Concerning Dependence Space. Fundamenta Informaticae 16/3–4, 1992, pp. 275–287.
101. Ohrn A., Vinterbo S., Szymański P., Komorowski J.: Modelling Cardiac Patient Set Residuals Using Rough Sets. Proceedings of the AMIA Annual Fall Symposium (formerly SCAMC), Nashville, TN, USA, October 25–29, 1997, pp. 203–207.
102. Ohrn A., Komorowski J., Skowron A., Synak P.: The Design and Implementation of a Knowledge Discovery Toolkit Based on Rough Sets - The Rosetta System, L. Polkowski, A. Skowron (Eds.), Rough Sets in Knowledge Discovery 1, Methodology and Applications, Physica-Verlag, Heidelberg, 1998, pp. 376–399.
103. Orłowska E.: A logic of Indiscernibility Relations A. Skowron (Ed.), Computation Theory, Lecture Notes in Computer Science 208, 1985, pp. 177–186.
104. Orłowska E.: Information Algebras, Lecture Notes in Computer Science 936, 1995, pp. 55–65.
105. Pagliani P.: From Concept Lattices to Approximation Spaces, Algebraic Structures of Some Spaces of Partial Objects. Fundamenta Informaticae 18/1, 1993, pp. 1–25.
106. Pal S.K., Skowron A. (Eds.): Rough-Fuzzy Hybridization A New Trend in Decision Making, Springer–Verlag, 1999.
107. Paszek P., Wakulicz-Deja A.: Optimization Diagnose in Progressive Encephalopathy Applying the Rough Set Theory, Proceedings of the Fourth European Congress on Intelligent Techniques and Soft Computing, Aachen, Germany, September 2–5, 1996, vol. 1, pp. 192–196.
108. Pawlak Z.: Rough Sets. International Journal of Computer and Information Science 11, 1982, pp. 341–356.

109. Pawlak Z.: Rough Relations, Bulletin of the Polish Academy of Sciences, Technical Sciences vol. 34 (9-10), 1986, pp. 587–590.
110. Pawlak Z.: Rough Sets. Theoretical Aspects of Reasoning about Data, Kluwer Academic Publishers, Dordrecht, 1991.
111. Pawlak Z., Skowron A.: Rough Membership Functions, M. Fedrizzi, J. Kacprzyk, R. R. Yager (Eds.), Advances in the Dempster-Shafer Theory of Evidence, John Wiley and Sons, New York, 1994, pp. 251–271.
112. Pawlak Z., Słowiński K., Słowiński R.: Rough Classification of Patients After Highly Selected Vagotomy for Duodenal Ulcer, Journal of Man–Machine Studies 24, 1986, pp. 413–433.
113. Peters J.F., Han L., Ramanna S.: Approximate Time Rough Software Cost Decision System: Multicriteria Decision-Making Approach, Proceedings of the 11th International Symposium on Foundations of Intelligent Systems, ISMIS'99, Warsaw, Poland, June 8-11, 1999, Lecture Notes in Artificial Intelligence 1609, Springer-Verlag, 1999, pp. 556–564.
114. Piasta Z., Lenarcik A., Tsumoto S.: Machine Discovery in Databases with Probabilistic Rough Classifiers. S. Tsumoto (Ed.): Bulletin of International Rough Set Society 1/2, 1997, pp. 51–57.
115. Polkowski L.: Mathematical Morphology of Rough Sets. Bulletin of the Polish Academy of Sciences Mathematics 41/3, 1993, pp. 241–273.
116. Polkowski L., Skowron A.: Rough Mereology, Lecture Notes in Artificial Intelligence 869, Springer-Verlag, Berlin 1994, pp. 85–94.
117. Polkowski L., Skowron A.: Rough Mereology: A New Paradigm for Approximate Reasoning, International Journal of Approximate Reasoning, Vol. 15, No 4, 1996, pp. 333–365.
118. Polkowski L., Skowron A.: Towards Adaptive Calculus of Granules, Proceedings of FUZZ-IEEE'98 International Conference, Anchorage, Alaska, USA, May 5–9 1998, pp. 111–116.
119. Polkowski L., Skowron A. (Eds.): Rough Sets in Knowledge Discovery 1: Methodology and Applications. Physica-Verlag, Heidelberg, 1998.
120. Polkowski L., Skowron A. (Eds.): Rough Sets in Knowledge Discovery 2: Applications, Case Studies and Software Systems. Physica-Verlag, Heidelberg, 1998.
121. Polkowski L., Skowron A., Komorowski J.: Towards a Rough Mereology-Based Logic for Approximate Solution Synthesis, Part 1. Studia Logica 58/1, 1997, pp. 143–184.
122. Polkowski L., Skowron A., Żytkow J.M.: Tolerance Based Rough Sets, T.Y.Lin, A.M.Wildberger (Eds.), Soft Computing Simulation Councils, San Diego 1995, pp. 55–58.
123. Pomykała J. A.: Approximation Operations in Approximation Space, Bulletin of the Polish Academy of Sciences, Mathematics, 35, 1987, pp. 653–662.
124. Pomykała J. A.: On Definability in the Nondeterministic Information System. Bulletin of the Polish Academy of Sciences, Mathematics, 36, 1988, pp. 193–210.
125. Quinlan J.R.: Learning Logical Definitions from Relations, Machine Learning, 5, 1990, pp. 239–266.
126. Raś Z.W.: Cooperative Knowledge-Based Systems. Journal of the Intelligent Automation and Soft Computing 2/2, 1996, pp. 193–202.
127. Raś Z.W.: Collaboration Control in Distributed Knowledge-Based Systems. Information Sciences 96/3-4, 1997, pp. 193–205.
128. Raś Z.W., Skowron A. (Eds.): Proceedings of the Tenth International Symposium on Methodologies for Intelligent Systems, Foundations of Intelligent Systems

(ISMIS'97), October 15-18, 1997, Charlotte, NC, USA, Lecture Notes in Artificial Intelligence 1325, Springer-Verlag, Berlin, pp. 1–630.
129. Rasiowa H., Skowron A.: Approximation Logic. In: Proceedings of Mathematical Methods of Specification and Synthesis of Software Systems Conference, Akademie Verlag 31, 1985, Berlin pp. 123–139.
130. Rauszer C.: Knowledge Representation Systems for Groups of Agents, J. Wolenski (Ed.), Philosophical Logic in Poland, Kluwer Academic Publishers, Dordrecht, 1994, pp. 217–238.
131. Schalkoff R.: Pattern Recognition: Statistical, Structural and Neural Approaches, Wiley, 1992.
132. Schreider J.A.: Equality, Resemblance and Order, Mir Publishers, Moscow, 1975.
133. Siromoney A.: A Rough Set Perspective of Inductive Logic Programming, L. De Raedt, S. Muggleton (Eds.), Proceedings of the IJCAI-97 Workshop on Frontiers of Inductive Logic Programming, Nagoya, Japan, August 1997, pp. 111–113.
134. Siromoney A., Inoue K.: A Framework for Rough Set Inductive Logic Programming - the gRS-ILP Model, Pacific Rim Knowledge Acquisition Workshop (PKAW98), Singapore, November 1998, pp. 201–217.
135. Siromoney A., Inoue K.: The gRS-ILP Model and Motifs in Strings. The Seventh International Workshop on Rough Sets, Fuzzy Sets, Data Mining, and Granular-Soft Computing (RSFDGrC'99), Ube, Yamaguchi, Japan November 9–11, Lecture Notes in Artificial Intelligence 1711, 1999.
136. Skowron A.: Data Filtration: A Rough Set Approach, W. Ziarko (Ed.), Rough Sets, Fuzzy Sets and Knowledge Discovery, Springer-Verlag, Berlin 1994, pp. 108–118.
137. Skowron A.: Extracting Laws from Decision Tables. Computational Intelligence 11/2, 1995, pp. 371–388.
138. Skowron A., Grzymała-Busse J.: From Rough Set Theory to Evidence Theory. R.R. Yager, M. Fedrizzi, and J. Kacprzyk (Eds.), Advances in the Dempster-Shafer Theory of Evidence, John Wiley and Sons, New York, 1994, pp. 193–236.
139. Skowron A., Nguyen H.S.: Boolean Resoning Scheme with Some Applications in Data Mining. 3rd European Conference of Principles and Practice of Knowledge Discovery in Databases, September 15-18, 1999, Prague, Czech Republic, Lecture Notes in Artificial Intelligence 1704, 1999, pp. 107–115.
140. Skowron A., Polkowski L.: Synthesis of Decision Systems from Data Tables, T. Y. Lin, N. Cercone (Eds.), Rough Sets and Data Mining Analysis of Imprecise Data, Kluwer Academic Publishers, Dordrecht, 1997, pp. 259–299.
141. Skowron A., Polkowski L., Komorowski J.: Learning Tolerance Relations by Boolean Descriptors: Automatic Feature Extraction from Data Tables, Proceedings of the Fourth International Workshop on Rough Sets, Fuzzy Sets, and Machine Discovery, November 6-8, 1996, Tokyo, Japan, pp. 11–17.
142. Skowron A., Polkowski L.: Rough Mereological Foundations for Design, Analysis, Synthesis and Control in Distributive Systems. Information Sciences 104/1–2, 1998, pp. 129–156.
143. Skowron A, Rauszer C.: The Discernibility Matrices and Functions in Information Systems, R. Słowiński (Ed.), Intelligent Decision Support. Handbook of Applications and Advances of Rough Sets Theory, Kluwer Academic Publishers, Dordrecht, 1992, pp. 331–362.
144. Skowron A., Stepaniuk J.: Towards an Approximation Theory of Discrete Problems, Fundamenta Informaticae 15(2), 1991, pp. 187–208.

145. Skowron A., Stepaniuk J.: Searching for Classifiers. M. De Glas, D. Gabbay (Eds.), Proceedings of the First World Conference on the Fundamentals of Artificial Intelligence (WOCFAI'91), July 1–5, 1991, Angkor, Paris pp. 447–460.
146. Skowron A., Stepaniuk J.: Intelligent Systems Based on Rough Set Approach. Foundations of Computing and Decision Sciences 18/3–4, 1993, pp. 343–360.
147. Skowron A., Stepaniuk J.: Approximations of Relations, W. Ziarko (Ed.), Rough Sets, Fuzzy Sets and Knowledge Discovery, Springer Verlag, London Berlin 1994, pp. 161–166 see also: Institute of Computer Science, Warsaw University of Technology, ICS Research Report 20/94 1994.
148. Skowron A., Stepaniuk J.: Generalized Approximation Spaces, Proceedings of the Third International Workshop on Rough Sets and Soft Computing, San Jose, November 10–12, 1994, pp. 156–163.
149. Skowron A., Stepaniuk J.: Generalized Approximation Spaces, T.Y.Lin, A.M.Wildberger (Eds.), Soft Computing, Simulation Councils, San Diego 1995, pp. 18–21 see also: Institute of Computer Science, Warsaw University of Technology, ICS Research Report 41/94 1994.
150. Skowron A., Stepaniuk J.: Decision Rules Based on Discernibility Matrices and Decision Matrices, T.Y.Lin, A.M.Wildberger (Eds.), Soft Computing, Simulation Councils, San Diego 1995, pp. 6–9 see also Institute of Computer Science, Warsaw University of Technology, ICS Research Report 40/94 1994.
151. Skowron A., Stepaniuk J.: Tolerance Approximation Spaces, Fundamenta Informaticae, 27, 1996, pp. 245–253.
152. Skowron A., Stepaniuk J.: Information Reduction Based on Constructive Neighborhood Systems, P.P. Wang (Ed.): Proceedings of the Fifth International Workshop on Rough Sets and Soft Computing (RSSC'97) at Third Annual Joint Conference on Information Sciences (JCIS'97). Duke University, Durham, NC, USA, Rough Set & Computer Science 3, March 1–5, 1997, pp. 158–160.
153. Skowron A., Stepaniuk J.: Constructive Information Granules, Proceedings of the 15th IMACS World Congress on Scientific Computation, Modelling and Applied Mathematics, August 24–29, 1997, Berlin, Germany, vol. 4 Artificial Intelligence and Computer Science, pp. 625–630.
154. Skowron A., Stepaniuk J.: Information Granules and Approximation Spaces, Proceedings of Seventh International Conference on Information Processing and Management of Uncertainty in Knowledge-Based Systems, Paris, France, July 6–10 1998, pp. 354–361.
155. Skowron A., Stepaniuk J.: Towards Discovery of Information Granules, 3rd European Conference of Principles and Practice of Knowledge Discovery in Databases, September 15–18, 1999, Prague, Czech Republic, Lecture Notes in Artificial Intelligence 1704, Springer-Verlag, 1999, pp. 542–547.
156. Skowron A., Stepaniuk J.: Information Granules in Distributed Environment, New Directions in Rough Sets, Data Mining, and Granular-Soft Computing (RSFD-GrC'99), Ube, Yamaguchi, Japan November 9–11, Lecture Notes in Artificial Intelligence 1711, Springer-Verlag, 1999, pp. 357–365.
157. Skowron A., Stepaniuk J.: Concept Approximation and Information Granules, International Journal of Intelligent Systems, submitted.
158. Skowron A., Suraj Z.: A Parallel Algorithm for Real–Time Decision Making, A Rough Set Approach. Journal of Intelligent Information Systems 7, 1996, pp. 5–28.
159. Słowiński K.: Rough Classification of HSV Patients, Słowiński R. (Ed.), Intelligent Decision Support - Handbook of Applications and Advances of the Rough Sets Theory. Kluwer Academic Publishers, Dordrecht, 1992, pp. 77–93.

160. Słowiński K., Słowiński R., Stefanowski J., Rough Sets Approach to Analysis of Data from Peritoneal Lavage in Acute Pancreatitis, Medical Informatics 13/3, 1988, pp. 143–159.
161. Słowiński K., Stefanowski J.: Multistage Rough Set Analysis of Therapeutic Experience with Acute Pancreatitis, L. Polkowski, A. Skowron (Eds.), Rough Sets in Knowledge Discovery 2. Applications, Case Studies and Software Systems, Physica-Verlag, Heidelberg, 1998, pp. 272–294.
162. Słowiński R. (Ed.): Intelligent Decision Support – Handbook of Applications and Advances of the Rough Sets Theory. Kluwer Academic Publishers, Dordrecht, 1992.
163. Słowiński R.: A Generalization of the Indiscernibility Relation for Rough Sets Analysis of Quantitative Information. Revista di Matematica per le Scienze Economiche e Sociali 15/1, 1992, pp. 65–78.
164. Słowiński R.: Strict and Weak Indiscernibility of Objects Described by Quantitative Attributes with Overlapping Norms, Foundations of Computing and Decision Sciences, Vol. 18, 1993, pp. 361–369.
165. Słowiński R., Stefanowski J.: Software Implementation of the Rough Set Theory, L. Polkowski, A. Skowron (Eds.), Rough Sets in Knowledge Discovery 2. Applications, Case Studies and Software Systems, Physica-Verlag, Heidelberg, 1998, pp. 581–586.
166. Słowiński R., Vanderpooten D.: Similarity Relation as a Basis for Rough Approximations. Warsaw University of Technology, Institute of Computer Science Research Report 53, 1995.
167. Stanfill C., Waltz D.: Toward Memory-Based Reasoning, Communications of the ACM 29, 1986, pp. 1213–1228.
168. Stefanowski J., Słowiński K.: Rough Set Theory and Rule Induction Techniques for Discovery of Attribute Dependencies in Medical Information Systems, Lecture Notes in Artificial Intelligence 1263, Springer-Verlag, 1997, pp. 36–46.
169. Stepaniuk J.: Elementary Approximation Theory. Bulletin of the Polish Academy of Sciences Tech. 38/1–12, 1990, pp. 121–128.
170. Stepaniuk J.: Approximation Logic of Programs. Bulletin of the Polish Academy of Sciences Tech. 38/1–12, 1990, pp. 129–138.
171. Stepaniuk J.: Applications of Finite Models Properties in Approximation and Algorithmic Logics. Fundamenta Informaticae 14/1, 1991, pp. 91–108.
172. Stepaniuk J.: Methods of Approximate Reasoning for Discrete Problems. Ph.D. Dissertation, Warsaw University, 1992.
173. Stepaniuk J.: Decision Rules for Consistent Decision Tables. Proceedings of the Polish–English Meeting on Information Systems, Bialystok, Poland, September 22, 1993, pp. 76–86.
174. Stepaniuk J.: Decision Rules for Decision Tables. Bulletin of the Polish Academy of Sciences Tech. 42/3, 1994, pp. 457–469.
175. Stepaniuk J.: Discernibility and Decision Matrices (in Polish). R. Kulikowski, L. Bogdan (Eds.), Wspomaganie Decyzji, Systemy Eksperckie, Institute of System Analysis PAS, Warsaw, Poland, 1995, pp. 440–443.
176. Stepaniuk J.: Properties and Applications of Rough Relations, Proceedings of the Fifth International Workshop on Intelligent Information Systems, Deblin, Poland, June 2–5, 1996, Institute od Computer Science, Polish Academy of Sciences, Warsaw, 1996, pp. 136–141 see also Institute of Computer Science, Warsaw University of Technology, ICS Research Report 26/96, 1996.

177. Stepaniuk J.: Similarity Based Rough Sets and Learning, Tsumoto S., Kobayashi, S., Yokomori, T., Tanaka, H. (Eds.), Proceedings of the Fourth International Workshop on Rough Sets, Fuzzy Sets and Machine Discovery (RSFD'96), Tokyo November 6–8 1996, pp. 18–22.
178. Stepaniuk J.: Rough Sets, First Order Logic and Attribute Construction. Proceedings of the Sixth International Conference, Information Processing and Management of Uncertainty in Knowledge–Based Systems (IPMU'96), July 1–5, 1996, Granada, Spain, 2, pp. 887–890.
179. Stepaniuk J.: Attribute Discovery and Rough Sets, Principles of Data Mining and Knowledge Discovery, First European Symposium, PKDD97, Trondheim, Norway, June 1997, Lecture Notes in Artificial Intelligence 1263, Springer Verlag, pp. 145–155.
180. Stepaniuk J.: Rough Sets Similarity Based Learning. Proceedings of the Fifth European Congress on Intelligent Techniques and Soft Computing, September 8–12, Aachen, Germany, Verlag Mainz, 1997, pp. 1634–1638.
181. Stepaniuk J.: Conflict Analysis and Groups of Agents. Proceedings of the Poster Session at Tenth International Symposium on Methodologies for Intelligent Systems (ISMIS'97), October 15–18, 1997, Charlotte, USA, pp. 174–185.
182. Stepaniuk J.: Approximation Spaces, Reducts and Representatives, L. Polkowski, A. Skowron (Eds.), Rough Sets in Knowledge Discovery 2. Applications, Case Studies and Software Systems, Physica-Verlag, Heidelberg, 1998, pp. 109–126.
183. Stepaniuk J.: Rough Relations and Logics, L. Polkowski, A. Skowron (Eds.), Rough Sets in Knowledge Discovery 1. Methodology and Applications, Physica-Verlag, Heidelberg 1998, pp. 248–260.
184. Stepaniuk J.: Approximation Spaces in Extensions of Rough Set Theory, Proceedings of the International Conference on Rough Sets and Current Trends in Computing, Warsaw, Poland, June 22–26, 1998, Lecture Notes in Artificial Intelligence 1424, pp. 290–297.
185. Stepaniuk J.: Optimizations of Rough Set Model, Fundamenta Informaticae Vol. 36 (2-3), October-November 1998, pp. 265–283.
186. Stepaniuk J.: Rough Set Data Mining of Diabetes Data, Proceedings of the 11th International Symposium on Foundations of Intelligent Systems, Warsaw, Poland, June 8-11, 1999, Lecture Notes in Artificial Intelligence 1609, Springer-Verlag, pp. 457–465.
187. Stepaniuk J.: Rough Sets and Relational Learning, Proceedings of the Seventh European Congress on Intelligent Techniques and Soft Computing, September 13–16, Aachen, Germany, Verlag Mainz, 1999, CD-ROM, 6 pages.
188. Stepaniuk J., Krętowski M.: Decision System Based on Tolerance Rough Sets, Proceedings of the Fourth International Workshop on Intelligent Information Systems, Augustow, Poland, June 5-9, 1995, Institute od Computer Science, Polish Academy of Sciences, Warsaw 1995, pp. 62–73 see also Institute of Computer Science, Warsaw University of Technology, ICS Research Report 36/95 1995.
189. Stepaniuk J., Maj M.: Data Transformation and Rough Sets, PKDD98, Nantes, France, September, 1998, Lecture Notes in Artificial Intelligence 1510, Springer-Verlag, pp. 441–449.
190. Stepaniuk J., Tyszkiewicz J.: Probabilistic Properties of Approximation Problems. Bulletin of the Polish Academy of Sciences Tech. 39/3, 1991, pp. 535–555.
191. Stepaniuk J., Urban M., Baszun-Stepaniuk E.: The Application of Rough Set Based Data Mining Technique in the Prognostication of the Diabetic Nephropathy Prevalence, Proceedings of the Seventh International Workshop on Intelligent

Information Systems, Malbork, Poland, June 15–19, 1998, Institute od Computer Science, Polish Academy of Sciences, Warsaw 1998, pp. 388–391.
192. Ślęzak D.: Approximate Reducts in Decision Tables, Proceedings of the Six International Conference on Information Processing and Management of Uncertainty in Knowledge-Based Systems, Granada, Spain, July 1–5, 1996, pp. 1159–1164.
193. Suraj Z.: Discovery of Concurrent Data Models from Experimental Tables, A Rough Set Approach. Fundamenta Informaticae 28/3–4, 1996, pp. 353–376.
194. Świniarski R.: Rough Set Expert System for On-Line Prediction of Volleyball Game Progress for US Olympic Team. B.D. Czejdo, I.I. Est, B. Shirazi, B. Trousse (Eds.), Proceedings of the Third Biennial European Joint Conference on Engineering Systems Design Analysis, July 1–4, 1996, Montpellier, France, pp. 15–20.
195. Tentush I.: On Minimal Absorbent Sets for some Types of Tolerance Relations, Bulletin of the Polish Academy of Sciences, Technical Sciences 43/1, 1995, pp. 79–88.
196. Torgo L.: Controlled Redundancy in Incremental Rule Learning, Lecture Notes in Artificial Intelligence 667, 1993, pp. 185–195.
197. Tsumoto S., Tanaka H.: PRIMEROSE, Probabilistic Rule Induction Method Based on Rough Set Resampling Methods. Computational Intelligence: An International Journal 11/2, 1995, pp. 389–405.
198. Tsumoto S., Tanaka H.: Machine Discovery of Functional Components of Proteins from Amino–Acid Sequences Based on Rough Sets Change of Representation. Journal of the Intelligent Automation and Soft Computing 2/2, 1996, pp. 169–180.
199. Tsumoto S.: Extraction of Experts Decision Process from Clinical Databases Using Rough Set Model, PKDD97, Trondheim, Norway, June 1997, Lecture Notes in Artificial Intelligence 1263, Springer Verlag, pp. 58–67.
200. Tsumoto S.: Formalization and Induction of Medical Expert System Rules Based on Rough Set Theory, L. Polkowski, A. Skowron (Eds.), Rough Sets in Knowledge Discovery 2. Applications, Case Studies and Software Systems, Physica-Verlag, Heidelberg, 1998, pp. 307–323.
201. Tsumoto S., Ziarko W.: The Application of Rough Sets - Based Data Mining Technique to Differential Diagnosis of Meningoencephalitis, Proceedings of the 9th International Symposium, Foundations of Intelligent Systems, Zakopane, Poland, 9–13 June, 1996, Lecture Notes in Artificial Intelligence 1079, pp. 438–447.
202. Tversky A.: Features of Similarity. Psychological Review 84/4, 1997, pp. 327–352.
203. Yao Y. Y.: On Generalizing Pawlak Approximation Operators, Proceedings of the International Conference on Rough Sets and Current Trends in Computing, Warsaw, Poland, June 22–26, 1998, Lecture Notes in Artificial Intelligence 1424, pp. 298–307.
204. Yao Y. Y., Lin T. Y.: Generalization of Rough Sets Using Modal Logic. Intelligent Automation and Soft Computing 2, 1996, pp. 103–120.
205. Yao Y. Y., Wong S. K. M., Lin T. Y.: A Review of Rough Set Models, T. Y. Lin, N. Cercone (Eds.), Rough Sets and Data Mining Analysis of Imprecise Data, Kluwer Academic Publishers, 1997, pp. 47–75.
206. Yao Y. Y., Zhong N.: An Analysis of Quantitative Measures Associated with Rules, Proceedings of The Third Pacific-Asia Conference on Knowledge Discovery and Data Mining, Beijing, China, April 26–28, 1999, Lecture Notes in Artificial Intelligence 1574, pp. 479–488.

207. Vakarelov D.: A Modal Logic for Similarity Relations in Pawlak Knowledge Representation Systems. Fundamenta Informaticae 15, 1991, pp. 61–79.
208. Vakarelov D.: Rough Polyadic Modal Logics, Journal of Applied Non-Classical Logics, vol. 1(1), 1991, pp. 9–36.
209. Vakarelov D.: Information Systems, Similarity Relations and Modal Logic, E. Orłowska (Ed.), Incomplete Information: Rough Set Analysis, Physica Verlag, Heidelberg, 1998, pp. 492–550.
210. Urban M., Baszun-Stepaniuk E., Stepaniuk J.: Application of the Rough Set Theory in the Prognostication of the Diabetic Nephropathy Prevalence. Preliminary Communication Endokrynologia, Diabetologia i Choroby Przemiany Materii Wieku Rozwojowego 1998, 4, 2, pp. 107–112.
211. Wakulicz–Deja A., Paszek P.: Diagnose Progressive Encephalopathy Applying the Rough Set Theory. International Journal of Medical Informatics 46, 1997, pp. 119–127.
212. Wasilewska A.: Linguistically Definable Concepts and Dependencies. Journal of Symbolic Logic 54/2, 1989, pp. 671–672.
213. Weiss S.M., Kulikowski C.A.: Computer Systems that Learn: Classification and Prediction Methods from Statistics, Neural Networks, Machine Learning and Expert Systems, Morgan Kaufmann, San Mateo, CA, 1991.
214. Wilson D. A., Martinez T. R.: Improved Heterogeneous Distance Functions, Journal of Artificial Intelligence Research, Vol. 6, 1997, pp. 1–34.
215. Wong S.K.M., Ziarko W., Ye L.W.: Comparision of Rough Set and Statistical Methods in Inductive Learning. Journal of Man–Machine Studies 24, 1986, pp. 53–72.
216. Wong S.K.M.: A Rough-Set Model for Reasoning about Knowledge, L. Polkowski, A. Skowron (Eds.), Rough Sets in Knowledge Discovery 1. Methodology and Applications, Physica–Verlag, Heidelberg, 1998, pp. 276–285.
217. Woolery L., Grzymała-Busse J.W.: Machine learning for an Expert System to Predict Preterm Birth Risk. Journal of the American Medical Informatics Association 1, 1994, pp. 439–446.
218. Wybraniec-Skardowska U.: On a Generalization of Approximation Space, Bulletin of the Polish Academy of Sciences, Mathematics, 37, 1989, pp. 51–61.
219. Zadeh L. A.: Similarity Relations and Fuzzy Orderings. Information Sciences 3 1971, pp. 177–200.
220. Zadeh L.A.: Fuzzy Logic = Computing with Words, IEEE Trans. on Fuzzy Systems Vol. 4, 1996, pp. 103–111.
221. Zadeh L.A.: Toward a Theory of Fuzzy Information Granulation and Its Certainty in Human Reasoning and Fuzzy Logic, Fuzzy Sets and Systems Vol. 90, 1997, pp. 111–127.
222. Zadeh L.A., Kacprzyk J. (Eds.): Computing with Words in Information/Intelligent Systems 1. Foundations, Physica-Verlag, Heidelberg, 1999.
223. Zadeh L.A., Kacprzyk J. (Eds.): Computing with Words in Information/Intelligent Systems 2. Applications, Physica-Verlag, Heidelberg, 1999.
224. Ziarko W.: The Discovery, Analysis and Representation of Data Dependencies in Databases, G. Piatetsky–Shapiro, W.J. Frawley (Eds.), Knowledge Discovery in Databases, AAAI Press/MIT Press, 1991, pp. 177–195.
225. Ziarko W.: Variable Precision Rough Sets Model, Journal of Computer and Systems Sciences, Vol. 46, No. 1, 1993, pp. 39–59.
226. Ziarko W., Shan N.: KDD–R: A Comprehensive System for Knowledge Discovery in Databases Using Rough Sets, Proceedings of the Third International Workshop

on Rough Sets and Soft Computing, San Jose, November 10–12, 1994, pp. 164–173.
227. Żakowski W.: On a Concept of Rough Sets. Demonstratio Mathematica XV, 1982, pp. 1129–1133.
228. Żytkow J.M., Zembowicz R.: Database Exploration in Search of Regularities, Journal of Intelligent Information Systems 2, 1993, pp. 39–81.

a_1	a_2	a_3	a_4	a_5	a_6	a_7	a_8	a_9	a_{10}	a_{11}	a_{12}
f	12	5	no	KIT_IIT	yes	no	7.28	yes	3 − 97	no	no
f	13	4	no	KIT_IIT	yes	no	8.69	no	3 − 97	no	no
f	11	5	yes	KIT	yes	no	9.6	no	3 − 97	yes	no
m	13	5	no	KIT_IIT	yes	no	8.6	yes	3 − 97	no	no
f	14	6	no	KIT_IIT	yes	no	7.68	yes	3 − 97	no	no
m	14	4	no	KIT	yes	yes	9	no	3 − 97	no	no
m	9	9	no	KIT_IIT	no	no	7.4	no	3 − 97	yes	no
m	16	2	no	KIT	yes	no	9	yes	3 − 97	no	no
f	7	12	yes	KIT_IIT	yes	no	8.06	no	3 − 97	yes	no
f	11	5	no	KIT_IIT	yes	no	11.61	no	3 − 97	yes	no
m	11	5	no	KIT	no	no	8.32	no	3 − 97	no	no
m	9	6	no	KIT	no	no	10.5	no	3 − 97	yes	yes
f	12	7	no	KIT	yes	no	8.26	no	3 − 97	no	no
f	10	5	no	KIT	no	no	7.96	no	< 3	no	no
f	9	6	no	KIT_IIT	yes	yes	9.44	no	3 − 97	no	no
f	12	7	no	KIT	yes	no	8.05	no	3 − 97	no	no
m	9	5	no	KIT	yes	no	8.48	no	3 − 97	no	no
f	13	5	no	KIT	yes	no	8.68	no	3 − 97	no	no
f	12	7	no	KIT_IIT	no	yes	9.42	no	3 − 97	no	no
f	12	6	yes	KIT_IIT	yes	yes	9.61	no	3 − 97	no	no
m	13	5	no	KIT_IIT	no	no	8.02	yes	3 − 97	no	no
f	12	6	no	KIT_IIT	yes	no	7.53	no	3 − 97	yes	no
m	12	6	no	KIT	yes	yes	8.36	yes	3 − 97	yes	no
m	11	4	no	KIT_IIT	yes	no	6.8	yes	3 − 97	no	no
f	11	8	yes	KIT	yes	no	11	yes	3 − 97	yes	yes
f	14	4	no	KIT_IIT	yes	no	8.47	no	> 97	yes	no
f	16	3	no	KIT_IIT	no	no	8.56	no	3 − 97	no	no
m	7	13	no	KIT	yes	no	9.07	yes	3 − 97	no	no
f	10	7	no	KIT_IIT	yes	no	10.07	no	3 − 97	yes	no
f	2	5	no	KIT	yes	no	7.15	no	< 3	no	no
f	5	5	no	KIT	yes	no	7.35	no	3 − 97	yes	no
f	3	11	no	KIT_IIT	no	no	8.88	no	3 − 97	no	no
m	11	7	yes	KIT	yes	no	8.83	yes	3 − 97	yes	yes
f	10	6	no	KIT	no	no	10.43	no	> 97	yes	yes
f	13	4	yes	KIT_IIT	yes	no	10.83	no	< 3	yes	yes
m	14	4	no	KIT	yes	no	7.07	yes	3 − 97	no	no
m	13	4	no	KIT_IIT	yes	no	8.83	no	3 − 97	yes	yes
m	4	5	no	KIT_IIT	yes	no	4.3	no	3 − 97	no	no
m	13	6	yes	KIT	no	no	6.9	no	3 − 97	no	no
f	12	7	no	KIT_IIT	yes	yes	11.02	yes	3 − 97	no	no
f	11	6	no	KIT_IIT	yes	no	6.95	no	> 97	yes	no
m	12	4	no	KIT_IIT	no	no	9.33	no	< 3	yes	no
f	11	4	no	KIT_IIT	yes	no	11.14	no	3 − 97	yes	yes
m	12	7	no	KIT	no	no	6.72	no	3 − 97	no	no
m	7	7	no	KIT_IIT	yes	yes	8.59	no	3 − 97	yes	no
m	10	5	no	KIT_IIT	yes	no	8	no	3 − 97	no	no
f	10	6	yes	KIT	yes	no	10.35	no	3 − 97	yes	yes
m	10	7	no	KIT	yes	no	7.55	no	< 3	no	no
f	14	7	no	KIT	yes	no	6	no	> 97	no	no
f	5	5	no	KIT	yes	no	5.6	no	3 − 97	no	no
m	7	7	no	KIT	yes	no	10.8	no	3 − 97	yes	yes
f	10	8	no	KIT	yes	no	7.8	no	> 97	yes	no
f	13	7	yes	KIT	yes	no	5	no	3 − 97	yes	no
f	13	5	no	KIT	yes	no	7.48	yes	< 3	no	no
f	14	5	no	KIT_IIT	no	no	6.97	no	3 − 97	no	no
m	12	6	no	KIT	yes	no	10.35	yes	3 − 97	yes	yes

Table 44. Diabetes Mellitus Data - Class "Yes"

a_1	a_2	a_3	a_4	a_5	a_6	a_7	a_8	a_9	a_{10}	a_{11}	a_{12}
m	1	4	no	KIT	yes	no	10	no	3 − 97	no	no
m	15	5	yes	KIT	yes	no	6.65	no	3 − 97	no	no
f	17	3	yes	KIT_IIT	no	no	7.03	no	3 − 97	no	no
m	8	4	no	KIT	yes	no	7.46	no	3 − 97	no	no
m	6	6	yes	KIT	yes	no	10.08	no	3 − 97	no	no
f	8	5	no	KIT	yes	no	9.52	no	3 − 97	no	no
f	14	4	no	KIT	yes	yes	5.9	no	3 − 97	no	no
m	16	4	no	KIT_IIT	no	no	7.18	no	3 − 97	no	no
m	15	7	no	KIT_IIT	yes	no	7.96	yes	3 − 97	no	no
m	12	5	no	KIT	no	no	12.73	no	3 − 97	no	no
m	15	7	no	KIT	yes	yes	8.2	yes	3 − 97	no	no
m	9	4	yes	KIT_IIT	yes	no	9.79	no	3 − 97	yes	no
f	16	3	no	KIT	no	yes	6.86	no	3 − 97	no	no
f	2	4	no	KIT	no	no	8.64	no	3 − 97	no	no
m	15	5	no	KIT_IIT	yes	no	5.8	no	3 − 97	no	no
m	10	4	yes	KIT_IIT	no	no	8.13	no	3 − 97	no	no
m	4	12	no	KIT_IIT	no	no	8.09	no	3 − 97	no	no
m	8	5	yes	KIT_IIT	no	no	7.47	no	3 − 97	yes	no
m	1	4	yes	KIT_IIT	yes	no	8.38	no	3 − 97	no	no
m	7	4	no	KIT	no	no	7.91	no	< 3	yes	no
m	9	5	no	KIT	yes	yes	6.64	no	3 − 97	no	no
f	7	5	yes	KIT_IIT	yes	no	7.51	no	3 − 97	no	no
m	5	4	no	KIT	no	no	7.43	no	3 − 97	no	no
f	5	6	no	KIT	yes	no	8.52	no	< 3	no	yes
f	10	4	no	KIT_IIT	no	no	8.75	yes	3 − 97	no	no
m	6	7	yes	KIT	yes	no	8.13	no	3 − 97	no	no
m	9	4	no	KIT	yes	no	7.9	no	3 − 97	no	no
m	11	6	yes	KIT_IIT	no	no	8.94	no	3 − 97	no	no
f	7	7	yes	KIT	no	no	7.54	no	3 − 97	no	no
m	10	5	no	KIT_IIT	yes	no	8.28	no	< 3	yes	yes
f	6	8	no	KIT	no	no	7.16	no	3 − 97	no	no
m	10	6	no	KIT	yes	no	9.75	no	3 − 97	no	no
m	8	4	yes	KIT_IIT	yes	no	8.58	no	3 − 97	yes	no
m	4	7	yes	KIT	no	no	9.2	no	3 − 97	yes	yes
m	14	5	yes	KIT	yes	no	7.57	no	3 − 97	no	no
m	11	7	no	KIT	yes	yes	6.21	no	3 − 97	no	no
m	5	5	yes	KIT	yes	no	7.67	no	3 − 97	no	no
f	5	7	yes	KIT	yes	no	7.8	no	3 − 97	no	yes
f	2	8	no	KIT	yes	no	8.13	no	3 − 97	no	no
m	13	6	no	KIT	no	no	9.35	no	3 − 97	no	no
f	14	7	no	KIT	no	no	8.4	no	3 − 97	no	no
f	8	11	no	KIT	yes	no	11.6	no	3 − 97	yes	yes
m	4	7	yes	KIT	yes	no	7.5	no	3 − 97	no	no
f	11	7	no	KIT	yes	no	5.46	yes	3 − 97	no	no
m	3	4	no	KIT	yes	no	7.15	no	3 − 97	yes	no
m	16	6	no	KIT	yes	no	7.87	no	3 − 97	yes	no
m	4	13	no	KIT_IIT	no	no	6.74	yes	3 − 97	no	no
m	2	8	yes	KIT	yes	no	10.7	no	3 − 97	yes	yes
m	13	7	no	KIT_IIT	yes	yes	8.92	yes	3 − 97	no	no
m	2	5	no	KIT	yes	no	7.78	no	3 − 97	no	no
f	4	5	no	KIT	yes	no	8.1	no	3 − 97	yes	no

Table 45. Diabetes Mellitus Data - Class "No"

Chapter 5

Various Approaches to Reasoning with Frequency Based Decision Reducts: A Survey

Dominik Ślęzak

Warsaw University
Banacha 2, 02-097 Warsaw, Poland
Polish-Japanese Institute of Information Technology
Koszykowa 86, 02-008 Warsaw, Poland
email:slezak@mimuw.edu.pl

Abstract: Various aspects of reduct approximations are discussed. In particular, we show how to use them to develop flexible tools for analysis of strongly inconsistent and/or noisy data tables. A special attention is paid to the notion of a rough membership decision reduct — a feature subset (almost) preserving the frequency based information about *conditions→decision* dependencies. Approximate criteria of preserving such a kind of information under attribute reduction are considered. These criteria are specified by using distances between frequency distributions and information measures related to different ways of interpreting rough membership based knowledge.

1 Introduction

Rough set theory ([20]) provides, in particular, tools for expressing inexact dependencies within data. These tools can be used in applications to both knowledge representation and decision support under uncertainty. Given a data table of records of some descriptions and/or measurements concerning available objects, rough set methods enable to extract causal connections between corresponding features and/or states. These connections can be applied to inductive reasoning about new, so far unseen cases, in a way well understandable for the user. Above advantages, as well as very effective computational framework for extraction of the most interesting dependencies from real-life data (see e.g. [7], [18], [49]), cause a rapid development of applications of rough sets to more and more scientific fields and practical tasks (see e.g. [11], [14], [19], [21], [22], [25], [41], [44], [45]).

Rough set approach is based on generalization of the relation of being an element of a set. Calculations over basic indiscernibility relations attach to each considered property lower and upper approximations, corresponding to elements which satisfy this property with certainty and which possibly satisfy it, respectively ([20]). Then, a subset of objects with considered property is called rough with respect to a given knowledge about object indiscernibility, if its lower approximation is not equal to the upper one. The set theoretic difference between the upper and lower approximations is called the boundary region of a given subset of objects.

In applications, properties of objects are often stated by means of a distinguished feature, called the decision attribute. The main task then is to approximate decision classes, i.e., object sub-domains corresponding to particular values of decision, in terms of indiscernibility relation induced by the rest of features, called conditional attributes. The need of considering such a model can be related, e.g., to the task of decision classification (decision forecasting) of new objects under information provided for them by conditional features. In such a case, operations on indiscernibility classes and approximations of decision classes lead to rough set based decision rules enabling to reason about new objects by analogy with dependencies extracted from already known data.

Rough sets adapted from other approaches some fundamental principles potentially improving effectiveness of decision rules concerned with the above task. One of such paradigms is the Minimum Description Length Principle (abbreviated as the MDL-Principle). It states, roughly speaking, that rules of the most simplified construction, which (almost) preserve consistency with data, are likely to classify so far unseen objects with the lowest risk of error. The MDL-Principle, adapted from statistics ([12], [27], [28]), perfectly fits the main ideas of rough sets, related to providing possibly simple and readable knowledge representation (see e.g. [20] or [31]). Experiments confirm that reasonable MDL's usage decreases classification error. Moreover, it enables to shorten decision rules and neglect features being the source of redundant information during further reasoning.

The process of extraction of optimal decision rules or decision reducts – subsets gathering groups of most relevant conditional features – bases on the balance between two aspects of the MDL-Principle, i.e., description simplicity and degree of preserving decision information under reduction of attributes or descriptors (see e.g. [18], [49]). While the first factor is reflected, usually, by searching for rules consisting of possibly small number of descriptors or reducts of possibly small cardinality, the second of above mentioned factors relies substantially on the way of understanding decision information itself. Original method provided by the theory of rough sets is based on preserving lower and upper approximations of considered concepts. Namely, given the family (lattice) of indiscernibility relations induced by each particular subset of conditional attributes, one can search for possibly small decision reducts allowing to induce (almost) the same rough set boundaries of decision classes as those before the feature reduction started. For the purpose of dealing with decision rules, characteristics based on generalized decision functions have been proposed ([35]). It enables to attach to particular objects and the whole indiscernibility equivalence classes the sets of possible decision values, leading to classifiers based on generalized inconsistent rules.

Nowadays, real-life applications compel us to consider more and more sophisticated tools for inconsistent decision analysis. One of proposed directions for research in this area is based on the process of boundary regions tuning ([51]). It results with tools for dealing with noisy data within a common rough set framework. Another possibility refers to analysis of boundary data by using a more detailed representation. It leads to enrichment of basic rough set notions by appropriate adaptation of constructs known from other approaches to data analysis, e.g., by introducing in [23] the notion of a rough membership function — an analogy of conditional frequencies considered in non-parametric statistics — or by comparing rough set approximations with fundamental notions of Dempster-Shafer theory in [34].

Taking into account different understanding of inconsistent decision information, one must remember about computational complexity of the problems of extracting optimal inexact (in any sense) decision rules and/or reducts from data. In [35] it is shown that a great advantage of classical approach lays in the correspondence between rough set optimization problems and the task of finding minimal prime implicants in boolean reasoning ([2], [30]). Starting from this observation, a common computational core and methodological canon for effective solving of fundamental tasks concerned with rough set approach to data analysis has been developed (for further references see Chapter by Bazan et al. in this Volume). Thus, we are going to pay a special attention to adapting the above, so called, discernibility representation to novel methods of handling inconsistent dependencies between conditional and decision attributes.

New approaches to knowledge extraction and representation are often referred to probabilistic tools, according to wide studied foundations of probability theory and statistics (cf. [24], [46]). Studies on comparison with statistical mechanisms have been also following the development of rough sets (started by [47] and continued in different directions, e.g. in [3]) We focus on frequency based methodology concerned with above mentioned rough membership functions (cf. [21], [22], [23], [45]). We begin from re-formulation of the notion of a decision reduct as preserving *conditions→decision* frequency distributions (compare e.g. with [23], [37], [38], [39]). Proposed definitions and presented results enable us to take advantages of both mentioned boolean based discernibility framework and analogies with probabilistic terminology. In particular, it gives opportunity of applying appropriately modified rough set search algorithms to computational problems related to the notions of probabilistic conditional independence and Markov boundary ([24]) — often useful in practical approaches based on probability theory and statistics (see e.g. [41]).

A disadvantage of dealing with frequencies in their strict form is that they are not robust to dynamic changes of data. On the other hand, by embedding rough membership distributions into Cartesian product of the space of real values, one can approximate the criterion for preserving frequency based information in terms of different distance measures and approximation thresholds (compare e.g. with [38], [43]). Such a similarity based approach is partially analogous to generalizations of rough set notions by defining indiscernibility as a tolerance relation (for further references see Chapters by Stepaniuk and Nguyen Sinh Hoa in this Volume). Presented study on approximate rough membership decision reducts may be thus regarded as complementary to tolerance based techniques of dealing with, e.g., real-valued or vaguely defined conditional attributes (cf. Chapters by Stepaniuk, Nguyen Sinh Hoa and Bazan et al. in this Volume).

Yet another way of understanding approximation of frequencies is related to the question what kind of information we actually need to keep. For example, one can observe that knowledge necessary for reconstruction of the values of generalized decision function is much poorer than in case of rough membership representation. Thus, if one is interested in basing, e.g., the scheme of new case classification on generalized decisions, then it is enough to search for optimal rules and reducts preserving just a relevant part of the whole frequency based information. In the same way we can try to model reasoning and classification strategies referring to other decision functions, defined by neglecting these aspects of frequencies which are redundant with respect to a given approach (compare with [40], [42]).

We present a possibly wide framework for mentioned approximations, providing discernibility characteristics for decision reducts connected with particular methodologies. We also discuss relationships between these characteristics and reduct approximations based on different information measures, like discernibility measure ([19]), conditional entropy ([3], [5], [9]) or the family of normalized information measures ([40], [42]). Measure based approaches enable to define and extract from data subsets of conditions which approximately keep a fixed kind of global *conditions→decision* information, computed as the weighted mean value of local performances of particular patterns. It seems to be especially promising in view of applications to extraction of effective decision rules robust to noise − patters of relatively small support have negligible influence on global measure quantities, so they become unimportant in view of the average approximate preserving requirements. Additional gain is that − unlike in case of previously mentioned techniques − reducts approximately keeping a given global information level generate inexact, relatively shorter and stronger local decision rules, even for consistent decision tables.

Usually, approximate decision reducts based on information measures are not provided with characterizations derivable by means of (approximate) boolean reasoning. Obviously, it is not a disadvantage unless generated rules lead to substantial errors in classification of new cases. Misclassifications may happen when situations neglected because of low support in training data turn out to occur frequently for so far unseen objects. To avoid such a risk − concerned, in a sense, with insufficient satisfaction of statistical assumptions − some methods balancing between classical discernibility and average information measures are required. Similarly as, e.g., in the field of normalized decision measures themselves (compare with [42]), studies on appropriate average discernibility characteristics are very difficult ([43]), providing solutions rather for special cases so far. Nevertheless, presented results give some insight on possibilities of further development of rough set based methodological framework in view of approximate classification and knowledge representation.

The material is organized as follows: In Section 2 we recall basic rough set notions concerned with analysis of (inconsistent) decision tables. Some attention is paid to discernibility and approximate discernibility characteristics for introduced constructs. Section 3 deals with different possibilities of usage of frequency based information for − among others − new case classification by inexact decision rules. In Section 4 we discuss different kinds of approximate rough membership decision reducts, related to weakened requirements of preserving frequency based information under feature reduction, as well as to re-formulation of these requirements for different strategies for reasoning with frequencies. Section 5 is devoted to approximations relying on different kinds of decision information measures and to methods based on average approximate discernibility characteristics. Section 6 concludes the paper with directions for further research on the considered topics.

We base substantially on results presented in our study so far (cf. [41], [42], [43]). Gathered together, these results imply a number of novel corollaries pointing to new directions for further theoretical research and applications.

2 Preliminaries

2.1 Rules and reducts

The main paradigm of rough set theory ([20]) states that the universe of known objects is assumed to be the only source of knowledge about a domain specified by our needs. Data based reasoning is then concerned with the analysis of dependencies between features labeling known cases with values from some pre-defined domain. Let us represent any sample of known data as an information system $\mathbf{A} = (U, A)$, where each attribute $a \in A$ is identified with a function $a : U \to V_a$ from the universe of objects U into the set V_a being such a domain for a.

If analysis is aimed at predicting values of a distinguished decision attribute under information provided by conditional attributes, then it is simpler to represent data as a decision table $\mathbf{A} = (U, A, d)$. Values $v_d \in V_d$ of $d \notin A$ correspond then to mutually disjoint decision classes of the form $d^{-1}(v_d)$. To represent *conditions→decision* dependencies, one can use propositional language over atomic formulas $"a = v_a"$, $a \in A \cup \{d\}$, $v_a \in V_a$. For a given decision table $\mathbf{A} = (U, A, d)$, we say that $\mathbf{A}$ satisfies the exact decision rule

$$\bigwedge_{a \in B} (a = v_a) \Rightarrow (d = v_d) \tag{1}$$

iff for any object $u \in U$ such that for each $a \in B$ the equality $a(u) = v_a$ holds, we have $d(u) = v_d$. Given such a decision rule, one can reason about new cases by analogy. Namely, if a new object $u_{new} \notin U$ is observed to have values $v_a \in V_a$ on corresponding conditional attributes $a \in B$ occurring in the left side of implication (1), then we are going to classify it as belonging to the decision class of $v_d \in V_d$.

Specification of V_a for each particular $a \in A \cup \{d\}$ depends on many factors, like preferences of the user, chosen algorithmic approach, or the nature of data itself. Elements of a given V_a can take the form of intervals for numeric $a \in A \cup \{d\}$ (cf. Chapter by Bazan et al. in this Volume), or, for example, subsets of values in case of symbolic conditions (cf. [15]). Then we should rather write, e.g., $"x < a \leq y"$ or $"a \in \{x, .., y\}"$, respectively, instead of simple descriptor

"$a = v_a$". Thus, let us use a more universal notation, which enables to rewrite any rule of the form (1) as

$$\bigwedge_{a \in B} (a, v_a) \Rightarrow (d, v_d) \qquad (2)$$

where generalized descriptors (a, v_a), $a \in B \cup \{d\}$, are understood as "$a = v_a$", "$a \in v_a$" or whatever else, depending on the context. If, for instance, a given attribute $a \in B$ had originally numerical values, then the value of a on any new object $u_{new} \notin U$ is to be observed as numerical as well. If now domain V_a of a is built from intervals, then u_{new} satisfies atomic formula "$a \in v_a$", iff the numerical value of u_{new} on a falls into the interval v_a.

To express dependencies at a higher level than that of decision rules, let us define, for any $B \subseteq A$, an information function labeling each object $u \in U$ with

$$Inf_B(u) = \prod_{a \in B} \{\langle a, a(u) \rangle\} \qquad (3)$$

Elements of the set

$$V_B^U = \{Inf_B(u) : u \in U\} \qquad (4)$$

of all supported information patterns on B correspond to equivalence classes of indiscernibility relation

$$IND(B) = \{(u_1, u_2) \in U \times U : Inf_B(u_1) = Inf_B(u_2)\} \qquad (5)$$

Definition 1. Given decision table $\mathbf{A} = (U, A, d)$, we say that $B \subseteq A$ exactly defines an attribute $d \notin A$ in $\mathbf{A}$, iff

$$IND(B) \subseteq IND(\{d\}) \qquad (6)$$

A subset $B \subseteq A$ which exactly defines d is called an exact decision reduct, iff it has no proper subset which defines d.

Condition (6) is equivalent to saying that for each $w_B \in V_B^U$ there exists $v_d \in V_d$ such that

$$(B, w_B) \Rightarrow (d, v_d) \qquad (7)$$

where we write "(B, w_B)" instead of "$\bigwedge_{a \in B} (a, v_a)$" for $w_B = \prod_{a \in B} (\langle a, v_a \rangle)$. Thus, we can say that each exact decision reduct is a minimal (in sense of inclusion) subset $B \subseteq A$, which generates the bunch of decision rules of the form (7), for all $w_B \in V_B^U$.

One can see that the property of defining d is closed with respect to supersets, in the following sense:

Definition 2. We say that a property $P : 2^A \to \{0, 1\}$ is closed with respect to supersets, iff for each $C \subseteq B \subseteq A$ we have the implication

$$(P(C) = 1) \Rightarrow (P(B) = 1) \qquad (8)$$

Analogously, P is called closed with respect to subsets, iff for each $C \subseteq B \subseteq A$ we have the implication

$$(P(B) = 1) \Rightarrow (P(C) = 1) \qquad (9)$$

An example of property satisfying the second of the above tendencies is applicability to new cases. Indeed, for any $C \subseteq B$, for each $w_B \in V_B^U$ we have implication

$$(w_B \in V_B^U) \Rightarrow (w_B^{\downarrow C} \in V_C^U) \qquad (10)$$

Thus, if a given subset $B \subseteq A$ is applicable to a given $u_{new} \notin U$, then any $C \subseteq B$ is applicable to u_{new} as well. As a result, we obtain the principle of balance between two considered properties, which substantially agrees with the idea of MDL-Principle. According to this, optimal subsets of conditional attributes can be regarded as decision reducts with a minimal number of attributes – which is a commonly used criterion for extraction of both exact and approximate reducts – or those with minimal number $|V_B^U|$ of generated decision rules – which is a proposal being recently developed e.g. in [49] (see also Chapter by Bazan et al. in this Volume).

2.2 Inconsistent data

The balance between applicability and the property of keeping decision information is very simple: According to (10), we are likely to base on possibly short decision rules or possibly small decision reducts. The situation changes a bit if minimal decision rules are still too complicated. Then, one would like to generate shorter rules which do not point at any decision class in a completely exact way. Such a tendency, being just an alternative for consistent data, becomes necessity in case of lack of complete specification of decision classes in terms of conditional attributes. Then, we must rely on some representation of initial inconsistency, to be able to measure its dynamics with respect to the reduction of information.

Theory of rough sets deals with such a requirement by generalizing the notion of being an element of a set. Calculations over basic indiscernibility relations attach to each considered property lower and upper approximations of objects, corresponding to elements which satisfy this property with certainty and which possibly satisfy it, respectively:

Definition 3. ([20]) For any decision table $\mathbf{A} = (U, A, d)$, a subset $B \subseteq A$ and a subset $X \subseteq U$ (being, e.g., a decision class $X = d^{-1}(v_d)$ for some $v_d \in V_d$), we define the lower and the upper approximation of X by

$$\begin{array}{l} LOW_B(X) = \bigcup_{w_B: Inf_B^{-1}(w_B) \subseteq X} Inf_B^{-1}(w_B) \\ UPP_B(X) = \bigcup_{w_B: Inf_B^{-1}(w_B) \cap X \neq \emptyset} Inf_B^{-1}(w_B) \end{array} \qquad (11)$$

respectively. The boundary of X with respect to B is then defined as the set theoretic difference

$$Bound_B(X) = UPP_B(X) - LOW_B(X) \qquad (12)$$

We say that a given $X \subseteq U$ is roughly defined with respect to the decision table $\mathbf{A} = (U, A, d)$ and a subset $B \subseteq A$ iff $Bound_B(X) \neq \emptyset$.

The above methodology enables to express inexact way of defining concepts under conditional knowledge given in terms of indiscernibility. Now, one can generalize the notion of an exact decision reduct as follows:

Definition 4. (compare with [35]) For any decision table $\mathbf{A} = (U, A, d)$, we say that a subset $B \subseteq A$ roughly defines d in $\mathbf{A}$ iff for each decision value $v_d \in V_d$ we have

$$Bound_B(d^{-1}(v_d)) = Bound_A(d^{-1}(v_d)) \tag{13}$$

A subset B which roughly defines d is called a rough decision reduct iff none of its proper subsets satisfies the above property.

Representing knowledge via decision rules requires attaching to each information pattern over considered conditional attributes a local decision information. As an example, let us consider the notion of a generalized decision function $\partial_{d/B} : V_B^U \to 2^{V_d}$, which, for any subset $B \subseteq A$, labels information patterns with possible decision values

$$\partial_{d/B}(w_B) = \{d(u) : u \in Inf_B^{-1}(w_B)\} \tag{14}$$

For each $w_B \in V_B^U$, the subset $\partial_{d/B}(w_B) \subseteq V_d$ contains exactly these decision values which occur in the indiscernibility class of w_B. Thus, in view of propositional logic, the implication

$$(B, w_B) \Rightarrow \vee_{v_d \in \partial_{d/B}(w_B)}(d, v_d) \tag{15}$$

is in some sense optimal with respect to $\mathbf{A}$. Namely, disjunction in its right side is the minimal one which keeps consistency of the rule (15) with $\mathbf{A}$. In other words, any implication with (B, w_B) as predecessor and the disjunction of a smaller number of atomic elements than $|\partial_{d/B}(w_B)|$ as the successor is not true in the model $\mathbf{A}$.

The generalized decision function enables us to reconsider the notion of a decision reduct for inconsistent data.

Definition 5. For any decision table $\mathbf{A} = (U, A, d)$, we say that a subset $B \subseteq A$ ∂-defines d, iff for any $w_A \in V_A^U$ there is equality

$$\partial_{d/B}\left(w_A^{\downarrow B}\right) = \partial_{d/A}(w_A) \tag{16}$$

A subset $B \subseteq A$ which ∂-defines d is called a ∂-decision reduct iff none of its proper subsets has this property.

Proposition 6. *(compare with [35]) For any decision table, notions of a rough decision and a ∂-decision reduct are equivalent.*

Definition 5 describes possibly minimal subsets of conditions which preserve information generated by the generalized decision function. One can see that condition (16) is formulated in a way relative to the whole set of attributes. Obviously, given the structure of the generalized decision, we could introduce the notion of a reduct as, e.g., a subset which assigns to its information patterns appropriately small sets of possible decision values. Then, however, the problem would arise in case when even the whole of A do not satisfy such a prior criterion. Thus, it is always better to understand optimal subsets of conditions as those not decreasing information provided before feature reduction.

From equivalence provided by Proposition 6 or characteristics given by Proposition 10, one can derive the following fact:

Proposition 7. *For any decision table, the property of ∂-defining decision is closed with respect to supersets.*

2.3 Discernibility characteristics

There is analogy between subsets of conditions defining decision and implicants in boolean reasoning (see [2], [30]). One can show it in two ways. First, let us interpret attributes as boolean variables, taking logical value 1 iff information about a given attribute's value is provided. Then we can use notation "$\bigwedge_{a \in B} a \Rightarrow d$", or simply "$B \Rightarrow d$" for the whole bunch of decision rules generated by B. If we have information about attributes $a \in B$, i.e., the predecessor of implication $B \Rightarrow d$ has logical value 1, then by firing an appropriate decision rule we can uniquely specify decision value. It gives us the wanted semantics, since information about B implies information about d, i.e. boolean variable d has logical value 1 as well.

The second way of expressing the mentioned correspondence is based on the so called discernibility characteristics for exact decision reducts. According to this way, one can see that a subset $B \subseteq A$ is an exact decision reduct iff it is minimal in sense of inclusion and such that for each pair $u, u' \in U$ we have

$$d(u) \neq d(u') \Rightarrow Inf_B(u) \neq Inf_B(u') \tag{17}$$

Condition (17) can be illustrated by using the structure of $|U| \times |U|$ discernibility matrix $M(\mathbf{A}) = (c_{ij})_{i,j=1,..,|U|}$ ([35]), where for each $i, j = 1, .., |U|$ we put

$$c_{ij} = \begin{cases} \{a \in A : a(u_i) \neq a(u_j)\} & if\ d(u_i) \neq d(u_j) \\ \emptyset & otherwise \end{cases} \tag{18}$$

Then it is quite obvious that subset $B \subseteq A$ is an exact decision reduct, iff it is minimal in sense of inclusion, such that for any $i, j = 1, .., |U|$ we have implication

$$c_{ij} \neq \emptyset \Rightarrow B \cap c_{ij} \neq \emptyset \tag{19}$$

Given such a structure, we can finally consider boolean discernibility function

$$f_{\mathbf{A}}(\bar{a}_1, .., \bar{a}_{|A|}) = \bigwedge_{1 \leq i < j \leq |U|} \{\bigvee_{a \in c_{ij}} \bar{a}\} \tag{20}$$

where boolean variables $\bar{a}_1, .., \bar{a}_{|A|}$ correspond to attributes $a_1, \ldots, a_{|A|}$, respectively, and where disjunction $\bigvee_{a \in \emptyset} \bar{a}$ over empty set of variables is assumed to have logical value of truth. In [35] it is shown that the set of all prime implicants ([2]) of $f_{\mathbf{A}}$ determines the set of all exact decision reducts for $\mathbf{A}$. Thus, one can conclude that the notion of discernibility is a key for referring exact *conditions*→*decision* dependencies to paradigms of boolean reasoning.

Besides an illustration of the above correspondence, discernibility matrices have provided the starting point for development of algorithms searching for reducts (for further references see Chapter by Bazan et al. in this Volume). Although nowadays majority of implementations relies on discernibility rather in an implicit way, it still remains a fundamental tool for comparing different modifications of classical rough set notions. In our study we are going to pay a special attention to the notion of discernibility, using, however, a bit changed variant of $M(\mathbf{A})$. One can derive, e.g. from characteristics provided by the function $f_{\mathbf{A}}$, that when searching for minimal decision reducts, i.e. minimal prime implicants of $f_{\mathbf{A}}$, there is no need to repeat in (20) disjunctions corresponding to discernibility sets for different pairs of objects and that we can replace the matrix $M(\mathbf{A})$ with the discernibility table

$$T(\mathbf{A}) = \{B \subseteq A : B \neq \emptyset \wedge \exists_{i,j}(B = c_{ij})\} \tag{21}$$

or – after application of so called absorption law, which states that there is no need to keep any $B \subseteq A$ such that at least one of its proper subsets already belongs to $T(\mathbf{A})$ – with reduced table

$$T'(\mathbf{A}) = \{B \in T(\mathbf{A}) : \neg\exists_{i,j}(c_{ij} \subset B)\} \tag{22}$$

The above derivations can be concluded as follows:

Proposition 8. *Given a decision table* $\mathbf{A} = (U, A, d)$, *a subset* $B \subseteq A$ *exactly defines* d *iff it intersects with each element of discernibility table* $T'(\mathbf{A})$ *or, equivalently, iff it corresponds to a boolean implicant of the function*

$$f'_{\mathbf{A}}(\bar{a}_1, .., \bar{a}_{|A|}) = \bigwedge_{B \in T'(\mathbf{A})} \{\bigvee_{a \in B} \bar{a}\} \tag{23}$$

Discernibility tables are far more similar to structures used in current implementations of discernibility based algorithms than matrices $M(\mathbf{A})$. In fact, some optimization concerning, e.g., operations on information patterns or usage of $T'(\mathbf{A})$ in further computations, lead to satisfactory performance (cf. Chapter by Bazan et al. in this Volume). Such effective tools for finding minimal exact decision reducts are extremally valuable in view of the following result possible to be derived from the above relationships.

Theorem 9. *([35]) The problem of finding minimal exact decision reduct is NP-hard.*

The above result suggests to look at complexity connected with the size of discernibility matrices or tables from another perspective. It turns out that one cannot escape from potentially exponential time of calculations, unless he decides to apply artificial intelligence based heuristics working on reasonably filtered (or sampled) structures allowing to extract information represented within $M(\mathbf{A})$,

$T(\mathbf{A})$ or $T'(\mathbf{A})$. Since practically all proposed types approximate decision reducts remain NP-hard with respect to the problem formulated analogously as in the above theorem, it is crucial to be able to express them within the same rough set framework, based on already mentioned algorithmic tools, in order to search for reasonably sub-optimal solutions in a relatively short time.

Minimal generalized decision reducts can be searched for in the same way as exact ones. It is just enough to replace the original decision attribute d in a given $\mathbf{A} = (U, A, d)$ with the generalized decision attribute $\partial_{d/A} : U \to 2^{V_d}$ such that

$$\partial_{d/A}(u) = \partial_{d/A}(Inf_A(u)) \qquad (24)$$

According to the following well–known fact, all further calculations can be performed on the so obtained consistent decision table $\mathbf{A}_\partial = (U, A, \partial_{d/A})$.

Proposition 10. *([30]) A subset $B \subseteq A$ is a ∂-decision reduct for a given decision table $\mathbf{A} = (U, A, d)$ iff it is a decision reduct for the consistent decision table $\mathbf{A}_\partial = (U, A, \partial_{d/A})$, where the decision attribute $\partial_{d/A}$ is defined by the formula (24).*

Corollary 11. *Given a decision table $\mathbf{A} = (U, A, d)$, a subset $B \subseteq A$ ∂-defines d iff it intersects with each element of the discernibility table $T_\partial(\mathbf{A})$ given by the formula*

$$T_\partial(\mathbf{A}) = \{B \subseteq A : B \neq \emptyset \wedge \exists_{i,j}(B = c_{ij}^\partial)\} \qquad (25)$$

where

$$c_{ij}^\partial = \begin{cases} \{a \in A : a(u_i) \neq a(u_j)\} & \text{if } \partial_{d/A}(u_i) \neq \partial_{d/A}(u_j) \\ \emptyset & otherwise \end{cases} \qquad (26)$$

Remark. Discernibility table (25) can be reduced to $T'_\partial(\mathbf{A})$ by application of the same absorption law as proposed for $T(\mathbf{A})$ in (22). We will implicitly assume possibility of applying this law to analogous discernibility characteristics, formulated from now on.

Example 1. Consider the decision table $\mathbf{A} = (U, A, d)$ given in Fig. 1. One can see that the universe splits into 11 equivalence classes of $IND(A)$, corresponding to information patterns $w_1, .., w_{11} \in V_A^U$ (compare with Fig. 2). Values of the generalized decision function are equal to

$$\partial_{d/A}(w_i) = \begin{cases} \{v_1\} & for \ i = 1, 2, 3, 4, 9 \\ \{v_2\} & for \ i = 6, 8, 11 \\ \{v_1, v_2\} & for \ i = 5, 7, 10 \end{cases} \qquad (27)$$

Since the table $T_\partial(\mathbf{A})$ consists here of almost all possible subsets of A (all such subsets are presented in Fig. 4, the third column) we present it after applying the absorption law. Then we obtain a reduced table

$$T'_\partial(\mathbf{A}) = \{\{a_1, a_4\}, \{a_2\}, \{a_3\}, \{a_1, a_5\}, \{a_4, a_5\}\} \qquad (28)$$

which leads to ∂-decision reducts

$$\{a_1, a_2, a_3, a_4\}, \{a_1, a_2, a_3, a_5\}, \{a_2, a_3, a_4, a_5\} \qquad (29)$$

It can be also easily seen that above subsets are indeed minimal ones which induce the same lower and upper approximations of all decision classes as the whole of A.

A	a_1	a_2	a_3	a_4	a_5	d
u_1	0	0	0	0	0	v_1
u_2	0	0	1	0	0	v_1
u_3	0	0	1	1	0	v_1
u_4	0	1	0	1	0	v_1
u_5	0	1	1	0	0	v_1
u_6	0	1	1	0	0	v_1
u_7	0	1	1	0	0	v_2
u_8	0	1	1	1	1	v_2
u_9	1	0	0	1	0	v_1
u_{10}	1	0	0	1	0	v_1
u_{11}	1	0	0	1	0	v_2
u_{12}	1	0	0	1	0	v_2
u_{13}	1	0	1	0	1	v_2
u_{14}	1	0	1	1	0	v_1
u_{15}	1	0	1	1	0	v_1
u_{16}	1	1	0	0	0	v_1
u_{17}	1	1	0	0	0	v_1
u_{18}	1	1	0	0	0	v_1
u_{19}	1	1	0	0	0	v_2
u_{20}	1	1	1	1	1	v_2

Fig. 1. Example of decision table

As one could expect, also in case of ∂-decision reducts the time complexity of finding optimal subsets of conditions is potentially exponential.

Theorem 12. *The problem of finding minimal ∂-decision reducts is NP-hard.*

Thus, a special attention should be paid to effective implementation of above mentioned absorption laws reducing the size of discernibility tables, as well as on fast heuristics working on such structures.

2.4 Approximate discernibility

Discernibility characteristics provided by (17) for exact decision reducts can be re-formulated to state that a given $B \subseteq A$ is an exact decision reduct, iff it is minimal in the sense of inclusion and such that for any pair $u, u' \in U$ we have

$$d(u) \neq d(u') \land Inf_A(u) \neq Inf_A(u') \Rightarrow Inf_B(u) \neq Inf_B(u') \qquad (30)$$

Indeed, for consistent decision tables, where for all pairs $u, u' \in U$

$$d(u) \neq d(u') \Rightarrow Inf_A(u) \neq Inf_A(u') \qquad (31)$$

holds, implication (30) is equivalent to (17). One can realize, however, that it leads to the notion of discernibility based reduct for not necessarily consistent tables.

Definition 13. Given a decision table $\mathbf{A} = (U, A, d)$, we say that a subset $B \subseteq A$ exactly discerns d iff for each pair $u, u' \in U$ condition (30) is satisfied. A subset $B \subseteq A$ which exactly discerns d is called an exactly discerning decision reduct iff none of its proper subsets satisfies this condition.

Exactly discerning decision reducts provide us with a useful generalization. Regardless of whether a given table is consistent or not, we can still use the same structure of discernibility table as introduced in Section 2.3.

Corollary 14. *Given a (possibly inconsistent) decision table* $\mathbf{A} = (U, A, d)$, *a subset* $B \subseteq A$ *exactly discerns* d *iff it intersects with each element of the discernibility table* $T(\mathbf{A})$ *given by the formula (21)*.

Basing on discerning decision reducts, one must still remember about the original meaning of the notion of a reduct, understood as a subset which preserves the information about decision, provided by the whole set of conditions. Obviously, one could argue that Definition 13 fulfills this requirement in terms of discernibility information. However, such a kind of information is not applicable directly in constructing decision rules for new object classification.

Let us note that $T(\mathbf{A})$ for the decision table $\mathbf{A} = (U, A, d)$ presented in Fig. 1 is exactly the same as $T_\partial(\mathbf{A})$ computed in Example 1. In particular, it agrees with the following result.

Proposition 15. *For each decision table* $\mathbf{A} = (U, A, d)$, *the inclusion* $T_\partial(\mathbf{A}) \subseteq T(\mathbf{A})$ *holds.*

Corollary 16. *Each subset which exactly discerns the decision ∂-defines it as well.*

Usually, exactly discerning reducts turn out to be much longer than ∂-decision ones, because they have to discern much more object pairs. Together with mentioned lack of inconsistency interpretation necessary for new case classification, it does not prognoze well for their applications. On the other hand, for consistent decision tables, both types are equivalent to the notion of an exact decision reduct. Then, if one wants to generate any kind of shorter inconsistent decision rules, it is necessary to learn how to approximate decision information induced by particular subsets of conditions – and it turns out that the easiest initial way of introducing the notion of approximate decision reduct is via approximate discernibility operations.

Let us consider discernibility measure

$$N_\mathbf{A}(B) = \frac{|\{(u,u') \in U \times U : (d(u) \neq d(u')) \wedge (Inf_B(u) \neq Inf_B(u'))\}|}{|\{(u,u') \in U \times U : (d(u) \neq d(u')) \wedge (Inf_A(u) \neq Inf_A(u'))\}|} \quad (32)$$

labeling subsets $B \subseteq A$ with the ratio of object pairs discerned by B, which are:

1. necessary to be discerned because of belonging to different decision classes,
2. possible to be discerned by the whole of A.

It is easy to notice that a given $B \subseteq A$ exactly discerns d iff equality $N_\mathbf{A}(B) = 1$ holds. Moreover, one can consider the following notion.

Definition 17. (cf. [19]) Given a decision table $\mathbf{A} = (U, A, d)$ and an approximation threshold $\varepsilon \in [0, 1)$, we say that a subset $B \subseteq A$ ε-discerns d iff we have the inequality

$$N_\mathbf{A}(B) \geq 1 - \varepsilon \qquad (33)$$

A subset $B \subseteq A$ which ε-discerns d is called a ε-discerning decision reduct iff none of its proper subsets satisfies the analogous inequality.

The main aim of considering approximate reducts is to reduce the number of conditions while not loosing much of determinism with respect to the initial table. Above, the loss of determinism is understood in terms of the fraction of object pairs, which are wrongly left as not discerned. The following result shows that for a fixed approximation threshold $\varepsilon \in [0, 1)$ the search for minimal ε-discerning decision reducts is potentially not easier than in the exact case.

Theorem 18. (cf. [19]) For each fixed approximation threshold $\varepsilon \in [0, 1)$, the problem of finding a minimal ε-discerning decision reduct is NP-hard.

On the other hand, it seems that by intelligent tuning of ε we are more likely to obtain interesting results via appropriate heuristics. The reader is referred to [16] for some study on possible design of such heuristics, aimed there at extraction of approximate association rules from data (see also [19]).

2.5 Plausible discernibility functions

The notion of ε-discerning decision reduct relates to approximate boolean reasoning. For any non-empty $B \subseteq A$, let us define discernibility assignment

$$n(B) = \frac{|\{(u, u') \in U \times U : d(u) \neq d(u') \wedge B = c_{ij}\}|}{|\{(u, u') \in U \times U : d(u) \neq d(u') \wedge Inf_A(u) \neq Inf_A(u')\}|} \qquad (34)$$

Then it can be easily observed that discernibility measure can be equivalently rewritten by using the formula

$$N_\mathbf{A}(B) = \sum_{C \subseteq A : B \cap C \neq \emptyset} n(C) \qquad (35)$$

In fact, such an interpretation of $N_\mathbf{A}$ draws a new kind of analogy between rough sets and Dempster-Shafer theory ([29]). Namely, formula (34) can be treated as attaching to elements of 2^A basic probability assignments summing to 1. Under such a construction, formula (35) becomes to play the role of plausibility measure ([29], see also [34]).

Provided with discernibility characteristics for ∂-decision reducts, we can consider their approximations in the above sense as well. Just like before, let us rely on analogy with Dempster-Shafer theory and put, for any non-empty $B \subseteq A$, basic ∂-discernibility assignment

$$n_\partial(B) = \frac{|\{(u,u') \in U \times U : (\partial_{d/A}(u) \neq \partial_{d/A}(u')) \wedge (B = c_{ij}^\partial)\}|}{|\{(u,u') \in U \times U : (\partial_{d/A}(u) \neq \partial_{d/A}(u')) \wedge (Inf_A(u) \neq Inf_A(u'))\}|} \quad (36)$$

Definition 19. Given a decision table $\mathbf{A} = (U, A, d)$, we say that a subset $B \subseteq A$ ε-discerns $\partial_{d/A}$ iff it intersects with subsets $C \subseteq A$ of total weight

$$N_{\partial/\mathbf{A}}(B) = \sum_{C \subseteq A : B \cap C \neq \emptyset} n_\partial(C) \quad (37)$$

not less than $1-\varepsilon$. A subset $B \subseteq A$ which ε-discerns $\partial_{d/A}$ is called a ε-discerning ∂-decision reduct iff none of its proper subsets satisfies the above inequality.

Theorem 20. *For each fixed approximation threshold $\varepsilon \in [0,1)$, the problem of finding a minimal ε-discerning ∂-decision reduct is NP-hard.*

Example 2. In Fig. 2 we list subsets of conditions with positive discernibility and/or ∂-discernibility assignments, computed for the decision table from Fig. 1. In Fig. 3 we present subsets, which are the most interesting in view of requirements of Definitions 17 and 19. The second column contains the interval ranges of thresholds $\varepsilon \in [0,1)$, for which particular subsets are ε-discerning decision reducts. Similarly, the third column corresponds to the notion of ε-discerning ∂-decision reduct.

One can see that in both cases the most accurate subset (starting to be a reduct for the smallest threshold) is $\{a_2, a_3, a_4\}$. However, while increasing ε, it turns out that optimal, two-element solutions to problems of finding a minimal ε-discerning decision and ∂-decision reduct become to differ — $\{a_2, a_3, a_4\}$ becomes to be reducible to $\{a_2, a_3\}$ and to $\{a_2, a_4\}$, respectively.

The above example, particularly Fig. 2, shows that calculations on discernibility assignments are potentially more expensive than in case of exactly discerning or ∂-decision reducts. Indeed, instead of purely logical characteristics leading to analysis of discernibility tables $T'(\mathbf{A})$ or $T'_\partial(\mathbf{A})$, one has to operate over potentially all feature subsets, weighted additionally with some numerical values. Thus, further development of approximate heuristics, taking advantages of relationship to both approximate boolean reasoning and Dempster-Shafer theory is required.

3 Rough membership distributions

3.1 Conditional frequencies

In a consistent decision table $\mathbf{A} = (U, A, d)$, where each class of relation $IND(A)$ drops into one of decision classes, decision rules enable deterministic classification

$B \subseteq A$	$n(B)$	$n_\partial(B)$
$\{a_2\}$	1/82	3/117
$\{a_3\}$	4/82	8/117
$\{a_1, a_2\}$	3/82	8/117
$\{a_1, a_3\}$	7/82	4/117
$\{a_1, a_4\}$	3/82	8/117
$\{a_1, a_5\}$	1/82	1/117
$\{a_2, a_3\}$	1/82	3/117
$\{a_2, a_4\}$	9/82	3/117
$\{a_2, a_5\}$	3/82	3/117
$\{a_3, a_4\}$	1/82	3/117
$\{a_3, a_5\}$	1/82	1/117
$\{a_4, a_5\}$	4/82	5/117
$\{a_1, a_2, a_3\}$	1/82	4/117
$\{a_1, a_2, a_4\}$	2/82	6/117

$B \subseteq A$	$n(B)$	$n_\partial(B)$
$\{a_1, a_2, a_5\}$	5/82	6/117
$\{a_1, a_3, a_4\}$	2/82	4/117
$\{a_1, a_3, a_5\}$	2/82	2/117
$\{a_1, a_4, a_5\}$	3/82	4/117
$\{a_2, a_3, a_4\}$	2/82	8/117
$\{a_2, a_3, a_5\}$	5/82	8/117
$\{a_2, a_4, a_5\}$	1/82	1/117
$\{a_3, a_4, a_5\}$	5/82	8/117
$\{a_1, a_2, a_3, a_4\}$	8/82	4/117
$\{a_1, a_2, a_3, a_5\}$	1/82	4/117
$\{a_1, a_2, a_4, a_5\}$	1/82	1/117
$\{a_1, a_3, a_4, a_5\}$	3/82	4/117
$\{a_2, a_3, a_4, a_5\}$	1/82	1/117
$\{a_1, a_2, a_3, a_4, a_5\}$	2/82	2/117

Fig. 2. Positive basic discernibility and ∂-discernibility assignments.

$B \subseteq A$	"ε-interval"	"(∂, ε)-interval"
$\{a_1, a_3\}$	[18/82, 36/82)	[15/117, 49/117)
$\{a_2, a_3\}$	[11/82, 36/82)	[18/117, 49/117)
$\{a_2, a_4\}$	[15/82, 35/82)	[16/117, 52/117)
$\{a_1, a_2, a_3\}$	[4/82, 11/82)	[5/117, 15/117)
$\{a_1, a_2, a_4\}$	[5/82, 15/82)	[9/117, 16/117)
$\{a_1, a_2, a_5\}$	[5/82, 15/82)	[11/117, 25/117)
$\{a_1, a_3, a_4\}$	[4/82, 13/82)	[6/117, 15/117)
$\{a_1, a_3, a_5\}$	[10/82, 18/82)	[6/117, 15/117)
$\{a_1, a_4, a_5\}$	[6/82, 15/82)	[4/117, 26/117)
$\{a_2, a_3, a_4\}$	[1/82, 11/82)	[1/117, 16/117)
$\{a_2, a_3, a_5\}$	[3/82, 11/82)	[8/117, 18/117)
$\{a_3, a_4, a_5\}$	[4/82, 13/82)	[11/117, 21/117)

Fig. 3. Threshold ranges of satisfaction of conditions for being an ε-discerning decision and ε-discerning ∂-decision reduct.

of objects. Obviously, for new cases it is not always true that such a reasoning by analogy provides the proper result. However, at least inside known universe it is surely valid. Moreover, according to, e.g., the MDL-Principle, the search of minimal rules potentially decreases the chance of wrong classification of new objects. The situation changes if we are about to deal with non-deterministic dependencies among attributes. In case of inconsistent decision tables, we must specify the way of dealing with uncertainty – basing, e.g., on generalized decision functions – appropriately adjusted to one's understanding of reasoning with data.

Such a way of reasoning with inconsistent data is not the only possibility. To provide a wider range of reasoning strategies, let us focus on conditional representation based on rough membership functions $\mu_{d/B} : V_d \times V_B^U \to [0,1]$, defined, for any fixed $B \subseteq A$, by the formula

$$\mu_{d/B}(v_d/w_B) = \frac{|Inf_B^{-1}(w_B) \cap d^{-1}(v_d)|}{|Inf_B^{-1}(w_B)|} \qquad (38)$$

where the number of objects with the pattern value w_B on B and the decision value v_d is divided by the number of objects with the pattern value w_B on B (compare with [23]). The value of $\mu_{d/B}(v_d/w_B)$ is, actually, the conditional frequency of occurrence of the given $v_d \in V_d$ under the condition $w_B \in V_B^U$ on $B \subseteq A$. It enables to consider approximate decision rules for any combination $w_B \in V_B^U$, $v_d \in V_d$, by putting

$$(B, w_B) \Rightarrow_{\mu_{d/B}(v_d/w_B)} (d, v_d) \qquad (39)$$

where $\mu_{d/B}(v_d/w_B)$ is understood as the rule's confidence (see e.g. [1]).

While specifying a set of useful decision rules, we need not to consider all combinations of conditional information patterns and decision values. Quite a reasonable approach is to choose, for any given $w_B \in V_B^U$, the decision value with the highest conditional frequency of occurrence, i.e., such $v_d \in V_d$ that

$$\mu_{d/B}(v_d/w_B) = \max_{k=1,..,|V_d|} \mu_{d/B}(v_k/w_B) \qquad (40)$$

which enables to assign to each pattern over $B \subseteq A$ a decision value which minimizes the risk of wrong classification.

From the formal point of view, we have to take into account the situation where there are more than one decision values satisfying (40). As a consequence, we obtain the majority decision function $m_{d/B} : V_B^U \to 2^{V_d}$ such that

$$m_{d/B}(w_B) = \{v_d \in V_d : \mu_{d/B}(v_d/w_B) = \max_k \mu_{d/B}(v_k/w_B)\} \qquad (41)$$

The function $m_{d/B}$, for any given $w_B \in V_B^U$, corresponds to the generalized majority decision rule

$$(B, w_B) \Rightarrow_\Lambda \left[\vee_{v_d \in m_{d/B}(w_B)} (d, v_d)\right] \qquad (42)$$

where

$$\Lambda = |m_{d/B}(w_B)| \cdot \max_k \mu_{d/B}(v_k/w_B) \qquad (43)$$

is a frequency based chance that decision value of an object fitting information pattern w_B belongs to $m_{d/B}(w_B)$.

Definition 21. Given a decision table $\mathbf{A} = (U, A, d)$, we say that a subset $B \subseteq A$ m-defines d iff

$$m_{d/B}\left(w_A^{\downarrow B}\right) = m_{d/A}(w_A) \qquad (44)$$

for any $w_A \in V_A^U$. A subset B which m-defines d is called an m-decision reduct, iff no proper subset of it m-defines d.

Obviously, one can imagine many other ways of constructing sets of possible answers using conditional frequencies. For example, let us present yet another method, being a combination of considered above. For any $B \subseteq A$, we define the function $m_{d/B \geq 0.7} : V_B^U \to 2^{V_d}$ by putting

$$m_{d/B \geq 0.7}(w_B) = \begin{cases} m_{d/B}(w_B) & if \ \max_k \mu_{d/B}(v_k/w_B) \geq 0.7 \\ \partial_{d/B}(w_B) & otherwise \end{cases} \quad (45)$$

and reconsider the notion of decision reduct in terms of comparing subsets of decision values generated by $m_{\geq 0.7}$. Definition of $m_{\geq 0.7}$-decision reduct is completely analogous to Definitions 5 and 21. Moreover, m-decision and $m_{\geq 0.7}$-decision reducts have analogous discernibility characteristics as in case of generalized decision functions. Let us present it as an example for m:

Proposition 22. *A subset $B \subseteq A$ is an m-decision reduct for a given decision table $\mathbf{A} = (U, A, d)$ iff it is a decision reduct of the consistent decision table $\mathbf{A}_m = (U, A, m_{d/A})$, where the decision attribute $m_{d/A}$ is defined analogously as in case of the function ∂ in the formula (24).*

Corollary 23. *Given a decision table $\mathbf{A} = (U, A, d)$, a subset $B \subseteq A$ m-defines d iff it intersects with each element of the discernibility table $T_m(\mathbf{A})$ given by the formula*

$$T_m(\mathbf{A}) = \{B \subseteq A : B \neq \emptyset \wedge \exists_{i,j}(B = c_{ij}^m)\} \quad (46)$$

where

$$c_{ij}^m = \begin{cases} \{a \in A : a(u_i) \neq a(u_j)\} & if \ m_{d/A}(u_i) \neq m_{d/A}(u_j) \\ \emptyset & otherwise \end{cases} \quad (47)$$

Besides practical gains from the above corollary, one can observe that such a discernibility characteristics for any kind of defining decision implies its closeness with respect to supersets.

Corollary 24. *For any decision table, the property of m-defining decision is closed with respect to supersets.*

It enables us to look at the process of finding minimal m-decision reducts as related to the MDL-Principle similarly as in case of exact reducts.

Example 3. Let us rewrite the decision table presented in Fig. 1 by focusing on A-based information patterns, i.e. $IND(A)$-classes, instead of on objects. In Fig. 4 each information pattern w_i, $i = 1,..,11$, is labeled with object support "#", i.e. the number of objects satisfying it. Moreover, values of functions $\partial_{d/A}$, $m_{d/A}$, $m_{d/A \geq 0.7}$, as well as values of rough membership function for each pattern are given. After an application of absorption, the reduced discernibility table for m becomes to have the form

$$T'_m(\mathbf{A}) = \{\{a_1, a_2\}, \{a_1, a_4\}, \{a_1, a_5\}, \{a_2, a_4\}, \{a_2, a_5\}, \{a_3\}, \{a_4, a_5\}\} \quad (48)$$

and the table $T'_{m \geq 0.7}(\mathbf{A})$ turns out to be equal to $T'_\partial(\mathbf{A})$ given by (28). The obtained m-decision and $m_{\geq 0.7}$-decision reducts are presented – together with ∂-decision ones – in Fig. 5.

V_A^U	#	a_1	a_2	a_3	a_4	a_5	∂	m	$m_{\geq 0.7}$	$\mu_{d/A}(v_i/\cdot), i=1,2$
w_1	1	0	0	0	0	0	$\{v_1\}$	$\{v_1\}$	$\{v_1\}$	$(1,0)$
w_2	1	0	0	1	0	0	$\{v_1\}$	$\{v_1\}$	$\{v_1\}$	$(1,0)$
w_3	1	0	0	1	1	0	$\{v_1\}$	$\{v_1\}$	$\{v_1\}$	$(1,0)$
w_4	1	0	1	0	1	0	$\{v_1\}$	$\{v_1\}$	$\{v_1\}$	$(1,0)$
w_5	3	0	1	1	0	0	$\{v_1,v_2\}$	$\{v_1\}$	$\{v_1,v_2\}$	$(2/3,1/3)$
w_6	1	0	1	1	1	1	$\{v_2\}$	$\{v_2\}$	$\{v_2\}$	$(0,1)$
w_7	4	1	0	0	1	0	$\{v_1,v_2\}$	$\{v_1,v_2\}$	$\{v_1,v_2\}$	$(1/2,1/2)$
w_8	1	1	0	1	0	1	$\{v_2\}$	$\{v_2\}$	$\{v_2\}$	$(0,1)$
w_9	2	1	0	1	1	0	$\{v_1\}$	$\{v_1\}$	$\{v_1\}$	$(1,0)$
w_{10}	4	1	1	0	0	0	$\{v_1,v_2\}$	$\{v_1\}$	$\{v_1\}$	$(3/4,1/4)$
w_{11}	1	1	1	1	1	1	$\{v_2\}$	$\{v_2\}$	$\{v_2\}$	$(0,1)$

Fig. 4. Decision table in terms of information patterns on A, with their object supports, set values of considered decision functions and values of rough membership function.

$B \subseteq A$	∂-reduct	m-reduct	$m_{\geq 0.7}$-reduct
$\{a_1,a_2,a_3,a_4\}$	+	+	+
$\{a_1,a_2,a_3,a_5\}$	+	+	+
$\{a_1,a_3,a_4,a_5\}$	−	+	−
$\{a_2,a_3,a_4,a_5\}$	+	+	+

Fig. 5. Generalized, majority and $m_{\geq 0.7}$-decision reducts.

According to Proposition 22, it is also possible to consider notions of ε-discerning m-decision (and also $m_{\geq 0.7}$-decision, according to a similar result) reducts, understood analogously as in Definition 19 for function ∂.

3.2 Frequency based decision reducts

In previous section we discussed different notions of decision reducts, corresponding to different ways of interpreting frequency based information. Now, let us consider possibilities of reasoning with frequencies in a more straightforward manner.

Given a fixed linear ordering $V_d = \langle v_1, ..., v_{|V_d|} \rangle$ let us apply the notion of the rough membership distribution function $\mu_{d/B} : V_B^U \to \triangle_{|V_d|-1}$, such that

$$\mu_{d/B}(w_B) = \langle \mu_{d/B}(v_1/w_B), ..., \mu_{d/B}(v_{|V_d|}/w_B) \rangle \qquad (49)$$

where

$$\triangle_{|V_d|-1} = \{s = \langle s_1, ..., s_{|V_d|} \rangle : s_1 \geq 0, .., s_{|V_d|} \geq 0, s_1 + .. + s_{|V_d|} = 1\} \qquad (50)$$

denotes ($|V_d|$-1)-dimensional simplex. Then, each vector $\mu_{d/B}(w_B)$ is said to induce the random μ-decision rule

$$(B, w_B) \Rightarrow \vee_{k=1,..,|V_d|} (d, v_k)_{\mu_{d/B}(v_k/w_B)} \qquad (51)$$

which attaches to $w_B \in V_B^U$ information concerning degrees $\mu_{d/B}(v_k/w_B)$ of hitting of $Inf_B^{-1}(w_B)$ into particular decision classes (see also [42]).

Rough membership distributions can be regarded as frequency based source of statistical estimation of joint probabilistic distributions ([46]). From this point of view, the following notion is very important, as closely related to probabilistic conditional independence ([24], [39]).

Definition 25. Given a decision table $\mathbf{A} = (U, A, d)$, we say that $B \subseteq A$ μ-defines attribute $d \notin A$ in $\mathbf{A}$, iff for each $w_A \in V_A^U$ we have

$$\mu_{d/B}\left(w_A^{\downarrow B}\right) = \mu_{d/A}(w_A) \qquad (52)$$

A subset $B \subseteq A$ which μ-defines d is called a μ-decision reduct iff it has no proper subset which μ-defines d.

The correspondence between Definition 25 and probability theory can be described as follows (cf. [24]):

1. A subset $B \subseteq A$ which μ-defines d can be regarded as such that d is conditionally independent on $A \setminus B$ under B — if we treat $\mathbf{A}$ as the generator of the product probabilistic space over $A \cup \{d\}$, then condition (52) coincides with such an independence statement.
2. A subset $B \subseteq A$ which is a μ-decision reduct can be regarded as the Markov boundary of d with respect to A.

Such analogy seems to be very important for applications based on probability and statistics, which use as basic the notion of conditional independence (like, e.g., the task of extracting optimal Bayesian nets from data in [39]). Development of studies on extraction of μ-decision reducts ([37], [38], [42]) by adaptation of rough set based algorithmic framework enables to use it to support computational aspects of applications related to Markov boundaries.

For consistent decision tables Definitions 1 and 25 are equivalent. Indeed, in such case, the distribution $\mu_{d/A}(w_A)$ corresponds for each $w_A \in V_A^U$ to a unique vertex of $\triangle_{|V_d|-1}$ and thus it must be also the case for $\mu_{d/B}\left(w_A^{\downarrow B}\right)$, if equality (52) is satisfied. In particular it implies that:

Theorem 26. *The problem of finding minimal μ-decision reduct is NP-hard.*

The result presented below implies that the search for minimal μ-decision reducts can be performed just like in consistent case of exact decision reducts again.

Proposition 27. *([38]) Given a decision table $\mathbf{A} = (U, A, d)$, a subset $B \subseteq A$ and $w_B \in V_B^U$, let us put*

$$V_A^U(w_B) = \left\{w_A \in V_A^U : w_A^{\downarrow B} = w_B\right\} \qquad (53)$$

as the set of all extensions of w_B on V_A^U. Then, for each $v_d \in V_d$, we have inequalities

$$\min_{w_A \in V_A^U(w_B)} \mu_{d/A}(v_d/w_A) \leq \mu_{d/B}(v_d/w_B) \leq \max_{w_A \in V_A^U(w_B)} \mu_{d/A}(v_d/w_A) \qquad (54)$$

Given (54), it is easy to realize what follows:

Proposition 28. *([38]) For a decision table $\mathbf{A} = (U, A, d)$, a subset $B \subseteq A$ is a μ-decision reduct iff it is a decision reduct of the consistent decision table $\mathbf{A}_\mu = (U, A, \mu_{d/A})$, where the decision attribute d is replaced by the rough membership distribution attribute $\mu_{d/A} : U \to \triangle_{|V_d|-1}$, defined by the formula*

$$\mu_{d/A}(u) = \mu_{d/A}(Inf_A(u)) \tag{55}$$

Corollary 29. *For a decision table $\mathbf{A} = (U, A, d)$, a subset $B \subseteq A$ μ-defines d iff it intersects with each element of the discernibility table $T_\mu(\mathbf{A})$ given by the formula*

$$T_\mu(\mathbf{A}) = \{B \subseteq A : B \neq \emptyset \wedge \exists_{i,j}(B = c_{ij}^\mu)\} \tag{56}$$

where

$$c_{ij}^\mu = \begin{cases} \{a \in A : a(u_i) \neq a(u_j)\} & \text{if } \mu_{d/A}(u_i) \neq \mu_{d/A}(u_j) \\ \emptyset & \text{otherwise} \end{cases} \tag{57}$$

Just like previously, calculations can be equivalently performed over the reduced table

$$T'_\mu(\mathbf{A}) = \{B \in T_\mu(\mathbf{A}) : \neg \exists_{i,j}(c_{ij}^\mu \subset B)\} \tag{58}$$

Corollary 30. *For any decision table, the property of μ-defining decision is closed with respect to supersets.*

Relation of μ-decision reducts to previous generalizations of exact decision reducts onto inconsistent data is the following:

Proposition 31. *For any decision table $\mathbf{A} = (U, A, d)$ we have inclusions*

$$T_\partial(\mathbf{A}), T_m(\mathbf{A}), T_{m \geq 0.7}(\mathbf{A}) \subseteq T_\mu(\mathbf{A}) \subseteq T(\mathbf{A}) \tag{59}$$

Corollary 32. *Each subset which exactly discerns decision μ-defines it as well. Each subset which μ-defines decision ∂-defines, m-defines and $m_{\geq 0.7}$-defines it.*

Example 4. For the decision table presented in Fig. 4 – where the last two columns can be regarded as storing values of μ-decision attribute defined by (55) – the reduced table $T'_\mu(\mathbf{A})$ is equal to $T'_\partial(\mathbf{A})$ and $T'_{m \geq 0.7}(\mathbf{A})$. However, for real-life data with more decision classes μ-discernibility tables are usually much longer than in case of ∂, $m_{\geq 0.7}$ or m.

3.3 Principles of weighted classification

Equipped with methods for extracting minimal μ-decision reducts from data, let us come back to the task of classification of new objects. Each μ-decision reduct $B \subseteq A$ generates μ-decision rules of the form (51). Given a new object $u_{new} \notin U$ which agrees with values of an information pattern $w_B \in V_B^U$ over B, we are likely to attach to it rough membership distribution $\mu_{d/B}(w_B)$ as expressing chances of dropping of u_{new} into particular decision classes. Equality (52) assures us that if u_{new} agreed also with some extension $w_A \in V_A^U(w_B)$ of

w_B, then it would be attached with the same random distribution $\mu_{d/A}(w_A)$, while reasoning relying on the whole A. Thus, the idea of searching for possibly minimal μ-decision reducts is to generate the same frequency based answers for cases which would be classified in terms of A and, in parallel, to increase chances of applicability to the rest of yet unknown domain.

If we base classification of new objects on just one μ-decision reduct, then at most one rule can be fired for a given u_{new} and reasoning is finished by providing the user with distribution of chances. The situation changes when we choose for classification more than one subset of conditions, i.e. when we have at disposal classification scheme $S \subseteq 2^A$ of the form

$$S = \{R_1, .., R_{|S|}\} \qquad (60)$$

where subsets $R_i \subseteq A$, $i = 1, .., |S|$ are, e.g., minimal μ-decision reducts. In such a case the chance that a new object u_{new} will fit at least one element of S becomes higher but, on the other hand, u_{new} can fire more than one μ-decision rule and the global answer must be somehow negotiated.

Literature provides many techniques of such negotiations (see Chapter by Bazan et al. in this Volume for references). Here we propose an approach which seems to fit frequency based reasoning in view of basic intuitions as well as analogies with other probabilistic methods. However, although our proposal is good to compare different approximations of reducts in terms of a common classification scheme, further experimental research is needed to find out actually best framework for negotiations among frequency decision rules.

In general, negotiations should take into account the strength of decision rules. One of the most common ways of defining such a strength for any kind of rule with predecessor of the form "(B, w_B)" concerns the number of supporting objects, represented in a relative way by the prior frequency

$$\mu_B(w_B) = \frac{\left|Inf_B^{-1}(w_B)\right|}{|U|} \qquad (61)$$

It seems to reflect a potential importance of a given rule since new examples are supposed to fit it with probability estimated just by $\mu_B(w_B)$. Moreover, such a quantity relates to the belief attached to a rule, because larger support provides more representative sample in a statistical sense (compare with [46]). Obviously, such an argumentation might contradict with real-life situations when talking about statistically representative samples is often a very controversial subject. Nevertheless, quantities of the form (61) remain the most straightforward way of expressing weights of decision rules, especially these based on frequency information.

Given classification scheme of the form (60) consisting of μ-decision reducts, given a new object $u_{new} \notin U$ with information $w_B \in \prod_{a \in B} V_a$ provided on a subset $B \subseteq A$, we are going to weight its chances of belonging to particular decision classes by the formula

$$W_S(v_d/w_B) = \sum_{i: w_B^{\downarrow R_i} \in V_{R_i}^U} \mu_{R_i}\left(w_B^{\downarrow R_i}\right) \cdot \mu_{d/R_i}\left(v_d/w_B^{\downarrow R_i}\right) \qquad (62)$$

and answer, e.g., with the strongest decision value, given by

$$d(u_{new}/w_B) = \max_{k=1,..,|V_d|} W_S(v_k/w_B) \qquad (63)$$

Actually, we can try to perform such a weighted classification also by using ∂-decision, m-decision or, e.g., $m_{\geq 0.7}$-decision. The only thing we must do is to express decision value sets induced by them as distributions. A simple idea is to represent sets as normalized characteristic functions. According to this, let us consider for example the generalized decision function and put

$$\partial^\mu_{d/B}(w_B) = \left\langle \partial^\mu_{d/B}(v_1/w_B), .., \partial^\mu_{d/B}(v_{|V_d|}/w_B) \right\rangle \qquad (64)$$

where, for $k = 1, .., |V_d|$, we put

$$\partial^\mu_{d/B}(v_k/w_B) = \begin{cases} |\partial_{d/B}(w_B)|^{-1} & if\ v_k \in \partial_{d/B}(w_B) \\ 0 & otherwise \end{cases} \qquad (65)$$

One can see that condition (16) in Definition 5 of ∂-decision reduct can be equivalently formulated in terms of $\partial^\mu_{d/B}$ instead of $\partial_{d/B}$. Thus, we do not change the sense of the notion of ∂-decision reduct by comparing generalized decision distributions of the form (64) instead of original decision value sets. Obviously, the same trick can be used also for functions m and $m_{\geq 0.7}$. So, we can proceed with classification based on normalized weights of different origin.

It is worth remembering – while classifying new cases with respect to different kinds of reducts within a one scheme – that for a given new case each fired decision rule should be of the same type as a reduct which has generated it. Thus, let us use the notation

$$S = \{\langle R_i, kind(i) \rangle\}_{i=1,..,|S|} \qquad (66)$$

where function $kind$ labels subsets of conditions with abbreviations of corresponding reduct types, like "∂", "m", etc.

Example 5. From the previous example we know that for the decision table presented in Fig. 1 the set of μ-decision reducts equals the sets of ∂-decision and $m_{\geq 0.7}$-decision reducts. For a better illustration, let us treat each of elements as a reduct of another kind and – according to the table presented in Fig. 5 – consider four-element classification scheme

$$S = \{\langle B_1, \partial \rangle, \langle B_2, \mu \rangle, \langle B_3, m \rangle, \langle B_4, m_{\geq 0.7} \rangle\} \qquad (67)$$

where

$$\begin{array}{ll} B_1 = \{a_1, a_2, a_3, a_4\} & B_2 = \{a_1, a_2, a_3, a_5\} \\ B_3 = \{a_1, a_3, a_4, a_5\} & B_4 = \{a_2, a_3, a_4, a_5\} \end{array} \qquad (68)$$

Assume that information about a new object $u_{new} \notin U$ is provided over the whole of A, as

$$w_A = (\langle a_1, 0 \rangle, \langle a_2, 1 \rangle, \langle a_3, 1 \rangle, \langle a_4, 1 \rangle, \langle a_5, 0 \rangle) \qquad (69)$$

One can see that only first three elements of S induce decision rules possible to be fired for u_{new}. Thus, the formula for weighting decision classes takes the form of

$$W_S(v_k/w_A) = \mu_{B_1}(0,1,1,1) \cdot \partial^\mu_{d/B_1}(v_k/0,1,1,1) \\ + \mu_{B_2}(0,1,1,0) \cdot \mu_{d/B_2}(v_k/0,1,1,0) \\ + \mu_{B_3}(0,1,1,0) \cdot m^\mu_{d/B_3}(v_k/0,1,1,0) \quad (70)$$

for $k = 1,2$, where values of normalized majority distribution m^μ_{d/B_3} are defined analogously as it was done in (64) for generalized decision. As a result we obtain that

$$W_S(v_1/w_A) = \frac{1}{20} \cdot 0 + \frac{3}{20} \cdot \frac{2}{3} + \frac{1}{20} \cdot 1 = \frac{3}{20} \quad (71)$$

is a bit higher than

$$W_S(v_2/w_A) = \frac{1}{20} \cdot 1 + \frac{3}{20} \cdot \frac{1}{3} + \frac{1}{20} \cdot 0 = \frac{2}{20} \quad (72)$$

However, one can also base a classification scheme on approximate decision reducts and, e.g., replace in (67) pair $\langle B_1, \partial \rangle$ with $\langle B'_1, \partial \rangle$ for $B'_1 = \{a_2, a_3, a_4\}$ known as a ε-discerning ∂-decision reduct of a high quality (see Fig. 3). Then, after repeating calculations under such a change, both decision weights become to be equal to 3/20.

The proposed classification scheme enables to handle different kinds of decision rules within the same framework. It agrees with recent tendency of generalization of fundamental rough set notions by their parameterizing, appropriately for particular applications (see e.g. [7], [33], [50]). Here, parameters of a model of classification refer to choosing different kinds of reducts as corresponding to different kinds of expressing inexact decision information. Tuning coefficients specific for particular reduct types – like, e.g., the choice of threshold 0.7 in case of function $m_{\geq 0.7}$ – is also possible. Since knowledge about data is usually insufficient for deciding which configuration of coefficients is optimal for particular data, there arises the need of adaptive way of searching for models being most robust to noise, fluctuations or dynamic changes in data (see e.g. [7], [13], [50]). It leads to the search through the space of parameterized models, evaluated due to results of classification based on optimal decision reducts derived under particular specifications of above factors and coefficients.

As a partial conclusion here, let us note that one can treat the following study on different kinds of reducts as developing more reasonable handlers aiming at parameterized description of the most effective models for inexact knowledge representation. Such handlers are possible to be combined with other aspects of rough set applications, like, e.g., these devoted to specifying the way of information synthesis itself ([7]), within a unified adaptive system. A general parameterized classification model can be, for instance, described as

$$S = \left\{ \langle R_i, \overrightarrow{kind}(i), \alpha_i \rangle \right\}_{i=1,...,|S|} \quad (73)$$

where $\overrightarrow{kind}$ is completed with additional generation parameters (like 0.7 in case of $m_{\geq 0.7}$). Coefficients $\alpha_i \in [0,1]$ are prior classification weights of rules induced by particular $R_i \subseteq A$. It corresponds to calculation of decision beliefs according to formula

$$W_S(v_d/w_B) = \sum_{i: w_B^{\downarrow R_i} \in V_{R_i}^U} \alpha_i \cdot \mu_{R_i}\left(w_B^{\downarrow R_i}\right) \cdot kind(i)_{d/R_i}^{\mu}\left(v_d/w_B^{\downarrow R_i}\right) \quad (74)$$

4 Approximations of frequency based information

4.1 Normalized distance measures

Talking about frequencies, we stated so far notions of decision reduct as strictly preserving particular types of information. On the other hand, we mentioned that a kind of approximation of criteria for considered m-decision or $m_{\geq 0.7}$-decision reducts are possible due to their discernibility characteristics, like it was proposed for ∂-decision in Section 2.4. According to Corollary 29, the same approach can be applied to μ-decision reducts. Then, we would obtain possibility of searching for subsets of conditions preserving information induced by rough membership functions almost everywhere, along known universe.

A certain difference between approaches to decision approximation based on frequencies and their interpretations – described so far by set functions ∂, m or $m_{\geq 0.7}$ – lays in the fact that these first correspond to very detailed information, too accurate to deal with dynamical data changes or other problems mentioned at the and of previous section. Indeed, if one wants to design a classification algorithm being dynamically reconfigured by new cases, then it must respond to deviations of values of rough membership distributions. Intuitively, small deviations should not have impact on the overall outcomes of weighted classification. However, the scheme of classification based on minimal μ-decision reducts may react in a very unstable way due to the fact any new case can influence the satisfaction of requirement of the accurate frequency distribution preserving.

One can see that relaxation of criteria for μ-decision reducts by applying so far proposed approximate discernibility approach does not solve the above problem. However, in case of rough membership distributions we have much more possibilities, since they can fill practically the whole subspace determined by $(|V_d|-1)$-dimensional simplex $\triangle_{|V_d|-1}$. It enables to introduce the whole class of intuitive approximations parameterized by the choice of the way of measuring the distance between distributions and by adjusting levels of thresholds up to which we agree to regard close distributions as practically indistinguishable.

Definition 33. Given an integer $n \geq 2$ and a simplex $\triangle_{n-1}$, we call a function $\rho : \triangle_{n-1} \times \triangle_{n-1} \to [0,1]$ an n-dimensional normalized distance measure iff

1. It satisfies usual metric conditions for distance measures and, moreover, it is normalized in the sense that its value for each pair of simplex vertices is equal to 1.

2. Its value does not depend on any permutation of coordinates, i.e., given any $s, s' \in \Delta_{n-1}$ and a permutation σ which defines simplex elements of the form

$$\sigma(s) = \langle s_{\sigma(1)}, .., s_{\sigma(n)} \rangle \qquad (75)$$

we have the equality $\rho(s, s') = \rho(\sigma(s), \sigma(s'))$.

We are going to talk about normalized distance measures without specifying their dimension, when it is anyway known in a given context.

Definition 34. Given a decision table $\mathbf{A} = (U, A, d)$, a normalized distance measure $\rho : \Delta_{|V_d|-1} \times \Delta_{|V_d|-1} \to [0, 1]$ and an approximation threshold $\varepsilon \in [0, 1)$, we say that a subset $B \subseteq A$ ($\rho \leq \varepsilon$)-approximately μ-defines d iff

$$\rho\left(\mu_{d/B}\left(w_A^{\downarrow B}\right), \mu_{d/A}(w_A)\right) \leq \varepsilon \qquad (76)$$

for each $w_A \in V_A^U$. Any subset $B \subseteq A$ satisfying the above condition is called a ($\rho \leq \varepsilon$)-approximating μ-decision reduct, iff none of its proper subsets ($\rho \leq \varepsilon$)-approximately μ-defines d.

Above approximation of original criterion (52) concerns, actually, specification of the tolerance relation, responsible for differing negligible noises or deviations from situations which do require consideration while the process of preserving frequency based information about decision under conditional feature reduction. Such an approach remains in a partial analogy with techniques concerning tolerance based reducts and rules (see e.g. [15], [36]), where similarity measures and thresholds are tuned to define indiscernibility classes being most accurate for description of decision values.

Considering thresholds $\varepsilon \in [0, 1)$ makes the search for relevant subsets of conditions to be more flexible, because of possibility of tuning the balance between their cardinality and degree of approximation. Still, one should remember that in view of both computational and methodological advantages it would be good to provide the above notions with some discernibility characteristics. From this point of view, the most optimistic hypothesis would concern the correspondence between Definition 34 and the following.

Definition 35. Given a decision table $\mathbf{A} = (U, A, d)$, a normalized distance measure ρ and $\varepsilon \in [0, 1)$, we say that a subset $B \subseteq A$ ($\rho > \varepsilon$)-discerns $\mu_{d/A}$ iff for any pair of information patterns $u, u' \in U$, we have the implication

$$\rho\left(\mu_{d/A}(u), \mu_{d/A}(u')\right) > \varepsilon \Rightarrow Inf_B(u) \neq Inf_B(u') \qquad (77)$$

A subset $B \subseteq A$ satisfying the above condition is called a ($\rho > \varepsilon$)-discerning μ-decision reduct iff none of its proper subsets ($\rho > \varepsilon$)-discerns $\mu_{d/A}$.

Normalized character of ρ assures us that both ($\rho \leq \varepsilon$)-approximating and ($\rho > \varepsilon$)-discerning μ-decision reducts coincide with exact decision reducts over any consistent decision table. Thus, we have the following:

Theorem 36. *([43]) Given a fixed $\varepsilon \in [0,1)$ and a normalized distance measure ρ, the problems of finding minimal $(\rho \leq \varepsilon)$-approximating and $(\rho > \varepsilon)$-discerning μ-decision reduct are NP-hard.*

On the other hand, at least in that second case, we can operate with discernibility table representation of the form

$$T_\mu^{\rho>\varepsilon}(\mathbf{A}) = \{B \subseteq A : B \neq \emptyset \wedge \exists_{i,j}(B = c_{ij}^{\rho>\varepsilon})\} \tag{78}$$

where

$$c_{ij}^{\rho>\varepsilon} = \begin{cases} \{a \in A : a(u_i) \neq a(u_j)\} & if \ \rho(\mu_{d/A}(u_i), \mu_{d/A}(u_j)) > \varepsilon \\ \emptyset & otherwise \end{cases} \tag{79}$$

to develop searching heuristics for suboptimal solutions of the problem concerning minimal $(\rho > \varepsilon)$-discerning μ-decision reducts. Similarly as in the case of ε-discerning decision reducts, such a search process may be simplified by a potential decrease of the number of object pairs necessary to be discerned under the increase of $\varepsilon \in [0,1)$. Additional gain, however, is that now it leads directly to a decrease of the number of sets in corresponding discernibility table. Thus, in many situations, appropriate tuning of ε enables us to search for reasonably sub-optimal solutions over structures of substantially smaller size.

The choice of ρ and ε is crucial for drawing the correspondence between mechanisms of approximation and discernibility, expressed, respectively, by Definitions 34 and 35. To give an example of possible relationship, let us consider a normalized distance measure Max defined by putting for each two elements $s, s' \in \Delta_{|V_d|-1}$ the value

$$Max(s, s') = \max_{k=1,\ldots,|V_d|} |s_k - s'_k| \tag{80}$$

Inequalities (54) from Proposition 27 allow for a proof of the correspondence presented below.

Proposition 37. *([43]) Given a decision table $\mathbf{A} = (U, A, d)$, any subset $B \subseteq A$ which $(Max > \varepsilon)$-discerns $\mu_{d/A}$ $(Max \leq \varepsilon)$-approximately μ-defines d as well. On the other hand, each $B \subseteq A$ which does not $(Max > 2 \cdot \varepsilon)$-discerns $\mu_{d/A}$ cannot $(Max \leq \varepsilon)$-approximately μ-define d.*

The above result gives just an approximate relationship between notions presented in Definitions 34 and 35. However, it enables to develop heuristics working over discernibility tables $T_\mu^{Max>\varepsilon}(\mathbf{A})$, defined like in (78) which search for reasonably sub-optimal solutions of the problem of finding minimal $(Max \leq \varepsilon)$-approximating μ-decision reducts.

Example 6. For any $\varepsilon \in [0,1)$, the discernibility table $T_\mu^{Max>\varepsilon}(\mathbf{A})$ corresponding to the decision table presented in Fig. 1 contains attribute subsets $\{a_1, a_5\}$, $\{a_2, a_5\}$, $\{a_3, a_5\}$ and $\{a_4, a_5\}$, necessary to discern pairs of patterns with rough membership distributions being different vertices of the decision simplex Δ_1. In Fig. 6, one can find pattern pairs necessary and satisfactory to be additionally considered for different approximation thresholds $\varepsilon \in (0,1)$, to obtain $(Max \leq \varepsilon)$-approximating μ-decision reducts.

In Fig. 7 we present reduced discernibility tables and $(Max \leq \varepsilon)$-discerning μ-decision reducts for particular approximation thresholds $\varepsilon \in (0,1)$. It turns out that $\{a_1, a_2, a_3, a_4\}$ is the most "stable" subset of conditions, being a $(Max > \varepsilon)$-discerning μ-decision reduct for any $\varepsilon \in [0,1)$.

Attribute subsets being optimal in terms of cardinality for higher thresholds are $\{a_1, a_3, a_5\}$ and $\{a_5\}$. First of them is a $(Max > \varepsilon)$-discerning μ-decision reduct for $\varepsilon \in [1/3, 1/2)$ and one can see that it $(Max \leq \varepsilon)$-approximately μ-defines d for any $\varepsilon \geq 1/5$. Subset $\{a_5\}$ becomes to be enormously valuable in terms of discernibility for thresholds not less than $1/2$. It turns out that it $(Max \leq \varepsilon)$-approximately μ-defines d for any $\varepsilon \geq 6/17$.

$\{w_i, w_j\}$	$Max(\mu_{d/A}(w_i), \mu_{d/A}(w_j))$	$\{a \in A : w_i^{\downarrow\{a\}} \neq w_j^{\downarrow\{a\}}\}$
$\{w_1, w_7\}$	$1/2$	$\{a_1, a_4\}$
$\{w_2, w_5\}$	$1/3$	$\{a_2\}$
$\{w_4, w_7\}$	$1/2$	$\{a_1, a_2\}$
$\{w_7, w_9\}$	$1/2$	$\{a_3\}$

Fig. 6. Pairs of information patterns necessary and satisfactory to be additionally considered for positive thresholds.

approximation threshold	elements of discernibility table
$0 \leq \varepsilon < 1/3$	$\{a_1, a_4\}, \{a_2\}, \{a_3\}, \{a_1, a_5\}, \{a_4, a_5\}$
$1/3 \leq \varepsilon < 1/2$	$\{a_1, a_2\}, \{a_1, a_4\}, \{a_1, a_5\}, \{a_2, a_5\}, \{a_3\}, \{a_4, a_5\}$
$1/2 \leq \varepsilon < 1$	$\{a_1, a_5\}, \{a_2, a_5\}, \{a_3, a_5\}, \{a_4, a_5\}$
approximation threshold	approximate μ-decision reducts
$0 \leq \varepsilon < 1/3$	$\{a_1, a_2, a_3, a_4\}, \{a_1, a_2, a_3, a_5\}, \{a_2, a_3, a_4, a_5\}$
$1/3 \leq \varepsilon < 1/2$	$\{a_1, a_2, a_3, a_4\}, \{a_1, a_3, a_5\}, \{a_2, a_3, a_4, a_5\}$
$1/2 \leq \varepsilon < 1$	$\{a_1, a_2, a_3, a_4\}, \{a_5\}$

Fig. 7. Discernibility tables and $(Max > \varepsilon)$-discerning μ-decision reducts for different thresholds $\varepsilon \in [0,1)$.

4.2 Normalized decision functions

In Section 3.3 we discussed how to use decision set based reducts during the process of negotiable classification. It turned out that the easiest way was to interpret the values of ∂, m and $m_{\geq 0.7}$-decision functions as uniform distributions of chances over engaged decision classes. One might regard it as an example of interpreting frequency based information due to parameters concerning the process of searching for most accurate inconsistency representation or, simply, due to intuition of a given user.

Labeling information patterns with distributions of chances provides quite a universal tool for comparing the performance of different approaches to reasoning. Rough membership functions corresponding to particular information patterns express the most accurate knowledge about *conditions→decision* dependencies and can be successfully applied to the construction of inexact classifiers. However, if one wants to reason under another interpretation of conditional information, he is able to model it by "forgetting" a part of frequency based knowledge which is redundant with respect to a given reasoning strategy. Below we argue that such an approach enables to use frequencies in many different ways, aimed at data cleaning and preprocessing.

Definition 38. Given an integer $n \geq 2$ and a simplex $\triangle_{n-1}$, we call $\phi : \triangle_{n-1} \to \triangle_{n-1}$, with values represented as

$$\phi(s) = \langle \phi(s)_1, ..., \phi(s)_n \rangle \tag{81}$$

an n-dimensional normalized decision function iff it satisfies the logical consistency assumption

$$\forall_{k=1,..,n} [(s_k = 0) \Rightarrow (\phi(s)_k = 0)] \tag{82}$$

the monotonic consistency assumption

$$\forall_{k,l=1,..,n} [(s_k \leq s_l) \Rightarrow (\phi(s)_k \leq \phi(s)_l)] \tag{83}$$

and the commutative consistency assumption which states that for any permutation σ of the form (75) we have

$$\phi(\sigma(s)) = \sigma(\phi(s)) \tag{84}$$

Given a decision table $\mathbf{A} = (U, A, d)$, a function $\phi \in NDF(|V_d|)$ and a subset $B \subseteq A$, we define the $\phi_{d/B}$-decision function $\phi_{d/B} : V_B^U \to \triangle_{|V_d|-1}$, with values represented as

$$\phi_{d/B}(w_B) = \langle \phi_{d/B}(v_1/w_B), ..., \phi_{d/B}(v_{|V_d|}/w_B) \rangle \tag{85}$$

by combination $\phi_{d/B} = \phi \circ \mu_{d/B}$, i.e., by putting for each $k = 1,..,|V_d|$

$$\phi_{d/B}(v_k/w_B) = (\phi(\mu_{d/B}(w_B)))_k \tag{86}$$

We will denote the family of n-dimensional normalized decision functions as $NDF(n)$, i.e., for any $\phi : \triangle_{n-1} \to \triangle_{n-1}$, we will write $\phi \in NDF(n)$, iff it satisfies (82), (83) and (84). We will talk about normalized decision functions $\phi \in NDF$, without specifying dimension when it is anyway known in a given context.

Assumptions (82) and (83) mean, respectively, that we cannot attach a positive chance to non-supported events and that the chances provided by reasoning strategy cannot contradict the chances derived from an information source. These assumptions imply a number of results concerning efficient search for condition subsets preserving decision information in terms of particular NDF-functions. Assumption (84) expresses the lack of any kind of differences in treatment of particular decision classes. We can say that all conditions of the above definition agree with the idea that we are not allowed to base on any additional knowledge besides that possible to be derived straightly from data.

Definition 38 cannot refer to all methods of adjusting numerical values to conditioned weights of particular events. Still, the expressive power of the above characteristics enables to describe a reasonably large group of strategies of reasoning with provided conditional frequency based information. An obvious remark here is that normalized ∂-decision m-decision and $m_{\geq 0.7}$-decision functions defined, respectively, by putting

$$(\partial^\mu(s))_k = \begin{cases} |\{s_l : s_l > 0\}|^{-1} & if \ s_k > 0 \\ 0 & otherwise \end{cases}$$
$$(m^\mu(s))_k = \begin{cases} |\{s_l : s_l = \max_m s_m\}|^{-1} & if \ s_k = \max_m s_m \\ 0 & otherwise \end{cases} \quad (87)$$
$$\left(m^\mu_{\geq 0.7}(s)\right)_k = \begin{cases} (m^\mu(s))_k & if \ max_l(s_l) \geq 0.7 \\ (\partial^\mu(s))_k & if \ max_l(s_l) < 0.7 \end{cases}$$

for $k = 1,..,|V_d|$, correspond to distribution based interpretations of the form (64) considered in Section 3.3. NDF-family contains, however, also more sophisticated functions enabling to model different strategies of dealing with noise and redundant dependencies.

As an example, let us consider the function $\mu^\varepsilon : \Delta_{n-1} \to \Delta_{n-1}$ such that for each given $s \in \Delta_{n-1}$ it takes the form of

$$(\mu^\varepsilon(s))_k = \begin{cases} s_k/\left(1 - \sum_{l:s_l < \varepsilon} s_l\right) & if \ s_k \geq \varepsilon \\ 0 & otherwise \end{cases} \quad (88)$$

if $\max_l s_l \geq \varepsilon$ and equals to $\partial^\mu(s)$ otherwise. One can see that μ^ε distributes frequencies of less supported decision classes onto these stronger ones. It prunes a bit classification rules keeping, on the other hand, relative proportions between strong decision classes.

Example 7. For decision table from Fig. 1, let us put $\varepsilon = 0.3$ and enrich it with values of $\mu^{0.3}_{d/A}$ as shown in Fig. 8.

Obviously, while talking about classification of new cases, we still want to rely on MDL-Principle.

Definition 39. Given a decision table $\mathbf{A} = (U, A, d)$ and a normalized decision function $\phi \in NDF$, we say that a subset $B \subseteq A$ ϕ-defines d iff for each $w_A \in V_A^U$ there is the equality

$$\phi_{d/B}\left(w_A^{\downarrow B}\right) = \phi_{d/A}(w_A) \quad (89)$$

V_A^U	#	a_1	a_2	a_3	a_4	a_5	$\mu_{d/A}(v_i/\cdot), i=1,2$	$\mu_{d/A}^{0.3}(v_i/\cdot), i=1,2$
w_1	1	0	0	0	0	0	$(1,0)$	$(1,0)$
w_2	1	0	0	1	0	0	$(1,0)$	$(1,0)$
w_3	1	0	0	1	1	0	$(1,0)$	$(1,0)$
w_4	1	0	1	0	1	0	$(1,0)$	$(1,0)$
w_5	3	0	1	1	0	0	$(2/3,1/3)$	$(2/3,1/3)$
w_6	1	0	1	1	1	1	$(0,1)$	$(0,1)$
w_7	4	1	0	0	1	0	$(1/2,1/2)$	$(1/2,1/2)$
w_8	1	1	0	1	0	1	$(0,1)$	$(0,1)$
w_9	2	1	0	1	1	0	$(1,0)$	$(1,0)$
w_{10}	4	1	1	0	0	0	$(3/4,1/4)$	$(1,0)$
w_{11}	1	1	1	1	1	1	$(0,1)$	$(0,1)$

Fig. 8. Decision table with distributions $\mu_{d/A}$ and $\mu_{d/A}^{0.3}$.

A subset $B \subseteq A$ which ϕ-defines d is called a ϕ-decision reduct iff none of its proper subsets satisfies analogous property.

In general, conditions (82) and (83) imply that the above is yet another generalization of the consistent case, i.e., that for consistent decision tables exact decision reducts and ϕ-decision reducts coincide, for any $\phi \in NDF$. Moreover, for the identity function $\phi = id$, the above is equivalent to the notion of μ-decision reduct.

Theorem 40. *([42]) For any normalized decision function $\phi \in NDF$, the problem of finding minimal ϕ-decision reduct is NP-hard.*

The following result, generalizing Proposition 27, enables to reconsider for all NDF-functions practically all techniques concerning rough membership distributions described so far.

Proposition 41. *([42]) Given $\mathbf{A} = (U, A, d)$, $B \subseteq A$ and $\phi \in NDF$, for any $w_A \in V_A^U$ we have inequalities*

$$\min_{w_A \in V_A^U(w_B)} \phi_{d/A}(v_d/w_A) \leq \phi_{d/B}(v_d/w_B) \leq \max_{w_A \in V_A^U(w_B)} \phi_{d/A}(v_d/w_A) \tag{90}$$

Proposition 42. *([42]) A subset $B \subseteq A$ is a ϕ-decision reduct for a given decision table $\mathbf{A} = (U, A, d)$ iff it is decision reduct for the consistent decision table $\mathbf{A}_\phi = (U, A, \phi_{d/A})$, where decision attribute $\phi_{d/A} : U \to \triangle_{|V_d|-1}$ is defined by the formula*

$$\phi_{d/A}(u) = \phi_{d/A}(Inf_A(u)) \tag{91}$$

Corollary 43. *Given a normalized decision function ϕ, and decision table $\mathbf{A} = (U, A, d)$, a subset $B \subseteq A$ ϕ-defines d iff it intersects with each element of the discernibility table $T_\phi(\mathbf{A})$ given by the formula*

$$T_\phi(\mathbf{A}) = \{B \subseteq A : B \neq \emptyset \wedge \exists_{i,j}(B = c_{ij}^\phi)\} \tag{92}$$

where
$$c_{ij}^{\phi} = \begin{cases} \{a \in A : a(u_i) \neq a(u_j)\} & if \ \phi_{d/A}(u_i) \neq \phi_{d/A}(u_j) \\ \emptyset & otherwise \end{cases} \quad (93)$$

Corollary 44. *For any decision table, the property of ϕ-defining decision is closed with respect to supersets.*

By introducing normalized decision functions as derivable from rough membership distributions, we provide ourselves with the following relationship.

Proposition 45. *Given $\mathbf{A} = (U, A, d)$ and $\phi \in NDF$, we have $T_\phi(\mathbf{A}) \subseteq T_\mu(\mathbf{A})$.*

Corollary 46. *Given $\phi \in NDF$, for any decision table, each subset which μ-defines decision ϕ-defines it as well.*

While classifying new cases, we can still rely on appropriately understood decision rules, referring to different kinds of normalized decision functions. Just like in a special case of $\phi = id$ before, each vector $\phi_{d/B}(w_B)$ is said to induce a random ϕ-decision rule

$$(B, w_B) \Rightarrow \vee_{k=1,\ldots,|V_d|} (d, v_k)_{\phi_{d/B}(v_k/w_B)} \quad (94)$$

which attaches to $w_B \in V_B^U$ information concerning degrees $\phi_{d/B}(v_k/w_B)$ of ϕ-based membership of $Inf_B^{-1}(w_B)$ into particular decision classes (see [42]).

Example 8. Let us reconsider classification scheme S stated by (67) and classify a new object $u_{new} \notin U$ provided with the following information:

$$w'_A = (\langle a_1, 1 \rangle, \langle a_2, 1 \rangle, \langle a_3, 0 \rangle, \langle a_4, 1 \rangle, \langle a_5, 0 \rangle) \quad (95)$$

Then, the formula for weighting decision classes takes the form of

$$\begin{aligned} W_S(v_k/w'_A) = & \mu_{B_2}(1,1,0,0) \cdot \mu_{d/B_2}(v_k/1,1,0,0) \\ & + \mu_{B_3}(1,0,1,0) \cdot m^{\mu}_{d/B_3}(v_k/1,0,1,0) \\ & + \mu_{B_4}(1,0,1,0) \cdot m^{\mu}_{d/B_4 \geq 0.7}(v_k/1,0,1,0) \end{aligned} \quad (96)$$

for $k = 1, 2$. So, we obtain that decision weights for u_{new} are equal to

$$W_S(v_1/w'_A) = \frac{4}{20} \cdot \frac{3}{4} + \frac{4}{20} \cdot \frac{1}{2} + \frac{1}{20} \cdot 1 = \frac{3}{10} \quad (97)$$

and

$$W_S(v_2/w'_A) = \frac{4}{20} \cdot \frac{1}{4} + \frac{4}{20} \cdot \frac{1}{2} + \frac{1}{20} \cdot 0 = \frac{3}{20} \quad (98)$$

Now, let us note that for considered decision table $\mu^{0.3}$-decision reducts coincide with μ-decision ones and modify S by replacing "μ" with "$\mu^{0.3}$" for its second subset of conditions B_2. For such an S', one can see that quantity

$$W_{S'}(v_1/w'_A) = \frac{4}{20} \cdot 1 + \frac{4}{20} \cdot \frac{1}{2} + \frac{1}{20} \cdot 1 = \frac{7}{20} \quad (99)$$

becomes to be more clearly higher than

$$W_{S'}(v_2/w'_A) = \frac{4}{20} \cdot 0 + \frac{4}{20} \cdot \frac{1}{2} + \frac{1}{20} \cdot 0 = \frac{1}{10} \quad (100)$$

4.3 Possible combinations of two approaches

Approximation methods considered in Sections 4.1 and 4.2 can be regarded as complementary to each other. First of them enables to omit the need of discerning almost equal frequency distributions, where the word "almost" corresponds to setting up certain tolerance parameters. Then, by tuning approximation threshold we can balance with both the time of computations and exactness of obtained results. In the second case, Proposition 45 provides us with potential possibility of relaxing these pairs of distributions which become to be equal under interpretation corresponding to a given normalized decision function strategy.

Let us say a few words about possible combination of these two processes.

Definition 47. Let a decision table $\mathbf{A} = (U, A, d)$, a normalized decision function $\phi \in NDF$, a normalized distance measure ρ and $\varepsilon \in [0, 1)$ be given.

1. We say that a subset $B \subseteq A$ $(\rho \leq \varepsilon)$-approximately ϕ-defines d iff for each $w_A \in V_A^U$ we have

$$\rho\left(\phi_{d/B}\left(w_A^{\downarrow B}\right), \phi_{d/A}(w_A)\right) \leq \varepsilon \qquad (101)$$

A subset B which satisfies (101) is called a $(\rho \leq \varepsilon)$-approximating ϕ-decision reduct iff none of its proper subsets $(\rho \leq \varepsilon)$-approximately ϕ-defines d.

2. We say that $B \subseteq A$ $(\rho > \varepsilon)$-discerns $\phi_{d/A}$ iff for each pair $u, u' \in U$ we have the implication

$$\rho\left(\phi_{d/A}(u), \phi_{d/A}(u')\right) > \varepsilon \Rightarrow Inf_B(u) \neq Inf_B(u') \qquad (102)$$

A subset B which satisfies the above condition is called a $(\rho > \varepsilon)$-discerning ϕ-decision reduct iff none of its proper subsets $(\rho > \varepsilon)$-discerns $\phi_{d/A}$.

Basing on Proposition 41, let us give an example of a generalization of previous results to the currently considered combination of approaches.

Proposition 48. *Given a decision table* $\mathbf{A} = (U, A, d)$, *any subset* $B \subseteq A$ *which* $(Max > \varepsilon)$-*discerns* $\phi_{d/A}$ $(Max \leq \varepsilon)$-*approximately* ϕ-*defines* d *as well. On the other hand, each* $B \subseteq A$ *which does not* $(Max > 2 \cdot \varepsilon)$-*discerns* $\phi_{d/A}$ *cannot* $(Max \leq \varepsilon)$-*approximately* ϕ-*define* d.

Corollary 49. *Let a normalized decision function* $\phi \in NDF$, *an approximation threshold* $\varepsilon \in [0, 1)$ *and a decision table* $\mathbf{A} = (U, A, d)$ *be given. A satisfactory condition for a subset* $B \subseteq A$ *to* $(Max \leq \varepsilon)$-*approximate* ϕ-*defining* d *is to that* B *intersect each element of the discernibility table*

$$T_\phi^{Max > \varepsilon}(\mathbf{A}) = \{B \subseteq A : B \neq \emptyset \wedge \exists_{i,j}(B = c_{ij}^{Max(\phi) > \varepsilon})\} \qquad (103)$$

where

$$c_{ij}^{Max(\phi) > \varepsilon} = \begin{cases} \{a \in A : a(u_i) \neq a(u_j)\} & \text{if } Max(\phi_{d/A}(u_i), \phi_{d/A}(u_j)) > \varepsilon \\ \emptyset & \text{otherwise} \end{cases} \qquad (104)$$

Similarly, a necessary condition for $(Max \leq \varepsilon)$-*approximate* ϕ-*defining* d *is to intersect each element of the discernibility table* $T_\phi^{Max > 2\varepsilon}(\mathbf{A})$ *defined analogously.*

Let us illustrate the above correspondence by considering the function $\mu^{0.3} \in NDF$.

Example 9. Analysis of dynamics of $(Max > \varepsilon)$-discerning $\mu^{0.3}$-decision reducts can be based on the table in Fig. 6, enriched additionally by the pattern pair $\{w_7, w_{10}\}$ corresponding to the discernibility set $\{a_2, a_4\}$ and the distance

$$Max(\mu_{d/A}^{0.3}(w_7), \mu_{d/A}^{0.3}(w_{10})) = 1/2 \qquad (105)$$

Thus, the sets of reducts for thresholds $\varepsilon \in [0, 1/3)$ and $\varepsilon \in [1/2, 1)$ remain unchanged with respect to $(Max > \varepsilon)$-discerning μ-decision reducts (Fig. 7). For $\varepsilon \in [1/3, 1/2)$ a new discernibility matrix takes the form of

$$T_{\mu^{0.3}}^{Max>\varepsilon}(\mathbf{A}) = T_{\mu}^{Max>\varepsilon}(\mathbf{A}) \cup \{\{a_2, a_4\}\} \qquad (106)$$

In particular, it implies that the subset $\{a_1, a_3, a_5\}$ looses its meaning in comparison with previous situation.

Also in case of such a combination of approximations presented reduct modifications are exact decision reducts for consistent decision tables. Thus, in particular, we obtain possibly most general complexity result concluding this part of our study.

Theorem 50. *For a given decision table, a given $\varepsilon \in [0, 1)$, a normalized distance measure $\rho : \Delta_{|V_d|-1} \times \Delta_{|V_d|-1} \to [0, 1]$ and a normalized decision function $\phi \in NDF$, the problems of finding minimal $(\rho \leq \varepsilon)$-approximating and $(\rho > \varepsilon)$-discerning ϕ-decision reducts are NP-hard.*

Once more, the above characteristics suggests to apply heuristics for searching for attributes optimal in the above sense. What was relatively easy to compute in the above example, may be impossible in a straightforward way in case of real-life data.

5 Measures of information

5.1 Average encoding of frequency based dependencies

Approximations studied in Section 4 enable to relax conditions on the notion of μ-decision reduct appropriately for particular data. They can be used for development of a more flexible version of rough set based MDL-Principle, where one can tune approximation thresholds $\varepsilon \in [0, 1)$ or formulas of functions $\phi \in NDF$ to search for smaller and smaller sets of relevant features keeping initial frequency based information up to a fixed degree.

In our opinion, above provided tools can be very useful for strongly inconsistent decision tables, especially under assumption about occurrence of noises related to conditional frequencies. However, for (almost) consistent tables it suddenly turns out that all notions of approximate decision reducts introduced so far become to be (almost) just equivalent to the exact case.

An exception to the above equivalence is the notion of ε-discerning decision reduct introduced in Section 2.4. The reason is that there approximation threshold is related not to definition of the object (pattern) pair discernibility criterion itself, but to requirement of sufficiently frequent satisfaction of this criterion. It enabled to express mentioned notion in terms of discernibility measure given by formula (32). Obviously, it is possible to generalize the idea of approximate discernibility in different ways, starting, e.g., with a proposal concerning ε-discerning ∂-decision reducts given in Definition 19. One could even search for minimal subsets of conditions in the process parameterized by two kinds of approximation thresholds $\varepsilon_0, \varepsilon_1 \in [0,1)$, where:

1. ε_0 would relate to discernibility criterion (102) in Definition 47, and
2. ε_1 — to the approximate performance over tables of weighted subsets of conditional features, just like it was proposed in Section 2.4.

Obtained scheme would be then a bit complicated, usually inexpressible in terms of any measure similar to (32), which is, actually, an analogy of rough set discernibility measures for extraction of decision trees from data (see e.g. [14]). On the other hand, it would anyway satisfy all so far considered requirements concerning flexibility of dealing with approximate discernibility.

Such a sophisticated approximation of the notion of μ-decision reduct is still possible to be based on other kinds of information measures, not (straightly) related to discernibility. These measures enable to cope with requirements concerning both differences between conditional frequency distributions and prior weights of occurrence of particular conditional patterns in data, by usage of a common approximation threshold.

Let us start from the most known information measure, called a conditional entropy ([5], [9]). For more advantages of entropy as concerned with rough set based applications the reader is referred to [3], [4], [41]. Besides basic properties listed below, we would like to note that conditional information measure of entropy is the one related to statistical concept of MDL-Principle in the most straight way ([12]).

Conditional entropy is defined by the formula

$$H_{\mathbf{A}}(d/B) = \sum_{w_B \in V_B^U} \mu_B(w_B) h_{d/B}(w_B) \qquad (107)$$

which labels each $B \subseteq A$ with average amount of conditioned uncertainties

$$h_{d/B}(w_B) = - \sum_{v_d \in \partial_{d/B}(w_B)} \mu_{d/B}(v_d/w_B) \log \mu_{d/B}(v_d/w_B) \qquad (108)$$

concerning random μ-decision rules induced by particular information patterns (see [5], [9]). In such a form, entropy $H_{\mathbf{A}}(d/B)$ is usually interpreted as a measure of the lack of information (chaos) keeping concerning d under indiscernibility relation induced by $B \subseteq A$. This intuition was used in many approaches developed within rough sets (see e.g. [3], [41]) or other fields (see [9] for further references).

In view of the purpose of approximate preserving of rough membership based information the following result is crucial:

Proposition 51. *([5]) For any* $\mathbf{A} = (U, A, d)$ *and a subset* $B \subseteq A$ *we have*

$$H_{\mathbf{A}}(d/B) \geq H_{\mathbf{A}}(d/A) \tag{109}$$

where equality holds iff B μ-*defines* d *in* $\mathbf{A}$.

Definition 52. Given $\varepsilon \in [0, 1)$ and $\mathbf{A} = (U, A, d)$, we say that a subset $B \subseteq A$ ($\triangle H \leq \varepsilon$)-approximately μ-defines d iff

$$H_{\mathbf{A}}(d/B) + \log(1 - \varepsilon) \leq H_{\mathbf{A}}(d/A) \tag{110}$$

A subset $B \subseteq A$ which satisfies the above inequality is called a ($\triangle H \leq \varepsilon$)-approximating μ-decision reduct iff there is no proper subset of B which ($\triangle H \leq \varepsilon$)-approximately μ-defines d.

Remark. In equality (110), we use coefficient "$\log(1 - \varepsilon)$" instead of "ε" to provide all introduced measure-oriented approximations of decision reduct with the same, normalized approximation range. Otherwise, we would have to deal with thresholds belonging to $[0, +\infty)$.

Proposition 51 provides the above introduced notion with useful characteristics:

Corollary 53. *Notions of* μ-*decision and* ($\triangle H \leq 0$)-*approximating* μ-*decision reduct are equivalent.*

Corollary 54. *For any decision table, the property of* ($\triangle H \leq \varepsilon$)-*approximate* μ-*defining decision is closed with respect to supersets.*

Monotonicity of entropy provided by Proposition 51 is very useful while searching for ($\triangle H \leq \varepsilon$)-approximating μ-decision reducts being optimal in terms of the MDL-Principle. It enables to keep thinking about optimal solutions in sense of the balance between quality and applicability of reducts. Although now even in consistent case ($\triangle H \leq \varepsilon$)-approximating μ-decision reducts are potentially shorter than exact ones, we still cannot run away from the following complexity characteristics, similarly as in case of ε-discerning decision reducts.

Theorem 55. *([41]) For any fixed* $\varepsilon \in [0, 1)$, *the problem of finding minimal* ($\triangle H \leq \varepsilon$)-*approximating* μ-*decision reduct is NP-hard.*

Another measure which can be used here is a normalized μ-information measure which labels subsets $B \subseteq A$ with quantities

$$E_{\mathbf{A}}(d/B) = \sum_{w_B \in V_B^U} \mu_B(w_B) e_{d/B}(w_B) \tag{111}$$

where

$$e_{d/B}(w_B) = \sum_{v_d \in V_d} \mu_{d/B}^2(v_d/w_B) \tag{112}$$

is the quantity of total probability that randomly chosen $u \in Inf_B^{-1}(w_B)$ — which has a chance $\mu_{d/B}(v_d/w_B)$ to belong to decision class of each particular

$v_d \in V_d$ – will be actually classified there by random μ-decision rule of the form (51) – what, in this case, is of probability $\mu_{d/B}(v_d/w_B)$ as well.

We can say that any $B \subseteq A$ satisfies a kind of approximate implication $B \Rightarrow_{E_\mathbf{A}(d/B)} d$, where the value of $E_\mathbf{A}(d/B)$ may be interpreted as the expected chance that a randomly chosen object $u \in U$ will be properly classified by using μ-decision rules (compare with [42]). Thus, it is reasonable to search for subsets of A which almost keep such understood efficiency of classification.

Definition 56. Let an approximation threshold $\varepsilon \in [0,1)$ and $\mathbf{A} = (U, A, d)$ be given. We say that a subset $B \subseteq A$ ($\triangle E \leq \varepsilon$)-approximately μ-defines d iff

$$E_\mathbf{A}(d/B) + \varepsilon \geq E_\mathbf{A}(d/A) \tag{113}$$

A subset $B \subseteq A$ which satisfies the above inequality is called an ($\triangle E \leq \varepsilon$)-approximating μ-decision reduct iff there is no proper subset of B which ($\triangle E \leq \varepsilon$)-approximately μ-defines d.

Proposition 57. *For any $\mathbf{A} = (U, A, d)$ and a subset $B \subseteq A$ we have*

$$E_\mathbf{A}(d/B) \leq E_\mathbf{A}(d/A) \tag{114}$$

where equality holds iff B μ-defines d in $\mathbf{A}$.

Corollary 58. *Notions of μ-decision and ($\triangle E \leq 0$)-approximating μ-decision reduct are equivalent.*

Corollary 59. *For any decision table, the property of ($\triangle E \leq \varepsilon$)-approximate μ-defining decision is closed with respect to supersets.*

Let us compare introduced above notions of reducts with ε-discernibility mechanism introduced in Section 2.4.

Example 10. The table in Fig. 9 contains approximate values of so far considered decision information measures for subsets of attributes pointed as interesting in previous examples, over decision table from Fig. 1. Last three columns include approximation thresholds for which considered subsets begin to ε-approximately, or ($\triangle * \leq \varepsilon$)-approximately μ-define d, where "$*$" denotes "H" and "E", respectively. The way of calculating these thresholds depends on specifications of Definitions 17, 52 and 56.

Again, the following implies that for real-life data it would be difficult to find minimal ($\triangle E \leq \varepsilon$)-approximating μ-decision reducts.

Theorem 60. *([42]) For any fixed $\varepsilon \in [0,1)$, the problem of finding minimal ($\triangle E \leq \varepsilon$)-approximating μ-decision reduct is NP-hard.*

Thus, for all considered information measures, it would be good to refer to previously mentioned algorithmic tools for extraction of decision reducts from data. The following result gives possibility of discernibility based search for suboptimal solutions of described optimization problems.

$B \subseteq A$	$N_\mathbf{A}$	$H_\mathbf{A}$	$E_\mathbf{A}$	$1 - N_\mathbf{A}$	$1 - \exp(\triangle H_\mathbf{A})$	$\triangle E_\mathbf{A}$
$\{a_5\}$	0.451	0.669	0.694	0.549	0.56	0.064
$\{a_1, a_3\}$	0.785	0.957	0.579	0.215	0.665	0.179
$\{a_2, a_3\}$	0.866	0.847	0.6	0.134	0.625	0.158
$\{a_2, a_4\}$	0.817	0.879	0.581	0.183	0.637	0.177
$\{a_1, a_3, a_5\}$	0.878	0.562	0.745	0.122	0.522	0.013
$\{a_2, a_3, a_4\}$	0.988	0.6	0.708	0.012	0.535	0.05
$\{a_1, a_2, a_3, a_4\}$	1	0.5	0.758	0	0	0

Fig. 9. Examples of values of different proposed information measures.

Proposition 61. *([41]) Let $\varepsilon \in [0,1)$ and a decision table $\mathbf{A} = (U, A, d)$ be given. Consider the normalized Euclidean distance measure Euc, defined, for any $s, s' \in \triangle_{|V_d|-1}$, by*

$$Euc(s, s') = \sqrt{\frac{1}{2} \sum_{k=1,..,|V_d|} (s_k - s'_k)^2} \qquad (115)$$

and the Kullback-Leibner measure ([9]) of directed divergence

$$KL(s, s') = \sum_{i=1}^{n} s_i \log\left(\frac{s_i}{s'_i}\right) \qquad (116)$$

Then any $B \subseteq A$ which $(Euc > \varepsilon)$-discerns $\mu_{d/A}$ μ-defines d $(\triangle E \leq \varepsilon^2)$-approximately. Similarly, if B $(KL > \varepsilon)$-discerns $\mu_{d/A}$, then it $(\triangle H \leq \varepsilon)$-approximately μ-defines d.

Corollary 62. *Given a decision table $\mathbf{A} = (U, A, d)$, a subset $B \subseteq A$ which intersects with each element of the discernibility table $T_\mu^{Euc>\varepsilon}(\mathbf{A})$ μ-defines d $(\triangle E \leq \varepsilon^2)$-approximately. Similarly, if B intersects with each element of table built analogously for distance KL, then it $(\triangle H \leq \varepsilon)$-approximately μ-defines d.*

Such characteristics enable to apply already mentioned powerful rough set algorithms basing in different ways on the idea of discernibility ([1], [18], [49]) to approximate solving of the above problems in a relatively short time. Obviously, it is difficult to think about too high quality understood in terms of comparison of cardinalities of minimal subsets intersecting with, e.g., $T_\mu^{Euc>\varepsilon}(\mathbf{A})$ and cardinalities of actually minimal subsets satisfying requirements of Definition 56. However, it is all we can do by referring to strict boolean reasoning during the analysis of measures which rely on prior weights substantially.

Actually, there exist rough set algorithms possible to be adapted to the search for, e.g., minimal $(\triangle E \leq \varepsilon)$-approximating μ-decision reducts in a more straightforward way, without necessity of using discernibility characteristics at all (see e.g. [48]). On the other hand, in some situations it is better to keep methodological relationship with discernibility, regardless of the fact whether it is needed for algorithmic implementation or not.

Example 11. Let us note that in any decision table with two decision classes considered distance measures *Max* and *Euc* are equivalent. Thus, basing on Example 6, we obtain that subsets of conditions contained in the lower part of table in Fig. 7 are simultaneously $(Euc > \varepsilon)$-discerning μ-decision reducts, for corresponding approximation thresholds. Just like before, let us focus on subsets $\{a_1, a_2, a_3, a_4\}$, $\{a_1, a_3, a_5\}$ and $\{a_5\}$. For the first of them, equalities

$$E_{\mathbf{A}}(d/A) = E_{\mathbf{A}}(d/\{a_1, a_2, a_3, a_4\}) = E_{\mathbf{A}}(d/\{a_2, a_3, a_4\}) + 1/20 \qquad (117)$$

imply that for any $\varepsilon \geq 1/20$ it is not a minimal set $(\Delta E \leq \varepsilon)$-approximately μ-defining d any more.

According to Proposition 61 subset $\{a_1, a_3, a_5\}$ is known as a $(\Delta E \leq \varepsilon)$-approximately μ-defining d for any $\varepsilon \geq 1/9$. Again, however, it is not an optimal set with respect to MDL-Principle at this approximation level, since also the third of considered sets satisfies inequality

$$E_{\mathbf{A}}(d/\{a_5\}) \geq E_{\mathbf{A}}(d/A) - 1/9 \qquad (118)$$

Similar disadvantages of considered strategy of approximation can be observed in case of the entropy measure and the Kullback-Leibner distance.

5.2 Average distance measures

Basing notion of approximate decision reduct on information measures has its obvious advantages with respect to searching for conditions generating valuable decision rules in average sense. It may turn out to be especially appropriate while analyzing efficiency of proposed scheme of weighted classification in view of statistical principles. What we mean is that if a given decision table $\mathbf{A} = (U, A, d)$ is representative in a statistical sense (see [46]), then we can assume that frequencies of occurrence of particular value combinations $w_B \in V_B^U$ of new cases on a given $B \subseteq A$ are going to be approximately equal to quantities $\mu_B(w_B)$ computed within $\mathbf{A}$. In such a case, one can indeed regard, e.g., measure $E_{\mathbf{A}}(d/B)$ as the expected chance that $\mu_{d/B}$-based classification of new cases will be proper. On the other hand, in data mining applications, we usually cannot afford to assume such a kind of validity of $\mathbf{A}$ as a statistical sample. In fact, in many situations treating a given decision table as a sample with respect to larger hypothetical universe does not make a lot of sense. Then, more exact comparison of performance of particular configurations, like in case of approaches based on, e.g., $(\rho > \varepsilon)$-discernibility seems to be better, since it focuses on each of information patterns in a similar degree.

Nevertheless, one can consider some hybrid method, which would enable to take the best from two above approaches. Thus, in this section we try to discuss possibilities of referring a kind of average $(\rho > \varepsilon)$-discernibility and $(\rho \leq \varepsilon)$-approximation to notions based on information measures.

Let us focus on rough membership distributions and go back to the notion of $(\rho \leq \varepsilon)$-approximating μ-decision reduct, for a given normalized distance measure ρ. Let us introduce global σ-average ρ-deviation measure labeling each $B \subseteq A$ with quantity

$$\sigma_{\mathbf{A}}^2(\rho(B:A)) = \sum_{w_B \in V_B^U} \mu_B(w_B)\, \sigma_{d/B}^2(\rho(w_B:A)) \tag{119}$$

where, for each particular $w_B \in V_B^U$,

$$\sigma_{d/B}^2(\rho(w_B:A)) = \sum_{w_A \in V_A^U(w_B)} \mu_{B/A}(w_A/w_B)\, \rho^2\left(\mu_{d/B}(w_B), \mu_{d/A}(w_A)\right) \tag{120}$$

is a local average deviation of rough membership distribution induced by w_B from distributions corresponding to particular extensions $w_A \in V_A^U(w_B)$, weighted by conditional frequencies of the form

$$\mu_{A/B}(w_A/w_B) = \frac{|Inf_A^{-1}(w_A)|}{|Inf_B^{-1}(w_B)|} \tag{121}$$

which express the chance that objects satisfying condition (B, w_B) also satisfy its extension (A, w_i). One can also rewrite formula (119) as

$$\sigma_{\mathbf{A}}^2(\rho(B:A)) = \frac{1}{|U|} \sum_{u \in U} \rho^2\left(\mu_{d/B}(u), \mu_{d/A}(u)\right) \tag{122}$$

what suggests to consider the following notion.

Definition 63. Given a decision table $\mathbf{A} = (U, A, d)$, a normalized distance measure ρ and an approximation threshold $\varepsilon \in [0,1)$, we say that a subset $B \subseteq A$ $(\sigma^2(\rho) \leq \varepsilon)$-approximately μ-defines d iff $\sigma_{\mathbf{A}}^2(\rho(B:A)) \leq \varepsilon$. If, moreover, none of proper subsets of B satisfies the above inequality, then we say that B is a $(\sigma^2(\rho) \leq \varepsilon)$-approximating μ-decision reduct.

An obvious fact is that such an average modification of the notion of $(\rho \leq \varepsilon)$-approximating μ-decision reduct enables to search for potentially smaller subsets of conditions by neglecting deviations of relatively low support.

Proposition 64. *Given a decision table $\mathbf{A} = (U, A, d)$, a normalized distance measure ρ and an approximation threshold $\varepsilon \in [0,1)$, each subset $B \subseteq A$ which $(\rho \leq \varepsilon)$-approximately μ-defines d does it also $(\sigma^2(\rho) \leq \varepsilon^2)$-approximately.*

Far more astonishing result connects $\sigma_{\mathbf{A}}^2$ with normalized μ-information measure $E_{\mathbf{A}}$.

Proposition 65. *([43]) Given a decision table $\mathbf{A} = (U, A, d)$ and a normalized Euclidean distance measure Euc, for each subset $B \subseteq A$ we have the equality*

$$\sigma_{\mathbf{A}}^2(Euc(B:A)) = E_{\mathbf{A}}(d/A) - E_{\mathbf{A}}(d/B) \tag{123}$$

In particular, it implies all properties of $E_{\mathbf{A}}$ presented in previous section. Moreover, simple derivation leads to the following fact:

Proposition 66. *For any decision table and any approximation threshold $\varepsilon \in [0,1)$, the property of being $(\sigma(\rho) \leq \varepsilon)$-approximating μ-decision reduct is closed with respect to supersets.*

Finally, we can use equality (123) to conclude what follows:

Corollary 67. *For any $\varepsilon \in [0,1)$, notions of $(\triangle E \leq \varepsilon)$-approximating and $(\sigma^2(Euc) \leq \varepsilon)$-approximating μ-decision reduct are equivalent.*

In a similar way we can try to state some average modification of the notion of $(\rho > \varepsilon)$-discerning μ-decision reduct. Again, we propose to rely on tools provided by probability and statistics and regard the following formula as analogy of statistical measures related to the idea of χ^2-test ([8]). For a given $w_B \in V_B$ let us put local χ^2-average ρ-indiscernibility measure

$$\chi^2_{\mathbf{A}}\left(\rho(w_B : V^U_A(w_B))\right) =$$
$$= \sum_{w_i, w_j \in V^U_A(w_B): i<j} \mu_{A/B}(w_i/w_B) \mu_{A/B}(w_j/w_B) \rho^2\left(\mu_{d/A}(w_i), \mu_{d/A}(w_j)\right) \quad (124)$$

Such a quantity tells us how much we pay for restricting to $B \subseteq A$ locally, within $Inf_B^{-1}(w_B) \subseteq U$ in terms of decision indiscernibility. By putting for $B \subseteq A$ a global χ^2-average ρ-indiscernibility measure

$$\chi^2_{\mathbf{A}}\left(\rho(B:A)\right) = \sum_{w_B \in V_B} \mu_B(w_B) \chi^2_{\mathbf{A}}\left(\rho(w_B : V^U_A(w_B))\right) \quad (125)$$

we can thus say how large is the above indiscernibility globally and consider the following notion:

Definition 68. *Given a decision table $\mathbf{A} = (U, A, d)$, a normalized distance measure ρ and a threshold $\varepsilon \in [0,1)$, we say that a subset $B \subseteq A$ $(\chi^2(\rho)) \leq \varepsilon$-discerns $\mu_{d/A}$ iff the inequality $\chi^2_{\mathbf{A}}(\rho(B:A)) \leq \varepsilon$ holds. Moreover, B which $(\chi^2(\rho)) \leq \varepsilon$-discerns $\mu_{d/A}$ is called a $(\chi^2(\rho)) \leq \varepsilon$-discerning μ-decision reduct iff none of its proper subsets satisfies an analogous inequality.*

A natural question is whether the correspondence between ε-discernibility and ε-approximation can happen in an average sense. Usually, even for distance measures for which results analogous to Proposition 37 hold, it is quite difficult to re-formulate them properly. One of exceptions is the normalized Euclidean distance measure again:

Proposition 69. *([43]) Given a decision table $\mathbf{A} = (U, A, d)$, for each subset $B \subseteq A$ we have the equality*

$$E_{\mathbf{A}}(d/A) - E_{\mathbf{A}}(d/B) = \chi^2_{\mathbf{A}}(Euc(B:A)) \quad (126)$$

Corollary 70. *For each decision table $\mathbf{A} = (U, A, d)$ and any threshold $\varepsilon \in [0,1)$, the property of $(\chi^2(Euc)) \leq \varepsilon$-discerning $\mu_{d/A}$ is closed with respect to supersets.*

Corollary 71. *For any decision table and threshold $\varepsilon \in [0,1)$, notions of $(\triangle E \leq \varepsilon)$-approximating, $(\sigma^2(Euc) \leq \varepsilon)$-approximating and $(\chi^2(Euc)) \leq \varepsilon$-discerning μ-decision reduct are equivalent.*

Equality (126) enables to express the notion of ε-average μ-decision reduct in terms of average Euc-based discernibility. In fact, one could try to take the advantage of such a characteristics in purpose of providing alternative interpretation of measure $E_\mathbf{A}$ as related to computing with weighted discernibility tables. The most straightforward would be to rewrite formula for measure $\chi_\mathbf{A}^2(Euc(B:A))$ to obtain the sum of local quantities indexed by pairs of information patterns $w_i, w_j \in V_A^U$, like it was done in Section 2.4 for ε-discerning decision reducts. Such a re-formulation can be presented as

$$\chi_\mathbf{A}^2(Euc(B:A)) = \sum_{1\leq i<j\leq |V_A^U|: w_i^{\downarrow B}=w_j^{\downarrow B}} \frac{1}{\mu_B(w_i^{\downarrow B})} \cdot avg_{Euc}(i,j) \qquad (127)$$

where

$$avg_{Euc}(i,j) = \mu_A(w_i)\mu_A(w_j)Euc^2(\mu_{d/A}(w_i), \mu_{d/A}(w_j)) \qquad (128)$$

depends just on information concerning particular pairs of information patterns on A. Unfortunately, the whole formula (127) relates to each given $B \subseteq A$ in a way disabling direct application of weighted discernibility approach. In [43] we deal with this problem as follows: First, we set up some approximation threshold $\varepsilon_0 \in [0,1)$ responsible for filtering pairs of patterns with similar rough membership distributions – and continue with discernibility table $T_\mu^{Euc>\varepsilon_0}(\mathbf{A})$ defined like in Corollary 62. Second, we label each its element $B \subseteq A$ with assignment

$$n_\chi^{Euc>\varepsilon_0}(B) = \sum_{i,j: B=\{a\in A: w_i^{\downarrow\{a\}}=w_j^{\downarrow\{a\}}\}} avg_\chi(i,j) \qquad (129)$$

where

$$avg_\chi(i,j) = \frac{\mu_A(w_i)\mu_A(w_j)}{\mu_A(w_i)+\mu_A(w_j)} \qquad (130)$$

is chosen to approximate quantity of χ^2-measure by inequality

$$\chi_\mathbf{A}^2(Euc(B:A)) \leq \varepsilon_0^2 + \sum_{1\leq i<j\leq |V_A^U|: Euc(\mu_{d/A}(w_i),\mu_{d/A}(w_j))>\varepsilon_0} avg_\chi(i,j) \qquad (131)$$

For weighted discernibility table

$$W_\chi T_\mu^{Euc>\varepsilon_0}(\mathbf{A}) = \{\langle B, n_\chi^{Euc>\varepsilon_0}(B)\rangle : B \in T_\mu^{Euc>\varepsilon_0}(\mathbf{A})\} \qquad (132)$$

we can fix one more threshold $\varepsilon_1 \in [0,1)$ and search for minimal subsets $B \subseteq A$ such that quantity

$$N_\chi^{Euc>\varepsilon_0}(B) = \sum_{C\subseteq A: B\cap C\neq \emptyset} n_\chi^{Euc>\varepsilon_0}(C) \qquad (133)$$

is not less than ε_1. Such a technique combines advantages of both tuning ε_0 to reduce the size of discernibility table and ε_1 – to obtain more flexible approximation of the notion of $(\Delta E \leq \varepsilon)$-approximating μ-decision reduct (for $\varepsilon \approx \varepsilon_0^2 + \varepsilon_1$) than in case of approach provided by Corollary 62.

Example 12. Fig. 10 consists of attribute subsets $B \subseteq A$ with positive assignments $n_\chi^{Euc>1/3}(B)$ computed for decision table from Fig. 1.

One can see that for relatively large ε_0 – here equal to $1/3$ – the size of weighted discernibility table may be a bit decreased with respect, e.g., to those from Fig. 2. Still, some further approximations, by applying some approximate absorption laws, are needed to reduce the above structure for effective calculations.

$B \subseteq A$	$n_\chi^{Euc>1/3}(B)$	$B \subseteq A$	$n_\chi^{Euc>1/3}(B)$
$\{a_3\}$	1/15	$\{a_1, a_3, a_5\}$	1/20
$\{a_1, a_2\}$	1/25	$\{a_1, a_4, a_5\}$	1/16
$\{a_1, a_3\}$	1/25	$\{a_2, a_3, a_5\}$	2/25
$\{a_1, a_4\}$	1/25	$\{a_2, a_4, a_5\}$	1/40
$\{a_1, a_5\}$	1/40	$\{a_3, a_4, a_5\}$	2/25
$\{a_2, a_5\}$	7/120	$\{a_1, a_2, a_3, a_5\}$	1/25
$\{a_3, a_5\}$	1/40	$\{a_1, a_2, a_4, a_5\}$	1/40
$\{a_4, a_5\}$	17/240	$\{a_1, a_3, a_4, a_5\}$	1/25
$\{a_1, a_2, a_5\}$	23/240	$\{a_2, a_3, a_4, a_5\}$	1/40
$\{a_1, a_3, a_4\}$	1/25	$\{a_1, a_2, a_3, a_4, a_5\}$	1/20

Fig. 10. Positive $n_\chi^{Euc>1/3}$-assignments.

5.3 Normalized information measures

In Section 5.1 we introduced notions of local and global normalized μ-information measure, as referring to expected chance of proper classification based on random μ-decision rules. Now, let us reconsider those quantities for the whole family of normalized decision functions introduced in Section 4.2.

Definition 72. Given a decision table $\mathbf{A} = (U, A, d)$, a normalized decision function ϕ and a subset $B \subseteq A$, we define a local ϕ-information measure $e_{\phi/B}(d/\cdot)$: $V_B^U \to [0,1]$ by putting, for each $w_B \in V_B^U$,

$$e_{\phi/B}(d/w_B) = \sum_{v_d \in V_d} \mu_{d/B}(v_d/w_B) \phi_{d/B}(v_d/w_B) \qquad (134)$$

Given such a measure, we can attach to any $w_B \in V_B^U$ an average ϕ-decision rule

$$(B, w_B) \Rightarrow_{e_{\phi/B}(d/w_B)} d \qquad (135)$$

where $e_{\phi/B}(d/w_B)$ is treated as the probability that a randomly chosen object $u \in U$ which fits information pattern w_B over B will be properly classified by the corresponding random ϕ-decision rule of the form (94).

Proposition 73. *([42]) Given a decision table* $\mathbf{A} = (U, A, d)$, *a subset* $B \subseteq A$ *and an information pattern* $w_B \in V_B^U$, *for each normalized decision function* ϕ *we have inequalities*

$$e_{\partial/B}(d/w_B) \leq e_{\phi/B}(d/w_B) \leq e_{m/B}(d/w_B) \tag{136}$$

where $e_{\partial/B}(d/w_B)$ *and* $e_{m/B}(d/w_B)$ *are given by (137) and (138), respectively.*

Let us examine the nature of above marginal local measures. In case of normalized generalized decision function ∂, we obtain that

$$e_{\partial/B}(d/w_B) = |\partial_{d/B}(w_B)|^{-1} \tag{137}$$

One can see that it is indeed the probability that for randomly chosen $u \in Inf_B^{-1}(w_B)$ we guess its decision value $d(u) \in \partial_{d/B}(w_B)$ by uniformly random choice among elements of $\partial_{d/B}(w_B)$. In case of normalized majority decision function m, we obtain that

$$e_{m/B}(d/w_B) = \max_{k=1,..,|V_d|} \mu_{d/B}(v_k/w_B) \tag{138}$$

For simplicity, let us assume that for a given $w_B \in V_B^U$ there is only one decision value reaching the maximum, e.g., $v_1 \in V_d$ such that $\mu_{d/B}(v_1/w_B) = 0.7$. Then, according to definition of m, there is $\phi_{d/B}(v_1/w_B) = 1$, so all objects $u \in Inf_B^{-1}(w_B)$ will be classified as belonging to decision class of v_1. Such a classification will be appropriate in 70%, at least within the universe of known cases.

We could say that reasoning with generalized decision based strategy takes the minimal and reasoning with majority strategy – the maximal amount of information provided by conditional frequencies. Indeed, it meets with the intuition that by reasoning with the set of possible decision values we obtain the most vague answer and, on the other hand, reasoning with the most frequently occurring decision values gives the sharpest output. In application to new cases, however, vagueness of generalized decision might occur to be much less risky than permanent counting on one decision class, what is the case in majority reasoning.

Analogously, we can generalize global normalized μ-information measure.

Definition 74. *Given a decision table* $\mathbf{A} = (U, A, d)$ *and a normalized decision function* ϕ, *we define a global* ϕ-*information measure* $E_\phi(d/\cdot) : 2^A \to [0, 1]$ *as*

$$E_{\phi/\mathbf{A}}(d/B) = \sum_{w_B \in V_B^U} \mu_B(w_B) e_{\phi/B}(d/w_B) \tag{139}$$

Given (139), we can say that any $B \subseteq A$ satisfies approximate boolean implication

$$B \Rightarrow_{E_{\phi/\mathbf{A}}(d/B)} d \tag{140}$$

where $E_{\phi/\mathbf{A}}(d/B)$ is interpreted as the expected chance that a randomly chosen object $u \in U$ will be properly classified by random ϕ-decision rule of the form (94) corresponding to distribution $\phi_{d/B}(Inf_B(u))$.

The following result is analogous to Proposition 73. Again, it may be treated as an argument for using the majority strategy of reasoning, as providing the highest expected chance of proper classification. It is, however, worth remembering that it corresponds to a simplified situation, when we classify new cases under implicit assumption that our decision table is representative for a given problem in statistical sense and that a domain updated by new cases is going to keep frequency based dependencies in somehow stable way.

Proposition 75. *([42]) Given a decision table* $\mathbf{A} = (U, A, d)$ *and a fixed* $B \subseteq A$, *for each normalized decision function ϕ we have inequalities*

$$E_{\partial/\mathbf{A}}(d/B) \leq E_{\phi/\mathbf{A}}(d/B) \leq E_{m/\mathbf{A}}(d/B) \tag{141}$$

where quantities $E_{\partial/\mathbf{A}}(d/B)$ *and* $E_{m/\mathbf{A}}(d/B)$ *correspond to generalized and majority decision functions, respectively.*

Given Definition 74, let us focus on the task of finding optimal subsets of attributes for classification based on the scheme analogous to that described for rough membership distributions in Section 3.3.

Definition 76. Given a normalized decision function ϕ, a threshold $\varepsilon \in [0, 1)$ and a decision table $\mathbf{A} = (U, A, d)$, we say that a subset $B \subseteq A$ ($\Delta E \leq \varepsilon$)-approximately ϕ-defines d iff

$$E_{\phi/\mathbf{A}}(d/B) + \varepsilon \geq E_{\phi/\mathbf{A}}(d/A) \tag{142}$$

A subset $B \subseteq A$ which satisfies the above inequality is called an ($\Delta E \leq \varepsilon$)-approximating ϕ-decision reduct iff there is no proper subset of B which ($\Delta E \leq \varepsilon$)-approximately ϕ-defines d.

According to the above, we obtain two parallel parameters for tuning the search for optimal conditions for classification of new objects. First of them, just like before, refers to the degree up to which we are likely to neglect the decrease of information provided by the smaller subsets $B \subseteq A$ with respect to the whole A. It refers to the following relationship:

Proposition 77. *Given any decision table, a threshold $\varepsilon \in [0,1)$ and a function $\phi \in NDF$, each subset of conditions which ϕ-defines decision ϕ-defines it ($\Delta E \leq \varepsilon$)-approximately as well.*

The second parameter corresponds to the choice of normalized decision function responsible for the strategy of classification, i.e., for the way of understanding the above information. The following result states that even in this situation, regardless of the choice of approximation degree $\varepsilon \in [0, 1)$ and function $\phi \in NDF$ we remain in the same complexity class of optimization problems:

Theorem 78. *([42]) For each normalized decision function $\phi \in NDF$ and for any approximation threshold $\varepsilon \in [0,1)$, the problem of finding minimal ($\Delta E \leq \varepsilon$)-approximating ϕ-decision reduct is NP-hard.*

The above finally confirms us that it is impossible to run away from complexity of MDL-based optimization, under any kind of non-trivial approximation of the notion of decision reduct. In other words, it gives the feeling that regardless of the way of approximate preserving information we should not expect the existence of fast deterministic algorithms extracting optimal solutions from data. Thus, again, the main attention should be paid to adapting different rough set based techniques applying artificial intelligence to searching for, e.g., minimal decision reducts.

From this point of view, further research is needed to draw more general correspondence between introduced modifications and fundamental principles of rough sets, crucial for efficiency of so far developed heuristics. In view of material gathered in Section 5, a good example here is the question, for which functions $\phi \in NDF$ the property of $(\triangle E \leq \varepsilon)$-approximate ϕ-defining decision is closed with respect to supersets. Here we illustrate different situations which may occur for particular normalized decision functions.

Proposition 79. *For any decision table* $\mathbf{A} = (U, A, d)$ *and a subset* $B \subseteq A$ *we have:*

1. $E_{\partial/\mathbf{A}}(d/B) \leq E_{\partial/\mathbf{A}}(d/A)$, *where equality holds, iff* B ∂-*defines* d *in* $\mathbf{A}$.
2. $E_{m/\mathbf{A}}(d/B) \leq E_{m/\mathbf{A}}(d/A)$, *where the property of* m-*defining decision just implies equality, without equivalence analogous to the above one.*
3. *No analogous inequality in case of function* $m/_{0.7}\partial$ − *just implication of equality* $E_{m/_{0.7}\partial/\mathbf{A}}(d/B) = E_{m/_{0.7}\partial/\mathbf{A}}(d/A)$ *by the fact of* $m/_{0.7}\partial$-*defining decision by* B *(what is nothing new in view of Proposition 77).*

Corollary 80. *For* $\varepsilon = 0$, *notions of* ∂-*decision and* $(\triangle E \leq \varepsilon)$-*approximating* ∂-*decision reduct are equivalent. However,* m-*decision and* $m/_{0.7}\partial$-*decision reducts may be larger than* $(\triangle E \leq \varepsilon)$-*approximating* m-*decision and* $m/_{0.7}\partial$-*decision reducts, respectively.*

Corollary 81. *For any decision table, the properties of* $(\triangle E \leq \varepsilon)$-*approximate* ∂-*defining and* m-*defining decision are closed with respect to supersets. This is, however, not the case for the property of* $m/_{0.7}\partial$-*defining decision.*

Further question is whether all such functions can be provided with average (or at least strict) discernibility characteristics similar to that discussed in Section 5.2. Our current state of knowledge does not allow for extractions of sub-families of reasoning strategies which perform well in the above sense. The only thing we know is that not all normalized decision functions do it.

Example 13. In Fig. 11 we include subsets which occurred in previous examples, considered now in view of their values for normalized measures corresponding to different NDF-functions. It illustrates, in particular, boundaries provided by (141) in Proposition 75.

By analyzing the column of measure $E_{m/_{0.7}\partial/\mathbf{A}}$ one can observe that it is not monotonic with respect to inclusion, so indeed we cannot say that property of $(\triangle E \leq \varepsilon)$-approximate $m/_{0.7}\partial$-defining decision is always closed with respect supersets (the same can be seen for $\mu^{0.3}$).

It also turns out that subset $\{a_5\}$ is a $(\triangle E \leq \varepsilon)$-approximating m-decision reduct for $\varepsilon = 0$, although it is not a m-decision reduct.

$B \subseteq A$	∂	$m_{\geq 0.7}$	μ	$\mu^{0.3}$	m
$\{a_5\}$	0.575	0.8	0.694	0.8	0.8
$\{a_1, a_3\}$	0.55	0.55	0.579	0.579	0.65
$\{a_2, a_3\}$	0.5	0.65	0.6	0.66	0.7
$\{a_2, a_4\}$	0.5	0.65	0.581	0.667	0.7
$\{a_1, a_3, a_5\}$	0.675	0.75	0.745	0.763	0.8
$\{a_2, a_3, a_4\}$	0.675	0.725	0.708	0.733	0.75
$\{a_1, a_2, a_3, a_4\}$	0.725	0.775	0.758	0.783	0.8

Fig. 11. Examples of values of different proposed information measures.

Despite the above mentioned open problems, necessary to be solved for better orientation in advantages and disadvantages of particular techniques of approximation of the notion of exact decision reduct, we can still try to get some intuition by performing experiments with data classification. Thus, as a conclusion, let us present an example of weighted classification by using ϕ-decision rules generated from $(\triangle E \leq \varepsilon)$-approximating ϕ-decision reducts, for different thresholds $\varepsilon \in [0, 1)$ and normalized decision functions $\phi \in NDF$.

Example 14. Let us consider a reduced classification scheme

$$S'' = \{\langle\{a_2, a_3, a_4\}, \partial\rangle, \langle\{a_5\}, m\rangle\} \tag{143}$$

where the first subset turns out to be a $(\triangle E \leq \varepsilon)$-approximating ∂-decision reduct for $\varepsilon \in [0.05, 0.225)$ and the second was pointed in previous example as $(\triangle E \leq \varepsilon)$-approximating m-decision reduct for $\varepsilon = 0$. Assume that information about a new object $u_{new} \notin U$ is provided over the whole A, as

$$w_A = (\langle a_1, 0\rangle, \langle a_2, 1\rangle, \langle a_3, 1\rangle, \langle a_4, 1\rangle, \langle a_5, 0\rangle) \tag{144}$$

just like in Example 5. Formula for weighting decision classes takes the form of

$$W_S(v_k/w_A) = \mu_{a_2,a_3,a_4}(1,1,1) \cdot \partial^{\mu}_{d/a_2,a_3,a_4}(v_k/1,1,1) + \mu_{a_5}(0) \cdot m^{\mu}_{d/a_5}(v_k/0) \tag{145}$$

for $k = 1, 2$. As a result we obtain that

$$W_{S''}(v_1/w_A) = \frac{2}{20} \cdot 0 + \frac{17}{20} \cdot \frac{13}{17} = \frac{13}{20} \tag{146}$$

is more than twice higher than

$$W_{S''}(v_2/w_A) = \frac{2}{20} \cdot 1 + \frac{17}{20} \cdot \frac{4}{17} = \frac{6}{20} \tag{147}$$

Let us also reconsider the case of a new object corresponding to conditional information

$$w'_A = (\langle a_1, 1\rangle, \langle a_2, 1\rangle, \langle a_3, 0\rangle, \langle a_4, 1\rangle, \langle a_5, 0\rangle) \qquad (148)$$

analyzed in Example 8. Here, the difference between

$$W_{S''}(v_1/w'_A) = \frac{1}{20} \cdot 1 + \frac{17}{20} \cdot \frac{13}{17} = \frac{14}{20} \qquad (149)$$

and

$$W_{S''}(v_2/w'_A) = \frac{1}{20} \cdot 0 + \frac{17}{20} \cdot \frac{4}{17} = \frac{4}{20} \qquad (150)$$

becomes even more visible.

6 Directions for further research

We compared different kinds of approximations to the notion of a decision reduct, known from previous study ([19], [37], [38], [39], [40]). In view of the need of representation of inconsistent information, we took as the starting point the notion of rough membership decision reduct, which is strongly related to the constructs of conditional independence and Markov boundary known from probability theory and statistics (see e.g. [24]). Approximate criteria of keeping frequency based decision information under conditional feature reduction were considered with respect to distances between frequency distributions and information measures related to different strategies of reasoning and classification under uncertainty.

Presented results provided computational complexity characteristics for optimization problems corresponding to particular approximations. Methodology of efficient solving of those problems were based on (approximate) discernibility operations, similar to those developed and implemented in case of original notion of exact (generalized) decision reduct (cf. Chapter by Bazan et al. in this Volume). It enables to implement and test algorithms searching for optimal approximate reducts of different types over real-life data.

Introduced notions were illustrated by examples of application of a common weighted scheme for classification of new cases. Development of adaptive methods searching for optimal classification models is possible due to a wide bunch of provided approximation parameters (compare, e.g., with [7], [50]). Still, further theoretical study is needed for setting the most appropriate ranges for tuning these parameters.

Acknowledgements Paper supported by the National Research Committee grant 8T11C02417 and ESPRIT project 20288 CRIT-2.

References

1. Bazan, J.: A Comparison of Dynamic and non-Dynamic Rough Set Methods for Extracting Laws from Decision Tables. In: L. Polkowski, A. Skowron (eds.), Rough Sets in Knowledge Discovery, Heidelberg, Physica-Verlag (1998) pp. 321–365.

2. Brown, E.M.: Boolean Reasoning. Kluwer Academic Publishers, Dordrecht (1990).
3. Duentsch, I., Gediga, G.: Uncertainty measures of rough set prediction. Artificial Intelligence, **106** (1998) pp. 77–107.
4. Duentsch, I., Gediga, G.: Rough set data analysis. Encyclopedia of Computer Science and Technology, Marcel Dekker (2000) to appear.
5. Gallager, R.G.: Information Theory and Reliable Communication. John Wiley & Sons, New York (1968).
6. Garey, M.R., Johnson, D.S.: Computers and Intractability: A Guide to the Theory of NP-Completeness. Freeman and Company, San Francisco (1979).
7. Góra, G., Ejdys, P.: System enabling synthesis of complex objects on rough mereology (documentation of MAS project based on rough mereology version 2.0). Manuscipt, ESPRIT project 20288 CRIT-2.
8. Iman, R.L.: A data-based approach to statistics. Duxbury Press, Wadsworth, Inc. (1994).
9. Kapur, J.N., Kesavan, H.K.: Entropy Optimization Principles with Applications. Academic Press (1992).
10. Komorowski, J., Polkowski, L., Skowron, A.: Towards a rough mereology–based logic for approximate solution synthesis Part 1. Studia Logica **58/1** (1997) 143–184.
11. Lenarcik, A., Piasta, Z.: Rough Classifiers Sensitive to Costs Varying from Object to Object. In: L. Polkowski, A. Skowron (eds.), Proceedings of the First International Conference on Rough Sets and Current Trends in Computing (RSCTC'98), June 22–26, Warsaw, Poland, Springer-Verlag, Berlin Heidelberg (1998), pp. 222–230.
12. Li, M., Vitanyi, P.: An Introduction to Kolmogorov Complexity and Its Applications. Springer-Verlag (1997).
13. Michalewicz, Z.: Genetic Algorithms + Data Structures = Evolution Programs. Springer-Verlag (1994).
14. Nguyen, H.S.: Efficient SQL-Querying Method for Data Mining in Large Data Bases. In: Proc. of Sixteenth International Joint Conference on Artificial Intelligence (IJCAI'99), Stockholm, Sweden (1999) pp. 806-811.
15. Nguyen, H.S., Nguyen, S.H.: Pattern extraction from data. Fundamenta Informaticae **34** (1998) pp. 129–144.
16. Nguyen, H.S., Nguyen, S.H.: Rough Sets and Association Rule Generation. Fundamenta Informaticae **40/4** (2000) pp. 383–405.
17. Nguyen, H.S., Skowron, A.: Quantization of real value attributes: Rough set and boolean reasoning approach. Bulletin of International Rough Set Society 1/1 (1996) pp. 5–16.
18. Nguyen, S.H., Skowron, A., Synak, P.: Discovery of data patterns with applications to decomposition and classification problems. In: L. Polkowski, A. Skowron (eds.), Rough Sets in Knowledge Discovery **2**, Physica-Verlag, Heidelberg (1998), pp. 55–97.
19. Nguyen, H.S., Ślęzak, D.: Approximate Reducts and Association Rules Correspondence and Complexity Results. In: N. Zhong, A. Skowron and S. Ohsuga (eds.), Proc. of the Seventh International Workshop on New Directions in Rough Sets, Data Mining, and Granular-Soft Computing (RSFDGrC'99), Yamaguchi, Japan, Lecture Notes in Artificial Intelligence **1711** (1999), pp. 137–145.
20. Pawlak, Z.: Rough sets – Theoretical aspects of reasoning about data. Kluwer Academic Publishers, Dordrecht (1991).

21. Pawlak, Z.: Rough Modus Ponens. In: Proceedings of the Seventh International Conference on Information Processing and Management of Uncertainty in Knowledge-Based Systems (IPMU'98), July 6–10, Paris, France (1998).
22. Pawlak, Z.: Decision rules, Bayes' rule and rough sets. In: N. Zhong, A. Skowron, S. Ohsuga (eds.), New Directions in Rough Sets, Data Mining and Granular-Soft Computing, LNAI **1711**, Springer Verlag, Berlin (1999) pp. 1–9.
23. Pawlak, Z., Skowron, A.: Rough membership functions. In: R.R. Yaeger, M. Fedrizzi, and J. Kacprzyk (eds.), Advances in the Dempster Shafer Theory of Evidence, John Wiley & Sons, Inc., New York, Chichester, Brisbane, Toronto, Singapore (1994), pp. 251–271.
24. Pearl, J.: Probabilistic Reasoning in Intelligent Systems: Networks of Plausible Inference. Morgan Kaufmann (1988).
25. Piasta, Z., Lenarcik, A., Tsumoto S.: Machine discovery in databases with probabilistic rough classifiers. In: S. Tsumoto, S. Kobayashi, T. Yokomori, H. Tanaka and A. Nakamura (eds.), Proceedings of The Fourth International Workshop on Rough Sets, Fuzzy Sets, and Machine Discovery (RSFD'96), November 6–8, The University of Tokyo (1996) pp. 353–359.
26. Polkowski, L., Skowron, A.: Rough mereology and analytical morphology: New developments in rough set theory. In: M. de Glass and Z. Pawlak (eds.), Proceedings of the Second World Conference on Fundamentals of Artificial Intelligence (WOCFAI'95), Angkor, Paris (1995) pp. 343–354.
27. Rissanen, J.: Modeling by the shortest data description. Authomatica, **14** (1978) pp. 465–471.
28. Rissanen, J.: Minimum-description-length principle. In: S. Kotz, N.L. Johnson (eds.), Encyclopedia of Statistical Sciences, New York, Wiley (1985) pp. 523–527.
29. Shafer, G.: A mathematical theory of evidence. Princeton, NJ: Princeton University Press.
30. Skowron, A.: Boolean reasoning for decision rules generation. In: Proc. of the 7-th International Symposium ISMIS'93, Trondheim, Norway, 1993; J. Komorowski, Z. Ras (eds.), Lecture Notes in Artificial Intelligence, **689**, Springer-Verlag (1993) pp. 295–305.
31. Skowron, A.: Extracting laws from decision tables. Computational Intelligence 11/2 (1995) pp. 371–388.
32. Skowron, A.: Synthesis of adaptive decision systems from experimental data. In: A. Aamodt, J. Komorowski (eds.), Proc. of the Fifth Scandinavian Conference on Artificial Intelligence (SCAI'95), May 1995, Trondheim, Norway, IOS Press, Amsterdam (1995) pp. 220–238.
33. Skowron, A.: Rough Sets in KDD. In preparation.
34. Skowron, A., Grzymala-Busse, J.: From rough set theory to evidence theory. In: R.R. Yaeger, M. Fedrizzi, and J. Kacprzyk (eds.), Advances in the Dempster Shafer Theory of Evidence, John Wiley & Sons, Inc., New York, Chichester, Brisbane, Toronto, Singapore (1994), pp. 193–236.
35. Skowron, A., Rauszer, C.: The discernibility matrices and functions in information systems. In: R. Słowiński (ed.), Intelligent Decision Support. Handbook of Applications and Advances of the Rough Set Theory, Kluwer Academic Publishers, Dordrecht (1992), pp. 311–362.
36. Skowron, A., Stepaniuk, J.: Tolerance approximation spaces. Fundamenta Informaticae 27/2-3 (1996) pp. 245–253.
37. Ślęzak, D.: Approximate reducts in decision tables. In: Proceedings of the Sixth International Conference, Information Processing and Management of Uncertainty

in Knowledge-Based Systems (IPMU'96), July 1–5, Granada, Spain (1996), **3**, pp. 1159–1164.
38. Ślęzak, D.: Searching for frequential reducts in decision tables with uncertain objects. In: L. Polkowski, A. Skowron (eds.), Proceedings of the First International Conference on Rough Sets and Current Trends in Computing (RSCTC'98), Warsaw, Poland, June 22–26, 1998, Lecture Notes in Artificial Intelligence 1424, Springer-Verlag, Heidelberg (1998), pp. 52–59.
39. Ślęzak, D.: Searching for dynamic reducts in inconsistent decision tables. In: Proceedings of the Seventh International Conference on Information Processing and Management of Uncertainty in Knowledge-Based Systems (IPMU'98), July 6–10, Paris, France (1998), **2**, pp. 1362–1369.
40. Ślęzak, D.: Decision information functions for inconsistent decision tables analysis. In: Proceedings of the Seventh European Congress on Intelligent Techniques & Soft Computing, September 13–16, Aachen, Germany (1999) p. 127.
41. Ślęzak, D.: Foundations of Entropy-Based Bayesian Networks: Theoretical Results & Rough Set-Based Extraction from Data. Submitted to IPMU'2000 (2000).
42. Ślęzak, D.: Normalized decision functions and measures for inconsistent decision tables analysis. To appear in Fundamenta Informaticae (2000).
43. Ślęzak, D.: Different approaches to reasoning with frequency-based decision reducts. In preparation (2000).
44. Ślęzak, D., Wróblewski, J.: Classification algorithms based on linear combinations of features. In: Proceedings of the Third European Conference PKDD'99, Praga, Czech Republik, 1999, LNAI **1704**, Springer, Heidelberg (1999) pp. 548–553.
45. Tsumoto, S.: Automated Induction of Medical Expert System Rules from Clinical Databases based on Rough Set Theory. Information Sciences **112** (1998) pp. 67–84.
46. Vapnik, V.N.: Statistical Learning Theory. John Wiley & Sons, Inc. (1999).
47. Wong, S.K.M., Ziarko, W., Li Ye, R.: Comparison of rough-set and statistical methods in inductive learning. International Journal of Man-Machine Studies, 24 (1986) pp. 53–72.
48. Wróblewski, J.: Theoretical Foundations of Order-Based Genetic Algorithms. Fundamenta Informaticae, **28** (3, 4), IOS Press (1996) pp. 423–430.
49. Wróblewski, J.: Genetic algorithms in decomposition and classification problems. In: L. Polkowski and A. Skowron (eds.), Rough Sets in Knowledge Discovery 2: Applications, Case Studies and Software Systems, Physica-Verlag, Heidelberg (1998), pp. 471–487.
50. Wróblewski, J.: Analyzing relational databases using rough set based methods. Submitted to IPMU'2000 (2000).
51. Ziarko, W.: Variable Precision Rough Set Model. Journal of Computer and System Sciences, 40 (1993) pp. 39–59.

PART 3:

METHODS AND APPLICATIONS: REGULAR PATTERN EXTRACTION, CONCURRENCY

Chapter 6

Regularity Analysis and its Applications in Data Mining

Sinh Hoa Nguyen

[1] Institute of Computer Sciences Warsaw University
 Banacha 2 02-097 Warsaw Poland
[2] Polish-Japanese Institute of Information Technology
 Koszykowa 86 02-008 Warsaw Poland
 email: hoa@mimuw.edu.pl

Abstract: Knowledge discovery is concerned with extraction of useful information from databases ([21]). One of the basic tasks of knowledge discovery and data mining is to synthesize the description of some subsets (concepts) of entities contained in databases. The patterns and/or rules extracted from data are used as basic tools for concept description. In this Chapter we propose a certain framework for approximating concepts. Our approach emphasizes extracting regularities from data. In this Chapter the following problems are investigated: (1) issues concerning the languages used to represent patterns; (2) computational complexity of problems in approximating concepts; (3) methods of identifying optimal patterns. Data regularity is a useful tool not only for concept description. It is also indispensable for various applications like classification or decomposition. In this Chapter we present also the applications of data regularity to three basic problems of data mining: classification, data description and data decomposition.

1 Introduction

Knowledge discovery is concerned with extraction of useful information from databases ([21]). Data mining is a main step in the multi-stage process of knowledge discovery. Data mining concentrates on exploration and analysis, by automatic or semiautomatic means, of large quantities of data in order to discover meaningful patterns and rules.

One of the basic tasks of knowledge discovery and data mining is to synthesize the description of some subsets (concepts) of entities contained in databases. The patterns and/or rules extracted from data are used as basic tools for concept description.

Usually, the nature of the investigated phenomenon suggests certain primitive concepts, i.e. sets of entities which are building blocks to form more complex concepts and describe properties. The following three problems (see [76]) are main parts of any process of synthesizing the descriptions of concepts:

(i) Determining primitive concepts from which approximations of more complex concepts are assembled. The problem is to construct relevant primitive concepts to assure that the synthesized concepts have high quality.
(ii) Defining the measures of closeness between constructed and target concepts. The problem is to define proper measures to assure that the optimal concepts constructed according to these measures approximate the target concepts well.
(iii) Synthesizing complex concepts from primitive ones. Having the primitive concepts, the problem is to define the proper synthesizing operation, which would assure that the complex concepts constructed from them are meaningful and useful.

A pattern is often used as a way to specify the primitive concept. The first stage of determining primitive concepts involves pattern extraction. To make the process of pattern extraction meaningful and efficient we need to solve the following three problems:

(i) To find a language for pattern representation. This aspect may be critical. The chosen language should be expressive enough to render all the essential concepts and data structures. On the other hand, too expressive a language may cause too high complexity of the process of inference or/and overfitting.
(ii) To determine a meaningful function to evaluate the quality of a pattern. The choice of a function measuring the description quality is also a difficult problem. Having a part of data set, the function should sufficiently estimate the closeness of approximate description to the real concept.
(iii) To find methods of identifying optimal patterns. Such problems may turn out to be computationally hard. The reason may be intrinsic, like hardness in a complexity class, or a sheer size of the data.

In this Chapter we propose a certain framework for approximating concepts. Our approach emphasizes extracting regularities from data. For a given concept C, a regularity is a group of features contained in data satisfied by a large number of examples from C. Regular features of data are used to express the (primitive) concepts. Usually, regularity is a pattern expressed in some logical language.

The results presented in this Chapter can be categorized into the following four groups:

- issues concerning the languages used to represent patterns;
- computational complexity of problems in approximating concepts;
- methods of identifying optimal patterns;
- experiments.

The following is their overview:

1. Languages
The conjunctive propositional language is often used in data mining. Patterns are defined as conjunctions of expressions (descriptors) of the form ($attribute = value$). This language is simple and in many cases sufficient for

applications. However the essential problem related to inductive learning is that we need to approximate a real concept from incomplete data. Therefore the description of a concept needs to be general enough. In numerous practical applications it is not sufficient to provide a satisfactory concept approximation only on a given data set. In inductive reasoning for instance, we also need a high approximation quality on extensions of that data set. To deal with this, one could extend the classical language to a more sophisticated and general one. The main advantage of these new languages is their expressive power which allows to describe real concepts more precisely. However, new languages should be sufficiently simple, so that patterns expressed in them can still be induced efficiently. We consider the following languages to express primitive concepts:

(i) *Simple descriptor language*: it consists of expressions of the form $(a = v_a)$, called *descriptors*, where a is an attribute and v_a is a value from a's domain $\mathcal{DOM}(a)$.

(ii) *Generalized descriptor language*: it consists of expressions of the forms $(a \in V_a)$ or $(a \in [v_1, v_2])$, where a is an attribute. The form $(a \in V_a)$ is often used for defining the concepts in symbolic domains, the value set V_a is some subset of the domain $\mathcal{DOM}(a)$. However, the form $(a \in [v_1, v_2])$ is used to define the concepts in numeric domains. In this case, $[v_1, v_2]$ is a real interval contained in the domain $\mathcal{DOM}(a)$.

(iii) *Neighborhood language*: it consists of descriptors which are characteristic functions of neighborhoods defined by "optimal" similarity (tolerance) relations found from data. Intuitively, for an object x_0 and similarity relation $\mathcal{R}$, a descriptor defined by the pair $(x_0, \mathcal{R})$ describes the set of objects x related with x_0 ($\langle x_0, x \rangle \in \mathcal{R}$). In Section 7 we explain precisely what we mean by optimal similarity relations.

Patterns defined by conjunctions of descriptors expressed in (generalized) descriptor language are called *templates*. The relational language is more sophisticated, patterns are defined by tolerance relations. These patterns are called *relational patterns*.

2. Computational complexity

The problem of finding relevant primitive concepts is often computationaly hard. Certain new results of this Chapter justify this claim. We present five new results:

The first result concerns the *optimal template problem* (see Subsection 2.3). We prove that, for a given positive integer k, the problem of finding the k primitive concepts defined by descriptors of the form $(a = v_a)$, such that the complex concept defined by their conjunction is supported by the population of the largest size, is NP-hard. The result is stated in Theorem 2 (Subsection 2.3, Section 2).

The second result is about the *optimal cut set problem* (see Subsection 4.2). The problem is to find a minimal set of primitive concepts of the form $a \in [v_1, v_2]$, necessary for construction of concepts, which describe the entire data set under the constrains requiring that discernibility be preserved. The problem has been investigated in [58], where is was shown that the problem is NP-hard if the number of features in the data set is an input parameter of the problem. In this Chapter we show that the problem of finding the optimal cut set is NP-hard if the number of attributes is a constant larger than 1 (see Theorem 8, Subsection 4.2, Section 4).

The third result concerns the *optimal set of cuts*. We consider here another optimization criterion. The set of cuts is *optimal* if it discerns all the objects by different decisions and the number of "regions" defined by these cuts is minimal. We show in Theorem 9 (Subsection 4.2, Section 4) that the optimization problem defined by this criterion is NP-hard.

The next result is about the *optimal family of partitions problem*. The motivation is to find a minimal set of primitive concepts of the form $(a \in V_a)$, which describe the entire data set under the constrains requiring that discernibility be preserved. A *partition* for a given attribute a is obtained by dividing the domain of a into disjoint subsets. A family of partitions is optimal if the number of subsets defined by these partitions is minimal. In Theorem 10 (Subsection 4.3, Section 4) we show the problem to find an optimal family of partitions to be NP-hard.

The last result concerns the *optimal binary partition problem*. For a given attribute a, a *binary partition* divides $a's$ domain into two disjoint subsets. A binary partition is optimal if it discerns a maximum number of objects from different decision classes. We show that this problem is NP-hard (see Theorem 12, Subsection 5.2, Section 5).

3a. Template extraction

The next part of this Chapter is concerned with developing methods to extract regularities hidden in data sets. We consider two kinds of regularities: regularities defined by templates and regularities defined in a relational language. The process of extracting regularities should find templates of high quality. The following two techniques are investigated in this Chapter.

We present a universal scheme for template extraction. It concerns the problem of feature selection (see [37]). The universality of the presented methods is shown by the fact that the templates generated using our algorithms can be applied as efficient tools for various data mining tasks like classification, description or decomposition. Moreover the universality of this approach is shown by the fact that it allows to extract not only simple patterns but also generalized ones. Details of these methods are presented in Section 3.

The other method concerns techniques of decision trees, which are based on splitting the labeled data in recursive manner to obtain subsets of objects with the same label. The result of splitting is a decision tree with internal nodes representing splitting functions; the leaves represent decision classes. Many systems have been investigated for decision tree induction [9, 70, 71], but the common drawback of these methods is the inefficient of splitting of a symbolic domain. We present a new approach based on value grouping (see Subsection 5.2). The method of constructing decision trees is discussed in Section 5.

3b. Extraction of relational patterns

We use similarity (tolerance) relations defined by a neighborhood language to extract primitive concepts. Similarity is often used as a tool for "information granulation". In many papers, the similarity relation has been studied as an extension of the indiscernibility relation used in rough set approach [62]. Instead of an equivalence relation, different generalizations have been proposed ([13], [24], [84], [35]). Similarity relations are usually defined as conjunctions of local similarity relations $\mathcal{R}_a$ of attributes. We consider in this Chapter a wide class of tolerance relations defined by different combinations of local similarities. The properties of tolerance relations and methods of searching for optimal tolerance relations are included in the third part of this Chapter. Properties of such relations are studied. They turn out to have simple and interesting geometrical interpretations, allowing to develop effective heuristics to search for sub-optimal relations in different similarity families. We present methods to extract relevant similarity relations from data rather than assuming apriori their definitions (see e.g. [19]). These similarity relations are extracted in optimization processes from some parameterized families of relations. The notions and properties of these similarity families are discussed in Section 7. In Section 8 we present methods for similarity extraction. The advantage of our method is its universality i.e. one can extract different kinds of similarity relations, in particular, the relations from some basic families (see Subsection 8.2, Section 8).

4. Results of experiments

Data regularity is a useful tool not only for concept description. It is also indispensable for various applications like classification or decomposition. All the heuristics discussed in the Chapter have been implemented. The experiments have demonstrated amply that our methods are not only efficient with respect to time but also of high quality in different application domains. In Section 10 we present experimental results related to three problems: classification, data description and data decomposition.

The *classification task* consists of two sub-tasks: examining the features of a new object and assigning it to one of the predefined set of decision classes. As tools for classification we use templates, decision trees and tolerance classes. As compared with standard rules, the generalized ones show fundamental advantages. They help to reduce the set of classifiers and improve radically the recognition ratio as well as the classification quality. The new decision tree method has been compared with the methods known before. It often provides a better accuracy of classification. The quality of classifiers defined by tolerance classes is compared to that of classifiers induced by means of k-NN methods [33]. In general, the distance extracted from data gives better results than distance measures taken independently from data by means of k-NN methods.

The second problem is *data description*. The problem is to construct some logical expression to describe a given concept. The quality of description is evaluated by simplicity of description and closeness of an approximating concept to the target concept. Templates are used as primitive concepts. Our experiments have demonstrated that the proposed methods can be used to generate a simple description of the concept and to approximate the concept with high confidence.

Often we need to deal with very large data bases, and no data mining or learning methods can be used directly. In this case the large data set should be partitioned into smaller ones. Data partition is the main task of *decomposition problem*. The goal is to divide data table into regular sub-tables. Experiments have shown that one can obtain "regular" sub-tables using proper templates; usually they contain a smaller number of decision classes than the original one. The regularity of sub-tables is often essential in further stages of the learning process.

Organization of the Chapter

The Chapter consists of eleven sections. In the first five sections $(2 \div 6)$ the results about the concept of a template are discussed. In the next three sections $(7 \div 9)$, the problems concerning relational patterns are investigated. The experimental results and final remarks are presented in the last two sections.

The basic notions related to templates are provided in Section 2. In this section we show hardness of the optimal template problem. In Section 3 we present methods for template extraction using the sequential searching approach. In Section 4 data segmentation is presented as a technique for template set extraction. We present in this section two results concerning the hardness of optimal cut set problems. In Section 5 we present a new method for decision tree construction. In this section we prove the hardness of the value grouping problem. We show how to apply templates to solve some data mining tasks in Section 6.

Basic notions related to relational patterns are discussed in Section 7. In this section we present also properties and structures of some families of tolerance relation. In Section 8 a general scheme for tolerance extraction is discussed. We discuss strategies for extraction of three different types of tolerance relations: local, global and categorical. Section 9 is dedicated to applications of tolerance relations in data mining.

In Section 10 the experimental results are reported. We discuss the quality of algorithms solving three basic problems: classification, data description and data decomposition. Section 11 consists of final remarks and main direction of future research.

2 Template preliminaries

This section introduces concepts and methodology of regularity analysis in terms of *templates*. To understand basic issues for template approach and to appreciate different methods for template generation, we should consider several of their aspects. The four major ones are: (1) What kind of measures should be used to determine the best templates or what are the criteria for template evaluation?

(2) What is the computational complexity of optimal template problems? (3) How to extract the (semi-)optimal template or the optimal template set? (4) How to apply the extracted templates in the data mining problems? We propose answers to these questions in this section.

2.1 Basic notions

One of the first decision to be made while developing a knowledge representation system is choosing data structures to represent primitive notions. One of them is the *attribute-value representation*. In this model data are represented by an *information system* $\mathbf{S} = (U, A)$, where U is a finite nonempty set of objects, A is a finite nonempty set of attributes (see [62]). Every attribute a is a functions $a : U \to \mathcal{DOM}(a)$, where $\mathcal{DOM}(a)$ is a finite set of values of the attribute a, called the *domain* of a. Every object x in an information system $\mathbf{S}$ is characterized by its *information vector*:

$$Inf_\mathbf{S}(x) = \{(a, a(x)) : a \in A\}.$$

A decision table is a special information system $\mathbf{S} = (U, A \cup \{d\})$, where $d \notin A$ is a special attribute called *decision*. Attributes in A are called *condition attributes*. One can assume that $d : U \to \mathcal{DOM}(d) = \{1, ..., m\}$. For any $i \in \mathcal{DOM}(d)$, the set $C_i = \{x \in U : d(u) = i\}$ is called the ith *decision class*.

We distinguish two attribute categories: numeric and symbolic. An attribute a is *numeric* (or *continuous*) if its domain $\mathcal{DOM}(a)$ can be ordered linearly (e.g. age of patients, income of customers). Otherwise, attribute is called *symbolic* (or *nominal*) (e.g. jobs of customers, colors).

We can describe relationships between objects by their attribute values. With respect to an attribute $a \in A$, a *indiscernibility relation* $IND(a)$ is defined by:

$$xIND(a)y \iff a(x) = a(y).$$

This definition can be extended for any subset $B \subseteq A$ as follows:

$$xIND(B)y \iff \forall_{a \in B}(a(x) = a(y)).$$

That is, two objects are considered to be *indiscernible* by the attributes in B, if and only if they have the same value for every attribute in B. The reflexivity, symmetry and transivity of $IND(B)$ follows from the properties of the relation $=$ between attribute values. Therefore $IND(B)$ is an equivalence relation.

For every object $x \in U$ and $B \subseteq A$ the equivalence class $[x]_{IND(B)}$ is defined by: $[x]_{IND(B)} = y \in U : xIND(B)y$.

Let $\mathbf{S} = (U, A)$ be an information system. Any clause of the form $D = (a \in V_a)$ is called the *descriptor* and the value set V_a is called the *range* of D. If a is a numeric attribute, we restrict the ranges of descriptors for a to real intervals, that means $V_a = [v_1, v_2] \subseteq \mathcal{DOM}(a) \subseteq \mathbb{R}$. In case of symbolic attributes, V_a can be any non-empty subset $V_a \subseteq \mathcal{DOM}(a)$. The *volume* of a given descriptor $D = (a \in V_a)$ is defined by

$$Volume(D) = |V_a \cap a(U)|$$

where $|V_a \cap a(U)|$ is the number of values occurring on the attribute a or the cardinality[3] of the set $(V_a \cap a(U))$. By $Prob(D)$, where $D = (a \in V_a)$, we denote the *hitting probability* of the set V_a i.e.

$$Prob(D) = \frac{|V_a|}{|\mathcal{DOM}(a)|}$$

If a is a numeric attribute then

$$Prob(D) = \frac{v_2 - v_1}{\max(\mathcal{DOM}(a)) - \min(\mathcal{DOM}(a))}$$

where $\max(\mathcal{DOM}(a))$, $\min(\mathcal{DOM}(a))$ denote the maximum and minimum value of domain $\mathcal{DOM}(a)$, respectively.

Any propositional formula $T = \bigwedge_{a \in B} (a \in V_a)$ (i.e. any conjunction of descriptors defined over attributes from $B \subseteq A$) is called *a template* of **S**. Template T is *simple*, if any descriptor of T has range of one element. Templates with descriptors consisting of more than one element are called *generalized*. For any $X \subseteq U$, the set of objects $\{x \in X : \forall_{a \in B}\, a(x) \in V_a\}$ from X satisfying T is denoted by $[T]_X$. Denote also $support_X(T) = |[T]_X|$.

Any template $T = D_1 \wedge ... \wedge D_k$, where $D_i = (a_i \in V_i)$, is characterized by the following parameters:

1. $length(T)$, which is the number of descriptors occurring in T;
2. $support(T) = support_U(T)$, which is the number of objects in U satisfying T;
3. $applength(T)$, which is $\sum_{1 \leq i \leq k} \frac{1}{Volume(D_i)}$ called the *approximated length* of the generalized template T.

Functions *applength* and *length* coincide for simple templates.

In some applications we are interested in templates "well matching" some chosen decision classes. Such templates are called *decision templates*. The template associated with one decision class is called *decision rule*. The precision of any decision template (e.g. $T \to (d \in V)$) is estimated by its *confidence ratio*. The confidence of decision rule $T \to (d = i)$ (associated with decision class C_i) is defined as

$$confidence_{C_i}(T) = \frac{support_{C_i}(T)}{support(T)}$$

2.2 Evaluation measures

In this section we address the issue of "template quality". It should be evaluated taking into account an individual application. In general, templates with high quality have the following properties:

(i) They are supported by a "large" number of objects;

---

[3] We denote the cardinality of a set X by $|X|$.

(ii) They are of high "specificity", i.e. they should be described by sufficiently "large" number of descriptors and the set of values $\mathcal{DOM}(a)$ in any descriptor $(a \in V_a)$ is to be "small".

Moreover, decision templates (decision rules) used for classification tasks should be of high *predictive accuracy*. Decision templates (decision rules) of high quality should have the following properties:

(i) They should be supported by "large" number of objects;
(ii) They should have high predictive accuracy.

Template quality For a given template T, its support is defined by function $support(T)$ and its length by $applength(T)$. Any quality measure is a combination of these functions. We use one of the following functions:

$$quality(T) = support(T) + applength(T) \qquad (1)$$
$$quality(T) = support(T) \cdot applength(T) \qquad (2)$$

Decision template quality In case of decision templates, one should take into consideration the confidence ratios. Let T be a decision template associated with the decision class C_i. The quality of T can be measured by one of the following functions:

$$quality(T) = confidence(T) \qquad (3)$$
$$quality(T) = support_{C_i}(T) \cdot applength(T) \qquad (4)$$
$$quality(T) = support_{C_i}(T) \cdot confidence(T) \qquad (5)$$

2.3 Computational complexity

In this section we present some results on computational complexity of the template problems. We will show that an optimization problem for *simple templates* is NP-hard.

We consider two problems. The first problem concerns the complexity of finding a template of the *fixed length* and with the *maximal support*. We will see that the support and the length play the symmetric roles. A similar result also holds for the *maximal length* and the *fixed support*.

The second problem concerns the complexity of finding the template of *maximal quality* measured by a combination of the template *support* and the template *length*.

Maximal support problem The subject of this section is to estimate the computational complexity of the problem to find the template with the *maximum support*. The template is L-*optimal* if the number of objects matching it is maximal among templates with the length equal to a given number L. We show that the template decision problem is NP-complete and the corresponding optimization problem is NP-hard.

A template decision problem is defined as follows.

> **PROBLEM:** Template Support (TS)
> **Instance:** *Information system* $\mathbf{S} = (A, U)$, *and positive integers* S, L
> **Question:** *Is there a template T with the length equal to L and the support at least S?*

The corresponding optimization problem is defined as follows.

> **PROBLEM:** Optimal Template Support (OTS)
> **Input:** *Information system* $\mathbf{S} = (A, U)$, *and positive integer* L
> **Output:** *Find a template T with the length L and the maximal support.*

Below we list some NP-complete problems used to show NP-completeness of the Template Support Problem.

> **PROBLEM:** Balanced Complete Bipartite Sub-graph (BCBS)[23]
> **Instance:** *Bipartite undirected graph* $G = (V_1 \cup V_2, E)$ *and a positive integer* $K \leq \min(|V_1|, |V_2|)$
> **Question:** *Are there two subsets $U_1 \subseteq V_1, U_2 \subseteq V_2$ satisfying $|U_1| = |U_2| = K$ and $\{u, v\} \in E$ for any $u \in U_1, v \in U_2$?*

The BCBS problem is NP-complete [23]. We consider a modified version of BCBS problem called the *Complete Bipartite Sub-graph* (CBS) problem.

> **PROBLEM:** Complete Bipartite Sub-graph (CBS)
> **Instance:** *Bipartite undirected graph* $G = (V_1 \cup V_2, E)$ *and positive integers* $K_1 \leq |V_1|, K_2 \leq |V_2|$
> **Question:** *Are there two subsets $U_1 \subseteq V_1, U_2 \subseteq V_2$ such that $|U_1| = K_1, |U_2| \geq K_2$ and $\{u, v\} \in E$ for any $u \in U_1, v \in U_2$?*

Lemma 1. *The CBS problem is NP-complete.*

Proof:
To see that CBS $\in$ NP, notice that a non-deterministic algorithm needs only to guess the subsets $U_1 \subseteq V_1$ and $U_2 \subseteq V_2$ with $|U_1| = K_1, |U_2| \geq K_2$ and to check in polynomial time if the sub-graph defined on $U_1 \cup U_2$ is complete, i.e. if $u \in U_1, v \in U_2$ implies $\{u, v\} \in E$.

Let $G|_{U_1 \cup U_2}$ denote the sub-graph of G being the restriction of G to $U_1 \cup U_2$. We say that the sub-graph $G|_{U_1 \cup U_2}$ has the size (K_1, K_2) if $|U_1| = K_1, |U_2| = K_2$.

We will transform BCBS to CBS. Let a graph G along with an integer K be an instance of BCBS. For CBS, we consider the same graph G with parameters $K_1 = K_2 = K$. It is clear that the graph G has a complete sub-graph $G|_{U_1 \cup U_2}$ such that $|U_1| = K_1$ and $|U_2| \geq K_2$ if and only if it contains a complete sub-graph $G|_{U_1 \cup U_2}$, where $|U_1| = K_1$ and $|U_2| = K_2$. We obtain in this way a polynomial reduction of BCBS to CBS. ∎

Lemma 2. *TS and CBS are polynomially equivalent.*

Proof:
First we show that TS is polynomially reducible to CBS. Let an information system $\mathbf{S} = (U, A)$ and positive integers $L \leq |A|, F \leq |U|$ be given as an arbitrary instance of the TS, where L denotes the length of template to be found out and matched by at least F objects. We shall construct a bipartite graph $G = (V_1 \cup V_2, E)$ and parameters K_1, K_2 such that G has a complete sub-graph of the size (K_1, K_2) if and only if there exists in $\mathbf{S}$ a template with the length L matched by at least F objects. $G = (V_1 \cup V_2, E)$ is constructed as follows: V_1 is the set of objects U and V_2 is the set of all attribute values. Formally, the vertex sets of graph G are defined by

$$V_1 = U \text{ and } V_2 = \{(a = v) : a \in A, v \in V_a\}$$

Any vertex $u \in V_1$ is connected with vertex $(a = v) \in V_2$ iff $a(u) = v$.

We recall that a template is a descriptor conjunction of the form

$$T = \bigwedge_{a \in B} (a = v_a), \text{ where } B \subseteq A$$

Hence every template can be treated as a subset of V_2. One can observe that if T is a template of length L and support F and $U_1 \subseteq V_1$ is a set of objects matching T then the sub-graph $G|_{U_1 \cup T}$ is a complete bipartite graph with $|U_1| = F$ and $|T| = L$. Conversely, any complete bipartite sub-graph $G|_{U_1 \cup T}$, where $U_1 \subseteq V_1, |U_1| = F$ and $T \subseteq V_2, |T| = L$ defines exactly one template T with length L and support S. An illustration of the graph G and a complete sub-graph $G|_{U_1 \cup T}$ is shown in Figure 1. Solid lines represent edges of the graph G and bold lines represent edges of the complete sub-graph $G|_{U_1 \cup T}$ defining the template T where U_1 is the set of objects matching it.

We conclude that the graph G has a complete sub-graph of size (S, L) if and only if an information system $\mathbf{S}$ has a template T with the length L and the support F. Graph G can be constructed in polynomial time from an information system $\mathbf{S}$.

We show that CBS can be transformed polynomially into TS. We assume that a bipartite graph $G = (V_1 \cup V_2, E)$ and positive integers K_1, K_2 are given as an instance of CBS. We shall construct an information system $\mathbf{S}$ and parameters F, L such that the system $\mathbf{S}$ has a template of the length L and the support at least equal to F if and only if there is a complete sub-graph of G with the size

S	a	b	c
u_1	1	Y	0
u_2	2	N	0
u_3	2	N	0
u_4	3	Y	0
u_5	2	N	1

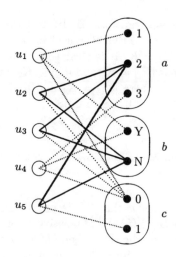

Fig. 1. The bipartite graph generated from the table and the sub-graph corresponding to the template: $(a = 2) \wedge (b = N)$

(K_1, K_2). First, we define the information system $\mathbf{S} = (U, A)$. The object set U is equal to the vertex set V_1 and attribute set A is equal to the vertex set V_2. An attribute $a \in A$ is a function $a : U \to \mathcal{DOM}(a)$ defined by

$$a(u) = \begin{cases} 0 & \text{if } (u, a) \in E \\ v_{a,u} & \text{otherwise} \end{cases}$$

For a given object u and a given attribute a, the value $v_{a,u}$ is defined as follows. Let $U_a \subseteq V_1$ be the set of all vertices not connected with the vertex a. We assume that $|U_a| = m$ and vertices from U_a are ordered by $u_{a,1}, u_{a,2}, ..., u_{a,m}$. Hence if $(u, a) \notin E$ then $u = u_{a,i}$ for some $i \in \{1, ..., m\}$. We take in this case $v_{a,u} = i$ (i.e. $a(u) = i$). In Figure 2 we give an example of a bipartite graph G and the corresponding information system $\mathbf{S}$. One can observe that the information system $\mathbf{S}$ can be constructed in polynomial time from a bipartite graph G. We can also see that every template T with support greater than 1 is of the form $T = \bigwedge \{(a = 0) : \text{for some } a \in A\}$. Therefore it determines exactly one bipartite sub-graph $G|_{U_1 \cup U_2}$, where $U_1 \subseteq V_1$ is the set of objects matching the template T and $U_2 \subseteq V_2$ is a set of attributes occurring in T, i.e. $U_2 = \{a : a \text{ occurs in } T\}$. Hence the table $\mathbf{S} = (U, A)$ with the parameters $F = K_1$, $L = K_2$ is the corresponding instance for TS. We obtain in this way the polynomial transformation of CBS into TS. ∎

Theorem 3. *[56] TS is NP-complete.* □

Proof:

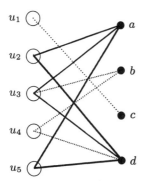

S	a	b	c	d
u_1	1	1	0	1
u_2	0	2	1	0
u_3	0	0	2	0
u_4	2	0	3	0
u_5	0	3	4	0

Fig. 2. The table constructed from the bipartite graph and the corresponding template: $(a = 0) \wedge (d = 0)$.

NP-completeness of TS results from NP-completeness of CBS. ∎

Now we observe that TS is not harder than the OTS, what along with the fact that TS is NP-complete, constitutes a proof that the considered optimization problem is NP-hard.

Theorem 4. *[56] If $P \neq NP$ then OTS is NP-hard.* □

Proof:
Suppose $P[\mathbf{S}, L]$ is a procedure that calculates for an information system $\mathbf{S}$ a template of length L with the maximal support. Then the decision problem TS with the instance: $\mathbf{S}$ - information system, L - template length, F - template support could be solved as follows: We call subroutine $S[\mathbf{S}, L]$ to compute the template T of the length L and with the maximal support. If $support(T) < S$ then the answer for the decision problem is negative, i.e. there is no template of length L and support at least S. Otherwise (i.e. $support(T) \geq S$), the answer for the decision problem is positive, i.e. there exists a template of length L and support at least S.

Hence TS could be solved in polynomial time if $S[\mathbf{S}, L]$ were a polynomial time subroutine for the OTS. From the NP-completeness of TS it follows that OTS is NP-hard and cannot be solved by any polynomial time algorithm unless $P = NP$. ∎

We can observe that the problem CBS is symmetric in the sense that if we exchange the roles of parameters K_1, K_2 we obtain again an NP-complete problem. Hence the TS problem with exchanged roles of support and length of a template is NP-complete, too. We obtain therefore the following two results:

Corollary 5. *Given an information system* $\mathbf{S} = (A, U)$ *and positive integers* S, L. *The decision problem of checking if there exists a template* T *of length equal to* L *and with support at least* S *is NP-complete.*

Corollary 6. *Given an information system* $\mathbf{S} = (A, U)$ *and positive integer* L. *The optimization searching problem for a template* T *(if any) of length* L *and the maximal support is NP-hard.*

Maximal quality problem We consider in this section the problem of *maximal quality*. For every template T, the quality is a function of $support(T)$ and $length(T)$. The template is *optimal* if its quality is maximal. Template Quality Problem can be formulated as the following decision problem:

PROBLEM: Template Quality Problem (TQP)
Instance: *An information system* $\mathbf{S} = (U, A)$, *an integer* K
Question: *Does there exist a template for* $\mathbf{S}$ *with the quality larger than* K?

One can show that TQP with the quality measure defined by

$$quality(T) = support(T) + length(T)$$

can be solved in polynomial time. This follows from the fact that TQP is equivalent to the Maximal Size Complete Bipartite Sub-graph problem (MSBS), which can be solved in polynomial time (see [23]). The (MSBS) problem is defined as follows:

PROBLEM: Maximal Size Complete Bipartite Sub-graph (MSCBS)
Instance: *Bipartite undirected graph* $G = (V_1 \cup V_2, E)$ *and a positive integer* $K_1 \leq |V_1|, K_2 \leq |V_2|$
Output: *Two subsets* $U_1 \subseteq V_1, U_2 \subseteq V_2$ *satisfying* $\{u, v\} \in E$ *for any* $u \in U_1, v \in U_2$ *with maximal* $|U_1| + |U_2|$.

The polynomial reduction TQP to MSCBS is identical to the construction in Theorem 2.

If the template quality is defined by the formula:

$$quality(T) = support(T) \times length(T)$$

then problem is not known to be NP-hard. We conjecture that it and corresponding optimization problem:

> **PROBLEM:** Optimal Template Quality Problem (OTQP)
> **Instance:** *An information system* $\mathbf{S} = (U, A)$
> **Output:** *A template T of the best quality (i.e. with maximal $support(T) \times length(T)$).*

are both NP-hard.

We have shown that the problem of searching for the optimal template is in general hard from the computational complexity point of view. The theoretical results justify us to look for efficient methods finding approximations of a template or a set of such templates. In the next section we will discuss how to apply the standard data mining techniques to the template extraction problem.

3 Sequential selection methods

The sequential searching approach is the first template extraction method discussed in this section. According to this method, a template is generated by choosing sequentially descriptors and a template set is constructed by selecting templates sequentially. First we discuss some methods for a single template extraction and then methods for template set extraction.

3.1 Basic generation schemes

What are basic methods for descriptor selection? One can start with the *empty descriptor set* and extend it by adding the most relevant descriptors from the original set. One can also begin with the original *full set* of descriptors and remove irrelevant descriptors from it. The former strategy is called the *sequential forward* strategy and the latter scheme is called the *sequential backward* strategy. Every strategy can be classified as *deterministic* or *random* depending on a method used for descriptor selection. A deterministic strategy always chooses the descriptor with the *optimal fitness*, whereas the random selector chooses templates according to some probability distribution.

Sequential forward generation This scheme starts with the *empty* template T. The template T is extended by adding one descriptor at a time that well fits the existing template. Let T_i be a temporary template obtained in the i-th step of construction. A new descriptor is selected according to the function $fitness_{T_i}$. For the temporary template T_i, the fitness of any descriptor D measured relatively to T_i reflects its potential ability to create a new template $T_{i+1} = T_i \wedge D$ of high quality. In the **Deterministic Sequential Forward Generation** (in short DSFG), the template T_i is extended by a descriptor with the value of maximum fitness. In general DSFG detects a template of high quality. The drawback of this method is that it can be stuck in a local extreme (the best template at the moment) and will never leave it. To avoid this situation, we consider the second scheme called **Random Sequential Forward Generation** (RSFG in short).

In RSFG scheme, descriptors are chosen randomly according to probability distribution defined by a fitness function.

The idea of DSFG is presented in the scheme below

ALGORITHM: (DSFG SCHEME)

1. $i := 0$; $T_i = \emptyset$; $T_{best} = \emptyset$.
2. **while** $(A \neq \emptyset)$
3.   choose an attribute $a \in A$ and a corresponding value set $V_a \subseteq \mathcal{DOM}(a)$ such that $(a \in V_a)$ is the best descriptor according to $fitness_{T_i}(.)$;
4. $T_i := T_i \wedge (a \in V_a)$; $i := i + 1$;
5.   **if** $(quality(T_{best}) < quality(T_i))$ **then** $T_{best} := T_i$;
6.   remove the attribute a from the attribute set A;
7. **endwhile**.
8. **return** the template T_{best};

To avoid the above mentioned drawback, the random scheme chooses a descriptor according to a probability distribution. Let P be a set of descriptors, which can be used to extend T_i. Let p_0 be a descriptor in P. In the simplest case, the distribution function $Prob$ can be defined by $Prob(p_0) = \frac{fitness_{T_i}(p_0))}{\sum_{p \in P} fitness_{T_i}(p)}$

A random version (RSFG) can be obtained from a DSFG. In Step 3, instead of choosing the best descriptor, we choose a descriptor according to the distribution function $Prob$.

Sequential backward generation The sequential backward method uses top-down strategy rather than bottom-up strategy used for DSFG. Starting from a *full template* T, (which is often defined by an information vector $Inf_\mathbf{S}(x)$ for some object x), the algorithm finds irrelevant descriptors and removes one descriptor at a time. After a removal the quality of a new template is estimated. The descriptor p is irrelevant if the template T without this descriptor is of better quality. Attributes are selected according to the *fitness function*, similarly as in DSFG.

Assume the full template T_{full} is of the form $T_{full} = \bigwedge_{1 \leq i \leq k}(a_i \in V_i)$, where k is a number of attributes in a data table and V_i are fixed subsets of $\mathcal{DOM}(a_i)$. The scheme below describes in details of Deterministic Sequential Backward Generation (DSBG):

> **ALGORITHM:** (DSBG SCHEME)
>
> 1. $i := k$; $T_i = T_{full}$; $T_{best} = T_{full}$.
> 2. **repeat**
> 3.   choose the descriptor $p \in T_i$ with the smallest fitness;
> 4. $T_i := T_i \backslash p$; $i := i - 1$;
> 5.   **if** $(quality(T_{best}) < quality(T_i))$ **then** $T_{best} := T_i$;
> 6. **until** $(i = 0)$ **or** $(quality(T_{best})$ is fixed);
> 7. **return** the template T_{best};

Analogously to the forward strategy, descriptor to be removed can also be chosen randomly. If $fitness_{T_i}(p_0)$ is the fitness of descriptor p_0 according to the template T_i (it increases, if a descriptor p_0 is relevant), then the distribution function for irrelevant descriptor elimination can be defined as follows:
$Prob(p_0) = 1 - \frac{fitness_{T_i}(p_0))}{\sum_{p \in T_i} fitness_{T_i}(p)}$.

For any algorithm based on sequential schemes we assume that three parameters are fixed: *estimation of the descriptor fitness, estimation of the template quality* and the *method of searching for the best descriptor* $(a \in V_a)$ of a given attribute a.

The template quality measures are discussed in Section 2.2, section 2. In the next section, we discuss the descriptor fitness function and a new approach for generalized descriptor $(a \in V_a)$ extraction.

3.2 Searching for optimal descriptors

We start from the notion of a function estimating the fitness of a descriptor relatively to a given template T. On can treat a descriptor $(a \in V_a)$ as a descriptor of "good quality" if the template $T \wedge (a \in V_a)$ has large support, although the set V_a is small.

Hence the fitness function is defined using the following two parameters:

- the number of objects supporting the template $T \wedge (a \in V_a)$
- the cardinality of the set V_a.

The fitness function should be proportional to the first parameter and inversely proportional to the second parameter. It can be defined by the following formula:
$$fitness_T(a \in V_a) = support_{[T]_U}(a \in V_a) \cdot Prob^{-1}(V_a)$$

The descriptor $a \in V_a$ is *optimal* if $fitness(a \in V_a)$ is maximal.

Now we can discuss searching heuristics for an optimal descriptor. Let $\mathbf{S} = (U, A \cup \{d\})$ be a given decision table [62] and let $a \in A$ be a fixed attribute. We consider the searching problem for range V_a of the descriptor $(a \in V_a)$.

Let $R_a \subseteq \mathcal{DOM}(a) \times \mathcal{DOM}(a)$ be a given similarity relation (we assume R_a to be reflexive). We define the *similarity class* of v by $[v]_{R_a} = \{u \in \mathcal{DOM}(a) : vR_au\}$ for any value $v \in \mathcal{DOM}(a)$. The similarity classes establish a covering of

$\mathcal{DOM}(a)$. In our case, the range V_a is computed by taking the similarity class $[v^*]_{R_a}$ of the properly chosen *generator* v^*.

The searching algorithm for the optimal range V_a often takes time propositional to the square of the number of objects, so it is not efficient for larger data tables. We propose in this section some algorithms returning the semi-optimal descriptors in time $O(n \log n)$, where n is the number of values in $\mathcal{DOM}(a)$.

In the next sections we present two main types of searching strategies for optimal descriptors. They are called *global* and *local strategies*.

We recall that for any value $v \in \mathcal{DOM}(a)$, $[a = v]_X$ denotes a set of objects x from X satisfying the template $(a(x) = v)$ and $support(a = v)$ denotes the cardinality of set $[a = v]_U$. We introduce the following notation:

For a given decision class C_i, we denote by $P_{a=v}(C_i) = \frac{support_{C_i}(a=v)}{support(a=v)}$ the conditional probability of the event that the randomly chosen object from $[a = v]_U$ belongs to the decision class C_i. The following formula

$$\mu_a(v) = \langle P_{a=v}(C_1), ..., P_{a=v}(C_m) \rangle$$

defines a probabilistic distribution on $[a = v]_U$.

Global strategies Let $v_{\max}^a$ be the most frequent value from $\mathcal{DOM}(a)$ occurring in **S**. For any similarity relation R_a, we compute the range V_a as follows:

- if a is a symbolic attribute then $V_a = [v_{\max}^a]_{R_a}$;
- if a is a numerical attribute then V_a is the largest real interval $[v_1, v_2]$ such, that $v_{\max}^a \in V_a$ and $\forall_{v \in a(U)} (v \in [v_1, v_2]) \Rightarrow (v_{\max}^a \ R_a \ v)$.

1. **Similar Frequency (SF).**

   The strategy based on frequency is used in *unsupervised learning model*. We use frequencies of values to determine the similarity of values. For a given real number $\varepsilon \in [0, 1]$, the parameterized similarity relation, called *similar frequency*, $R_a^\varepsilon \subseteq \mathcal{DOM}(a) \times \mathcal{DOM}(a)$ is defined by

   $$\forall_{u,v \in \mathcal{DOM}(a)} \left(u R_a^\varepsilon v \Leftrightarrow \frac{|support(a=u) - support(a=v)|}{support(a=u)} \leq \varepsilon \right)$$

   The similar frequency is a useful tool for extracting *regular blocks* of objects with similar values on condition attributes. The advantage of the method is a possibility to operate without any knowledge about decision classes.

2. **Similar Distribution relatively to Decisions (SDD)** .

   This strategy belongs to the *supervised learning* methods. The similarity relation R_a^ε is defined by distribution $\mu_a(v)$ of decision classes i.e.

   $$\forall_{u,v \in \mathcal{DOM}(a)} (u R_a^\varepsilon v \Leftrightarrow distance(\mu_a(u), \mu_a(v)) \leq \epsilon)$$

   To measure the *distance* between distribution vectors, one can use any distance function defined in m-dimensional space $\mathbb{R}^m$ (e.g. the *Euclidean distance*).

   The **SDD** strategy is an important tool for extracting similar objects with respect to defined by them distribution among decision classes.

3. **Similar Predominant Decision Class (SPDC)** .
   The Similar Predominant Decision Class techniques are used for extracting *decision rules*. Let us consider the distribution vector

   $$\mu_a(v) = \langle P_{a=v}(C_1), ..., P_{a=v}(C_m) \rangle$$

   The decision class C_i is called *predominant* for the set $[a = v]_U$ if $P_{a=v}(C_i)$ is maximal in $\mu_a(v)$. We denote by $C^*_{a=v}$ the predominant decision class for $[a = v]_U$.
   The definition of similarity relation R_a^ϵ is also based on the distribution vectors with the following modification:

   $$\forall_{u,v \in \mathcal{DOM}(a)} (u R_a^\epsilon v \Leftrightarrow distance\,(P_{a=u}(C^*_{a=u}), P_{a=v}(C^*_{a=v})) \leq \epsilon)$$

   The similar predominant decision class strategy is useful for extracting decision rules associated with one decision class.

Local strategies In this section we discuss some techniques called *local strategies* to search for templates related to a single object. Let $x \in U$. The purpose is to find a good template T_x that covers a given object $x \in U$ and simultaneously is supported by as many as possible other objects with the same decision as x in the system **S**. Local techniques often use the similarity of a given object to its neighbors. The range V_a of the best descriptor will be defined analogously to the definition of V_a in the previous section with the following modification: instead of $v_{\max}$, the value $a(x)$ will be chosen as the generator of the condition set V_a.

For a given attribute $a \in A$, let R_a be a predefined similarity on $\mathcal{DOM}(a)$. If a is a symbolic attribute, then $V_a = [a(x)]_{R_a}$. If a is a numerical attribute, the set V_a is a maximal real interval $[v_1, v_2]$ such that $a(x) \in V_a$ and $\forall_{v \in \mathcal{DOM}(a)} (v \in [v_1, v_2]) \Rightarrow (a(x)\ R_a\ v)$.

The local versions of **SDD** and **SPDC** strategies, presented in the previous section, are denoted by **L-SDD** and **L-SPDC**. The experimental results (reported in Section 10.3) show that the local strategies can be used to extract decision rules of high quality.

One can observe that all proposed algorithms can be implemented by sorting attribute values in time $O(kn \log n)$, where n is the number of objects and k is the number of attributes in the data set.

In the next section we present methods for optimal template set generation.

3.3 Template set generation

The approach is to cover a given data set by a minimal family of independent templates. The idea is adopted from a heuristic for *minimal set covering problem*. We assume that every template has associated a quality function reflecting its covering power. The algorithm describes a greedy process, which in every iteration step finds the template with the maximal quality. This process is performed until all objects are covered. Every template is constructed by some generator from the training table. The problem is to find for a set of generators such that the template set extracted using them covers the whole considered data and is of a small size.

The generator x is "good" if the local template T_x is of high quality. The set of generators is optimal if the template set generated from them covers the whole data set and is of the minimal cardinality. However, the searching problem for the *minimal covering set* is NP-hard. Below we present some heuristics for this problem.

The general covering scheme can be presented as a greedy process searching in every iteration step for template with the best quality. This process is continued until all objects are covered.

The general searching scheme is presented bellow:

ALGORITHM: (MINIMAL COVERING BY LOCAL TEMPLATES)

Input: A decision table $\mathbf{S} = (U, A \cup d)$.
Output: A (semi-)minimal covering of U i.e. such a set $\mathbf{T_S}$ of templates that
$\forall_{x \in U} \exists_{T \in \mathbf{T_S}} x \in [T]_U$.
1. $S := \emptyset$;
2. Find a local template T_x of the best quality, where $x \in U$;
3. $S := S \cup \{T\}$;
4. Remove from U all objects matching T;
5. If $U \neq \emptyset$ then **go to** Step 2, else **stop**.

Now, the main problem is to find the optimal set of generators. High quality generators should have the following properties:

(i) Generator must be *typical*, that means it is *similar* (*close*) to many objects from the data set.
(ii) In case of decision templates, generator should be similar to many objects with the same decision and it should be different from many objects with different decisions.
(iii) Generator should be *representative*: that means the intersection of supporting sets of different generators must be as small as possible.

In our approach, the generators are selected using the weight functions reflecting the quality of descriptors.

Let $dist : U \times U \to \mathbb{R}_0^+$ be a given distance function defined on the object space U (e.g. the normal Euclidean or Hamming function). The function $weight : U \to \mathbb{R}$ can be defined using one of the following:

1. *Average Distance*:
$$weight(x) = \frac{1}{n} \sum_{y \neq x} dist(x,y)$$

2. *Static Distance*:
$$weight(x) = \sum_{d(y)=d(x)} dist(x,y) - \sum_{d(z) \neq d(x)} dist(x,z)$$

3. *Dynamic Distance*:

$$weight(x) = dist(x, G) = \min_{y \in G}\{dist(x, y)\}$$

where $G = \{x_1, ..., x_i\}$ is the set of generators in the current step.

4 Segmentation based method

The main idea is adopted from data discretization. Discretization is usually seen as pre-processing of other data mining tasks. Its role is to reduce the continuous attribute domains [22, 52]. Discretization approach, therefore, is applicable only for numeric attributes. In this work the technique is adopted and generalized to two kinds of data: numeric and symbolic. We call this new approach *data segmentation*.

Segmentation approach is used for the *supervised learning model*, when we have to deal with a decision attribute. Let a decision table $\mathbf{S} = (U, A \cup d)$ be given. Let $card(A) = k$. If an attribute $a \in A$ is numeric, any object $x \in U$ can be interpreted as a point in k-dimensional space $\mathbb{R}^k$. The coordinates of a point x are defined by an information vector

$$Inf_\mathbf{S}(x) = \{(a, a(x)) : a \in A\}$$

The main idea of the segmentation approach is to divide domains of attributes into disjoint intervals by "cuts". These cuts create a disjoint "rectangle" regions in space $\mathbb{R}^k$ (see Figure 3).

The aim is to search for a minimal set of "cuts" (or minimal set of regions) such that any region defined by these cuts contains objects from one decision class only. Segmentation of numeric data (considered as data discretization) is a subject of many papers (see e.g. [14, 16, 52, 63]). In this work, we present some new methods, which are applicable to symbolic attributes too. Instead of partitioning of the attribute domain into real intervals, we consider the problem of partitioning of the symbolic domain into groups of similar values. The new techniques for mixed data are presented in Section 4.3.

4.1 Basic notions

Let $\mathbf{S} = (U, A \cup \{d\})$ be a decision table where $U = \{x_1, x_2, \ldots, x_n\}$, $A = \{a_1, ..., a_k\}$ and $d : U \to \{1, ..., r\}$. Any pair (a, c) where $a \in A$ and $c \in \mathbb{R}$ will be called a *cut on* V_a. Any set of cuts: $\{(a, c_1^a), (a, c_2^a), \ldots, (a, c_{k_a}^a)\} \subset A \times \mathbb{R}$ on $V_a = [l_a, r_a) \subset \mathbb{R}$ (for $a \in A$) uniquely defines a partition $\mathbf{P}_a$ of V_a into sub-intervals i.e. $\mathbf{P}_a = \{[c_0^a, c_1^a), [c_1^a, c_2^a), \ldots, [c_{k_a}^a, c_{k_a+1}^a)\}$ where $l_a = c_0^a < c_1^a < c_2^a < \ldots < c_{k_a}^a < c_{k_a+1}^a = r_a$ and $V_a = [c_0^a, c_1^a) \cup [c_1^a, c_2^a) \cup \ldots \cup [c_{k_a}^a, c_{k_a+1}^a)$.

Therefore, any set of cuts $\mathbf{P} = \bigcup_{a \in A} \mathbf{P}_a$ defines from $\mathbf{S} = (U, A \cup \{d\})$ a new decision table $\mathbf{S}^{\mathbf{P}} = (U, A^{\mathbf{P}} \cup \{d\})$ called $\mathbf{P}$-*segmentation of* $\mathbf{S}$, where $A^{\mathbf{P}} = \{a^{\mathbf{P}} : a \in A\}$ and $a^{\mathbf{P}}(x) = i \Leftrightarrow a(x) \in [c_i^a, c_{i+1}^a)$ for $x \in U$ and $i \in \{0, .., k_a\}$.

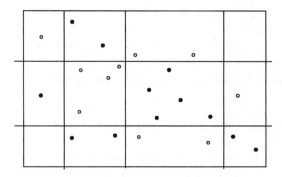

Fig. 3. Regions defined by a set of cuts.

Two sets of cuts $\mathbf{P}', \mathbf{P}$ are equivalent, i.e. $\mathbf{P}' \equiv_\mathbf{S} \mathbf{P}$, iff $\mathbf{S}^\mathbf{P} = \mathbf{S}^{\mathbf{P}'}$. The equivalence relation $\equiv_\mathbf{S}$ has a finite number of equivalence classes. In the sequel we will not discern between equivalent families of partitions.

Let ∂_A and $\partial_A \mathbf{P}$ be generalized decisions of $\mathbf{S}$ and $\mathbf{S}^\mathbf{P}$, respectively. We say that the set of cuts $\mathbf{P}$ is $\mathbf{S}$-*consistent* if $\partial_A = \partial_A \mathbf{P}$. The $\mathbf{S}$-consistent set of cuts $\mathbf{P}^{irr}$ is $\mathbf{S}$-*irreducible* if $\mathbf{P}$ is not $\mathbf{S}$-consistent for any $\mathbf{P} \subset \mathbf{P}^{irr}$. The $\mathbf{S}$-consistent set of cuts $\mathbf{P}^{opt}$ is $\mathbf{S}$-*optimal* if $card(\mathbf{P}^{opt}) \leq card(\mathbf{P})$ for any $\mathbf{S}$-consistent set of cuts $\mathbf{P}$.

4.2 Complexity of segmentation problems

Let us consider the problem of finding optimal set of cuts for a given decision table. If the number of attributes in the decision table S is equal to 1, then the discretization problem is solved in polynomial time (actually in $O(n \log n)$, where n is a number of objects in decision tables). It has been shown (see [58]) that for an arbitrary decision table the problem of discretization is NP-hard. This result is related to decision tables with arbitrary number of attributes. We improve this result by showing that if the number of attributes is limited to 2, the optimal discretization problem still remains hard.

Theorem 7. *[58] The decision problem of checking if for a given decision table* $\mathbf{S} = (U, A \cup \{d\})$, *where* $card(A) = 2$, *and an integer* k *there exists an irreducible set of cuts* $\mathbf{P}$ *in* $\mathbf{S}$ *such that* $card(\mathbf{P}) < k$ *is NP-complete. The problem of finding for optimal set of cuts* $\mathbf{P}$ *in a given decision table* $\mathbf{S}$ *is NP-hard.* □

We consider a special consistent decision table $\mathbf{S} = (U, \{a, b\} \cup \{d\})$ with two real value condition attributes a, b and binary decision $d : U \to \{0, 1\}$. Such a decision table is a representation of the set of points $S = \{(a(u_i), b(u_i)) : u_i \in U\}$ of the plane $\mathbb{R}^2$ partitioned into two disjoint categories $S = S_1 \cup S_2$. We represent a partition by assigning black and white colors to points. Any cut (a, c) on a (or (b,c) on b), where $c \in \mathbb{R}$, can be represented by a vertical (or horizontal) line. The set of cuts is **A-consistent** if the set of lines representing them defines a partition of the plane into regions in such a way that if any two points are in the same region then they have the same color. Such a set of lines is said to be *consistent*. The discretization problem (**k-D2**) for a decision table with two condition attributes can be defined as follows:

PROBLEM: k-D2 – Discretization in $\mathbb{R}^2$ by k cuts.
Input: *Set S of points $P_1, ..., P_n$ of the plane, partitioned into two disjoint categories S_1, S_2 and a natural number k.*
Question: *Is there a consistent set of at most k lines?*

We also consider the corresponding optimization problem:

PROBLEM: OD2 – Optimal Discretization in $\mathbb{R}^2$.
Input: *Set S of points $P_1, ..., P_n$ of the plane, partitioned into two disjoint categories S_1, S_2 and a natural number T.*
Output: *An A-optimal set of cuts.*

Theorem 8. *[15] The decision problem k-D2 is NP-complete and the optimization version of this problem is NP-hard.* □

Proof:
An instance I of Set Cover consists of $S = \{u_1, u_2, ..., u_n\}$, $\mathcal{F} = \{S_1, S_2, ..., S_m\}$, where $S_j \subseteq S$ and $\bigcup_{i=1}^{m} S_i = S$, and an integer K and the question is if there are K sets from $\mathcal{F}$ whose sum contains all elements of S. We need to construct an instance I' of k-D2 such that I has a positive answer iff I' has a positive answer. The construction of I' is quite similar to the construction described in the previous section. We start by building a grid-line structure consisting of vertical and horizontal strips. The regions are in rows labeled by $y_{u_1}, ..., y_{u_n}$ and columns labeled by $x_{S_1}, ..., x_{S_m}, x_{u_1}, ..., x_{u_n}$ (see Figure 4). In the first step, for any element $u_i \in S$ we define a family $\mathcal{F}_i = \{S_{i_1}, S_{i_2}, ..., S_{i_{m_i}}\}$ of all subsets containing the element u_i.

If $\mathcal{F}_i$ consists of exactly $m_i \leq m$ subsets, then subdivide the row y_{u_i} into m_i strips, corresponding to the subsets from $\mathcal{F}_i$. For each $S_j \in \mathcal{F}_i$ place one pair of black and white points in the strip labeled by $u_i \in S_j$ inside a region (x_{u_i}, y_{u_j}) and the second pair in the column labeled by x_{S_j} (see Figure 4). In each region (x_{u_i}, y_{u_j}) add a special point in the top left corner with a color different from the color of the point on the top right corner. This point is introduced to force at least one vertical line across a region. Place the configuration R_{u_i} for u_i in the region labeled by (x_{u_i}, y_{u_i}). Examples of R_{u_1} and R_{u_2} where $\mathcal{F}_{u_1} = \{S_1, S_2, S_4, S_5\}$ and $\mathcal{F}_{u_2} = \{S_1, S_3, S_4\}$, are depicted in the Figure 4.

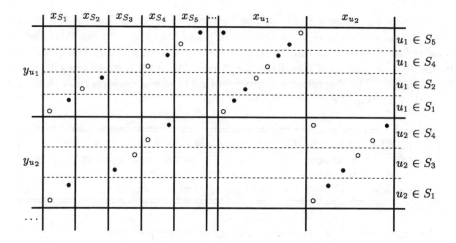

Fig. 4. Construction of configurations R_{u_1} and R_{u_2} where $\mathcal{F}_{u_1} = \{S_1, S_2, S_4, S_5\}$ and $\mathcal{F}_{u_2} = \{S_1, S_3, S_4\}$

The configuration R_{u_i} requires at least m_i lines to be separated, among them at least one vertical. Thus, the whole construction for u_i requires at least $m_i + 1$ lines. Let I' be an instance of k-D2 defined by the set of all points forcing the grid and all configurations R_{u_i} with $K = k + \sum_{i=1}^{n} m_i + (2n + m + 2)$ as the number, where the last component $(2n + m + 2)$ is the number of lines defining the grid. If there is a covering of S by k subsets $S_{j_1}, S_{j_2}, ..., S_{j_k}$, then we can construct K lines that separate well the set of points, namely $(2n + m + 2)$ grid lines, k vertical lines in columns corresponding to $S_{j_1}, S_{j_2}, ..., S_{j_k}$ and m_i lines for the each element u_i $(i = 1, ..n)$.

On the other hand, let us assume that there are K lines separating the points from instance I'. We show that there exists a covering of S by k subsets. There is a set of lines such that for any $i \in \{1, ..., n\}$ there are exactly m_i lines passing across the configuration R_{u_i} (i.e. the region labeled by (x_{u_i}, y_{u_i})), among them exactly one vertical line. Hence, there are at most k vertical lines on rows labeled by $x_{S_1}, ..., x_{S_m}$. These lines determine k subsets which cover the whole S. ∎

Next we will consider the discretization problem that minimizes the *number of homogeneous regions* defined by a set of cuts. We will show that the new problem is NP-hard too.

The new descretization problem is called Optimal Splitting. We will show that the Optimal Splitting problem is NP-hard, even when the number of attributes is limited to 2. The Optimal Splitting problem is defined as follows:

> **PROBLEM: 2-OS** Optimal Splitting in $\mathbb{R}^2$
> **Input:** *A set S of points $P_1, ..., P_n$ in the plane $\mathbb{R}^2$, partitioned into two disjoint categories S_1, S_2 and a natural number T.*
> **Question:** *Is there a consistent set of lines such that the partition of the plane into regions defined by them consists of at most T regions?*

Theorem 9. *[15] 2-OS is NP-complete.* □

Proof:
It is clear that 2-OS is in NP. The NP-hardness part of the proof is done by reducing 3SAT to 2-OS (cf. [23]).

Let $\Phi = C_1 \wedge ... \wedge C_k$ be an instance of 3SAT. We construct an instance I_Φ of 2-OS such that Φ is satisfiable iff there is a sufficiently small consistent set of lines for I_Φ. The description of I_Φ will specify a set of points S, which will be partitioned into two subsets of white and black points. A pair of points with equal horizontal coordinates is said to be *vertical*, similarly, a pair of points with equal vertical coordinates is *horizontal*. If a configuration of points includes a pair of horizontal points p_1 and p_2 of different colors, then any consistent set of lines will include a vertical line L separating p_1 and p_2, which will be in the vertical strip with p_1 and p_2 on its boundaries. Such a strip is referred to as a *forcing strip*, and the line L as *forced* by points p_1 and p_2. Horizontal forcing strips and forced lines are defined similarly. The instance I_Φ has an underlying grid-like structure consisting of vertical and horizontal forcing strips. The rectangular regions inside the structure and consisting of points outside the strips are referred to as *f-rectangles* of the grid. The f-rectangles are arranged into rows and columns.

For each propositional variable p occurring in C use one special row and one special column of rectangles. In the f-rectangle that is at the intersection of the row and column place configuration R_p as depicted on Figure 5.

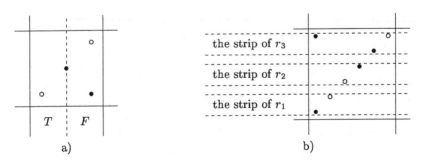

Fig. 5. a) Configuration R_p b) Configuration R_C

Notice that R_p requires at least one horizontal and one vertical line to separate the white from the black points. If only one such vertical line occurs in a consistent set of lines, then it separates either the left or the right white point from the central black one, what we interpret as an assignment of the value *true* or *false* to p, accordingly.

For each clause C in Φ use one special row and one special column of f-rectangles. Let C be of the form $C = r_1 \vee r_2 \vee r_3$, where the variables in the literals r_i are all different. Subdivide the row into three strips corresponding to the literals. For each such r_i place one black and one white points, of distinct vertical and horizontal coordinates, inside its strip in the column of the variable of r_i, in the 'true' vertical strip if $r_i = p$, and in the 'false' strip if $r_i = \neg p$. These two points are referred to as *configuration* $R_{C,i}$. In the region of the intersection of the row and column of C place configuration R_C as depicted in Figure 5. Notice that R_C requires at least three lines to separate the white from the black points, and among them at least one vertical. An example of a fragment of this construction is depicted in Figure 6. Column x_{p_i} and row x_{p_i} correspond to variable p_i, row y_C corresponds to clause C.

Let the underlying grid of f-rectangles be minimal to accommodate this construction. Add horizontal rows of f-rectangles, their number greater by 1 then the size of Φ. Suppose conceptually that a consistent set of lines W includes exactly one vertical and one horizontal line per each R_p, and exactly one vertical and two horizontal lines per each R_C, let L_1 be the set of all these lines. There is also the set L_2 of lines inside the forcing strips, precisely one line per each strip. We have $W = L_1 \cup L_2$. Let the number of horizontal lines in W be equal to l_h nad vertical to l_v. That many lines create $T = (l_h - 1) \cdot (l_v - 1)$ regions, and this number is the last component of I_Φ.

Next we show the correctness of the reduction. Suppose first that Φ is satisfiable, let us fix a satisfying assignment of logical values to the variables of Φ. The consistent set of lines is determined as follows. Place one line into each forcing strip. For each variable p place one vertical and one horizontal line to separate points in R_p, the vertical line determined by the logical value assigned to p. Each configuration R_C is handled as follows. Let C be of the form $C = r_1 \vee r_2 \vee r_3$. Since C is satisfied, at least one $R_{C,i}$, say $R_{C,1}$, is separated by the vertical line that separates also R_p, where p is the variable of r_1. Place two horizontal lines to separate the remaining $R_{C,2}$ nad $R_{C,3}$. They also separates two pairs of points in R_C. Add one vertical line to complete separation of the points in R_C. All this means that there is a consistent set of lines which creates T regions.

On the other hand, suppose that there is a consistent set of lines for I_Φ, which determines at most T regions. The number T was defined in such a way that two lines must separate each R_p and three lines each R_C, in the latter case at least one of them vertical. Notice that a horizontal line contributes fewer regions than a vertical one because the grid of splitting strips contains much more rows than columns. Hence one vertical line and two horizontal lines separate each R_C, because changing horizontal to vertical would increase the number of regions beyond T. It follows that for each clause $C = r_1 \vee r_2 \vee r_3$ at least one $R_{C,i}$ is separated by a vertical line of R_p, where p is the variable of r_i, and this yields a satisfying truth assignment. ∎

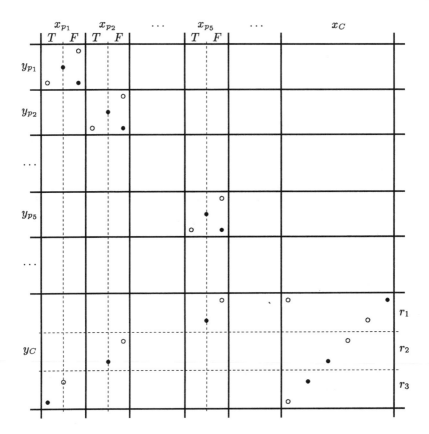

Fig. 6. Construction of configurations R_{p_i} and R_C for $C = p_1 \wedge \neg p_2 \wedge \neg p_5$

4.3 Symbolic value partition

In case of symbolic value attributes the problem of finding partitions of value sets into a "small" number of subsets is more complicated than for the numeric attributes. Once again we apply Boolean reasoning approach to construct a partition of symbolic value sets into small number of subsets.

Let $\mathbf{S} = (U, A \cup \{d\})$ be a decision table where

$$A = \left\{ a_i : U \to V_{a_i} = \{v_1^{a_i}, ..., v_{n_i}^{a_i}\} \right\}$$

for $i \in \{1, ..., k\}$. Any function $P_{a_i} : V_{a_i} \to \{1, ..., m_i\}$ (where $m_i \leq n_i$) is called a *partition of* V_{a_i}. The *rank of* P_{a_i} is the value $rank(P_{a_i}) = card(P_{a_i}(V_{a_i}))$. The function P_{a_i} defines a new *partition attribute* $b_i = P_{a_i} \circ a_i$ i.e. $b_i(u) = P_{a_i}(a_i(u))$ for any object $u \in U$.

The family of partitions $\{P_a\}_{a \in B}$ is $B - consistent$ iff

$$\forall_{u,v \in U} \left[d(u) \neq d(v) \wedge (u,v) \notin IND(B) \right] \Rightarrow \exists_{a \in B} \left[P_a(a(u)) \neq P_a(a(v)) \right] \quad (6)$$

where $IND(B)$ is indiscernibility relation defined in Section 2.1. It means that if two objects u, v are discerned by B and d, then they are discerned by the partition attribute defined by $\{P_a\}_{a \in B}$. We consider the following optimization problem called *the symbolic value partition problem*:

PROBLEM: Symbolic Value Partition Problem
Input *For a given decision table* $\mathbf{S} = (U, A \cup \{d\})$, *a set of nominal attributes* $B \subseteq A$ *and an integer* K.
Output *Check if there is a* B − *consistent family of partitions with* $\sum_{a \in B} rank(P_a) \leq K$.

We show that this problem is NP-complete.

Theorem 10. *The decision problem of checking if for a given decision table* $\mathbf{S} = (U, A \cup \{d\})$ *and an integer* K, *there exists a consistent family of partitions* $\{\mathbf{P_a}\}(a \in A)$ *such that* $\sum_{a \in A} rank(P_a) \leq k$ *is NP-complete. The problem of finding for a consistent family of partition with minimal* $\sum_{a \in A} rank(P_a)$ *is NP-hard.* □

Proof:
It is clear that the considered decision problem is in NP. We will show that this problem is harder than Minimal Graph Covering problem (i.e. the problem of finding minimal set of vertices which cover all the edges in a given graph). Let us consider an instance of Minimal Graph Covering problem, which consists of graph $\mathbf{G} = (V, E)$ and integer $k \leq |V|$. The problem is to check if there is a set of vertices $V' \subseteq V$ such that $|V'| = k$ and for any edge $e = (u, v) \in E$ either $u \in V'$ or $v \in V'$.

Let graph $G = (V, E)$ be an instance of Minimal Vertex Cover Problem, where $V = \{v_1, v_2, ...v_n\}$ and $E = \{e_1, e_2, ...e_m\}$. We assume the every edge e_i is represented by two-element set of vertices i.e. $e_i = \{v_{i_1}, v_{i_2}\}$. We construct the corresponding decision table $\mathbb{S}(\mathbb{G}) = (\mathbb{U}, \mathbb{A} \cup \{\})$ for Symbolic Value Partition Problem as follows:

1. The set U consists of m objects corresponding to m edges of the graph G and one additional object i.e. $U = \{u^*\} \cup \{u_{e_1}, u_{e_2}, ..., u_{e_m}\}$
2. The set A consists of n attributes corresponding to n vertices

$$A = \{a_{v_1}, a_{v_2}, ..., a_{v_n}\}$$

. The value of attribute $a \in A$ over the object $u \in U$ is defined by:

$$a_{v_j}(u_{e_i}) = \begin{cases} 1 & \text{if } v_j \in e_i \\ 0 & \text{otherwise} \end{cases} \quad \text{and} \quad a_{v_j}(u^*) = 0$$

for $j \in \{1, ..., n\}$ and $i \in \{1, .., m\}$.

3. The decision attribute is define by

$$d(u_{e_i}) = 1 \text{ and } d(u^*) = 0$$

for $i \in \{1,..,m\}$.

This construction can be done in polynomial time. An example given in Figure 7.

Any subset $V' = \{v_{i_1},...,v_{i_k}\}$ of vertices defines the family of partitions $\mathbf{P}_{V'} = \{\mathbf{P}_{a_{v_i}} : v_i \in V\}$ of values of attributes from $\mathbb{S}(\mathbf{G})$ as follows:

1. $\mathbf{P}_{a_{v_i}}(0) = \mathbf{P}_{a_{v_i}}(1) = 0$ if $v_i \notin V'$; (i.e. this attribute can be omitted without loss of discernibility between objects)
2. $\mathbf{P}_{a_{v_i}}(0) = 0$; $\mathbf{P}_{a_{v_i}}(1) = 1$ if $v_i \in V'$; (i.e this attribute is chosen to keep the discernibility between objects)

Notice that $rank(\mathbf{P}_{V'}) = n + k$.

We show that V' covers all edges of $\mathbf{G}$ if and only if $\mathbf{P}_{V'}$ is a consistent family of partition for $\mathbb{S}(\mathbf{G})$. Indeed, the family of partitions $\mathbf{P}_{V'}$ is consistent with $\mathbb{S}(\mathbf{G})$ if and only if it discerns u^* from objects u_{e_j} for $j = 1,..,m$. Thus for any object u_{e_j} there exists an attribute a_{v_i} such that $\mathbf{P}_{a_{v_i}}(u_{e_j}) \neq \mathbf{P}_{a_{v_i}}(0) = \mathbf{P}_{a_{v_i}}(u^*)$, i.e. $a_{v_i}(u_{e_j}) = 1$. The last condition means that the vertex v_i is adjacent to the edge e_j.

Example We illustrate the proof of Theorem 10 by the graph $\mathbf{G} = (V, E)$ with five vertices $V = \{v_1, v_2, v_3, v_4, v_5\}$ and six edges $E = \{e_1, e_2, e_3, e_4, e_5, e_6\}$. The decision table $\mathbb{S}(\mathbf{G})$ consists of five conditional attributes $\{a_{v_1}, a_{v_2}, a_{v_3}, a_{v_4}, a_{v_5}\}$, decision d and 7 objects $\{u^*, u_{e_1}, u_{e_2}, u_{e_3}, u_{e_4}, u_{e_5}, u_{e_6}\}$. Decision table $\mathbb{S}(\mathbf{G})$ constructed from the graph $\mathbf{G}$ is presented as follows:

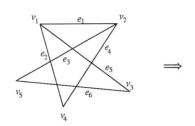

$\mathbb{S}(\mathbf{G})$	a_{v_1}	a_{v_2}	a_{v_3}	a_{v_4}	a_{v_5}	d
u^*	0	0	0	0	0	0
u_{e_1}	1	1	0	0	0	1
u_{e_2}	1	0	0	1	0	1
u_{e_3}	0	1	0	0	1	1
u_{e_4}	0	1	0	1	0	1
u_{e_5}	1	0	1	0	1	1
u_{e_6}	0	0	1	0	1	1

Fig. 7. An illustrative example for proof of NP-hardness of Symbolic Value Partition Problem

■

This concept is useful when we want to reduce domains of attributes with large cardinalities. The discretization problem can be derived from the partition problem by adding the monotonicity condition for family $\{P_a\}_{a \in A}$ such that

$$\forall_{v_1, v_2 \in V_a} [v_1 \leq v_2 \Rightarrow P_a(v_1) \leq P_a(v_2)]$$

We distinguish two approaches for solving this problem, namely the *local partition method* and the *global partition method*. The first approach is based on grouping the values of each attribute independently, whereas the second approach is based on grouping of attribute values simultaneously for all attributes.

Local partition method For any fixed attribute $a \in A$, we want to find such a partition P_a that preserves consistency condition (6) for the attribute a (i.e. $B = \{a\}$).

For any partition P_a the equivalence relation $\approx_{P_a}$ is defined by: $v_1 \approx_{P_a} v_2 \Leftrightarrow P_a(v_1) = P_a(v_2)$ for all $v_1, v_2 \in V_a$. We consider the relation $\mathbf{UNI}_a$ defined on V_a by

$$v_1 \mathbf{UNI}_a v_2 \Leftrightarrow \forall_{u, u' \in U} (a(u) = v_1 \wedge a(u') = v_2) \Rightarrow d(u) = d(u') \quad (7)$$

The relation $\mathbf{UNI}_a$ defined by (7) is an equivalence relation. The relevant properties of $\mathbf{UNI}_a$ are as follows:

Proposition 11. *If P_a is a-consistent then $\approx_{P_a} \subseteq \mathbf{UNI}_a$. The equivalence relation $\mathbf{UNI}_a$ defines a minimal a−consistent partition on a.*

Global partition method We consider the discernibility matrix [75] of the decision table $\mathbf{S}$: $\mathbf{M}(\mathbf{S}) = [m_{i,j}]_{i,j=1}^n$ where $m_{i,j}$ is the set of all attributes having different values on objects u_i, u_j i.e. $m_{i,j} = \{a \in A : a(u_i) \neq a(u_j)\}$. Observe that if we want to discern between objects u_i and u_j we have to keep one of the attributes from $m_{i,j}$. We would like to have the following property: *for any two objects u_i, u_j there exists an attribute $a \in m_{i,j}$ such that the values $a(u_i), a(u_j)$ are discerned by P_a*. Hence instead of cuts in case of numeric values (defined by pairs (a_i, c_j)), one can discern objects by triples $\left(a_i, v_{i_1}^{a_i}, v_{i_2}^{a_i}\right)$ called chains, where $a_i \in A$ for $i = 1, ..., k$ and $i_1, i_2 \in \{1, ..., n_i\}$.

One can build a new decision table $\mathbf{S}^+ = (U^+, A^+ \cup \{d^+\})$ (analogously to the table $\mathbf{S}^*$ (see Section 3.2)) assuming $U^+ = U^*$; $d^+ = d^*$ and $A^+ = \{(a, v_1, v_2) : (a \in A) \wedge (v_1, v_2 \in V_a)\}$. Now one can apply to A^+ the Johnson's heuristic to find a minimal set of chains discerning all pairs of objects from different decision classes [18].

Our problem can be solved by an efficient heuristics of graph coloring. The "*graph $k-$colorability*" problem is formulated as follows:

PROBLEM: Graph $k-$ Colorability Problem
  Input: *Graph $G = (V, E)$, positive integer $k \leq |V|$.*
  Output: **1** *if G is $k-$colorable, (i.e. if there exist a function $f : V \rightarrow \{1, ..., k\}$ such that $f(v) \neq f(v')$ whenever $(v, v') \in E$) and **0** otherwise.*

This problem is solvable in polynomial time for $k = 2$, but is NP-complete for all $k \geq 3$. However, similarly to discretization, one can apply some efficient heuristic searching for optimal graph coloring determining optimal partitions of attribute value sets.

For any attribute a_i in a semi-minimal set X of chains returned from the above heuristic (e.g. Johnson's heuristic) we construct graph $\Gamma_{a_i} = \langle V_{a_i}, E_{a_i} \rangle$, where E_{a_i} is the set of all chains in X of the attribute a_i. Any coloring of all graphs Γ_{a_i} defines an A-consistent partition of value sets. Hence heuristics searching for minimal graph coloring return also sub-optimal partitions of attribute value sets.

One can see that the constructed Boolean formula has $O(knl^2)$ variables and $O(n^2)$ clauses, where l is the maximal value of $card(V_a)$ for $a \in A$. When prime implicants have been constructed a heuristic for graph coloring can be applied.

A	a	b	d
u_1	a_1	b_1	0
u_2	a_1	b_2	0
u_3	a_2	b_3	0
u_4	a_3	b_1	0
u_5	a_1	b_4	1
u_6	a_2	b_2	1
u_7	a_2	b_1	1
u_8	a_4	b_2	1
u_9	a_3	b_4	1
u_{10}	a_2	b_5	1

M(A)	u_1	u_2	u_3	u_4
u_5	$b_{b_4}^{b_1}$	$b_{b_4}^{b_2}$	$a_{a_2}^{a_1}, b_{b_4}^{b_3}$	$a_{a_3}^{a_1}, b_{b_4}^{b_1}$
u_6	$a_{a_2}^{a_1}, b_{b_2}^{b_1}$	$a_{a_2}^{a_1}$	$b_{b_3}^{b_2}$	$a_{a_3}^{a_2}, b_{b_2}^{b_1}$
u_7	$a_{a_2}^{a_1}$	$a_{a_2}^{a_1}, b_{b_2}^{b_1}$	$b_{b_3}^{b_1}$	$a_{a_3}^{a_2}$
u_8	$a_{a_4}^{a_1}, b_{b_2}^{b_1}$	$a_{a_4}^{a_1}$	$a_{a_4}^{a_2}, b_{b_3}^{b_2}$	$a_{a_4}^{a_3}, b_{b_2}^{b_1}$
u_9	$a_{a_3}^{a_1}, b_{b_4}^{b_1}$	$a_{a_3}^{a_1}, b_{b_4}^{b_2}$	$a_{a_3}^{a_2}, b_{b_4}^{b_3}$	$b_{b_4}^{b_1}$
u_{10}	$a_{a_2}^{a_1}, b_{b_5}^{b_1}$	$a_{a_2}^{a_1}, b_{b_5}^{b_2}$	$b_{b_5}^{b_3}$	$a_{a_3}^{a_2}, b_{b_5}^{b_1}$

Fig. 8. The decision table and the corresponding discernibility matrix.

Example Let us consider the decision table presented in Figure 8 and a reduced form of its discernibility matrix.

Firstly, from the Boolean function f_S with Boolean variables of the form $a_{v_1}^{v_2}$ (corresponding to the chain (a, v_1, v_2) described in Section 5.2) we find the shortest prime implicant:

$$a_{a_2}^{a_1} \wedge a_{a_3}^{a_2} \wedge a_{a_4}^{a_1} \wedge a_{a_4}^{a_3} \wedge b_{a_4}^{a_1} \wedge b_{a_4}^{a_2} \wedge b_{a_3}^{a_2} \wedge b_{a_3}^{a_1} \wedge b_{a_5}^{a_3},$$

which can be represented by graphs (Figure 9). Next we apply a heuristic to color vertices of those graphs as it is shown in Figure 9. The colors are corresponding to the partitions:

$$P_{\mathbf{a}}(a_1) = P_{\mathbf{a}}(a_3) = 1; \quad P_{\mathbf{a}}(a_2) = P_{\mathbf{a}}(a_4) = 2$$
$$P_{\mathbf{b}}(b_1) = P_{\mathbf{b}}(b_2) = P_{\mathbf{b}}(b_5) = 1; P_{\mathbf{b}}(b_3) = P_{\mathbf{b}}(b_4) = 2$$

and at the same time one can construct the new decision table (Figure 8).

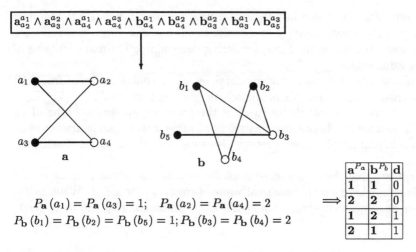

Fig. 9. Coloring of attribute value graphs and the reduced table.

5 Decision Tree Methods

Decision trees are data structures used, in particular, for classification tasks. The main idea of decision tree induction is to split training samples into smaller ones in a top-down manner until some terminating condition holds. In our consideration, the splitting process stops if all (or almost all) objects associated with a given node belong to the same decision class. There is a number of methods for decision tree induction such as ID3 [70], C4.5 [71], MD [52], CART [9], which are based on information entropy measure or other similarity measures. These methods have some disadvantages, namely they are not applicable for symbolic attributes with large domains.

We will present in this section a new method, called *MDG-method* [4], which is suitable for both kinds of data: symbolic and numeric.

5.1 Preliminaries

A decision tree is a labeled tree in which internal nodes have associated with them certain *test functions*, edges starting from the same node are labeled by values of the test function labeling this node, and external nodes (leaves) represent decision classes. Hence every path in a decision tree represents *decision rule* saying that an object with the specific feature values belongs to a certain class. Special techniques are used in order to minimize the size of decision tree (e.g. the number of leaves) and to extract the function fitting well to all kinds of attribute domains. Below we present some known test functions:

1. A function in any internal node is of the form "*attribute*" and edges going out from a node are labeled by values of this attribute. The degrees of nodes are then equal to the cardinality of the attribute domain. This concept

---

[4] Maximal Discernibility by Grouping

is often used for attributes with small domains. For continuous attributes or attributes with a large domain, we need some preprocessing tools likes *discretization* or *value grouping* to reduce attribute domains.

2. Test functions of the form "*attribute = value*" are defined by a pair consisting of the attribute and the value from the domain of this attribute. These test functions have one of two possible values: *False* and *True* (or 0 and 1). This kind of functions is suitable for symbolic attributes.
3. Test functions of the form "*attribute < value*". These test functions have also one of two values: *False* and *True*. This kind of functions is suitable for numeric and continuous attributes.
4. Test functions can have more complicated form called *oblique hyperplanes* or just hyperplanes, for short. This is a generalization of the third form by taking a linear combination of several attributes (i.e. $w_1 \cdot f_1 + ... + w_k \cdot f_k > value$, where $f_1, ..., f_k$ are attributes and $w_1, ..., w_k$ are coefficients to be found) instead of the single attribute. Geometrically, any inequality $w_1 \cdot f_1 + ... + w_k \cdot f_k > w$ defines a binary partition of multi-dimensional space into two half-spaces. One can generalize the attributes defined by hyperplanes to attributes defined by high-dimensional surfaces represented by linear combinations of single attributes and possible multiplications of attributes. This kind of test functions is can be applied for numeric and continuous attributes.

Decision trees are often constructed by a "top-down" approach. Starting from first node, called *the root*, containing all training examples, the constructing algorithm is looking for a feature to label this node with and partition the set of examples into subsets (related to values of the chosen test function). Then some new nodes arise holding these subsets of examples and one can continue the construction for new nodes until some stop condition is satisfied. The task for every step of the decision tree construction can be described as follows: For a given set of examples $E = \{e_1, ..., e_n\}$ with every example pre-classified to one of the decision classes $C_1, ..., C_d$, we would like to find among candidates to be split the one that does the best job of separating the examples into groups where a single class predominates.

Construction methods differ by the choice of splitter forms, the definition of quality measures of splitters and the strategy of searching.

Two test functions widely used in literature are of the form "*attribute = value*" for symbolic attributes and "*attribute < value*" for numeric attributes.

5.2 Decision tree construction - MDG method

In this section we consider a new method for symbolic domain partition. We start from some properties of optimal partition problem.

Let $\mathbf{S} = (U, A \cup \{d\})$ be a decision table. For a fixed attribute a and an object set $Z \subseteq U$, we define the *discernibility degree* of two disjoint sets V_1 and V_2 of values from V_a (denoted by $Disc(V_1, V_2|Z)$) by

$$Disc(V_1, V_2|Z) = |\{(u_1, u_2) \in Z^2 : d(u_1) \neq d(u_2) \land (a(u_1), a(u_2)) \in V_1 \times V_2\}|$$

Let P be an arbitrary partition of V_a. For any two objects $u_1, u_2 \in U$ we say that the pair of objects (u_1, u_2) is *discerned* by P if $[d(u_1) \neq d(u_2)] \wedge [P(a(u_1)) \neq P(a(u_2))]$. The *discernibility degree* $Disc(P|Z)$ of partition P over the set of objects $Z \subset U$ is defined as the number of pairs of objects from Z discerned by P, i.e.

$$Disc(P|Z) = |\{(u_1, u_2) \in Z \times Z : (u_1, u_2) \text{is discerned by P}\}|$$

In this section, similarly to the decision tree approach, we consider an optimization problem called the *Binary Optimal Partition* (BOP) problem which is described as follows:

PROBLEM: Binary Optimal Partition (BOP)
Input: *A set of objects Z and an attribute a.*
Output: *A binary partition P (i.e. $rank(P) = 2$) of V_a such that $Disc(P, Z)$ is maximal.*

We will show that BOP problem is NP-hard with respect to the size of V_a. The proof will also suggest some natural heuristics to search for optimal partitions.

Complexity of binary partition problem Let $Z \subseteq U$ and $\sigma \in V_d$. Denote by $s(V, \delta | Z)$, the number of objects u from $Z \cap C_\sigma$ such that $a(u) \in V \subseteq V_a$, i.e., $s(V, \delta | Z) = |\{u \in Z : (a(u) \in V) \wedge (d(u) = \delta)\}|$. In case of $rank(P) = 2$ (without loss of generality one can assume $P: V_a \to \{1, 2\}$) and $V_d = \{0, 1\}$, the discernibility degree of P is expressed by

$$Disc(P|Z) = \sum_{i \in V_1; j \in V_2} [s(i, 0|Z) \cdot s(j, 1|Z) + s(i, 1|Z) \cdot s(j, 0|Z)] \qquad (8)$$

where $V_1 = P^{-1}(1)$, $V_2 = P^{-1}(2)$ and $s(v, \delta | Z) = s(\{v\}, \delta | Z)$. In this section we fix the set of objects Z to be equal to U and for simplicity of notation, Z will be omitted in the functions described above.

To prove the NP-hardness of BOP problem we consider the corresponding decision problem called the *Binary Partition problem* which is defined as follows

---
PROBLEM: Binary Partition (BP)
Input: *A value set $V = \{v_1, \ldots, v_n\}$, two functions: $s_0, s_1 : V \to \mathbf{N}$ and a positive integer K.*
Question: *Is there a binary partition of V into two disjoint subsets $P(V) = \{V_1, V_2\}$ such that the discernibility degree of P defined by*

$$Disc(P) = \sum_{i \in V_1, j \in V_2} [s_0(i) \cdot s_1(j) + s_1(i) \cdot s_0(j)]$$

is satisfying the inequality $Disc(P) \geq K$?

If the BP problem is NP–complete then the BOP problem is NP–hard.

Theorem 12. *[55] The binary partition problem is NP-complete.* □

Proof:
It is clear that BP $\in$ NP. The NP-completeness of BP problem can be shown by a polynomial transformation from *Set Partition Problem* (SPP) defined as a problem of checking, for a given finite set of positive integers $S = \{n_1, n_2, ..., n_k\}$, if there is a partition of S into two disjoint subsets S_1 and S_2 such that $\sum_{i \in S_1} n_i = \sum_{j \in S_2} n_j$.

It is known that the SPP is NP-complete [23]. We will show that SPP is polynomially transformable to BP. Let $S = \{n_1, n_2, ..., n_k\}$ be an instance of SPP. The corresponding instance of BP problem is specified as follows:

- $V = \{1, 2, ..., k\}$;
- $s_0(i) = s_1(i) = n_i$ for $i = 1, .., k$;
- $K = \frac{1}{2} \left(\sum_{i \in V} n_i \right)^2$

For any partition P of the set V_a into two disjoint subsets V_1 and V_2 the discernibility degree of P can be expressed by:

$$Disc(P) = \sum_{i \in V_1; j \in V_2} [s_0(i) \cdot s_1(j) + s_1(i) \cdot s_0(j)] =$$

$$= \sum_{i \in V_1; j \in V_2} 2n_i n_j = 2 \cdot \sum_{i \in V_1} n_i \cdot \sum_{j \in V_2} n_j =$$

$$\leq \frac{1}{2} \left(\sum_{i \in V_1} n_i + \sum_{j \in V_2} n_j \right)^2 = \frac{1}{2} \left(\sum_{i \in V} n_i \right)^2 = K$$

i.e. for any partition P we have the inequality $Disc(P) \leq K$ and the equality holds iff $\sum_{i \in V_1} n_i = \sum_{j \in V_2} n_j$. Hence P is a good partition of V (into V_1 and V_2) for BP problem iff it defines a good partition of S (into $S_1 = \{n_i\}_{i \in V_1}$ and $S_2 = \{n_j\}_{j \in V_2}$) for SPP problem. Therefore BP problem is NP-complete and BOP problem is NP hard. ∎

Grouping algorithms Now we are ready to describe some approximate solutions for BOP problem. These heuristics are quite similar to 2-mean clustering algorithms, but instead of the Euclidean measure we use the discernibility degree function. First at all, let us explore several properties of the $Disc$ function.

The values of function $Disc(V_1, V_2)$ can be computed using the function s, namely

$$Disc(V_1, V_2) = \sum_{\delta_1 \neq \delta_2} s(V_1, \delta_1) \cdot s(V_2, \delta_2)$$

It is easy to observe that $s(V, \delta) = \sum_{v \in V} s(v, \delta)$, hence the function $Disc$ is additive function, i.e.:

$$Disc(V_1 \cup V_2, V_3) = Disc(V_1, V_3) + Disc(V_2, V_3) \text{ if } V_1 \cap V_2 = \emptyset$$
$$Disc(V_1, V_2 \cup V_3) = Disc(V_1, V_2) + Disc(V_1, V_3) \text{ if } V_2 \cap V_3 = \emptyset$$

The first binary partition algorithm called *grouping by minimizing discernibility* begins with the family of singletons $\mathbf{V}_a = \{\{v_1\}, ..., \{v_m\}\}$, and for any $V \in \mathbf{V}_a$, a vector $s(V) = [s(V, 0), s(V, 1), ..., s(V, r)]$ of its occurrences in decision classes $C_0, C_1, ..., C_r$ is associated. In every step we look for two nearest sets V_1, V_2 from $\mathbf{V}_a$ with respect to the function $Disc(V_1, V_2)$ and then we replace them by a new set $V = V_1 \cup V_2$ with occurrence vector $s(V) = s(V_1) + s(V_2)$. The algorithm is continued until $\mathbf{V}_a$ contains two sets and then suboptimal binary partition is stored in $\mathbf{V}_a$.

We illustrate the effect of this heuristics for the decision table presented in the Example from Figure 8 (Section 4.3). Let us consider the attribute **b** with five symbolic values $V_{\mathbf{b}} = \{b_1, b_2, b_3, b_4, b_5\}$. We depict in Figure 10 the consecutive steps of our method. The leftmost picture is of a labeled graph constructed directly from the decision table from Figure 8. Its edges are labeled by discernibility degree between values. We group the values b_4 and b_5 because they are connected by the edge with smallest discernibility degree. The result of the first grouping step is presented in the center picture. One can see the new vertex labeled by $\{b_4, b_5\}$ instead of old vertices b_4 and b_5. We obtain the semi–optimal partition $P_{\mathbf{b}} = \{\{b_1, b_4, b_5\}, \{b_2, b_3\}\}$ (see Figure 10).

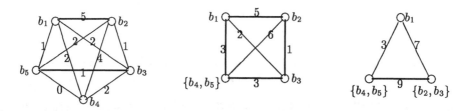

Fig. 10. An example of grouping algorithm by minimizing discernibility

The second techniques is called *grouping by maximizing discernibility*. The algorithm also begins with the family of singletons $\mathbf{V}_a = \{\{v_1\}, ..., \{v_m\}\}$, but first we look for two singletons with the largest discernibility degree to create kernels of two groups, let us denote them by $V_1 = \{v_1\}$ and $V_2 = \{v_2\}$. For any symbolic value $v_i \notin V_1 \cup V_2$ we compare discernibility degrees $Dics(\{v_i\}, V_1)$ and $Dics(\{v_i\}, V_2)$ and attach v_i to the group with smaller discernibility degree with respect to v_i. This process ends when all values from V_a have been exhausted.

This method is illustrated in Figure 11. First, we choose the vertices b_1 and b_2 as kernels of groups, because they are connected by the edge with maximum discernibility degree. As a result we obtain the partition $P_b = \{\{b_1, b_4, b_5\}, \{b_2, b_3\}\}$. One can see that for this data, the results of two heuristics are the same.

[h]

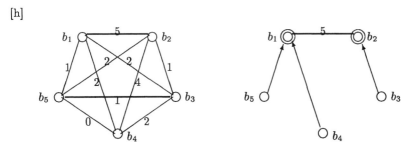

Fig. 11. Example of grouping algorithm by maximizing discernibility

Decision tree construction algorithm In [53] we have presented an efficient method for decision tree construction called the *MD algorithm*. It has been developed for continuous (numeric) data only. In this section we extend the MD algorithm to data with mixed attributes, i.e. both numeric and symbolic, by using binary partition algorithms presented in the previous Sections.

Given a decision table $\mathbf{S} = (U, A \cup \{d\})$, we assume that there are two types of attributes: numeric and symbolic. We also assume that the type of attributes is given by a predefined function $type : A \to \{N, S\}$ where

$$type(a) = \begin{cases} N \text{ if } a \text{ is a Numeric attribute} \\ S \text{ if } a \text{ is a Symbolic attribute} \end{cases}$$

We can use grouping methods to generate a decision tree. The structure of the decision tree is defined as follows: internal nodes of a tree are labelled by logical test functions of the form $T : U \to \{True, False\}$ and external nodes (leaves) are labelled by decision values.

In this paper we consider two kinds of tests related to attribute types. In the case of symbolic attributes $a_j \in A$ we use *test functions defined by partition*:

$$T(u) = True \iff [a_j(u) \in V]$$

where $V \subset V_{a_j}$. For numeric attributes $a_i \in A$ we use *test functions defined by discretization*:

$$T(u) = True \iff [a_i(u) \leq c] \iff [a_i(u) \in (-\infty; c)]$$

where c is a *cut* in V_{a_i}. Below we present the process of a decision tree constructing. We additionally use some object sets to label nodes of the decision tree. This third kind of labels will be removed at the end of the construction process. The algorithm is presented in as follows.

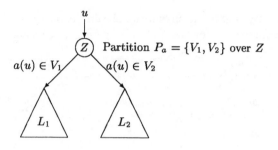

Fig. 12. The decision tree

ALGORITHM: (MDG – DECISION TREE)

Create a decision tree with one node labelled by the object set U;
for each leaf L labelled by an object set Z do
begin
  for each attribute $a \in A$ **do**
    if $type(a) = S$ **then** search for the optimal binary partition $P_a = \{V_1, V_2\}$ of V_a;
    if $type(a) = N$ **then** search for the optimal cut c in V_a and set $P_a = \{V_1, V_2\}$ where $V_1 = (-\infty; c)$ and $V_2 = (c; \infty)$;
  choose an attribute a so that $Disc(P_a|Z)$ is maximal and label the current node by the formula $[a \in V_1]$;
  for $i := 1$ to 2 **do**
  $$Z_i = \{u \in Z : a(u) \in V_i\}; \quad L_i = \begin{cases} v_d & \text{if } d(Z_i) = \{v_d\} \\ Z_i & \text{if } card(d(Z_i)) \geq 2 \end{cases}$$
  create two successors of the current node and **label** them by L_1 and L_2;
end;
if all leaves are labelled by decisions **then** STOP **else** goto Step 2;

---

6 Applications

Templates are patterns reflecting regularities occurring in data. They are useful constructions for many data mining tasks such as classification, description or decomposition. In this section we will discuss how to apply templates extracted from data to solve basic data mining problems.

6.1 General scheme for description of decision classes

We outline a general searching scheme for regularities in data based on templates. However, our main goal is to find a set of rules binding attributes and decisions i.e. approximate decision rules. To this end we follow the steps described below. Suppose that a decision table **S** is given. We are interested in the description of its ith decision class by a set of decision rules, i.e. by the decision algorithm for this class.

Step 1: Produce a set of templates covering the decision class, i.e. templates belonging to one decision class. Algorithms for template generation can be adopted to this new kind of a template: one can simply change the formula for the template fitness (see Section 2.2).

Step 2: Combine the templates obtained in the previous step into groups and apply the operations of generalization and/or contraction. One can use one of these operators in an adaptive way to obtain decision algorithm of better quality. Repeat Step 2 until the quality of obtained decision algorithm is sufficiently good.

Step 3: If the quality of the decision algorithm is not satisfactory then repeat the construction process from Step 1 else use the algorithm (maybe after some post-processing) as the approximate definition of the i-th decision class.

A strategy for choosing templates can also depend on the estimation of how promising these templates can be for the construction of the decision class approximation by application of different operators like grouping, generalization, contraction. The grouping procedures are executed after templates have been chosen. In this step the following principles should be observed:

(i) Two templates covering almost the same objects from the class and almost disjoint on objects not belonging to the class should be separated by grouping procedures;

(ii) The family of intersections of different templates in one group should not be "close" to the partition of the decision class into one element sets.

Groups of templates are received as the results of these procedures. Different approximate coverings of the decision class are constructed by applying generalization to these groups. Next, the grouping procedures are executed again as a pre-processing for contraction. The process continues until a description of the decision class with a sufficient quality is constructed; otherwise, the construction process is estimated as unsuccessful and it is redone starting from some previous construction level by applying another grouping, generalization or contraction strategies. The generalization operator may be understood in the simplest cases as the union of objects matching one of the templates, alternatively as a minimal template including all the templates. The contraction, in the simplest case, can be defined as the intersection of the templates. For both operators, one may take into account e.g. weights attached to the attributes: the "less important" attribute can be generalized or fixed etc.

The quality of decision algorithm obtained using this method depends on the quality of the decision class description returned by the algorithm, and also on its complexity - we would like to produce rules as simple as possible.

6.2 Data decompostion

Decomposition is a problem of partitioning a large data table into smaller ones. One can use templates extracted from data to partition data table into blocks of objects with common (or similar) features. Every block consists of objects matching one template. We consider two decomposition schemes: decomposition by a template tree and decomposition by a template set.

Decomposition by Template Tree The main goal is to construct a decomposition tree. Decomposition tree is a binary tree, whose every internal node is labeled by some template and external node (leaf) is associated with a set of objects matching all templates in a path from the root to a given leaf. Let **S** be a data table. The algorithm for decomposition tree construction can be presented as follows:

ALGORITHM: (DECOMPOSITION BY TEMPLATE TREE)

Step 1 Find the best (generalized) template T in **S**.
Step 2 Divide A onto two sub-tables: A_1 containing all objects satisfying T, $A_2 = A - A_1$.
Step 3 If obtained sub-tables are of acceptable size (in the sense of rough set methods) then stop else repeat 1–3 for all "too large" sub-tables.

This algorithm produces a binary tree of sub-tables with corresponding sets of decision rules for sub-tables in the leaves of the tree. Any non-leaf vertex of the obtained tree refers to a template.

Decomposition by minimal set covering Our goal is to search for a partition consisting of sub-tables of feasible sizes. It means that these sub-tables should not be too large to be analyzed by existing algorithms and at the same time they should be not too small for assuring generality of decision rules extracted from them. We also optimize (minimize) the number of generated sub-tables. In addition, we want to reach sub-tables with some degree of a regularity.

Our idea is based on searching for a set of sub-tables covering the domain. Next we choose from it an optimal cover for our domain. The "optimal" cover set can be defined using different criteria, for example, the cover set is optimal if it is of minimal cardinality.

To construct a minimal set of sub-tables covering a domain, one can use algorithms presented in Section 3.3.

6.3 Classification problem

The classification is one of the most important tasks in data mining. Using knowledge extracted from training data set, the problem is to classify a new case to proper decision class. In this section we present some approaches to the classification problem.

Classification by the binary decomposition tree Suppose we have a binary tree created in the process of decomposition (BDT method) as described in Section 6.2. Let x be a new object and $A(T)$ be a sub-table containing all objects matching T, we evaluate x starting from the root of the tree as follows:

ALGORITHM: (CLASSIFICATION BY BDT)

Step 1: If x matches template T found for A then go to sub-tree related to $A(T)$ else go to sub-tree related to $A(\neg T)$.
Step 2: If x is at the leaf of the tree then go to 3 else repeat $1-2$ substituting $A(T)$ (or $A(\neg T)$) for A, respectively.
Step 3: Apply decision rules calculated [44],[71], [7] for sub-table attached to the leaf to classify x.

This algorithm uses a binary decision tree, however it should not be confused with C4.5 [71], ID3 [41] or other algorithms using decision trees. The difference is that the above algorithm splits the object domain (universe) into sub-domains and for a new case we search for the most similar (from the point of view of the templates) sub-domain. Then rough set methods, C4.5, etc., may be used for the classification of this new case relatively to the matched sub-domain.

Classification by covering sub-domains Another approach for new object classification is based on decision rules extracted from sub-tables covering the domain. Any sub-table from a cover set is defined by a template matching it. Assume $\{T_1, T_2, ..., T_m\}$ is a set of templates defining the cover set, then a new object x can by classified as follows:

ALGORITHM: (CLASSIFICATION BY COVERING SUB-DOMAINS)

Step 1: Use well known methods, from e.g. [44],[71],[6],[48], to generate decision rules for any sub-table from the cover set.
Step 2: Classify x to proper sub-tables by matching it to templates from $\{T_1, T_2, ..., T_m\}$.
Step 3: Use decision rules of sub-tables found in Step 2 to classify x.
step 4: If x is classified to more then two decision classes, use voting strategy to classify x.

Classification by decision templates The above methods are of two-stage classification type. A new object is first classified to proper sub-domains. Then it is classified using the rules extracted from these sub-domains. These methods are helpful when the investigated domain is large and there is a need to decompose the domain into smaller sub–domains.

Another method can be based on direct classification using templates learned from the whole data domain. We use decision templates to yield classification. The classification process consists of two phases. At first, we generate all decision templates, that create a description of decision classes. The classification process is presented in the scheme below:

ALGORITHM: (CLASSIFICATION BY DECISION TEMPLATES)

Step 1: Generate a decision set **T** of templates, which covers the training table.
Step 2: Classify a new object x using templates chosen from **T**. In conflicting cases, the fitness function is used to resolve conflicts.

The are four problems to be solved in any classification process: (1) Which algorithm should be used for a single template generation?, (2) Which algorithms should be used for a template set generation? (3) Which templates should be chosen (from the generated template set) for new object classification, and (4) How to resolve conflicts?

One can use methods presented in Sections 3.2, 3.2 for the decision template extraction; the methods presented in Section 3.3 are suitable for template set generation. To classify a new object, one can adopt two strategies: classification by matching templates and classification by closest templates. In the former case, a new object x is classified using templates matched by x and in the latter case, x is classified by templates closest to x. The distance function of a given object x to the template T can be defined, for example, by

1. *Pseudo-Hamming distance*:

$$HDist(x,T) = 1 - \frac{|\{D_i \in T : x \text{ satisfying } D_i\}|}{length(T)}$$

2. *Pseudo-Euclidean distance*:

$$EDist(x,T) = \sqrt{\sum_{D \in T} d^2(x,D)}$$

where $d(x,D)$ is a distance of an object x to a descriptor D.

For any object x and a descriptor $D = (a \in V_a)$, where $V_a = [v_1, v_2]$, the distance $d(v, [v_1, v_2])$ is defined by:

$$d(v, [v_1, v_2]) = \begin{cases} v_1 - v & \text{if } v \leq v_1, \\ 0 & \text{if } v_1 \leq v \leq v_2, \\ v - v_2 & \text{otherwise.} \end{cases}$$

If V_a is a sub-set of a symbolic domain, the $d(v, V_a)$ is defined by:

$$d(v, V_a) = \begin{cases} 0 & \text{if } v \in V_a, \\ 1 & \text{otherwise.} \end{cases}$$

The next problem is how to assign x to the proper decision class. Below we present some functions estimating the *weight* of a decision class. Object x is classified to the decision class with the *maximal weight*. For a template T and an object x, let $T(x) =$ **true** denote the fact that object x is matching the template T.

- **Majority voting**
  The majority voting is one of the simplest strategies. The function measuring the weight of decision class c is defined by
  $$fitness(c) = |\{T : (T(x) = \text{true}) \wedge (d(T) = c)\}|$$
  where $T(x) =$ **true** denotes that x matches to T. According to the defined weight function, object x is classified to a decision class most frequently appearing in templates matching x.

- **Total fitness strategy**
  Contrary to the majority strategy, we take into consideration not only the number of templates, but their *weight* too. As the first example of template weight, we consider its fitness function. The *fitness* function maps every template T to the number of objects supporting it. Let **T** be a set of templates, that cover a training table. To any template T we associate a decision $d(T)$ and the number $fitness(T)$ of supporting objects. Having a fitness function for templates, we can define a *weight* for decision class. Weight of decision class c is a function reflecting a number of objects from this decision class, that are similar to x. Weight of the class c can be defined by formula:
  $$weight(c) = \sum_{T(x)=\text{true},d(T)=c} fitness(T)$$

- **Average accuracy strategy**
  In the previous section a weight of a template has been defined as the number of objects supporting the template. In this section we present another weight function defined by the template *accuracy*. For every decision template T associated with decision c, the $confidence(T)$ is defined by $\frac{fitness_c(T)}{fitness(T)}$, where $fitness_c(T)$ is the number of objects from decision class c supporting T and $fitness(T)$ is the number of all objects supporting T.
  Let NN be the set of templates nearest (best matching) to the considered object x. Hence a weight of the decision class c is defined as the average accuracy of all the templates related to c, that belong to the neighborhood of x. The weight of class c is defined by formula:
  $$weight(c) = \frac{\sum_{T(x)=\text{true},d(T)=c} accuracy(T)}{|\{T : (T(x) = \text{true}) \wedge (d(T) = c)\}|}$$

- **Max-min accuracy strategy**
  Having accuracy function as the weight of a template, we can consider other functions to measure weight of a decision class. The weight of the decision

class c can be defined as the *minimal confidence* of templates related to c, that is
$$weight(c) = \min_{T(x)=\mathbf{true}, d(T)=c} \{confidence(T)\}.$$

7 Relational patterns

In this section we consider a new kind of patterns defined by binary relations defined over objects of the data set. These patterns are called *relational patterns*. We consider the patterns defined by *tolerance relations*, i.e., reflexive and symmetric (or only reflexive) relations. A tolerance relation in the literature is often viewed as an extension of the indiscernibility relation used in rough set approach (see [13, 25, 34, 40, 84]). The tolerance relations are usually defined by means of conjunction of local similarity relations $\mathcal{R}_a$ of attributes. We consider a wider class of tolerance relations defined by different combinations of local similarities. Properties of tolerance relations and searching methods for these relations are included in this section. These relations turn out to have simple and interesting geometrical interpretations, helping to develop effective heuristics to search in different similarity families for sub-optimal relations. We present methods for extracting relevant similarity relations from data rather than assuming their definitions a priori (see e.g. [19]). These similarity relations are extracted from some parameterized families of relations by tuning parameters to optimal ones. We start by presenting basic notions and properties of these tolerance families.

7.1 Basic notions

Let $\mathbf{S} = (U, A \cup \{d\})$ be given a decision table.

A relation $\tau \subseteq U \times U$ is called a *tolerance (similarity) relation*, if

- $\forall_{x \in U} \ \langle x, x \rangle \in \tau$ (*reflexivity*);
- $\forall_{x,y \in U} \ (\langle x, y \rangle \in \tau \Rightarrow \langle y, x \rangle \in \tau)$ (*symmetry*).

We also consider weak tolerance relations, i.e. reflexive relations. In the sequel we omit prefix "weak" when it follows from the context that the relation is tolerance or weak tolerance.

If $\mathbf{S} = (U, A \cup \{d\})$ is a given decision table and $A = \{a_1, \ldots, a_k\}$ then any object x is described by an information vector $Inf_A(x) = \langle a_1(x), a_2(x), \ldots a_k(x) \rangle$. Let us consider information space $INF_A = \prod_{a \in A} V_a$, where V_a is the domain of the attribute $a \in A$. Assume that τ_A is a similarity relation defined on INF_A. Hence a similarity relation in $U \times U$ can be defined by

$$\forall_{x,y \in U} \ \{\langle x, y \rangle \in \tau \Leftrightarrow \langle Inf_A(x), Inf_A(y) \rangle \in \tau_A\}.$$

For any $x \in U$ the *similarity class* $[x]_\tau$ is defined by

$$[x]_\tau = \{y \in U : \langle x, y \rangle \in \tau\}.$$

We say that the similarity relation τ *identifies* objects x and y if $\langle x,y \rangle \in \tau$; otherwise we say that it *discerns* them.

One can define the *lower approximation* and the *upper approximation* of any subset $X \subseteq U$ with respect τ to by

$$\underline{\tau(X)} = \{x \in U : [x]_\tau \subseteq X\}; \quad \overline{\tau(X)} = \{x \in U : [x]_\tau \cap X \neq \emptyset\},$$

respectively.

Let $\delta_a : U \times U \to \Re^+ \cup \{0\}$ be a function defined for any attribute a. The function δ_a is a *similarity measure* if it satisfies the following conditions:

- the value of $\delta_a(x,y)$ depends on the values $a(x)$ and $a(y)$ only;
- $\delta_a(x,x) = 0$;
- $\delta_a(x,y) = \delta_a(y,x)$

The *parametric local relation* $\tau_a(\varepsilon_a)$ can be defined by

$$\langle x,y \rangle \in \tau_a(\varepsilon_a) \Leftrightarrow \delta_a(x,y) < \varepsilon_a$$

where ε_a is a threshold and δ_a is a similarity measure.

Hence, a *global similarity relation* $\tau \subseteq U \times U$ can be defined by

$$\langle x,y \rangle \in \tau \Leftrightarrow \Psi_R(\delta_{a_1}(x,y), \delta_{a_2}(x,y), \ldots, \delta_{a_k}(x,y)) = \textbf{true} \tag{9}$$

where $\Psi(\xi_1, \xi_2, ..., \xi_k)$ is an open formula of the first order logic and Ψ_R is its realization in a relational structure of real numbers such that

$$\Psi_R(0, 0, \ldots, 0) = \textbf{true}$$

By C_k we denote the set $\{\langle r_1, r_2, ..., r_k \rangle \in R^k : 0 \leq r_i \text{ for } i = 1, ..., k\}$. Any relation τ defined by (9) determines a subset $\overline{\tau} \subseteq C_k$ equal to

$$\{\langle r_1, r_2, ..., r_k \rangle \in C_k : \Psi_R(r_1, r_2, ..., r_k) = \textbf{true}\}.$$

One can define different similarity relations using different formulas

$$\Psi(\xi_1, \xi_2, ..., \xi_k)$$

We list four important similarity relation classes. Using them one can define a rich family of other relations.

1. *Max form*:
$$\langle x,y \rangle \in \tau_1(\varepsilon) \Leftrightarrow \max_{a_i \in A}\{\delta_{a_i}(x,y)\} \leq \varepsilon$$

2. *Sum form*:
$$\langle x,y \rangle \in \tau_2(\varepsilon_1, ..., \varepsilon_k) \Leftrightarrow$$

3. *Conjunction form*
$$\bigwedge_{a_i \in A} [\delta_{a_i}(x,y) \leq \varepsilon_i]$$

4. *Parameterized sum*:

$$\langle x, y \rangle \in \tau_3(w_1, ..., w_k, w) \Leftrightarrow \sum_{a_i \in A} w_i \cdot \delta_{a_i}(x, y) + w \leq 0$$

5. *Product form*:

$$\langle x, y \rangle \in \tau_7(w) \Leftrightarrow \prod_{a_i \in A} \delta_{a_i}(x, y) \leq w$$

where $\delta_{a_i}(x, y)$ is a predefined similarity measure for $i = 1, ..., k$ and ε_i, ε, w_i, w are real numbers, called *parameters*.

A similarity relation $\tau \subseteq U \times U$ is *consistent* with a decision table $\mathbf{S} = (U, A \cup \{d\})$ if

$$\langle x, y \rangle \in \tau \Rightarrow (d(x) = d(y)) \vee (\langle x, y \rangle \in IND(A))$$

for any objects $x, y \in U$.

The similarity relation is *inconsistent* if it is not consistent.

One can see that if a similarity relation is consistent with the decision table $\mathbf{S}$ then it contains only pairs of objects with the same decision, however *inconsistent* similarity may contain discernible pairs of objects with different decisions.

The relation τ is *optimal* in the family $\mathcal{T}$ for a given $\mathbf{S}$ if τ contains the maximum number of pairs of objects among similarity relations from $\mathcal{T}$ consistent with $\mathbf{S}$.

A similarity relation $\tau \subseteq U \times U$ is U_1-*consistent*, where $U_1 \subseteq U$ if

$$\langle x, y \rangle \in \tau \Rightarrow (d(x) = d(y)) \vee (\langle x, y \rangle \in IND(A))$$

for any objects $x \in U_1, y \in U$. We denote by τ_{U_1} the U_1-*consistent* similarity relation.

7.2 Similarity measures

We discuss in this section examples of distance functions that are often used to measure similarity between two values of a given attribute in the training set.

For a decision table $\mathbf{S} = (U, A \cup \{d\})$, we consider similarity relations constructed from some *similarity measures* predefined for attribute values. We list some basic similarity measures used to define the most important similarity relation.

We distinguish two kinds of similarity measures: for attributes with numeric values and for attributes with symbolic values.

Similarity measure $\delta_a : U \times U \to \mathbb{R}^+ \cup \{0\}$ for a numeric attribute $a \in A$ can be defined as follows. Let $x, y \in U$. The range of attribute a, denoted by $range(a)$, is defined by $range(a) = max(a) - min(a)$, where $max(a)$ and $min(a)$ are the maximum and minimum values, respectively, observed in the training set for the attribute a.

1. (*The linear difference*)
$$\delta_a(x,y) = |a(x) - a(y)|.$$
Linear difference can be normalized by
$$\delta_a(x,y) = \frac{|a(x) - a(y)|}{range(a)}.$$

2. (*The square difference*)
$$\delta_a(x,y) = [a(x) - a(y)]^2.$$
Square difference can be normalized by
$$\delta_a(x,y) = \left[\frac{a(x) - a(y)}{range(a)}\right]^2.$$

3. (*The constant dimension difference*)
$$\delta_a(x,y) = [a(x) - a(y)]^r$$
where r is a positive integer. The constant dimension difference can be normalized by
$$\delta_a(x,y) = \left[\frac{a(x) - a(y)}{range(a)}\right]^r$$

In the symbolic attribute case, a variant of δ_a is obtained by stipulating
$$\delta_a(x,y) = \begin{cases} 1 & a(x) = a(y) \\ 0 & otherwise \end{cases}$$

A more sophisticated variant can be defined by
$$\delta_a(x,y) = \sum_{k \in V_d} \frac{||C_k \cap [x]_{IND(a)}| - |C_k \cap [y]_{IND(a)}||}{|C_k|},$$
where C_k denotes the k-th decision class and V_d the set of decision values.

Using these similarity measures to general similarity relation forms 9 presented in Section 7.1 one can define various similarity relations. Below we present some of the most important ones. These relations are motivated by standard distance functions often used in machine learning systems. The name of a relation is derived from the corresponding distance function.

1. (*Chebychev relation*) Applying the *linear difference* measure to the *max form* we obtain the Chebychev relation
$$\langle x,y \rangle \in \tau(\varepsilon) \Leftrightarrow \max_{a_i \in A} \{|a(x) - a(y)|\} \leq \varepsilon.$$

The *generalized Chebychev relation* can be obtained by setting *linear difference* measure to the *conjunction form*
$$\langle x,y \rangle \in \tau(\varepsilon_1, ..., \varepsilon_k) \Leftrightarrow \bigwedge_{a_i \in A} [|a_i(x) - a_i(y)| \leq \varepsilon_i].$$

2. (*Mahattan / city-block*) Applying the *linear difference* measure to the *sum form* we obtain the Mahattan relation

$$\langle x,y\rangle \in \tau(w) \Leftrightarrow \sum_{a_i \in A} |a(x) - a(y)| + w \leq 0.$$

The *generalized Mahattan relation* can be obtained by setting *linear difference* measure to the *parametrized sum form*

$$\langle x,y\rangle \in \tau(w_1,...,w_k,w) \Leftrightarrow \sum_{a_i \in A} w_i \cdot |a(x) - a(y)| + w \leq 0.$$

3. (*Euclidian relation*) Applying the *square difference* measure to the *sum form* we obtain the Euclidian relation

$$\langle x,y\rangle \in \tau(w) \Leftrightarrow \sum_{a_i \in A} [a(x) - a(y)]^2 + w \leq 0.$$

The *generalized Euclidian relation* can be obtained by setting the *square difference* measure to the *parametrized sum form*

$$\langle x,y\rangle \in \tau_4(w_1,...,w_k,w) \Leftrightarrow \sum_{a_i \in A} w_i \cdot [a(x) - a(y)]^2 + w \leq 0.$$

4. (*Minkowsky relation*) Applying the *higher dimension difference* measure to the *sum form* we obtain the Minkowsky relation

$$\langle x,y\rangle \in \tau_4(w) \Leftrightarrow \sum_{a_i \in A} [a(x) - a(y)]^r + w \leq 0.$$

The *generalized Minkowsky relation* can be obtained by setting the *constant dimension difference* measure to the *parameterized sum form*

$$\langle x,y\rangle \in \tau_4(w_1,...,w_k,w) \Leftrightarrow \sum_{a_i \in A} w_i \cdot [a(x) - a(y)]^r + w \leq 0.$$

7.3 Approximate space. Geometrical interpretation

In this section we show that the families of similarity relations proposed in Section 7.1 have clear geometrical interpretations, i.e., they can be represented in a straightforward way by subsets of a real affine space $\mathbb{R}^k$. Knowledge about similarity relation representation plays an important role in learning processes. It helps to search for an approximate description of the corresponding subset of real affine space $\mathbb{R}^k$ in order to find a semi-optimal similarity relation τ.

For a decision table $\mathbf{S} = (U, A \cup \{d\})$ with k condition attributes and a set $\{\delta_{a_i}\}_{a_i \in A}$ of predefined similarity measures we build the similarity table $\mathbf{B} = (U', A' \cup \{D\})$ constructed from decision table $\mathbf{S}$ and the set $\{\delta_{a_i}\}_{a_i \in A}$ of similarity measures. Every object u of the table $\mathbf{B}$ can be represented by point $p(u) = \left[a'_1(u),...,a'_k(u)\right] \in \mathbb{R}^k$ of one of two categories "white" or "black".

Point $p(u) \in \mathbb{R}^k$ is "white" iff $\{u_0 \in U' : p(u_0) = p(u)\}$ is non-empty and it consists of objects with the decision D equal to 0 only; otherwise $p(u)$ is "black". Below we present geometric interpretations of some standard similarity relations. We take the functions: $\delta_{a_i}(x,y) = |a_i(x) - a_i(y)|$ for any attribute $a_i \in A$ to define similarity of values from $\mathcal{DOM}(a)$.

Let us consider, as an example, a table with two attributes representing the quantity of vitamin A and C in apples and pears. We want to extract the

Vit.A	Vit.C	Fruit	Vit.A	Vit.C	Fruit
1.0	0.6	Apple	2.0	0.7	Pear
1.75	0.4	Apple	2.0	1.1	Pear
1.3	0.1	Apple	1.9	0.95	Pear
0.8	0.2	Apple	2.0	0.95	Pear
1.1	0.7	Apple	2.3	1.2	Pear
1.3	0.6	Apple	2.5	1.15	Pear
0.9	0.5	Apple	2.7	1.0	Pear
1.6	0.6	Apple	2.9	1.1	Pear
1.4	0.15	Apple	2.8	0.9	Pear
1.0	0.1	Apple	3.0	1.05	Pear

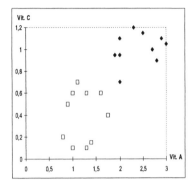

Table 1. Apple and pear data.

similarities of fruits of one category. The data about apples and pears are shown in Figure 13.

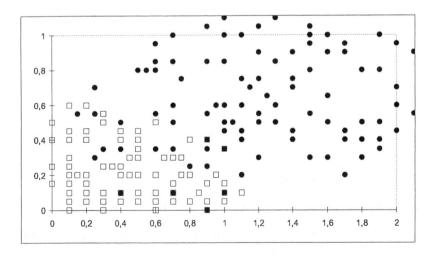

Fig. 13. A visualization of object pair space.

We consider geometric interpretations of tolerance relations in the space of pairs of objects from the fruit table.

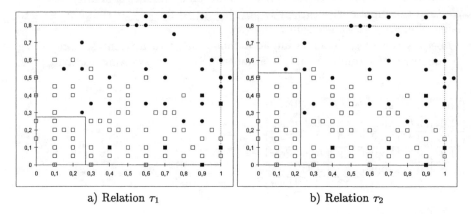

a) Relation τ_1 b) Relation τ_2

Fig. 14. Interpretation of similarity relations τ_1 and τ_2

1. The first similarity relation class consists of relation in *max form*

$$\langle x, y \rangle \in \tau_1(\varepsilon) \Leftrightarrow \max_{a_i \in A} \{\delta_{a_i}(x, y)\} \leq \varepsilon$$

where ε is a non-negative real number. The relation $\tau_1(\varepsilon)$ defines the following subset of $\mathbb{R}^k$:

$$\overline{\tau_1(\varepsilon)} = \{(r_1, ..., r_k) \in C_k : 0 \leq r_i \leq \varepsilon \text{ for } i = 1, ..., k\}.$$

Hence $\overline{\tau_1(\varepsilon)}$ is a *hypercube* with edges of length ε. This hypercube is attached to the origin O of axes (Figure 14a).
By $\mathcal{T}_1$ we denote the family of all similarity relations of the form $\tau_1(\varepsilon)$ where ε is a positive real.

2. The second similarity relation class, called the *descriptor conjunction*, consists of relations in *conjunctive form*

$$\langle x, y \rangle \in \tau_2(\varepsilon_1, ..., \varepsilon_k) \Leftrightarrow \bigwedge_{a_i \in A} [\delta_{a_i}(x, y) \leq \varepsilon_i]$$

where $\varepsilon_1, ..., \varepsilon_k$ are non-negative real numbers. The relation $\tau_2(\varepsilon_1, ..., \varepsilon_k)$ defines the following subset $\overline{\tau_2(\varepsilon_1, ..., \varepsilon_k)} \subseteq \mathbb{R}^k$:

$$\overline{\tau_2(\varepsilon_1, ..., \varepsilon_k)} = \{(r_1, ..., r_k) \in C_k : 0 \leq r_i \leq \varepsilon_i \text{ for } i = 1, ..., k\}$$

$\overline{\tau_2(\varepsilon_1, ..., \varepsilon_k)}$ is an *interval* in R^k with boundaries $\varepsilon_1, \varepsilon_2, ..., \varepsilon_k$. It is attached to the origin O of axes (Figure 14b).
By $\mathcal{T}_2$ we denote the family of all similarity relations of the form $\tau_2(\varepsilon_1, ..., \varepsilon_k)$, where $\varepsilon_1, ..., \varepsilon_k$ are positive real numbers.

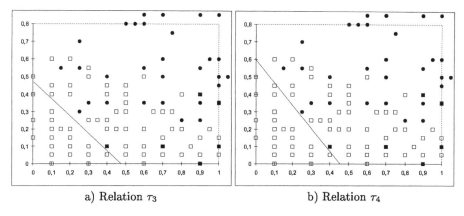

Fig. 15. The interpretation of similarity relations τ_3 and τ_4

3. The third similarity relation class τ_3, called the *linear combination*, consists of relations in *sum form*

$$\langle x, y \rangle \in \tau_3(w) \Leftrightarrow \sum_{a_i \in A} \delta_{a_i}(x, y) + w \leq 0$$

where w is a real number. The relation $\tau_3(w)$ defines the following subset of $\mathbb{R}^k$

$$\overline{\tau_3(w)} = \left\{ (r_1, ..., r_k) \in C_k : \sum_{i=1}^{k} r_i + w \leq 0 \right\}$$

Hence $\overline{\tau_3(w)}$ is a region in C_k under the hyperplane $H: \sum_{i=1}^{k} x_i + w = 0$ (Figure 15a). By $\mathcal{T}_3$ we denote the family of all similarity relations of the form $\tau_3(w)$.

4. The next similarity relation class τ_4, called the *parameterized linear combination*, consists of relations in a *parameterized sum form*

$$\langle x, y \rangle \in \tau_4(w_1, ..., w_k, w) \Leftrightarrow \sum_{a_i \in A} w_i \cdot \delta_{a_i}(x, y) + w \leq 0$$

where $w_1, ..., w_k, w$ are real numbers. The relation $\tau_4(w_1, ..., w_k, w)$ defines the following subset of $\mathbb{R}^k$

$$\overline{\tau_3(w_1, ..., w_k, w)} = \left\{ (r_1, ..., r_k) \in C_k : \sum_{i=1}^{k} w_i \cdot r_i + w \leq 0 \right\}$$

Hence $\overline{\tau_4(w_1, ..., w_k, w)}$ is a region in C_k under the hyperplane $H: \sum_{i=1}^{k} w_i \cdot x_i + w = 0$ (Figure 15a). By $\mathcal{T}_4$ we denote the family of all similarity relations of the form $\tau_4(w_1, ..., w_k, w)$.

5. The last class of similarity relations consists of relations in the *product form*

$$\langle x,y \rangle \in \tau_5(w) \Leftrightarrow \prod_{a_i \in A} \delta_{a_i}(x,y) \leq w,$$

where w is a non-negative real number. The set $\overline{\tau_5(w)}$ is equal to

$$\{(r_1, ..., r_k) \in C_k : r_1 \cdot ... \cdot r_k \leq w\}$$

Hence it is a region in C_k bounded by *hyperboloid* (Figure 16.b).

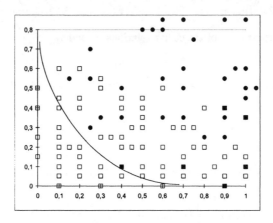

Fig. 16. Interpretation of tolerance relation τ_5

8 Similarity relation learning

For a given general parameterized formula defining a similarity relation class we would like to extract the relevant values of parameters occurring in this formula and defining the best similarity relation between objects from the training table.

8.1 Searching methods

We discuss three methods to construct similarity relations from a given decision table $\mathbf{S} = (U, A \cup \{d\})$, called *global*, *local*, and *categorical*. The choice among these methods depends on the application problem at hand.

The global method extracts from the whole space of pairs of object $U \times U$ the optimal similarity relation which describes similarity between objects from one decision class. This method gives a simple description of the similarity relation. The relation computed by the global method determines global similarities of all pairs of objects. Notice that using the global method one cannot generate a relevant similarity related to a given object or a given group of objects.

The second strategy of similarity relation extraction, called the *local method*, is a searching strategy for the optimal similarity relations τ_x where $x \in U$. We restrict the searching space to $\{x\} \times U$ while constructing the similarity relation τ_x. The similarity relation τ_x is optimal for a given object x if τ_x discerns x from all the objects with decisions different from $d(x)$ and at the same time it identifies x with the maximum number of objects with the same decision $d(x)$. Similarity relations generated by local methods describe well the similarity of a given object to another but they do not describe the global similarities for the decision table.

The *categorical method* extracts a similarity relation optimal with respect to a given decision class. The tolerance relation τ_C is optimal for a given decision class C if it discerns the objects from C from all the objects not belonging to C and at the same time τ_C identifies the maximum number of pairs of objects from C. For construction of τ_C one should examine the set $C \times U$ only. This method is more expensive than the local method with respect to time and space complexity but the similarity relation generated by this method describes well the similarities of groups of objects characteristic for a given decision.

Global method Let $\mathbf{S} = (U, A \cup \{d\})$ be a decision table and let δ_a be the similarity measure for any attribute $a \in A$. The problem of extracting a tolerance relation from a given class of parameterized similarity relations is a searching problem for the parameters such that the similarity relation defined by these parameters (returned as output from the searching process) is optimal. Our goal is to find a global similarity relation that *discerns* between all pairs of objects with different decisions and *identifies* (makes similar) maximum number of pairs of objects with the same decision.

In the first stage of construction, we define the new decision table called the *similarity table*, which consists of information about the object similarity calculated from similarity measures. The *universe* of the similarity table is defined by the set of all pairs of objects from table $\mathbf{S}$ and the *attribute values* are the values of the similarity measure function for pairs of objects. The new table has a binary decision. The decision value for any pair of objects is equal to 0 if its objects have the same decision in the original table $\mathbf{S}$, and 1 otherwise. Formal definition of the similarity table $\mathbf{B}$ from table $\mathbf{S}$ and the set of similarity measures $\{\delta_a\}_{a \in A}$ is presented below

$\mathbf{B} = (U', A' \cup \{D\})$, where $U' = U \times U$
$A' = \left\{ a' : U' \to \mathbb{R}^+ : a'(\langle x, y \rangle) = \delta_a(x, y) \right\}$,
$D(x, y) = \begin{cases} 0 \text{ if } d(x) = d(y) \\ 1 \text{ otherwise.} \end{cases}$

The searching problem for the optimal similarity relation of table $\mathbf{S}$ among relations from a given class of similarity relations can be considered as the problem of decision rule extraction from the decision table $\mathbf{B}$. We are interested in decision rules describing the decision class of $\mathbf{B}$ with decision 0, i.e., the class associated with pairs of objects of the table **S** with the same decision. Our goal is to find a rule of the form $\Psi\left(a_1'(u), a_2'(u), \ldots, a_k'(u)\right) \Rightarrow (D=0)$ with a large support, i.e. satisfied by as many as possible objects $u \in U'$.

Local method The local method constructs similarity relations relative to objects. We assume only the reflexivity of relations, so we consider weak–tolerance (similarity) relations. Let $\mathbf{S} = (U, A \cup \{d\})$ be a decision table and let δ_a be a similarity measure for any attribute $a \in A$. For a given object x the local method extracts the similarity relation τ_x optimal with respect to x. The goal is to find a tolerance relation discerning x from all the objects with decisions different from $d(x)$ and identifying the maximum number of objects with decision $d(x)$. Analogously to the global method we construct a new decision table which contains information about the similarity of the object x to other objects. The new table $\mathbf{B}_x$ is defined from **S** and the set of similarity measures $\{\delta_a\}_{a \in A}$ by

$\mathbf{B}_x = (U', A' \cup \{D\})$ where $U' = \{x\} \times U$;
$A' = \left\{a' : U' \to \mathbb{R}^+ : a'(\langle x, y \rangle) = \delta_a(x, y)\right\}$;
$D(x, y) = \begin{cases} 0 \text{ if } d(x) = d(y) \\ 1 \text{ otherwise.} \end{cases}$

The searching problem for the optimal similarity relation relatively to the given object x and a table **S** can be transformed to the problem of decision rule extraction from the decision table $\mathbf{B}_x$. Again our goal is to find a rule of the form

$$\Psi\left(a_1'(u), a_2'(u), \ldots, a_k'(u)\right) \Rightarrow (D=0)$$

satisfied by as many objects $u \in U'$ as possible. One can see that using this method we do not consider all the pairs of objects but only the pairs containing a given object x. Hence the size of the table $\mathbf{B}_x$ is linear relatively to the size of table **S**. Therefore the local method needs less memory than the global method.

Categorical method The last method extracts the optimal similarity relation relatively to the given decision class C (C-*optimal similarity relation*, in short). We assume only the reflexivity of relations. For this purpose we construct the similarity table **B** that contains information about the similarity of objects from decision class C to objects from U. Given the decision table $\mathbf{S} = (U, A \cup \{d\})$ the similarity table $\mathbf{B}_C$ is defined as follows:

$\mathbf{B}_C = (U', A' \cup \{D\})$, where $U' = C \times U$;
$A' = \left\{a' : U' \to \mathbb{R}^+ : a'(\langle x, y \rangle) = \delta_a(x, y)\right\}$;
$D(x, y) = \begin{cases} 0 \text{ if } d(x) = d(y) \\ 1 \text{ otherwise.} \end{cases}$

The searching problem for C-optimal similarity relation for table **S** can be transformed to the problem of decision rule extraction from the decision table $\mathbf{B}_C$. Our goal is to find the rule of the form $\Psi\left(a_1'(u), a_2'(u), \ldots, a_k'(u)\right) \Rightarrow (D=0)$ satisfied by as many objects $u \in U'$ as possible.

8.2 Algorithms to generate similarity relations

In Section 7.1 we presented five parameterized similarity relation classes. These classes can be divided into two groups: classes characterized by one parameter (*max form*, *sum form*, *product form*) and classes characterized by k parameters (conjunction form, parameterized sum form). The main goal for every class is to find the optimal (semi-optimal) similarity relation. It means that we have to find, for a similarity relation class, the proper collection of parameters, i.e. defining the best relation. One can show that, for a group of one parameter relations, the problem of searching for the optimal relation can be solved in polynomial time. However, the problem of searching for the optimal similarity relation for a second group seems to be hard. The time complexity of the exhaustive searching problem for optimal similarity relation parametrized by k parameters for a set of n objects is of order $\theta(n^k)$ because we have to test all the possible values of parameter vector, where the number of possible values for one parameter is usually $\theta(n)$. Below we present some heuristics for extraction of the similarities in the second group. The geometrical interpretation of similarity relation classes shows to be very useful for all constructed algorithms.

Conjunctive form Let us consider first the similarity relation class consisting of relations in *conjunctive form*, i.e.

$$\langle x,y \rangle \in \tau(\varepsilon_1,...,\varepsilon_k) \Leftrightarrow \bigwedge_{a_i \in A}[\delta_{a_i}(x,y) \leq \varepsilon_i]$$

Any such similarity relation defines in the real affine space $\mathbb{R}^k$ the set

$$\overline{\tau(\varepsilon_1,...,\varepsilon_k)} = \{(r_1,...,r_k) \in C^k : 0 \leq r_i \leq \varepsilon_i \text{ for } i=1,...,k\} \quad (10)$$

Details of the algorithm are presented below

ALGORITHM: (CONJUNCTIVE FORM)

Input: The set of points from $V_{a'_1} \times V_{a'_2} \times ... \times V_{a'_k} \subseteq \mathbb{R}^k$ labeled "white" or "black".
Output: The set $\{\varepsilon_i \in V_{a'_i} : i = 1..k\}$ of parameters for a (semi-)optimal interval
begin
  sort values of lists $V_{a'_1}, ..., V_{a'_k}$ in an increasing order;
  $\varepsilon_1 = 0, \varepsilon_2 = \infty, ..., \varepsilon_k = \infty$;
  repeat
    $\varepsilon_1 = v$, where v is the first element from the list $V_{a'_1}$ such that there exists a black point p with $a'_1(p) = v$;
    for ($i \in \{2, .., k\}$)
      set $\varepsilon_i = v_i$, where $v_i = \min\{\varepsilon_i, a'_i(p)\}$ so, that the interval $I(\varepsilon_1, ..., \varepsilon_k)$ with the modified parameters ε_1 and ε_i still contains only "white" points from $\mathbb{R}^k$;
      m_i = number of "white" points in the new interval $I(\varepsilon_1, ..., \varepsilon_k)$;
    endfor
    choose the parameter ε_{i_0} and the value v_{i_0} corresponding to the interval $I(\varepsilon_1, ..., \varepsilon_k)$ with the maximum number m_{i_0} of "white" points;
    $\varepsilon_{i_0} = v_{i_0}$;
    $V_{a'_1} = V_{a'_1} \setminus \{v\}$;
  until ($V_{a'_1} = \emptyset$)
  among generated intervals, choose the interval $I(\varepsilon_1, ..., \varepsilon_k)$ with the maximum number of "white" points from $\mathbb{R}^k$. In this way we obtain semi-optimal parameters $\{\varepsilon_1, ..., \varepsilon_k\}$.
end (Algorithm)

For given $\varepsilon_1, ..., \varepsilon_k$, the set (10) is included in the interval $I(\varepsilon_1, ..., \varepsilon_k)$ from $\mathbb{R}^k$. Our goal is to find parameters $\varepsilon_1, ..., \varepsilon_k$ such that the interval $I(\varepsilon_1, ..., \varepsilon_k)$ consists of "white" points (x, y) with the same decision for x and y, at the same time, of as many as possible of them. Starting from the empty interval of the form $I(0, \infty, ..., \infty)$ we gradually augment one chosen parameter, for example ε_1, and at the same time decrease one of the remaining parameter so, that the interval $I(\varepsilon_1, ..., \varepsilon_k)$ still consists of points of one kind and it contains as many as possible points. The idea of the algorithm is illustrated in Figure 17. In this example we show the two-dimensional intervals with parameters $\varepsilon_1, \varepsilon_2$. In every step of the algorithm we augment the first parameter ε_1 and decrease the second parameter ε_2 to obtain the new interval $I(\varepsilon_1, \varepsilon_2)$.

Linear combination form Let us consider relation in a *parameterized linear combination form* defined by the formula

$$[\langle x, y \rangle \in \tau(w_1, ..., w_k, w)] \Leftrightarrow \sum_{i=1}^{k} w_i \cdot \delta_{a_i}(x, y) + w \leq 0$$

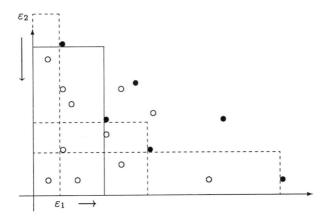

Fig. 17. Interpretation of an algorithm searching for an optimal interval $I(\epsilon_1, \epsilon_2)$.

Obviously, the parameterized linear combination belongs to the second similarity group. This relation defines the following set:

$$\tau(w_1, ..., w_k, w) = \left\{ (r_1, ..., r_k) \in C_k : \sum_{i=1}^{k} w_i \cdot r_i + w \leq 0 \right\}$$

For given parameters $w_1, ..., w_k, w$ the formula (8.2) describes the set of points with positive coordinates that are below the hyperplane $H = \sum_{i=1}^{k} w_i \cdot x_i + w = 0$. This hyperplane is determined by $(k+1)$ parameters. We are interested in hyperplanes having non-negative intersections with all axes of the space R^k. Hence $w_i > 0$ for any i and $w < 0$. Any hyperplane divides the space into two half-spaces. Our goal is to find a hyperplane H such that the half-space below the hyperplane H contains only "white" points and the number of these points is as large as possible. Searching for the optimal hyperplane H, we randomly choose a hyperplane $H = \sum_{i=1}^{k} w_i \cdot x_i + w$. After that we try to rotate this hyperplane by fixing k parameters, for example $w, w_1, ..., w_{j-1}, w_{j+1}, ..., w_k$ and modifying only one parameter w_j. We would like to find a value of w_j such that the modified hyperplane determines a new partition of set of objects. From the equation of hyperplane H we have $w_j = \frac{-\sum_{i \neq j} w_i \cdot x_i - w}{x_j}$. Any point $p_0 = [x_1^0, x_2^0 ... x_k^0] \in C^k$ is below H iff $H(p_0) < 0$, i.e. $w_j < \frac{-\sum_{i \neq j} w_i \cdot x_i^0 - w}{x_j^0}$ and it is above H iff $w_j > \frac{-\sum_{i \neq j} w_i \cdot x_i^0 - w}{x_j^0}$.

Let $S_j(p_0) = \frac{-\sum_{i \neq j} w_i \cdot x_i^0 - w}{x_j^0}$, where $p_0 = [x_1^0, x_2^0 ... x_k^0]$. We construct a set

$$S = \left\{ S_j(p_0) : p_0 = [a_1^{'}(u), ..., a_k^{'}(u)] \text{ for any } u \in U^{'} \right\} \quad (11)$$

Any value $w_j > 0$ chosen from S determines a new hyperplane defining a new partition of the set of points. For any defined hyperplane we move it until all the points below hyperplane are "white". Among constructed hyperplanes we choose the best. The idea of the algorithm is illustrated in Figure 18. In our example we show two-dimensional hyperplane (straight line) defined by parameters w, w_1, w_2. In every step of the algorithm, we first rotate the initially chosen line by modifying w_1 and assuming w, w_2 are fixed then translate it to the "good" position to obtain the new hyperplane $H(w, w_1, w_2)$.

The algorithm is presented below:

ALGORITHM: (LINEAR COMBINATION)

Input: The set of labeled points of the space $\mathbb{R}^k$.
Output: The set of parameters $\{w, w_1, ..., w_k\}$ of the semi-optimal hyperplane
  $H(w, w_1, ..., w_k)$.
begin
  $H(w, w_1, ..., w_k) =$ randomly chosen hyperplane;
  for (any $j = 1..k$)
    construct the set S defined in (11) and sort S in increasing order;
    for (any positive $v \in S$)
      $w_j = v$;
      move $H(w, w_1, ..., w_k)$ to a good position i.e. with all "white" points
        below it and calculate the number of these white points. The fitness
        of the hyperplane is equal to this number.
    endfor
  endfor
  among good hyperplanes we choose a hyperplane of the maximal fitness.
end

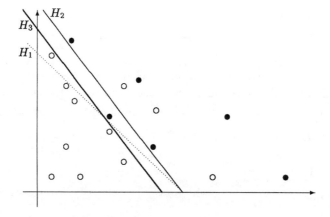

Fig. 18. H_1 - a randomly chosen hyperplane; H_2 - after rotation; H_3 - after translation.

Product form The last similarity relation class, which we will discuss is a class defined by relations in *product form*, i.e.

$$\langle x,y \rangle \in \tau(w) \Leftrightarrow \prod_{a_i \in A} \delta_{a_i}(x,y) \leq w$$

The considered relation belongs to the first similarity relation group, because it has associated one parameter w only.

The relation in the product form defines in real affine space $\mathbb{R}^k$ the set

$$\overline{\tau(\varepsilon)} = \{(r_1, ..., r_k) \in C^k : \Pi_{1 \leq i \leq k} r_i \leq w\} \tag{12}$$

The $\overline{\tau(w)}$ represents the part of R^k below *hyperboloid* $F : \prod_{1 \leq i \leq k} r_i = w$. Analogously to the case of hyperplane, the hyperboloid F is optimal if it satisfies the following two conditions:

1. The part of $\mathbb{R}^k$ below F contains only "white" points.
2. The number of these points is maximum.

The first condition holds if all the "black" points are situated above F, that means, for every "black" point $P = [p_1, p_2, ..., p_k]$ the following inequality holds: $\prod_{1 \leq i \leq k} p_i \geq w$.

The second condition is satisfied if F is placed as far as possible with respect to the origin O, that means w is the maximum possible value.

From these observations we can conclude that F is the optimal hyperboloid if its parameter w satisfies the equation:

$$w = min \prod_{1 \leq i \leq k} p_i : \text{ for all "black" points } P = [p_1, p_2, ..., p_k]$$

The details of the algorithm can be presented in the scheme below:

ALGORITHM: (PRODUCT FORM RELATION)

Input: The set S of n labeled points in the space $\mathbb{R}^k$.
Output: The value of parameter w of the optimal hyperboloid $F(w)$.
begin
  for every "black" point $P = [p_1, p_2, ..., p_k]$ in input set S compute the value
    $v_P = \prod_{1 \leq i \leq k} p_i$;
  $w = min \{v_P : P \text{ is a "black" point in } S\}$;
end

The first step takes times $O(n \cdot k)$ to count all products. The second step takes $O(n)$ time to find the minimum product value. Therefore the time complexity of the algorithm is $O(n \cdot k)$.

9 Applications

9.1 Data clustering

The goal of the clustering problem is to group objects that are classified as similar by a similarity relation. The question is how to combine objects into "homogenous" clusters, that is, clusters contain as many objects from one decision class and as few objects from another classes as possible. The quality of a cluster depends on the number of objects from a given decision class and the number of objects from other decision classes that belong to the cluster. A cluster is better if the former number is larger and the latter number is smaller. We will show how we can construct the clusters of good quality by heuristics based on similarity relations. We consider two cases of similarity relations: *consistent* and *inconsistent*.

First we focus on application of *consistent* tolerance relations to the clustering problem. Let $\mathbf{S} = (U, A \cup \{d\})$ be a decision table. For this table one can compute a *consistent* similarity relation τ (by *global method*) or similarity relation family $\{\tau_{x_0}\}$ for every object $x_0 \in U$ (by *local method*) or tolerance family $\{\tau_D\}$ for every decision class D (by *categorical method*). For the similarity τ we define transitive closure τ^* by

$$\tau^0 = \{(x,y) \subseteq U \times U : (x,y) \in \tau\}$$
$$\tau^k = \{(x,y) \subseteq U \times U : \exists_z (x,z) \in \tau^{k-1} \wedge (z,y) \in \tau\}$$
$$\tau^* = \bigcup_{k \geq 0} \tau^k$$

The x_0-transitive closure $\tau_{x_0}^*$ of the relative similarity τ_{x_0} is defined by

$$\tau_{x_0}^0 = \{(x_0,y) \subseteq \{x_0\} \times U : (x_0,y) \in \tau\}$$
$$\tau_{x_0}^k = \{(x_0,y) \subseteq \{x_0\} \times U : \exists_z (x_0,z) \in \tau_{x_0}^{k-1} \wedge (z,y) \in \tau_z\}$$
$$\tau_{x_0}^* = \bigcup_{k \geq 0} \tau_{x_0}^k$$

The D-transitive closure τ_D^* of τ_D is defined by

$$\tau_D^0 = \{(x,y) \subseteq D \times U : (x,y) \in \tau\}$$
$$\tau_D^k = \{(x,y) : \exists_z (x,z) \in \tau_D^{k-1} \wedge (z,y) \in \tau_{D_z}, \text{ for } x \in C, z \in D_z\}$$
$$\tau_D^* = \bigcup_{k \geq 0} \tau_D^k$$

where D_z is a decision class containing z.

The cluster C can be defined as the object set $[x]_{\tau^*}$ (or $[x]_{\tau_x^*}$ or $[x]_{\tau_D^*}$) for some object $x \in U$. The object x is called a *generator* of the cluster C.

The clusters of the universe U can be constructed in a straightforward way by the following algorithm.

ALGORITHM: (CLUSTERING I)

repeat
    Choose randomly an object $x \in U$;
    $C_i = [x]_{\tau^*}$ (or $[x]_{\tau_x^*}, [x]_{\tau_{d(x)}^*}$),
    $U = U \setminus C_i$;
    $i = i + 1$;
until $U = \emptyset$

Clusters determined by the algorithm are disjoint and they contain objects with the same decision. We can observe also that clusters generated by consistent similarity relations do not depend on the generator.

In a more general case we have to deal with inconsistent similarity relations. Recall that a similarity relation is inconsistent if it contains pairs of objects with different decisions. Consider cluster defined by $C = [x]_{\tau^k}$ (or $[x]_{\tau_x^k}$ or $[x]_{\tau_D^k}$, where $x \in D$) for some generator $x \in U$ and some positive number k. One can see that cluster C may contain objects with different decisions. The quality of a cluster in this case depends on the generator x and on the number k. A good generator corresponds to the object x defining $[x]_{\tau^k}$ ($[x]_{\tau_x^k}, [x]_{\tau_D^k}$) with good quality among all the objects from U. We extend its similarity class by successive iteration of similarity $\tau^k (k = 1, 2...)$. After every step of extension we examine the quality of the new class. If the quality of the cluster rapidly decreases we stop the process of extension with the current value of the parameter k, otherwise we continue the process. The improved method of searching for good clusters is presented in the Algorithm *Clustering II*.

ALGORITHM: (CLUSTERING II)

repeat
    1. Find $x \in U$ with the best quality of $[x]_\tau$ ($[x]_{\tau_x}, [x]_{\tau_{d(x)}}$);
    2. Find $C_i = [x]_{\tau^k}$ (or $[x]_{\tau_x^k}, [x]_{\tau_{d(x)}^k}$), where $x \in U$ is an object obtained from previous step;
        The number k in Step 2 is computed as follows: for $k = 1, 2...$ we construct the cluster $[x]_{\tau^k}$ as an extension of the cluster $[x]_{\tau^{k-1}}$ and investigate the quality of the obtained cluster.
    3. $U = U \setminus C_i$; $i = i + 1$;
until $U = \emptyset$

9.2 Data classification

In this section we discuss classification methods using tolerance relations extracted from data. The classifiers are defined by a family **C** of tolerance classes $[x]_{\tau_x}$), where x is an object in U and τ_x is a tolerance relation related to x. We use for classification the family **C** covering the training table.

General classification scheme can be described as follows:

Phase 1 Generate a family **C** of tolerance classes to cover the training set.

Phase 2 Classify a new object using some tolerance classes chosen from **C**. In conflicting cases, use the *fitness function* to resolve conflicts.

The main problems are: (1) How to construct the classifier set? and (2) How to classify a new object using extracted classifiers?

To construct the tolerance relation, one can use the heuristics presented in Section 8. Having tolerance relation τ (or family of local tolerance relations τ_x for all $x \in U$), the family of tolerance classes is defined. In the case of a classification task, one can use a small set of tolerance class covering the whole data universe.

The next problem is related to a choice of the relevant tolerance classes among extracted classes in the classification process. There are two strategies: *classification by the matched tolerance classes* and *classification be the closest neighbors*. In case of the first strategy, object x is classified by objects of tolerance classes that x belongs to. In the other case, x is classified by closest classes. The idea is similar to k-NN approach. The main difference is the method for the distance measuring. The distance is defined by tolerance relations extracted from data. For example, let us consider two distance functions constructed according to tolerance relations from two basic families: *conjunction of local relations* and *square combination*.

Let x_0 be an object of the universe U and let τ_{x_0} be a tolerance relation related to x_0. If τ_{x_0} is in conjunctive form then $(x, y) \in \tau_{x_0}$ iff

$$(|a_1(x) - a_1(y)| \leq \epsilon_1) \wedge (|a_2(x) - a_2(y)| \leq \epsilon_2) \wedge ... \wedge (|a_k(x) - a_k(y)| \leq \epsilon_k).$$

The distance $HD(x, x_0)$ is defined by

$$d(x, x_0) = 1 - \frac{|\{a_i : |a_i(x) - a_i(x_0)| \leq \epsilon_i\}|}{k}$$

If τ_{x_0} is in square combination form then $(x, y) \in \tau_{x_0}$ iff

$$(w_1 \cdot [a_1(x) - a_1(y)]^2 + w_2 \cdot [a_2(x) - a_2(y)]^2 + ... + w_k \cdot [a_k(x) - a_k(y)]^2 \leq w_0.$$

The distance $ED(x, [x_0]_{\tau_{x_0}})$ is defined by

$$ED(x, x_0) = d(x, x_0) - w_0$$

where

$$d(x, x_0) = (w_1 \cdot [a_1(x) - a_1(x_0)]^2 + w_2 \cdot [a_2(x) - a_2(x_0)]^2 + ... + w_k \cdot [a_k(x) - a_k(x_0)]^2.$$

10 Results of experiments

The main objective of this section is an experimental evaluation of methods presented in the dissertation. We consider three applications in our experiments, namely *classification*, *data description* and *data decomposition*. The quality of algorithms for these applications is estimated using different criteria. The classifying algorithms are characterized by a *prediction accuracy*. The algorithms for data description are characterized by the *size* and the *preciseness* of a given pattern set used for concept approximation. Good decomposition algorithms should show the ability of partitioning large domain into regular sub-domains.

We consider three methods for the classifier generation related to *decision trees*, *template induction* and *tolerance class inferring*. The decision trees are generated by *DTG* algorithm (see section 5, Section 5.2). The quality of the *DTG* tree is compared with the well known decision tree algorithms from the literature like, ID3 [70], C4.5 [71], and MD [52].

The second classification method is based on template set. Templates are used as generalized decision rules in classification tasks. These generalized rules show some advantages in comparison with widely used simple rules. The recognition ratio of testing data radically increases without decreasing the classification accuracy.

We also present a new approach for unseen case classification. The idea has been adopted from case based reasoning area [33], where a new object is classified using nearest neighbors with a distance function often predefined independently from data. The main difference of our approach is that we use the distance functions extracted from data. The discovered templates or tolerance relations describe not only the similarity relations in data, they also define the distance functions relevant for data. We report the efficiency of new approaches by comparing their quality with well known methods like k-NN [5] in case based reasoning.

The next aim of our experiments was to estimate the quality of the proposed methods for concept description. We consider two sub-problems: *description of decision classes* and *description of data universe*. In description problems, the complex concepts are synthesized from primitive ones. In our experiments, the primitive concepts are defined by templates or neighborhoods defined by some tolerance relations extracted from data. The target concept is defined by disjunction of some primitives concepts. The description is good if it is simple and it approximates concepts with high precision. The first factor is often estimated by the number of primitive concepts occurring in the description and the second factor is evaluated by the closeness of the constructed concept and the target concept. The experimental results show that our approaches are characterized by results of high quality for description problems with respect to both criteria.

The last aim of our experiments was to check the quality of proposed methods for data decomposition. For a given number s, the problem is to find a partition of data table into regular sub-tables with support closest to s. Regularity of data table in our experiments is estimated by the number of decision classes contained in a sub-table and/or the cardinality of a predominate class contained in that sub-table. A sub-table is regular if it contains a few classes and/or the predominate class is large according to the sub-table size.

10.1 Data sets

We describe the data sets used in our experiments and some testing models used for decision algorithms.

The performance of different algorithms presented in this dissertation has been evaluated on both artificial and real-world domains. Artificial domains are useful because they allow us to vary parameters, understand the specific problems that algorithms exhibit, and test conjectures. Real-world domains are useful because they come from real-world problems that we do not always understand and are therefore actual problems on which we would like to improve performance. All real-world data sets used are from the UC Irvine repository [45], which contains over 100 data sets mostly contributed by researchers in the field of machine learning.

We chose for experiments the following real-world domains:

Australian credit screening (australian) There are 690 instances from an Australian credit company. The task is to determine whether to give a credit card to an applicant. The data set was first used in Quinlan (1987). There are six continuous features and nine nominal ones.

Breast cancer Wisconsin (breast) There are 699 instances collected from Dr. Wolberg's clinical cases at the University of Wisconsin (Wolberg & Mangasarian 1990, Zhang 1992a). These were collected over a period of two and a half years, and the problem is to determine whether the tumors were benign or malignant based on data for each cancer patient. There are ten features identifying the disease symptoms.

Heart disease (heart) There are 270 instances from Dr. Detrano. The task is to distinguish the presence or absence of heart disease in patients. There are seven nominal features and six continuous. The features include: age, sex, chest pain type, cholesterol, fasting blood sugar, resting ECG, max heart rate, etc.

Iris plants (iris) There are 150 instances. The problem is to distinguish three iris families: setosa, virgince and versicolor. Data table has four continuous features.

Lymphography (lympho) There are 148 instances. The task is to diagnose stadiums of lymphography disease. There are four classes and 18 features, among which 15 are nominal and three continuous.

Indian diabetes (diabetes) There are 768 instances from the National Institute of Diabetes and Kidney Diseases. The task is to determine whether the patient shows signs of diabetes according to World Health Organization criteria. There are eight features, all are continuous.

Satellite image dataset (satimage) The original data was generated from data purchased from NASA by the Australian Centre for Remote Sensing. The database consists of the multi-spectral values of pixels in 3×3 neighborhoods in a satellite image. The classification is associated with the central pixel in each neighborhood. The aim is to predict this classification, given the multi-spectral values. There are 36 (4 spectral bands × 9 pixels in neighborhood) features, all are numeric in the range 0 to 255. The data table has 6 decision classes. The training table contains 2000 objects.

Soybean (soybean) There are 683 instances in this data set. The task is to diagnose soybean diseases. There are 19 classes and 35 discrete features describing leaf properties and various abnormalities.

Tic-tac-toe game (tic-tac-toe) There are 958 instances corresponding to the legal tic-tac-toe endgame boards. There are 9 symbolic attributes, each corresponding to one tic-tac-toe square. The values of attributes are x, o, b: x - player x has taken, o - player o has taken, b - blank. There are two decision classes: 1 if x wins and 0, otherwise. The task is to discover the winning situation for player x.

We choose for experiments the following artificial domains: Monk's problems. The Monk's problems are three artificial problems that allow comparison of algorithms. In the given domain, robots have six different nominal features as follows:

- **HeadShape** $\in \{round, square, octagon\}$,
- **BodyShape** $\in \{round, square, octagon\}$,
- **IsSmiling** $\in \{yes, no\}$,
- **Holding** $\in \{sword, balloon, flag\}$,
- **JacketColor** $\in \{red, yellow, green, blue\}$,
- **HasTie** $\in \{yes, no\}$.

The three problems are as follows:

Monk1 The standard training set contains instances of the following target concept:

$$(\textbf{HeadShape} = \textbf{BodyShape}) \vee (\textbf{JacketColor} = \text{'red'})$$

The training table consists of 124 objects, among which 62 objects belong to the concept and 124 do not. The testing table contains 432 objects, among them half belongs to the concept and half do not.

Monk2 The standard training set contains instances from the following target concept: **Exactly two of the features have their first value.** For example the object ⟨ '<u>round</u>', 'square', '<u>yes</u>', 'flag', 'yellow', 'no'⟩ belongs to the target concept.

The training set contains 169 objects, among them 105 belong to the concept and 64 do not. However the testing table contains 432 objects, among which, 190 belong to the concept and 242 do not.

Monk3 The standard training set contains instances of the following target concept:

$$(\textbf{JacketColor} =' green' \wedge \textbf{Holding} =' sword') \vee$$
$$(\textbf{JacketColor} =' blue' \wedge \textbf{BodyShape} =' octagon')$$

There are 122 objects in the training table, among them 62 objects belong to the concept and 60 not belong to it. The testing table contains 432 objects, among them 204 belong to the concept and 228 not belong to it.

The task is to find the rules which distinguish the objects belonging to the concept from such ones not belonging to the concept.

Table 2 provides a summary of characteristics of the data sets. The table contains information about the object number and the attribute number in the data table including the number of symbolic and numeric attributes. The last column records distribution of predominate decision classes in the decision tables.

No.	Data sets	Train size	Features All	Features Sym	Features Num	No. classes	Predominate class
1	Australian	690	6	0	6	2	52.31%
2	Breast	286	10	0	10	2	70.27%
3	Diabetes	768	8	0	8	2	65.10%
4	Heart	270	13	7	6	2	55.55%
5	Iris	150	4	0	4	3	33.33%
6	Lympho	148	18	15	3	4	54.72%
7	SatImage	4435	36	0	36	6	24.17%
8	Soybean(S)	47	35	0	35	4	36.17%
9	Tic-tac-toe	958	9	9	0	2	43.63%
10	Monk1	124	6	6	0	2	50.00%
11	Monk2	169	6	6	0	2	62.13%
12	Monk3	122	6	6	0	2	50.81%

Table 2. Characteristics of data sets used in experiments

10.2 Testing methods

We have adopted two methods used in Machine Learning for decision algorithm testing.

The first method is called "*train-and-test*". The original data table is split into a *training table* and a *testing table*. The knowledge is extracted from the training table and then it is checked in the testing table. For example, if N is the number of objects of the testing table and N_1 the number of objects from this data table properly recognized by decision algorithms then the classification accuracy is defined by $q = \frac{N_1}{N} 100\%$. Quality of decision algorithms can be also defined by a fraction of quantity of properly recognized objects to quantity of recognized ones. In Monk's problems and Satellite image analysis, the "train-and-test" method is applied for quality evaluation.

The second method is known in Machine Learning as the *m-fold cross validation* $(CV - m)$, where m is an arbitrary positive integer. In $CV - m$ approach we divide a given data table into m equal parts. Every part is tested via the decision algorithm achieved from $(m-1)$ remaining parts. The classification quality of the learning algorithm is equal to the average value of all classification qualities i.e. $q = \frac{1}{n}\sum_{i=1}^{m} n_i$ where n_i is a number of properly recognized objects in the ith test. In the rest of data sets (except Monks and SatImage), we used m-fold-cross validation for quality evaluation.

10.3 Classification quality

In this section we present the quality of classifiers generated by decision algorithms presented in the paper. They are related to three techniques: *decision trees, template set* and *tolerance classes*. The efficiency of our algorithms is reported by comparing the classification accuracy of presented algorithms with the corresponding algorithms in the literature. The best results of classification over 12 data sets from some papers (see [6], [5], [43]) has been selected. For better comparison we present these results in Table 3.

Data sets	Train size	Test size	Accuracy (%)	Methods
Australian	690	CV-10	86.90	Cal5
Breast	699	CV-10	78.00	Assistant-86
Diabetes	768	CV-10	77.70	Logdisc
Heart	270	CV-10	83.60	NaiveBayes
Iris	150	CV-10	96.67	MD-H, ID3
Lympho	148	CV-10	85.00	RSES-lib
Soybean(S)	47	CV-10	98.00	MD
Tic-tac-toe	958	CV-10	98.00	MD
SatImage	4435	2000	90.60	k-NN
Monk1	124	432	100.00	RSES-lib
Monk2	169	432	100.00	Backpropagation 1
Monk3	122	432	100.00	Assistant-86

Table 3. The best classification results selected from the literature.

The presented algorithms can be classified to the following groups:

- Statistical methods: Cal5, Logdisc, NaiveBayes [42], k-NN [5].
- Neural network based methods: Backpropagation 1 [42].
- Decision tree based methods: Assistant-86 [42], ID3 [71], MD [52].
- Decision rule based methods: RSES-lib [7] (rough set methods).

Classification by the decision tree We present the result related to the algorithm DTG (see section 5, Section 5.2) for decision tree generation. A new aspect of our method comparing to the other approaches is the way the symbolic domain is split. The DTG divides the symbolic domain in two sub-domains such that the partition in the best way discerns objects from different decision classes. The other approaches, like for example C4.5, divide the symbolic domain into as many sub-domains as is the cardinality of this domain. If a considered domain is large, the tree generated by presented methods becomes too large and decision rules extracted from such decision trees are not general enough. The advantage of DTG tree is its simple structure and the high quality of classification. One can compare the quality of method DTG with some known methods reported in the literature (see Table 4).

Data sets	Train size	Test size	ID3	C4.5	MD	DTG
Australian	690	CV-5	78.26	**85.36**	83.69	84.49
Breast	699	CV-5	62.07	**71.00**	69.95	69.95
Diabetes	768	CV-5	66.23	70.84	71.09	**76.17**
Heart	270	CV-5	77.78	77.04	77.04	**81.11**
Iris	150	CV-5	**96.67**	94.67	95.33	**96.67**
Lympho	148	CV-5	73.33	77.01	71.93	**82.02**
Soybean(S)	47	CV-5	**100**	95.56	100.00	98.00
Tic-tac-toe	958	CV-5	84.38	84.02	**97.70**	**97.70**
Monk1	124	432	81.25	75.70	**100.00**	93.05
Monk2	169	432	69.91	65.00	**97.70**	83.33
Monk3	122	432	90.28	97.20	93.51	94.00
Average			80.01	81.21	85.54	**86.95**

Table 4. The classification quality of decision tree methods.

Classification by set of templates Another classification method is by using template set. Classifiers are defined by decision templates covering the training table.

We compare the quality of two kinds of templates: *simple* and *generalized*. Simple templates are constructed by L-SF algorithm (section 3, Section 3.2). Generalized templates are generated by L-SDD algorithm (section 3, Section 3.2). Taking into account the low time complexity, we have chosen the Dynamic Weight strategy (presented in section 3, Section 3.3) for rule set generation and the Max Support strategy (presented in section 6, Section 6.3) for conflict resolving.

The quality of any classification algorithm is characterized by the *accuracy* and the *recognition ratio*. The accuracy is measured by a fraction of the number of good classified objects to the number of objects recognized by a given decision algorithm. The recognition ratio is defined by the ratio of the number of recognized objects to the number of all objects in the testing table.

One of the most important aspect in the classification process is related to choosing proper templates for new object classification. We investigate two strategies: classification by *matched templates* and classification by *closest templates*.

Classification by matched templates A new object x is classified according to a set of templates matching x. The classification strategy is presented by the following scheme:

ALGORITHM: (CLASSIFICATION BY MATCHED TEMPLATES)

Phase 1 Generate a decision templates set.
Phase 2 Classify a new object x using the template set satisfied by x. The fitness function is used to resolve conflicts.

The main advantage of this strategy is its high accuracy ratio in the recognized region. In Table 5 we present the classification quality of simple and generalized templates. The results are obtained by two algorithms L-SF (simple templates) and L-SDD (generalized templates) using a 10-cross validation model.

Data sets	Simple templates		Generalized templates	
	Acc.(%)	Rec.(%)	Acc.(%)	Rec.(%)
Australian	**92.94**	62.31	88.60	**90.50**
Breast	**76.75**	60.00	**75.71**	**96.15**
Diabetes	**78.35**	70.84	75.27	**82.50**
Heart	**92.00**	70.67	90.11	**83.30**
Iris	100.00	96.67	96.67	**100.00**
Lympho	94.00	58.36	91.70	**82.80**
Soybean(S)	**100.00**	72.72	**100.00**	**91.11**
Tic-tac-toe	**100.00**	78.48	99.04	**97.59**
SatImage	**89.09**	52.00	88.34	**94.40**
Monk1	87.00	62.09	**97.43**	**94.35**
Monk2	**81.00**	65.00	76.78	**90.00**
Monk3	93.00	64.80	**94.00**	**97.00**

Table 5. Classification quality of satisfied template based strategy(**Acc.**: accuracy of classification; **Rec.**: Recognition ratio;).

Comparing with the results presented in Table 3, one can see that the classification received by satisfied templates gives high accuracy in recognized part. In Australian credit data the accuracy is improved from 86.90% to 92.94%. The interesting results are obtained also for Heart disease data, where the accuracy is increased from 83.60% to 92.00%. Usually simple templates give a higher classification accuracy. The advantages of generalized templates is shown by the fact that its recognition ratio is substantially improved and at the same time the classification accuracy is comparable to the simple ones. The interesting results appeared in two data sets: Monk1 and Monk2. In these data, the generalized templates show much better recognition ratio (94.43 % instead 62.09% for Monk1 and 97.00% instead 64.80% for Monk2) and simultaneously, it shows a better accuracy (97.35 % instead of 87.00% for Monk1 and 94.00% instead of 93.00% for Monk2).

Now we report the classification quality in individual decision classes. For many data sets the average accuracy might be low, but it shows a very high accuracy in some decision classes. In Table 6 we present the results related to generalized templates whose quality are presented in Table 5.

Data sets	Accuracy(%)	Accuracy in decision classes(%)
Australian	90.11	**95.1** 86.9
Breast	75.11	**78.9** 64.5
Diabetes	75.27	65.9 **77.5**
Heart	90.11	**100** 83.3
Iris	96.70	**100** 90 **100**
Lympho	91.70	**92.3** 90 **100**
Soybean(S)	100.00	**100 100**
Tic-Tac-Toe	100.00	**100 100**
Sat. Image	88.34	**97.4 96.34** 80.5 60.13 **91.46%** 83.97

Table 6. Accuracy in decision classes.

In Figure 19 one can compare the accuracy ratio of simple and generalized classifiers. These results are compared with the best ones presented in Table 3.

In Table 7 one can compare the number of simple templates and generalized templates used in the classification process. One can substantially reduce the size of the classifier set using generalized templates. Especially, to cover the Australian credit data, one can use only 2 generalized templates instead of 214 simple templates. The last column presents the reduction ratio, which is measured as a fraction of the number of simple classifiers to the number of generalized classifiers. In general, one can reduce the classifier set from 2 to 10 times.

Classification by closest templates Classification by matching rules has as its main advantage the high accuracy in the recognized space of objects. It is a useful tool in some application domains e.g. medicine or market analysis, where preciseness is most important. In many other applications we need to obtain as reliable answers for all the questions (cases) as we can. A natural idea is to modify the *matching* strategy to classification by *closeness*. This idea is adopted from case based reasoning (CBR) [33]. A new object is classified using closest objects in the training data table. The problem is how to measure the distance between objects. The distance function in CBR application is often predefined independently from data. There are two fundamental differences with our approach: the classifier set is reduced to a template set covering the training

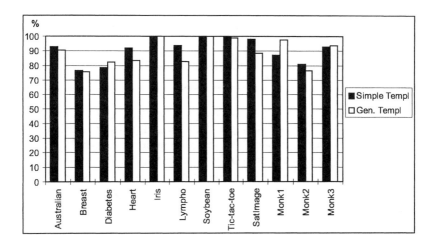

Fig. 19. Classification accuracy of the strategy based on matched templates.

Data sets	No simple classifiers	No gen. classifiers	Reduct ratio
Australian	214	2	107
Breast	67	43	1.55
Diabetes	315	189	1.66
Heart	84	8	10.5
Iris	7	4	1.75
Lympho	55	23	2.39
Soybean(S)	7	5	1.4
Tic-tac-toe	122	43	2.83
SatImage	2258	432	5.22
Monk1	64	19	3.36
Monk2	102	32	3.18
Monk3	63	22	2.86

Table 7. The numbers of simple and generalized templates used in classification process.

table and the distance function is defined by the "distance" to the extracted templates, i.e. to the object sets represented by templates. The classification process is based on the following scheme:

> **ALGORITHM:** (CLASSIFICATION BY CLOSEST TEMPLATES)
>
> **Phase 1** Generate a decision templates set.
> **Phase 2** Classify a new object x using template set closest to x. Use the fitness function to resolve conflicts.

We again adopt L-SF, L-SDD algorithms for template generation. The second phase is somewhat modified. We use two functions: pseudo-Hamming and pseudo-Euclidean (presented in section 6, Section 6.3) for distance measuring.

In Table 8 we present the classification quality based on "closeness" strategy.

The classification by distance function always gives 100% recognition ratio. Comparing the quality of simple and generalized templates one can observe that the generalized templates give better accuracy than simple ones.

Compared to the k-NN method, our approaches show the following advantages: the classification accuracy is improved and the classifier set is reduced. The second feature allows us to reduce the time needed in the classification phase.

Data sets	k-NN	Simple templates	Generalized templates
Australian	**86.00**	82.17	85.50
Breast	74.30	70.96	**74.46**
Diabetes	68.70	61.56	**74.20**
Heart	82.00	75.33	**86.10**
Iris	95.10	94.67	**96.67**
Lympho	77.00	80.38	**81.70**
Soybean(S)	-	**100.00**	91.11
Tic-tac-toe	98.10	**99.47**	98.12
SatImage	**90.60**	81.70	86.02
Monk1	-	87.90	**92.93**
Monk2	53.5	**77.54**	78.12
Monk3	-	83.86	**92.12**
Average	80.58	80.42	**84.54**

Table 8. Classification accuracy by simple and generalized templates (100% recognition). Closeness is defined by the pseudo–Hamming distance.

Figure 20 depicts the reduction ratio of classifier set size in simple and generalized template methods with respect to the k-NN method. For template set **T** and the set of training objects U, the reduction ratio is measured by $\frac{card(U)}{card(\mathbf{T})}$.

Now we analyze the quality of classification algorithm using pseudo-Euclidean measure.

In Table 9 we present the quality of the same classifier set, but with pseudo-Euclidean function used for distance measuring. One can observe that the latter measure shows to be more suitable than the pseudo-Hamming distance for some

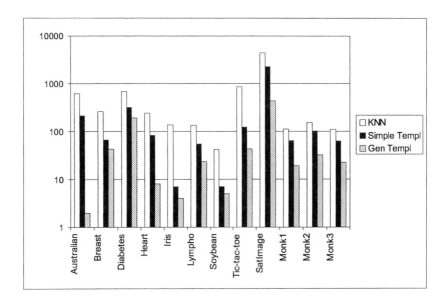

Fig. 20. Classifier set reduction by simple and generalized template methods in comparison with respect to the k-NN method.

data sets: Australian credit data, Breast or Heart disease.

The two distance measures are compared in Figure 21. For classification, we have used the same classifier set (generalized templates) but two different distance functions: pseudo-Hammimng and pseudo-Euclidean measure.

Classification based on tolerance relation The next classification method is based on tolerance relations extracted from data.

General classification scheme can be described as follows:

ALGORITHM: (CLASSIFICATION BY TOLERANCE CLASSES)

Phase 1 Generate a family C of tolerance classes to cover the training set.
Phase 2 Classify a new object using some tolerance classes chosen from C. Use the *fitness function* to resolve conflicts.

We compare here the quality of two kinds of tolerance relation: tolerances in *conjunctive form* and tolerances in *combination form*. There is a number of tolerance relations that can be defined by a conjunction of local relations. We concentrate here on the relation defined by $(x,y) \in R$ iff $(|a_1(x) - a_1(y)| \leq \epsilon_1) \wedge (|a_2(x) - a_2(y)| \leq \epsilon_2) \wedge ... \wedge (|a_k(x) - a_k(y)| \leq \epsilon_k)$. If a is a symbolic attribute, the value $|a(x) - a(y)|$ is defined by

Data sets	k-NN	Simple templates	Generalized templates
Australian	86.00	84.20	**87.53**
Breast	74.30	72.39	**75.86**
Diabetes	68.70	63.12	**70.06**
Heart	82.00	86.00	**87.01**
Iris	95.10	90.00	**96.67**
Lympho	77.00	**83.27**	81.70
Soybean(S)	-	**97.77**	93.33
Tic-tac-toe	98.10	**99.37**	97.80
SatImage	**90.60**	82.70	87.22
Monk1	-	80.65	**97.85**
Monk2	53.5%	71.06	71.06
Monk3	-	81.48	**92.12**
Average	80.58	81.34	**83.87**

Table 9. Classification accuracy by simple and generalized templates (100% recognition). Closeness is defined by the pseudo-Euclidean distance.

$$|a(x) - a(y)| = \begin{cases} 0 \text{ if a(x)=a(y)} \\ 1 \text{ otherwise.} \end{cases}$$

Analogously, there are different relations that can be defined in the combination form. We focus on the one defined by $(x,y) \in R$ iff $(w_1 \cdot [a_1(x) - a_1(y)]^2 + w_2 \cdot [a_2(x) - a_2(y)]^2 + ... + w_k \cdot [a_k(x) - a_k(y)]^2 \leq w_0$.

To generate the relations in the *conjunctive form*, we use the Conjunctive Form Algorithm presented in section 8, Section 8.2. The Combination Form Algorithm (presented in section 8, Section 8.2) is adopted to generate the relation being in *combination form*. Because of the low time and space complexity, we choose the local method (see section 8, Section 8.1) for tolerance extraction and the Max Support strategy (presented in section 6, Section 6.3) is adopted for conflict resolving.

In the classification process we need also to choose among the extracted classifiers the relevant ones for classifying a given new object. We consider two strategies for solving this problem: classification by tolerance classes containing a new object and classification by classes closest to a new object. Below we present the results related to these strategies.

Classification by matched tolerance classes One important aspect occurring in any algorithm constructing tolerance classes is a *homogeneous ratio* ρ of tolerance classes used for covering the training set. For a tolerance class $[x]_{\tau_x}$, the homogeneous ratio is defined by $\rho([x]_{\tau_x}) = \frac{card\{y \in [x]_{\tau_x} : d(y) = d(x)\}}{card\{[x]_{\tau_x}\}}$. The classifier is called *precise* if its homogeneous ratio $\rho = 100\%$, and it is called *approximate* if $\rho < 100\%$. The classifiers with high homogeneous ratio show usually a high

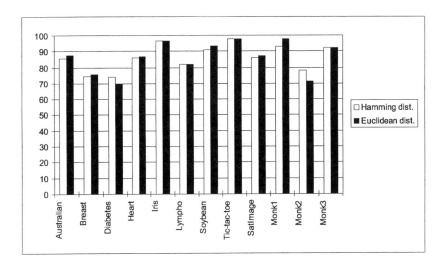

Fig. 21. The classification quality using the Hamming distance and the Euclidean distance functions.

classification accuracy but they show a low recognition ratio. In Table 10 we present the quality of classifiers defined by relation in conjunctive form. Now one can compare the recognition ratio of precise and approximate classifiers.

The quality of classifiers defined by conjunctive form relation is presented in Table 11. The relations in combination form show advantage for 5 data sets: Australian credits, Lymphography, Heart and Monk1 and Monk3. For the rest of data tables the conjunctive form appears to be better.

Classification by closest tolerance classes We adopt two functions for distance measuring: HD according to Hamming distance and ED according to Euclidean distance presented in section 9, Section 9.2. The former function is applied to relations in the conjunctive form and the latter is applied to relations in the combination form. A new object is classified using objects in the closest tolerance class.

In Table 12 we present the quality of conjunctive and combination form relations obtained using "closeness" strategy. In these experiments the approximate classifiers are used.

10.4 Data description quality

In this section we discuss the efficiency of template approaches for regularity extraction and data description. We distinguish two description tasks: *data universe description* and *decision class description*. In both cases the target concept is defined by disjunction of primitive concepts, which are defined by templates. We use two kinds of templates: simple and generalized templates for primitive

Data sets	Precise classifiers		Approx. classifiers		
	Acc.(%)	Rec.(%)	ρ	Acc.(%)	Rec.(%)
Australian	87.47	65.07	0.80	85.25	91.15
Breast	77.30	74.86	0.70	75.20	91.24
Diabetes	75.70	70.10	0.65	74.47	82.81
Heart	83.70	80.33	0.75	82.10	90.5
Iris	96.67	95.50	0.90	96.67	100.00
Lympho	77.50	60.01	0.70	76.50	85.50
Soybean(S)	98.00	70.00	0.85	97.75	90.25
Tic-tac-toe	98.10	92.20	0.90	96.65	91.20
SatImage	87.65	70.30	0.75	85.25	87.00
Monk1	91.20	70.64	0.85	95.93	90.00
Monk2	87.00	75.00	0.80	85.00	82.33
Monk3	95.00	65.00	0.85	91.00	85.12

Table 10. The quality of matched tolerance class strategy. Classifiers are defined by conjunctive form relation (**Acc.**: accuracy of classification; **Rec.**: Recognition percent).

Data sets	Precise classifiers		Approximate classifiers		
	Acc.(%)	Rec.(%)	ρ	Acc.(%)	Rec.(%)
Australian	**88.47**	70.07	0.8	87.25	**91.15**
Breast	**76.20**	80.86	0.7	75.20	**90.20**
Diabetes	74.70	73.10	65	**75.20**	**85.50**
Heart	**84.70**	82.33	0.8	82.15	**87.5**
Iris	95.00	91.50	90	**95.70**	**100.00**
Lympho	**83.50**	75.01	0.75	82.50	**85.50**
Soybean(S)	95.00	70.00	0.85	**96.75**	**90.25**
Tic-tac-toe	95.10	85.20	0.90	**96.65**	**91.20**
SatImage	**85.65**	72.30	0.85	83.25	**85.00**
Monk1	**95.20**	74.64	0.8	94.93	**92.00**
Monk2	**85.00**	79.00	0.85	89.10	**87.33**
Monk3	94.00	72.00	0.85	**95.00**	80.12

Table 11. The quality of "matching" strategy. Classifiers are defined by conjunctive form relation.(**Acc.**: accuracy of classification; **Rec.**: Recognition percent)

concept expression. To analyze the quality of simple templates, the attribute domains are discretized using the algorithm presented in [61].

Data universe description

Template quality First we analyze the quality of single templates used for primitive concept description. A template is of good quality if the concept it describes is regular. The regularity of a concept is often estimated by its *support*

Data sets	k-NN	Conjunctive form	Combination form
Australian	**86.00**	80.72	82.46
Breast	**74.30**	71.6	74.00
Diabetes	68.70	68.03	**74.30**
Heart	82.00	79.62	**83.40**
Iris	95.10	**96.67**	95.70
Lympho	77.00	81.01	**83.70**
Soybean(S)	**98.00**	97.53	93.11
Tic-tac-toe	98.10	**98.43**	97.00
SatImage	**90.10**	85.65	83.90

Table 12. The quality of the "closest" strategy. Classifiers are defined by conjunction and combination form relation.

and *length*. Another criterion for template evaluation is the support and the homogeneous degree. For a template T, let $[T]_U$ be a set of objects satisfied by T in table U, and let C_i be a decision predominate class among those one contained in $[T]_U$. The homogeneous degree $\rho(T)$ is defined by $\frac{card([T]_{C_i})}{card([T]_U)}$.

In experiments we adopt two algorithms for template generation: Similar Frequency (SF) (for simple template generation) and Similar Distribution of Decision Classes (SDD) (for generalized template generation), see Section 3.2, section 3. The quality of templates in construction process is estimated by the following function

$$quality(T) = support(T) \cdot length(T)$$

In Table 13 we present the support and the length of the best simple and generalized templates.

The best templates cover 25-30 % objects of data set on the average. In the case of Diabetes data, the best templates cover 40 % of data set. As one can expect, the generalized templates cover more objects than simple ones. Simultaneously, experiments have showed that the generalized templates preserve the homogeneous factor estimated for simple templates. In Table 14, we present the distribution of decision classes contained in these templates. Comparing it with Table 13, one can observe the following fact: the best generalized template covers nearly two times more objects, simultaneously its homogeneous ratio is comparable to that of the simple template.

Quality of data description Now we discuss the quality of description of data. A description is good if it is simple, that means the description is defined by a small number of templates. In experiments, the quality of description is estimated by the number of templates used in description and the average homogeneous degrees of these templates. In Table 15 we present these two factors of descriptions for simple and generalized templates.

These experiments demonstrate that the descriptions defined by template approach are simple. The number of templates used for data description often

Data sets	Data size No.Obj × No.Attr	Simple templates Supp. × Length	Gen. templates Supp. × Length
Australian	690 × 6	134 × 3	260 × 2
Breast	286 × 10	146 × 3	162 × 5
Diabetes	768 × 8	302 × 2	311 × 4
Heart	270 × 13	64 × 3	135 × 2
Iris	150 × 4	57 × 2	57 × 2
Lympho	148 × 18	91 × 4	70 × 5
SatImage	4435 × 36	1661 × 5	1661 × 5
Soybean(S)	47 × 35	13 × 5	13 × 5
Tic-tac-toe	958 × 9	155 × 2	256 × 4
Monk1	124 × 6	34 × 2	87 × 2
Monk2	169 × 6	44 × 2	86 × 3
Monk3	122 × 6	35 × 2	44 × 3

Table 13. The support and the length of the best simple and generalized templates.

Data sets	No. classes	Dec. class distribution	
		Best simple template	Best gen. template
Australian	2	[**93.3** 6.7]	[**92.7** 7.3]
Breast	2	[**82.9** 17.1]	[**82.1** 17.9]
Diabetes	2	[**67.2** 32.8]	[**82.6** 17.4]
Heart	2	[**87.5** 12.5]	[**77.1** 22.9]
Iris	3	[0 **100.0** 0]	[0 **100.0** 0]
Lympho	4	[2.2 **53.9** 43.9 0]	[0 42.85 **57.15** 0]
SatImage	6	[**50.0** 25.9 0 0.8 15.9 8.0]	[**50.0** 25.9 0 0.8 15.9 8.0]
Soybean(S)	4	[0 0 0 **100.0**]	[0 0 0 **100.0**]
Tic-tac-toe	2	[25.8 **74.2**]	[17.9 **82.1**]
Monk1	2	[**61.8** 38.2]	[**58.6** 41.4]
Monk2	2	[**56.8** 43.2]	[**61.6** 38.4]
Monk3	2	[**57.1** 42.9]	[2.3 **97.7**]

Table 14. The distribution of decision classes contained in the best simple and generalized templates.

is of order 3-5% of the data size. Moreover, the descriptions demonstrate a high regularity which is defined by homogeneous degree. The next interesting observation is related to three data sets: Australian, Soybean and Monk3. The first data table contains two decision classes. Using only two generalized templates, one can express these decision classes with average accuracy 85.70%. The second data set Soybean contains four decision classes. Using only four templates, one can describe these classes with 100% accuracy. The last data set is Monk3, which consists of three classes. Experiments show that three templates are sufficient for describing all classes and to approximate these classes with 93.79% accuracy.

Data sets	No. objects	Simple template		Gen. template	
		D. size	A.H. (%)	D. size	A.H. (%)
Australian	690	20	80.00	2	**85.70**
Breast	286	13	77.30	10	76.44
Diabetes	768	19	70.07	10	72.13
Heart	270	28	87.08	9	85.84
Iris	150	6	90.60	5	97.26
Lympho	148	7	60.90	10	83.41
SatImage	4435	38	62.04	23	65.54
Soybean(S)	47	6	78.60	4	**100.00**
Tic-tac-toe	958	21	70.97	4	71.56
Monk1	124	4	57.05	5	79.39
Monk2	169	4	65.39	2	62.75
Monk3	122	5	54.56	3	**93.79**

Table 15. The quality of data description: **D. size** is a number of templates necessary for data covering; **A. H.** is the average homogeneous degree of such templates.

Decision class description A decision class is approximated by disjunction of templates it covers, analogously to data universe description. Therefore the quality of description depends on the quality of a singular decision template. Below we analyze the quality of decision templates generated for a given decision class.

Decision template quality The quality of a decision template $T \to (d = d_0)$ can be estimated by a *support* and a *confidence*. The support of a decision template is the number of objects in universe satisfying the expression T (i.e. $card([T]_U)$). Let C be a decision class related to decision d_0. Hence confidence is defined by $\frac{card([T]_C)}{card([T]_U)}$. The template is good if both the factors of support and confidence are high.

For decision template generation, we use two algorithms: Local-Maximal Frequency (L-SF) for simple template generation and Local- Similar Distribution of Decision Classes (L-SDD) for generalized template generation. In any data table, we choose the predominant decision class and next try to approximate it. We present the quality of the best template related to the predominant decision class in Table 16.

One can observe that for some data sets the L-SDD algorithm generates really efficient rules. For example in *Australian credit data*, we have extracted the rule with confidence equal to 92.69%, which covers 72.22% objects in a given decision class. In data related to *Breast cancer disease*, we have obtained a rule with 84.13% confidence, which covers 51% of the decision class and in *Diabetes table* a rule with confidence 84.13% and covering ratio of 73.54%.

Quality of decision class description The quality of decision class description is evaluated by the size of template set covering the decision class and the

Data sets	Largest class	Simple templates		Gen. templates	
		Support	Conf.(%)	Support	Conf.(%)
Australian	360	64	100.00	260	92.69
Breast	149	22	77.30	145	84.13
Diabetes	499	44	100.00	367	85.46
Heart	150	46	100.00	57	93.98
Iris	50	39	100.00	39	100.00
Lympho	80	15	98.00	17	98.00
SatImage	1072	38	62.04	23	65.54
Soybean(S)	17	13	100.00	13	100.00
Tic-tac-toe	418	65	100.00	89	100.00
Monk1	62	6	100.00	16	100.00
Monk2	105	16	100.00	14	100.00
Monk3	62	13	100.00	27	100.00

Table 16. The quality of the best decision template.

preciseness of description. The preciseness is estimated in our experiment by the following function. Let $\mathbf{T}_{C_i}$ be a template set covering the decision class C_i. Then the preciseness of covering is defined by $\frac{card(C_i)}{card([\bigcup_{T \in \mathbf{T}_{C_i}} [T]_U])}$.

In Table 17 we present the quality of description defined by simple templates and generalized templates, respectively. The results are related to predominant decision classes.

Data sets	Simple templates		Gen. templates	
	Descr. size	Precise(%)	Descr. size	Precise(%)
Australian	117	100.00	21	95.65
Breast	47	99.35	34	99.20
Diabetes	209	100.00	13	90.25
Heart	48	100.00	40	100.00
Iris	1	100.00	1	100.00
Lympho	24	100.00	10	96.95
SatImage	38	62.04	23	65.54
Soybean(S)	1	100.00	1	100.00
Tic-tac-toe	25	100.00	8	100.00
Monk1	36	100.00	10	96.95
Monk2	51	100.00	17	98.03
Monk3	29	100.00	12	95.83

Table 17. The quality of predominant class approximation.

One can observe by comparing with description defined by simple templates that the description defined by generalized templates is of a smaller size in almost all cases. Interesting results have been received for two data sets: Australian and Diabetes. In the former table, the description size is reduced from 117 to 21 (72%), preserving still preciseness (95%) and in the last table one can reduce the size of description from 209 to 13 (94%) preserving preciseness (90.25%).

10.5 Decomposition quality

The goal of decomposition is to divide a given large data table into smaller sub-tables, on which another data mining or machine learning methods cannot effectively operate. For a given number s, the problem is how to partition the data table into a small number of regular blocks with size closed to s. We use for decomposition two algorithms SF and SDD for simple and generalized templates, respectively. We have chosen for experiment the SatImage data containing 4435 objects, 36 condition attributes and 6 decision classes. In Figure 22 we present the relationship between size of sub-tables and average number of decision classes contained in the sub-tables. Tuning the size of a basic block, one can obtain the sub-domains with high homogeneity. We consider for example the behaviour of simple templates. In Figure 22, if the size is within the range [100-150] then the simple templates define the sub-tables containing no more than 2 classes (on the average). If support is increasing to 300, sub-tables contain usually no more than 4 decision classes. The number of classes contained in sub-tables approaches 5 if support increases to 700.

We present the relationship between the size and the number of sub-tables generated by the decomposition algorithm in Figure 23. On the axis X we represent sizes of sub-tables and on the axis Y the number of sub-tables needed to cover the entire data universe. One can obtain the suitable number of sub-domains by tuning the size of basic blocks. For a given size the number of sub-domains constructed by generalized templates is always smaller than the number of sub-domains constructed by simple templates. As Figure 23 demonstrates if the size of basic blocks is greater than 600, one can use less then 50 sub-domains (in case of simple templates) and 30 sub-domains (in case of generalized templates) for the entire domain description.

11 Conclusions

In this section we summarize the research contributions of the paper and we present some directions for future work.

11.1 Summary

We have discussed a general methodology for regularity extraction from data. Data regularities are expressed as patterns described by some formulas. Two types of patterns have been investigated in the paper. The first type is related to patterns defined in the propositional language (templates), and the second type

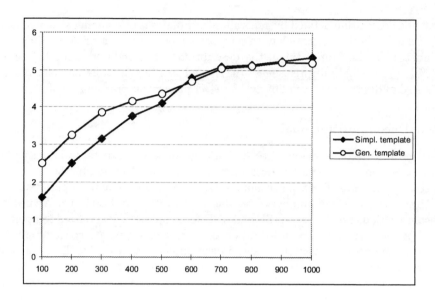

Fig. 22. Relationship of the size of sub-tables to the number of decision classes contained in them.

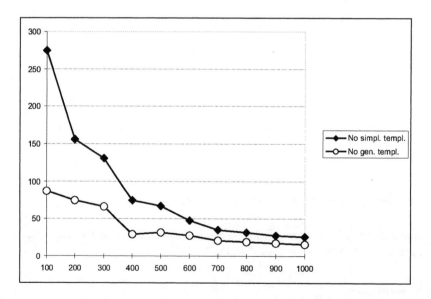

Fig. 23. Relationship of the size of sub-tables to the number of decomposed sub-tables.

is related to patterns defined by a tolerance (similarity) relation extracted from data (relational pattern). We have developed algorithms for pattern extraction and presented the framework for the pattern extraction problem. The following aspects of pattern extraction process were among the topics of consideration: (1) the language for pattern expression; (2) the evaluation function of the pattern quality; (3) the computational complexity of the optimal pattern searching problem; (4) the methods for semi-optimal pattern generation.

We have generalized the language for template expression. Instead of the descriptors of the form $(attribute = value)$ we have considered generalized descriptors of the form $(attribute \in ValueSet)$.

We have discussed some criteria for optimal template estimation. The problems of searching for the optimal template and the optimal template set have been shown to be hard for some evaluation functions. We have obtained theoretical hardness justification of the computational complexity of problems related to *optimal template searching*, *optimal cut set construction* and *optimal partition construction*.

The main goal of the paper was to develop regularity extraction methods. The large part of the paper is dedicated to the methods for optimal template (template set) generation. In the paper some new approaches have been investigated. The first approach is based on feature selection. We presented universal scheme for extracting both kinds of templates: decision templates and templates without a decision. These algorithms allow to extract generalized templates.

We have also discussed a new approach to construct decision trees. Comparing with well known decision tree methods, the approach has shown advantages: it generates trees with a simple structure and a high quality of classification.

More sophisticated patterns can be defined by some binary relations over the objects of data universe. The problem addressed in the paper is how to extract similarity relations from data. Analogously to the case of template patterns, we have discussed the framework for similarity extraction problem. We have considered the structures and properties of some basic families of similarity relations. These properties were used for semi-optimal relation extraction. We presented in the paper a universal scheme for optimal similarity construction, and efficient heuristics for computing semi-optimal solutions for some basic similarity families.

Almost all the methods presented in the paper have been implemented and patterns generated by these methods were used to solve specific basic problems in data mining: namely *classification*, *data description* and *decomposition*. The experimental results show the efficiency of our methods (with respect to computational time and their quality) in many applications.

11.2 Future work

There are some open problems that could be of interest.

The first problem is concerned with the complexity of the template problem assuming other optimalization criteria e.g $quality(T) = support(T) \times length(T)$.

The complexity of the problem of searching for the optimal similarity in a given parameterized family of relations is still open.

The second problem is concerned with the development of methods for template and similarity relation extraction in case of dynamically changing data. That could mean for example that the data universe is changing with time. The methods allowing adaptive learning of patterns from such data are of great important for many applications.

Yet another problem that is important for any classification problem is how to resolve conflicts occurring in the classification process. The popular method is to use predefined fitness functions for conflict resolving. The question is, how to learn the fitness from data? One can expect that such functions can improve the classification quality.

Acknowledgements

The contents of the paper is adopted from my Ph.D thesis with the same title. The dissertation was completed at Warsaw University in July 1999.

Thanks go first to my advisor, dr hab. Bogdan S. Chlebus, for his good advice, support and guidance.

Special thanks to prof. Andrzej Skowron, who has introduced me to the field of data mining and provided a constant guidance and inspiration. Without his advice and criticisms this work would not have been possible.

Thanks to all the members of the Algorithms and Complexity Group for providing a supportive and friendly environment.

Thanks to all the colleagues of the Mathematical Logic Group for unconditional support. Special thanks to prof. Lech Polkowski and Marcin Szczuka for reading this paper and suggesting improvements.

Some results presented here were obtained during joint research. I am very grateful to Bogdan Chlebus, H. Son Nguyen, Andrzej Skowron, Piotr Synak and Jakub Wróblewski for allowing me to include our joint work.

Thanks to my husband H. Son Nguyen for always encouraging me to do my best, and for his unconditional support and assistance.

The paper is supported by the Polish State Committee for Scientific Research - grants No. 8T11C03614 and the Research Program of the European Union - ESPRIT-CRIT2 No. 20288.

References

1. Agrawal R., Imielinski T., Suami A, 1993. *Mining Association Rules Between Sets of Items in Large Datatabes*, ACM SIGMOD. Conference on Management of Data, Washington, D.C., pp. 207-216.
2. Agrawal R., Mannila H., Srikant R., Toivonen H., Verkamo A.I., 1996. *Fast Discovery of Association Rules*. In: V.M. Fayad, G.Piatetsky Shapiro, P. Smyth, R. Uthurusamy, (Eds.), Advanced in Knowledge Discovery and Data Mining, AAAI/MIT Press, pp. 307-328.

3. Agrawal R., Piatetsky-Shapiro G., (Eds.), 1998. *Proc. of the Four International Conference on Knowledge Discovery and Data Mining*, August 27-31, AAAI PRESS, Menlo Park, California.
4. Anzai Y., (Ed.), 1992. *Pattern Recognition and Machine Learning*. Academic Press, inc. Harcourt Brace Jovanovich, Publishers.
5. Bay S. D., 1998. *Combining Nearest Neighbor Classifiers Through Multiply Feature Subsets*. In: Proc. of the International Conference on Machine Learning. Morgan Kaufmann Publishers. Madison.
6. Bazan J., 1998. *A Comparison of Dynamic Non-dynamic Rough Set Methods for Extracting Laws from Decision Tables*. In: Polkowski and Skowron [67], pp. 321–365.
7. Bazan J. G., 1998. *Metody Wnioskowań Aproksymacyjnych dla Syntezy Algorytmów Decyzyjnych*. PhD Dissertation, Warsaw University.
8. Bezdek J., 1996. *A Sampler of Non-Neural Fuzzy Models for Clustering and Classification*. Manuscript of Tutorial at the Fourth European Congress on Intelligent Techniques and Soft Computing, September 2-5.
9. Breiman L., Friedman J. H., Olshen R. A., Stone, P. J., 1984. *Classification and Regression Trees*. Belmont, CA: Wadsworth International Group.
10. Brown F. M. 1990. *Boolean Reasoning*. Kluwer Academic Publishers, Dordrecht.
11. Bouchon–Meunier B., Delgado M., Verdegay J.L., Vila M.A., Yager R.R., 1996. *Proceedings of the Sixth International Conference on Information Processing Management of Uncertainty in Knowledge–Based Systems (IPMU'96)*. July 1-5, Granada, Spain **1–3**, pp. 1–1546.
12. Cattaneo G., 1997. *Generalized Rough Sets*. Preclusivity Fuzzy-Intuitionistic (BZ) Lattices, Studia Logica 58, pp. 47–77.
13. Cattaneo G., 1998. *Abstract Approximation Spaces for Rough Theories*. In: Polkowski and Skowron [67], pp. 59–98.
14. Chiu D.K.Y., Cheung B., Wong, A.K.C., 1990. *Information Synthesis Based on Hierarchical Entropy Discretization*. Journal of Experimental and Theoretical Artificial Intelligence **2**, pp. 117–129.
15. Chlebus B. S., Nguyen S. Hoa, 1998. *On Finding Optimal Discretizations for Two Attributes*. In: Polkowski and Skowron [69], pp. 537–544.
16. Chmielewski, M. R., Grzymala-Busse, J. W., 1994. *Global Discretization of Attributes as Preprocessing for Machine Learning*. Proc. of the III International Workshop on RSSC94, November, pp. 294–301.
17. Clark P., Niblett R., 1989. *The CN2 Induction Algorithm*. Machine Learning, 3, pp. 261–284.
18. Cormen T. H., Leiserson C. E., Rivest R. L. *The Set Covering Problem*. Introduction to Algorithms. The MIT Press, Cambridge, Massachusetts, London, England, pp. 974–978.
19. Everitt B. S., (Ed.), 1993. *Cluster Analysis*. Reprinted 1998 by Arnold, a member of the Hodder Headline Group.
20. Fayyad U.M., 1991. *On the Induction of Decision Trees for Multiple Concept Learning*. PhD Dissertation, the University of Michigan.
21. Fayyad U., Piatetsky-Shapiro G., (Eds.), 1996. *Advances in Knowledge Discovery and Data Mining*. MIT/AAAI Press.
22. Fayyad U. M., Irani K.B., 1993. *Multi-Interval Discretization of Continuous-Valued Attributes for Classification Learning*. In: Proc. of the 13th International Joint Conference on Artificial Intelligence, Morgan Kaufmann, pp. 1022-1027.

23. Garey M.R., Johnson D.S., 1979. *Computers and Interactability. A Guide to the Theory of NP-Completeness.* W.H. Freeman and Company New York.
24. Greco S., Matarazzo B., Słowiński R., 1995. *Rough Set Approach to Multi-Attribute Choice Ranking Problems.* Institute of Computer Science, Warsaw University of Technology, ICS Research Report **38/95**; see also, G. Fandel, T. Gal (Eds.), *Multiple Criteria Decision Making*, Proc. of 12th International Conference in Hagen, Springer–Verlag, Berlin, pp. 318–329.
25. Greco S. Matarazzo B. Słowiński R., 1998. *Rough Approximation of a Preference Relation in a Pairwise Comparison Table.* In: Polkowski and Skowron [67] pp. 13–36.
26. Greco S. Matarazzo B. , Słowiński R. 1998. *A New Rough Set Approach to Multicriteria and Multiattribute Classification.* In: Polkowski and Skowron [69], pp. 60–67.
27. Greco S., Matarazzo B., Sowiski R. 1999. *On Joint Use of Indicernibility, Similarity and Dominance in Rough Approximation of Decision Classes.* Proc. of Fifth International Conference on Integrating Technology and Human Decisions: Global Bridges into the 21th Century. Athens, Greece, July 4-7.
28. Heath D., Kasif S., Salzberg S., 1993. *Induction of Oblique Decision Trees.* Proc. of 13th International Joint Conference on Artificial Intelligence. Chambery, France, pp. 1002-1007.
29. Hu X., Cercone N., 1995. *Rough Set Similarity Based Learning from Databases.* Proc. of the First International Conference of Knowledge Discovery and Data Mining. August 20-21, Montreal, Canada, pp. 162-167.
30. Komorowski J., Ras Z.W., (Eds.), 1993. *Proceedings of the Seventh International Symposium on Methodologies for Intelligent Systems (ISMIS'93)*, Trondheim, Norway, June 15–18, Lecture Notes in Computer Science **689**, Springer–Verlag, Berlin.
31. Krawiec K., Słowiński R., Vanderpooten D., 1996. *Construction of Rough Classifires Based on Application of a Similarity Relation.* Proc. of the Fourth International Workshop on Rough Set, Fuzzy Set and Machine Discovery. November 6-8, Tokyo, Japan, pp. 23–30.
32. Krętowski M., Stepaniuk J., 1996. *Selection of Objects and Attributes a Tolerance Rough Set Approach.* Proc. of the Ninth International Symposium on Methodologies for Intelligent Systems . June 9-13, Zakopane, Poland, pp. 169–180.
33. Lenz M., Bartsch-Sporl B., Hans-Dieter Burkhard, Wess S., (Eds.), 1998. *Case–Based Reasoning Technology: From Fundation to Application.* LNAI **1400**, Springer–Verlag, Berlin.
34. Lin T.Y., 1989. *Neighbourhood System and Approximation in Database and Knowledge Base Systems.* Proc. of the Fourth International Symposium on Methodologies of Intelligent System.
35. Lin T.Y., 1998. *Granular Computing on Binary Relations I.* In: Polkowski and Skowron [67], pp. 107–121.
36. Lin T.Y., 1989. *Granular Computing on Binary Relations II.* In: Polkowski and Skowron [67], pp. 122–140.
37. Liu H., Motoda H., (Eds.), 1998. *Feature Selection for Knowledge Discovery and Data Mining.* Kluwer Academic Publishers.
38. Mannila H., Toivonen H., Verkamo A. I., 1994. *Efficient Algorithms for Discovering Association Rules.* In: U. Fayyad and R. Uthurusamy (Eds.): AAAI Workshop on Knowledge Discovery in Databases, Seattle, WA, pp. 181–192.

39. Marcus S., 1994. *Tolerance Rough Sets, Čech Topologies, Learning Process*, Bull. Polish Academy. Ser. Sci. Tech., Vol. **42**, No. **3**, pp. 471–487.
40. Maritz P., 1996. *Pawlak and Topological Rough Sets in Terms of Multi-Functions*. Glasnik Matematicki **31/51**, pp. 159–178.
41. Michalski R. S., Mozetic I., Hong J., Lavrac H. (1986). *The Multi-Purpose Incremental Learning System AQ15 and Its Testing Application to Three Medical Domains*. Proc. of the Fifth National Conference on AI, Filadelfia, Morgan-Kaufmann, pp. 1041–1045.
42. Michie, D., Spiegelhanter, D.J. and Taylor, C.C., (Eds.), 1994. *Machine Learning, Neural and Statistical Classification*. Great Britain: Ellis Horwood.
43. Mitchell T. M., (Ed.), 1997. *Machine Learning*. The McGraw-Hill Companies, Inc.
44. Mollestad T., Skowron A., 1996. *A Rough Set Framework for Data Mining of Propositional Default Rules*. Proc. ISMIS-96, Zakopane, Poland, June, pp. 448–457.
45. Murthy S., Aha D., 1996. UCI Repository of Machine Learning Data Tables. http://www/ics.uci.edu/mlearn.
46. Murthy S., Kasif S., Saltzberg S., Beigel R., 1993. *OC1: Randomized Induction of Oblique Decision Trees*. Proc. of the Eleventh National Conference on AI, July, pp. 322–327.
47. Murthy S., Kasif S., Saltzberg S., 1994. *A System for Induction of Oblique Decision Trees*, In Proc. of Sixth International Machine Learning Workshop, Ithaca N.Y., Morgan Kaufmann.
48. Nguyen S. Hoa, Nguyen H. Son, 1996. *Some Efficient Algorithms for Rough Set Methods*. In: Bouchon–Meunier, Delgado, Verdegay, Vila, and Yager [11], pp. 1451–1456.
49. Nguyen S. Hoa, Nguyen T. Trung, Skowron A., Synak P., 1996. *Knowledge Discovery by Rough Set Methods*. Proc. of the International Conference on Information Systems Analysis and Synthesis, July 22-26, Orlando, USA, pp. 26–33.
50. Nguyen S. Hoa., Polkowski L., Skowron A., Synak P., Wróblewski J.,1996. *Searching for Approximate Description of Decision Classes*. Proc. of the Fourth International Workshop on Rough Sets, Fuzzy Sets, and Machine Discovery, November 6-8, Tokyo, Japan, pp. 153–161.
51. Nguyen S. Hoa., Skowron A., Synak P., 1996. *Rough Sets in Data Mining: Approximate Description of Decision Classes*. Proc. of the fourth European Congress on Intelligent Techniques and Soft Computing, Aachen, Germany, September 2-5, pp. 149–153.
52. Nguyen S. Hoa, Skowron A., 1997. *Searching for Relational Patterns in Data*. Proc. of the First European Symposium on Principles of Data Mining and Knowledge Discovery, June 25-27, Trondheim, Norway, pp. 265–276.
53. Nguyen S. Hoa, Nguyen H. Son, 1998. *Pattern Extraction from Data*. Fundamenta Informaticae **34**, 1998, pp. 129–144.
54. Nguyen S. Hoa, Nguyen H. Son, 1998. *The Decomposition Problem in Multi-Agent Systems*. In: J. Komorowski, A. Skowron, I. Duntsch (Eds.): Proc. of the ECAI'98 Workshop on Synthesis of Intelligent Agent Systems from Experimental Data, Brighton, UK.
55. Nguyen S. Hoa, Nguyen H. Son, 1998. *Pattern Extraction from Data*. Proc. of the Seventh International Conference on Information Processing and Management of Uncertainty in Knowledge-based Systems (IPMU'98), Paris, France, July 6-10, pp. 1346–1353.

56. Nguyen S. Hoa, Skowron A., Synak P., 1998. *Discovery of Data Patterns with Applications to Decomposition and Classification Problems.* In: Polkowski and Skowron [67], pp. 55–97.
57. Nguyen S. Hoa, 1999. *Discovery of Generalized Patterns.* Proc. of the Eleventh International Symposium on Methodologies for Intelligent Systems (ISMIS'99), Warsaw, Poland, June 8–12, Lecture Notes in Computer Science **1609**, Springer-Verlag, Berlin, pp. 574–582.
58. Nguyen H. Son, Skowron A., 1995. *Quantization of Real Value Attributes: Rough Set and Boolean Reasoning Approach.* Proc. of the Second Annual Joint Conference on Information Sciences, Wrightsville Beach, NC, USA, September 28 - October 1, pp. 34–37.
59. Nguyen H. Son, Nguyen S. Hoa, Skowron A., 1996. *Searching for Features Defined by Hyperplanes.* In: Z. W. Raś, M. Michalewicz (Eds.), Proc. of the IX International Symposium on Methodologies for Information Systems ISMIS'96, June 1996, Zakopane, Poland. Lecture Notes in AI **1079**, Berlin, Springer-Verlag, pp. 366–375.
60. Nguyen H. Son, 1997. *Discretization of Real Value Attributes: Boolean Reasoning Approach.* PhD Dissertation, Warsaw University.
61. Nguyen H. Son, Nguyen S. Hoa, 1998. *Discretization methods in data mining.* In: Polkowski and Skowron [67], pp. 451–482 .
62. Pawlak Z., 1991. Rough Sets. *Theoretical Aspects of Reasoning about Data*, Kluwer Academic Publishers, Dordrecht.
63. Pfahringer B., 1995. *Compression-Based Discretization of Continuous Attributes.* In A.Prieditis, S.Russell, (Eds.), Proc. of the Twelfth International Conference on Machine Learning, Morgan Kaufmann.
64. Piatetsky-Shapiro G., 1991. *Discovery, Analysis and Presentation of Strong Rules.* In: Piatetsky-Shapiro G. and Frawley W.J. (Eds.): Knowledge Discovery in Databases, AAAI/MIT, pp. 229 – 247.
65. Polkowski L., Skowron A., Żytkow J., 1995. *Tolerance Based Rough Sets.* In: Soft Computing, T.Y.Lin, A.M. Wildberger (Eds.), San Diego, Simulation Council, Inc., pp. 55-58.
66. Polkowski L., Skowron A., 1996. *Rough Mereological Approach to Knowledge-based Distributed AI.* In: J.K. Lee, J. Liebowitz, Y. M. Chae (Eds.): Critical Technology. Proc. of the Third World Congress on Expert Systems, pp. 774-781, Seoul, Cognisant Communication Corporation, New York.
67. Polkowski L., Skowron A., (Eds.), 1998. *Rough Sets in Knowledge Discovery 1: Methodology and Applications.* Physica-Verlag, Heidelberg.
68. Polkowski L., Skowron A., (Eds.), 1998. *Rough Sets in Knowledge Discovery 2: Methodology and Applications.* Physica-Verlag, Heidelberg.
69. Polkowski L. Skowron A. (Eds.), 1998, *Proceedings of the First International Conference on Rough Sets and Soft Computing (RSCTC'98).* Warszawa, Poland, June 22-27, Springer-Verlag, LNAI **1424**.
70. Quinlan, J.R., 1986. *Induction of Decision Trees.* In: Machine Learning **1**, pp. 81–106.
71. Quinlan J. R., 1993. *C4.5: Programs for Machine Learning.* Morgan Kaufmann, San Mateo, CA.
72. Stefanowski J. 1998. *On Rough Set Based Approaches to Induction of Decision Rules.* In: Polkowski and Skowron [67], pp. 500–529.
73. Stepaniuk J., 1996. *Similarity Based Rough Sets and Learning.* Proc. of the Fourth International Workshop on Rough Sets, Fuzzy Sets and Machine Discovery,

November 6-8, Tokyo, Japan, pp. 18-22.
74. Skowron A., Polkowski L., Komorowski J., 1996. *Learning Tolerance Relation by Boolean Descriptions: Automatic Feature Extraction from Data Tabes.* Proc. of the Fourth International Workshop on Rough Set, Fuzzy Set and Machine Discovery . November 6-8, Tokio, Japan, pp. 11-17.
75. Skowron A., Rauszer C., 1992. *The Discernibility Matrices and Functions in Information Systems.* In: R. Słowiński (Eds.): Intelligent Decision Support. Handbook of Applications and Advances of the Rough Sets Theory, Kluwer, Dordrecht, pp. 331-362.
76. Skowron A., Pal S. K., (Eds.), 1998. *Rough-Fuzzy Hybridization: A New Trend in Decision Making.* Springer-Verlag. Singapore, pp. 1-97.
77. Skowron A., Stepaniuk J., 1996. *Tolerance Approximation Spaces.* In: Fundamenta Informaticae, August, **27**(2-3), pp. 245-253.
78. Słowiński R., (Ed.), 1992, *Intelligent Decision Support – Handbook of Applications and Advances of the Rough Sets Theory.* Kluwer Academic Publishers, Dordrecht.
79. Słowiński R., 1993. *Rough Set Learning of Preferential Attitude in Multi-Criteria Decision Making.* In: Komorowski and Ras [30], pp. 642-651.
80. Słowiński R., 1994. *Handling Various Types of Uncertainty in the Rough Set Approach.* In: Ziarko [95], pp. 366-376.
81. Słowiński R., 1994. *Rough Set Analysis of Multi-Attribute Decision Problems.* In: Ziarko [95], pp. 136-143.
82. Słowiński R., 1995. *Rough Set Approach to Decision Analysis.* AI Expert **10**, pp. 18-25.
83. Słowiński R., Vanderpooten D., 1995. *Similarity Relation as a Basis for Rough Approximations.* In: P. Wang (Ed.): Advances in Machine Intelligence & Soft Computing, Bookwrights, Raleigh NC (1997) pp. 17-33.
84. Słowiński R., Vanderpooten D., 1997. *A Generalized Definition of Rough Approximations Based on Similarity.* IEEE Trans. on Data and Knowledge Engineering.
85. Stepaniuk J., 1996. *Similarity Based Rough Sets and Learning.* In: Tsumoto, Kobayashi, Yokomori, Tanaka, and Nakamura [89], pp. 18-22.
86. Stepaniuk J., 1998. *Approximation Spaces, Reducts and Representatives.* In: Polkowski and Skowron [67], pp. 109-126.
87. Tentush I., 1995. *On Minimal Absorbent Sets for some Types of Tolerance Relations,* Bulletin of the Polish Academy of Sciences, **43**(1), pp. 79-88.
88. Toivonen H., Klemettinen M., Ronkainen P., Hatonen P., Mannila H., 1995. *Pruning and Grouping Discovered Association Rules.* In: Mlnet: Familiarisation Workshop on Statistics, Machine Learning and Knowledge Discovery in Databases , Heraklion, Crete, April, pp. 47-52.
89. Tsumoto S., Kobayashi S., Yokomori T., Tanaka H., Nakamura A., (Eds.), 1996. *Proceedings of the Fourth International Workshop on Rough Sets, Fuzzy Sets, and Machine Discovery (RSFD'96).* The University of Tokyo, November 6-8.
90. Uthurusamy H., Fayyad V.M., Spangler S., 1991. *Learning Useful Rules from Inconclusive Data.* In: Piatetsky-Shapiro G. and Frawley W.J. (Eds.): Knowledge Discovery in Databases, AAAI/MIT, pp. 141-157.
91. Yao Y.Y., 1998. *Generalized Rough Set Models.* In: Polkowski and Skowron [67], pp. 286-318.
92. Yao Y.Y., Wong S.K.M., Lin T.Y., 1997. *A Review of Rough Set Models.* In: T.Y. Lin, N.Cercone (Eds.): Rough Sets and Data Mining Analysis of Imprecise Data, Kluwer Academic Publishers, pp. 47-75.

93. Westphal C., Blaxton T., (Eds.), 1998. *Data Mining Solution.* Wiley Computer Publishing.
94. Zadeh L.A., 1997. *Toward a Theory of Fuzzy Information Granulation and Its Certainty in Human Reasoning and Fuzzy Logic.* Fuzzy Sets and Systems **90**, pp. 111–127.
95. Ziarko W., (Ed.), 1994. *Rough Sets, Fuzzy Sets and Knowledge Discovery*, Workshops in Computing, Springer Verlag & British Computer Society.
96. Ziarko W., 1998. *Rough Sets as a Methodology for Data Mining.* In: Polkowski and Skowron [67], pp. 554–576.
97. Żytkow J., Baker J., 1991. *Interactive Mining of Regularities in Data Bases.* In: Piatetsky-Shapiro, W. J. Frawley, (Eds.): Knowledge Discovery in Databases, Menlo Park, CA: AAAI Press, pp. 31–53.

Chapter 7

Rough Set Methods for the Synthesis and Analysis of Concurrent Processes

Zbigniew Suraj

Institute of Mathematics
Pedagogical University of Rzeszów
Rejtana 16A, 35-310 Rzeszów, Poland
e-mail: zsuraj@univ.rzeszow.pl

Abstract. In this Chapter rough set methods for the modeling of concurrent processes are considered. The research is motivated by the problems coming from the domains such as, for example: knowledge discovery systems, data mining, control design, decomposition of information systems, object identification in real-time. this Chapter includes, in particular, the description of automatic methods for the modeling and analysis of concurrent systems specified by information systems. In this Chapter the following problems are considered:

1. The synthesis problem of concurrent systems specified by information systems.
2. The problem of discovering concurrent data models from experimental tables.
3. The re-engineering problem for cooperative information systems.
4. The real-time decision making problem.
5. The control design problem for discrete event systems.

Rough set theory, Boolean reasoning, theory of Petri nets as well as self-implemented computer tools are used for this purpose. The methods presented in the paper as well as further investigations of interconnections between rough set theory and concurrency may stimulate the development of both theoretical and practical research related to the areas mentioned above.

Keywords: Rough sets, concurrent models, Petri nets, knowledge discovery, data mining, system decomposition, control design, real-time systems, cooperative information systems, computer based tools.

1 Introduction

The aim of this Chapter is to present an unified approach to the synthesis problem as well as the analysis problem of concurrent systems based on the rough set philosophy [166], Boolean reasoning [34] and Petri nets [187].

The synthesis problem is the problem of synthesizing a concurrent system model from observations or specification of processes running in a given concurrent system. However, the analysis problem consists in revealing important information about the structure and dynamic behaviour of the modeled system. This information can then be used to evaluate the modeled system and suggest improvements or changes. Both the problems have been discussed in the literature for various types of formalisms, among others: parallel programs ([109], [205]), COSY-expressions [94], Petri nets ([106], [59], [154], [138], [19], [54], [6], [142]), but the approach presented here for these problems is new.

The research is motivated by the problems coming from the domains such as, for example: knowledge discovery systems ([189], [328], [63], [192], [193], [45]), data mining ([63], [115], [45], [130]), control design ([229], [175], [50], [320], [115], [81]), decomposition of information systems ([303], [226], [237], [46], [298], [164]), object identification in real-time ([229], [93], [108], [58], [264], [247]).

In this Chapter the following problems are considered:

1. The synthesis problem of concurrent systems specified by information systems [168].
2. The problem of discovering concurrent data models from experimental tables.
3. The re-engineering problem for cooperative information systems.
4. The real-time decision making problem.
5. The control design problem for discrete event systems.

Rough set theory, Boolean reasoning, theory of Petri nets as well as self-implemented computer tools are used for this purpose.

Rough set theory is a relative new mathematical and Artificial Intelligence (AI) ([313], [232], [225], [141]) technique introduced by Z. Pawlak [166] in the early 1980s. It lays on the crossroads of fuzzy sets ([315], [323], [293], [176]), theory of evidence ([230], [238], [314]), neural networks ([304], [86], [212]), genetic algorithms ([77], [133]), chaotic systems([76], [37], [162]), Petri nets ([187], [186], [206], [100]) and many others branches of AI.

Rough set theory has found many interesting real-life applications in medicine, banking, industry and others (see e.g. [168], [260], [115], [192], [193], [194], [161], [45], [141], [163]).

The rough set theory is dealing with incomplete or imprecise data. A major feature of rough set theory in terms of practical applications is the classification of empirical data and subsequent decision making. In general, in order to understand and use the data it is necessary to derive underlying knowledge about the data, i.e., what it represents. Such knowledge can be represented in many forms. In the rough set theory, the information systems [168] and the rules are the most common form of representing knowledge. In this Chapter, the rules are made of two parts: a condition part and an action part. They will be represented using the following condition-action form: *if* (conditions) *then* (action), where 'conditions' is a list of conditions linked by logical operators (and, or) and 'action' is an action to be taken if the condition part evaluates to true.

In this Chapter the processes of complex systems are specified by information systems or the rules extracted from such systems.

Main advantages of knowledge rule representation are high readability and modularity. Rules can be added and deleted quite easily. The knowledge represented in this way is thus highly modifiable. The disadvantage of knowledge rule representation is in the difficulty of capturing and revealing an overall picture of the problem, as in the case of using a graph representation.

Significant efforts have been spent in the attempt to automate the inference process leading to the identification of objects with particular attention to real-time systems. The major part of approaches to reasoning problem solving, based on Artificial Intelligence methodologies, very often concentrate on logical formalism. They are usually able to provide, in quite clear way, a formal notion of the concept of reasoning problem and of reasoning solution. The major problem with logical approaches is that emphasis is usually given more to declarative definitions than to procedural aspects, because of the intrinsic declarative nature of any logical framework. Here for the description of problems mentioned above and their solutions, we shall use the methods and the techniques proposed by both the rough set theory as well as the Petri net theory.

In this Chapter we shall treat complex systems and their components uniformly as *concurrent processes*. A process is here an object that is designed for a possible continuous interaction with its user, which can be another process. An interaction can be an input or output of a value, which we treat abstractly as a *communication*. In between two subsequent communications the process usually engages in some *internal actions*. These proceed autonomously at a certain speed and are not visible to the user. However, as a result of such internal actions the process behaviour may appear *nondeterministic* to the user. Concurrency arises because there can be more than one user and inside the process more than one active subprocess. The behaviour of a process is unsatisfactory for its user if it does not communicate as desired. The reason can be that the process stops too early (*deadlock*) or that it engages in an infinite loop of internal actions (*divergence*). Thus most processes are designed to communicate arbitrarily long without any danger of deadlock or divergence.

Since the behaviour of concurrent processes is in general very complicated, it is not surprising that their description has been approached from rather various points of view. In particular, Petri nets ([187], [186], [206], [142]), algebraic process terms ([122], [88], [131], [89], [87], [132], [8], [95]) and logical formulas of temporal or ordinary predicate logic ([190], [17], [191], [29], [60], [26]) have been used.

Petri nets are used to describe processes as concurrent and interacting machines with all details of their operational behaviour.

We could take as our machine model *automata* in the sense of classical automata theory [203] also known as *transition systems* [103]. Automata are fine except that they cannot represent situations where parts of a machine work independently or concurrently. Since we are after such a representation, we use Petri nets instead.

An attractive alternative to Petri nets are *event structures* introduced in [153] and further developed by many researchers ([310], [311], [132], [33], [306], [308]). Event structures are more abstract than nets because they do not record states, only events. But in order to forget about states, event structures must not contain cycles. This yields infinite event structures even in cases where finite (but cyclic) nets suffice. We prefer whenever possible finite objects, and thus

Petri nets, as our basic machine-like view of process. Besides, there are also the special-purpose specification languages: protocol description languages such as ESTELLE [36] or LOTOS [28], real-time languages such as ESTEREL [18] or LUSTRE [79].

Another alternative to Petri nets could be Harel's formalism of *state charts* [83]. Like nets, state charts extend the classical notion of an automaton, but that extension models only fully synchronous parallelism. Since we aim here at modeling asynchronous parallelism with synchronous communication, state charts cannot replace Petri nets for our machine-like view of processes.

In the late seventies, *temporal logic* [60] was proposed as a query language for the specification of reactive and distributed systems; a few years later, *model-checking* was introduced as a technique for the verification of arbitrary temporal properties. The temporal logics used in the verification of systems are usually state-based, but also action-based logics have been suggested, especially in connection with process algebras. Temporal logics can be *linear-time* and *branching-time*; linear-time logics are interpreted on the single computations of a system, while branching-time logics are interpreted on the tree of all its possible computations. The most popular linear and branching-time temporal logics are LTL (*linear-time propositional temporal logic*) and CTL (*computation tree logic*). Many important properties of concurrent systems, like deadlock-freedom, reachability, liveness (in the Petri net sense) can be expressed in LTL or in CTL (often in both).

One may regret such a diversity, but we think that these different descriptions can be seen as expressing complementary views of concurrent processes, each one serving its own purpose. It is worth to point that in [160] a theory of concurrent processes where three different semantic description methods mentioned above in one uniform framework have been presented.

In this Chapter as a model for concurrency Petri nets are chosen. Our choice of Petri nets is motivated additionally by the following:

1. Petri nets are good models for describing concurrent, nondeterministic and asynchronous activities with in addition well-developed net theory and formal analysis techniques.
2. Petri nets are based on a simple extension of the concepts of state and transition known from automata. The extension is that in nets both states and transitions are distributed over several places. This allows an explicit distinction between concurrency and sequentiality.
3. Petri nets have a graphical representation that visualizes the basic concepts about processes like sequentiality, choice, concurrency and synchronization.
4. Since Petri nets allow cycles, a large class of processes can be represented by finite nets.
5. Petri nets are the tool which can support not only the specification activity, but also the evolution of the behaviour of the system starting from the model provided by the specification.
6. One of the further reasons for utility of Petri nets in modeling systems follows from the fact that different interpretations may be given to places, transitions and tokens. Places may represent conditions, transitions may represent

actions or events, and tokens may indicate that the conditions on which they reside are true; places may also represent states, transitions may represent state changes and tokens may represent objects, such as jobs or resources that change state.
7. Petri nets allow the construction of models amenable both for correctness and efficiency analysis.
8. Petri nets can be implemented using many different techniques (hardware, microprogrammed, software).
9. Petri nets may be applied to a broad range of systems, including information systems, operating systems, databases, communication protocols, computer hardware architectures, security systems, manufacturing systems, defense command and control, business processes, banking systems, chemical processes, nuclear waste systems and telecommunications.
10. Moreover, the possibility of using Petri nets in Artificial Intelligence, in particular also in rough set applications, is recently receiving a lot of attention ([121], [296], [251], [252], [253], [254], [255], [177], [257], [258], [259], [278], [279], [281], [178], [180], [283], [284], [285], [286], [183], [182]).

Indeed, in the last few years, we assisted in a growing interest in using Petri net models as the basis for the construction of knowledge representation and reasoning frameworks. However, attention essentially focused on tasks, in principle, different from problems solving mentioned above, as it is possible to notice in the following review of the most important works concerning the application of Petri net models to Artificial Intelligence and to the rough set theory.

An important feature of rough sets as well as Petri nets is that the theories are followed by practical implementations of toolkits that support the user in the construction of a model and its analysis in a natural and effective way. We use our *ROSEPEN* computer tools (described shortly in section 8). They are running on *IBM PC* microcomputers under *MS-DOS* operating system.

In [51], Deng and Chang introduce the *G-net model*; such a model is a Petri net-like model intended to be a uniform framework for representing declarative and procedural knowledge. They compare their model with classical knowledge representation formalisms like *frames* and *semantic networks* and propose, among others things, some interesting algorithms for inheritance and classification which are typical kinds of inference in these formalisms.

Production rules are perhaps the AI formalism most studied from a Petri net point of view. In [295], Valette and Bako discuss a way of modeling production rules trying to exploit the efficient RETE algorithm [71] in their framework, while Giordana and Saitta [75] use *Predicate/Transition Nets (Pr/T-nets)*, proposed in [74] as an extension to the classical model by Petri, for modeling production rules; classical problems related to production rules are discussed in this framework and place invariant analysis techniques [107] are proposed for validation of the set of rules and for deducing conclusions from a given set of facts.

The application of Petri net models to the knowledge base verification task is also one of the most popular kind of applications; the major part of approaches address this problem in the context of rule base system as shown in ([150], [149], [117], [82]), however the application of Petri net analysis seems feasible also for validating different kind of models as, for instance, causal models (see [196] for the description of a possible approach).

Other kinds of application have also been proposed. Lin et al. [111] discuss how to perform logical inference on a set of Horn clauses modeled as a Petri net, by using transition invariant analysis techniques of the net model and they also show how this kind of analysis is equivalent to perform resolution on the set of clauses.

In [145], Murata et al. investigate a Petri net framework for logic programming applicable to reasoning in the presence of inconsistent knowledge; they show how the framework of *paraconsistent logic programming* introduced in [25] can be captured using Pr/T-nets.

Planning problems are faced using Pr/T-nets in [144] and [319]; in particular, the task of plan synthesis is addressed in [319] by applying some results in [146] concerning the computation of the solution of a logic program modeled through a Pr/T-net.

The representation of uncertainty in Petri nets model has also been a topic of big interest with different proposal of *Fuzzy Petri Nets* ([116], [43]) and with the integration of Petri nets and Possibilistic Logic [38].

In [196] Portinale shows how Petri nets can be used as models for diagnostic knowledge representation and reasoning.

The basic problems solution such as e.g.: the synthesis problem, decomposition problems, re-engineering problems, identification problems are actually a neglected area in frameworks trying to apply Petri nets to rough set tasks. However, some attempts to address the problem of approximate reasoning using fuzzy and rough fuzzy Petri nets can be found in ([179], [180], [183], [182]) and more recently in [184].

A short review of rough set results in the Petri net context, which are connected with the issues discussed in this Chapter, is described below.

In [180], Peters proposes a new class of Petri nets: roughly fuzzy Petri nets in which new models of clock representation systems are discussed. Some applications are pointed to aimed at extracting rough fuzzy approximations from universes of fuzzy objects.

An approach to designing software quality decision systems is provided in the paper [181]. The decision rule extracted from a given software quality decision system make it possible for a designer to evaluate the consistency and correctness of assessments of software characteristics for a particular system. The basic structure of a software quality decision system is represented by a rough Petri net model.

In [185], the authors present an approach to the design of approximate time rough controllers. The clocks used to measure durations required to achieve controller objectives are modeled as approximate time windows.

The paper [182] considers the construction of Petri nets to simulate the computation performed by decision systems. It is strongly connected with the papers [259] and [183].

In the paper [184] the construction of Petri nets to simulate conditional computation in various forms of systems is presented. Two families of guards on transitions are introduced. The first family of guards provide a basis for transitions with a form of continuous enabling, whereas the second one of guards is derived from approximations of our knowledge of input values.

This Chapter is organized as follows.

Section 2 presents the basic notions and notations from both rough set theory as well as Petri nets.

Section 3 contains two approaches to the synthesis problem of concurrent systems. In the first approach we assume that the specification of a given concurrent system is encoded in data table represented by an information system S. Next we construct for a given information system S its concurrent model, i.e., a marked Petri net (N_S, M_S) with the following property: the set of all reachable markings from the initial marking of the net (the reachability set) $R(N_S, M_S)$ defines an extension S' of S created by adding to S all new global states corresponding to markings from $R(N_S, M_S)$. All (new) global states in S' are consistent with all rules true in S. Moreover, S' is the largest extension of S with that property. In the second approach, upon introducing a new notion of a dynamic information system we propose a method for constructing an elementary net system equivalent to a given dynamic information system in the sense that the related transition systems are isomorphic. This provides a method for synthesis of a concurrent system specified by a dynamic information system.

Section 4 provides a method for discovery of concurrent data models. The method is based on a decomposition of a given data table into so called components linked by some connections (communications) which allow to preserve some constraints. Any component represents in a sense the strongest functional module of the system. The connections between components represent constraints which must be satisfied when these functional modules coexist in the system. The components together with the connections define a so called covering of the system. Finally, the coverings of the system are used to construct its concurrent model in the form of a Petri net.

On the base of the synthesis method described in section 3, we present in section 5 an approach to re-engineering of a cooperative information system when the specification for the system is being changed e.g. by adding new requirements. Whenever the costs of unchanged parts of the system are known then it is also possible to estimate the global cost of the re-engineering.

An approach to real-time decision making is discussed in section 6. We consider decision tables with the values of conditional attributes (conditions) measured by sensors. These sensors produce outputs after an unknown but finite number of time units. We construct an algorithm for computing a highly parallel program represented by a Petri net from a given decision table. The constructed net allows to identify objects in decision tables to an extent which makes appropriate decisions possible. The outputs from sensors are propagated through the net with maximal speed. This is done by an appropriate implementation of all rules true in a given decision table.

Section 7 deals with an application of rough set methods in control design. The main stages of our approach are: the control specification by decision tables, generation of rules from the specification of the system behaviour, and converting rules set into a concurrent program represented in the form of a Petri net.

In the last section we describe the computer tools based on rough set methods and Petri nets for the synthesis and analysis of concurrent systems discovered from data tables.

2 Preliminaries of Rough Sets and Petri Nets

We assume that the reader is acquainted with the basic notions of rough set theory, like information systems, decision tables, indiscernibility relations, discernibility matrices, information functions, reducts, rules, etc., and also with the basic notions of Petri net theory, like firing rule, reachability marking, liveness, boundedness. This section just fixes some notation.

2.1 Rough Sets

2.1.1 Information Systems and Decision Tables

Information systems (sometimes called data tables, attribute-value systems, condition-action tables, knowledge representation systems etc.) are used for representing knowledge. The notion of an information system presented here is due to Z. Pawlak and was investigated by several authors (see e.g. the bibliography in [168], [193]). Among research topics related to information systems are: rough set theory, problems of knowledge representation, problems of knowledge reduction, dependencies in knowledge bases. Rough sets have been introduced [166] as a tool to deal with inexact, uncertain or vague knowledge in artificial intelligence applications.

In this subsection, we will discuss some terminology associated with information systems.

An *information system* is a pair $S = (U, A)$, where U - is a non-empty, finite set called the *universe*, A - is a non-empty, finite set of *attributes*, i.e., $a : U \to V_a$ for $a \in A$, where V_a is called the *value set* of a.

Elements of U are called *objects* and interpreted as e.g. cases, states, patients, observations. Attributes are interpreted as features, variables, processes, characteristic conditions etc.

The information system can be represented as a finite data table, in which the *columns* are labeled by attributes, the *rows* by objects and on the position corresponding to the row u and column a the value $a(u)$ appears. Each row in the table describes the information about some object in S.

In the paper attributes are meant to denote the processes of the system, the values of attributes are understood as local states of processes and objects are interpreted as global states of the system.

The set $V = \bigcup_{a \in A} V_a$ is said to be the *domain* of A. A function $cost : A \to \Re_+ \cup \{0\}$, where $\Re_+$ is the set of non-negative real numbers, is called the *cost of attributes*. In the following the cost of attributes will be represented by a vector, i.e., $cost = (cost(a_1), ..., cost(a_m))$, where $a_1, ..., a_m \in A$ and $m =$card(A). We assume that the attributes in the set A of attributes have been ordered in some way. For $S = (U, A)$, a system $S' = (U', A')$ such that $U \subseteq U'$, $A' = \{a' : a \in A\}$, $a'(u) = a(u)$ for $u \in U$ and $V_a = V_{a'}$ for $a \in A$ will be called an $U'-extension$ of S (or an extension of S, in short). S is then called a *restriction* of S'. If $S = (U, A)$ then $S' = (U, B)$ such that $A \subseteq B$ will be referred to as a B-*extension* of S. S is also called a *subsystem of S'*.

Example 1. [172] Let us consider an information system $S = (U, A)$ such that $U = \{u_1, u_2, u_3\}$, $A = \{a, b, c\}$, the values of the attributes are defined as in Table 1 and the cost of attributes is equal to $cost(a) = 1.5$, $cost(b) = 0.4$, and $cost(c) = 3$.

U/A	a	b	c
u_1	1	1	0
u_2	0	2	0
u_3	0	0	2

Table 1. An example of an information system

This information system we can treat as a specification of system behaviour concerning distributed traffic signals control presented in Figure 1.

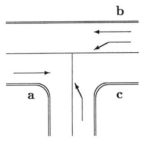

Fig. 1. T-intersection

In this case we assume that attributes a, b, and c denote the traffic signals, objects labeled by u_1, u_2, u_3 denote the possible states of the observed system, whereas entries of the table 0, 1 and 2 denote colours of the traffic lights, red, green and green arrow, respectively.

In a given information system, in general, we are not able to distinguish all single objects (using attributes of the system). Namely, different objects can have the same values on considered attributes. Hence, any set of attributes divides the universe U into some classes which establish a partition [168] of the set of all objects U. It is defined in the following way.

Let $S = (U, A)$ be an information system. With any subset of attributes $B \subseteq A$ we associate a binary relation $ind(B)$, called an *indiscernibility relation*, which is defined by $ind(B) = \{(u, u') \in U \times U$ for every $a \in B, a(u) = a(u')\}$.

Notice that $ind(B)$ is an equivalence relation and $ind(B) = \bigcap_{a \in B} ind(a)$, where by $ind(a)$ we denote $ind(\{a\})$.

If $u\ ind(B)\ u'$, then we say that the objects u and u' are indiscernible with respect to attributes from B. In other words, we cannot distinguish u from u' in terms of attributes in B.

Any information system $S = (U, A)$ determines an *information function*

$$Inf_A : U \to P(A \times V)$$

defined by $Inf_A(u) = \{(a, a(u)) : a \in A\}$, where $V = \bigcup_{a \in A} V_a$ and $P(X)$ denotes the powerset of X. The set $\{Inf_A(u) : u \in U\}$ is denoted by $INF(S)$.

Hence, $u\ ind(A)\ u'$ if and only if $Inf_A(u) = Inf_A(u')$.

The values of an information function will be sometimes represented by vectors of the form $(v_1, ..., v_m), v_i \in V_a$, for $i = 1, ..., m$, where $m = \text{card}(A)$. Such vectors are called *information vectors* (over V and A).

In the paper we also consider a special case of information systems called *decision tables* [168]. In any decision table with a set of attributes a partition of that set into conditions and decisions is given. For our purposes it will be sufficient to consider decision tables with one decision only because one can always transform a decision table with more than one decision into a decision table with exactly one decision by simple coding. One can interpret a decision attribute as a sort of a classification of the universe of objects given by an expert (decision-maker, operator, physician, etc.). We adopt the following definition:

A *decision table* is any information system of the form $S = (U, A \cup \{d\})$, where $d \notin A$ is a distinguished attribute called *decision*. The elements of A are called *conditional attributes (conditions)*. The cardinality of the image

$$d(U) = \{k : d(u) = k \text{ for some } u \in U\}$$

is called the *rank* of d and is denoted by $r(d)$. We assume that the set V_d of values of the decision d is equal to $\{1, ..., r(d)\}$. Let us observe that the decision d determines a partition $\{X_1, ..., X_{r(d)}\}$ of the universe U, where $X_k = \{u \in U : d(u) = k\}$ for $1 \leq k \leq r(d)$. The set X_i is called the i-th *decision class of S*.

Any decision table $S = (U, A \cup \{d\})$ can be represented by a data table with the number of rows equal to the cardinality of the universe U and the number of columns equal to the cardinality of the set $A \cup \{d\}$. On the position corresponding to the row u and column a the value $a(u)$ appears.

Let $S = (U, A)$ be an information system, where $A = \{a_1, ..., a_m\}$. Pairs (a, v) with $a \in A, v \in V$ are called *descriptors*. By $\text{DESC}(A, V)$ we denote the set of all descriptors over A and V. Instead of (a, v) we also write $a = v$ or a_v.

The set of *terms* over A and V is the least set containing descriptors (over A and V) and closed with respect to classical propositional connectives: $\neg$ (negation), $\vee$ (disjunction), and $\wedge$ (conjunction), i.e., if τ, τ' are terms over A and V then $\neg \tau, (\tau \vee \tau'), (\tau \wedge \tau')$ are terms over A and V.

The meaning $\| \tau \|_S$ (or in short $\| \tau \|$) of a term τ in S is defined inductively as follows:

$$\| (a, v) \| = \{u \in U : a(u) = v\} \text{ for } a \in A \text{ and } v \in V_a;$$

$$\| \tau \vee \tau' \| = \| \tau \| \cup \| \tau' \|;$$

$$\| \tau \wedge \tau' \| = \| \tau \| \cap \| \tau' \|;$$

$$\| \neg \tau \| = U - \| \tau \|.$$

Two terms τ and τ' are equivalent, $\tau \Leftrightarrow \tau'$, if and only if $\| \tau \| = \| \tau' \|$. In particular we have: $\neg(a = v) \Leftrightarrow \bigvee \{a = v' : v' \neq v \text{ and } v' \in V_a\}$.

2.1.2 Set Approximations

Some subsets (classes) of objects in an information system cannot be distinguished in terms of the available attributes. They can only be roughly (approximately) defined. The idea of *rough sets* consists of the approximation of a set by a pair of sets, called the *lower* and the *upper* approximation of this set.

Let $S = (U, A)$ be an information system, $B \subseteq A$ be a set of attributes, and let $X \subseteq U$ denotes a set of objects. Then sets $\{u \in U : [u]_B \subseteq X\}$ and $\{u \in U : [u]_B \cap X \neq \emptyset\}$ are called the *B-lower* and the *B-upper approximation* of X in S, and are denoted by $\underline{B}X$ and $\overline{B}X$, respectively.

The set $BN_B(X) = \overline{B}X - \underline{B}X$, will be called the *B-boundary* of X. When $B = A$ we write also $BN_S(X)$ instead of $BN_A(X)$.

Sets which are unions of some classes of the indiscernibility relation $ind(B)$ are called *definable* by B. The set X is *B-definable* if and only if $\underline{B}X = \overline{B}X$.

The set $\underline{B}X$ is the set of all elements of U which can be with certainty classified as elements of X, given the knowledge represented by attributes from B, $\overline{B}X$ is the set of elements of U which can be possibly classified as elements of X, employing the knowledge represented by attributes from B, and $BN_B(X)$ is the set of elements which can be classified neither to X nor to $-X$ given knowledge B.

If $X_1, ..., X_{r(d)}$ are decision classes of S then the set

$$\underline{B}X_1 \cup ... \cup \underline{B}X_{r(d)}$$

is called the *B-positive region* of S and denoted by $POS_B(d)$.

If $C \subseteq A$ then the set $POS_B(C)$ is defined as $POS_B(d)$ where $d(u) = \{a(u) : a \in C\}$ for $u \in U$ is an attribute representing the set C of attributes and called the *B-positive region of C* in S.

The B-positive region of C in S contains all objects in U which can be classified perfectly without error into distinct classes defined by $ind(C)$, based only on information in relation $ind(B)$.

2.1.3 Dependencies in Information Systems

Some attributes in an information system may depend on one another. Change of a given attribute may cause changes of other attributes in some nonlinear way. Rough sets determine a degree of attributes' dependency and their significance. In the indiscernibility relation, dependency of attributes is one of the important features of information systems.

In this subsection we recall some notions related to dependencies in information systems and we introduce a new notion connected with an equivalence of attribute sets in information systems. The last notion plays an important role in reduction of rule sets generated from a given information system.

Let $S = (U, A)$ be an information system and let $B, C \subseteq A$. We say that the *set C depends* on B in S in *degree k* ($0 \leq k \leq 1$), symbolically $B \xrightarrow[S,k]{} C$, if $k = \frac{card(POS_B(C))}{card(U)}$, where $POS_B(C)$ is the *B-positive region* of C in S.

If $k = 1$ we write $B \xrightarrow[S]{} C$ instead of $B \xrightarrow[S,k]{} C$ and we say that C *totally dependent* on B in S. In this case $B \xrightarrow[S]{} C$ means that $ind(B) \subseteq ind(C)$. If the right hand side of a dependency consists of one attribute only, we say the dependency is *elementary*.

It is easy to see that a simple property given below is true.

Proposition 1. *Let $S = (U, A)$ be an information system and let $B, C, D \subseteq A$. If $B \xrightarrow[S]{} C$ and $B \xrightarrow[S]{} D$ then $B \xrightarrow[S]{} C \cup D$.*

If the set C is totally dependent on B and vice versa, then we say that C and B are *equivalent* in S. In this case $B \xleftrightarrow[S]{} C$ means that $ind(B) = ind(C)$.

A pair (B, C) of subsets of A is called *a strong component* of S if and only if the following conditions are satisfied:

1) $B \cap C = \emptyset$,
2) B and C are equivalent in S,
3) B and C are minimal (with respect to $\subseteq$) in A.

By SCOMP(S) we denote the set of all strong components of S.

The condition 3 of the definition of the strong component of S we can also formulate in the following way:
B and C are minimal (with respect to $\subseteq$) in A if and only if:

1) if $(B, C) \in$ SCOMP(S) then $(B - \{a\}, C) \notin$ SCOMP(S) for any $a \in B$,
2) if $(B, C) \in$ SCOMP(S) then $(B, C - \{a\}) \notin$ SCOMP(S) for any $a \in C$.

2.1.4 Rules in Information Systems

One of the important applications of rough sets is the generation of decision rules for a given information system for classification of known objects, or prediction of classes for new objects unseen during design. Using an original or reduced decision table, one can find the rules classifying objects through determining the decision attributes value based on condition attributes values.

Rules express some of the relationships between values of the attributes described in the information systems. This subsection contains the definition of rules as well as other related concepts.

Let $S = (U, A)$ be an information system and let $B \subset A$. For every $a \notin B$ we define a function $d_a^B : U \to P(V_a)$ such that

$$d_a^B(u) = \{v \in V_a : \text{ there exists } u' \in U \; u' \, ind(B) \, u \text{ and } a(u') = v\},$$

where $P(V_a)$ denotes the powerset of V_a. Hence, $d_a^B(u)$ is the set of all the values of the attribute a on objects indiscernible with u by attributes from B. If the set

$d_a^B(u)$ has only one element, this means that the value $a(u)$ is uniquely defined by the values of attributes from B on u.

A *rule* over A and V is any expression of the following form:

$$(1) \quad a_{i_1} = v_{i_1} \vee ... \vee a_{i_r} = v_{i_r} \Rightarrow a_p = v_p$$

where $a_p, a_{i_j} \in A, v_p, v_{i_j} \in V_{a_{i_j}}$ for $j = 1, ..., r$.

A rule of the form (1) is called *trivial* if $a_p = v_p$ appears also on the left hand side of the rule. The rule (1) is *true in* S (or in short: is *true*) if

$$\emptyset \neq \| a_{i_1} = v_{i_1} \wedge ... \wedge a_{i_r} = v_{i_r} \| \subseteq \| a_p = v_p \|$$

The fact that the rule (1) is true in S is denoted in the following way:

$$(2) \quad a_{i_1} = v_{i_1} \wedge ... \wedge a_{i_r} = v_{i_r} \overrightarrow{S} a_p = v_p.$$

In the case (2) we also shall say that the values (local states) $v_{i_1}, ..., v_{i_r}, v_p$ of processes $a_{i_1}, ..., a_{i_r}, a_p$ can *coexist* in S. By $D(S)$ we denote the set of all rules true in S.

Let $S = (U, A \cup \{d\})$ be a decision table. A rule $\tau \Rightarrow d = v$, where τ is a term over A and V, d is a decision of S, $v \in V_d$, is called a *decision rule*.

Let $R \subseteq D(S)$. An information vector $\mathbf{v} = (\mathbf{v}_1, ..., \mathbf{v}_m)$ is *consistent* with R if and only if for any rule $a_{i_1} = v_{i_1} \wedge ... \wedge a_{i_r} = v_{i_r} \overrightarrow{S} a_p = v_p$ in R if $\mathbf{v}_{i_j} = v_{i_j}$ for $j = 1, ..., r$ then $v_p = \mathbf{v_p}$. The set of all information vectors consistent with R is denoted by $\mathrm{CON}(R)$.

Let $S = (U, A)$ be an information system, $D(S)$ - the set of all rules true in S, c - the cost of attributes in S, and let x be a rule of the form: $a_{i_1} = v_{i_1} \wedge ... \wedge a_{i_r} = v_{i_r} \overrightarrow{S} a_p = v_p$. The *cost of a rule* x is called the sum of costs of attributes on the left hand side of the rule. The cost of a rule x is denoted by $\mathrm{COST}(x)$. Similarly, the *cost of a subset of rules* is called the sum of the costs of rules in the subset. It is assumed that the cost of the empty set is equal to 0.

Remark. 2.1 By the cost of a subset of rules one can also understand the sum of the costs of rules in the subset assuming that each descriptor appearing multi-times in rules from the subset is taken into consideration only once.

Let $S' = (U', A')$ be a U'-extension of $S = (U, A)$. We say that S' is a *consistent extension* of S if and only if $D(S) \subseteq D(S')$. S' is a *maximal* consistent extension of S if and only if S' is a consistent extension of S and for any consistent extension S'' of S' we have $D(S'') = D(S')$.

In the paper we apply the Boolean reasoning approach to the rule generation [239].

The Boolean reasoning approach [34], due to G. Boole, is a general problem solving method consisting of the following steps: (i) construction of a Boolean function corresponding to a given problem; (ii) computation of prime implicants of the Boolean function; (iii) interpretation of prime implicants leading to the solution of the problem.

It turns out that this method can be also applied to the generation of rules with certainty coefficients [240]. Using this approach one can also generate the rule sets being outputs from some algorithms known in machine learning, like AQ-algorithms ([126], [250]).

2.1.5 Discernibility Matrix

Frequently discernibility of objects is more interesting than the specific values of attributes. In these situations an information system may be represented as a discernibility matrix. Skowron and Rauszer [249] have introduced two notions, namely the *discernibility matrix* and the *discernibility function*, which will help to compute minimal forms of rules with respect to the number of attributes on the left hand side of the rules. With these two notions, we can store the differences between the attributes of each pair of objects in a matrix called *discernibility matrix*. The discernibility matrix contains fewer data than those of an information system but holds all needed information used to check whether a set of attributes is a reduct (see: subsection 2.1.6).

Let $S = (U, A)$ be an information system, and let us assume that $U = \{u_1, ..., u_n\}$, and $A = \{a_1, ..., a_m\}$. By $M(S)$ we denote an $n \times n$ matrix (c_{ij}), called the *discernibility matrix* of S, such that $c_{ij} = \{a \in A : a(u_i) \neq a(u_j)\}$ for $i, j = 1, ..., n$.

Intuitively an entry c_{ij} consists of all the attributes discerning objects u_i and u_j. Since $M(S)$ is symmetric and $c_{ii} = \emptyset$ for $i = 1, ..., n$, $M(S)$ can be represented using only elements in the lower triangular part of $M(S)$, i.e., for $1 \leq j < i \leq n$.

With every discernibility matrix $M(S)$ one can uniquely associate a *discernibility function* $f_{M(S)}$, defined as follows.

A *discernibility function* $f_{M(S)}$ for an information system S is a Boolean function of m propositional variables $a_1^*, ..., a_m^*$ (where $a_i \in A$ for $i = 1, ..., m$) defined as the conjunction of all expressions $\bigvee c_{ij}^*$, where $\bigvee c_{ij}^*$ is the disjunction of all elements of $c_{ij}^* = \{a^* : a \in c_{ij}\}$, where $1 \leq j < i \leq n$ and $c_{ij} \neq \emptyset$. In the sequel we write a instead of a^*.

2.1.6 Reduction of Attributes

In many applications classification of objects is one of the most frequently encountered tasks. Classification can be considered as a process of determining a unique class for a given object. A given set of objects, characterized by the set of condition and decision attributes, can be classified into a disjoint family of classes based on the values of decision attributes. Each class can be determined in terms of features of corresponding condition attributes belonging to a class. If a given set of objects with a given set of attributes is classifiable, classification may be possibly achieved by some subset of attributes. Frequently only a few important attributes are sufficient to classify objects. This is consistent with human perception and classification ability based on intelligent attention, and selection of most important features of objects.

Some attributes in an information system may be redundant and can be eliminated without losing essential classificatory information. The process of finding a smaller set of attributes (than the original one) with the same or close classificatory power as the original set is called *attribute reduction*. As a result the original larger information system may be reduced to a smaller system containing attributes.

Rough set allows us to determine for a given information system the most important attributes from a classificatory point of view. A *reduct* is the essential part of an information system (related to a subset of attributes) which can discern all objects discernible by the original information system. A *core* is a common part of all reducts. Core and reduct are fundamental rough sets concepts which can be used for knowledge reduction.

Let $S = (U, A)$ be an information system. Any minimal subset $B \subseteq A$ such that $ind(B) = ind(A)$ is called a *reduct* in the information system S [168]. The set of all reducts in S is denoted by RED(S). The intersection of all the reducts in S is called the *core* of S.

Proposition 2 gives an important property which enables us to compute all reducts of S.

Proposition 2. *[249] Let $S = (U, A)$ be an information system, and let $f_{M(S)}$ be a discernibility function for S. Then the set of all prime implicants [305] of the function $f_{M(S)}$ determines the set RED(S) of all reducts of S, i.e., $a_{i_1} \wedge ... \wedge a_{i_k}$ is a prime implicant of $f_{M(S)}$ if and only if $\{a_{i_1}, ..., a_{i_k}\} \in RED(S)$.*

2.1.7 Relationships between Dependencies in Information System and Reducts

In the following propositions the important relationships between the reducts and the dependencies are given.

Proposition 3. *[168] Let $S = (U, A)$ be an information system and $B \in RED(S)$. If $A - B \neq \emptyset$ then $B \overrightarrow{S} A - B$.*

Proposition 4. *[168] If $B \overrightarrow{S} C$ then $B \overrightarrow{S} C'$, for every $\emptyset \neq C' \subseteq C$. In particular, $B \overrightarrow{S} C$ implies $B \overrightarrow{S} \{a\}$, for every $a \in C$.*

Proposition 5. *[168] Let $B \in RED(S)$. Then attributes in the reduct B are pairwise independent, i.e., neither $\{a\} \overrightarrow{S} \{a'\}$ nor $\{a'\} \overrightarrow{S} \{a\}$ holds, for any $a, a' \in B, a \neq a'$.*

Below we present a procedure for computing reducts [249].

PROCEDURE for computing RED(S):

Input: An information system S.
Output: The set of all reducts in S.

Step 1. Compute the discernibility matrix for the system S.

Step 2. Compute the discernibility function $f_{M(S)}$ associated with the discernibility matrix $M(S)$.

Step 3. Compute the minimal disjunctive normal form of the discernibility function $f_{M(S)}$ (The normal form of the function yields all the reducts).

The finding of minimal (with respect to cardinality) reducts for an information system is a combinatorial NP-hard computational problem [249]. Finding the reducts can be considered similarly as finding the minimal disjunctive normal form for a logical expression given in the conjunctive normal form [168]. In general the number of reducts of a given information system can be exponential with respect to the number of attributes (i.e., any information system S has at most m over $[m/2]$ reducts, where $m=\text{card}(A)$). Nevertheless, existing procedures for reduct computation are efficient in many applications and for more complex cases one can apply some efficient heuristics (see e.g. [15], [151], [242], [248], [152]).

Example 2. Applying the above procedure for the information system S from Example 1, we obtain the following discernibility matrix $M(S)$ presented in Table 2 and discernibility function $f_{M(S)}$ presented below:

$$f_{M(S)}(a,b,c) = (a \vee b) \wedge (a \vee b \vee c) \wedge (b \vee c).$$

U	u_1	u_2	u_3
u_1			
u_2	a,b		
u_3	a,b,c	b,c	

Table 2. The discernibility matrix $M(S)$ for the information system S from Example 1

We consider non-empty entries of the table (see: Table 2), i.e., $a,b;\ b,c$ and a,b,c; next a,b,c are treated as Boolean variables and the disjunctions $a \vee b;\ b \vee c$ and $a \vee b \vee c$ are constructed from these entries; finally, we take the conjunction of all the computed disjunctions to obtain the discernibility function corresponding to $M(S)$.

After reduction (using the absorption laws) we get the following minimal disjunctive normal form of the discernibility function $f_{M(S)}(a,b,c) = (a \wedge c) \vee b$.

There are two reducts: $R_1 = \{a,c\}$ and $R_2 = \{b\}$ of the system. Thus $\text{RED}(S) = \{R_1, R_2\}$.

Example 3 illustrates how to find all dependencies among attributes using Propositions 3 and 4.

Example 3. Let us consider again the information system S from Example 1. By Proposition 3 we have for the system S the dependencies: $\{a,c\} \overrightarrow{S} \{b\}$ and $\{b\} \overrightarrow{S} \{a,c\}$. Next, by Proposition 4 we get the following elementary dependencies: $\{a,c\} \overrightarrow{S} \{b\}, \{b\} \overrightarrow{S} \{a\}, \{b\} \overrightarrow{S} \{c\}$.

Proposition 6. Let $S = (U, A)$ be an information system, and $B, C \subseteq A$. The sets B and C are equivalent in S if and only if the following conditions are satisfied:

1. For each $b \in B$ C is $\{b\}$-reduct of S [168].
2. For each $c \in C$ B is $\{c\}$-reduct of S.

From Proposition 2.6 a method follows for computing the set of all strong components of a given information system $S = (U, A)$. It can be obtained by the following procedure:

PROCEDURE for computing SCOMP(S):

Input: An information system $S = (U, A)$ and the set RED(S) of all reducts of S.

Output: Strong components of S, i.e., the set SCOMP(S).

Step 1. For each $R \in RED(S)$ compute minimal subsets $B \subseteq R$ such that $B \underset{S}{\rightarrow} \{a\}$, for any $a \in A - R$, i.e., for each subsystem $S' = (U, R \cup \{a\})$ of S with $a \in A - R$ compute the discernibility function $f_{M(S')}$. In this step we compute the so called $\{a\}$ - reducts of R, for $a \in A - R$.

Step 2. For each $R \in RED(S)$ compute minimal subsets $C \subseteq A - R$ such that $C \underset{S}{\rightarrow} \{b\}$, for any $b \in R$, i.e., for each subsystem $S'' = (U, (A-R) \cup \{b\})$ of S with $b \in R$ compute the discernibility function $f_{M(S'')}$. In this step we compute $\{b\}$ - reducts of $A - R$, for $b \in R$.

Step 3. Choose all pairs (B,C) such that $B \cap C = \emptyset$ and $B \underset{S}{\rightarrow} \{a\}$ was computed in Step 1 for any $a \in C$ and $C \underset{S}{\rightarrow} \{b\}$ was computed in Step 2 for any $b \in B$.

2.1.8 Minimal Rules in Information Systems

Now we present a method for generating the minimal form of rules (i.e., rules with a minimal number of descriptors on the left hand side) in information systems and decision tables. The method is based on the idea of Boolean reasoning [34] applied to discernibility matrices defined in [249] and modified here for our purposes. In fact, this subsection presents the first step in the construction of a concurrent model of knowledge embedded either in a given information system or a given decision table which in the next sections are considered.

Let $S = (U, A \cup \{a^*\})$ be an information system and $a^* \notin A$. We are looking for all minimal rules (i.e., with minimal left hand sides) in S of the form: $a_{i_1} = v_{i_1} \wedge ... \wedge a_{i_r} = v_{i_r} \underset{S}{\Rightarrow} a = v$, where $a \in A \cup \{a^*\}, v \in V_a, a_{i_j} \in A$ and $v_{i_j} \in V_{a_{i_j}}$ for $j = 1, ..., r$.

The above rules express functional dependencies between the values of the attributes of S. These rules are computed from systems of the form $S' = (U, B \cup \{a\})$ where $B \subset A$ and $a \in A - B$ or $a = a^*$.

First, for every $v \in V_a, u_l \in U$ such that $d_a^B(u_l) = \{v\}$ a modification $M(S'; a, v, u_l)$ of the discernibility matrix is computed from $M(S')$.

By $M(S'; a, v, u_l) = (c_{ij}^*)$ (or M, in short) we denote the matrix obtained from $M(S')$ in the following way:

> **if** $i = l$ **then** $c_{ij}^* = \emptyset$;
> **if** $c_{lj} \neq \emptyset$ **and** $d_a^B(u_j) \neq \{v\}$ **then** $c_{lj}^* = c_{lj} \cap B$
> **else** $c_{lj}^* = \emptyset$.

Next, we compute the discernibility function f_M and the prime implicants [305] of f_M taking into account the non-empty entries of the matrix M (when all entries c_{ij}^* are empty we assume f_M to be always true).

Finally, every prime implicant $a_{i_1} \wedge ... \wedge a_{i_r}$ of f_M determines a rule $a_{i_1} = v_{i_1} \wedge ... \wedge a_{i_r} = v_{i_r} \Rightarrow_S a = v$, where $a_{i_j}(u_l) = v_{i_j}$ for $j = 1, ..., r$, $a(u_l) = v$.

Let $S = (U, A)$ be an information system. In the following we shall apply the above method for every $R \in \text{RED}(S)$. First we construct all rules corresponding to non-trivial dependencies between the values of attributes from R and $A - R$ and next all rules corresponding to non-trivial dependencies between the values of attributes within a reduct R. These two steps are realized as follows:

(i) For every reduct $R \in \text{RED}(S)$, $R \subset A$ and for every $a \in A - R$ we consider the system $S' = (U, R \cup \{a\})$. For every $v \in V_a$, $u_l \in U$ such that $d_a^R(u_l) = \{v\}$ we construct the discernibility matrix $M(S'; a, v, u_l)$, next the discernibility function f_M and the set of all rules corresponding to prime implicants of f_M.

(ii) For every reduct $R \in \text{RED}(S)$ with $\text{card}(R) > 1$ and for every $a \in R$ we consider the system $S'' = (U, B \cup \{a\})$, where $B = R - \{a\}$. For every $v \in V_a$, $u_l \in U$ such that $d_a^B(u_l) = \{v\}$ we construct the discernibility matrix $M(S''; a, v, u_l)$, then the discernibility function f_M and the set of all rules corresponding to prime implicants of f_M.

The set of all rules constructed in this way for a given $R \in \text{RED}(S)$ is denoted by $\text{OPT}(S, R)$. We put $\text{OPT}(S) = \bigcup\{\text{OPT}(S, R) : R \in \text{RED}(S)\}$.

Remark. 2.2 Let $S = (U, A \cup \{d\})$ be a decision table and $d \notin A$. In this case we assume that the attribute a^* appearing in the method for generating the minimal form of rules is replaced by the decision d. The minimal rules computed in the above way now express functional dependencies between the values of the conditional attributes of S as well as functional dependencies between the values of the conditional attributes of S and the values of the decision attribute of S. These rules are computed from systems of the form $S' = (U, B \cup \{a\})$ where $B \subset A$ and $a \in A - B$ or $a = d$.

The set of all rules constructed in the above way for any $a \in A \cup \{d\}$ is denoted by $\text{OPT}(S, a)$.

We put $\text{OPT}(S) = \bigcup\{\text{OPT}(S, a) : a \in A \cup \{d\}\}$.

For any $a \in A \cup \{d\}$ and $u \in U$ we take $B = A$, if $a = d$; $B = (A \cup \{d\}) - \{a\}$ otherwise and we take $v = a(u)$. We compute all minimal rules true in $S' = (U, B \cup \{a\})$ of the form $\tau \Rightarrow a = v$, where τ is a term in conjunctive form over B and $V_B = \bigcup_{a \in B} V_a$, with a minimal number of descriptors in any conjunct. To obtain all possible functional dependencies between the attribute values it is necessary to repeat this process for all possible values of a and for all remaining attributes from $A \cup \{d\}$.

Let us observe that if $a_{i_1} = v_{i_1} \wedge ... \wedge a_{i_r} = v_{i_r} \overrightarrow{\underset{S}{}} a_p = v_p$ is a rule from OPT(S), then $U \cap \| a_{i_1} = v_{i_1} \wedge ... \wedge a_{i_r} = v_{i_r} \|_S \neq \emptyset$.

Proposition 7. *[169] Let $S=(U, A)$ be an information system, $R \in RED(S)$, and $R \subset A$. Let $f_{M(S')}$ be a relative discernibility function for the system $S' = (U, R \cup \{a^*\})$ where $a^* \in A - R$. Then all prime implicants of the function $f_{M(S')}$ correspond to all $\{a^*\}$ - reducts of S'.*

Now we are ready to present a very simple procedure for computing an extension S' of a given information system S. Let OPT(S) be the set of all rules constructed as described above.

PROCEDURE for computing an extension S' of S:

Input: An information system $S = (U, A)$ and the set OPT(S) of rules.

Output: An extension S' of S.

Step 1. Compute all admissible global states of S, i.e., the cartesian product of the value sets for all attributes a from A.

Step 2. Verify using the set OPT(S) of rules which admissible global states of S are consistent with rules true in S.

The next example illustrates how to find all non-trivial dependencies between the values of attributes in a given information system. At the end of example we give information about an extension of the information system.

Example 4. Let us consider the information system S from Example 1 and the discernibility matrix for S presented in Table 2. We compute the set of rules corresponding to non-trivial dependencies between the values of attributes from the reduct R_1 of S with b (i.e., those outside of this reduct) as well as the set of rules corresponding to non-trivial dependencies between the values of attributes within the reduct of that system. In both cases we apply the method presented above.

Let us start by computing the rules corresponding to non-trivial dependencies between the values of attributes from the reduct $R_1 = \{a, c\}$ of S with b.

We have the following subsystem $S_1 = (U, B \cup \{b\})$, where $B = R_1$, from which we compute the rules mentioned above:

U/B	a	c	b	d_b^B
u_1	1	0	1	$\{1\}$
u_2	0	0	2	$\{2\}$
u_3	0	2	0	$\{0\}$

Table 3. The subsystem $S_1 = (U, B \cup \{b\}$ with the function d_b^B, where $B = \{a, c\}$

In the table the values of the function d_b^B are also given. The discernibility matrix $M(S_1; b, v, u_l)$ where $v \in V_b$, $u_l \in U$, $l = 1, 2, 3$, obtained from $M(S_1)$ in the above way is presented in Table 4.

U	u_1	u_2	u_3
u_1		a	a,c
u_2	a		c
u_3	a,c	c	

Table 4. The discernibility matrix $M(S_1;b,v,u_l)$ for the matrix $M(S_1)$

The discernibility functions corresponding to the values of the function d_b^B are the following:

Case 1. For $d_b^B(u_1) = \{1\}$: $a \wedge (a \vee c) = a$.

We consider non-empty entries of the column labeled by u_1 (see: Table 4), i.e., a and a,c; next a,c are treated as Boolean variables and the disjunctions a and $a \vee c$ are constructed from these entries; finally, we take the conjunction of all the computed disjunctions to obtain the discernibility function corresponding to $M(S_1;b,v,u_l)$.

Case 2. For $d_b^B(u_2) = \{2\}$: $a \wedge c$.
Case 3. For $d_b^B(u_3) = \{0\}$: $(a \vee c) \wedge c = c$.

Hence we obtain the following rules: $a_1 \overrightarrow{S} b_1$, $a_0 \wedge c_0 \overrightarrow{S} b_2$, $c_2 \overrightarrow{S} b_0$.

Now we compute the rules corresponding to all non-trivial dependencies between the values of attributes within the reduct R_1.

We have the following two subsystems $(U, C \cup \{c\})$, $(U, D \cup \{a\})$ of S, where $C = \{a\}$, and $D = \{c\}$, from which we compute the rules mentioned above:

U/C	a	c	d_c^C
u_1	1	0	$\{0\}$
u_2	0	0	$\{0,2\}$
u_3	0	2	$\{0,2\}$

Table 5. The subsystem $(U, C \cup \{c\})$ with the function d_c^C, where $C = \{a\}$

U/D	c	a	d_a^D
u_1	0	1	$\{0,1\}$
u_2	0	0	$\{0,1\}$
u_3	2	0	$\{0\}$

Table 6. The subsystem $(U, D \cup \{a\})$ with the function d_a^D, where $D = \{c\}$

In the tables the values of the functions d_c^C, and d_a^D are also given.

Discernibility functions corresponding to the values of these functions are the following:

Table 5. For $d_c^C(u_1) = \{0\}$: a.

Table 6. For $d_a^D(u_3) = \{0\}$: c.

Hence we obtain the following rules:

From Table 5: $a_1 \overrightarrow{\underset{S}{\Rightarrow}} c_0$.

From Table 6: $c_2 \overrightarrow{\underset{S}{\Rightarrow}} a_0$.

Finally, the set of rules corresponding to all non-trivial dependencies between the values of attributes within the reduct R_1 has the form: $a_1 \overrightarrow{\underset{S}{\Rightarrow}} c_0$, $c_2 \overrightarrow{\underset{S}{\Rightarrow}} a_0$.

Eventually, we obtain the set OPT(S, R_1) of rules corresponding to all non-trivial dependencies for the reduct R_1 in the considered information system S:

$$a_1 \overrightarrow{\underset{S}{\Rightarrow}} b_1,\ a_0 \wedge c_0 \overrightarrow{\underset{S}{\Rightarrow}} b_2,\ c_2 \overrightarrow{\underset{S}{\Rightarrow}} b_0,\ a_1 \overrightarrow{\underset{S}{\Rightarrow}} c_0,\ c_2 \overrightarrow{\underset{S}{\Rightarrow}} a_0.$$

In a similar way one can compute the set OPT(S, R_2) of rules corresponding to all non-trivial dependencies for the reduct R_2 in the system S. This set consists of one kind of rules, i.e., the rules corresponding to all non-trivial dependencies between the values of attributes from R_2 with a, c of the form:

$$b_1 \overrightarrow{\underset{S}{\Rightarrow}} a_1,\ b_0 \vee b_2 \overrightarrow{\underset{S}{\Rightarrow}} a_0,\ b_1 \vee b_2 \overrightarrow{\underset{S}{\Rightarrow}} c_0,\ b_0 \overrightarrow{\underset{S}{\Rightarrow}} c_2,$$

whereas the second set of rules corresponding to all non-trivial dependencies between the values of attributes within the reduct R_2 is empty, because this reduct has only one element.

The set OPT(S) of all rules constructed in this way for the information system S of Example 1 is the union of sets OPT(S, R_1) and OPT(S, R_2).

There are five rules for the reduct R_1. The costs of these rules are equal to 3, 1.5, 4.5, 1.5, 3, respectively. Consequently, the cost of the set OPT(S,R_1) of rules is equal to 13.5. And, the cost of the set OPT(S,R_2) of rules equals 2.4.

Remark. 2.3 Taking into consideration the fact that on the left hand side in rules from the set R_1 the descriptors a_1 and a_2 appear twice, the cost of the set OPT(S,R_1) of rules is equal to 9, for the same reason the cost of the set OPT(S,R_2) of rules equals 1.2.

It is easy to verify that in this case the extension S' of the system S computed by using our procedure presented above is the same as the original one.

Remark. 2.4 The rules from the set OPT(S) constructed for the information system S of Example 2.1 explain behaviour of the system from Figure 1.

Remark. 2.5 Our approach to rule generation is based on procedures for the computation of reduct sets. It is known that in general the reduct set can be of exponential complexity with respect to the number of attributes. Nevertheless, there are several methodologies allowing to deal with this problem in practical applications. Among them are the feature extraction techniques or clustering methods known in pattern recognition [147] and machine learning [126], allowing to reduce the number of attributes or objects so that the rules can be efficiently generated from them. Another approach is suggested in [15]. It leads to the computation of only so called the most stable reducts from the reduct set in a sampling process of a given decision table. The rules are produced from these stable reducts only. This last technique can be treated as relevant feature extraction from a given set of features. The result of the above techniques applied to a given information system is estimated as successful if rules can be efficiently generated from the resulting compressed information system by the Boolean reasoning method and if the quality of the classification of unseen objects by these rules is sufficiently high.

In the paper we assume that the information systems which create inputs for our procedures satisfy those conditions.

2.1.9 Generalized Decisions and Relative Reducts in Decision Tables

If $S = (U, A \cup \{d\})$ is a decision table then we define a function

$$\delta_A : U \to P(\{1, ..., r(d)\}),$$

called the *generalized decision* in S, by

$$\delta_A(u) = \{i : \text{there exists } u' \in U \; u' \text{ ind}(A) u \text{ and } d(u) = i\}.$$

A decision table S is called *consistent (deterministic)* if card $(\delta_A(u)) = 1$ for any $u \in U$, otherwise S is *inconsistent (non-deterministic)*.

A subset B of the set A of attributes of a decision table $S = (U, A \cup \{d\})$ is a *relative reduct* of S if and only if B is a minimal set with the following property: $\delta_B = \delta_A$. The set of all relative reducts in S is denoted by RED(S, d).

2.1.10 Rules in Inconsistent Decision Tables

Let us now consider inconsistent decision tables. One can transform an arbitrary inconsistent decision table $S = (U, A \cup \{d\})$ into a consistent decision table $S_\delta = (U, A \cup \{\delta_A\})$ where $\delta_A : U \to P(\{1, ..., r(d)\})$ is the generalized decision in S. It is easy to see that S_δ is a consistent decision table. Hence one can apply to S_δ the methods for rule synthesis (see: subsection 2.1.8).

2.1.11 Rough Set Methods for Data Reduction and Extraction of Rules

Reducts offer the same classificational possibilities as the whole system, but they require a smaller set of attributes. However, as this approach is not always sufficient, there exist additional tools, like e.g. dynamic reducts and rules [15].

The underlying idea of dynamic reducts stems from the observation that reducts generated from an information system are not stable in the sense that they are sensitive to changes introduced to the information system by removing a randomly chosen set of objects. The notion of a dynamic reduct encompasses stable reducts, i.e., reducts that are most frequent in random samples created by subtables of the given decision table [15]. We show here how dynamic reducts can be computed from reducts and how dynamic rules can be generated from dynamic reducts. Dynamic reducts have shown their utility in various experiments with data sets of various kinds e.g. marked data [90], monk's problems [129], hand-written digit recognition [14] or medical data [15].

The quality of unseen object classification by rules generated from dynamic reducts increases especially when data are very noisy, like market data [90]. In all the tested cases we have obtained a substantial reduction of the rule set without decreasing the quality of the classification of unseen objects. The results of experiments with dynamic reducts show that attributes from these reducts can be treated as relevant features [15].

To capture the fact that some reducts are chaotic, we consider random samples forming subtables of a given decision table $S = (U, A \cup \{d\})$. We shall call a *subtable* of S any information system $S' = (U', A \cup \{d\})$ such that $U' \subseteq U$.

Let F be a family of subtables of S and let ε be a real number from the unit interval $[0,1]$.

The set $\text{DR}_\varepsilon(S, F)$ of (F, ε)-*dynamic reducts* is defined by

$$\text{DR}_\varepsilon(S, F) = \left\{ C \in \text{RED}(S, d) : \frac{|\{S' \in F : C \in \text{RED}(S', d)\}|}{|F|} \geq 1 - \varepsilon \right\}$$

The number $\frac{|\{S' \in F : C \in \text{RED}(S', d)\}|}{|F|}$ is called the *stability coefficient* of C relative to F, for $C \in \text{RED}(S', d)$.

We present one of the existing techniques for computing dynamic reducts [15]. Experiments with different data sets have shown that this type of dynamic reducts allows to generate rules with better quality of classification of new objects than the other methods. The method consists in the following.

Step 1: A random set of subtables is taken from the given table; for example,

10 samples of the size of 90% of the decision table,
10 samples of the size of 80% of the decision table,
10 samples of the size of 70% of the decision table,
10 samples of the size of 60% of the decision table,
10 samples of the size of 50% of the decision table.

Step 2: Reducts for all of these tables are calculated; for example, reducts for any of the 50 randomly chosen tables.

Step 3: Reducts with stability coefficients higher than a fixed threshold are extracted.

The reducts selected in step 3 are regarded as true dynamic reducts.

One can compute dynamic reducts using approximations of reducts [15] instead of reducts.

If a set of dynamic reducts (with stability coefficients greater than a given threshold) has been computed, it is necessary to decide how to compute the set of rules. There exist several methods for this purpose. The first is based on the (F, ε)-dynamic core of S, i.e., on the set $\bigcup \text{DR}_\varepsilon(S, F)$. We apply the methods based on Boolean reasoning presented in ([174], [240]) to generate rules (with a minimal number of descriptors) from conditional attributes belonging to the dynamic core. The second method is based on the rule set construction for any chosen dynamic reduct. The final rule set is equal to the union of all these sets.

In our experiments we have received somewhat better results of tests by applying the second method. If an unseen object has to be classified, it is first matched against all the rules from the constructed rule set. Next the final decision is predicted by applying some strategy predicting the final decision from all the "votes" of the rules. The simplest strategy we have tested was the majority voting, i.e., the final decision is the one supported by the majority of the rules. The proper decision can be also predicted by an application of fuzzy methods.

The idea of dynamic reducts can be adjusted to a new method of dynamic rule computation. From a given data table a random set of subtables is chosen. For example:

10 samples of the size of 90% of the decision table,
10 samples of the size of 80% of the decision table,
10 samples of the size of 70% of the decision table,
10 samples of the size of 60% of the decision table,
10 samples of the size of 50% of the decision table.

Thus we receive 50 new decision tables. Then the rule sets for all these tables are calculated. In the next step a rule memory is constructed where all rule sets are stored. Intuitively, a dynamic rule appears in all (or almost all) experimental subtables. The rules can be also computed from the so-called local reducts used to generate rules with a minimal number of descriptors [174].

Several experiments performed with different data tables (see: [15], [14]) show that our strategies for decision algorithm synthesis increase the quality of unseen object classification or/and allow to reduce the number of rules without decreasing classification quality.

2.2 Petri Nets

In this Chapter classical Petri nets ([186], [206], [142]) are used as a tool for representing and analyzing the knowledge represented by an information system (see: sections 3, 4 and 5) as well as a tool for computing a highly parallel program from a given decision table (see: sections 6 and 7). After modeling an information system by a Petri net, many desirable properties of the system can be revealed by analyzing properties of the constructed Petri net (see: section 8, and also e.g. [277]).

There are several analysis techniques for Petri nets, including reachability trees and net invariants ([186], [206], [142]). Some reduction rules have also been investigated to reduce the complexity of nets prior to the analysis phase. However, these analysis techniques were initially developed for classical Petri nets [186] and do not apply to high-level Petri nets (especially to coloured Petri nets) [100], which are characterized by a great diversity of linear functions that are associated to their arcs. Therefore, unlike analysis algorithms for classical Petri nets that use integer matrices, analysis algorithms for high-level Petri nets need to manipulate matrices composed by linear functions. This fact introduces high complexities in the development and execution of these algorithms. As a consequence, a necessary step must be taken to convert the high-level net of the complex system model into a low-level net expressed in terms of classical Petri nets (cf. [82], p. 256). Based on the above observations and the fact that we have got computer tools for the design and analysis of Petri net models ([277], [291]). In this Chapter we use classical Petri nets.

At first, we recall some basic concepts and notation from Petri net theory.

2.2.1 Structure and Dynamics

A *Petri net* contains two types of nodes, *circles P* (places) and *bars T* (transitions). The relationship between the nodes is defined by two sets of relations α, defines the relationship between places and transitions, and β defines the relationship between transitions and places. The relations between nodes are represented by directed arcs. A Petri net N is defined as a quadruple $N = (P, T, \alpha, \beta)$. Such Petri nets are called *ordinary*.

A *marking m* of a Petri net is an assignment of black dots (tokens) to the places of the net for specifying the state of the system being modeled with a Petri net.

In classical Petri nets, a token represents a typeless fragment of information. Tokens are used to define the execution of a Petri net. Places represent storage for input or for output. Transitions represent activities (transformations) which transform input into output.

The number of tokens in a place p_i is denoted by m_i and then $m = (m_1, \ldots, m_l)$, where l is the total number of places of the net. The initial distribution of tokens among the places is called the *initial marking* and is denoted by M. A Petri net N with a marking M is called a *marked Petri net* and it is denoted by (N, M). In the paper we only use nets in that all markings are binary, i.e., $m(p) \in \{0, 1\}$ for any place p. *Input* and *output places* of a transition are those which are initial nodes of an incoming or terminal nodes of an outgoing arc of the transition, respectively. In a similar way we define *input* and *output transitions* of a place.

The dynamic behaviour of the system is represented by the *firing* of the corresponding transition, and the evolution of the system is represented by a *firing sequence* of transitions. We assume that nets constructed in sections 3, 4 and 5 of the paper act according to the following *transition (firing) rule*:

(1) A transition t is *enabled* if and only if each input place p of t is marked by at least one token.

(2) A transition can fire only if it is enabled.

(3) When a transition t fires, a token is removed from each input place p of t, and t adds a token to each output place p' of t.

A sample (marked, ordinary) Petri net is given in Figure 2.

Fig. 2. Sample Petri Net

Place p_1 is an input place for the transition t. The presence of a token in place p_1 enables the transition t. When the transition t fires, it removes the token from place p$_1$, and adds a token to its output place p_2.

A marking m' is said to be *reachable* from a marking M if there exists a sequence of firings that transforms M to m'. The set of all possible markings reachable from M in a net N is called the *M-reachability set* of N, and it is denoted by $R(N, M)$.

A Petri net N is said to be *live* if, no matter what marking has been reached from initial marking M, it is possible to ultimately fire any transition of the net by progressing through some further firing sequence; otherwise a net N is said to be *not live*. This means that a live Petri net guarantees deadlock-free operation, no matter what firing sequence is chosen. A marking m' reachable from initial marking M in a net N is said to be *dead* if and only if there is no transition in N enabled by m'.

2.2.2 Petri Nets with Priorities

We also recall an extension of the Petri net model. Petri nets with *priorities* have been suggested by Hack [78]. Priorities can be associated with transitions so if t and t' are both enabled, then the transition with the highest priority will fire first. In the paper we only use ordinary Petri nets with priorities.

Remark. 2.6 In the remaining sections (except section 6) of this Chapter we shall assume that there are in the constructed nets the transitions with three different priorities: 0, 1 and 2. All transitions identifying the values of attributes have priority 0, the transitions representing rules have priority 1, and all the remaining (i.e., the reset transitions transforming a current marking of the net to its initial marking) - 2.

For more detailed information about Petri nets we refer the reader to: [186], [206], [266], [142], [100], [97], [300], [50], [227], [11], [69], [27], [302], [209], [210].

3 The Synthesis Problem of Concurrent Systems

The aim of this section is to present an approach to the synthesis problem of concurrent systems based on the rough set philosophy [168] and Boolean reasoning [34]. We consider two versions of the problem. They are discussed in the next subsections.

The idea of concurrent system representation by information systems is due to Professor Z. Pawlak [169]. Some relationships of information systems and rough set theory with the synthesis problem have been recently discussed in ([169], [172]).

3.1 The Synthesis Problem of Concurrent Systems Specified by Information Systems

3.1.1 Introduction

In this subsection we present the first version of the synthesis problem that informally can be formulated as follows.

Synthesis Problem 1 (SP1). Let $A = \{a_1, a_2, ..., a_m\}$ be a non-empty, finite set of *processes*. With every process $a \in A$ we associate a finite set V_a of its *local states*. We assume that behaviour of such a process system is presented by a designer in a form of a table. Each row in the table includes record of states of processes from A, and each record is labeled by an element from the set U of *global states* of the system. In the following a pair (U, A) is denoted by S.

The problem we can formulate is as follows. Construct for a given information system S its concurrent model, i.e., a marked Petri net (N_S, M_S) with the following property: the reachability set $R(N_S, M_S)$ defines an extension S' of S created by adding to S all new global states corresponding to markings from $R(N_S, M_S)$. All (new) global states in S' are consistent with all rules true in S. Moreover, S' is the largest extension of S with that property.

Our considerations are based on the notion of processes independence. We apply the definition of the total independence of processes which is a modification of the independence definition used in [169]. The main idea of the total independence of two sets B and C of processes can be explained as follows: two sets B and C of processes are totally independent in a given information system S if and only if in S the set of local states of processes from B (from C) does not uniquely determine the set of local states of processes from C (from B). This property can be formulated by applying the partial dependency and rule notions [168]. The total independency of processes allows us to obtain our main result, i.e., an algorithm for solving SP1. Our method for constructing a Petri net consists of two phases. In the first phase, all dependencies between the local states of processes in the system are extracted from the given set of global states. In the second phase, a Petri net corresponding to these dependencies is being built. In this subsection, we use ordinary Petri nets with priorities for modeling target systems.

The synthesis algorithm proposed in this subsection is easier to understand and simpler to implement than presented in [252]. Some new relationships of the knowledge represented by information systems and its Petri net models are here also considered. Constructed nets make it possible to evaluate the behaviour of process systems specified by information systems, and to trace computations in rules derived from information systems.

The main ideas discussed in the subsection are illustrated by simple examples to avoid a very complex description of nets.

3.1.2 Computing Concurrent Data Models from Information Systems

We base our considerations about independency of processes on the notions of dependency and partial dependency of sets of attributes in an information system S. The set of attributes C depends in S on the set of attributes B in S if one can compute the values of attributes from C knowing the values of attributes from B. The set of attributes C depends in S partially in degree k $(0 \le k < 1)$ on the set of attributes B in S if the B-positive region of C in S consists of k % of global states in S.

A set of processes $B \subseteq A$ in a given information system $S = (U, A)$ is called *partially independent* in S if there is no partition of B into sets C and D such that D is dependent on C in S. We show that maximal partially independent sets in S are exactly reducts in S. In this way we have a method for computing maximal partially independent sets (in S) based on methods of reducts computing [249].

We say that a set $B \subseteq A$ is a *totally independent set of processes in* $S=(U, A)$ if there is no partition of B into C and D such that D depends on C in S in the degree $0 < k \le 1$.

In the following we show a method for computing maximal totally independent sets of processes in $S = (U, A)$. These are all totally independent maximal subsets of reducts in S.

3.1.3 Dependencies in Information System and Independence of Processes

Now we present two basic notions related to independency of processes.

Let $S = (U, A)$ be an information system (of processes) and let $\emptyset \ne B \subseteq A$. The set B of processes is called *totally independent* in S if and only if $\text{card}(B) = 1$ or there is no partition of B into C, D such that $C \xrightarrow[S,k]{} D$, where $k > 0$.

Let $S = (U, A)$ be an information system (of processes) and let $\emptyset \ne B \subseteq A$. The set B of processes is called *partially independent* in S if and only if $\text{card}(B) = 1$ or there is no partition of B into C, D such that $C \xrightarrow[S]{} D$.

One can prove from the above definitions the following properties.

Proposition 8. *If B is a totally independent set of processes in S and $\emptyset \ne B' \subseteq B$ then B' is also totally independent set of processes in S.*

Proposition 9. *B is a totally independent set of processes in S if and only if $\text{card}(B) = 1$ or $B - \{a\} \xrightarrow[S,0]{} \{a\}$ for any $a \in B$.*

Proposition 10. *B is a partially independent set of processes in S if and only if $\text{card}(B)=1$ or B consists of B-indispensable [168] attributes in S only.*

3.1.4 Reducts as Maximal Partially Independent Sets of Processes

We have the following relationship between the partially independent sets of processes and reducts:

Proposition 11. *B is a maximal partially independent set of processes in S if and only if $B \in RED(S)$, where $RED(S)$ denotes the set of all reducts in S.*

In order to compute the partially independent parts of a given information system one can execute the procedure for generating reducts (see: subsection 2.1.7).

3.1.5 Maximal Totally Independent Sets of Processes

In the previous subsection we have discussed the problem of construction of the family of partially independent sets of processes and a relationship between these sets and reducts. Now we are interested in a construction of all maximal totally independent sets of processes.

From the definition of totally independent sets of processes in a given information system S it follows that for an arbitrary totally independent set B in S there is a reduct $C \in \text{RED}(S)$ such that $B \subseteq C$. Hence to find all maximal totally independent sets of processes it is enough to find for every $C \in \text{RED}(S)$ all maximal independent subsets of C.

To find all maximal totally independent sets of processes in $S = (U, A)$ it is enough to perform the following steps:

Step 1. **T**:= RED(S); **I**:= $\{\{a_1\}, ..., \{a_m\}\}$;

Step 2. **if** (**T** is empty) **then goto** Step 4
   else begin
     CHOOSE_A_SET $B \in$ **T**;
     T:= **T**−$\{B\}$
   end;

Step 3. **if** card(B) ≤ 1 **then goto** Step 2;
$L := 0$;
for every $a \in B$ **do**
 if $B - \{a\} \overrightarrow{S,k} \{a\}$ for some $k > 0$ **then**
 T := **T** $\cup \{B - \{a\}\}$
 else $L := L + 1$;
 if $L = \text{card}(B)$ **then** **I** := **I** $\cup \{B\}$;
 goto Step 2;

Step 4. The maximal sets in **I** (with respect to the inclusion $\subseteq$) are maximal totally independent sets in S.

Let OPT(S) be the set of all rules of the form (1) $a_{i_1} = v_{i_1} \wedge ... \wedge a_{i_r} = v_{i_r} \overrightarrow{S} a = v$, with the left hand side in minimal form (see: subsection 2.1.8). If γ is in the form (1) then by $L(\gamma)$ we denote the set $\{a_{i_1}, ..., a_{i_r}\}$. It is easy to see that one can take in the first line of Step 1 the instruction **T** := $\{L(\gamma) : \gamma \in \text{OPT}(S)\}$ instead of **T**:= RED(S). In this way we obtain more efficient version of the presented method. The time and space complexity of the discussed problem is, in general, exponential because of the complexity of RED(S) computing. Nevertheless, existing procedures and heuristics help us to compute all maximal independent sets for many practical applications.

At the end let us note the following characterization of reducts being maximal totally independent set of processes:

Proposition 12. Let $S = (U, A)$ be an information system and let $C \in RED(S)$ with $card(C) > 1$. C is a maximal totally independent set of processes in S if and only if $card(d_a^C(u)) > 1$ for every $u \in U$ and $a \in C$.

Now we are ready to formulate formally the first synthesis problem of concurrent systems specified by information systems as follows.

The Synthesis Problem 1:

Let $S = (U, A)$ be an information system. Construct a marked Petri net (N_S, M_S) such that $INF(N_S, M_S) = CON(D(S))$, where:

(i) $INF(N_S, M_S)$ denotes the set $\{v(M)/B : M \in R(N_S, M_S)\}$ and B is the set of all places of nets N_S except a so-called *the start places* corresponding to all processes appearing in reducts of a given information system S (see: subsection 3.1.7, **Construction of a Net Representing Rules**), and
(ii) $CON(D(S))$ is the set of all information vectors consistent with all rules true in S.

3.1.6 Initial Transformations of Rules

Now we present a new method for transforming rules representing a given information system into a Petri net (cf. [256]). First initial transformations of rules are performed. There are two kinds of rules (see: Figure 3).

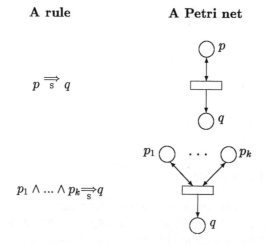

Fig. 3. Illustration of initial transformation rules; $p, p_1,...,p_k, q$ are descriptors in S

3.1.7 Transformation of Rules into Petri Nets

In this subsection a method for transforming rules representing an information system (with respect to any reduct of a given system) into a Petri net is presented. It is consists of five basic steps which are briefly discussed below:

1. A net representing all attributes in a reduct of a given information system is constructed.
2. The net obtained in the first step is extended by adding the elements (places, transitions and arcs) of the net induced by the rules determined by:
   - all non-trivial dependencies between the values of attributes not in a reduct and those within a reduct of the information system,
   - all non-trivial dependencies between the values of attributes within a reduct of the information system.
3. The net obtained in the second step is extended by adding places and arcs representing constraints between the values of attributes which must be satisfied when these values coexist in the system.
4. The net obtained in the third step is extended by adding transitions which after firing lead from a current marking to the initial marking of the net.
5. Priorities to all transitions of the result net are attached.

This method is repeated for all reducts of the information system. Finally, the obtained nets are merged on common elements. Another transformation method of information systems into Petri nets has been presented in the paper [256]. The approach presented in this paper makes the appropriate construction of a net much more readable. Moreover, one can compare better our approach with that presented in [169].

Construction of a Net Representing Rules

Let us assume that all rules of a given information system have the form presented in Figure 3. In this case we should execute five stages described below.

Stage 1. The construction of a net representing a set of all attributes from a reduct of a given information system.

Remark. 3.1 The remaining attributes, i.e., attributes not in a given reduct, are represented in a net by a set of places. Each attribute is represented by a set of places consisting of as many places as many values has that attribute.

Let $S = (U, A)$ be an information system, $R \subset A$ is a reduct of S, $a \in R$ and $V_a = \{a_1, ..., a_{k(a)}\}$.

A net corresponding to any attribute $a \in R$ has the structure and the initial marking presented in Figure 4.

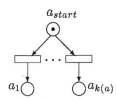

Fig. 4. The net representation of an attribute $a \in R$ with the initial marking

The place a_{start} is called the *start place* of the attribute a. The remaining places of the net represent possible values of the attribute a. The transitions of the net represent the process of identifying the local states of the process a. Let us observe that only the start place a_{start} is marked in the initial marking of the net. In the following example the first stage of the method for the reduct R_1 of the information system S from Example 1 is illustrated.

Example 5. Consider again the information system from Example 1. The attributes $a, c \in R_1$ are represented by nets shown in Figure 5. The initial marking of these nets presented in the figure is corresponding to the start marking. This marking is not represented in Table 1.

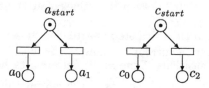

Fig. 5. Nets representing the attributes $a, c \in R_1$

Stage 2. The construction of a net representing the rules corresponding to:
  • all non-trivial dependencies between the values of attributes not in a reduct and those within a reduct of the information system, i.e., belonging to a reduct R of S and the remaining attributes of S (the attributes from $A - R$),
  • all dependencies between the values of attributes within a reduct R of the information system S.

This stage of our construction is explained by two examples presented below.

Example 6. Let us consider at first the following rules obtained in Example 4 for the information system from Example 1:

$$c_2 \underset{S}{\Rightarrow} b_0, \quad a_1 \underset{S}{\Rightarrow} b_1, \quad a_0 \wedge c_0 \underset{S}{\Rightarrow} b_2.$$

These rules correspond to the dependencies between the values of attributes a, c within the reduct R_1 of S and that outside of the reduct, i.e., the attribute b.

A net representation of the above rules obtained by an application of our initial transformation of rules and the net from Figure 5 is illustrated in Figure 6. In this figure the places b_0, b_1, and b_2 represent the values of the attribute b.

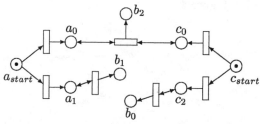

Fig. 6. Net representation of rules from Example 6

In the following example the second step of our construction mentioned in the second stage is illustrated.

Example 7. Let us consider the rules corresponding to all dependencies between the values of attributes within a reduct R_1 of the information system S from Example 1. In this case we have the following rules:

$$a_1 \underset{S}{\Rightarrow} c_0, \quad c_2 \Rightarrow a_0.$$

A net representation of the above rules obtained by an application of our initial transformation of rules and the net from Figure 5 is shown in Figure 7.

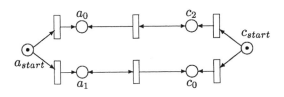

Fig. 7. Net representation of rules from Example 7

Stage 3. The construction of additional places and arcs representing constraints between the values of attributes which must be satisfied when these values coexist in the system.

In order to model e.g. the dependencies between admissible values of an attribute a and an attribute b in a given system S the marked place labeled by b_{sem} with appropriate arcs (as is shown in Figure 8) is introduced. This place ensures that x fires only if y has not been fired yet. It means that there is no non-trivial dependency between values of a_i and b_j represented in the net by places with the same names, respectively. In other words, there is no rule in S of this form: $a_i \underset{S}{\Rightarrow} b_j$. This construction is executed for any pair of inadmissible values of attributes in S.

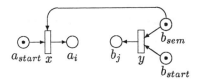

Fig. 8. A net representation of constraints between values of attributes in an information system

Example 8. Figure 9 shows the construction described in Stage 3 for attributes a and c from the reduct R_1 of the information system S presented in Example 1.

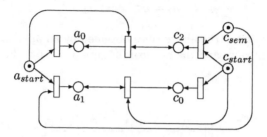

Fig. 9. A net representation of constraints between local states of processes a and c from Example 8

Stage 4. The construction of so called *reset transitions*, i.e., transitions which after firing lead from a current marking to the initial marking of the net.

The net obtained in Stage 3 is extended by the reset transitions. Each reset transition corresponds to one object of a given information system.

This construction is explained in the following example.

Example 9. Let us consider again the information system S from Example 1 and the net from Figure 6. This net should be extended to three reset transitions more corresponding to objects of S. For the reason of better clarity of the picture only one reset transition t_{reset} corresponding to the first object of the information system S from Example 1 is shown in Figure 10.

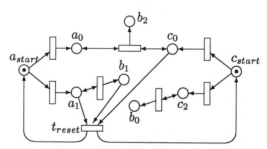

Fig. 10. The net with reset transition t_{reset} from Example 9

Stage 5. The attachment of priorities to each transition of the net obtained in Stage 4.

We assume that there are only transitions with three different priorities: 0, 1 and 2. All transitions identifying the values of attributes have priority 0, the transitions representing rules have priority 1, and all the remaining (i.e., the reset transitions) - 2.

Example 10. Figure 11 shows the net from Figure 10 with priorities. If we now combine the nets from Figures 7, 9 and 11, we obtain the net (N_{R_1}, M_{R_1}) corresponding to the reduct R_1 of the information system from Example 1. In the similar way one can construct the net (N_{R_2}, M_{R_2}) for the reduct R_2. The result, i.e., the net (N_S, M_S) is obtained by composing (N_{R_1}, M_{R_1}) and (N_{R_2}, M_{R_2}). In general, the net (N_S, M_S) is obtained by composing all nets (N_R, M_R) for $R \in \text{RED}(S)$.

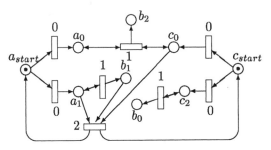

Fig. 11. The net with priorities from Example 10

Remark. 3.2 One can verify that the reachability set of the result net is consistent with all rules valid in the information system S.

Remark. 3.3 On the basis of Petri net approach it is possible to understand better the structure of rules which are true in a given information system. It is also possible to represent in a graphical way the dynamic interactions between values attributes from that system.

3.1.8 Solution Properties of the Synthesis Problem 1

In this subsection we show a relationship between the reachability set for (N_S, M_S) and the set of all information vectors consistent with rules true in S.

The method described in the previous subsection for constructing from an arbitrary information system S its concurrent model in the form of a marked Petri net (N_S, M_S) has the property which we formulate in the form of theorems presented below.

Theorem 13. *Let $S = (U, A)$ be an information system, (N_S, M_S) a marked Petri net constructed by means our method presented in the previous section. Then $INF(N_S, M_S) = CON(D(S))$, where $INF(N_S, M_S)$ denotes the set $\{v(M)/B : M \in R(N_S, M_S)\}$, B is the set of all places of nets N_S except the start places corresponding to all attributes appearing in reducts of S, and $CON(D(S))$ is the set of all information vectors consistent with all rules true in S.*

Sketch of the proof: First let us observe that

(1) $CON(D(S)) = CON(OPT(S))$.

In our construction of a Petri net we eliminate all markings defining information vectors which are inconsistent with CON($OPT(S)$) and only those marking. Hence

(2) INF(N_S, M_S) = CON($OPT(S)$).

From (1) and (2) we obtain our thesis.

Let $U \subset U'$, U' be a set of the same cardinality as INF(N_S, M_S) and f a bijection between U' and INF(N_S, M_S) such that $f(u) = (a_1(u), ..., a_m(u))$ for $u \in U$. We assume that $A = \{a_1, ..., a_m\}$ and $A' = \{a'_i : a_i \in A\}$. By EXT($S$) we denote the information system (U', A') such that $a'_i(u) = (f(u))_i$ for $u \in U'$ and $i = 1, 2, ..., m$.

Theorem 14. *EXT(S) is the largest consistent extension of S.*

Theorem 15. *Let S be an information system and (N_S, M_S) its concurrent model constructed by using our method. Then the net N_S is live at the initial marking M_S if and only if INF(S)=INF(EXT(S)).*

Theorem 16. *Let S be an information system, (N_S, M_S) its concurrent model constructed by using our method and EXT(S) be the largest consistent extension of S. Then*

1. INF(S) is the proper subset of INF(EXT(S)) if and only if the net N_S is not live at the initial marking M_S.

2. The set of all dead markings of the net N_S corresponds to the set of new global states consistent with the knowledge represented by S.

3.2 The Synthesis Problem of Concurrent Systems Specified by Dynamic Information Systems

Introducing a new notion of a dynamic information system we propose a method for constructing an elementary net system equivalent to a given dynamic information system in the sense that the related transition systems are isomorphic. This provides a method for synthesis of a concurrent system specified by a dynamic information system.

Now we consider the second version of the synthesis problem that informally can be formulated as follows.

Synthesis Problem 2 (SP2). Let $A = \{a_1, ..., a_m\}$ be a non-empty finite set of processes and V a finite set of its local states. We assume that the behaviour of such a process system is presented by a designer in a form of two integrated subtables denoted by S and TS, respectively. Each row in the first subtable includes the record of local states of processes from A, and each record is labeled by an element from the set U of global states of the system, whereas the second subtable represents a transition system. Columns of the second subtable are labeled by events, rows, analogously as for the underlying system, by objects of interest and entries of the subtable for a given row (state) are follower states of that state. The subtables S and TS consist of a so-called *a dynamic information* *system* denoted by DS. In addition we assume that the first row in a subtable S represents the initial state of a given transition system TS.

The problem is: For a given dynamic information system DS with its transition system TS, find (if there exists) a concurrent model in the form of an elementary net system N [294] with the property: the transition system TS is isomorphic to the transition system associated with the constructed elementary net system N.

This problem has been solved in the literature for various types of nets ranging from elementary net systems to classical Petri nets [6]. It has also been shown that the synthesis problem for elementary net systems is NP-complete [5].

In this subsection we propose two methods for solving to this problem based on rough set approach. In the first method we assume that the table representing a given dynamic information system contains all possible state combinations, i.e., the table contains the whole knowledge about the observed behaviour of the system. In the second one only a part of possible observations is contained in the table, i.e., they contain partial knowledge about the system behaviour only. In the subsection we discuss both approaches.

A method for constructing from a given dynamic information system DS its concurrent model in the form of an elementary net system N is our main result. The elementary net system obtained from a given transition system TS by applying our method has the following property: a given transition system TS is isomorphic to the transition system associated with the constructed net system N. The set of all global states of DS is consistent with all rules true in the underlying information system S of DS. The set of all global states of DS represents the largest extension of S consistent with the knowledge represented by S.

Our method for constructing a Petri net model consists of two phases. In the first phase, all dependencies between processes in the system are extracted from the given set of global states, the extension of the system is computed and, if necessary, a modification of the given transition system is done. In the second phase, an elementary net system corresponding to the computed extension of the given dynamic information system is built by employing a method solving the synthesis problem of Petri nets presented in [54].

A designer of concurrent systems can draw Petri nets directly from a specification in a natural language. We propose a method which allows automatically to generate an appropriate Petri net from a specification given by a dynamic information system and/or rules. This kind of specification can be more convenient for the designers of concurrent systems than drawing directly nets especially when they are large. The designer of concurrent systems applying our method is concentrated on a specification of local processes dependencies in global states. These dependencies are represented by an information system ([168], [174], [240], [252], [253], [255]). The computing process of the solution is iterative. In a successive step the constructed so far net is automatically redesigned when some new dependencies are discovered and added to a specification. The nets produced automatically by application of our method can be simplified by an application of some reduction procedures. This problem is out of scope of this Chapter. We expect that our method can be applied as a convenient tool for the synthesis of larger systems ([4], [231]).

We illustrate our ideas by an example of traffic signal control from subsection 2.1.

It is still worth to mention that discovering relations between observed data is the main objective of the machine discovery area (cf. [327]). Our main result can be interpreted as a construction method of all global states consistent with knowledge represented by the underlying system S of DS (i.e., with all rules true in S). For example, checking if a given global state is consistent with S is equivalent to checking if this state is reachable from the initial state of the net system N representing DS. It seems that our approach can be applied for synthesis and analysis of knowledge structure by means of its concurrent models.

3.2.1 Transition Systems

Transition systems create a simple and powerful formalism for explaining the operational behaviour of models of concurrency. This subsection contains basic notions and notations connected with transition systems that will be necessary for understanding of our main result.

A *transition system* is a quadruple $TS = (S, E, T, s_0)$, where S is a non-empty set of *states*, E is a set of *events*, $T \subseteq S \times E \times S$ is the *transition relation*, $s_0 \in S$ is the *initial state*.

A transition system can be pictorially represented as a rooted edge-labeled directed graph. Its nodes and its directed arcs represent states and state transitions, respectively. As different state transitions may be caused by equal events, different arcs may be labeled by equal symbols. If $(s, e, s') \in T$ then a transition system TS can go from s to s' as a result of the event e occurring at s.

Example 11. In Figure 12 a transition system is shown, where the initial state is indicated by an extra arrow without source and label.

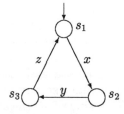

Fig. 12. An example of a transition system

An isomorphism between transition systems is defined in the following way: Let $TS = (S, E, T, s_0)$ and $TS' = (S', E', T', s'_0)$ be two transition systems.

A bijection $f : S \to S'$ is an *isomorphism* from TS to TS' (denoted $f : TS \to TS'$) if and only if the following two conditions are satisfied:

(i) $f(s_0) = s'_0$

(ii) $(s, e, s') \in T$ if and only if $(f(s), e, f(s')) \in T'$.

Two transition systems TS and TS' are called *isomorphic* (denoted $TS \simeq TS'$) if and only if there exists an isomorphism $f : TS \to TS'$.

It is worth to observe that we demand that the set of events of E from a transition system TS coincide with the set of events of E' from TS'.

Let $TS = (S, E, T, s_0)$ be a transition system. We say that the event e has *concession in* the state s (is *enabled at s*) if there exists a state s' such that $(s, e, s') \in T$.

The notion of regions, introduced in [59] is important for this Chapter.

Let $TS = (S, E, T, s_0)$ be a transition system. A set R of states of TS is a *region* of TS if and only if for equally labeled arcs (s, e, s') and (s_1, e, s_1') holds:

if $s \in R$ and $s' \notin R$ then $s_1 \in R$ and $s_1' \notin R$, and
if $s \notin R$ and $s' \in R$ then $s_1 \notin R$ and $s_1' \in R$.

$\emptyset$ and S are called *trivial regions* of TS. By R_{TS} we denote the set of all non-trivial regions of TS.

Let $TS = (S, E, T, s_0)$ be a transition system.
For $e \in E$,
$^{\bullet}e = \{R \in R_{TS} : \text{there exists } (s, e, s') \in T \ s \in R \text{ and } s' \notin R\}$
is called the *pre − region* of e,
$e^{\bullet} = \{R \in R_{TS} : \text{there exists } (s, e, s') \in T \ s \notin R \text{ and } s' \in R\}$
is called the *post − region* of e.

Example 12. For the transition system shown in Figure 12, $X = \{s_1\}, Y = \{s_2\}$ and $Z = \{s_3\}$ are regions, and $^{\bullet}x = \{X\}, y^{\bullet} = \{Z\}$.

3.2.2 Elementary Net Systems

In this subsection we recall basic notions connected with the basic system model of net theory, called *elementary net system* [294].

In net theory, models of concurrent systems are based on objects called nets which specify the local states and local transitions and the relationships between them.

A triple $N = (S, T, F)$ is called a *net* if and only if:

(i) S and T are disjoint sets (the elements of S are called $S − elements$, the elements of T are called $T − elements$).

(ii) $F \subseteq (S \times T) \cup (T \times S)$ is a binary relation, called the *flow relation*.

(iii) For each $x \in S \cup T$ there exists $y \in S \cup T$ such that $(x, y) \in F$ or $(y, x) \in F$.

In the following the S-elements will be called *conditions* and the T-elements will be called *events*. Moreover, we use B to denote the set of conditions and E to denote the set of events; consequently a net will be denoted as the triple (B, E, F).

Let $N = (B, E, F)$ be a net. For $x \in B \cup E, ^{\bullet}x = \{y : (y, x) \in F\}$ is called the *preset* of x, $x^{\bullet} = \{y : (x, y) \in F\}$ is called the *postset* of x.

The element $x \in B \cup E$ is called *isolated* if and only if $^{\bullet}x \cup x^{\bullet} = \emptyset$.

It is worth to observe that the condition (iii) in the net definition states that we do not permit isolated elements in considered nets.

The net $N = (B, E, F)$ is called *simple* if and only if distinct elements do not have the same pre- and postset, i.e., for each $x \in B \cup E$ the following condition is satisfied:

if $^\bullet x = ^\bullet y$ and $x^\bullet = y^\bullet$ then $x = y$.

A quadruple $\mathbf{N} = (B, E, F, c_0)$ is called an *elementary net system* if and only if:

(i) $N = (B, E, F)$ is a simple net without isolated elements, called the *underlying net* of $\mathbf{N}$ and denoted by $N_{\mathbf{N}}$,

(ii) $c_0 \subseteq B$ is the *initial state*.

In diagrams the conditions will be drawn as circles, the events as boxes and elements of the flow relations as directed arcs. The initial state will be indicated by marking (with small black dots) the elements of the initial state.

Example 13. An elementary net system shown in Figure 13 has three conditions X, Y, Z, and three events x, y, z. Its initial state is $\{X\}$. The preset of x is equal to $\{X\}$, and the postset of y is $\{Z\}$.

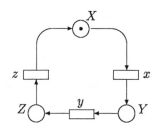

Fig. 13. An elementary net system

From now on we will often refer to elementary net systems as just net systems. The dynamics of a net system are straightforward. The states of a net system consist of a set of conditions that hold concurrently. The system can go from a state to a state through the occurrence of an event. An event can occur at a case if and only if all its pre-conditions (i.e., conditions in its preset) hold and none of its post-conditions (i.e., conditions in its postset) hold at the state. When an event occurs then all its pre-conditions cease to hold and all its post-conditions begin to hold. Formally, the dynamics of a net system is described by the so-called *transition relation* of that net system.

Let $N = (B, E, F)$ be a net. Then $tr_N \subseteq P(B) \times E \times P(B)$ is the *transition relation* of N defined as follows: $(c, e, c') \in tr_N$ if and only if $c - c' = ^\bullet e$ and $c' - c = e^\bullet$.

Let $\mathbf{N} = (B, E, F, c_0)$ be a net system.

(i) $C_{\mathbf{N}}$ is the *state space* of $\mathbf{N}$ and it is the smallest subset of $P(B)$ containing c_0 which satisfies the condition: if $(c, e, c') \in tr_{N_{\mathbf{N}}}$ and $c \in C_{\mathbf{N}}$ then $c' \in C_{\mathbf{N}}$.

(ii) tr_N is the *transition relation* of $\mathbf{N}$ and it is tr_{N_N} restricted to $C_N \times E \times C_N$.
(iii) E_N is the set of active events of $\mathbf{N}$ and it is the subset of E given by $E_N = \{e :$ there exists $(c, e, c') \in tr_N\}$.

It is possible to associate a transition system with a net system to explain its operational behaviour.

Let $\mathbf{N} = (B, E, F, c_0)$ be a net system. Then the transition system $TS_N = (C_N, E_N, tr_N, c_0)$ is called the *transition system associated with* $\mathbf{N}$.

A transition system TS is an *abstract transition system* if and only if there exists a net system N such that $TS \simeq TS_N$.

Example 14. The state space of the net system presented in Figure 13 is $\{\{X\}, \{Y\}, \{Z\}\}$. It is easy to verify that the transition system associated with the net system of Figure 13 is isomorphic with the transition system shown in Figure 12.

3.2.3 Dynamic Information Systems

Now we introduce the notion of a *dynamic information system* which plays a central role in this subsection.

A *dynamic information system* is a quintuple $DS = (U, A, E, T, u_0)$, where:
(i) $S = (U, A)$ is an information system called the *underlying system* of DS,
(ii) $TS = (U, E, T, u_0)$ is a transition system.

Dynamic information systems will be presented in the form of two integrated subtables. The first subtable represents the underlying system, whereas the second one the transition system. Columns of the second subtable are labeled by events, rows, analogously as for the underlying system, by objects of interest and entries of the subtable for a given row (state) are follower states of that state. The first row in the first subtable represents the initial state of a given transition system. Both subtables have the same number of rows, but the number of columns is different.

Example 15. In Table 7 an example is shown of a dynamic information system $DS = (U, A, E, T, u_0)$ such that its underlying system is represented by Table 1, whereas the transition system can be represented by the similar graph as in Figure 12. In this case the initial state of the system is represented by u_1. We show also that, for instance in the state u_2 the event y has concession and when it occurs then a new state u_3 of DS appears.

U/A	a	b	c	U/E	x	y	z
u_1	1	1	0		u_2		
u_2	0	2	0			u_3	
u_3	0	0	2				u_1

Table 7. A dynamic information system

Now we are ready to formulate formally the second synthesis problem of concurrent systems specified by dynamic information systems as follows.

The Synthesis Problem 2:

Let $DS = (U, A, E, T, u_0)$ be a dynamic information system. Is a given transition system $TS = (U, E, T, u_0)$ an abstract transition system? If yes, construct a net system N satisfying $TS \simeq TS_N$.

3.2.4 The Solution of the Synthesis Problem 2

In this subsection we present a solution of the synthesis problem stated in above.

The First Approach

A solution method of the problem is based on the approach proposed in [54]. Now we describe shortly their approach connected with a procedure to decide whether or not a given transition system TS is an abstract transition system. In the positive case, the procedure provides a net system whose transition system is isomorphic to TS.

Since every condition corresponds to a region and every region generates a potential condition we can construct a net system from a transition system, using only generated conditions.

Let $DS = (U, A, E, T, u_0)$ be a dynamic information system, let $TS = (U, E, T, u_0)$ be the transition system of DS, and let m be a set of regions of TS. Then the *m-generated net system* is $\boldsymbol{N}_m^{TS} = (m, E, F, c_0)$ where for each region $R \in m$ and each event $e \in E$ the following conditions are satisfied:

(i) $(R, e) \in F$ if and only if $R \in {}^\bullet e$,
(ii) $(e, R) \in F$ if and only if $R \in e^\bullet$,
(iii) $R \in c_0$ if and only if $u_0 \in R$.

Example 16. The transition system from Example 3.2 with the regions X, Y, Z of Example 12 generates the net system shown in Figure 13.

We can now formulate the synthesis problem in the following way: Given a transition system TS, construct the net system generated by the regions of TS. If the transition system associated with this net system is isomorphic to TS, then the net system is a basic solution to the synthesis problem and the procedure is finished. In the opposite case, there exists no a net system which corresponds to TS and so TS is no abstract transition system. This fact follows from the following

Theorem 17. *[54] A transition system TS is an abstract transition system if and only if $TS \simeq TS_{N_m^{TS}}$, where m denotes the set of all regions of TS.*

Example 17. The transition system TS from Example 3.2 is an abstract transition system. The transition system associated with the net system from Figure 13 is shown in Figure 14. It is isomorphic to TS.

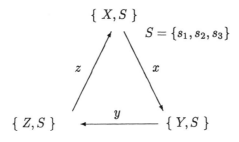

Fig. 14. The transition system associated with the net system from Figure 13

To decide if two graphs are isomorphic is in general a non-trivial problem. Fortunately, the procedure proposed above, decides this problem very easily since there exists at most one isomorphism transforming a given transition system TS onto a transition system associated with a net system generated by the regions of TS. It follows from the following proposition, which is reformulated to our formalism:

Proposition 18. *[54] Let $DS = (U, A, E, T, u_0)$ be a dynamic information system, let $TS = (U, E, T, u_0)$ be its transition system, and let m denote the set of all regions of TS. Then there is exactly one isomorphism f from TS to $TS_{N_m^{TS}}$, where N_m^{TS} denotes the m-generated net system which is defined as follows: $f(s) = \{R \in m : s \in R\}$.*

The Second Approach

Now we describe shortly a solution to the synthesis problem 2 based on the second approach, i.e., we assume that a given data table DS consists of only partial knowledge about the system behaviour. Thus, we at first compute an extension DS' of the data table DS, i.e., the system in which the set of global states of DS' is consistent with all rules true in the underlying information system S of DS as well as the set of global states of DS' represents the largest extension of S consistent with the knowledge represented by S. Next, for finding a solution of the synthesis problem in the form of a net system we use the method described in the previous section. The idea of our method is presented by example and a very simple procedure given below.

Example 18. Let us consider an example of a decision table $S = (U, A \cup \{d\})$ defined by the data table presented in Table 8.

U/A	a	b	c	a'	b'	c'	d
u_1	1	1	0	0	2	0	1
u_2	1	1	0	1	1	0	0
u_3	1	1	0	0	0	2	0
u_4	0	2	0	0	0	2	1
u_5	0	2	0	0	2	0	0
u_6	0	2	0	1	1	0	0
u_7	0	0	2	1	1	0	1
u_8	0	0	2	0	0	2	0
u_9	0	0	2	0	2	0	0

Table 8. An example of a decision table S

In the example we have $U = \{u_1, u_2, ..., u_9\}$, $A = \{a, b, c, a', b', c'\}$. The decision is denoted by d. The possible values of attributes (conditions and the decision) from $A \cup \{d\}$ are equal to 0, 1 or 2. This data table has been constructed on the basis of the dynamic information system $DS = (U, A, E, T, u_0)$ from Example 15. Table 8 contains all possible pairs of global states from the underlying system of DS. The value of decision d is equal to 1 if and only if there exists an event $e \in E$ such that $(u, e, u') \in T$. Thus, this decision table we can treat as a description of the characteristic function of the transition relation T. For the decision table S we obtain the following discernibility matrix $M(S)$ presented in Table 9.

Now we compute the set of rules corresponding to non-trivial dependencies between the values of conditions and the decision values. In this case we also apply the method for generating the minimal form of rules presented in the subsection 2.1.8. Let us start by computing the decision rules corresponding to the conditions $A = \{a, b, c, a', b', c'\}$ and the decision d. We have the decision table $S = (U, A \cup \{d\})$ from which we compute the decision rules mentioned below. In the table the values of the function d_d^A are also given. The discernibility matrix $M(S; d, v, u_l)$ where $v \in V_d$, $u_l \in U$, $l = 1, 2, ..., 9$, obtained from $M(S)$ in the above way is presented in Table 11.

Discernibility functions corresponding to the values of the function d_d^A after reduction (using the absorption laws) are the following:

Case 1. For $d_d^A(u_1) = \{1\}$:
$a \wedge a' \wedge c' \vee b \wedge a' \wedge c' \vee a \wedge b' \vee b \wedge b'$.

Case 2. For $d_d^A(u_2) = \{0\}$:
$a \wedge a' \vee b \wedge a' \vee c \wedge a' \vee a \wedge b' \vee b \wedge b' \vee c \wedge b'$.

Case 3. For $d_d^A(u_3) = \{0\}$:
$a \wedge b' \vee a \wedge c' \vee b \wedge b' \vee b \wedge c'$.

Case 4. For $d_d^A(u_4) = \{1\}$:

$b \wedge c' \vee b \wedge b' \vee a \wedge c \wedge b' \vee a \wedge c \wedge c'$.

Case 5. For $d_d^A(u_5) = \{0\}$:
$a \wedge b' \vee b \wedge b' \vee b \wedge c' \vee a \wedge c \wedge c' \vee a \wedge a' \wedge c'$.

Case 6. For $d_d^A(u_6) = \{0\}$:
$b \wedge a' \vee b \wedge b' \vee b \wedge c' \vee c \wedge b' \vee c \wedge a' \vee a \wedge c \wedge c'$.

Case 7. For $d_d^A(u_7) = \{1\}$:
$b \wedge a' \vee b \wedge b' \vee c \wedge a' \vee c \wedge b'$.

Case 8. For $d_d^A(u_8) = \{0\}$:
$b \wedge a' \vee b \wedge b' \vee b \wedge c' \vee c \wedge a' \vee c \wedge b' \vee c \wedge c'$.

Case 9. For $d_d^A(u_9) = \{0\}$:
$b \wedge a' \vee c \wedge a' \vee a \wedge b' \vee a \wedge a' \wedge c' \vee b \wedge b' \vee c \wedge b'$.

Hence we obtain the following decision rules:

$a_1 \wedge a_0' \wedge c_0' \vee b_1 \wedge a_0' \wedge c_0' \vee a_1 \wedge b_2' \vee b_1 \wedge b_2' \underset{s}{\Rightarrow} d_1$,

$b_2 \wedge c_2' \vee b_2 \wedge b_0' \vee a_0 \wedge c_0 \wedge b_0' \vee a_0 \wedge c_0 \wedge c_2' \underset{s}{\Rightarrow} d_1$,

$b_0 \wedge a_1' \vee b_0 \wedge b_1' \vee c_2 \wedge a_1' \vee c_2 \wedge b_1' \underset{s}{\Rightarrow} d_1$,

$a_1 \wedge a_1' \vee b_1 \wedge a_1' \vee c_0 \wedge a_1' \vee a_1 \wedge b_1' \vee b_1 \wedge b_1' \vee c_0 \wedge b_1' \underset{s}{\Rightarrow} d_0$,

$a_1 \wedge b_0' \vee a_1 \wedge c_2' \vee b_1 \wedge b_0' \vee b_1 \wedge c_2' \underset{s}{\Rightarrow} d_0$,

U	u_1	u_2	u_3	u_4	u_5	u_6	u_7	u_8	u_9
u_1									
u_2	a',b',d								
u_3	b',c',d	a',b',c'							
u_4	a,b,b',c'	a,b,a',b',c',d	a,b,d						
u_5	a,b,d	a,b,a',b'	a,b,b',c'	b',c',d					
u_6	a,b,a',b',d	a,b	a,b,a',b',c'	a',b',c',d	a',b'				
u_7	a,b,c,a',b'	a,b,c,d	a,b,c,a',b',c',d	b,c,a',b',c'	b,c,a',b',d	b,c,d			
u_8	a,b,c,b',c',d	a,b,c,a',b',c'	a,b,c	b,c,d	b,c,b',c'	b,c,a',b',c'	a',b',c',d		
u_9	a,b,c,d	a,b,c,b',c'	a,b,c b',c'	b,c,b',c',d	b,c	b,c,a',b'	a',b',d	b',c'	

Table 9. The discernibility matrix $M(S)$ for the decision table S from Example 18

U/A	a	b	c	a'	b'	c'	d	d_d^A
u_1	1	1	0	0	2	0	1	{1}
u_2	1	1	0	1	1	0	0	{0}
u_3	1	1	0	0	0	2	0	{0}
u_4	0	2	0	0	0	2	1	{1}
u_5	0	2	0	0	2	0	0	{0}
u_6	0	2	0	1	1	0	0	{0}
u_7	0	0	2	1	1	0	1	{1}
u_8	0	0	2	0	0	2	0	{0}
u_9	0	0	2	0	2	0	0	{0}

Table 10. The decision table from Example 18 with the function d_d^A

U	u_1	u_2	u_3	u_4	u_5	u_6	u_7	u_8	u_9
u_1									
u_2	a',b'								
u_3	b',c'								
u_4		a,b,a',b',c'	a,b						
u_5	a,b			b',c'					
u_6	a,b,a',b'			a',b',c'					
u_7		a,b,c	a,b,c,a',b',c'		b,c,a',b'	b,c			
u_8	a,b,c,b',c'			b,c			a',b',c'		
u_9	$a,b,c,$			b,c,b',c'			$a',b',$		

Table 11. The discernibility matrix $M(S;d,v,u_l)$ for the $M(S)$

$$a_0 \wedge b_2' \vee b_2 \wedge b_2' \vee b_2 \wedge c_0' \vee a_0 \wedge c_0 \wedge c_0' \vee a_0 \wedge a_0' \wedge c_0' \underset{s}{\Rightarrow} d_0,$$

$$b_2 \wedge a_1' \vee b_2 \wedge b_1' \vee b_2 \wedge c_0' \vee c_0 \wedge b_1' \vee c_0 \wedge a_1' \vee a_0 \wedge c_0 \wedge c_0' \underset{s}{\Rightarrow} d_0,$$

$$b_0 \wedge a_0' \vee b_0 \wedge b_0' \vee b_0 \wedge c_2' \vee c_2 \wedge a_0' \vee c_2 \wedge b_0' \vee c_2 \wedge c_2' \underset{s}{\Rightarrow} d_0,$$

$$b_0 \wedge a_0' \vee c_2 \wedge a_0' \vee a_0 \wedge b_2' \vee a_0 \wedge a_0' \wedge c_0' \vee b_0 \wedge b_2' \vee c_2 \wedge b_2' \underset{s}{\Rightarrow} d_0.$$

These decision rules allow us to verify which global states of the dynamic information system DS from Example 15 are in the transition relation T of DS.

Let $DS = (U, A, E, T, u_0)$ be a dynamic information system and $S = (U, A)$ its underlying system. Sometimes, it is possible that an extension of the underlying system S of DS contains new global states consistent with the knowledge represented by S, i.e., with the all rules from the set OPT(S). We can obtain the extension of the system S by applying the procedure for computing an extension S' of S described in subsection 2.1.8. Thus, the method for finding the decision rules in a given dynamic information system presented in the above example allows us to extend the transition relation T of DS to a new transition relation T'. In consequence, we obtain a new dynamic information system $DS' = (U', A, E', T', u_0)$ called an *extension of the dynamic information system DS*, where $S' = (U', A)$ is an extension of S, E', is a set of events, $E \subset E'$, and T' is the extension of the transition relation T, $T' \subseteq U' \times E' \times U'$. Further, for constructing from a dynamic information system DS' with its transition system $TS' = (U', E', T', u_0)$ describing the behaviour of DS' a concurrent model in the form of an elementary net system we can proceed analogously to the method presented above.

Now we are ready to present a very simple procedure for computing an extension $DS' = (U', A, E', T', u_0)$ of a given dynamic information system $DS = (U, A, E, T, u_0)$.

PROCEDURE for computing an extension DS' of DS:

Input: A dynamic information system $DS = (U, A, E, T, u_0)$ with its underlying system $S = (U, A)$.

Output: An extension DS' of the system DS.

Step 1. Construct the decision table $S' = (U', A \cup \{d\})$ with the function d_d^A in the way described in Example 3.14.

Step 2. Compute the discernibility matrix $M(S')$.

Step 3. Compute the discernibility matrix $M(S'; d, v, u_l)$ where $v \in V_d$, $u_l \in U'$, $l = 1, 2, ..., card(U')$ for the $M(S')$.

Step 4. Compute the discernibility functions corresponding to the values of the function d_d^A in the way described in subsection 2.1.8.

Step 5. Compute the decision rules true in S', i.e., the set $D'(S')$ of rules corresponding to non-trivial functional dependencies between the values of conditions and the decision values from the decision table S'.

Step 6. Compute an extension $S'' = (U', A)$ of the underlying system S of DS using procedure described in subsection 2.1.8.

Step 7. Compute an extension T' of the transition relation T using the decision rules obtained in Step 5 in the following way:

1. Construct all possible pairs of global states of S, i.e., a set $U \times U$,
2. Verify using the set of decision rules obtained in Step 5 which pairs of global states of S are consistent with these rules, i.e., execute instructions
   (i) $T' := \emptyset$; $E' := \emptyset$.
   (ii) For every pair $(u, u') \in U \times U$ do
       if $(u, u') \in U \times U$ and an information vector v corresponding to a pair (u, u') is consistent with $D'(S')$ then add (u, e, u') to T' and e to E'.

Step 8. Construct the extension $DS' = (U', A, E', T', u_0)$.

It is easy to verify that the extension DS' of the dynamic information system DS from Example 15 computed by using our procedure presented above is the same as the system DS (see: Example 4 and Example 18). Thus, the net system for the extension DS' is identical as for the system DS (see: Figure 13).

3.3 Summary

The contribution of this section is the presentation of an approach to modeling complex systems specified by information systems with Petri nets. The application of Petri nets to represent the given information systems and the modified definition of these systems enable us:

• to represent in an elegant and visual way the dependencies between local states of processes in the system,

• to simulate and analyze computations performed by the systems,

• to observe concurrent and/or sequential processes of the systems,

• to discover in a simple way new dependencies between local states of processes being in the systems.

Moreover, drawing Petri nets by hand one can produce very compact solutions for problems solved rather by small nets. For large models some automatic methods could be accepted even if the nets produced by them are not so compact or small. Comparing the presented examples it is possible to see that our method for solving SP1 can also produce solutions close to those obtained by designers.

We have also presented a method for solving SP2. Our solution is based on a construction of a solution of the synthesis problem of Petri nets discussed in [54]. We have proposed a solution of the synthesis problem of a net system from a dynamic information system. It is also possible to solve this problem for finite place/transition Petri nets, since that finite self-loop-free place/transition nets are equivalent to vector addition systems, introduced by Karp and Miller [102]. The solution of our problem for place/transitions Petri nets is also simple to obtain.

The methods presented in the section allow to generate automatically from an arbitrary (dynamic) information system its concurrent model in the form of a net. We have implemented a program on *IBM PC* generating a net model of the system specified by a (dynamic) information system. The resulting net can be analyzed by *PN-tools* [277].

4 Discovery of Concurrent Data Models

4.1 Introduction

The main objective of machine discovery is the determination of relations between data and of data models.

Machine discovery has elaborated methods for the discovery of relations among observed data (cf. [327]) and of data models. Recent research shows that these methods can be applied among others in knowledge discovery in databases

[189], scientific discovery [235], concept discovery [148], automated data analysis [104] and discovery in mathematics ([44], [9]).

Large databases can be a source of useful knowledge. Yet this knowledge is implicit in the data. It must be mined and expressed in a concise, useful form (e.g. in the form of rules, conceptual hierarchies, statistical patterns, equations, and the like). Automation of knowledge discovery is important because databases are growing in a size and a number, and the standard data analysis techniques are not designed for exploration of huge hypotheses spaces.

Decomposition is the breakdown of a complex system into smaller, relatively independent subsystems. It is the main tool available to simplify the construction of complex man-made systems. System decomposition problems are an essential part of the system analysis and design process. However, despite the importance of system decomposition, there is no general approach for accomplishing it. Rather, decomposition relies on an analyst's experience and expertise.

The system decomposition methods are applied in many areas (see e.g. in the field of system engineering and large-scale systems ([303], [226], [164], [46], [237]), in the logic synthesis [298], in Petri net theory [20].

The aim of this section is to present an approach to the decomposition of information systems. Decomposition is a means for discovery of data models represented by concurrent systems from experimental tables. Experimental data are represented by information systems, and Petri nets are models for concurrency.

Decomposition of large experimental data tables can be treated as one of the fundamental tools in data mining. It is usually imposed by the high computational complexity of the search for relations between data on one hand and/or the structure of the process of data models discovery on the other.

Our approach to the decomposition of information systems consists of three levels. First we show how experimental data tables are represented by information systems. Next we discuss how any information system S can be decomposed (with respect to any of its reduct) into components linked by some connections which allow to preserve some constraints. Any component represents in a sense the strongest functional module of the system. The connections between components represent constraints which must be satisfied when these functional modules coexist in the system. The connections are represented by some special rules. The components together with the connections define a so called covering of S. Finally, we use the coverings of the information system S to construct its concurrent model in the form of a marked Petri net (N_S, M_S) with the property described in subsection 3.1.8.

The behaviour of the constructed concurrent systems models is consistent with data tables from which they are extracted; their properties (like their invariants) can be considered as higher level laws of experimental data. From these invariants some new forms of laws can be deduced to express e.g. relationships between different components of the system.

In the section we investigate decomposition problems which can now be roughly defined as follows:

Component Extraction Problem:
Let S be an information system. Compute all components of S.

Covering Problem:
Let S be an information system. Compute the family of all coverings of S.

Our approach can be applied for automatic feature extraction (see: subsection 4.3) and for control design of systems represented by experimental data tables (see: section 7).

4.2 Decomposition of Information Systems

We present in this subsection concepts and notation related to the decomposition of information systems as well as a method for constructing components and coverings of a given information system with respect to its reducts.

Let $S = (U, A)$ be an information system. An information system S is said to be *covered with constraints* C (or C-covered, in short) *by information systems* $S_1 = (U_1, A_1), \ldots, S_k = (U_k, A_k)$, if $\text{INF}(S') = \{Inf_{A_1}(u_1) \cup \ldots \cup Inf_{A_k}(u_k) : Inf_{A_1}(u_1) \cup \ldots \cup Inf_{A_k}(u_k) \in \text{CON}(C) \text{ and } u_i \in U_i \text{ for } i = 1, \ldots, k\}$, where S' is a maximal consistent extension of S and C is a set of rules.

The pair $(\{S_1, \ldots, S_k\}, C)$ is called a C - *covering* of S (or a covering of S, in short). The sets $S_1, \ldots, S_k$ are its *components* and C is the set of *constraints* (connections, or communications).

Example 19. Let us consider the information system S from Example 1. It is easy to see that the information systems $S_1 = (U_1, A_1)$, $S_2 = (U_2, A_2)$ and $S_3 = (U_3, A_3)$ represented by Table 12, 13, 14, respectively, and the empty set of constraints C yield a C-coverings of S.

U_1/A_1	b	a
u_1	1	1
u_2	2	0
u_3	0	0

Table 12. The information system S_1

U_1/A_2	b	c
u_1	1	0
u_2	2	0
u_3	0	2

Table 13. The information system S_2

U_1/A_3	a
u_1	1
u_2	0

Table 14. The information system S_3

Remark. 4.1 More advanced example of a covering of a given information system is shown in [279].

From the definition of a covering the obvious proposition presented below follows.

Proposition 19. *Every information system has at least one covering.*

If $S = (U, A)$ then the system $S = (U', A')$ such that $U' \subseteq U$, $A' = \{a' : a \in B \subseteq A\}$, $a'(u) = a(u)$ for $u \in U'$ and $V_{a'} = V_a$ for $a \in A$ is said to be a *subsystem* of S.

Example 20. Every information system in Example 19 is a subsystem of the system S from Example 1.

Let $S = (U, A)$ be an information system and let $R \in \text{RED}(S)$. An information system $S' = (U', A')$ is a *normal component* of S (with respect to R) if and only if the following conditions are satisfied:

(i) S' is a subsystem of S,
(ii) $A' = B \cup C$ where B is a minimal (with respect to $\subseteq$) subset of R such that $B \underset{S}{\to} \{a\}$ for some $a \in A - R$, and C is the set of all attributes a with the above property.

The set of all normal components of S (with respect to R) is denoted by $\text{NCOMP}_R(S)$.

Remark. 4.2 From condition (ii) we have that sets B and C establish a partition of A'.

Example 21. Subsystems S_1 and S_2 are normal components of S (with respect to the reduct R_2) from Example 1, but the subsystem S_3 is not. A more detailed explanation of this fact is included in Example 22

The next proposition is a direct consequence of the definition of a normal component of an information system and Proposition 3

Proposition 20. *Every information system has at least one normal component (with respect to any of its reduct).*

Let $S' \in \text{NCOMP}_R(S)$ and $S' = (U', B_{S'} \cup C_{S'})$. By X_R we denote the set of all attributes which simultaneously occur in normal components of S (with respect to R) and in the reduct R, i.e., $X_R = \underset{S' \in \text{NCOMP}_R(S)}{\bigcup} B_{S'}$.

Let X_R be a set defined for S and R as above. We say that a subsystem $S' = (U', A')$ of S is to be a *degenerated component* of S (with respect to R) if and only if $A' = \{a\}$ for some $a \in R - X_R$. We denote this fact by $\{a\} \underset{S}{\to} \emptyset$ (the empty set).

In the sequel an component (with respect to a reduct) will be assumed to be either a normal component or a degenerated component (with respect to the reduct).

Proposition 21. *Let $S = (U, A)$ be an information system and let R be its reduct. Then the information system S consists of $\text{card}(R - X_R)$ degenerated components (with respect to R).*

Let $S = (U, A)$ be an information system, $R \in \text{RED}(S)$. We say that S is *R-decomposable into components* or that S is *C-coverable by components* (with respect to R) if and only if there exist components $S_1 = (U_1, B_1 \cup C_1), ..., S_k = (U_k, B_k \cup C_k)$ of S (with respect to R) with a set of communications C such that $B_1 \cup ... \cup B_k = R$ and $C_1 \cup ... \cup C_k = A - R$, yielding a *C-covering* of S.

The set of communications (connections) includes:

(i) rules corresponding to non-trivial dependencies between the values of attributes in B_i ($i = 1, ..., k$) called *internal communications* (the *internal connections*) within the component S_i of S,

(ii) rules corresponding to non-trivial dependencies between the values of attributes in a set B_i ($i = 1, ..., k$) and those in the set $A - A_i$, where $A_i = B_i \cup C_i$ called *external communications* (the *external connections*) with the outside of S_i.

From the above definition and from Proposition 3 follow the theorem and the proposition presented below.

Theorem 22. *Every information system is a C-covering (with respect to any its reduct), where C is the set of all internal and external communications of S.*

We obtained a constructive method of the information system (data table) decomposition into functional modules interconnected by external communications. One can observe a similarity of our data models to those used in general system theory and control design.

Proposition 23. *Let R be a reduct of an information system S. Then S has at least one C-covering (with respect to R), where C is the set of all internal and external communications of S.*

We denote by $\text{COVER}_R(S)$ the family of all C-covering of S (with respect to R), where C is the set of all internal and external communications of S.

4.2.1 Procedures for Computing Components and Coverings

Now we are ready to present a method for computing of the components of a given information system (with respect to its reduct).

All normal components of a given information system $S = (U, A)$ (with respect to a reduct $R \in \text{RED}(S)$) can be obtained by the following procedure:

PROCEDURE for computing $\text{NCOMP}_R(S)$:

Input: An information system $S = (U, A)$, a reduct $R \in \text{RED}(S)$.

Output: Normal components of S (with respect to R), i.e., the set $\text{NCOMP}_R(S)$ without degenerated components of S.

Step 1. Compute all dependencies of the form: $R \underset{S}{\rightarrow} \{a\}$, for any $a \in A - R$.

Step 2. Compute the discernibility function $f_{M(S')}$ for each subsystem $S' = (U, R \cup \{a\})$ of S with $a \in A - R$. In this step we compute the so called $\{a\}$-reducts of R, for $a \in A - R$ (see: Proposition 7).

Step 3. For all dependencies of the form $B \underset{S}{\to} \{a_{i_1}\}, ..., B \underset{S}{\to} \{a_{i_k}\}$, where B is any subset of R obtained in Step 2, construct a dependency $B \underset{S}{\to} C$, where $C = \{a_{i_1}\} \cup ... \cup \{a_{i_k}\}$. The set C is the maximal (with respect to $\subseteq$) subset in $A - R$ such that the dependency $B \underset{S}{\to} C$ is true. Now the subsystem $S'' = (U', B \cup C)$ of S defines a normal component of S (with respect to R).

The correctness of this method follows from Proposition 1, Proposition 7 and from the definition of components (with respect to a reduct).

One can see that the time and space complexity of the discussed problem is, in general, exponential because of the complexity of RED(S) computing.

Example 22. Applying the above procedure for the computation $\text{NCOMP}_R(S)$ for the information system S of Example 1 and its reducts $R \in \text{RED}(S)$ we get that the system has two normal components: $S_1 = (U_1, A_1)$ (with respect to the reduct $R_1 = \{a, c\}$ of the form: $A_1 = B_1 \cup C_1, B_1 = \{a, c\}, C_1 = \{b\}$, and $S_2 = (U_2, A_2)$ (with respect to the reduct $R_2 = \{b\}$) such that $A_2 = B_2 \cup C_2$, where $B_2 = \{b\}, C_2 = \{a, c\}$.

There are only internal communications in the component $B_1 = \{a, c\}$ of the system S (with respect to the reduct R_1) represented by rules corresponding to non-trivial dependencies between the values of attributes within the set B_1 of the form: $a_1 \underset{S}{\Rightarrow} c_0, c_2 \underset{S}{\Rightarrow} a_0$.

To compute a covering of an information system by its components (with respect to a reduct) it is sufficient to perform the following procedure.

PROCEDURE for computing $\text{COVER}_R(S)$:

Input: An information system $S = (U, A)$, a reduct $R \in \text{RED}(S)$.

Output: A family of coverings of S, i.e., $\text{COVER}_R(S)$.

Step 1. Compute all normal and degenerated components of S (with respect to R).

Step 2. Compute the set C of all external and internal communications of S.

Step 3. Choose those combinations of components which together with C yield a C-covering by components of S (with respect to R). This step is to be performed as long as new solutions are obtained.

Example 23. The information system S of Example 1 has one covering $(\{S_1\}, C)$ (with respect to the reduct R_1) as well as one covering $(\{S_2\}, C')$ (with respect to the reduct R_2), where S_1, and S_2 denote components of S (with respect to these reducts). The set C of communications in the first covering consists of two rules: $a_1 \underset{S}{\Rightarrow} c_0, c_2 \underset{S}{\Rightarrow} a_0$, and the set C' is equal to the empty set.

An example of a construction of a given information system from its components one can find in the paper [279]. That construction is based on the approach presented in the subsection 3.1.

From Theorem 3.1 and 3.2 and the definition of a C-covering of a given information system we obtain the theorem establishing an important property of the decomposition of information systems proposed in this subsection.

Corollary 24. *Let S' be a maximal consistent extension of an information system S constructed by the method presented in subsection 3.1. Then $\text{COVER}_R(S) = \text{COVER}_R(S')$, for any reduct R of S.*

4.3 Summary

Our method for decomposition of information systems can be applied for automatic feature extraction. The properties of the constructed concurrent systems (e.g. their invariants) can be interpreted as higher level laws of experimental data. New features can be also obtained by performing for a given decision table $S = (U, A \cup \{d\})$ the following steps:

Step 1. Extract from S a subtable S_i corresponding to the decision i, for any $i \in V_d$, i.e., $S_i = (U_i, A_i)$, where $U_i = \{u \in U : d(u) = i\}$, $A_i = \{a^i : a \in A\}$, and $a^i(u) = a(u)$ for $u \in U_i$.

Step 2. Compute the components of S_i for any $i \in V_d$.

Step 3. For a new object u compute the values of components defined on information included in $Inf_A(u)$ and check in what cases the computed values of components are matching $Inf_A(u)$ (i.e., they are included in $Inf_A(u)$). For any $i \in V_d$ compute the ratio $n_i(u)$ of the number of components matching $Inf_A(u)$ to all components of S_i. The simplest strategy classifies u to the decision class i_0, where $n_{i_0}(u) = \max_i n_i(u)$.

The decomposition method presented here has been implemented in $C++$ and preliminary tests are promising [285].

We also study some applications of our method in control design from experimental data tables (see: section 7).

The application of Petri nets to representing a given information system enable us:

• to represent in an elegant and visual way the dependencies between components in the system, and their dynamic interactions,

• to observe concurrent and sequential subsystems (components) of the system.

On the basis of Petri net approach it was possible to understand better the structure of those rules which are true in a given information system.

5 Re-engineering of Cooperative Information Systems

5.1 Introduction

The aim of this section is to present a methodology for re-engineering the cooperative information systems ([80], [96], [70], [101]) when the specification for the system is being changed e.g. by adding new requirements.

The approach presented in this section consists of three basic steps with reference to decomposition of an information system.

First we show how experimental data tables are represented by information systems (see: subsection 2.1).

Further we discuss how an information system S can be decomposed (with respect to any of its reduct) into functional components linked by some communications (communication lines) which allow to preserve some constraints. The components together with the communications define coverings of S (see: subsection 4.2).

Next, we use the coverings of the information system S to construct its cooperative information system in the form of a marked Petri net (N_S, M_S) with the property described in subsection 3.1.8.

Finally in this section, we present a methodology for re-engineering (reconstruction) of the synthesized cooperative information system when the specification for the system is being changed e.g. by adding new requirements. The methodology can be treated as an adaptive strategy in re-engineering of cooperative information systems in the following sense: it allows to produce a plan of re-engineering of a given system by specifying which parts (components and communications) can remain unchanged and which must be changed to satisfy new requirements. The new requirements can be expressed e.g. by adding new objects into the specification, new attributes or new attribute values.

Moreover, we discuss how one can estimate the cost of the re-engineering when e.g. the cost of added new components, communications or their modifications are known. The cost of the re-engineering is treated as the cost of adding new functional modules, communication lines or the cost of their modifications.

5.2 The Statement of Problems

In this subsection there are formulated three re-engineering problems which are discussed in the following.

Re-engineering Problem (RP):

Let S be an information system representing a specification of a cooperative information system, CIS(S) be a cooperative information system produced by the synthesis algorithms (see: subsection 3.1) from the information system S, and let S' be an information system representing a new specification.

Produce a plan (algorithm) of the re-engineering of a given cooperative information system CIS(S) for construction from CIS(S) of a new cooperative information system CIS(S') satisfying the new specification represented by an information system S'.

Re-engineering Cost Problem 1 (RCP1):

Suppose a specification represented by an information system S together with a cost vector of attributes, a cooperative information system $CIS(S)$ produced by the synthesis algorithm (see: subsection 3.1) from the information system S, a new specification given by an information system S' and a new cost vector of attributes from S' are given.

Compute a cost $COST(S)$ of the re-engineering of a given cooperative information system $CIS(S)$ to a system $CIS(S')$ satisfying the new specification represented by an information system S' (with respect to any reduct from both systems).

Re-engineering Cost Problem 2 (RCP2):

Suppose as for RCP1, and let a non-negative real number k be given.

Compute a cost $COST(S)$ of the re-engineering of a given cooperative information system $CIS(S)$ to a system $CIS(S')$ satisfying the new specification represented by an information system (S') (if any) less than a given threshold k.

The problem RCP1 concerns the case in which the cost is computed for any reduct and any covering of old and new system. The problem RCP2 deals with the case when the upper constraint for the cost is given.

5.3 Re-engineering Problem Solution

Now we present a methodology for re-engineering of the synthesized cooperative information system when the specification for the system is changing. Our methodology is formulated in the form of procedure given below:

PROCEDURE for solving RP:

Input: A specification represented by information system S, a cooperative information system $CIS(S)$ produced by the synthesis algorithm presented in subsection 3.1 from an information system S, and a new specification given by an information system S'.

Output: A plan of the re-engineering of a given system $CIS(S)$ satisfying a new specification represented by an information system S'.

Step 1. Compute all dependencies in the table S'.

Step 2. Compute all normal and degenerated components of S' (with respect to any reduct R in S'); use the procedure $COMP_R(S)$.

Step 3. Compute the set C of all external and internal communications of S'.

Step 4. Remain unchanged components of S and unchanged communications of S.

Step 5. Change components and communications of $CIS(S)$ which are different in $CIS(S')$ according to the new rules (representing new specification) and add completely new components and communications to $CIS(S)$.

Now we show an example of a re-engineering of the cooperative information system (with respect to its reducts).

Example 24. Let S denote the data table from Example 1, CIS(S) be the net representation of S constructed by using the method described in subsection 3.1. Let us consider a new specification represented by a data table S' defined as in Table 15.

U'/A'	a	b	c
u_1	1	1	0
u_2	0	2	0
u_3	0	0	2
u_4	0	1	2

Table 15. A data table S' representing a new specification of a system

It is easy to see that the system S' is U'-extension of S; a new object u_4 has been added to the table S. It is also possible to modify the table S on many other ways e.g. by adding attributes or by extending the set of attribute values. We restrict our considerations in the section to adding new objects.

First we should repeat calculations for the data table S'. In consequence we get two reducts of S': $R'_1 = \{b, c\}$ and $R'_2 = \{a, b\}$.

Next applying the method described in subsection 2.1.8 for generating rules with a minimal number of descriptors on their left hand sides, for the information system S' we obtain that:

1. The set of rules corresponding to non-trivial dependencies between the values of attributes from the reduct R'_1 of S' with a is as follows:
   $b_0 \vee b_2 \vee c_2 \underset{S'}{\Rightarrow} a_0$, $b_1 \wedge c_0 \underset{S'}{\Rightarrow} a_1$.
   The last rule is new for the reduct R'_1.

2. The set of rules corresponding to all non-trivial dependencies between the values of attributes within the reduct R'_1 has the form: $b_2 \underset{S'}{\Rightarrow} c_0$, $b_0 \underset{S'}{\Rightarrow} c_2$. The cost of the set OPT(S', R'_1) of rules is changing and now is equal to 8 (or 7.2, see: Remark 2.1).

3. The set of rules corresponding to non-trivial dependencies between the values of attributes from the reduct R'_2 of S with c is changing and now it has the form:
   $a_1 \vee b_2 \underset{S'}{\Rightarrow} c_0$, $b_0 \underset{S'}{\Rightarrow} c_2$, $a_0 \wedge b_1 \underset{S'}{\Rightarrow} c_2$.
   The last rule is also new for this reduct.

4. The set of rules corresponding to all non-trivial dependencies between the values of attributes within the reduct R'_2 is the following: $b_0 \vee b_2 \underset{S'}{\Rightarrow} a_0$, $a_1 \underset{S'}{\Rightarrow} b_1$. The cost of the set OPT(S', R_2) of rules is equal to 6.5 (or 4.2, see: Remark 2.1).

It is also worth observing that in comparison with the set of all rules for the information system S, the rules $c_2 \underset{S'}{\Rightarrow} b_0$, $b_1 \underset{S'}{\Rightarrow} a_1$, $b_1 \underset{S'}{\Rightarrow} c_0$, and $a_0 \wedge c_0 \underset{S'}{\Rightarrow} b_2$ have been deleted.

Applying the procedures for computing normal components and for computing coverings for the information system S' we get in consequence two coverings. One covering $(\{S_1'\}, C''')$ (with respect to the reduct R_1'), where S_1' denotes the normal component (with respect to R_1') of the system S' of the form: $(S_1' = (U_1', A_1'), A_1' = B_1' \cup C_1', B_1' = \{b, c\}, C_1' = \{a\}$, and the set C''' of communications includes only internal communications of the form: $b_2 \underset{S'}{\Rightarrow} c_0$, and $b_0 \underset{S'}{\Rightarrow} c_2$. The external communications for this component do not exist because in this case we have in the information system S' only one component S_1' (with respect to R_1'). Beside, we have one covering $(\{S_2'\}, C'''')$ (with respect to the reduct R_2'), where S_2' denotes the normal component (with respect to R_2') of the system S' of the form: $(S_2' = (U_2', A_2'), A_2' = B_2' \cup C_2', B_2' = \{a, b\}, C_2' = \{c\}$. The set C'''' of communications consists of the rules mentioned above at the item 4.

From the considerations presented above it follows that both coverings of the system S' should be changed by adding new rules and deleting not valid rules (see also: Example 4).

5.4 The Solution of Re-engineering Problems under Cost Constraints

In order to compute the cost of re-engineering of a given cooperative information system, firstly one more notion has to be defined.

In the following it is assumed that we are given: a specification represented by a data table S, a cooperative information system CIS(S) produced by the synthesis algorithm presented in subsection 3.1 (cf. also [256]) from the data table S, the set OPT(S) of all rules with a minimal number of descriptors on the left hand side of S, the cost of components from the data table S, a new specification given by a data table S', a new specification represented by a data table S', a cooperative information system CIS(S') produced by the synthesis algorithm presented in subsection 3.1 from the data table S', the set OPT(S') of all rules with a minimal number of descriptors on the left hand side of S', the cost of components from the data table S'.

Let R_{rej} denotes the set of all rules rejected from the set OPT(S) of rules by new specification S', and let R_{add} be the set of all rules added to the set OPT(S) of rules by new specification S'. The cost of re-engineering of cooperative information system CIS(S) is the sum of costs of rules from the union of sets R_{rej} and R_{add}. The cost of re-engineering of CIS(S) is denoted by COST(S).

Remark. 5.1 Sometimes, from practical point of view, it is more convenient to assume that the cost of rules x from R_{rej} should be computed as $w * \text{COST}(x)$, where w denotes a weight of the cost COST(x) of the rule x, and w is a real number from the unit interval (0,1].

Remark. 5.2 The union of sets R_{rej} and R_{add} from the above definition is equal to the symmetric difference of the sets OPT(S) and OPT(S') of rules.

Now a procedure for solving RCP1 can be formulated in the following way:

PROCEDURE for solving RCP1:

Input: As for RCP1 (see: subsection 5.2).

Output: A cost COST(S) of the re-engineering of a given system CIS(S) for construction from CIS(S) (with respect to each reduct R of S) of a new cooperative information system CIS(S') satisfying the new specification represented by the data table S' (with respect to each reduct R' of S').

Step 1. Solve RP for the data table S' (with respect to any reduct R and any reduct R').

Step 2. Compute for each two reducts R and R' a cost of changed components and communications of CIS(S) which are different in CIS(S') according to the new rules (representing a new specification with respect to R') and a cost of completely new components and communications to CIS(S) (with respect to R) and add them.

Using results of the procedure for solving RCP1, one can easily describe a procedure for solving RCP2.

PROCEDURE for solving RCP2:

Input: As for RCP1 (see: subsection 5.2), and a non-negative real number k.

Output: A cost COST(S) of the re-engineering of a given system CIS(S) for construction from CIS(S) (with respect to each reduct R of S) of a new cooperative information system CIS(S') satisfying the new specification S' (with respect to each reduct R' of S') less than a given threshold k.

Step 1. Solve RCP1.

Step 2. Choose solutions of RCP1 satisfying a condition: "The cost COST(S) of re-engineering of CIS(S) is less than k".

Example 25. Let S denote the data table from Example 1, OPT(S) be the set of all rules of S from Example 4, *cost* be the cost of components of S from Example 1, CIS(S) be a cooperative information system for S obtained by using the method presented in subsection 3.1, a new specification given by a data table S' from Example 24, OPT(S') be the set of all rules of S' from Example 24, let *cost'* be the cost of components of S'; we assume that *cost'* = *cost*, and CIS(S') be a cooperative information system produced by our synthesis algorithm presented in subsection 3.1 from the data table S'. Then we have $R_{rej} = \{b_1 \underset{s}{\Rightarrow} a_1, b_1 \underset{s}{\Rightarrow} c_0, c_2 \underset{s}{\Rightarrow} b_0, a_0 \wedge c_0 \underset{s}{\Rightarrow} b_2\}$ and $R_{add} = \{a_0 \wedge b_1 \underset{s'}{\Rightarrow} c_2, b_1 \wedge c_0 \underset{s'}{\Rightarrow} a_1\}$.

The cost COST(S) of re-engineering of CIS(S) is equal to 5.3+8.3 $*w$, (or 7.9 $*w$, see: Remark 2.1) and w belongs to (0,1].

5.5 Summary

In this section there is proposed an approach to re-engineering of the synthesized cooperative information systems when the specification for the system is being changed. It provides also information about the cost of the re-engineering. This method can be applied in the system technology domain. On the base of this method at the Pedagogical University of Rzeszów a computer system supporting the user in the process of reconstructing of cooperative information systems has been developed [285]. It suggests the user which parts of the system can be unchanged and how to change the remain parts when the specification for the system is being changed. If the costs of modules and connections are known the system computes also the cost of re-engineering.

6 Real-Time Decision Making

6.1 Introduction

In this section we assume that a decision table S representing experimental knowledge is given [168]. It consists of a number of rows labeled by elements from a set of objects U, which contain the results of sensor measurements represented by a value vector of conditional attributes (conditions) from A together with a decision d corresponding to this vector. Values of conditions are identified by sensors in a finite but unknown number of time units. In some applications the values of conditional attributes can be interpreted as states of local processes in a complex system and the decision value is related to the global state of that system ([169], [239]). Sometimes it is necessary to transform a given experimental decision table by taking into account other relevant features (new conditional attributes) instead of the original ones. This step is necessary when the decision algorithm constructed directly from the original decision table yields an inadequate classification of unseen objects or when the complexity of decision algorithm synthesis from the original decision table is too high. In this case some additional time is necessary to compute the values of new features after the results of sensor measurements are given. The input for our algorithm consists of a decision table (if necessary, pre-processed as described above, see also: subsection 2.1.11).

We shall construct a parallel network allowing to make a decision as soon as a sufficient number of attribute values is known as a result of measurements and conclusions drawn from the knowledge encoded in S [240]. In the section we formulate this problem and present its solution.

First of all we assume that the knowledge encoded in S is represented by rules automatically extracted from S.

We consider only rules true in S, i.e., any object u from U matching the left hand side of the rule also matches its right hand side and there is an object u matching the left hand side of the rule. We assume that the knowledge encoded in S is complete in the sense that unseen objects have attribute value vectors consistent with rules extracted from S. This assumption may be too restrictive, since the rules for classifying new objects should be generated from relevant features (attributes) only. In this case we propose to apply only rules generated by dynamic reducts [15].

The rule is active if the values of all the attributes on its left hand side have been measured. An active rule which is true in S can be used to predict the value of the attribute on its right hand side even if its value has not been measured yet.

Our algorithm should propagate information from sensors (attributes) to other attributes as soon as possible. This is the reason for generating rules in minimal form, i.e., with a minimal number of descriptors on its left hand side. We present a method for generating rules which are minimal and true in S.

The final step of our algorithm consists in an implementation of the generated rule set by means of a Petri net (synchronized by a clock). The constructed Petri net occurs to have the properties of the parallel network mentioned above. We prove that any of its computations leading to decision making has minimal length, i.e., no prediction of the proper decision based on the knowledge encoded in S and the measurement of attribute values is possible before the end of the computation.

Each step of a computation of the constructed Petri net consists of two phases. In the first phase a checking is performed to see if some new values of conditions have been identified by sensors and next, in the second phase, new information about values is transmitted through the net at high speed. The whole process is realized by an implementation of the rules true in a given decision table.

The problem of active rule selection based on a given set of conditional attributes with values known at each given moment of time has been studied by many authors (e.g. in [71]). In our case the activated rules can activate new conditions. An iteration of this process can eventually activate (indicate) the proper decision. Our algorithm can be viewed as a parallel implementation of this idea.

The aim of our approach is to create basic tools for real-time applications in such areas as real-time decision making by groups of intelligent robots (agents) [175], error detecting in distributed systems [40], navigation of intelligent mobile robots [47] and in general in real-time knowledge-based control systems [229]. In our opinion, the rough set approach seems to be very suitable for real-time decision making from incomplete or uncertain information. We shall discuss this topic in another paper.

In the present section, we use ordinary Petri nets with priorities [78] as a model of the target system.

6.2 An Example

In this subsection we give an example of a decision table as well as we compute for that table the discernibility matrix, the discernibility function, reducts and the minimal form of rules in the decision table.

Example 26. Let us consider an example of a decision table $S = (U, A \cup \{d\})$ defined by the data table presented in Table 16.

U/A	a	b	c	d
u_1	1	0	1	0
u_2	0	0	0	1
u_3	2	0	1	0
u_4	0	0	1	2
u_5	1	1	1	0

Table 16. An example of a decision table

In the example we have $U = \{u_1, u_2, u_3, u_4, u_5\}$, $A = \{a, b, c\}$.

The decision is denoted by d. The possible values of attributes (conditions and the decision) from $A \cup \{d\}$ are equal to 0,1 or 2.

For this decision table we obtain the following discernibility matrix $M(S)$ presented in Table 17 and discernibility function $f_{M(S)}$ presented below:

U	u_1	u_2	u_3	u_4	u_5
u_1					
u_2	a,c,d				
u_3	a	a,c,d			
u_4	a,d	c,d	a,d		
u_5	b	a,b,c,d	a,b	a,b,d	

Table 17. The discernibility matrix $M(S)$ for the decision table S

$$f_{M(S)}(a,b,c,d) = a \wedge b \wedge (a \vee b) \wedge (a \vee d) \wedge (c \vee d) \wedge (a \vee b \vee d) \wedge \\ \wedge (a \vee c \vee d) \wedge (a \vee b \vee c \vee d).$$

After simplification (using the absorption laws) we get the following minimal disjunctive normal form of the discernibility function

$$f_{M(S)}(a,b,c,d) = a \wedge b \wedge (c \vee d) = \\ = (a \wedge b \wedge c) \vee (a \wedge b \wedge d).$$

There are two reducts: $R_1 = \{a,b,c\}$ and $R_2 = \{a,b,d\}$ of this decision table. Thus $\text{RED}(S) = \{R_1, R_2\}$.

Now we compute the set of rules corresponding to non-trivial functional dependencies between the values of conditions and the decision values as well as the set of rules corresponding to functional dependencies between the values of conditions of that decision table. In both cases we apply the method presented in subsection 2.1.8 (see also: Remark 2.2).

Let us start by computing decision rules corresponding to conditions $A = \{a,b,c\}$ and the decision d.

We have the decision table $S = (U, A \cup \{d\})$ from which we compute decision rules mentioned above.

U/A	a	b	c	d	d_d^A
u_1	1	0	1	0	{0}
u_2	0	0	0	1	{1}
u_3	2	0	1	0	{0}
u_4	0	0	1	2	{2}
u_5	1	1	1	0	{0}

Table 18. The decision table from Table 16 with the function d_d^A

In the table the values of the function d_d^A are also given. The discernibility matrix $M(S; d, v, u_l)$, where $v \in V_d$, $u_l \in U, l = 1, 2, 3, 4, 5$, obtained from $M(S)$ in the above way is presented in Table 19.

U	u_1	u_2	u_3	u_4	u_5
u_1		a, c		a	
u_2	a, c		a, c	c	a, b, c
u_3		a, c		a	
u_4	a	c	a		a, b
u_5		a, b, c		a, b	

Table 19. The discernibility matrix $M(S; d, v, u_l)$ for the matrix $M(S)$

Discernibility functions corresponding to the values of the function d_d^A are the following:

Case 1. For $d_d^A(u_1) = \{0\} : (a \vee c) \wedge a = a$.
Case 2. For $d_d^A(u_2) = \{1\}: (a \vee c) \wedge (a \vee c) \wedge c \wedge (a \vee b \vee c) = c$.
Case 3. For $d_d^A(u_3) = \{0\} : (a \vee c) \wedge a = a$.
Case 4. For $d_d^A(u_4) = \{2\}: a \wedge c \wedge a \wedge (a \vee b) = a \wedge c$.
Case 5. For $d_d^A(u_5) = \{0\}: (a \vee b \vee c) \wedge (a \vee b) = a \vee b$.

Hence we obtain the following decision rules:

$$a_1 \vee a_2 \vee b_1 \underset{S}{\Rightarrow} d_0, \; c_0 \underset{S}{\Rightarrow} d_1, a_0 \wedge c_1 \underset{S}{\Rightarrow} d_2.$$

Now we compute the rules corresponding to non-trivial functional dependencies between condition values of the decision table.

We have the following three subsystems $(U, B \cup \{c\})$, $(U, C \cup \{b\})$, $(U, D \cup \{a\})$ of S, where $B = \{a, b\}, C = \{a, c\}$, and $D = \{b, c\}$, from which we compute the rules mentioned above:

U/B	a	b	c	d_c^B
u_1	1	0	1	{1}
u_2	0	0	0	{0,1}
u_3	2	0	1	{1}
u_4	0	0	1	{0,1}
u_5	1	1	1	{1}

Table 20. The subsystem $(U, B \cup \{c\})$ with the function d_c^B, where $B = \{a, b\}$

U/C	a	c	b	d_b^C
u_1	1	1	0	$\{0,1\}$
u_2	0	0	0	$\{0\}$
u_3	2	1	0	$\{0\}$
u_4	0	1	0	$\{0\}$
u_5	1	1	1	$\{0,1\}$

Table 21. The subsystem $(U, C \cup \{b\})$ with the function d_b^C, where $C = \{a,c\}$

U/D	b	c	a	d_a^D
u_1	1	1	0	$\{0,1,2\}$
u_2	0	0	0	$\{0\}$
u_3	0	1	2	$\{0,1,2\}$
u_4	0	1	0	$\{0,1,2\}$
u_5	1	1	1	$\{1\}$

Table 22. The subsystem $(U, D \cup \{a\})$ with the function d_a^D, where $D = \{b,c\}$.

In the tables the values of the functions d_c^B, d_b^C, and d_a^D are also given.
Discernibility functions corresponding to the values of these functions are the following:

Table 20 *Case 1.*
    For $d_c^B(u_1) = \{1\}$: a.
    Case 2.
    For $d_c^B(u_3) = \{1\}$: a.
    Case 3.
    For $d_c^B(u_5) = \{1\}$: $a \vee b$.

Table 21 *Case 1.*
    For $d_b^C(u_2) = \{0\}$: $a \vee c$.
    Case 2.
    For $d_b^C(u_3) = \{0\}$: a.
    Case 3.
    For $d_b^C(u_4) = \{0\}$: a.

Table 22 *Case 1.*
    For $d_a^D(u_2) = \{0\}$: $c \wedge c \wedge (b \vee c) = c$.
    Case 2.
    For $d_a^D(u_5) = \{1\}$: $(b \vee c) \wedge b \wedge b = b$.

Hence we obtain the following rules:

From Table 20: $a_1 \vee a_2 \vee b_1 \underset{s}{\Longrightarrow} c_1$.
From Table 21: $a_0 \vee a_2 \vee c_0 \underset{s}{\Longrightarrow} b_0$.
From Table 22: $c_0 \underset{s}{\Longrightarrow} a_0$, $b_1 \underset{s}{\Longrightarrow} a_1$.

Eventually, we obtain the set OPT(S) of rules corresponding to all non-trivial functional dependencies in the considered decision table S:

$$a_1 \vee a_2 \vee b_1 \underset{S}{\Rightarrow} d_0,\ c_0 \underset{S}{\Rightarrow} d_1,$$
$$a_0 \wedge c_1 \underset{S}{\Rightarrow} d_2,\ a_1 \vee a_2 \vee b_1 \underset{S}{\Rightarrow} c_1,$$
$$a_0 \vee a_2 \vee c_0 \underset{S}{\Rightarrow} b_0,\ c_0 \underset{S}{\Rightarrow} a_0,\ b_1 \underset{S}{\Rightarrow} a_1.$$

6.3 Attribute Value Propagation

We present in this subsection the definition of a computation set for an information system S. We shall show later how this set of computations can be simulated by a network formed by a Petri net corresponding to S.

Computations of S represent sequences of configurations. The configuration following a given one (different from terminal) in the sequence is constructed by applying new measurement results of attribute values or by applying the knowledge encoded in the rules from $D(S)$.

If inf $\in$ INF(S), where $S = (U, A \cup \{d\})$ is a decision table and $B \subseteq A$, then by inf | B we denote the restriction of inf to B, i.e., $\{(a,v) \in$ inf: $a \in B\}$. By PART_ INF(S) we denote the set of all restrictions of information vectors from S. Elements of PART_ INF(S) are called *configurations*.

The empty function inf| $\emptyset$ will be interpreted as the *starting configuration*.

We construct extensions of a configuration inf| $B \in$ PART_ INF(S) which are obtained by appending some pairs (a,v) to inf| B, where $a \in A - B$ and v is either the result of a new measurement or is calculated by applying all rules from $D(S)$ active in the configuration inf| B.

We introduce two transition relations on configurations which allow to define precisely the above idea.

inf T_M inf' if and only if inf $\subseteq$ inf' and inf' is consistent with $D(S)$,

inf T_R inf' if and only if inf' $=$ inf $\cup \{(a,v)$: there exists a rule r in $D(S)$ active in inf with the right hand side equal to $a = v\}$.

We write inf T inf' if and only if inf T_M inf' or inf T_R inf'. By COMP(S) we denote the set of all finite sequences inf(0), inf(1), ..., inf(k) of configurations such that:

(i) inf(0) is the starting configuration;
(ii) inf(0) T_M inf(1),
inf(1) T_R inf(2),
...
inf(2p-1) T_M inf(2p),
inf(2p) T_R inf(2p+1),
...
inf(k-1) T inf(k);
(iii) d occurs in inf(k) but not in inf(i) for $i < k$ (inf(k) is called the *terminal configuration*). The elements the of COMP(S) are called *S-computations* (or *computations* of S) and they are denoted by *com*.

6.4 More about Petri Net Model

In this subsection Petri nets are used as a tool for computing a highly parallel program from a given decision table. After modeling a decision table by a Petri net states are identified in the net to an extent allowing to take the appropriate decisions.

In the subsection, it is assumed Petri nets are governed by the following transition firing rules:

Rule 1. As in subsection 2.2.

Rule 2. Conditions 1 and 2 of Rule 1 remain unchanged. Condition 3 now has the form:

3'. Whenever a transition t fires, t only reads a copy of each of its tokens in its input places (tokens in input places of t remain as a result of the firing of t), and t adds a token to each output place p' of t, provided the marking of p' is empty. Otherwise, the marking of p' is unchanged [116].

The significance of the first firing rule is that more than one transition can be enabled at the same time (concurrent processing is possible). The second transition rule has been introduced to facilitate modeling rules where tokens represent conditions in a logical implication. In the case where the output place of a transition t represents conclusion of a rule, the firing of t preserves the information (conditions) that enabled the transition t.

We also assume that if several transitions are simultaneously enabled in the same marking (i.e., transitions are *concurrent*) then they can be fired by in one and the same step and the resulting marking is computed. If by application of Rule 1 (see: subsection 2.2) the firing of a transition t disables an enabled transition t', where $t' \neq t$ *(conflict)*, the transition to fire is chosen at random among enabled transitions.

In the Petri nets constructed in this section we distinguish two types of transitions. The first type (with firing Rule 1) will correspond to sensors and the second (with firing Rule 2) to transitions transmitting information about identified local states throughout the net.

In this section we use ordinary Petri nets with priorities. All markings are binary, i.e., $m(p) \in \{0,1\}$ for any place p. We shall assume that there are only transitions with two different priorities: 0 and 1. All transitions corresponding to decisions have priority 1, and all the remaining - 0.

We assume that in the nets considered in the section transitions can be fired at discrete moments of time: $0, 1, 2, \ldots$. This additional timer which synchronizes transition firing can be easily implemented by connecting the presented nets with a special synchronizing module simulating a time clock. This modification allows to restrict transition firing only to discrete moments of time pointed out by the synchronizing module. We also assume that the period of the synchronizing clock is fixed (equal to τ).

6.5 Transformation of Decision Tables into Petri Nets

Now we present a method for transforming rules representing a given decision table into a Petri net.

In subsection 6.6 we show that the constructed net simulates the set COMP(S) of computations. The definition of simulation is presented in subsection 6.6.

The method of a Petri net construction for a given decision table is described in three stages. First, a method for constructing a net representing the set of all conditions of a given decision table is described. Then, the net obtained in the first stage is extended by adding the elements (arcs and transitions) of the net defined by a set of rules created on the basis of all non-trivial functional dependencies between the condition attribute values of a given decision table. Finally, we add to the net obtained in the above two steps the transitions and arcs representing the rules corresponding to all non-trivial functional dependencies between the condition attribute values and the decision values represented by the table. Such an approach makes the appropriate construction of a net much more readable.

Let $S = (U, A \cup \{d\})$ be a decision table, and let $V_a = \{v_1, \cdots, v_{k(a)}\}$ be a finite set of values for $a \in A$.

***Stage* 1.** The construction of a net representing the set of all conditions of a given decision table.

A net corresponding to any condition $a \in A$ has the structure and the initial marking presented in Figure 15.

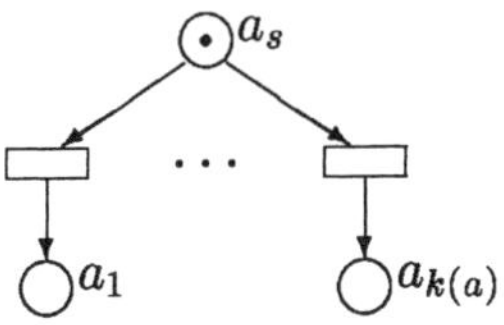

Fig. 15. The net corresponding to a condition with an initial marking

The place a_s is called the *start place* of the condition a. The remaining places of the net represent possible values of the condition a. The transitions of the net represent the process of identifying the values of the condition a. Let us observe that only the start place a_s is marked in the initial marking of the net. The transitions are labeled by values in the time interval $[0, +\infty)$ and a positive integer τ. This means that the time necessary to fire these transitions is $n \bullet \tau$, where n is unknown. Let us note that if the enabled transition corresponding to a distinguished condition is considered at any time $k \bullet \tau$ and there is no other enabled transition in the net, then that transition need not fire.

In the example which follows we illustrate the first stage of the method for the decision table from Example 6.2.

Example 27. Let us consider again the decision table from Example 6.2. The conditions a, b, and c are represented by the nets shown in Figure 16.

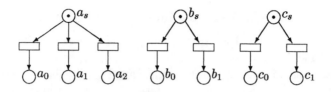

Fig. 16. Nets representing conditions a, b, and c

Stage 2. The construction of a net representing a set of rules corresponding to all non-trivial functional dependencies between the condition attribute values of a given decision table.

The rules in this case are of the form:

(1) $p_1 \wedge \cdots \wedge p_k \underset{S}{\Rightarrow} r$, where $k > 1$, or
(2) $p \underset{S}{\Rightarrow} r$.

By $p_1 \wedge \cdots \wedge p_k \vee q_1 \wedge \cdots \wedge q_l \underset{S}{\Rightarrow} r$ we shall denote the set of rules

$$p_1 \wedge \cdots \wedge p_k \underset{S}{\Rightarrow} r, \cdots, q_1 \wedge \cdots \wedge q_l \underset{S}{\Rightarrow} r.$$

The nets representing the rules (1) and (2) are illustrated in Figures 17 and 18, respectively.

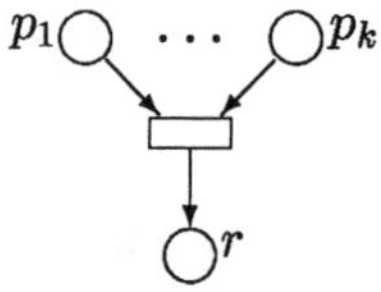

Fig. 17. Net representation of rules. Case (1): $p_1 \wedge \cdots \wedge p_k \underset{S}{\Rightarrow} r$, where $k > 1$

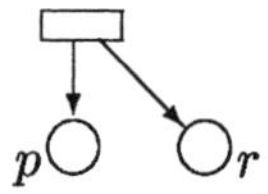

Fig. 18. Net representation of rules. Case (2): $p \underset{S}{\Rightarrow} r$

Example 28. Let us consider at first the following rules obtained in Example 6.2 for the decision table from the same example:

$$a_1 \vee a_2 \vee b_1 \underset{S}{\Rightarrow} c_1, \ a_0 \vee a_2 \vee c_0 \underset{S}{\Rightarrow} b_0, \ c_0 \underset{S}{\Rightarrow} a_0, \ b_1 \underset{S}{\Rightarrow} a_1.$$

In Figure 19 a net (N_1, M_1) representing the above rules is shown.

Stage 3. The construction of a net representing the rules corresponding to all non-trivial functional dependencies between the condition attribute values and the decision values represented by a given decision table.

In this case the rules have the form (1) or (2). The construction of a net representing this kind of rules is analogous to the case of the construction of a net for rules described in Stage 2. It is worth mentioning that the number of new transitions added in Stage 3 is equal to the number of rules corresponding to the decision of the given decision table, i.e., card(OPT(S,d)). Besides, to simplify the constructed nets, we do not draw the places corresponding to the values of the decision d of a given decision table.

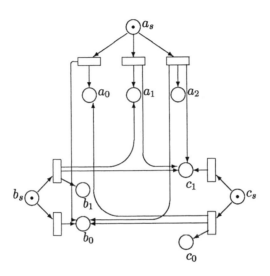

Fig. 19. Net representation of rules from Example 6.2

In the example which follows we illustrate the method described above.

Example 29. Let us now consider the rules corresponding to all non-trivial dependencies between the attribute values of the conditions a, b, c and the decision values d of the decision table from Example 6.2:

$$a_1 \vee a_2 \vee b_1 \underset{S}{\Longrightarrow} d_0, \quad c_0 \underset{S}{\Longrightarrow} d_1, \quad a_0 \wedge c_1 \underset{S}{\Longrightarrow} d_2.$$

We start from the nets constructed in Example 6.2 for the considered decision table. The resulting net $(N2, M2)$ of our construction is shown in Figure 20.

Now if we combine the nets (N_1, M_1) and (N_2, M_2) from Figures 19 and 20, we obtain the net (N_3, M_3) corresponding to the decision table from Example 6.2.

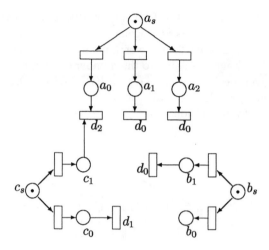

Fig. 20. Net representation of conditions a, b, c together with all non-trivial dependencies between them and the decision d

At the end of this section we describe how the resulting net (N_S, M_S) constructed as above should work. The net (N_S, M_S) starts from the initial marking in which any start place a_s for the condition a is marked by 1 and all the other places are marked by 0. (This marking describes the situation when there no value of the condition a is identified.)

Starting from the initial marking the net (N_S, M_S) performs the computation in steps described below.

T: TIME:=0;
A: **if** (there is an enabled transition for the decision)
   then fire it and go_ to RESTART;
   TIME:=TIME+τ;
B: Check if some values of conditions have been identified by sensors.
   if (the set of identified values of conditions is non-empty)
   then fire all transitions for corresponding conditions and put tokens in places representing those identified values;
C: **if** (there is an enabled transition for the decision)
   then fire it and go_ to RESTART;
D: fire all enabled transitions corresponding to rules and go_ to A;
   RESTART: Restart the initial marking of the net and go_ to T.

In the next subsection we shall study a fundamental property of the Petri net (N_S, M_S) constructed as above for a given decision table.

6.6 A Fundamental Property of the Petri Net for State Identification

The algorithm described in subsection 6.5 has a simple structure. Any finite computation of the constructed net can be divided into a finite number of segments. Each segment has two phases. In the first phase the values measured by sensors are caught by the net. In the second phase this information is transmitted throughout the net. This step is performed at very high speed by the part of the net which is an implementation of rules true in S.

We assume that outputs of sensors are measured at discrete moments of time τ, $\tau 2$, ... and the information propagation time throughout the part corresponding to rules is less than τ.

The computation ends when the decision is computed. After that the net is restarted. Our solution is minimal in the following sense: for an arbitrary finite computation the number of values of conditions identified by sensors necessary for decision taking is minimal. In consequence, the decision can be taken by the net after n units of time, each of length τ, and n is the smallest number with that property. Now we are going to present in a more formal way the computation property of the constructed nets described above.

The set P of places (different from "clock" places) in the net (N_S, M_S) constructed for a decision table S can be presented as the union of two disjoint sets: I and V, where $I = \{a_s : a \in A \text{ and } V = \bigcup_{a \in A}\{a = j : j \in V_a\}$.

Each marking m of the net (N_S, M_S) can be presented in a unique way as the union $m_I \cup m_V$ where m_I and m_V are the restrictions of m to I and V, respectively.

We now introduce two transition relations, namely, the input transition relation INPUT(S) and the message transition relation MT(S) of (N_S, M_S). Both are binary relations in the set of all markings m of (N_S, M_S) such that the information $X = \{(a, j) : m_V(a = j) = 1\}$ is consistent with any rule true in S, i.e., if $\bigwedge Y \underset{S}{\Rightarrow} a = j$, $y \subseteq X$ and a occurs in X, then $(a, j) \in X$.

We assume that m INPUT(S) m' if and only if for any $a \in A$ the following condition holds:

$m_I(a_s) = 1 \,\&\, m'_I(a_s) = 0 \,\&\, \exists! j \in V_a\, m'_V(a = j) = 1$
$\&\, m_V(a = j) = 0 \,\&\, \forall i \neq j\, m_V(a = j) = m'_V(a = j) \vee m_I(a_s) = m'_I(a_s)$
$\&\, \forall j \in V_a\, m'_V(a = j) = m_V(a = j)$

The meaning of these conditions is illustrated in Figures 21, 22 and 23.

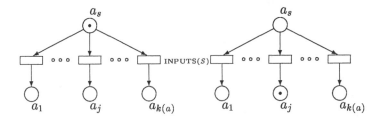

Fig. 21. Case 1: m INPUT(S) m' and $m \neq m'$

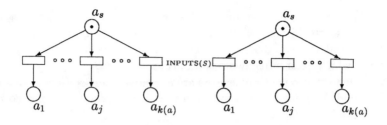

Fig. 22. Case 2: m INPUT(S) m' and $m = m'$

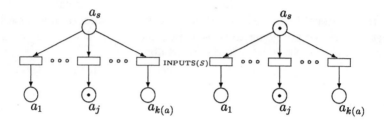

Fig. 23. Case 3: m INPUT(S) m' and $m = m'$

The second transition relation is defined by:

$m\mathrm{MT}(S)m' \Leftrightarrow m_I = m'_I \,\&\, \forall (a, j) \in \mathrm{DESC}(A, V) \forall X \subset \mathrm{DESC}(A, V)$

$\left(\forall x \in X(m_V(x) = 1) \,\&\, \left(\bigwedge X \underset{S}{\Rightarrow} a = j\right) \Rightarrow m'_V(a = j) = 1\right) \,\&\, \neg \exists X \subset \mathrm{DESC}(A, V)$

$\left(\forall x \in X(m_V(x) = 1) \,\&\, \left(\bigwedge X \underset{S}{\Rightarrow} a = j\right)\right) \Rightarrow (m'_V(a = j) = m_V(a = j)).$

Hence m MT(S) m' means that all differences between m' and m are due to some rules true in S.

A finite computation of (N_S, M_S) is any finite sequence $(m(0), ..., m(k))$ of its markings such that

(1) $m^{(0)}$ is the initial marking, i.e., $m_I^{(0)}(a_s) = 1$ for $a \in A$ and $m_V^{(0)}(x) = 0$ for $x \in V$;

(2) $m^{(k)}$ is a terminal marking, i.e., there exists an enabled transition corresponding to the decision;

(3) $m^{(i)}$ (INPUT(S);MT(S)) $m^{(i+1)}$ for $i = 0, ..., k - 1$, where denotes composition of binary relations.

It follows from our assumption that the transition from a marking $m^{(i)}$ to $m^{(i+1)}$ takes less than τ units of time.

In subsection 6.5 we have presented an algorithm for constructing a Petri net (N_S, M_S) corresponding to a given S. The constructed Petri net simulates the set COMP(S) of computations and now we are going to explain this in more detail.

We define a one-to-one coding function *code* mapping configurations of S to markings of (N_S, M_S):

code(inf) = m if and only if $[(a, v) \in$ inf if and only if the place labeled by $a = v$ is marked in $m]$ & for any a: $[a_S$ is marked in m if and only if a does not occur in inf].

The function code can be extended to sequences of configurations by code(inf,...) = (code(inf),...).

Theorem 25. *Let $S = (U, A \cup \{d\})$ be a given decision table and let (N_S, M_S) be the Petri net corresponding to S constructed by the algorithm presented above. Then we have $COMP(N_S, M_S) = \{code(com) : com \in COMP(S)\}$.*

Proof: One can verify that:

(i) inf T_M inf' if and only if code(inf) INPUT(S) code(inf'),
(ii) inf T_R inf' if and only if code(inf) MT(S) code(inf'),
(iii) (d, v) is in inf if and only if the place labeled by $d = v$ is marked in code(inf).

Hence by induction on the length of computations we obtain our theorem.

We also obtain the following property of (finite) computations of (N_S, M_S):

Theorem 26. *Let $(m^{(0)}, ..., m^{(k)})$ be an arbitrary finite computation of (N_S, M_S). Then*

1. *The decision rule $\bigwedge\{a = j : m_V^{(k)}(a = j) = 1 \ \& \ a \in A\} \Rightarrow d = r$, where d_r is the enabled transition in $m^{(k)}$, is true in S.*
2. *For any jjk and $r \bigwedge\{a = p : m_V^{(j)}(a = p) = 1 \ \& \ a \in A\} \Rightarrow d = r$ is not true in S.*

6.7 Summary

The presented algorithm enables a very fast identification of objects specified in a given decision table. Nevertheless, some parts of the constructed net can be redundant, i.e., unnecessary for taking proper decisions and slowing down the process of decision taking. The papers ([182], [183]) contain some extension of our results presented in this section. The contribution of the paper is the presentation of an approach to modeling real-time decision systems with rough fuzzy Petri nets [180]. The advantage to doing this derives from fact that computations performed by such a system can be simulated and analyzed. The basic problem for real-time knowledge based control algorithms [229] is related to the time/space trade-off. For example, if there is not enough memory to store the parallel network representing knowledge encoded in a given decision table, it is necessary to reduce it, which can lead to an increase of time necessary for making decisions. Another interesting problem arises when decision tables change with time and the constructed net ought to be modified by applying some strategies discovered during the process of changes. These are examples of problems which we would like to investigate applying the approach presented in the section.

7 An Application of Rough Set Methods in Control Design

7.1 Introduction

The designing of control for complex systems or devices is a very important research task from theoretical as well as practical point of view. A substantial effort has been made by rough set community to develop rough control methods (see e.g. [113], [114], [115], [139], [140], [141], [158], [167], [170], [173], [49], [137], [135], [136], [185], [292], [321], [81], [320]). The aim of the section is to demonstrate a methodology for the automatic control design from the specification of the discrete event system given in the form of a decision table [168].

In fact, in the present section we investigate a problem which informally can be stated as follows:

Control Design Problem (CDP):
Let S be a decision table representing a specification of a given discrete event control system.
Produce a plan (algorithm) of the design of a given discrete event control system.

An algorithm for computing a highly parallel program represented by a Petri net from a given decision table has been proposed in the paper [259] (see also: section 6). In this section we show an application of that algorithm after some modifications for CDP.

In our approach we use dependencies and rules describing interactions between different components of processes to design rule-based control of processes and taking into account a typical control loop. This allows to base the control design on deeper analysis of knowledge about processes represented in decision tables than in some other approaches based only on decision rules (see e.g. [115], [135], [170], [321]). For example we take into account that controlled process is composed out of interacting modules and the necessity of those modules coordination by the controller.

In the present section, we use ordinary Petri nets with priorities [186] as a model of the target system.

We illustrate our ideas by an example of control design for a very simple dosing tank presented in [81].

7.2 A Standard Scheme of a Controlled System

To make the section self-contained, the running example, adopted from [81], is shortly sketched. In Figure 24 we show a standard scheme of a controlled system. The system consists of two main elements: the plant and the controller. The plant is the system which is to be controlled by the controller, and the control is the system which imposes control to the plant. A typical control loop includes these two elements and it is closed by a flow of sensor signals and actor signals as depicted in Figure 24.

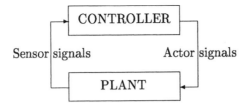

Fig. 24. Scheme of a control loop

7.3 The Control Model

The traditional approach to the control design of systems requires that control strategy to achieve a specific goal be known in advance and clearly described in the form of a control algorithm. In our approach the control strategy will be characterized by parameters of the process which are directly related to the imposed control strategy. The values of parameters define the observable states of the process. The proper control decision corresponds to each observable state. Hence, from the control viewpoint, a control process can be characterized by:

- the space of observable states determined by the state parameters,
- the space of control determined by control parameters.

In the rough set approach, the space of observable states is explicitly defined by a finite set of condition attributes, and the space of control is explicitly determined by a finite set of decision attributes. For our purposes it will be sufficient to consider decision tables with one decision only because one can always transform a decision table with more than one decision into a decision table with exactly one decision by simple coding. The knowledge about the control is represented by a set of decision rules. In such notation the rules represent dependencies between the set of conditions and the decision. Hence, a process control can be represented in the form of a decision table. The obtained decision table can be analyzed from many viewpoints. We shall analyze the decision table representing control in order to obtain:

1. Elimination of these condition attributes which do not influence the relations between the set of values of condition attributes and the decision.
2. Computing the dependencies between the set of values of condition attributes and the decision, and vice versa.
3. Reduction of redundant rules.
4. Checking whether there are no contradictory decision rules, i.e. rules in which different values of the decision correspond to the same values of condition attributes.

7.4 Illustrative Example

In this subsection we discuss how to pass from the informal description of the behavior of an exemplary plant to the synthesis of decision rules for the control of that plant, and next to a concurrent control representation (a control program) in the form a Petri net using rough set methods.

At first we discuss a simplified version of a plant presented in [81]. A scheme of the plant is illustrated in Figure 25. It consists of the dosing tank and two valves. The dosing tank can be filled and discharged by operating the valves denoted by A and B. These valves are operated by a human operator as well as by a controller. The level in the tank as well as the positions of the valves are indicated by switching sensors. It is assumed that a control ensures the following behavior:

1. Both valves can not be opened at the same time.
2. The tank can not be overflow.
3. The tank must be filled completely before it must be emptied completely.

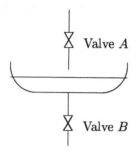

Fig. 25. A scheme of the dosing tank

It is easy to see that the observable state of the plant can be characterized by the following condition attributes:

oA - the valve A operated by the human operator,
oB - the valve B operated by the human operator,
vA - the valve A,
vB - the valve B,
tk - the tank.

The characteristic states of the control process (controller) are:

por - the initial state of controller,
fil - filling of the tank,
wai - waiting for opening vB,
emp - emptying of the tank, i.e., $V_{ctrl} = \{por, fil, wai, emp\}$.

We assume that:

1) the set of attribute values oA, oB, vA and vB are equal and they have the same elements, i.e., $V_{oA} = V_{oB} = V_{vA} = V_{vB} = \{c, o\}$, where:

c - denotes that the valve A (B) is closed,
o - denotes that the valve A (B) is open;

2) the set of attribute values tk consists of elements e, pf and f, where:

e - denotes that the tank is empty,
pf - denotes that the tank is partially emptying,
f - denotes that the tank is full, i.e., $V_{tk} = \{e, pf, f\}$.

The above description of the considered control of the plant in the form of a decision table $S = (U, A \cup \{d\})$ shown in Table 23, where the set of objects $U = \{u_1, u_2, u_3, u_4, u_5, u_6\}$, the set of condition attributes $A = \{oA, oB, vA, vB, tk\}$. The decision is denoted by $ctrl$.

7.5 Rough Sets Analysis

Now we present, using our illustrative example, the basic stages of the rough set analysis needed for our purposes.

7.5.1 Reduction of Parameters

Rough set methods offer possibility to reduce dispensable information.

From Proposition 2.2 a method follows for computing the set of all reducts of a given information system. For the decision table S presented in Table 23 we obtain the discernibility matrix $M(S)$ presented in Table 24 and the discernibility function $f_{M(S)}$ presented below:

U/A	oA	oB	vA	vB	tk	$ctrl$
u_1	c	c	c	c	e	por
u_2	o	c	o	c	e	fil
u_3	o	c	o	c	pf	fil
u_4	c	c	c	c	f	wai
u_5	c	o	c	o	f	emp
u_6	c	o	c	o	pf	emp

Table 23. An example of a decision table S

U	u_1	u_2	u_3	u_4	u_5	u_6
u_1						
u_2	$oA, vA, ctrl$					
u_3	$oA, vA, tk, ctrl$	tk				
u_4	$tk, ctrl$	$oA, vA, tk, ctrl$	$oA, vA, tk, ctrl$			
u_5	$oB, vB, tk, ctrl$	$oA, oB, vA, vB, tk, ctrl$	$oA, oB, vA, vB, tk, ctrl$	$oB, vB, ctrl$		
u_6	$oB, vB, tk, ctrl$	$oA, oB, vA, vB, tk, ctrl$	oA, oB, vA, vB, tk	$oB, vB, tk, ctrl$	tk	

Table 24. The discernibility matrix $M(S)$ for the decision table S

$f_{M(S)}(oA, oB, vA, vB, tk, ctrl) = tk \wedge (tk \vee ctrl) \wedge (oA \vee vA \vee ctrl) \wedge (oB \vee vB \vee ctrl) \wedge (oA \vee oB \vee vA \vee vB \vee ctrl) \wedge (oA \vee vA \vee tk \vee ctrl) \wedge (oB \vee vB \vee tk \vee ctrl) \wedge (oA \vee oB \vee vA \vee vB \vee tk \vee ctrl)$.

After simplification (using the absorption laws) we get the minimal disjunctive normal form of the discernibility function as follows:

$f_{M(S)}(oA, oB, vA, vB, tk, ctrl) = (oA \wedge oB \wedge tk) \vee (oA \wedge vB \wedge tk) \vee (oB \wedge vA \wedge tk) \vee (vA \wedge vB \wedge tk) \vee (tk \wedge ctrl)$.

There are five reducts: $R_1 = \{oA, oB, tk\}$, $R_2 = \{oA, vB, tk\}$, $R_3 = \{oB, vA, tk\}$, $R_4 = \{vA, vB, tk\}$ and $R_5 = \{tk, ctrl\}$ of the system. Thus $\text{RED}(S) = \{R_1, R_2, R_3, R_4, R_5\}$.

7.5.2 Extraction of Dependencies

Extraction of all strong components from a given data table allows us among others to delete the redundant rules from the set of rules representing knowledge included in the table.

By Proposition 2.3 we have for the system S the dependencies:

$$\{oA, oB, tk\} \overrightarrow{S} \{vA, vB, ctrl\},$$
$$\{oA, vB, tk\} \overrightarrow{S} \{vA, oB, ctrl\},$$
$$\{oB, vA, tk\} \overrightarrow{S} \{oA, vB, ctrl\},$$
$$\{vA, vB, tk\} \overrightarrow{S} \{oA, oB, ctrl\},$$
$$\{tk, ctrl\} \overrightarrow{S} \{oA, oB, vA, vB\}.$$

Next, by Proposition 2.4 we get the following elementary dependencies:

$$\{oA, oB, tk\} \overrightarrow{S} \{vA\}, \{oA, oB, tk\} \overrightarrow{S} \{vB\},$$
$$\{oA, oB, tk\} \overrightarrow{S} \{ctrl\}, \{oA, vB, tk\} \overrightarrow{S} \{vA\},$$
$$\{oA, vB, tk\} \overrightarrow{S} \{oB\}, \{oA, vB, tk\} \overrightarrow{S} \{ctrl\},$$
$$\{oB, vA, tk\} \overrightarrow{S} \{oA\}, \{oB, vA, tk\} \overrightarrow{S} \{vB\},$$
$$\{oB, vA, tk\} \overrightarrow{S} \{ctrl\}, \{vA, vB, tk\} \overrightarrow{S} \{oA\},$$
$$\{vA, vB, tk\} \overrightarrow{S} \{oB\}, \{vA, vB, tk\} \overrightarrow{S} \{ctrl\},$$
$$\{tk, ctrl\} \overrightarrow{S} \{oA\}, \{tk, ctrl\} \overrightarrow{S} \{oB\},$$
$$\{tk, ctrl\} \overrightarrow{S} \{vA\}, \{tk, ctrl\} \overrightarrow{S} \{vB\}.$$

Now, applying the procedure for computing of the strong components from the decision table S (see: subsection 2.1.7) we obtain six strong components in S of the form:

(B,C), (D,E), (F,G), (F,H), (F,I), (F,J) with $B = \{oA\}$, $C = \{vA\}$, $D = \{oB\}$, $E = \{vB\}$, $F = \{ctrl\}$, $G = \{oA, oB, tk\}$, $H = \{oA, vB, tk\}$, $I = \{oB, vA, tk\}$, $J = \{vA, vB, tk\}$.

7.5.3 Generation of Rules

In the subsection we describe how one can pass from a given decision table to the set of all rules corresponding to non-trivial functional dependencies in S. In our example the rules generated from the given decision table S represent knowledge about the control of the plant.

Let us consider the given decision table S and the discernibility matrix for S presented in Table 24. We compute the set of rules corresponding to non-trivial functional dependencies between the values of conditions and the decision values as well as the set of rules corresponding to functional dependencies between the values of conditions of that decision table. In both cases we apply the method presented in subsection 2.1.

Let us start by computing the decision rules corresponding to the conditions $A = \{oA, oB, vA, vB, tk\}$ and the decision $ctrl$.

We have the decision table $S = (U, A \cup \{ctrl\})$ from which we compute the decision rules mentioned above.

In Table 25 the values of the function d_{ctrl}^A are also given. The discernibility matrix $M(S; ctrl, v, u_l)$ where $v \in V_{ctrl}$, $u_l \in U$, $l = 1, 2, 3, 4, 5, 6$, obtained from $M(S)$ in the above way is presented in Table 26.

U/A	oA	oB	vA	vB	tk	ctrl	d_{ctrl}^A
u_1	c	c	c	c	e	por	$\{por\}$
u_2	o	c	o	c	e	fil	$\{fil\}$
u_3	o	c	o	c	pf	fil	$\{fil\}$
u_4	c	c	c	c	f	wai	$\{wai\}$
u_5	c	o	c	o	f	emp	$\{emp\}$
u_6	c	o	c	o	pf	emp	$\{emp\}$

Table 25. The decision table S with the function d_{ctrl}^A

U	u_1	u_2	u_3	u_4	u_5	u_6
u_1		oA, vA	oA, vA, tk	tk	oB, vB, tk	oB, vB, tk
u_2	oA, vA			oA, vA, tk	oA, oB, vA, vB, tk	oA, oB, vA, vB, tk
u_3	oA, vA, tk			oA, vA, tk	oA, oB, vA, vB, tk	oA, oB, vA, vB
u_4	tk	oA, vA, tk	oA, vA, tk		oB, vB	oB, vB, tk
u_5	oB, vB, tk	oA, oB, vA, vB, tk	oA, oB, vA, vB, tk	oB, vB		
u_6	oB, vB, tk	oA, oB, vA, vB, tk	oA, oB, vA, vB	oB, vB, tk		

Table 26. The discernibility matrix $M(S; ctrl, v, u_l)$ for the matrix $M(S)$

Discernibility functions corresponding to the values of the function d_{ctrl}^A are the following:

Case 1. For $d_{ctrl}^A(u_1) = \{por\}$: $(oA \lor vA) \land (oA \lor vA \lor tk) \land tk \land (oB \lor vB \lor tk) = (oA \lor vA) \land tk = oA \land tk \lor vA \land tk$.

Case 2. For $d_{ctrl}^A(u_2) = \{fil\}$: $(oA \lor vA) \land (oA \lor vA \lor tk) \land (oA \lor oB \lor vA \lor vB \lor tk) = oA \lor vA$.

Case 3. For $d_{ctrl}^A(u_3) = \{fil\}$: $(oA \lor vA \lor tk) \land (oA \lor oB \lor vA \lor vB \lor tk) \land (oA \lor oB \lor vA \lor vB) = oA \lor vA \lor oB \land tk \lor vB \land tk$.

Case 4. For $d_{ctrl}^A(u_4) = \{wai\}$: $tk \land (oA \lor vA \lor tk) \land (oB \lor vB) \land (oB \lor vB \lor tk) = tk \land oB \lor tk \land vB$.

Case 5. For $d^A_{ctrl}(u_5) = \{emp\} : (oB \lor vB \lor tk) \land (oA \lor oB \lor vA \lor vB \lor tk) \land (oB \lor vB) = oB \lor vB$.

Case 6. For $d^A_{ctrl}(u_5) = \{emp\} : (oB \lor vB \lor tk) \land (oA \lor oB \lor vA \lor vB \lor tk) \land (oA \lor oB \lor vA \lor vB) \land (oB \lor vB \lor tk) = (oB \lor vB \lor tk) \land (oA \lor oB \lor vA \lor vB) = oB \lor vB \lor oA \land tk \lor vA \land tk$.

Hence we obtain the following decision rules:

For *Case 1*: $oA(c) \land tk(e) \underset{S}{\Rightarrow} ctrl(por)$, $vA(c) \land tk(e) \underset{S}{\Rightarrow} ctrl(por)$.

For *Case 2*: $oA(o) \underset{S}{\Rightarrow} ctrl(fil), vA(o) \underset{S}{\Rightarrow} ctrl(fil)$.

For *Case 3*: $oA(o) \underset{S}{\Rightarrow} ctrl(fil), vA(o) \underset{S}{\Rightarrow} ctrl(fil), oB(c) \land tk(pf) \underset{S}{\Rightarrow} ctrl(fil)$, $vB(c) \land tk(pf) \underset{S}{\Rightarrow} ctrl(fil)$.

For *Case 4*: $vB(c) \land tk(f) \underset{S}{\Rightarrow} ctrl(wai), oB(c) \land tk(f) \underset{S}{\Rightarrow} ctrl(wai)$.

For *Case 5*: $oB(o) \underset{S}{\Rightarrow} ctrl(emp), vB(o) \underset{S}{\Rightarrow} ctrl(emp)$.

For *Case 6*: $oA(c) \land tk(pf) \underset{S}{\Rightarrow} ctrl(emp), vA(c) \land tk(pf) \underset{S}{\Rightarrow} ctrl(emp), oB(o) \underset{S}{\Rightarrow} ctrl(emp), vB(o) \underset{S}{\Rightarrow} ctrl(emp)$.

In order to obtain the remaining rules corresponding to all non-trivial functional dependencies between attribute values of the decision table it is sufficient to consider the following five subsystems $(U, B \cup \{tk\})$, $(U, C \cup \{vB\})$, $(U, D \cup \{vA\})$, $(U, E \cup \{oB\})$, $(U, F \cup \{oA\})$ of S, where $B = \{oA, oB, vA, vB, ctrl\}, C = \{oA, oB, vA, tk, ctrl\}, D = \{oA, oB, vB, tk, ctrl\}, E = \{oA, vA, vB, tk, ctrl\}$, and $F = \{oB, vA, vB, tk, ctrl\}$. Proceeding analogously as above, we obtain the following rules:

For the system $(U, B \cup \{tk\})$:

$ctrl(por) \underset{S}{\Rightarrow} tk(e)$, $ctrl(wai) \underset{S}{\Rightarrow} tk(f)$.

For the system $(U, C \cup \{vB\})$:
$oB(c) \underset{S}{\Rightarrow} vB(c), oB(o) \underset{S}{\Rightarrow} vB(o), tk(e) \underset{S}{\Rightarrow} vB(c), ctrl(por) \underset{S}{\Rightarrow} vB(c), ctrl(fil) \underset{S}{\Rightarrow} vB(c), ctrl(wai) \underset{S}{\Rightarrow} vB(c), ctrl(emp) \underset{S}{\Rightarrow} vB(o), oA(o) \underset{S}{\Rightarrow} vB(c), vA(o) \underset{S}{\Rightarrow} vB(c)$,
$oA(c) \land tk(pf) \underset{S}{\Rightarrow} vB(o), vA(c) \land tk(pf) \underset{S}{\Rightarrow} vB(o)$.

For the system $(U, D \cup \{vA\})$:
$oA(c) \underset{S}{\Rightarrow} vA(c), oA(o) \underset{S}{\Rightarrow} vA(o), ctrl(por) \underset{S}{\Rightarrow} vA(c), ctrl(fil) \underset{S}{\Rightarrow} vA(o), ctrl(wai) \underset{S}{\Rightarrow} vA(c), ctrl(emp) \underset{S}{\Rightarrow} vA(c), oB(c) \land tk(pf) \underset{S}{\Rightarrow} vA(o), vB(c) \land tk(pf) \underset{S}{\Rightarrow} vA(o)$,
$tk(f) \underset{S}{\Rightarrow} vA(c), oB(o) \underset{S}{\Rightarrow} vA(c), vB(o) \underset{S}{\Rightarrow} vA(c)$.

For the system $(U, E \cup \{oB\})$:
$vB(c) \underset{S}{\Rightarrow} oB(c), vB(o) \underset{S}{\Rightarrow} oB(o), tk(e) \underset{S}{\Rightarrow} oB(c), ctrl(por) \underset{S}{\Rightarrow} oB(c), ctrl(fil) \underset{S}{\Rightarrow}$

$oB(c), ctrl(wai) \underset{S}{\Rightarrow} oB(c), ctrl(emp) \underset{S}{\Rightarrow} oB(o), oA(o) \underset{S}{\Rightarrow} oB(c), vA(o) \underset{S}{\Rightarrow} oB(c),$
$oA(c) \wedge tk(pf) \underset{S}{\Rightarrow} oB(o), vA(c) \wedge tk(pf) \underset{S}{\Rightarrow} oB(o).$

For the system $(U, F \cup \{oA\})$:

$vA(c) \underset{S}{\Rightarrow} oA(c), vA(o) \underset{S}{\Rightarrow} oA(o), ctrl(por) \underset{S}{\Rightarrow} oA(c), ctrl(fil) \underset{S}{\Rightarrow} oA(o), ctrl(wai) \underset{S}{\Rightarrow} oA(c), ctrl(emp) \underset{S}{\Rightarrow} oA(c), oB(c) \wedge tk(pf) \underset{S}{\Rightarrow} oA(o), vB(c) \wedge tk(pf) \underset{S}{\Rightarrow} oA(o), tk(f) \underset{S}{\Rightarrow} oA(c), oB(o) \underset{S}{\Rightarrow} oA(c), vB(o) \underset{S}{\Rightarrow} oA(c).$

Eventually, we obtain the set OPT(S) of rules corresponding to all non-trivial functional dependencies in the considered decision table S:

$oA(c) \wedge tk(e) \underset{S}{\Rightarrow} ctrl(por), vA(c) \wedge tk(e) \underset{S}{\Rightarrow} ctrl(por), oA(o) \underset{S}{\Rightarrow} ctrl(fil), vA(o) \underset{S}{\Rightarrow} ctrl(fil),$

$oB(c) \wedge tk(pf) \underset{S}{\Rightarrow} ctrl(fil), vB(c) \wedge tk(pf) \underset{S}{\Rightarrow} ctrl(fil), vB(c) \wedge tk(f) \underset{S}{\Rightarrow} ctrl(wai),$
$oB(c) \wedge tk(f) \underset{S}{\Rightarrow} ctrl(wai), oB(o) \underset{S}{\Rightarrow} ctrl(emp), vB(o) \underset{S}{\Rightarrow} ctrl(emp),$
$oA(c) \wedge tk(pf) \underset{S}{\Rightarrow} ctrl(emp), vA(c) \wedge tk(pf) \underset{S}{\Rightarrow} ctrl(emp),$
$ctrl(por) \underset{S}{\Rightarrow} tk(e), ctrl(wai) \underset{S}{\Rightarrow} tk(f), ctrl(por) \underset{S}{\Rightarrow} vB(c), ctrl(fil) \underset{S}{\Rightarrow} vB(c),$
$ctrl(wai) \underset{S}{\Rightarrow} vB(c), ctrl(emp) \underset{S}{\Rightarrow} vB(o), ctrl(por) \underset{S}{\Rightarrow} vA(c), ctrl(fil) \underset{S}{\Rightarrow} vA(o),$
$ctrl(wai) \underset{S}{\Rightarrow} vA(c), ctrl(emp) \underset{S}{\Rightarrow} vA(c), ctrl(por) \underset{S}{\Rightarrow} oB(c), ctrl(fil) \underset{S}{\Rightarrow} oB(c),$
$ctrl(wai) \underset{S}{\Rightarrow} oB(c), ctrl(emp) \underset{S}{\Rightarrow} oB(o), ctrl(por) \underset{S}{\Rightarrow} oA(c), ctrl(fil) \underset{S}{\Rightarrow} oA(o),$
$ctrl(wai) \underset{S}{\Rightarrow} oA(c), ctrl(emp) \underset{S}{\Rightarrow} oA(c),$
$oB(c) \underset{S}{\Rightarrow} vB(c), oB(o) \underset{S}{\Rightarrow} vB(o), tk(e) \underset{S}{\Rightarrow} vB(c), oA(o) \underset{S}{\Rightarrow} vB(c), vA(o) \underset{S}{\Rightarrow} vB(c), oA(c) \wedge tk(pf) \underset{S}{\Rightarrow} vB(o), vA(c) \wedge tk(pf) \underset{S}{\Rightarrow} vB(o),$
$oA(c) \underset{S}{\Rightarrow} vA(c), oA(o) \underset{S}{\Rightarrow} vA(o), oB(c) \wedge tk(pf) \underset{S}{\Rightarrow} vA(o), vB(c) \wedge tk(pf) \underset{S}{\Rightarrow} vA(o), tk(f) \underset{S}{\Rightarrow} vA(c), oB(o) \underset{S}{\Rightarrow} vA(c), vB(o) \underset{S}{\Rightarrow} vA(c),$
$vB(c) \underset{S}{\Rightarrow} oB(c), vB(o) \underset{S}{\Rightarrow} oB(o), tk(e) \underset{S}{\Rightarrow} oB(c), oA(o) \underset{S}{\Rightarrow} oB(c), vA(o) \underset{S}{\Rightarrow} oB(c),$
$oA(c) \wedge tk(pf) \underset{S}{\Rightarrow} oB(o), vA(c) \wedge tk(pf) \underset{S}{\Rightarrow} oB(o),$
$vA(c) \underset{S}{\Rightarrow} oA(c), vA(o) \underset{S}{\Rightarrow} oA(o), oB(c) \wedge tk(pf) \underset{S}{\Rightarrow} oA(o), vB(c) \wedge tk(pf) \underset{S}{\Rightarrow} oA(o), tk(f) \underset{S}{\Rightarrow} oA(c), oB(o) \underset{S}{\Rightarrow} oA(c), vB(o) \underset{S}{\Rightarrow} oA(c).$

7.5.4 Reduction of Redundant Rules

Now we consider the rule reduction problem. In many cases we have the redundant sets of rules. If we find the strong components of a given decision table then we can use only some part of rules representing all non-trivial functional dependencies between values of attributes from those sets. This question we explain more precisely using an example presented below.

Let us consider again the set OPT(S) of rules corresponding to all non-trivial functional dependencies in the given decision table S.

Since the following sets of attributes $\{oA\}$ and $\{vA\}$, $\{oB\}$ and $\{vB\}$, $\{ctrl\}$ and $\{oA, oB, tk\}$, $\{ctrl\}$ and $\{oA, vB, tk\}$, $\{ctrl\}$ and $I=\{oB, vA, tk\}$, $\{ctrl\}$ and

$J=\{vA, vB, tk\}$ are equivalent, respectively, it is sufficient to take into account, for instance, the rules computed from the dependencies:

Case 1. $\{oA\} \overset{\leftrightarrow}{S} \{vA\}$; $\{oB\} \overset{\leftrightarrow}{S} \{vB\}$. (Strong components of S.)

Case 2. $\{vA, vB, tk\} \vec{S} \{ctrl\}$. (Sensor signals, see: Figure 24.)

Case 3. $\{ctrl\} \vec{S} \{vA, vB, tk\}$. (Actor signals, see: Figure 24.)

Case 4. $tk(e) \underset{S}{\Rightarrow} vB(c)$, $tk(f) \underset{S}{\Rightarrow} vA(c)$, $vA(c) \wedge tk(pf) \underset{S}{\Rightarrow} vB(o)$, $vB(c) \wedge tk(pf) \underset{S}{\Rightarrow} vA(o)$, $vB(o) \underset{S}{\Rightarrow} vA(c)$, $vA(o) \underset{S}{\Rightarrow} vB(c)$. (Internal dependencies.)

If we omit in the set OPT(S) of rules corresponding to all non-trivial functional dependencies in the considered decision table S in which on both left or right hand side appear descriptors with names oA, oB then we obtain the reduced set of rules without the redundant rules with respect to equivalent dependencies in the given data table S (see: Figure 26).

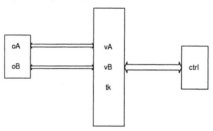

Fig. 26. Scheme of a reduction of redundant rules with respect to equivalent dependencies in S

It means that the system of generators presented below generates all non-trivial dependencies between attribute values of the given table S:

For *Case 1.* $oB(c) \underset{S}{\Rightarrow} vB(c)$, $oB(o) \underset{S}{\Rightarrow} vB(o)$, $vB(c) \underset{S}{\Rightarrow} oB(c)$, $vB(o) \underset{S}{\Rightarrow} oB(o)$, $oA(c) \underset{S}{\Rightarrow} vA(c)$, $oA(o) \underset{S}{\Rightarrow} vA(o)$, $vA(c) \underset{S}{\Rightarrow} oA(c)$, $vA(o) \underset{S}{\Rightarrow} oA(o)$.

For *Case 2.* $vA(c) \wedge tk(e) \underset{S}{\Rightarrow} ctrl(por)$, $vA(o) \underset{S}{\Rightarrow} ctrl(fil)$, $vB(c) \wedge tk(pf) \underset{S}{\Rightarrow} ctrl(fil)$, $vB(c) \wedge tk(f) \underset{S}{\Rightarrow} ctrl(wai)$, $vB(o) \underset{S}{\Rightarrow} ctrl(emp)$,
$vA(c) \wedge tk(pf) \underset{S}{\Rightarrow} ctrl(emp)$.

For *Case 3.* $ctrl(por) \underset{S}{\Rightarrow} tk(e)$, $ctrl(wai) \underset{S}{\Rightarrow} tk(f)$, $ctrl(por) \underset{S}{\Rightarrow} vB(c)$, $ctrl(fil) \underset{S}{\Rightarrow} vB(c)$, $ctrl(wai) \underset{S}{\Rightarrow} vB(c)$, $ctrl(emp) \underset{S}{\Rightarrow} vB(o)$, $ctrl(por) \underset{S}{\Rightarrow} vA(c)$, $ctrl(fil) \underset{S}{\Rightarrow} vA(o)$, $ctrl(wai) \underset{S}{\Rightarrow} vA(c)$, $ctrl(emp) \underset{S}{\Rightarrow} vA(c)$.

For *Case 4.* $tk(e) \underset{S}{\Rightarrow} vB(c)$, $tk(f) \underset{S}{\Rightarrow} vA(c)$, $vA(c) \wedge tk(pf) \underset{S}{\Rightarrow} vB(o)$, $vB(c) \wedge tk(pf) \underset{S}{\Rightarrow} vA(o)$, $vB(o) \underset{S}{\Rightarrow} vA(c)$, $vA(o) \underset{S}{\Rightarrow} vB(c)$.

Sometimes, the following situation can appear. If we have rules of the form $a(i) \underset{S}{\Rightarrow} b(j)$, $b(j) \underset{S}{\Rightarrow} c(k)$, and $a(i) \underset{S}{\Rightarrow} c(k)$ then the last rule is redundant, since the rule is a consequence of two remaining. In our case this situation does not appear.

By using strong components of the given decision table we have reduced the set of rules from 58 up to 30 ones.

7.6 More about Petri Net Model

Now we are giving the further information concerning some assumptions about Petri net model for the sake of the control design method presented in this section. In the following we assume that Petri nets are governed by the transition (firing) rules (Rule 1 and Rule 2) presented in section 6.

In the Petri nets constructed in the section we distinguish two types of transitions. The first type (with firing Rule 1) will correspond to sensors and the second (with firing Rule 2) to transitions transmitting information about identified local states throughout the net.

In the section we use ordinary Petri nets with priorities, although it could be also possible to use signal/event nets introduced in the paper [269]. We shall assume that there are only transitions with three different priorities: 0, 1, and 2. All transitions corresponding to decisions have the highest priority 2, transitions corresponding to dependencies between the values of conditional attributes 1, and all the remaining - 0.

7.7 The Solution to Control Design Problem

We present a procedure for transforming rules representing a given decision table into a Petri net.

Let $S = (U, A \cup d)$ be a given decision table, and let $R \in RED(S)$.

PROCEDURE for constructing a Petri net (N_S, M_S) from a given decision table S:

Input: The set of all minimal rules of S corresponding to a reduct R.

Output: A Petri net (N_S, M_S) such that any of its computation leading to decision making has the minimal length (see: section 6).

Step 1. Construct a net representing the set of all attributes from R of S and describing decision d of S (sensors).

Step 2. Extend the net obtained in Step 1 by adding the elements (arcs and transitions) of the net defined by the set of rules representing all non-trivial functional dependencies between the attribute values from R (a propagation of a new information throughout a net).

Step 3. Extend the net obtained in Step 2 by adding the elements of the net defined by the set of rules representing all non-trivial functional dependencies between the attribute values from R and the decision d (a change of a decision value).

Step 4. Extend the net obtained in Step 3 by adding the elements of the net defined by the set of rules representing all non-trivial functional dependencies between the decision value of d and those from R (a change of conditional values from R; to close a control loop).

Step 5. Implement all rules representing all non-deterministic dependencies between attribute values (local states) from R of S.

We present more details of our approach.

Let $S = (U, A \cup \{d\})$ be a decision table, and let $V_a = \{v(0), \cdots, v(k(a))\}$ be a finite set of values for $a \in A$. We assume also that the value sets of all attributes from A are ordered in some way.

In our illustrative example we can assume that the values of attributes will appear in the following order:
  (a) For attributes oA, oB, vA, vB: c, o, c, o,... .
  (b) For attribute tk: e, pf, f, pf, e, pf, f, pf,...
  (c) For decision $ctrl$: por, fil, wai, emp, por, fil, wai, emp,... .

Of course, such assumption has influence on the changes of global states in the system represented by the decision table S. This assumption is in our case quite natural. The values of the decision d will be represented by the set of single places identifying possible values of d. For example, in the section 6 it has been assumed that the attribute values from a given decision table appear in a non-deterministic order.

Step 1. The construction of a net representing an attribute a from R of S is shown in Figure 27.

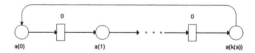

Fig. 27 The net corresponding to an attribute a with the priority 0

The place $a(0)$ is called the *start place* of the attribute a. The remaining places of the net represent possible values of the attribute a. The transitions with priority 0 of the net represent the process of identifying the values of the attribute a. Let us observe that only the start place $a(0)$ is marked in the initial marking of the net.

Let us consider again the given decision table S and its reduct $R_4 = \{vA, vB, tk\}$. The conditions vA, vB, tk and the decision $ctrl$ are represented by nets and places shown in Figure 28.

Steps 2 – 4. The construction of a net representing a set of all rules among:
  1) the attribute values of a reduct R of a given decision table,
  2) the attribute values from R and the decision d,
  3) the decision values of d and those from R.

The rules in this case are of the form:
(1) $p(j) \underset{S}{\Rightarrow} q(l)$, where $p, q \in A$, $j \in V_p$ and $l \in V_q$,
(2) $q(l) \underset{S}{\Rightarrow} p(j)$, where $p, q \in A$, $j \in V_p$ and $l \in V_q$,
(3) $p(j) \Longleftrightarrow_S q(l)$, where $p, q \in A$, $j \in V_p$ and $l \in V_q$,
(4) $p(1) \wedge \cdots \wedge p(k) \underset{S}{\Rightarrow} r$, where $k > 1$.

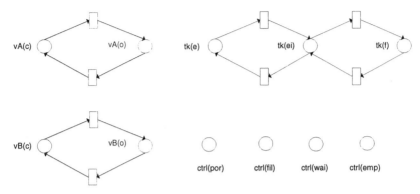

Fig. 28. Nets and places representing attributes vA, vB, tk, and $ctrl$

The nets representing the rules (1), (2), (3) and (4) are illustrated in Figures 29, 30, 31, 32, respectively.

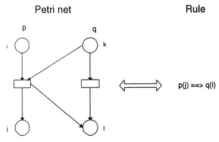

Fig. 29. Net representation of rules. Case (1): $p(j) \underset{S}{\Rightarrow} q(l)$, where $p, q \in A$, $j \in V_p$, $l \in V_q$

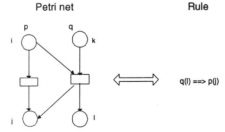

Fig. 30. Net representation of rules. Case (2): $q(l) \underset{S}{\Rightarrow} p(j)$, where $p, q \in A$, $j \in V_p$, $l \in V_q$

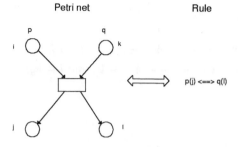

Fig. 31. Net representation of rules. Case (3): $p(j) \iff_S q(l)$, where $p, q \in A$, $j \in V_p$, $l \in V_q$

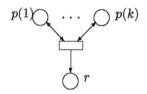

Fig. 32. Net representation of rules. Case (4): $p(1) \wedge \cdots \wedge p(k) \underset{S}{\Rightarrow} r$, where $k > 1$

For simplicity of the following pictures let us consider only two rules of the form: $vA(o) \underset{S}{\Rightarrow} vB(c), vB(o) \underset{S}{\Rightarrow} vA(c)$ and one rule of the form: $vB(c) \wedge tk(pf) \underset{S}{\Rightarrow} ctrl(fil)$ obtained for the reduct R_4 of the considered decision table S.

In Figure 33 a net representing the first two rules is shown, and in Figure 34 it is illustrated the construction of a net representing the third one. The net implementation of non-deterministic dependencies between the value $vA(c)$ and values $vB(o), vB(c)$ as well as between the value $vB(c)$ and values $vA(o), vA(c)$ is illustrated in Figure 35. The process vA can pass from the local state $vA(c)$ to the local state $vA(o)$ by firing the transition t_1 or the transition t_2. Analogously, the process vB can pass from the local state $vB(c)$ to the local state $vA(o)$ by firing the transition t_3 or the transition t_4. The choice of a transition in the both cases depends from the local state of the process vB and vA, respectively. In the case of non-deterministic dependencies between attribute values, one can also take into account in a net implementation the frequency for appearing of a given attribute value.

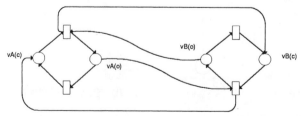

Fig. 33. Net representation of rules of the form: $vA(o) \underset{S}{\Rightarrow} vB(c), vB(o) \underset{S}{\Rightarrow} vA(c)$

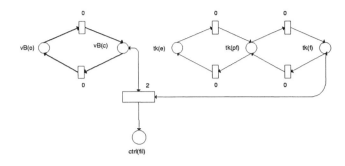

Fig. 34. Net representation of rules of the form: $vB(c) \wedge tk(pf) \underset{S}{\Rightarrow} ctrl(fil)$ with the priority 2

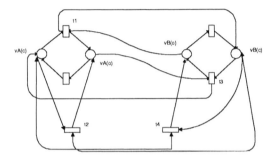

Fig. 35. Net representation of non-deterministic dependencies between the values $vA(c)$ and $vB(o)$, $vB(c)$ as well as $vB(c)$ and $vA(o)$, $vA(c)$

In order to construct the whole net representing all rules corresponding to non-trivial functional dependencies between the attribute values of the given decision table it is necessary to repeat our construction for all rules. This construction is omitted in the section.

7.8 Summary

In the section we have demonstrated a methodology for control design of discrete event systems specified by data tables. The presented method seems to be promising for automatic concurrent control design in case of problems which are difficult to model using standard mathematical methods. Our approach is based on rough set approach and Petri nets. It allows to design the control in automatic way. It can be especially applied when behavior requirements of the system are changed in real-time, because programming flexible automation devices by hand is a time-consuming and error-prone task.

Drawing Petri nets by hand one can produce very compact solutions for problems solved rather by small nets. For large models some automatic methods could be accepted even if the nets produced by them are not so compact or small. Comparing the presented example it is possible to see that our method for solving CDP can also produce solutions close to those obtained by designers [81].

Our approach allows also to estimate the cost of discrete event system reconstruction when the cost of changed parts of the system is known [283]. It is important to note that our methodology is supposed by software tools, namely the *Rosetta* system [159] for data analysis and rules extraction on the basis of the rough sets theory, as well as the *PN-tools* system for computer aided design and analysis of concurrent models [277]. Although our approach looks quite promising, as demonstrated by the example of dosing tank, more experiments with control processes is needed before the methodology attains its maturity. It will be important to develop methods for converting the received formal models (represented in our case by Petri nets) into the programmable logic controller [320].

8 Computer Tools for the Synthesis and Analysis of Concurrent Processes

8.1 Introduction

ROSEPEN is a computer system for automatic synthesis and analysis of concurrent processes discovered in data tables.

Features currently offered by the computational kernel of the system include algorithms for:

1. Automatic discovery of concurrent models from data tables.

2. Automatic generation of parallel programs for decision-making based on a given decision table.

3. Reconstruction of concurrent models when specifications for the systems are being changed.

4. Computation of the reconstruction costs, if the costs of components and communications of the systems can be estimated.

5. Design in interactive and/or semi-automatic mode of concurrent models.

6. Analysis and verification of the basic structural and dynamic properties of the model.

ROSEPEN is running on *IBM PC* microcomputers under *MS-DOS* operating system with *Windows* interface. The system is primarily intended to research and education goals.

ROSEPEN is developed at the Pedagogical University of Rzeszów. This system is based on the newest research results obtained on the crossroad of theory and application rough sets and/or Petri nets. The current version of the system, described in the next section, includes some separate modules which cooperate to one another.

8.2 System Architecture

ROSEPEN consists of two main components. The first component of this system is based on the rough set methods. It contains three main and two auxiliary programs.

1. *CAI* is a program which contains two logical modules: *Rule-Editor* and *Generator*.

1.1. *Rule-Editor* is a window-based graphical module for generating and editing rule sets in semi-automatic mode.

1.2. *Generator* is a set of two submodules for generating in automatic mode:

(i) From an arbitrary information system represented by rules extracted from the system, its concurrent models in the form of a marked Petri net with the following property: the reachability set of the net corresponds exactly to the set of global states (objects) consistent with all rules true in the given information system. The reachability set of the net represents the largest extension of the given information system consistent with the knowledge represented by this system.

(ii) The parallel algorithms (represented by Petri nets) from the given decision tables, which allow us to take the proper decision related to the identified global states (new objects). For more information see: [252], [259].

2. *RSC* is a program which contains three modules: *Constructor*, *Reconstructor* and *Classifier*.

2.1. *Constructor* is a module for automatic discovery of data models represented by concurrent systems from experimental data tables (information systems). The basic step of the construction consists of decomposition of any information system (with respect to any of its reduct) into components linked by some connections which allow to preserve some constraints. Any component represents in a sense the strongest functional module of the system. The connections between components represent constraints which must be satisfied when these functional modules coexist in the system. The components together with the connections defines a covering of the system. The coverings of the system are used to construction of its concurrent model in the form of a marked Petri net with the analogous property as it has been defined above for the result net obtained by *Generator* described above. For more information see: [258], [279].

2.2. *Reconstructor* is a module for supporting the process of reconstructing of the synthesized (e.g. by means the module *Constructor*) concurrent system when the specification for the system is being changed e.g. by adding new requirements (new objects, attributes or values of attributes). This module produces a plan of reconstruction of a given system by specifying which parts (components and/or connections) can remain unchanged and which must be changed to satisfy new requirements. It can make comparisons between the results obtained from different components and/or coverings. If the cost of components and connections can be estimated, the system computes also the cost of reconstruction.

For more information see: [278], [281].

2.3. *Classifier* is a module for classification of new objects with synthesized components of a system (computed by *Constructor*, see above) using a voting scheme. For more information see: [279].

3. *DIS* is a program which can be used for synthesis of concurrent models from observations or specification encoded in data table representing a dynamic information system. For more information see: [284].

There are also available two auxiliary programs for optimization of a set of rules by removing redundant rules, i.e., rules subsumed by other rules and for computing an extension of a given information system see: [255], [256].

The second component of *ROSEPEN* is a fully integrated software package (called *PN-tools*) that can support designers and analysts in editing, validation, analysis and execution of standard Petri net models, i.e., place/transition nets, self-modifying nets, nets with inhibitor arcs, priority nets, and generalized stochastic Petri nets. The nets can be structured hierarchically to facilitate reading and understanding. The package combines the following modules:

1. A graphical editor and a fully integrated simulator of nets.
2. A textual editor which is an interactive program for construction and editing of all kinds of Petri nets accepted by this package.
3. A net analyzer that allows to prove general system characteristics without any simulation experiments. The characteristics include among others liveness, boundedness, reachability, invariants and the basic structural information (conflicts, deadlocks and traps, state machine decomposition and covering). The net analyzer implements also a set of transformation rules by which it is possible to reduce the size and complexity of the net without changing important properties of the net. It is also possible to execute the performance analysis for generalized stochastic Petri net models. For more information see: [277], [276].

8.3 Input Data

The first component of *ROSEPEN* can handle rule sets and data tables which are provided in the form of text, i.e., saved as the ASCII files. The format of rules consists of *if-then* rules with a conjunctive antecedent, and it is according to the syntax of rules generated by *Rosetta* (see: [159]). *ROSEPEN* uses:

(i) rules corresponding to all non-trivial dependencies between the values of attributes from the reduct of a given information system with those outside of that reduct,

(ii) rules corresponding to all non-trivial dependencies between the values of attributes within the reduct of that system (see: [252], [259], [279], [281], [284]).

The data tables processed by *ROSEPEN* represent the decision tables in the standard form in the case of generating parallel algorithms from experimental data for real-time decision making ([258], [279]), and in the modified form for synthesis of concurrent systems specified by dynamic information systems. The input data in the form of rules can be either introduced by means internal rule editor or imported from text files. The data tables in the both cases are imported from text files.

The second component of *ROSEPEN* allows the user to input and to output the nets in many different textual formats. Those formats can be used as input to the grapical/textual editors and/or the analysis modules in the package. The graphical modules can generate a net format acceptable by many printers, plotters and text processors.

8.4 Output Knowledge

The first component of *ROSEPEN* generates concurrent data models represented by Petri nets, and written in the form of a text table [255]. This component can also cooperate with the second one of the system for further comprehensive analysis of the obtained net model. The reports of the analysis are written in the text file.

8.5 User

ROSEPEN is primarily intended to research and education goals.

8.6 Summary

ROSEPEN is a prototype system for discovery and comprehensive analysis of Petri net models.

9 Concluding Remarks

The paper includes, in particular, the description of automatic techniques for the modelling and analysis of concurrent systems specified by information systems. Petri nets have been chosen as a model for concurrency. The application of Petri nets to represent a given information system and a modified definition of these systems enables:

- to discover in a simple way new dependencies between local states of processes (and also components) being in the system,
- to represent in an elegant and visual way the dependencies between local states of processes (and also components) in the system,
- to observe concurrent and sequential subsystems of the system,
- to understand better the structure and dynamics of a given information system.

In section 3 we have formulated a method of the synthesis problem of concurrent systems specified by information systems as well as dynamic information systems.

We would like to investigate to what extent our method could be applied for automatic synthesis of parallel programs from examples ([231], [262]).

Moreover, to some extent, it is a matter of taste which of the modelling method of concurrent systems to use. Drawing Petri nets by hand one can produce very compact solutions for problems solved rather by small nets. For large models some automatic methods could be accepted even if the produced by them nets are not so compact or small. Comparing the presented examples it is possible to see that our method can also produce solutions close to those obtained by designers.

We have also proposed a solution of the synthesis problem of a net system from a dynamic information system. It is possible to solve this problem for finite place/transition Petri nets [102]. The solution of our problem for place/transitions Petri nets is also simple to obtain.

In section 4 a method for discovery of data models represented by concurrent systems from experimental tables has been considered. The method has been applied to automatic data models discovery from experimental tables with Petri nets as models for concurrency. This method can also be applied for automatic feature extraction. The properties of the constructed concurrent systems (e.g. their invariants) can be interpreted as higher level laws of experimental data. New features can be also obtained by performing for a given decision table a procedure presented in section 4. We have studied some applications of our method in control design from experimental data tables.

In section 5 an approach to reconstruction of the synthesized cooperative information systems under cost constraints when the specification for the system is being changed, has been proposed. This approach can be treated as an adaptive strategy in reconstructing of cooperative information systems under cost constraints. It allows to produce a plan of reconstruction of a given system by specifying which parts can remain unchanged and which must be changed to satisfy new requirements. It provides also information about the cost of the reconstruction. This method can be applied in the system technology domain.

The parallel algorithm presented in section 6 allows a very fast identification of objects specified in a given decision table. Nevertheless, some parts of the constructed net can be redundant, i.e. unnecessary for taking proper decisions and slowing down the process of decision taking.

The basic problem for real-time knowledge based control algorithms [229] is related to the time/space trade-off. For example, if there is not enough memory to store the parallel network representing knowledge encoded in a given decision table, it is necessary to reduce it, which can lead to an increase of time necessary for making decisions. Another interesting problem arises when decision tables change with time and the constructed net ought to be modified by applying some strategies discovered during the process of changes. These are examples of problems which we would like to investigate applying the approach presented in the paper.

In section 7 we have demonstrated a methodology for control design of discrete event systems specified by data tables. The presented method seems to be promising for automatic concurrent control design in case of problems which are difficult to model using standard mathematical methods. Our approach allows to design the control in automatic way. It can be especially applied when behavior requirements of the system are changed in real-time, because programming flexible automation devices by hand is a time-consuming and error-prone task.

It is important to note that the methodologies proposed in the paper are supposed by software tools, namely the *Rosetta* system [159] for data analysis and rules extraction on the basis of the rough sets theory, as well as the *PN-tools* system for computer aided design and analysis of concurrent models represented by Petri nets [277]. For the control design problem, it will be important to develop methods for converting the received formal models (represented in our case by Petri nets) into the programmable logic controller [320].

On the base of the methods presented in the paper at the Pedagogical University of Rzeszów a computer system supporting the user in the process of solving of the considered problems has been developed [285] (see also: section 8). All resulting nets obtaining from the computer system can be analyzed by *PN-tools* for the reason of the further analysis of Petri net model [277].

The methods presented in the paper as well as the further investigations of interconnections between rough set theory and concurrency may stimulate the development of both theoretical and practical research related to the areas such as, for example: knowledge discovery systems, control design, decomposition of information systems, object identification in real-time.

Acknowledgment

I am grateful to Professor A. Skowron for stimulating discussions and interesting suggestions about this work. I also wish to thank Professor J.F. Peters, Department of Electrical and Computer Engineering, University of Manitoba and my colleagues at the Institute of Mathematics of Warsaw University for discussions related to research on modeling real-time decision making systems and rough set methods. I also want to thank my colleagues from the Institute of Mathematics of Rzeszów University for their high quality programming. This work was partially supported by the grant #8 T11 C01 011 from the State Committee for Scientific Research (KBN) in Poland and by the ESPRIT project 20288 CRIT- 2.

References

1. Ajmone Marsan, M., Balbo, G., Conte, G.: *Performance Models of Multiprocessor Systems.* The MIT Press, Cambridge, 1986.
2. Ajmone Marsan, M., Balbo, G., Conte, G., Donatelli, S., Franceschinis, G.: *Modelling with Generalized Stochastic Petri Nets.* Wiley, New York, 1995.
3. Arnold, A.: *Finite Transition Systems.* Prentice-Hall, New York, 1994.
4. Baar, A., Cohen, P.R., Feigenbaum, E.A.: *The Handbook of Artificial Intelligence 4.* Addison-Wesley, Reading, 1989.
5. Badouel, E., Bernardinello, L., Darondeau, Ph.: *The synthesis problem for elementary net systems is NP-complete.* Inria Research Report 2558/1995. To appear in Theoretical Computer Science.
6. Badouel, E., Darondeau, Ph.: *Theory of Regions.* In: W. Reisig: *Elements of Distributed Algorithms.* Springer-Verlag, Berlin, 1998, 529-586.

7. Baccelli, F., Jean-Maria, A., Mitrani, I. (eds.): *Quantitative Methods in Parallel Systems*. Springer-Verlag, Berlin, 1995.
8. Baeten, J.C.M., Weijland, P.: *Process Algebra*. Cambridge University Press, 1990.
9. Bagai, R., Shanbhogue, V., Żytkow, J.M., Chou, S.C.: *Automatic Theorem Generation in Plane Geometry*. In: J. Komorowski, Z. Raś (eds.): Methodologies for Intelligent Systems, Springer-Verlag 1993, Berlin, 215-224.
10. Bandemer, H., Gottwald, S.: *Fuzzy Sets, Fuzzy Logic, Fuzzy Methods with Applications*. Wiley, New York, 1995.
11. Baumgarten, B.: *Petri-Netze. Grundlagen und Anwendungen*. Spektrum Akademischer Verlag, 1996.
12. Bause, F., Kritzinger, P.S.: *Stochastic Petri Nets. An Introduction to the Theory*. Vieweg, Braunschweig/Wiesbaden, 1996.
13. Bazan, J.: *A Comparison of Dynamic and Non-Dynamic Rough Set Methods for Extracting Laws from Decision Tables*. In: L. Polkowski, A. Skowron (eds.): Rough Sets in Knowledge Discovery 1. Methodology and Applications, Physica-Verlag 1998, Heidelberg, 321-365.
14. Bazan, J., Son, N.H., Trung, N.T., Skowron, A., Stepaniuk, J.: *Some Logic and Rough Set Applications for Classifying Objects*. Institute of Computer Science Research Report 38/94, Warsaw University of Technology, 1994, Poland.
15. Bazan, J., Skowron, A., Synak, P.: *Dynamic Reducts as a Tool for Extracting Laws from Decision Tables*. Proceedings of the Symposium on Methodologies for Intelligent Systems, Charlotte, NC, Lecture Notes in Artificial Intelligence Vol. 869, Springer-Verlag 1994, Berlin, 346-355.
16. Bednarczyk, M.A.: *Categories of Asynchronous Systems*. Ph.D. Thesis, University of Sussex, Report 1/88, School of Cognitive and Computing Sciences, University of Sussex.
17. Ben-Ari, M., Pnueli, A., Manna, Z.: *The temporal logic of branching time*. Acta Informatica 20 (1983), 207-226.
18. Berry, G., Gonthier, G.: *The Esterel synchronous programming language: Design, semantics, implementation*. Science of Computer Programming 19 (1992), 87-152.
19. Bernadinello, L.: *Synthesis of Net Systems*. Proceedings of the Application and Theory of Petri Nets, Lecture Notes in Computer Science Vol. 691, Springer-Verlag 1993, Berlin, 89-105.
20. Berthelot, G.: *Transformations and Decompositions of Nets*. In: W. Brauer, W. Reisig, G. Rozenberg (eds.): Petri Nets: Central Models and Their Properties. Advances in Petri Nets 1986 Part I, Lecture Notes in Computer Science Vol. 254, Springer-Verlag 1987, Berlin, 359-376.
21. Best, E.: *COSY: Its Relation to Nets and to CSP*. In: W. Brauer, W. Reisig, G. Rozenberg (eds.): Petri Nets: Applications and Relationships to Other Models of Concurrency. Advances in Petri Nets 1986 Part II, Lecture Notes in Computer Science Vol. 255, Springer-Verlag 1987, Berlin, 416-440.
22. Best, E.: *Semantik. Theorie sequentieller und paralleler Programmierung*. Vieweg, Braunschweig/Wiesbaden, 1995.
23. Best, E., Fernandez, C.: *Notations and Terminology on Petri Net Theory*. Arbeitspapiere der GMD 195, March 1987.
24. Best, E., Fernandez, C.: *Nonsequential Processes. A Petri Net View*. EATCS Monographs on Theoretical Computer Science, Vol. 13, Springer-Verlag, Berlin, 1988.
25. Blair, H.A., Subrahmanian, V.S.: *Paraconsistent logic programming*. Theoretical Computer Science 68-1 (1989), 135-154.

26. Bolc, L., Szałas, A. (eds.): *Time & Logic: a computational approach.* UCL Press, London, 1995.
27. Bolch, G., Greiner, S., de Meer, H.: *Queueing Networks and Markov Chains.* John Wiley, 1998.
28. Bolognesi, T., Brinksma, E.: *Introduction to the ISO specification language LOTOS.* Computer Networks and ISDN Systems, Vol. 14, 1987, 25-59.
29. Bradfield, J., Stirling, C.: *Verifying temporal properties of processes.* In: J.C.M. Baeten, J.W. Klop (eds.): Proceedings of CONCUR'90 Theories of Concurrency: Unification and Extension, Lecture Notes in Computer Science Vol. 458, Springer-Verlag 1990, 115-125.
30. Brams, G.W.: *Reséaux de Petri: Théorie et Pratique, Vol. 1: Théorie et Analyse, Vol. 2: Modélisation et Applications.* Edition Masson, 1983.
31. Brauer, W., Reisig, W., Rozenberg, G. (eds.): *Petri Nets: Central Models and Their Properties.* Advances in Petri Nets 1986 Part I, Lecture Notes in Computer Science Vol. 254, Springer-Verlag 1987, Berlin.
32. Brauer, W., Reisig, W., Rozenberg, G. (eds.): *Petri Nets: Applications and Relationships to Other Models of Concurrency.* Advances in Petri Nets 1986 Part II, Lecture Notes in Computer Science Vol. 255, Springer-Verlag 1987, Berlin.
33. Brookes, S.D., Hoare, C.A.R., Roscoe, A.W.: *A Theory of Communicating Sequential Processes.* Journal of the ACM, Vol. 31-3, 1984, 560-599.
34. Brown, E.M.: *Boolean Reasoning.* Kluwer Academic Publishers, Dordrecht, 1990.
35. Brownston, L., Farrell, R., Kant, E., Martin, N.: *Programming Expert Systems in OPS5. An Introduction to Rule-Based Programming.* Addison-Wesley, Reading, 1986.
36. Budkowski, S., Dembiński, P.: *An introduction to Estelle: a specification language for distributed systems.* Computer Networks and ISDN Systems 14 (1987), 3-24.
37. Cambel, A.B.: *Applied Chaos Theory: A Paradigm for Complexity.* Academic Press, San Diego, 1993.
38. Cardoso, J., Valette, R., Dubois, D.: *Petri nets with uncertain markings.* In: G. Rozenberg (ed.): Advances in Petri Nets, Lecture Notes in Computer Science Vol. 483, Springer-Verlag 1990, Berlin, 65-78.
39. Cardoso, J., Valette, R., Dubois, D.: *Fuzzy Petri nets: An overview.* Proceedings of the 13th IFAC World Congress, San Francisco, CA, USA, June 30 - July 5, 1996, 443-448.
40. Chaib-Draa, B., Moulin, B., Mandian, R., Millot, P.: *Trends in Distributed Artificial Intelligence.* Artificial Intelligence Review, Vol. 6-1, 1992, 35-66.
41. Chandrasekaran, B., Bhatnagar, R., Sharma, D.D.: *Real-Time Disturbance Control.* Communications of the ACM, Vol. 34-8, 1991, 32-47.
42. Chandy, K.M., Misra, J.: *Parallel Program Design: A Foundation.* Addison-Wesley, 1988.
43. Chen, S.M., Ke, J.S., Chang, J.F.: *Knowledge representation using fuzzy Petri nets.* IEEE Transactions on Knowledge and Data Engineering, Vol. 2-3, 1990, 311-319.
44. Chou, S.C.: *Mechanical Theorem Proving.* R. Reidel Publishing Company, Dordrecht, 1988.
45. Cios, K., Pedrycz, W., Świniarski, R.: *Data Mining. Methods for Knowledge Discovery.* Kluwer Academic Publishers, Dordrecht, 1998.
46. Courtois, P.J.: *On time and space decomposition of complex structures.* Communications of the ACM, Vol. 2-6, 1985, 590-603.

47. Crowley, J. L.: *Navigation for an Intelligent Mobile Robot.* IEEE Journal of Robotics and Automation RA-1 (1985), 31-41.
48. Czaja, L.: *Finite Processes in Cause-Effect Structures and Their Composition.* Information Processing Letters, Vol. 31-6, 1989, 305-310.
49. Czogała, E., Mrózek, A., Pawlak, Z.: *The idea of rough-fuzzy controller.* Fuzzy Sets and Systems 72 (1995), 61-63.
50. David, R., Alla, H.: *Petri Nets and Grafcet. Tools for modeling discrete events systems.* Prentice-Hall, New York, 1992.
51. Deng, Y., Chang, S.K.: *A G-net model for knowledge representation and reasoning.* IEEE Transactions on Knowledge and Data Engineering, KDE 2-3 (1990), 295-310.
52. Desel, J. (ed.): *Structures in Concurrency Theory.* Springer-Verlag, Berlin, 1995.
53. Desel, J., Esparza, J.: *Free Choice Petri Nets.* Cambridge University Press, 1995.
54. Desel, J., Reisig, W.: *The synthesis problem of Petri nets.* Acta Informatica 33-4 (1996), 297-315.
55. Desrochers, A.A., Al-Jaar, R.Y.: *Applications of Petri Nets in Manufacturing Systems - Modeling, Control, and Performance Analysis.* IEEE Control Systems Society Press, New York, 1995.
56. DiCesare, F., Harhalakis, G., Proth, J.M., Silva, M., Vernadat, F.B.: *Practice of Petri Nets in Manufacturing.* Chapman & Hall, London, 1993.
57. Diekert, V., Rozenberg, G. (eds.): *The book of traces.* Word Scientific, Singapore, 1995.
58. Edwards, K.: *Real-Time Structured Methods. Systems Analysis.* Wiley, New York, 1993.
59. Ehrenfeucht, A., Rozenberg, G.: *Partial 2-structures; Part II, State Space of Concurrent Systems.* Acta Informatica 27 (1990), 348-368.
60. Emerson, E.A.: *Temporal and Modal Logic.* In: J. van Leeuwen (ed.): Handbook of Theoretical Computer Science Vol. B: Formal Models and Semantics, Elsevier Science Publishers, Amsterdam, 1990, 995-1072.
61. *Extended Abstracts of International Workshop on Rough Sets: State of the art and perspectives*, Poznań-Kiekrz, Poland, 1992.
62. Fagin, R., Halpern, J.Y., Moses, Y., Vardi, M.Y.: *Reasoning about Knowledge.* The MIT Press, Cambridge, 1995.
63. Fayyad, U.M., Piatetsky-Shapiro, G., Smyth, P., Uthurusamy, R. (eds.): *Advances in Knowledge Discovery and Data Mining.* AAAI Press, Menlo Park, 1996.
64. Feldbrugge, F.: *Petri Net Tools.* In: G. Rozenberg (ed.): Advances in Petri nets 1985, Lecture Notes in Computer Science Vol. 222, Springer-Verlag 1986, Berlin, 203-223.
65. Feldbrugge, F.: *Petri net tool overview 1989.* In: G. Rozenberg (ed.): Advances in Petri Nets 1989, Lecture Notes in Computer Science Vol. 424, Springer-Verlag 1990, Berlin, 151-178.
66. Feldbrugge, F.: *Petri net tool overview 1992.* In: G. Rozenberg (ed.): Advances in Petri Nets 1993, Lecture Notes in Computer Science Vol. 674, Springer-Verlag 1992, Berlin, 169-209.
67. Feldbrugge, F., Jensen, K.: *Petri net tool overview 1986.* In: W. Brauer, W. Reisig, G. Rozenberg (eds.): Petri Nets: Applications and Relationships to Other Models of Concurrency. Advances in Petri Nets 1986 Part II, Lecture Notes in Computer Science Vol. 255, Springer-Verlag 1987, Berlin, 20-61.

68. Feldbrugge, F., Jensen, K.: *Computer tools for High-level Petri nets.* In: K. Jensen, G. Rozenberg (eds.): High-level Petri Nets. Theory and Application, Springer-Verlag 1991, Berlin, 691-717.
69. Fernandez, C., Merceron, A., Parada, V., Rozenberg, G., Solar, M. (eds.): *Advanced Course on Petri Nets.* Editorial de la Universidad de Santiago de Chile, 1996.
70. Fong, J., Huang, S.-M.: *Information Systems Reengineering.* Springer-Verlag, Berlin, 1997.
71. Forgy, C.L.: *Rete: A Fast Algorithm for the Many Pattern/Many Object Pattern Match Problem.* Artificial Intelligence, Vol. 19-1, 1982, 17-37.
72. Garcia, O.N., Chien, Y.-T. (eds.): *Knowledge-Based Systems: Fundamentals and Tools.* IEEE Computer Society Press, Los Alamitos, 1991.
73. Genesereth, M.R., Nilsson, N.J.: *Logical Foundations of Artificial Intelligence.* Morgan Kaufmann Publishers, Palo Alto, 1987.
74. Genrich, H.J., Lautenbach, K.: *System modeling with high-level Petri nets.* Theoretical Computer Science 13 (1981), 109-136.
75. Giordana, A., Saitta, L.: *Modeling production rules by means of Predicate/Transition Networks.* Information Sciences 35 (1985), 1-41.
76. Gleick, J.: *Chaos: Making a New Science.* Viking, 1987.
77. Goldberg, D.E.: *Genetic Algorithms in Search, Optimization, and Machine Learning.* Addison-Wesley, Reading, 1989.
78. Hack, M.: *Decidability Questions for Petri Nets.* Ph.D. Thesis, Department of Electrical Engineering, Massachusetts Institute of Technology, Cambridge, 1975.
79. Halbwachs, N., Caspi, P., Raymond, P., Pilaud, D.: *The synchronous dataflow programming language LUSTRE.* Proceedings of the IEEE, Vol. 79, 1991, 1305-1320.
80. Hammer, M., Stanton, S.A.: *The Reengineering Revolution.* Harper Collins Publishers, 1995.
81. Hanisch, H.M., Luder, A.: *A Signal Extension for Petri Nets and its Use in Controller Design.* Proceedings of the Workshop on Concurrency, Specification and Programming, Berlin, September, 1998.
82. Harhalakis, G., Vernadat, F.B.: *Petri nets for manufacturing information systems.* In: F. DiCesare, G. Harhalakis, J.M. Proth, M. Silva, F.B. Vernadat: *Practice of Petri Nets in Manufacturing.* Chapman & Hall, London, 1993, 185-290.
83. Harel, D.: *Statecharts: a visual formalism for complex systems.* Science of Computer Programming 8 (1987), 231-274.
84. Haverkort, B.R.: *Performance of Computer-Communication Systems.* John Wiley and Sons, 1998.
85. Haves-Roth, B.: *Architectural Foundations for Real-Time Performance in Intelligent Agents.* Journal of Real-Time Systems 2 (1990), 99-125.
86. Haykin, S.: *Neural Networks. A Comprehensive Foundation.* Macmillan, New York, 1994.
87. Hennessy, M.: *Algebraic Theory of Processes.* MIT Press, Cambridge, 1988.
88. Hoare, C.A.R.: *Communicating sequential processes.* Communications of the ACM 21 (1978), 666-677.
89. Hoare, C.A.R.: *Communicating Sequential Processes.* Prentice-Hall, London, 1985.
90. HRL: *Marked data.* Hughes Research Laboratories, personal communication, 1994.

91. Hura, G.S. (ed.): *Special Issue. Petri Nets and Related Graph Models.* Microelectronics and Reliability, Vol. 31-4, 1991, 559-816.
92. Hurley, R.B.: *Decision Tables in Software Engineering.* Van Nostrad Reinhold Company, New York, 1983.
93. *IEEE Transactions on Software Engineering, Special Issue: Specification and Analysis of Real-Time Systems*, Vol. 18-9, 1992.
94. Janicki, R.: *Transforming Sequential Systems into Concurrent Systems.* Theoretical Computer Science 36 (1985), 27-58.
95. Janicki, R., Lauer, P.E.: *Specification and Analysis of Concurrent Systems. The COSY Approach.* Springer-Verlag, Berlin, 1992.
96. Jarke, M., Jeusfeld, M.A., Peters, P., Szczurko, P.: *Cooperative Information Systems Engineering.* In: Z. Raś, M. Michalewicz (eds.): Proceedings of the 9th International Symposium on Methodologies for Intelligent Systems (ISMIS'96), Zakopane, Poland, June 9-13, 1996, Lecture Notes in Artificial Intelligence Vol. 1079, Springer-Verlag 1996, Berlin, 34-49.
97. Jensen, K.: *Coloured Petri Nets. Basic Concepts, Analysis Methods and Practical Use 1.* Springer-Verlag, Berlin, 1992.
98. Jensen, K.: *Coloured Petri Nets. Basic Concepts, Analysis Methods and Practical Use 2.* Springer-Verlag, Berlin, 1995.
99. Jensen, K.: *Coloured Petri Nets. Basic Concepts, Analysis Methods and Practical Use 3.* Springer-Verlag, Berlin, 1997.
100. Jensen, K., Rozenberg, G. (eds.): *High-level Petri nets. Theory and Application.* Springer-Verlag, Berlin, 1991.
101. Johson, P.M.: *Reengineering inspection.* Communications of the ACM 41-2 (1998), 49-52.
102. Karp, R.M., Miller, R.E.: *Parallel program schemata.* Journal of Computer and System Sciences 3-4 (1969), 147-195.
103. Keller, R.M.: *Formal verification of parallel programs.* Communications of the ACM 19 (1976), 371-384.
104. Kodratoff, Y., Michalski, R. (eds.): *Machine Learning Vol. 3.* Morgan Kaufmann Publishers, San Mateo, 1992.
105. Kotov, V.E.: *Petri Nets.* Nauka Publ. Company, Moskov, 1984, in Russian.
106. Krieg, B.: *Petrinetze und Zustandsgraphen.* IFI-Bericht B-29/77, Institut für Informatik, Universität Hamburg, 1977, Germany.
107. Lautenbach, K.: *Linear Algebraic Techniques for Place/Transition Nets.* In: W. Brauer, W. Reisig, G. Rozenberg (eds.): Petri Nets: Central Models and Their Properties. Advances in Petri Nets 1986 Part I, Lecture Notes in Computer Science Vol. 254, Springer-Verlag 1987, Berlin, 142-167.
108. Lee, Y.H., Krishna, C.M.: *Readings in Real-Time Systems.* IEEE Computer Society Press, Los Alamitos, 1993.
109. Lengauer, C., Hehner, E.C.R.: *A Methodology for Programming with Concurrency: An Informal Presentation.* Science of Computer Programming 2 (1982), 1-18.
110. Leszak, M., Eggert, H.: *Petri-Netz-Methoden und -Werkzeuge. Hilfsmittel zur Entwurfsspezifikation und -validation von Rechensystemen.* Springer-Verlag, Berlin, 1989.
111. Lin, C., Chaundhury, A., Whinston, A., Marinescu, D.C.: *Logical inference of Horn clauses in Petri net models.* IEEE Transactions on Knowledge and Data Engineering, KDE 5-3 (1993), 416-425.

112. Lin, T.Y. (ed.): *Proceedings of The Third International Workshop on Rough Sets and Soft Computing*, San Jose, CA, November 10-12, 1994.
113. Lin, T.Y.: *Rough - fuzzy controllers for complex systems*. In: P. P. Wang (ed.): Second Annual Joint Conference on Information Sciences (JCIS'95), Wrightsville Beach, North Carolina, USA, September 28 - October 1, 1995, 18-21.
114. Lin, T.Y.: *Fuzzy controllers: An integrated approach based on fuzzy logic, rough sets, and evolutionary computing*. In: T.Y. Lin (ed.): Proceedings of the Workshop on Rough Sets and Data Mining at 23rd Annual Computer Science Conference, Nashville, Tenessee, March 2, 1995, 48-56. Also in: [115], 123-138.
115. Lin, T.Y., Cercone, N. (eds.): *Rough Sets and Data Mining. Analysis for Imprecise Data*. Kluwer Academic Publishers, Dordrecht, 1997.
116. Looney, C.G.: *Fuzzy Petri Nets for Rule-Based Decision-making*. IEEE Transactions on Systems, Man, and Cybernetics 18-1 (1988), 178-183.
117. Lopez, B., Meseguer, P., Plaza, E.: *Knowledge based systems validation: a state of the art*. AI Communications 3-2 (1990), 58-72.
118. Manna, Z., Pnuelli, A.: *The Temporal Logic of Reactive and Concurrent Systems Vol. I: Specification*. Springer-Verlag, Berlin, 1992.
119. Manna, Z., Pnuelli, A.: *Temporal Verification of Reactive Systems Vol. II: Safety*. Springer-Verlag, Berlin, 1995.
120. Manson, P.R.: *Petri Net Theory, A Survey*. Technical Report No. 139, University of Cambridge, Cambridge, England, 1988.
121. Martinez, J., Muro, P.R., Silva, M., Smith, S.F., Villaroel, J.L.: *Merging artificial intelligence techniques and Petri nets for real-time schedulling and controll of production systems*. Proceedings of the 12th IMACS World Congress on Scientific Computation, Paris, 1988, 528-531.
122. Mazurkiewicz, A.: *Concurrent program schemes and their interpretations*. Technical Report DAIMI PB-78, Aarhus University, 1977.
123. Mazurkiewicz, A.: *Compositional Semantics of Pure Place/Transition systems*. In: G. Rozenberg (ed): Advances in Petri Nets 1985. Lecture Notes in Computer Science Vol. 222, Springer-Verlag 1986, Berlin, 307-330.
124. Mazurkiewicz, A.: *Trace Theory*. In: W. Brauer, W. Reisig, G. Rozenberg (eds.): Petri Nets: Applications and Relationships to Other Models of Concurrency. Advances in Petri Nets 1986 Part II, Lecture Notes in Computer Science Vol. 255, Springer-Verlag 1987, Berlin, 279-324.
125. Mazurkiewicz, A.: *Introduction to trace theory*, Chapter 1 of V. Diekert, G. Rozenberg (eds.): The book of traces. Word Scientific, Singapore, 1995.
126. Michalski, R., Carbonell, J.G., Mitchell, T.M. (eds.): *Machine Learning: An Artificial Intelligence Approach*, Vol. 1, Tioga/Morgan Publishers, Los Altos, 1983.
127. Michalski, R., Carbonell, J.G., Mitchell, T.M. (eds.): *Machine Learning: An Artificial Intelligence Approach*, Vol. 2, Tioga/Morgan Publishers, Los Altos, 1986.
128. Michalski, R., Kerschberg, L., Kaufman, K.A., Ribeiro, J.S.: *Mining for Knowledge in Databases: The INLEN Architecture, Initial Implementation and First Results*. Intelligent Information Systems: Integrating Artificial Intelligence and Database Technologies 1-1 (1992), 85-113.
129. Michalski, R., Wnęk, J.: *Constructive Induction: An Automated Improvement of Knowledge Representation Spaces for Machine Learning*. Proceedings of a Workshop on Intelligent Information Systems: Practical Aspects of Artificial Intelligence, Vol. II, Augustów, Poland, June 7-11, 1993, 188-236.
130. Michalski, R., Bratko, I., Kubat, M.(eds.): *Machine Learning and Data Mining. Methods and Applications*. Wiley, New York, 1998.

131. Milner, R.: *A calculus of communicated systems*. Lecture Notes in Computer Science, Vol. 92, Springer-Verlag, Berlin, 1980.
132. Milner, R.: *Communication and Concurrency*. Prentice-Hall, New York, 1989.
133. Mitchell, M.: *An Introduction to Genetic Algorithms*. MIT Press, Cambridge, 1996.
134. Molloy, M.K.: *Performance analysis using stochastic Petri nets*. IEEE Trans. Comput., C-31, No. 9, 1982, 913-917.
135. Mrózek, A.: *Rough Sets in Computer Implementation of Rule-Based Control of Industrial Processes*. In: R. Słowiński (ed.): Intelligent Decision Support: Handbook of Applications and Advances of Rough Sets Theory, Kluwer Academic Publishers 1992, Dordrecht, 19-31.
136. Mrózek, A., Płonka, L.: *Knowledge Representation in Fuzzy and Rough Controllers*. In: M. Dąbrowski, M. Michalewicz, Z.W. Raś (eds.): Proceedings of the Third International Workshop on Intelligent Information Systems, Wigry, Poland, June 6-10, 1994, Institute of Computer Science Polish Academy of Sciences, Warsaw, 324-337. Also in: Fundamenta Informaticae 30 (1997), 345-358.
137. Mrózek, A., Płonka, L., Kędziera, J.: *The methodology of rough controller synthesis*. Proceedings of the 5th IEEE International Conference on Fuzzy Systems (FUZZ-IEEE'96), New Orleans, Louisiana, September 8-11, 1996, 1135-1139.
138. Mukund, M.: *Petri Nets and Step Transition Systems*. International Journal of Foundations of Computer Science 3-4 (1992), 443-478.
139. Munakata, T.: *Rough control: Basic ideas and applications*. In: P.P. Wang (ed.): Second Annual Joint Conference on Information Sciences (JCIS'95), Wrightsville Beach, North Carolina, USA, September 28 - October 1, 1995, 340-343.
140. Munakata, T.: *Rough control: a perspective*. In: Lin, T.Y., Cercone, N. (eds.): Rough Sets and Data Mining. Analysis for Imprecise Data, Kluwer Academic Publishers 1997, Dordrecht, 77-88.
141. Munakata, T.: *Fundamentals of the New Artificial Intelligence*. Springer-Verlag, Berlin, 1998.
142. Murata, T.: *Petri Nets: Properties, Analysis and Applications*. Proceedings of the IEEE 77-4 (1989), 541-580.
143. Murata, T.: *Temporal uncertainty and fuzzy-timing high-level Petri nets*. Proceedings of the 17th International Conference on Applications and Theory of Petri Nets, Osaka, Japan, 1996, 10-28.
144. Murata, T., Nelson, P.C., Yim, J.: *A Predicate-Transition Net model for multiple agent planning*. Information Science 57-58 (1991), 361-384.
145. Murata, T., Subrahmanian, V.S., Wakayama, T.: *A Petri net model for reasoning in the presence of inconsistency*. IEEE Transactions on Knowledge and Data Engineering, KDE 3-3 (1991), 281-292.
146. Murata, T., Zhang, D.: *A Predicate-Transition Net model for parallel interpretation of logic programs*. IEEE Transactions on Software Engineering, SE 14-4 (1988), 481-497.
147. Nadler, M., Smith, E.P.: *Pattern Recognition Engineering*. Wiley, New York, 1993.
148. Natarajan, B.K.: *Machine Learning: A Theoretical Approach*. Morgan Kaufmann, San Mateo, 1991.
149. Nazareth, D.L.: *Investigating the applicability of Petri nets for rule-based system verification*. IEEE Transactions on Knowledge and Data Engineering, KDE 5-3 (1993), 402-415.

150. Nguyen, T.A., Perkins, W.A., Laffey, T.J., Pecora, D.: *Knowledge base validation.* AI Magazine, summer 1987, 67-75.
151. Nguyen, S.H., Skowron, A.: *Quantization of real value attributes.* Proceedings of the Second Joint Annual Conference on Information Sciences, Wrightsville Beach, NC, September 28 - October 1, 1995, 34-37.
152. Nguyen, S.H.: *Discretization of real-valued attributes: Boolean reasoning approach.* Ph.D. Thesis, Faculty of Mathematics, Computer Science and Mechanics, Warsaw University, 1997.
153. Nielsen, M., Plotkin, G.D., Winskel, G.: *Petri nets, event structures and domains, Part 1.* Theoretical Computer Science 13 (1981), 85-108.
154. Nielsen, M., Rozenberg, G., Thiagarajan, P.S.: *Elementary Transition Systems.* Theoretical Computer Science 96-1 (1992), 3-33.
155. Nielsen, M., Sassone, V., Winskel, G.: *Relationships Between Models of Concurrency.* In: J.W. de Bakker, W.P. de Roever, G. Rozenberg (eds.): A Decade of Concurrency, Lecture Notes in Computer Science Vol. 803, Springer-Verlag 1993, Berlin, 425-476.
156. Ochmański, E.: *Recognizable trace languages,* Chapter 6 of V. Diekert, G. Rozenberg (eds.): The book of traces. Word Scientific, Singapore, 1995.
157. Ochsenschläger, P., Prinoth, R.: *Modellierung verteilter Systeme. Konzeption, Formale Spezifikation und Verifikation mit Produktnetzen.* Vieweg, Braunschweig/Wiesbaden, 1995.
158. Øhrn, A.: *Rough logic control.* (Project), Technical Report, Knowledge Systems Group, Norwegian Institute of Technology, 1993.
159. Øhrn, A., Komorowski, J., Skowron, A., Synak, P.: *The Rosetta Software System.* In: L. Polkowski, A. Skowron (eds.): Rough Sets in Knowledge Discovery 2. Applications, Case Studies and Software Systems, Physica-Verlag 1998, Heidelberg, 572-575.
160. Olderog, E.R.: *Nets, Terms and Formulas.* Cambridge University Press, 1991.
161. Orłowska, E. (ed.): *Incomplete Information: Rough Set Analysis.* Physica-Verlag, Heidelberg, 1998.
162. Ott, E.: *Chaos in Dynamical Systems.* Cambridge University Press, 1993.
163. Pal, S.K., and Skowron, A. (eds.): *Rough-Fuzzy Hybridization. A New Trend in Decision Making,* Springer-Verlag, Singapore, 1999.
164. Paulson, D., Wand,Y.: *An Automated Approach to Information Systems Decomposition.* IEEE Transactions on Software Engineering 18-3 (1992), 174-189.
165. Pagnoni, A.: *Project Engineering. Computer-Oriented Planning and Operational Decision Making.*, Springer-Verlag, Berlin, 1990.
166. Pawlak, Z.: *Rough sets.* International Journal of Computer and Information Sciences, Vol. 11, 1982, 341-356.
167. Pawlak, Z.: *Decision Tables and Decision Algorithms.* Bulletin of the Polish Academy of Sciences 33-9,10 (1985), 487-494.
168. Pawlak, Z.: *Rough sets - theoretical aspects of reasoning about data.* Kluwer Academic Publishers, Dordrecht, 1991.
169. Pawlak, Z.: *Concurrent versus sequential the rough sets perspective.* Bulletin of the EATCS 48 (1992), 178-190.
170. Pawlak, Z.: *Rough real functions and rough controllers.* In: T.Y. Lin (ed.): Proceedings of the Workshop on Rough Sets and Data Mining at 23rd Annual Computer Science Conference, Nashville, Tenessee, March 2, 1995, 57-62. Also in: [115], 139-147.

171. Pawlak, Z.: *Rough Set Rudiments.* Institute of Computer Science Research Report 96, Warsaw University of Technology, 1996, Poland.
172. Pawlak, Z.: *Some Remarks on Explanation of Data and Specification of Processes Concurrent.* Bulletin of International Rough Set Society 1-1 (1997), 1-4.
173. Pawlak, Z., Munakata, T.: *Rough Control: Application of rough set theory to control.* Proceedings of the Fourth European Congress on Intelligent Techniques and Soft Computing (EUFIT'96) Vol. 1, Aachen, Germany, September 2-5, 1996, Verlag Mainz, 209-218.
174. Pawlak, Z., Skowron, A.: *A rough set approach for decision rules generation.* Institute of Computer Science Research Report 23/93, Warsaw University of Technology, 1993, Poland. Also in: Proceedings of the IJCAI'93 Workshop: The Management of Uncertainty in Artificial Intelligence, France, 1993.
175. Payton, D. W., Bihari, T. E.: *Intelligent Real-Time Control of Robotic Vehicles.* Communications of the ACM, Vol. 34-8, 1991, 48-63.
176. Pedrycz, W.: *Fuzzy Sets Engineering.* CRC Press, Boca Raton, FL, 1995.
177. Pedrycz, W., Gomide, F.: *A generalized fuzzy Petri net model.* IEEE Transactions on Fuzzy Systems 2-4 (1994), 295-301.
178. Pedrycz, W., Peters, J.F.: *Learning in fuzzy Petri nets.* In: Cardoso, J., Sandri, S. (eds.): Fuzzy Petri Nets, Physica-Verlag 1998, Heidelberg, in press.
179. Pedrycz, W., Peters, J.F., Ramanna, S., Furuhashi, T.: *From data to fuzzy Petri nets: generalized model and calibration abilities.* Proceedings of the Seventh International Fuzzy Systems Association World Congress (IFSA'97), Vol. III, 1997, 294-299.
180. Peters, J.F.: *Time and Clock Information Systems: Concepts and Roughly Fuzzy Petri Net Models.* In: L. Polkowski, A. Skowron (eds.): Rough Sets and Knowledge Discovery 2, Physica-Verlag 1998, Heidelberg, 385-417.
181. Peters, J.F., Ramanna, S.: *A rough sets approach to assessing software quality: concepts and rough Petri net models.* In: [163], 349-380.
182. Peters, J.F., Skowron, A., Suraj, Z., Pedrycz, W., Ramanna, S.: *Approximate Real-Time Decision Making: Concepts and Rough Fuzzy Petri Net Models.* International Journal of Intelligent Systems, **14**-4 (1998), 4-37.
183. Peters, J.F., Skowron, A., Suraj, Z., Ramanna, S., Paryzek, A.: *Modelling Real-Time Decision-Making Systems with Roughly Fuzzy Petri Nets.* In: J. Komorowski, A. Skowron (eds.): Proceedings of the 6th European Congress on Intelligent Techniques and Soft Computing (EUFIT'98), Aachen, Germany, September 7-10, 1998, 985-989.
184. Peters, J.F., Skowron, A., Suraj, Z., Ramanna, S.: *Guarded Transitions in Rough Petri Nets.* In: Proceedings of the 7th European Congress on Intelligent Techniques and Soft Computing (EUFIT'99), Aachen, Germany, September 13-16, 1999, Abstract (171p.), the full version of the paper on CD-ROM, BC3.
185. Peters, J.F., Ziaei, K., Ramanna, S.: *Approximate Time Rough Control: Concepts and Application to Satellite Attitude Control.* In: L. Polkowski, A. Skowron (eds.): Proceedings of the International Conference on Rough Sets and Current Trends in Computing (RSCTC'98), Warsaw, Poland, June, 1998, Lecture Notes in Artificial Intelligence Vol. 1424, 491-498.
186. Peterson, J.L.: *Petri Net Theory and the Modeling of Systems.* Prentice-Hall, Englewood Cliffs, 1981.
187. Petri, C.A.: *Kommunikation mit Automaten.* Schriften des IIM Nr. 2, Institut für Instrumentelle Mathematik, Bonn, 1962. Also in: Communication with Automata

(in English). Griffiss Air Force Base, New York Technical Report RADC-TR-65-377, Vol. 1, Suppl. 1, 1966.
188. *Petri Net Newsletter, Special Volume: Petri Net Tools. Overview 92.* Bonn, Germany: Gesellschaft für Informatik (GI), Special Interest Group on Petri Nets and Related System Models, No. 41, April 1992.
189. Piatetsky-Shapiro, G., Frawley, W. (eds.): *Knowledge discovery in databases.* The AAAI Press, Menlo Park, 1991.
190. Pnueli, A.: *The temporal logic of programs.* Proceedings of the 18th IEEE Symposium on Foundations of Computer Science, 1977, 46-57.
191. Pnueli, A.: *Applications of temporal logic to the specification and verification of reactive systems: a survey of current trends.* In: J.W. de Bakker, W.P. de Roever, G. Rozenberg (eds.): Current Trends in Concurrency, Lecture Notes in Computer Science Vol. 224, Springer-Verlag 1986, Berlin, 510-584.
192. Polkowski, L., Skowron, A. (eds.): *Rough Sets in Knowledge Discovery 1. Methodology and Applications.* Physica-Verlag, Heidelberg, 1998.
193. Polkowski, L., Skowron, A. (eds.): *Rough Sets in Knowledge Discovery 2. Applications, Case Studies and Software Systems.* Physica-Verlag, Heidelberg, 1998.
194. Polkowski, L., Skowron, A. (eds.): *Rough Sets and Current Trends in Computing. Proceedings of the First International Conference (RSCTC'98)*, Warsaw, Poland, June 1998, Lecture Notes in Artificial Intelligence Vol. 1424, Springer-Verlag, Berlin, 1998.
195. Popova, L.: *On Time Petri Nets.* Journal of Informormation Processing and Cybernetics (EIK), Vol. 27, No. 4, 1991, 227-244.
196. Portinale, L.: *Petri Net Models for Diagnostic Knowledge Representation and Reasoning.* Ph.D. Thesis, Computer Science Department, University of Torino, Italy, 1994.
197. Pratt, V.: *Modelling Concurrency with Partial Orders.* International Journal of Parallel Programming, Vol. 15, Plenum, 33-71.
198. *Proceedings of the International Workshop on Timed Petri Nets,* Torino, Italy, July 1-3, 1985.
199. *Proceedings of the International Workshop on Petri Nets and Performance Models,* Madison, WI, August 24-26, 1987.
200. *PNPM89: Petri Nets and Performance Models.* Proceedings of the 3rd International Workshop, Kyoto, Japan, 1989, IEEE Computer Society Press.
201. *PNPM91: Petri Nets and Performance Models.* Proceedings of the 4th International Workshop, Melbourne, Australia, 1991, IEEE Computer Society Press.
202. Proth, J.M., Xie, X.: *Petri Nets. A Tool for Design and Management of Manufacturing Systems.* Wiley, 1996.
203. Rabin, M.O., Scott, D.S.: *Finite automata and their decision problems.* IBM J. Research 3 (2), 1959.
204. Ramchandani, C.: *Analysis of Asynchronous Concurrent Systems by Timed Petri Nets.* Cambridge, Mass.: MIT, Dept. Electrical Engineering, Ph.D. Thesis, 1974. Also in: Cambridge, Mass.: MIT, Project MAC, Technical Report 120, Feb., 1974.
205. Reif, J.H. (ed.): *Synthesis of Parallel Algorithms.* Morgan Kaufmann Publishers, San Mateo, 1993.
206. Reisig, W.: *Petri Nets. An Introduction.* EATCS Monographs on Theoretical Computer Science, Vol. 4, Springer-Verlag, Berlin, 1985.
207. Reisig, W.: *A Primer in Petri Net Design.* Springer-Verlag, Berlin, 1992.
208. Reisig, W.: *Elements of Distributed Algorithms.* Springer-Verlag, Berlin, 1998.

209. Reisig, W., Rozenberg, G. (eds.): *Lectures on Petri Nets I: Basic Models.* Lecture Notes in Computer Science Vol. 1491, Springer-Verlag, Berlin, 1998.
210. Reisig, W., Rozenberg, G. (eds.): *Lectures on Petri Nets II: Applications.* Lecture Notes in Computer Science Vol. 1492, Springer-Verlag, Berlin, 1998.
211. Reutenauer, C.: *The Mathematics of Petri Nets.* Masson and Prentice Hall International (UK) Ltd, 1990.
212. Ripley, B.D.: *Pattern Recognition and Neural Networks.* Cambridge University Press, 1996.
213. Roscoe, A.W.: *The Theory and Practice of Concurrency.* Prentice-Hall, Englewood Cliffs, 1998.
214. Rozenberg, G. (ed.): *Advances in Petri Nets 1984.* Lecture Notes in Computer Science Vol. 188, Springer-Verlag 1985, Berlin.
215. Rozenberg, G. (ed): *Advances in Petri Nets 1985.* Lecture Notes in Computer Science Vol. 222, Springer-Verlag 1986, Berlin.
216. Rozenberg, G. (ed.): *Advances in Petri Nets 1987.* Lecture Notes in Computer Science Vol. 266, Springer-Verlag 1987, Berlin.
217. Rozenberg, G. (ed.): *Advances in Petri Nets 1988.* Lecture Notes in Computer Science Vol. 340, Springer-Verlag 1988, Berlin.
218. Rozenberg, G. (ed.): *Advances in Petri Nets 1989.* Lecture Notes in Computer Science Vol. 424, Springer-Verlag 1990, Berlin.
219. Rozenberg, G. (ed): *Advances in Petri Nets 1990.* Lecture Notes in Computer Science Vol. 483, Springer-Verlag 1991, Berlin.
220. Rozenberg, G. (ed.): *Advances in Petri Nets 1991.* Lecture Notes in Computer Science Vol. 524, Springer-Verlag 1991, Berlin.
221. Rozenberg, G. (ed.): *Advances in Petri Nets 1992.* Lecture Notes in Computer Science Vol. 609, Springer-Verlag 1992, Berlin.
222. Rozenberg, G. (ed.): *Advances in Petri Nets 1993.* Lecture Notes in Computer Science Vol. 674, Springer-Verlag 1993, Berlin.
223. Rozenberg, G., Thiagarajan, P.S.: *Petri nets: basic notions, structure, behaviour.* In: J.W. de Bakker, W.-P. de Roever, G. Rozenberg (eds.), Current Trends in Concurrency. Lecture Notes in Computer Science Vol. 224, Springer-Verlag 1986, Berlin, 585-668.
224. Rozenblyum, L.Y.: *Petri Nets.* Tekh. Kibern. (USRR), Vol. 21, No. 5, 1983, 12-40.
225. Russell, S., Norvig, P.: *Artificial Intelligence: Modern Approach.* Prentice-Hall, Englewood Cliffs, 1995.
226. Sage, A.P.: *Methodology for Large Scale Systems*, McGraw-Hill, New York, 1977.
227. Sahner, R.A., Trivedi, K.S., Puliafito, A.: *Performance and Reliability Analysis of Computer Systems: An Example-Based Approach Using the SHARPE Software Package.* Kluwer Academic Publishers, Dordrecht, 1995.
228. Sassone, V., Nielsen, M., Winskel, G.: *Models for Concurrency: Towards a Classification.* Theoretical Computer Science 170, Elsevier, 1996, 297-348.
229. Schoppers, M.: *Real-Time Knowledge-Based Control Systems.* Communications of the ACM, Vol. 34-8, 1991, 26-30.
230. Shafer, G.: *A mathematical theory of evidence.* Princeton University Press, 1976.
231. Shapiro, S.C., Eckroth, D.: *Encyclopedia of Artificial Intelligence Vol. 1.* Wiley, New York, 1987, 18-35.
232. Shapiro, S.C. (ed.): *Encyclopedia of Artifficial Intelligence Vol. 1,2.* Second Edition, Wiley, New York, 1992.
233. Shavlik, J.W., Dietterich, T.G. (eds.): *Readings in Machine Learning.* Morgan Kaufmann Publishers, San Mateo, 1990.

234. Shields, M.W.: *Concurrent machines*. Computer Journal, Vol. 28, Cambridge University Press, 1985, 449-465.
235. Shrager, J., Langley, P. (eds.): *Computational Models of Scientific Discovery and Theory Formation*. Morgan Kaufmann Publishers, San Mateo, 1990.
236. Silva, M.: *Las Redes de Petri*. En la Automática y la Informática. Edotorial AC, Madrit, 1985.
237. Simon, H.A.: *The Sciences of the Artificial*. 2nd edition, MIT Press, Cambridge, 1981.
238. Skowron, A.: *The rough set theory and evidence theory*. Fundamenta Informaticae 13 (1990), 245-262.
239. Skowron, A.: *Boolean reasoning for decision rules generation*. In: J. Komorowski, Z. Raś (eds.): Proceedings of the Seventh International Symposium (ISMIS'93), Trondheim, Norway, 1993, Lecture Notes in Artificial Intelligence Vol. 689, Springer-Verlag 1993, Berlin, 295-305.
240. Skowron, A.: *A synthesis of decision rules: applications of discernibility matrix properties*. Proceedings of the Conference Intelligent Information Systems, Augustów, Poland, June 7-11, 1993.
241. Skowron, A.: *Synthesis of adaptive decision systems from experimantal data (invited talk)*. In: A. Aamadt, J. Komorowski (eds.): Proceedings of the Fifth Scandinavian Conference on Artificial Intelligence (SCAI-95), IOS Press 1995, Amsterdam, 220-238.
242. Skowron, A.: *Extracting laws from decision tables: a rough set approach*. Computational Intelligence, Vol. 11-2, 1995, 371-388.
243. Skowron, A., Grzymała-Busse, J.: *From the rough set theory to the evidence theory*. Institute of Computer Science Research Report 8/91, Warsaw University of Technology, 1991, Poland. Also in: R.R. Yager, M. Federizzi, J. Kacprzyk (eds.): Advances in the Dempster-Shafer theory of evidence, Wiley, New York, 1993, 193-236.
244. Skowron, A., Peters, J. F., Suraj, Z.: *An Application of Rough Set Methods to Control Design*. In: H.D. Burkhard, L. Czaja, H.S. Nguyen, P. Starke (eds.): Proceedings of the Workshop on the Concurrency, Specification and Programming, Warsaw, Poland, 28-30 September, 1999, 214-235.
245. Skowron, A., Polkowski, L.: *Synthesis of decision systems from data tables*. Institute of Computer Science Research Report 64/95, Warsaw University of Technology, 1995, Poland.
246. Skowron, A., Polkowski, L.: *Rough mereology: A new paradigm for approximate reasoning*. Journal of Approximate Reasoning, Vol. 15-4, 1996, 333-365.
247. Skowron, A., Polkowski, L.: *Decision Algorithms: a Survey of Rough Set - Theoretical Methods*. Fundamenta Informaticae 30 (1997), 345-358.
248. Skowron, A., Polkowski, L., Komorowski, J.: *Learning tolerance relations by Boolean descriptors: automatic feature extraction from data tables*. Proceedings of the Fourth International Workshop on Rough Sets, Fuzzy Sets, and Machine Discovery (RSFD-96), Tokyo, Japan, November 6-8, 1996, 11-17.
249. Skowron, A., Rauszer, C.: *The discernibility matrices and functions in information systems*. In: R. Słowiński (ed.): Intelligent Decision Support: Handbook of Applications and Advances of Rough Sets Theory, Kluwer Academic Publishers 1992, Dordrecht, 331-362.
250. Skowron, A., Stepaniuk, J.: *Decision Rules Based on Discernibility Matrices and Decision Matrices*. In: T.Y. Lin (ed.), Proceedings of The Third International

Workshop on Rough Sets and Soft Computing, San Jose, CA, November 10-12, 1994, 156-163.
251. Skowron, A., Suraj, Z.: *A Rough Set Approach to Real-Time State Identification*. Bulletin of the EATCS 50 (June 1993), 264-275.
252. Skowron, A., Suraj, Z.: *Rough Sets and Concurrency*. Bulletin of the Polish Academy of Sciences 41-3 (1993), 237-254.
253. Skowron, A., Suraj, Z.: *Synthesis of concurrent systems specified by information systems. Part 1*. Institute of Computer Science Research Report 4/93, Warsaw University of Technology, 1993, Poland.
254. Skowron, A., Suraj, Z.: *A rough set approach to real-time state identification for decision making*. Institute of Computer Science Research Report 18/93, Warsaw University of Technology, 1993, Poland.
255. Skowron, A., Suraj, Z.: *Synthesis of concurrent systems specified by information systems. Part 2. Examples of synthesis*. Institute of Computer Science Research Report 38/93, Warsaw University of Technology, 1993, Poland.
256. Skowron, A., Suraj, Z.: *Synthesis of concurrent systems specified by information systems*. Institute of Computer Science Research Report 39/94, Warsaw University of Technology, 1994, Poland.
257. Skowron, A., Suraj, Z.: *Discovery of concurrent data models from experimental data tables: a rough set approach*. Institute of Computer Science Research Report 15/95, Warsaw University of Technology, 1995, Poland.
258. Skowron, A., Suraj, Z.: *Discovery of Concurrent Data Models from Experimental Tables: A Rough Set Approach*. In: Usama M. Fayyad, Ramasamy Uthurusamy (eds.): Proceedings of the First International Conference on Knowledge Discovery and Data Mining (KDD-95), Montreal, Canada, August 19-21, 1995, The AAAI Press, Menlo Park, 288-293.
259. Skowron, A., Suraj, Z.: *A Parallel Algorithm for Real-Time Decision Making: A Rough Set Approach*. Journal of Intelligent Information Systems 7, Kluwer Academic Publishers, Dordrecht, 1996, 5-28.
260. Słowiński, R. (ed.): *Intelligent Decision Support: Handbook of Applications and Advances of the Rough Sets Theory*. Kluwer Academic Publishers, Dordrecht, 1992.
261. Słowiński, R., Stefanowski, J.: *Rough-set reasoning about uncertain data*. In: W. Ziarko (ed.): Fundamenta Informaticae 27-2,3 (1996) (special issue), IOS Press, Amsterdam, 229-243.
262. Smith, D.R.: *The synthesis of LISP programs from examples: a survey*. In: A. Bierman, G. Guiho, Y. Kodratoff (eds.): Automatic Program Construction Techniques, Macmillan, New York, 1984, 307-324.
263. Snow, C.R.: *Concurrent Programming*. Cambridge University Press, 1992.
264. Stankovic, J.A., Ramamritham, K. (eds.): *Advances in Real-Time Systems*. IEEE Computer Society Press, Los Alamitos, 1993.
265. Starke, P.H.: *Petri-Netze*. VEB Deutscher Verlag der Wissenschaften, Berlin, 1980.
266. Starke, P.H.: *Petri-Netz-Maschine: A Software Tool for Analysis and Validation of Petri Nets*. Proceedings of the 2nd International Symposium on Systems Analysis and Simulation, Berlin, 1985, Pergamon, Oxford, 474-475.
267. Starke, P.H.: *On the mutual simulatability of different types of Petri nets*. In: K. Voss, H.J. Genrich, G. Rozenberg (eds.): Concurrency and Nets, Special Volume of Advances in Petri Nets, Lecture Notes in Computer Science, Springer-Verlag 1987, Berlin, 481-495.

268. Starke, P.H.: *Analyse von Petri-Netz-Modellen*. B.G. Teubner, Stuttgart, 1990.
269. Starke, P.H., Hanisch, H.M.: *Analysis of Signal/Event-Nets*. Proceedings of the 6th IEEE International Conference on Emerging Technologies and Factory Automation (ETFA'97), Los Angeles, USA, September, 1997, 253-257.
270. Starke, P.H.: *Signal/Event Net Analyzer, Version 0.6*. Humboldt University, Berlin, 1998.
271. Suraj, Z.: *A Graphical System for Petri Net Design and Simulation - GRAPH*. Petri Net Newsletter No. 35, Special Interest Group on Petri Nets and Related System Models, Gesellschaft für Informatik (GI), Germany, 1990, 32-36.
272. Suraj, Z.: *GRAPH*. Petri Net Newsletter No. 41. Special Volume: Petri Net Tools. Overwiew 92, Special Interest Group on Petri Nets and Related System Models, Gesellschaft für Informatik (GI), Germany, 1992, 20-21. Also in: Advances in Petri Nets 1993, Lecture Notes in Computer Science Vol. 674, Springer-Verlag 1993, Berlin, 187-188.
273. Suraj, Z.: *A System for the Design and Analysis of Petri Nets*. Institute of Computer Science Research Report 3/93, Warsaw University of Technology, 1993, Poland.
274. Suraj, Z.: *Tools for Generating and Analyzing Concurrent Models Specified by Information Systems*. In: T.Y. Lin (ed.): Proceedings of the Third International Workshop on Rough Sets and Soft Computing, San Jose, CA, November 10-12, 1994, 610-617.
275. Suraj, Z., Komarek, B.: *GRAF. Graphical System for Construction and Analysis of Petri Nets*. Akademicka Oficyna Wydawnicza PLJ, Warszawa, 1994, in Polish.
276. Suraj, Z: *Tools for Generating Concurrent Models Specified by Information Systems*. In: T.Y. Lin, A.M. Wildberger (eds.): Soft Computing, Simulation Councils, Inc., The Society for Computer Simulation, San Diego, CA, 1995, 107-110.
277. Suraj, Z.: *PN-tools: Environment for the Design and Analysis of Petri Nets*. Control and Cybernetics 24-2 (1995), Systems Research Institute of Polish Academy of Sciences, 199-222.
278. Suraj, Z.: *An Application of Rough Set Methods to Cooperative Information Systems Re-engineering*. In: S. Tsumoto, S. Kobayashi, T. Yokomori, H. Tanaka, A. Nakamura (eds.): Proceedings of the Fourth International Workshop on Rough Sets, Fuzzy Sets and Machine Discovery (RSFD-96), Tokyo, Japan, November 6-8, 1996, The Tokyo University Press, 364-371.
279. Suraj, Z.: *Discovery of Concurrent Data Models from Experimental Tables: A Rough Set Approach*. Fundamenta Informaticae 28-3,4 (1996), IOS Press, Amsterdam, 353-376.
280. Suraj, Z., Gąsior, T.: *Tools for the Design and Analysis of Concurrent Systems*. In: V. Tchaban (ed.): Proceedings of the 1st International Modelling School - Krym'96, Alushta, September 12-17, 1996, Ukraine, 79-81.
281. Suraj, Z.: *Reconstruction of Cooperative Information Systems under Cost Constraints: A Rough Set Approach*. In: P. Wang (ed.), Proceedings of the First International Workshop on Rough Sets and Soft Computing (RSSC-97), Relaigh, NC, USA, March 1-5, 1997, 364-371.
282. Suraj, Z.: *Computer Aided Design and Analysis of Concurrent Models Represented by Petri Nets*. In: K. Hrubina (ed.), Proceedings of the Scientific Conference with International Participation "Informatics and Mathematics", Presov, Slovakia, September 4-5, 1997, 65-70.
283. Suraj, Z.: *Reconstruction of Cooperative Information Systems under Cost Constraints: A Rough Set Approach*. Journal of Information Sciences 111 (1998),

Elsevier Science Publishers, 273-291.
284. Suraj, Z.: *The Synthesis Problem of Concurrent Systems Specified by Dynamic Information Systems.* In: L. Polkowski, A. Skowron (eds.): Rough Sets in Knowledge Discovery 2. Applications, Case Studies and Software Systems. Studies in Fuzziness and Soft Computing, Physica-Verlag 1998, Heidelberg, 418-448.
285. Suraj, Z.: *TAS: Tools for Analysis and Synthesis of Concurrent Processes using Rough Set Methods.* In: L. Polkowski, A. Skowron (eds.): Rough Sets in Knowledge Discovery 2. Applications, Case Studies and Software Systems. Studies in Fuzziness and Soft Computing, Physica-Verlag 1998, Heidelberg, 587-590.
286. Suraj, Z.: *Discovery of Communicating Agent Systems from Experimental Data: A Rough Set Approach.* In: J. Komorowski, I. Düntsch, A. Skowron (eds.): Proceedings of the 13th European Conference on Artifficial Intelligence (ECAI'98): Workshop on Synthesis of Intelligent Agent Systems from Experimental Data, Brighton, UK, August 24-28, 1998, 76-90.
287. Suraj, Z.: *ROSEPEN: Environment for the Synthesis and Analysis of Concurrent Processes based on Rough Set Methods and Petri Nets.* In: S. Tsumoto, Y.Y. Yao, M. Hadjimichael (eds.): Bulletin of International Rough Set Society, Vol. 2, No. 1, June, 1998, 37-39.
288. Suraj, Z., Szpyrka, M.: *Computer tools for the design and analysis of concurrent systems represented by Petri nets.* In: S. Paszczyński (ed.): Proceedings of 2nd International Conference on Computer Systems and Networks - designing, application, utilization, Rzeszów, Poland, June 25-26, 1998, Zeszyty Naukowe Wyższej Szkoły Informatyki i Zarządzania w Rzeszowie, Nr 6/1998, 163-176.
289. Suraj, Z.: *An Application of Rough Sets and Petri Nets to Controller Design.* In: S. Aoshima, L. Polkowski, M. Toho (eds.): Proceedings of the International Conference on Intelligent Techniques in Robotics, Control and Decision Making, February 22-23, 1999, Warsaw, Polish-Japanese Institute of Information Technology, 86-96.
290. Suraj, Z.: *Petri Nets and Rough Sets in Controller Design.* In: L. Portinale, R. Valette, D. Zhang (eds.): Proceedings of the Workshop on Application of Petri Nets to Intelligent System Development with 20th International Conference on Applications and Theory of Petri Nets, Williamsburg, USA, June 21-25, 1999.
291. Suraj, Z., Szpyrka, M.: *Petri Nets and PN-tools. Tools for the Construction and Analysis of Petri Nets*, Pedagogical University Publisher, Rzeszów, 1999, in Polish.
292. Szladow, A.J., Ziarko, W.P.: *Knowledge-based process control using rough sets.* In: R. Słowiński (ed.): Intelligent Decision Support: Handbook of Applications and Advances of Rough Sets Theory, Kluwer Academic Publishers 1992, Dordrecht, 49-60.
293. Terano, T., Asai, K., Sugeno, M.: *Fuzzy Systems: Theory and Its Applications.* Academic Press, San Diego, 1992.
294. Thiagarajan, P.S.: *Elementary Net Systems.* In: W. Brauer, W. Reisig, G. Rozenberg (eds.): Petri Nets: Central Models and Their Properties. Advances in Petri Nets 1986 Part I, Lecture Notes in Computer Science Vol. 254, Springer-Verlag 1987, Berlin, 26-59.
295. Valette, R., Bako, B.: *Software implementation of Petri nets and compilation of rule-based systems.* In: Advances in Petri Nets 1991. Lecture Notes in Computer Science Vol. 524, Springer-Verlag 1991, Berlin, 296-316.
296. Valette, R., Courvoisier, M.: *Petri nets and artificial intelligence.* Proceedings of International Workshop on Emerging Technologies for Factory Automation,

North Queensland, Australia, 1992. Also in: R. Żurawski, T. Dillon (eds.), Modern Tools for Manufacturing Systems, Elsevier, 1993, 385-405.
297. Valk, R.: *Self-Modifying Nets: a Natural Extension of Petri Nets*. In: G. Ausiello, C. Bohm (eds.): Automata, Languages and Programming, Lecture Notes in Computer Science Vol. 62, Springer-Verlag 1978, Berlin, 464-476.
298. Varma, D., Trachtenberg, E.A.: *Design automation tools for efficient implementation of logic functions by decomposition*. IEEE Transactions on CAD, Vol. 8-8, 1989, 901-916.
299. Vasiliev, V., Kuzmuk, V.V.: *Petri Nets, Parallel Algorithms and Models of Multiprocessor Systems*. Naukova Dumka, Kiev, USSR, 1990, in Russian.
300. Viswanadham, N., Narahari, Y.: *Performance Modeling of Automated Manufacturing Systems*, Prentice-Hall, Englewood Cliffs, 1992.
301. Voss, K., Genrich, H.J., Rozenberg, G. (eds.): *Concurrency and Nets*. Advances in Petri Nets, Springer-Verlag 1987, Berlin.
302. Wang, J.: *Timed Petri Nets: Theory and Application*. Kluwer Academic Publishers, Dordrecht, 1998.
303. Warfield, J.N.: *Societal Systems: Planning, Policy and Complexity*, Wiley, New York, 1976.
304. Wasserman, P.D.: *Advanced Methods in Neural Computing*. van Nostrand Reinhold, 1993.
305. Wegener, I.: *The complexity of Boolean functions*. Wiley and B.G. Teubner, Stuttgart, 1987.
306. Winkowski, J.: *An Algebraic Description of System Behaviours.*, Theoretical Computer Science, Vol. 21, 1982, 315-340.
307. Winkowski, J.: *An Algebraic Way of Defining the Behaviour of Place/Transition Petri Nets*. Petri Net Newsletter No. 33, Bonn, Special Interest Group on Petri Nets and Related System Models, August 1989.
308. Winkowski, J., Maggiolo-Schettini, A.: *An Algebra of Processes.*, Journal of Computer and System Sciences, Vol. 35-2, 1987, 206-228.
309. Winkowski, J.: *Concatenable Weighted Pomsets and Their Applications to Modelling Processes of Petri Nets.*, Fundamenta Informaticae, Vol. 28-3,4 (1996), 403-422.
310. Winskel, G.: *Events in computation*. Ph.D. Thesis, Depertment of Computer Science, University of Edinburg, 1980.
311. Winskel, G.: *Event structures*. In: W. Brauer, W. Reisig, G. Rozenberg (eds.): Petri Nets: Applications and Relationships to Other Models of Concurrency. Advances in Petri Nets 1986 Part II, Lecture Notes in Computer Science Vol. 255, Springer-Verlag 1987, Berlin, 325-392.
312. Winskel, G., Nielsen, M.: *Models for Concurrency*. In: S. Abramsky et al. (eds.), Hanbook of Logic in Computer Science Vol. 4, Oxford University Press, 1995, 1-148.
313. Winston, P.H.: *Artificial Intelligence*. Third Edition, Addison-Wesley, Reading, 1992.
314. Yager, R., Fedrizzi, M., Kacprzyk, J. (eds.): *Advances in the Dempster-Shafer Theory of Evidence*. Wiley, New York, 1994.
315. Zadeh, L.A.: *Fuzzy sets*. Information and Control 8 (1965), 338-353.
316. Zadeh, L.A.: *Fuzzy logic = computing with words*. IEEE Transactions on Fuzzy Systems 4-2 (1996), 103-111.
317. Zadeh, L.A.: *Toward a theory of fuzzy information granulation and its certainty in human reasoning and fuzzy logic*. Fuzzy Sets and Systems 90-2 (1997), 111-128.

318. Zembowicz, R., Żytkow, J.M.: *Discovery of Equations: Experimental Evaluation of Convergence.* Proceedings of the AAAI-92, The AAAI Press, Menlo Park, 1992, 70-75.
319. Zhang, D.: *Planning with Pr/T nets.* In: Proceedings of the IEEE International Conference on Robotics and Automation, Sacramento, CA, 1991, 769-775.
320. Zhou, M.C., DiCesare, F.: *Petri Net Synthesis for Discrete Event Control of Manufacturing Systems.* Kluwer Academic Publishers, Dordrecht, 1993.
321. Ziarko, W.P.: *Acquisition of control algorithms from operation data.* In: R. Słowiński (ed.): Intelligent Decision Support: Handbook of Applications and Advances of Rough Sets Theory, Kluwer Academic Publishers 1992, Dordrecht, 61-75.
322. Ziarko, W.P. (ed.): *Proceedings of the International Workshop on Rough Sets and Knowledge Discovery (RSKD'93)*, Banff, Canada, October 11-15, 1993.
323. Zimmermann, H.J.: *Fuzzy Set Theory and Its Applications.* Kluwer Academic Publishers, Boston, 1985.
324. Zuberek, W.M.: *Timed Petri Nets, Definitions, Properties, and Applications.* Microelektronics and Reliability, Vol. 31, No. 4, 1991, 627-644.
325. Zuse, K.: *Anwendungen von Petri-Netzen.* Vieweg, Braunschweig/Wiesbaden, 1982.
326. Żurawski, R., Zhou, M.C.: *Special issue on Petri nets in manufacturing.* IEEE Trans. on Industrial Electronics, Vol. 41-6, 1994.
327. Żytkow, J.: *Interactive mining of regularities in databases.* In: G. Piatetsky-Shapiro, W. Frawley (eds.), Knowledge discovery in databases, The AAAI Press, Menlo Park, 1991.
328. Żytkow, J.M. (ed.): *Proceedings of the ML-92 Workshop on Machine Discovery (MD-92)*, Aberdeen, Scotland, UK, July 4, 1992, NIAR Report 92-12, National Institute for Aviation Research, The Wichita State University, Wichita, Kansas, 1992, USA.
329. Żytkow, J.: *Introduction: Cognitive Autonomy in Machine Discovery*, Machine Learning, Vol. 12, 1993, Kluwer Academic Publishers, Boston, 7-16.
330. Żytkow, J., Zembowicz, R.: *Database Exploration in Search of Regularities*, Journal of Intelligent Information Systems, Vol. 2, 1993, Kluwer Academic Publishers, Boston, 39-81.

PART 4:

METHODS AND APPLICATIONS: ALGEBRAIC AND STATISTICAL ASPECTS, CONFLICTS, INCOMPLETENESS

Chapter 8

Conflict Analysis

Rafał Deja

Alta s.c.
ul. Tokarskiego 4/14, 40-749 Katowice, Poland
e-mail: rd@alta.pl

1 Introduction

Computer support for different human activities has grown up in the latest years. Actually the researchers in Artificial Intelligence benefit from this fact in many fields not considered some years ago. Conflict analysis is one of the fields whose importance is increasing nowadays as distributed systems of computers are starting to play a significant role in the society. The computer aided conflict analysis must be applied when intelligent machines (agents) interact. However this is only one from many different areas where a conflict can arise like business, government, political or military operations, labour–management negotiations etc.etc.

In this Chapter, we first examine nature of conflicts as we are formally defining the conflict situation model. Then we investigate a consensus problem and we discuss the methods of solving it. Finally, other problems related to resolving conflicts are analysed.

The introduced model enhances the one proposed and investigated by Pawlak in [24, 26, 27, 28]. The new model is based on information stored in the information tables and on constrains among agents' attributes. Any data table may form a Pawlak information system and constraints are usually given in the form of boolean formulas. Boolean reasoning has been found as the best tool for conflict analysis within this model.

2 Pawlak model

The simple model introduced by Pawlak [24] forms the basis for the model presented in this paper. In the Pawlak model, attitudes of agents to specific issues are depicted in the form of a table in which agents are represented by rows and issues by columns. The value assigned to each agent and to each attribute (issue) is in the set $\{-1, 0, 1\}$, where $-1, 0, 1$ mean, respectively, that an agent is *against*, *neutral* or *favourable* toward the issue.

Formally, any table described above can be represented as an *information system* defined as follows:

An information system is a pair $S = (U, A)$, where

U - is a nonempty, finite set called the universe; elements of U are called objects (here agents),

A - is a nonempty, finite set of attributes (issues).

Every attribute $a \in A$ is a map, $a : U \to V_a$, where the set V_a is the value set of a; elements of V_a are referred to as *opinions*, i.e. $a(x)$ is the opinion of an agent x about an issue a. The domain of each attribute (for the conflict analysis model) is restricted to three values only, i.e. $V_a = \{-1, 0, 1\}$, which means against, neutral, favourable, respectively.

Example 1. Let us first analyse a conflict between an employer and an employee. The example is taken from the author's observation and is used to present the defined notions rather then resolve a real conflict. Job attributes considered for the worker are compensation and work conditions. On the other hand, employers are interested in factory profit, good investment level and, maybe, worker's satisfaction. We can think about these attributes quite generally, for example, compensation can consist of the worker's salary and all his income but it also can include the repeated profit division like the social fund. Similarly worker's conditions include a modern and safe work place and in addition a nice team and development possibilities. One can easily find that these aspects contradict each other in this example. We analyse this problem more deeply in the sequel.

Let us choose the issues for the Pawlak model (agents are voting on):
a - increasing the employees' incomes,
b - improving the work conditions,
c - increase the factory profit by reducing the costs of work,
d - increase the level of investment.

Then, the information table: Table 1, where ag_1 is the employee and ag_2 is the employer, can describe the conflict situation.

	a	b	c	d
ag_1	1	1	-1	0
ag_2	-1	0	1	1

Table 1. The conflict situation in the Pawlak model

The tension of the conflict [28] in the described situation can be calculated as equal to 1.

Analysis of conflicts described by the Pawlak model is restricted to outermost conclusions like finding the most conflicting attributes or the coalitions of agents if more than two take part in the conflict [8].

Because in the Pawlak model the reason for the conflict cannot be determined, there is no way to specify the situation for avoiding the conflict. Moreover, we cannot be sure that the issues the agents vote represent the issues each agent takes care of. In the next section, we will define a model allowing for answering the following basic questions.

- What are the conflict reasons?
- How the consensus can be found?
- Is it possible to satisfy all the agents?

3 New model

In the Pawlak model, conflicts are presented at the outermost level. Some issues are chosen, and the agents are asked to specify their views: are they favourable, neutral or against. In the real world, views on the issues to vote are consequences of the decision taken, based on the local issues, the current state and some background knowledge. Therefore, the Pawlak model is enhanced here by adding to this model some local aspects of conflicts. The introduced model also gives a possibility to check if the issues to vote are chosen correctly, i.e., if the local issues determine the decisions.

3.1 Local states

The information about the local states U_{ag} of an agent ag can be represented in the form of an information table, creating the agent ag's information system $I_{ag} = (U_{ag}, A_{ag})$ where $a : U_{ag} \to V_a$ for any $a \in A_{ag}$ and V_a is the value set of attribute a. We assume:

$$V(ag) = \bigcup_{a \in A(ag)} V_a$$

Any local state $s \in U_{ag}$ is explicitly described by its *information vector* $InfA_{ag}(s)$, where $InfA_{ag}(s) = \{(a, a(s)) : a \in A_{ag}\}$. The set $\{InfA_{ag}(s) : s \in U_{ag}\}$ is denoted by $INF_{A_{ag}}$ and it is called the *information vector set of* ag. We assume that sets A_{ag} are pairwise disjoint, i.e., $A_{ag} \cap A_{ag'} = \emptyset$ for $ag \neq ag'$. This condition emphasises that any agent is describing the situation in its own way. The manner of understanding the *same world* by each agent can be completely different. Relationships among attributes of different agents will be defined by constraints as shown in Section 3.3.

Example 2 illustrates local states for the labour–management conflict.

Subjective evaluation of local states (similarity of states) Every agent has favourable (target) states in the set of local states, i.e., those states the agent wants to reach. In the information table of ag the states the agent ag cannot accept can also appear; being in such a state could mean a disaster for the agent. Actually, the agent evaluates each state. The *subjective evaluation* corresponds to an order (or partial order) of the states of the agent information table. We assume that the function e_{ag} called the *target function*, assigns an evaluation score to each state. An exemplary target function used in our examples is defined to be a function $e_{ag} : U_{ag} \to R[0,1]$. The states with score 1 are mostly preferred by the agent as target states, while the states with score 0 are not acceptable.

The current information vector of ag is a vector describing the current local state of the agent ag.

The state evaluation can also help us to find the state similarity (see e.g. [30] for references on similarity in rough set investigations). For any $\varepsilon > 0$ and $s \in U_{ag}$, we define ε-neighbourhood of s by:

$$\tau_{ag,\varepsilon}(s) = \{s\prime \in U_{ag} : |e_{ag}(s) - e_{ag}(s')| \leq \varepsilon\}$$

The family $\{\tau_{ag,\varepsilon}(s)\}_{s \in U_{ag}}$ defines a tolerance relation $\tau_{ag,\varepsilon}$ in $U_{ag} \times U_{ag}$ by $s\tau_{ag,\varepsilon}s\prime$ iff $s\prime \in \tau_{ag,\varepsilon}(s)$.

Example 2. Let us consider the situation described in Example 1. Ag consists of two agents: ag_1– the employee and ag_2–the employer.

Table 2 shows agent's ag_1 states, i.e., views on local issues (attributes) a, b and the state subjective evaluation.

a–compensation,
b–work conditions

local states	a	b	evaluation e_{ag1}
s1	2	2	1
s2 (current)	2	1	$\frac{2}{3}$
s3	1	2	$\frac{1}{3}$
s4	1	1	0
s5	2	0	0

Table 2. Local states of agent ag_1 with subjective evaluation

The agent ag_2 describes its view on local issues k, l, m where
k–factory profit,
l–level of investment,
m–worker's satisfaction.

local states	k	l	m	evaluation e_{ag2}
s1	2	2	2	1
s2	1	2	2	$\frac{2}{3}$
s3	1	1	2	$\frac{1}{3}$
s4	1	1	1	$\frac{1}{3}$
s5 (current)	2	0	1	0

Table 3. Local states of agent ag_2 with subjective evaluation

For simplicity, let us assume that attributes' domains for both agents are the same, and values belong to the set $V = \{0, 1, 2\}$. One can interpret the values from set V as *small*, *medium* and *high* levels, respectively. For example, the state s_1 of the agent ag_1 expresses a high level of compensation and high level of work conditions. Similarly, the state s_3 of the agent ag_2 means medium profit level with medium level of investment, while at the same time worker satisfaction is high.

Distance function A tolerance relation τ describes similarity of states according to the subjective evaluation. However, it is necessary to describe the state similarity according to differences between values of attributes. Similarity of states from U_{ag} can be often defined as follows.

We assume that for any $a \in A_{ag}$ there is a distance function:

$$d_a : U_{ag} \times U_{ag} \to R+$$

For example, $d_a(s, s\prime) = |a(s) - a(s\prime)|$ if $V_{ag}(a) \subseteq R$.

Next we define the distance function

$$d : U_{ag} \times U_{ag} \to R+ \text{ by } d(s, s\prime) = F(d_{a1}(s, s\prime), ..., d_{am}(s, s\prime))$$

where $A_{ag} = \{a1, ..., am\}$ and $F : R_+^m \to R+$ is a function like e.g.

$$F(r1, ..., rm) = \sqrt{r_1^2 + ... + r_m^2}$$

The function F depends on the problem and should be chosen reflecting the problem specificity.

Crucial for the negotiation process results and for ability to solve any conflict is agents' willingness to change the current state (possibly giving up some resources). This disposition is the basis to define *closeness* of states agents are ready to accept. Closeness is defined by a distance function in the following manner: two states s and s' are close iff $d(s, s\prime) < \varepsilon_{ag}$, where ε_{ag} is a given threshold for ag. Consequently, a closeness neighbourhood of the state s with a diameter ε_{ag} is defined by $\{s\prime : d(s, s\prime) < \varepsilon_{ag})\}$.

Example 3. Let $\varepsilon(ag_2) = \frac{2}{3}$. The example of closeness neighbourhood of the local state s_2 with the diameter $\frac{2}{3}$ is presented in Table 4.

We assume $F(v1, v2, v3) = \frac{1}{3}(v1 + v2 + v3)$ and

$$d(s2, s2_1) = F(d_k(s2, s2_1), d_l(s2, s2_1), d_m(s2, s2_1)).$$

Hance $d(s2, s2_1) = \frac{1}{3}(1 + 0 + 0) = \frac{1}{3}$.

local states	k	l	m
s2	1	2	2
$s2_1$	2	2	2
$s2_2$	1	1	2
$s2_3$	1	2	1

Table 4. The closeness of state 2 within the threshold $\frac{2}{3}$

Local set of goals (targets) The target function introduces a partial order in the set of local states so that one can find the maximal element(s) (with the highest evaluation) and the minimal one(s). Maximal elements can be interpreted as those, which are targets of the agent, i.e., the agent wants to reach them e.g. in a negotiation process. The agent ag's *set of goals (targets)* denoted by $T(ag)$ is defined as the set of target states of ag, which means

$$T(ag) = \{s \in U_{ag} : e_{ag}(s) > \mu_{ag}\}$$

and μ_{ag} is the boundary level, chosen by the agent ag - it is subjective which evaluation level is acceptable by the agent.

Example 4. In the considered situation, the minimal acceptable level of evaluation by the both agents will be, e.g., a score greater than $\frac{1}{3}$. Accordingly sets of goals of agents ag_1 and ag_2 are as follows: $T(ag_1) = \{s1, s2\}$ and $T(ag_2) = \{s1, s2\}$.

The set of goals can also be presented in the propositional form. The information table with scores is going to be converted to the decision table in which the decision 1 means that the state belongs to the set of goals, while 0 that it does not. Then the rules for decision 1 are found (for the method of rule generation see e.g. [29]).

The decision table of an agent ag_1 with the threshold $\frac{1}{3}$ is constructed and presented in Table 5.

states	a	b	decision d
s1	2	2	1
s2 (current)	2	1	1
s3	1	2	0
s4	1	1	0
s5	2	0	0

Table 5. The local set of goals - the decision table

Rule for d_1: $(a_2 \wedge b_2) \vee (a_2 \wedge b_1) \rightarrow d_1$
Rule for d_0: $(a_1 \wedge b_2) \vee (a_1 \wedge b_1) \vee b_0 \rightarrow d_0$

Remark. We consider rules minimal with respect to the number of descriptors on the left-hand side (see e.g. [20, 29] for references to decision rules generation), i.e., they can be used to specify the new decisions for the states not yet included in the table.

Remark. In the rest of the paper, the parentheses are omitted in boolean expressions, according to the rule that the conjunction operator binds more strongly than that of disjunction. Thus, the expression of the form $a \wedge b \vee g \wedge d$ is understood as $(a \wedge b) \vee (g \wedge d)$. Furthermore, the conjunction sign $\wedge$ will be omitted in long formulas. Boolean variables like a_2 are understood as $a = 2$.

3.2 Situation

Let us consider a set Ag consisting of n agents $ag_1,...,ag_n$. A *situation* of Ag is any element of the Cartesian product $S(Ag) = \prod_{i=1}^{n} INF^*(ag_i)$, where $INF*(ag_i)$ is the set of all possible information vectors of the agent ag_i, defined by

$$INF^*(ag) = \{f : A(ag) \to \bigcup_{a \in A(ag)} V_a(ag) : f(a) \in V_a(ag) \ for \ a \in A(ag)\}$$

The situation $\bar{s}(Ag) \in S(Ag)$ corresponding to the global state

$$\bar{s} = (s_1, ..., s_n) \in U_{ag_1} \times ... \times U_{ag_n}$$

is defined by $(Inf_{A_{ag_1}}(s_1), ..., Inf_{A_{ag_n}}(s_n))$.

Example 5. An example of a current situation is the one presented in Table 6.

	a	b	k	l	m
current	2	1	2	0	1

Table 6. Current situation

3.3 Constraints

Constraints are described by some dependencies among local states of agents. Without any dependencies, any agent could take the next state freely. If there is no influence of a given agent on states of other agents – there is no conflict at all. Dependencies among local states of agents come, e.g., from the bound on the number of resources (any kind of a resource may be considered, e.g. water on Golan Hills see [24] or an international position [21], everything that is essential for agents). Constraining relations are introduced to express which local states of agents can coexist in the (global) situation. More precisely, *constraints* are used to define a subset $S(Ag)$ of global situations.

Constraints restrict the set of possible situations to admissible situations satisfying constraints. We will consider only admissible situations (shortly, situations) in the rest of the paper.

Example 6. The following dependencies restrict the set of situations and are constraints in our example. Attribute names here stand for the variables corresponding to attribute values. Constants here have been taken experimentally to express relationships and to allow comparison of any two variables.

1. $a > 0$ (compensation must be medium at least)
2. $1.5 + k > a + l$ (division of profit – a very simple case, i.e., the company uses its current profit for all expenses)
3. $2.5 + m > a + b$ (workers' satisfaction comes from a good level of salary and work conditions)

Constraints above can be converted to propositional formulas ($f_{\varphi 1}$, $f_{\varphi 2}$ and $f_{\varphi 3}$), respectively. The conjunction of formulas $f_\varphi = f_{\varphi 1} \land f_{\varphi 2} \land f_{\varphi 3}$ defines all admissible situations in our example. Let us see how formulas $f_{\varphi 1}$, $f_{\varphi 2}$ and $f_{\varphi 3}$ are created.

The equation $a > 0$ yields the formula $f_{\varphi 1} = a_1 \lor a_2$. The next formula (from the equation $1.5 + k > a + l$) is much more complex:

$f_{\varphi 2} = k_0 a_0 l_0 \lor k_0 a_0 l_1 \lor k_0 a_1 l_0 \lor k_1 a_0 l_0 \lor k_1 a_0 l_1 \lor k_1 a_0 l_2 \lor k_1 a_1 l_0 \lor k_1 a_1 l_1$
$\lor k_1 a_2 l_0 \lor k_2 a_0 l_0 \lor k_2 a_0 l_1 \lor k_2 a_0 l_2 \lor k_2 a_1 l_0 \lor k_2 a_1 l_1 \lor k_2 a_1 l_2 \lor k_2 a_2 l_0 \lor k_2 a_2 l_1$

The formula $f_{\varphi 3}$ is created in a similar way:

$f_{\varphi 3} = m_0 a_0 b_0 \lor m_0 a_0 b_1 \lor m_0 a_0 b_2 \lor m_0 a_1 b_0 \lor m_0 a_2 b_0 \lor m_0 a_1 b_1 \lor m_1 a_0 b_0 \lor m_1 a_0 b_1 \lor m_1 a_0 b_2 \lor m_1 a_1 b_0 \lor m_1 a_2 b_0 \lor m_1 a_1 b_1 \lor m_1 a_1 b_2 \lor m_1 a_2 b_1 \lor m_2 a_0 b_0 \lor m_2 a_0 b_1 \lor m_2 a_0 b_2 \lor m_2 a_1 b_0 \lor m_2 a_2 b_0 \lor m_2 a_1 b_1 \lor m_2 a_1 b_2 \lor m_2 a_2 b_1 \lor m_2 a_{22}$

As already mentioned, constraints describe the situations that are admissible i.e. all local states can coexist in the admissible situation. For example, the situation $a = 2$, $b = 2$, $k = 2$, $l = 2$, $m = 2$ is not admissible because of constraints 2 and 3. The set of all admissible situations is described by the prime implicants of the boolean formula $f_\varphi = f_{\varphi 1} \land f_{\varphi 2} \land f_{\varphi 3}$.

3.4 Objective evaluation of situations

Agents could possibly not care about the global good. However, the real consensus (a non-conflicting situation) can be found only when the global good is taken into account by all participants [21]. Thus, the objective evaluation of situations is introduced to score on situations. More precisely the *quality function of the situations* is the function $q : S(Ag) \to R[0, 1]$ which assigns the evaluation score to each situation, where $S(Ag)$ is the set of all admissible situations. The score function specification can be as follows. An expert could give the score of some situations. Next, rules can be generated for different degrees of the score function value.

The set of situations satisfying a given level of quality t is defined by:

$$Score_{Ag}(t) = \{\bar{s} \in \prod_{ag \in Ag} U_{ag} : q(\bar{s}) \geq t\}$$

Example 7. Values of the function q and some admissible situations (these scored by an expert) of our example are presented in Table 7.

Situations	a	b	k	l	m	$q(S)$	decision
S1	1	1	2	2	1	$\frac{2}{3}$	1
S2	1	0	0	0	2	0	0
S3	1	0	1	0	2	0	0
S4	1	0	1	1	2	$\frac{1}{3}$	0
S5	1	0	2	0	2	0	0
S6	1	0	2	1	2	$\frac{1}{3}$	0
S7	1	0	2	2	2	$\frac{1}{3}$	0
S8	1	1	0	0	2	0	0
S9	1	1	1	0	2	0	0
S10	1	1	1	1	2	$\frac{1}{3}$	0
S11	1	1	2	0	2	$\frac{1}{3}$	0
S12	1	1	2	1	2	$\frac{1}{3}$	0
S13	1	1	2	2	2	$\frac{2}{3}$	1
S14	1	2	0	0	2	0	0
S15	1	2	1	0	2	0	0
S16	1	2	1	1	2	$\frac{1}{3}$	0
S17	1	2	2	0	2	0	0
S18	1	2	2	1	2	$\frac{2}{3}$	1
S19	1	2	0	0	1	0	0
S20	1	2	1	0	1	0	0
S21	1	2	1	1	1	$\frac{1}{3}$	0
S22	1	2	2	1	1	$\frac{2}{3}$	1
S23	2	1	2	1	2	$\frac{2}{3}$	1
S24	2	2	2	1	2	$\frac{2}{3}$	1
S25	1	2	2	2	1	1	1
S26 (current)	2	1	2	0	1	0	0

Table 7. Admissible situations with the quality score

Minimal rules for admissible situations with the quality score not lower than $\frac{2}{3}$ follow – these rules are going to be used in calculations in the next section:

$b_1 l_2 \lor b_2 l_2 \lor a_1 b_1 m_1 \lor a_1 k_2 m_1 \lor l_2 m_1 \lor b_2 k_2 l_1 \lor b_2 k_2 m_1 \lor k_2 {}_1 m_1 \lor a_2 l_1 \lor a_2 m_2 \lor a_2 b_2 \rightarrow q(S) \geq \frac{2}{3}$

3.5 Global preference function vs. objective situation evaluation

Though the situation is objectively evaluated by the quality function the influence of local preferences (defined by subjective evaluation) onto the global situation evaluation has to be outlined. One solution is to consider local preferences while looking for the consensus (Problem 5.2). The other solution is to express the global situation evaluation based on local preferences. For this purpose the global preference function is introduced in this section, which passes the local states evaluation onto the level of the global situation. The consensus reached based on the global preference function denotes the agreement between agents (found without an expert's help). Such a consensus is usually more stable but might be objectively worse than the one proposed by an expert – the agents may not take care about the global good.

Global preference function The global preference function for the admissible situation S corresponding to the global state $\bar{s}=(s_1,...,s_n)$ can be defined by:

$$p(\bar{s} = (s_1, s_2, ..., s_n)) = F(e_{ag_1}(s_1), e_{ag_2}(s_2), ..., e_{ag_n}(s_n))$$

where F is a suitable, chosen function e.g. $F(r_1, ..., r_m) = \sum_{i=1}^{m} r_i$.
Consequently the set of all acceptable global situations is defined by:

$$S_{accept_{Ag}}(t) = \{S : p(S) \geq t\}, \text{ where } t \text{ is a given threshold.}$$

Remark. We assume that agents express evaluations of local states in the same way (they use the same scale).

Remark. The function used in this Chapter is very simple, however the global evaluation can be described in any suitable form (also non–linear) like in the form of decision rules.

3.6 System with constraints

The multi–agent system, with local states for each agent defined and the global situations satisfying constraints, will be called *the system with constraints*. We denote our system with constraints by M_{Ag}.

4 Conflict definition

In Section 3.1-3.6 the system with constraints has been defined. In such systems, conflict can be defined on several different levels.

4.1 Local conflict

The agent ag is in the ε-*local conflict* in a state s iff s does not belong to the ε-neighbourhood of s', for any s' from the set of ag-targets where ε is a given threshold.

Local conflicts for an agent ag arise from the low level of subjective evaluation of the current state by ag. The value $Cl_{ag}(s)$, which can be treated as a degree of the local conflict for ag at $s \in U_{ag}$ is defined by

$$Cl_{ag}(s) = \begin{cases} f_{ag}(s) - \varepsilon, & \text{when } f_{ag}(s) > \varepsilon \\ 0, & \text{otherwise} \end{cases}$$

where ε is a given threshold. The function f_{ag} evaluates the distance from the state s to the set of targets of ag, i.e. $f_{ag}(s) = \min\{|e_{ag}(s) - e_{ag}(s\prime)| : s\prime \in T(ag)\}$, where $e_{ag}(s)$ is the subjective evaluation by ag at the local state s.

Example 8. We choose the threshold ε in our example to be equal 0, i.e., we want to obtain states without any local conflict. For the state s_2 of the agent ag_1, $Cl_{ag1}(s2) = 0$ – the current state belongs to the set of targets. However, $Cl_{ag2}(s5) = \frac{2}{3} - 0 = \frac{2}{3}$ i.e. the agent ag_2 is in a local conflict at s_5, the current state s_5 is not satisfactory for agent ag_2.

4.2 Global conflict

Global conflict can be measured by applying the global preference function or based on an expert evaluation. The difference lies in the way of considering the global good (see Section 3.5 for explanation).

Global conflict (based on an expert evaluation) A situation S is called t–*objectively conflicting* for Ag where t is a given threshold iff S does not belong to the set $Score_{Ag}(t)$. When the current situation is conflicting for Ag then agents from Ag are in the *objective global conflict*. The difference between the situation score and the given threshold can be treated as a global conflict degree, i.e.,

$$Cg_{ag}(S) = \begin{cases} t - q(S), & \text{when } t > q(S) \\ 0, & \text{otherwise} \end{cases}$$

where t is the given threshold and q is the quality function.

Example 9. In discussed example, let us take $t = \frac{2}{3}$. The current situation S26 is t–conflicting for ag_1, ag_2 and the global conflict factor is equal to $Cg(S26) = \frac{2}{3} - 0 = \frac{2}{3}$.

Global conflict (based on agents preferences) Consequently, a situation S is called $t-conflicting$ for Ag where t is a given threshold iff S does not belong to the set $S_{accept_{Ag}}(t)$. When the current situation is conflicting for Ag then agents from Ag are in the global conflict. The difference between the situation score and the given threshold can be treated as this kind of conflict degree, i.e.,

$$Cp_{ag}(S) = \begin{cases} t - p(S), \text{ when } t > p(S) \\ 0, \text{ otherwise} \end{cases}$$

where t is the given threshold and p is the global preference function.

5 Problems

The introduced above conflict model gives us possibility, first to understand and, then, to analyse different kinds of conflicts. Particularly, the most fundamental problem is widely investigated, that is, the possibility to achieve the consensus. As in everyday's life, the consensus can be found on several levels and under some conditions.

5.1 Consensus

The consensus problem can be defined as follows.

INPUT

The system with constraints M_{Ag} defined in Section 3.
t - an acceptable threshold of the objective global conflict for Ag.

OUTPUT

The set of all situations with eliminated global conflict i.e., $Cg_{Ag}(S') = 0$, where S' is any new, reconstructed situation. That means that the quality score of the new, reconstructed situation cannot be lower then the given threshold t.

ALGORITHM

The algorithm must analyse all admissible situations and find these with the quality score not lower than the given threshold t.

Finding the solution consists in retrieving the formula which describes the set $Score_{Ag}(t)$ and verifying it against constraints (not all admissible situations have to be considered by an expert). To do this, the information table with scored situations is converted into a decision table. We are looking for a formula (rule) describing the decision that the situation is not conflicting. How to create such a formula has been shown in Example 3.6. Finally, the formula f_N describing the consensus problem is as follows:

$f_N = f_C \wedge f_\varphi$

where f_C is the formula which describes the set $Score_{Ag}(t)$ and f_φ describes constraints.

One can find that changing the global situation does not solve all the problems. The quality of local states of agents is not considered – the local conflict can be even stronger then before. In the sections that fallow, we are going to analyse this problem more deeply and we will try to eliminate local conflicts as well.

Example 10. The formula f_C for our conflict with a delimiter $t=\frac{2}{3}$ has been created in Example 7 and the formula f_φ in Example 6. Calculations give us the following formula f_N in the normal disjunctive form. Each prime implicant denotes the proposal for a non-conflicting situation.

$f_N = a_1b_1k_2l_2m_0 \vee a_1b_1k_2l_2m_1 \vee a_1b_1k_2l_2m_2 \vee a_1b_2k_2l_2m_1 \vee a_1b_2k_2l_2m_2 \vee$
$a_1b_0k_2l_2m_1 \vee a_2b_0k_2l_1m_0 \vee a_2b_0k_2l_1m_1 \vee a_2b_1k_2l_1m_1 \vee a_2b_0k_2l_1m_2 \vee a_2b_1k_2l_1m_2 \vee$
$a_2b_2k_2l_1m_2 \vee a_2b_0k_1l_0m_2 \vee a_2b_1k_1l_0m_2 \vee a_2b_2k_1l_0m_2 \vee a_2b_0k_2l_0m_2 \vee a_2b_1k_2l_0m_2 \vee$
$a_2b_2k_2l_0m_2 \vee a_1b_1k_0l_0m_1 \vee a_1b_1k_1l_0m_1 \vee a_1b_1k_1l_1m_1 \vee a_1b_1k_2l_0m_1 \vee a_1b_1k_2l_1m_1 \vee$
$a_1b_0k_2l_0m_1 \vee a_1b_2k_2l_0m_1 \vee a_1b_0k_2l_1m_1 \vee a_1b_2k_2l_1m_1 \vee a_1b_2k_2l_1$

Remark. It could be noticed that not every resulting non–conflicting situation has been scored (considered) by an expert. The method for rule generation applied here allows for searching for the solution within the equivalence classes of decisions – we cannot request the expert to specify all admissible situations. Furthermore, a useful information about agents' behavior can be achieved in this way from the historical data (like previous conflicts).

Remark. All calculations here have been done with the program module created by the author. Without computer aid this kind of analysis would be practically impossible. More about calculations can be found in Section 8.

5.2 Consensus on local preferences

In this section a conflict analysis is proposed where local information tables and the set of local goals are taken into consideration.

INPUT

The system with constraints M_{Ag} defined in Section 3.
t - an acceptable threshold of the objective global conflict for Ag.

OUTPUT

All situations with the objective evaluation reduced to degree at most t, and without local conflict for any agent.

The problem in this section consists in looking for a better compromise: additionally it is required that any new situation is constructed in the way that all local states in this situation are favourable for the agents.

ALGORITHM

The algorithm is based on verification of global situations from $Score_{Ag}(t)$ with the local set of goals of agents and constraints. The problem is described by the formula f:

$$f = \bigwedge_{ag \in Ag} t(ag) \land f_C \land f_\varphi$$

where $t(ag)$ is the disjunction of targets of the agent ag, and $f_C \land f_\varphi$ is the formula investigated in Section 5.1 representing all admissible situations without the global conflict regarding the threshold t.

Situations, which can be found using this algorithm, are better then the previous one – local preferences are taken into account.

Example 11. The formula $f_C \land f_\varphi$ has been already constructed in the previous example. Formulas $t(ag_1)$ and $t(ag_2)$ are based on sets of goals of agents ag_1 and ag_2, respectively. Example 4 shows the way the formula $t(ag_1) = a_2b_2 \lor a_2b_1$ can be found. The formula $t(ag_2)$ is found in the same way: $t(ag_2) = l_2$.

Thus, the formula f is the following conjunction:

$f = (a_2b_2 \lor a_2b_1) \land l_2 \land (a_1b_1k_2l_2m_0 \lor a_1b_1k_2l_2m_1 \lor a_1b_1k_2l_2m_2 \lor a_1b_2k_2l_2m_1 \lor a_1b_2k_2l_2m_2 \lor a_1b_0k_2l_2m_1 \lor a_2b_0k_2l_1m_0 \lor a_2b_0k_2l_1m_1 \lor a_2b_1k_2l_1m_1 \lor a_2b_0k_2l_1m_2 \lor a_2b_1k_2l_1m_2 \lor a_2b_2k_2l_1m_2 \lor a_2b_0k_1l_0m_2 \lor a_2b_1k_1l_0m_2 \lor a_2b_2k_1l_0m_2 \lor a_2b_0k_2l_0m_2 \lor a_2b_1k_2l_0m_2 \lor a_2b_2k_2l_0m_2 \lor a_1b_1k_0l_0m_1 \lor a_1b_1k_1l_0m_1 \lor a_1b_1k_1l_1m_1 \lor a_1b_1k_2l_0m_1 \lor a_1b_1k_2l_1m_1 \lor a_1b_0k_2l_0m_1 \lor a_1b_2k_2l_0m_1 \lor a_1b_0k_2l_1m_1 \lor a_1b_2k_2l_1m_1 \lor a_1b_2k_2l_1m_2)$

Within the given data no solution for this problem can be found - the goals of the agents cannot coexist, so they are rejected by the constraints. We will look for the solution in the closeness neighbourhood of the local targets (in the local closeness).

5.3 Consensus on local closeness

INPUT

The system with constraints M_{Ag} defined in Section 3.
t - an acceptable threshold of the objective global conflict for Ag.
The closeness threshold α.

OUTPUT

All situations with the objective evaluation reduced to degree at most t, and without local conflict for any agent. The new situation can be constructed from the local states closeness, i.e., from the states having the distance from those in the information table less than α.

ALGORITHM

The algorithm is similar to the previous one, but each state from the set of goals of any agent is enlarged on the closeness. Precisely, the boolean formula f' defines the solution.

$$f' = \bigwedge_{ag \in Ag} t'(ag) \wedge f_C \wedge f_\varphi$$

where $t'(ag)$ is the formula which describes the agent ag's set of targets with closeness, that is each state from the set of targets and state closeness are considered. The formula f_C describes global situations from the set $Score_{Ag}(t)$, and f_φ stands for constraints.

Example 12. The formula f_φ has been defined in Example 6, and f_C in Example 7.

Let us consider closeness of the goals with the threshold $\alpha=\frac{1}{2}$, i.e. if $d(s,s') < \frac{1}{2}$, then a local state s' is close to s. Let the distance function be defined by

$$d(s,s\prime) = \frac{1}{card(A(ag))} \sum_{a \in A(ag)} |s(a) - s(a')|$$

For the agent ag_1, the closeness neighbourhood with the threshold $\frac{1}{2}$ is not giving new states. Thus, the formula for the local goals of this agent remains the same as one found in Example 4. The local set of goals for the agent ag_2 is shown in Table 8.

states	k	l	m	decision	order e_{ag2}
$s1$	2	2	2	1	1
$s1_1$	2	1	2	1	
$s1_2$	2	2	1	1	
$s2$	1	2	2	1	$\frac{2}{3}$
$s2_1$	0	2	2	1	
$s3$	1	1	2	0	$\frac{1}{3}$
$s3_1$	0	1	2	0	
$s3_2$	1	0	2	0	
$s4$	1	1	1	0	$\frac{1}{3}$
$s4_1$	0	1	1	0	
$s4_2$	1	0	1	0	
$s4_3$	1	1	0	0	
$s5$	2	0	1	0	0
$s5_1$	2	1	1	0	
$s5_2$	2	0	2	0	
$s5_3$	2	0	0	0	

Table 8. Local states of agent ag_2 with closeness

One can notice that the states from the closeness neighbourhood of s_1 can be the same as those from the closeness neighbourhood of the state s_5 while the states s_1 and s_5 have completely different evaluation values. We will take the upper boundary of the set of target states (as specified in Table 8 by the states with decision 1).

In order to find out the minimal rules for decision 1, the discernibility between the set of local goals and the other states has to be found. The discernibility matrix is presented in Table 9.

	$s1$	$s1_1$	$s1_2$	$s2$	$s2_1$
$s3$	k, l	k	k, l	l	k, l
$s3_1$	k, l	k	k, l, m	k, l	l
$s3_2$	k, l	k, l	k, l, m	l	k, l
$s4$	k, l, m	k, m	k, l	l, m	k, l, m
$s4_1$	k, l, m	k, m	k, l	k, l, m	l, m
$s4_2$	k, l, m	k, l, m	k, l	l, m	k, l, m
$s4_3$	k, l, m	k, m	k, l, m	l, m	k, l, m
$s5$	l, m	l, m	l	k, l, m	k, l, m
$s5_1$	l, m	m	l	k, l, m	k, l, m
$s5_2$	l	l	l, m	k, l	k, l
$s5_3$	l, m	l, m	l, m	k, l, m	k, l, m

Table 9. Discernibility matrix

Prime implicants for each considered state are as follows: s_1 : l, s_{1_1}: $k \wedge l \wedge m$, s_{1_2}: l, $_2$: l and s_{2_1}: l. These attributes are considered while generating the decision rules and consequently $t'(ag_2)$ is the formula as follows. We are always looking for the minimal rules to simplify the formula and speed up the computation.

$t'(ag_2) = l_2 \vee k_2 l_1 m_2 \vee l_2 \vee l_2 \vee l_2 = l_2 \vee k_2 l_1 m_2$ where

$t'(ag_1)$ has been found in Example 4, i.e., $t'(ag_1) = a_2 b_2 \vee a_2 b_1$. Thus the formula f' is as follows:

$f\prime = (a_2 b_2 \vee a_2 b_1) \wedge (l_2 \vee k_2 l_1 m_2) \wedge (a_1 b_1 k_2 l_2 m_0 \vee a_1 b_1 k_2 l_2 m_1 \vee a_1 b_1 k_2 l_2 m_2 \vee a_1 b_2 k_2 l_2 m_1 \vee a_1 b_2 k_2 l_2 m_2 \vee a_1 b_0 k_2 l_2 m_1 \vee a_2 b_0 k_2 l_1 m_0 \vee a_2 b_0 k_2 l_1 m_1 \vee a_2 b_1 k_2 l_1 m_1 \vee a_2 b_0 k_2 l_1 m_2 \vee a_2 b_1 k_2 l_1 m_2 \vee a_2 b_2 k_2 l_1 m_2 \vee a_2 b_0 k_1 l_0 m_2 \vee a_2 b_1 k_1 l_0 m_2 \vee a_2 b_2 k_1 l_0 m_2 \vee a_2 b_0 k_2 l_0 m_2 \vee a_2 b_1 k_2 l_0 m_2 \vee a_2 b_2 k_2 l_0 m_2 \vee a_1 b_1 k_0 l_0 m_1 \vee a_1 b_1 k_1 l_0 m_1 \vee a_1 b_1 k_1 l_1 m_1 \vee a_1 b_1 k_2 l_0 m_1 \vee a_1 b_1 k_2 l_1 m_1 \vee a_1 b_0 k_2 l_0 m_1 \vee a_1 b_2 k_2 l_0 m_1 \vee a_1 b_0 k_2 l_1 m_1 \vee a_1 b_2 k_2 l_1 m_1 \vee a_1 b_2 k_2 l_1 m_2)$

After reduction, we get:

$f' = a_2 b_1 k_2 l_1 m_2 \vee a_2 b_2 k_2 l_1 m_2$

Thus, situations presented in Table 10 are proposed as the solution in the conflict i.e. the consensus.

situation	a	b	k	l	m
S1	2	1	2	1	2
S2	2	2	2	1	2

Table 10. Not conflicting situations

5.4 Consensus based on acceptable situations

The consensus problem based on acceptable situations can be defined by:

INPUT

The system with constraints M_{Ag} defined in Section 3.
h – the acceptable level of the global preference function.

OUTPUT

The set of all acceptable situations with the threshold h i.e. for which the global preference function is greater than or equal h.

ALGORITHM

Our algorithm requires generating all admissible situations and scoring them by applying the global preference function. Then the set $S_{accept}(h)$ can easily be found and described by the left side of an appropriate decision formulas like in the consensus problem.

Another possible solution is to distribute the required threshold into every agent: $h \leq h_1 + ... + h_n$. Then for any agent, the description of local states satisfying h_i has to be found (formula $t(ag_i)$). The conjunction of components found in the previous step constructs the boolean formula, whose prime implicants form the solutions (non-conflicting situations). However the formula has to be verified against the constraints. For any different distribution the new formula must be created e.g. $f_1, ..., f_m$. The disjunction of these formulas $f = f_1 \vee ... \vee f_m$ describes the problem of finding all possible non-conflicting situations. Summarising, the whole formula is as follows:

$$f = \bigvee_{1 \leq i \leq m} f_i$$

where m is the number of possible distributions of the threshold h and f_i is is the formula describing the consensus problem in the i-th distribution:

$$f_i = \bigwedge_{ag \in Ag} t_i(ag) \wedge f_\varphi$$

where $t_i(ag)$ is the formula which describes the agent ag's set of targets with the given threshold in the i-th distribution and f_φ specifies the constraints.

Conflict level distribution The distribution of the global conflict level consists in passing the conflict from the global into the local level. The way of dividing the global conflict level should reflect the global preference function used in the given conflict. That is all agents features (e.g. agent importance) applied in the global preference function must be considered. Here in the paper both simple global preference function and the way of distribution are used (see Section 6).

5.5 Consensus considering both agents preferences and the global good

The consensus problem defined in the previous section is enlarged with the requirement to consider the global good too i.e.:

INPUT

> The system with constraints M_{Ag} defined in Section 3.
> t – an acceptable threshold of the objective global conflict for Ag.
> h – the acceptable level of the global preference function.

OUTPUT

All situations with the objective evaluation reduced to degree at most t, and belonging to the set $S_{accept}(h)$, i.e., for which the global preference function is greater than or equal h.

ALGORITHM

The first step of the algorithm requires calculating prime implicants of the boolean formula described in the previous section. The next step is to verify the resulting formula from the first step with the formula describing the global situations from $Score_{Ag}(t)$. The way the formula describing the $Score_{Ag}(t)$ is created was shown in Example 7.

6 Example

The simple example of three co-operative intelligent agents is presented in this section. The example recalls the previously defined notions in another type of situation and exemplifies algorithms for resolving the consensus problem based on agents preferences Problem 5.4 and 5.5.

6.1 Conflict subject

Agents ag_1, ag_2 and ag_3 have to paint their elements of the car with the colours accordingly: sea-green, violet and coral. The appropriate colours can be obtained by mixing red, green and blue components as follows. Sea-green colour can be obtained from 2 bottles of blue colour and 1 bottle of green, violet by mixing 2 bottles of blue, 2 bottles of red and one bottle of green and coral from 2 bottles of red and 2 of green. However there are only 4 bottles of red, 4 bottles of blue and 3 of green colour and a single bottle cannot be divided or shared between the agents. The deal is to paint the elements as best as possible i.e. with the appropriate colour or the closest possible shade.

Let us consider the following attributes:

ag_1: v, b (Italian agent: "verde", "blue");

ag_2: c, z, n (Polish agent: "czerwony", "zielony", "niebieski");

ag_3: r, g (English agent: "red", "green")

where r, c denote the number of bottles with red component taken by agents ag_2 and ag_3, respectively. Similarly, g, z, v denote the number of taken bottles with the green component and b, n with the blue one, respectively.

Local states of agents are presented in Table 11, Table 12 and Table 13.

v	b	e_{ag1}
≥ 1	≥ 2	1 (see-green)
≥ 1	1	$\frac{1}{2}$
0	≥ 1	$\frac{1}{4}$
≥ 0	0	0

Table 11. Agent ag_1's local states

c	z	n	e_{ag2}
≥ 2	≥ 1	≥ 2	1 (violet)
≥ 1	0	≥ 1	$\frac{3}{4}$ (light shade)
1	1	1	$\frac{3}{4}$ (dark shade)
≥ 1	≥ 1	0	$\frac{1}{2}$
0	≥ 1	≥ 1	$\frac{1}{2}$
0	≥ 1	0	0
≥ 1	0	0	0
0	0	≥ 1	0
0	0	0	0

Table 12. Agent ag_2's local states

States with a condition on attribute values (e.g. $\geq$) denote all states with values in the Cartesian product of attributes domains restricted to the given condition. Thus for example the state $c \geq 1$ and $z \geq 1$ and $n \geq 1$ represents all the states with values from the set $E = \{2,3,4\} \times \{1,2,3\} \times \{2,3,4\}$. The condition *greater then* comes from the assumption that agents can take from the stock more bottles then they really need.

r	g	e_{ag3}
≥ 2	≥ 2	1 (coral)
≥ 2	1	$\frac{3}{4}$ (orange)
1	≥ 1	$\frac{1}{2}$
≥ 1	0	0
0	≥ 1	0
0	0	0

Table 13. Agent ag_3's local states

Sets of targets for each agent are seperated with double lines (the boundary level is $\frac{1}{2}$).

The constraints are due to the limited number of bottles with components i.e., $r + c \leq 4$, $g + z + v \leq 3$, $b + n \leq 4$. They can be converted into the propositional formula:
$f_\varphi = (r_0c_0 \lor r_0c_1 \lor r_0\,c_2 \lor r_0c_3 \lor r_0c_4 \lor r_1c_0 \lor r_1c_1 \lor r_1c_2 \lor r_1c_3 \lor r_2c_0 \lor r_2c_1 \lor r_2c_2 \lor r_3c_0 \lor r_3c_1 \lor r_4c_0) \land (g_0z_0v_3 \lor g_0z_0v_2 \lor g_0z_0v_1 \lor 0z_0v_0 \lor g_0z_3v_0 \lor g_0z_2v_0 \lor g_0z_1v_0 \lor g_3z_0v_0 \lor g_2z_0v_0 \lor g_1z_0v_0 \lor g_0z_1v_2 \lor g_0z_2v_1 \lor g_1z_0v_2 \lor g_2z_0v_1 \lor g_1z_2v_0 \lor g_2z_1v_0 \lor g_0z_1v_1 \lor g_1z_0v_1 \lor g_1z_1v_0 \lor g_1z_1v_1) \land (b_0n_0 \lor b_0n_1 \lor b_0n_2 \lor b_0n_3 \lor b_0n_4 \lor b_1n_0 \lor b_1n_1 \lor b_1n_2 \lor b_1n_3 \lor b_2n_0 \lor b_2n_1 \lor b_2n_2 \lor b_3n_0 \lor b_3n_1 \lor b_4n_0)$

The current global situation (conflicting) is presented in Table 14.

	v	b	c	z	n	r	g
Sc	1	2	0	2	2	4	0

Table 14. The current global situation

6.2 Analysis

The value of the global preference function for the current situation presented in 14 is

$$p(Sc) = \sum_{i=1}^{3} e_{ag_i}(s_i) = 1 + \frac{1}{2} + 0 = \frac{3}{2}$$

where s_i is the agent ag_i's part of the situation Sc.

Let the threshold h be $2\frac{3}{4}$. Thus $Cp(Sc) = 2\frac{3}{4} - 1\frac{1}{2} = 1\frac{1}{4}$ and agents are in conflict.

Concerning the specified states evaluation we can distribute this threshold between agents in the following manner:
$(h_1(ag_1) = 1, h_1(ag_2) = 1, h_1(ag_3) = 1) \land ((h_2(ag_1) = 1, h_2(ag_2) = \frac{3}{4}, h_2(ag_3) = 1) \lor (h_2(ag_1) = 1, h_2(ag_2) = 1, h_2(ag_3) = \frac{3}{4}))$

where $h_i(ag)$ is a threshold in the i-th distribution for ag.

For each agent we have to find the formulas describing the states locally evaluated into 1. Additionally the formula describing the states evaluated by agents ag_2 and ag_3 into $\frac{3}{4}$ has to be found. The formulas are the left-hand sidse of minimal rules found from the agents' local tables. The decision is 1 for the states with score equal to the given threshold h. We obtained the following formulas:

$t_1(ag_1) = t_2(ag_1) = t_3(ag_1) = v_3b_4 \lor v_3b_3 \lor v_3b_2 \lor v_2b_4 \lor v_2b_3 \lor v_2b_2 \lor v_1b_4 \lor v_1b_3 \lor v_1b_2$

$t_1(ag_2) = t_3(ag_2) = c_4z_3n_4 \lor c_4z_3n_3 \lor c_4z_3n_2 \lor c_4z_2n_4 \lor c_4z_2n_3 \lor c_4z_2n_2 \lor c_4z_1n_4 \lor c_4z_1n_3 \lor c_4z_1n_2 \lor c_3z_3n_4 \lor c_3z_3n_3 \lor c_3z_3n_2 \lor c_3z_2n_4 \lor c_3z_2n_3 \lor c_3z_2n_2 \lor c_3z_1n_4 \lor c_3z_1n_3 \lor c_3z_1n_2 \lor c_2z_3n_4 \lor c_2z_3n_3 \lor c_2z_3n_2 \lor c_2z_2n_4 \lor c_2z_2n_3 \lor c_2z_2n_2 \lor c_2z_1n_4 \lor c_2z_1n_3 \lor c2z_1n_2$

$t_2(ag_2) = c_4z_0n_4 \lor c_4z_0n_3 \lor c_4z_0n_2 \lor c_3z_0n_4 \lor c_3z_0n_3 \lor c_3z_0n_2 \lor c_2z_0n_4 \lor c_2z_0n_3 \lor c_2z_0n_2 \lor c_4n_1 \lor c_3n_1 \lor c_2n_1 \lor c_1n_4 \lor c_1n_3 \lor c_1n_2 \lor c1n_1$

$t_1(ag_3) = t_2(ag_3) = r_4g_3 \lor r_3g_3 \lor r_2g_3 \lor r_4g_2 \lor r_3g_2 \lor r_2g_2$

$t_3(ag_3) = r_4g_1 \lor r_3g_1 \lor r_2g_1$

Thus the problem can be transformed into the problem of finding prime implicants of the following boolean formula f, where f_φ has been presented in Section 6.1.

$f = (t_1(ag_1) \land t_1(ag_2) \land t_1(ag_3) \land f_\varphi) \lor (t_2(ag_1) \land t_2(ag_2) \land t_2(ag_3) \land f_\varphi) \lor (t_3(ag_1) \land t_3(ag_2) \land t_3(ag_3) \land f_\varphi)$

Finally, prime implicants of the formula f form the solution i.e. non–conflicting situations. Table 15 presents all non-conflicting situations, where S1 comes from the second distribution and S2-S10 from the third.

	v	b	c	z	n	r	g
S1	1	2	2	1	2	2	1
S2	1	2	1	0	2	2	2
S3	1	3	1	0	1	2	2
S4	1	2	1	0	1	2	2
S5	1	3	2	0	1	2	2
S6	1	2	2	0	1	2	2
S7	1	2	1	0	2	3	2
S8	1	3	1	0	1	3	2
S9	1	2	1	0	1	3	2
S10	1	2	2	0	2	2	2

Table 15. Non-conflicting situation

7 Coalitions

Coalitions can be extracted by finding the relations among agents in current and/or historical data. Roughly speaking agents are in a coalition when their state evaluation on the same situation is similar (with respect to the given threshold). More precisely two agents coalition is a tolerance relation γ such that:

$$< ag, ag\prime > \in \gamma \Leftrightarrow D(ag, ag\prime) \leq t$$

where t is the given threshold and D is the distance function defined as follows in general:

$$D(ag, ag\prime) = F(f(e_{ag1}(S1), e_{ag2}(S1)), ..., f(e_{ag1}(Sm), e_{ag2}(Sm)))$$

where $e_{ag}(S)$ is agent's ag evaluation of the state from the situation S, m – the number of situations available, f and F chosen, suitable functions e.g.

$$D(ag, ag\prime) = \frac{1}{m} \sum_{i=1}^{m} |e_{ag}(S_i) - e_{ag\prime}(S_i)|$$

A coalition C is the set of agents such that

$$\forall (ag, ag\prime \in C) max D(ag, ag\prime) \leq t$$

Example 13. Let us consider five agents and 8 situations. We are going to find coalitions among the agents with the threshold t=7/8. In the table below (Table 16) we present only the evaluation of each agent state in the considered situation.

	e_{ag1}	e_{ag2}	e_{ag3}	e_{ag4}	e_{ag5}
S1	1	0	2	1	0
S2	0	0	1	2	1
S3	2	0	2	1	0
S4	0	1	2	2	2
S5	0	0	2	1	0
S6	1	1	2	1	0
S7	0	1	0	2	0
S8	1	2	1	2	1

Table 16. Local states evaluations of exemplary data

Now distances between agents must be calculated e.g.:

$$D(ag_1, ag_2) = \frac{1}{8}(|1-0|+|0-0|+|2-0|+|0-1|+|0-0|+|1-1|+|0-1|+|1-2|) = \frac{3}{4}$$

The distances can be presented in the distance table - Table 17.
Thus only one bigger coalition can be determined: $\{ag_1, ag_2, ag_5\}$.

	ag_1	ag_2	ag_3	ag_4	ag_5
ag_1					
ag_2	$\frac{6}{8}$				
ag_3	$\frac{7}{8}$	$\frac{11}{8}$			
ag_4	$\frac{9}{8}$	$\frac{6}{8}$	$\frac{8}{8}$		
ag_5	$\frac{7}{8}$	$\frac{5}{8}$	$\frac{8}{8}$	$\frac{8}{8}$	

Table 17. The distance table

8 Calculation strategies

Boolean calculations of formulas described in the previous section can be time consuming. In the consensus problem we have to verify the local goals $f_1, ..., f_n$ against the formula of admissible situations and/or constraints f. This usually yields long formulas looking like : $g = f_1 \wedge f_2 \wedge ... \wedge f_n \wedge f$. Calculating prime implicants of such formulas is usually a hard-computational problem. Therefore depending on the formula, some simple strategies or eventually quite complex heuristics must be used to resolve the problem in real time. The important fact which can be used in calculation strategies is that the result (if any) is the disjunction of selected components of the formula f. This last remark has been applied in the program module created by the author.

The discussed problems, especially consensus problem, can be treated as the numerical CSP problems as well (see e.g. [3, 4] or [32]). The entry points are the constraints, which in this case will not be transformed into the propositional form. The quality score of the global situation must be set to the already computed situation. If the score does not satisfy the threshold, a next solution has to be searched.

8.1 Simple strategies

Simple strategies can be based on the Boolean algebra rules. First, the absorption rule has to be considered when choosing the formulas to calculate the formulas conjunction – a shorter formula can strongly reduce the longer formula being an extension of the shorter one. In the case when the attribute domain of a given component is small, it may be worthy to replace that component with the disjunction of negations using de Morgan rules. Let us take for example the formula $f = a_0 b_2 \vee a_0 b_0 \vee a_0 b_1 \vee a_0 c_1$. Assuming $\overline{b_2} = b_0 \vee b_1$ the disjunction $a_0 b_0 \vee a_0 b_1 = a_0(b_0 \vee b_1)$ can be replaced with $a_0 \overline{b_2}$. Thus we obtain $f = a_0 b_2 \vee a_0 \overline{b_2} \vee a_0 c_1$ which can be reduced further using the absorption rule to $f = a_0$.

Example 14. Let us present an example of a simple strategy with the consensus problem. The problem is to find all prime implicants of the boolean formula $f_1 \wedge f_2 \wedge ... \wedge f_n \wedge f$. To make the calculations faster the shortest formula from $f_1, ..., f_n$ (e.g., f_1) should be chosen and matched against the formula f. The formula f is then reduced to the components satisfying the formula f_1. Next, the algorithm repeats looking for the shortest formula among $f_2,...,f_n$. The process ends when the whole formula f is eliminated (no solution can be found) or we have scanned through all the components of formulas $f_1,...,f_n$ - all prime implicants have been found.

Let us consider the calculation from Example 12. There are formulas $f_1 = a_2b_2 \lor a_2b_1$, $f_2 = l_2 \lor k_2l_1m_2$ and a formula that describes these admissible situations for which the quality value is greater than the given threshold, i.e.,

$f = a_1b_1k_2l_2m_0 \lor a_1b_1k_2l_2m_1 \lor a_1b_1k_2l_2m_2 \lor a_1b_2k_2l_2m_1 \lor a_1b_2k_2l_2m_2 \lor a_1b_0k_2l_2m_1 \lor a_2b_0k_2l_1m_0 \lor a_2b_0k_2l_1m_1 \lor a_2b_1k_2l_1m_1 \lor a_2b_0k_2l_1m_2 \lor a_2b_1k_2l_1m_2 \lor a_2b_2k_2l_1m_2 \lor a_2b_0k_1l_0m_2 \lor a_2b_1k_1l_0m_2 \lor a_2b_2k_1l_0m_2 \lor a_2b_0k_2l_0m_2 \lor a_2b_1k_2l_0m_2 \lor a_2b_2k_2l_0m_2 \lor a_1b_1k_0l_0m_1 \lor a_1b_1k_1l_0m_1 \lor a_1b_1k_1l_1m_1 \lor a_1b_1k_2l_0m_1 \lor a_1b_1k_2l_1m_1 \lor a_1b_0k_2l_0m_1 \lor a_1b_2k_2l_0m_1 \lor a_1b_0k_2l_1m_1 \lor a_1b_2k_2l_1m_1 \lor a_1b_2k_2l_1m_2.$

According to our strategy the formula f_1 is chosen. It simplifies the formula f to $f' = a_2b_1k_2l_1m_1 \lor a_2b_1k_2l_1m_2 \lor a_2b_2k_2l_1m_2 \lor a_2b_1k_1l_0m_2 \lor a_2b_2k_1l_0m_2 \lor a_2b_1k_2l_0m_2 \lor a_2b_2k_2l_0m_2$. Then the formula f_2 is applied and the result is obtained i.e. $f'' = a_2b_1k_2l_1m_2 \lor a_2b_2k_2l_1m_2$.

Another way of calculation, when looking for the best solution, is to choose the components of the formula f relative to the global evaluation (starting from the best score) and verify this component within the formulas $f_1,...,f_n$.

8.2 Agent clustering

The detection of coalitions can help us in resolving the consensus problem. Problems with many agents involved can be divided into smaller problems and solved separately. Furthermore, we can apply agents clustering based on the coalitions (for basics about cluster analysis see e.g. [10]). The hypothesis is that we can rid of some of the agents by replacing each coalition with one representative, and resolve equivalently the consensus problem between the new set of agents. (We permit the boundary level of information lost.) The proposals described above require redefining the conflict situation (including constraints). Another approach is to take advantage of coalition knowledge during calculation of consensus problem. This approach consists of the following steps:

1. Finding coalitions.
2. Choosing one agent from the coalition as the coalition representative.
3. Removing from the formula $g = f_1 \land f_2 \land ... \land f_n \land f$ (describing consensus problem) all formulas f_i of agents removed by clustering.
4. Calculating prime implicants of the new formula g'.
5. (Optional). Verifying formula g' within formulas f_i removed in the third step.

After the forth step we obtain the approximated solution. The local goals of agents reduced by clustering are not considered. Thus the 5th step is recommended to reach full solution. The algorithm presented here (even including the 5th point) usually reduces the number of conjunctions needed for finding the solution. This is due to reducing the longest formula f (representing constraints) first by most significant formulas. This reduction has also been proven experimentally. The calculation of all prime implicants (4400) of the constraints in Example 6 takes about 7 seconds (within the author program) while the calculation of longer formula resolving consensus problem of the same example takes about 1 second. In the last case some prime implicants of the constraints are not calculated at all – they are reduced by formulas of agents preferences.

Computer implementations of this approach can be reduced by applying the priority among agents. The formulas are matched against constraints in the priority order.

9 Agents moves

The conflict situation is usually flowing (unstable). Agents are changing their views on some issues in respond to changing the external situation and/or other agents' moves. The agents' willingness of changing the state (particularly of giving up some of resources) is the fundamental assumption in the negotiation process. On the other hand the strategy of any agent is to reach the preferred state i.e., the one from its target. Because of constraints, moving from one state to another can cause the other agents to change their states.

Considering acceptable situations, there are two general possibilities when an agent is going to change its state:

1. Improving one agent's state (by changing its current state to the one from the set of targets) does not force other agents into the states out from their set of targets. These moves avoid conflicts.
2. Improving one agent's state causes other agents (agent) to change their states into less preferred.

9.1 Tension

Transition relation for ag is a binary relation μ in $U_{ag} \times U_{ag}$ such that $<s, s\prime> \in \mu$ if s is the current state and s' is any state from the local states and $s \neq s\prime$. *Upward transition* H_{ag} is a relation in $U_{ag} \times U_{ag}$ such that:

$$<s, s\prime> \in H_{ag} \leftrightarrow e_{ag}(s\prime) > e_{ag}(s)$$

and $s, s' \in U_{ag}$. Accordingly, D_{ag} is the *downward transition* relation in $U_{ag} \times U_{ag}$:

$$<s, s\prime> \in D_{ag} \leftrightarrow e_{ag}(s\prime) < e_{ag}(s)$$

In a global situation S, where ag_1 is in the state s_{ag_1} and ag_2 is in s_{ag_2}, there is a tension between ag_1 and ag_2 if the move of ag_1 according to H_{ag_1} requires – due to constraints – the agent ag_2 move to s_{ag_2}' such that $(s_{ag_2}, s_{ag_2}\prime) \in D_{ag_2}$.

Example 15. Let us consider the example from Section 3.

	v	b	c	z	n	r	g	p
Sc	1	2	0	2	2	4	0	$1\frac{1}{2}$

Table 18. Conflicting situation

Assuming that in the situation from Table 18 the agent ag_3 is going to change its state into $r = 4, g = 1$, this move is only possible due to constraints when the agent ag_1 or ag_2 returns occupied resources i.e. $v = 0$ or $z = 0$. One can notice that releasing resources by ag_1 causes its state to become less preferred, while ag_2 stays in the same state evaluated . Thus in the sense of our definition, ag_3 is in tension with ag_1, while not in the tension with ag_2.

10 Conclusions

We have presented and discussed the extension of the Pawlak conflict model. The understanding of the underlying local states as well as constraints in the given situation is the basis for any analysis of our world. The local preferences as well as the evaluation of the global situation are observed as factors defining the strength of the conflict and can suggest the way to reach the consensus.

The fundamental consensus problem has been analysed in the paper. Then, Boolean reasoning has been successfully applied as a tool for solving presented problems. A program module created by the author allows to resolve much more complex conflicts than presented in the paper.

Acknowledgements

The author is grateful to Professor Andrzej Skowron. This work would have never been done without Professor Skowron's inspiration, help and patience.

References

1. Angur, M. (1996). A Hybrid Conjoint-Measurement and Bi-Criteria Model for a 2 Group Negotiation Problem. Socio-Economic Planning Sciences 30(3), pp. 195-206.
2. Avouris, M and Gasser, L (1992). Distributed Artificial Intelligence: Theory and Praxis. Boston, Mass.: Kluwer Academic.
3. Beringer, B. and De Backer, B. (1998). Combinatorial problem solving in Constraint Programming with cooperating Solvers. Logic Programming: Formal Methods and Practical Applications. C Beirle and L. Palmer editors, North Holland.
4. Botelho, S.S.C. (1998). A distributed scheme for task planning and negotiation in multi-robot systems. 13th ECAI. Edited by Henri Prade. Published by John Wiley & Sons, Ltd.
5. Brown, F. N. (1990). Boolean Reasoning, Kluwer, Dordrecht.

6. Bui, T. (1994). Software Architecture for Negotiator Support: Co-op and Negotiator. Computer-Assisted Negotiation and Mediation Symposium, Harvard Law School, Cambridge, MA.
7. Chmielewski, M. and Grzymała-Busse, J. (1992). Global Discretization of Continuous Attributes as Pre-processing for Inductive Learning. Department of Computer Science, University of Kansas, TR-92-7.
8. Deja, R. (1996). Conflict Analysis. Proceedings of the Fourth International Workshop on Rough Sets, Fuzzy Sets and Machine Discovery. The University of Tokyo, 6-8 November, pp. 118-124.
9. Deja, R. (1996). Conflict Model with Negotiation. Bulletin of the Polish Academy of Sciences, Technical Sciences, vol. 44, no. 4, pp. 475-498.
10. Everitt, B. (1980). Cluster Analysis. London, United Kingdom: Heinmann Educational Books, Second Edition.
11. Fang, L., Hipel, K.W. and Kilgour, D.M. (1993). Interactive Decision Making: the Graph Model for Conflict Resolution. Wiley, New York.
12. Fraser, N.M. and Hipel, K.W.(1984). Conflict Analysis: Models and Resolutions North-Holland, New York.
13. Fraser, N. M and Hipel, K. W. (1983). Dynamic modeling of the Cuba missile crisis. Journal of the Conflict Management and Peace Science 6 (2), 1-18.
14. Grzymała-Busse, J. (1992). LERS - a System for Learning from Examples Based on Rough Sets. In Słowiński R. [ed.] Intelligent Decision Support. Handbook of Applications and Advances of the Rough Sets Theory. Kluwer, 3-18.
15. Hipel, K.W. and Meiser, D.B.(1993). Conflict analysis methodology for modeling coalition formation in multilateral negotiations. Information and Decision Technologies.
16. Howard, N. (1975). Metagame analysis of business problems. INFOR 13, pp. 48-67.
17. Howard, N. and Shepanik, I. (1976). Boolean algorithms used in metagame analysis. Univeristy of Ottawa. Canada.
18. Kersten, G.E. and Szpakowicz, S. (1994). Negotiation in Distributed Artificial Intelligence: Drawing from Human Experiences, Proceedings of the 27th Hawaii International Conference on System Sciences. Volume IV, J.F. Nunamaker and R.H. Sprague, Jr. (eds.), Los Alamitos, CA: IEEE Computer Society Press (pp. 258-270).
19. Kersten, G.E., Rubin, S and Szpakowicz, S. (1994). Medical Decision Making in Negoplan. Moving Towards Expert Systems Globally in the 21st Century. Proceedings of the Second World Congress on Expert Systems, J. Liebovitz (ed.) Cambridge, MA: Macnillan (pp. 1130-1137).
20. Komorowski, J., Pawlak, Z., Polkowski, L., Skowron, A., (1999). Rough sets: A tutorial. in: S.K. Pal and A. Skowron (eds.), Rough fuzzy hybridization: A new trend in decision making, Springer-Verlag, Singapore, pp. 3-98.
21. Nęcki, Z. (1994). Negotiations in business. Professional School of Business Edition. (The book in Polish). Krakow 1994.
22. Nguyen S. H.; Skowron A., 1997, "Searching for Relational Pattern on Data", Proceedings of The First European Symposium on Principles of Data mining and Knowledge Discovery, Trondheim, Norway, June 25–27, pp. 265–276.
23. Pawlak, Z. (1981). Information Systems - Theoretical Foundations. (The book in Polish). PWN Warsaw.
24. Pawlak, Z. (1984). On Conflicts. Int. J. of Man-Machine Studies. 21, pp. 127-134.
25. Pawlak, Z. (1991). Rough Sets - Theoretical Aspects of Reasoning about Data. Kluwer Academic Publishers, Dordrecht.

26. Pawlak, Z. (1993). Anatomy of Conflicts. Bull. EATCS, 50, pp. 234-246.
27. Pawlak, Z. (1993). On Some Issues Connected with Conflict Analysis. Institute of Computer Science Reports, 37/93, Warsaw University of Technology.
28. Pawlak, Z. (1998). An Inquiry into Anatomy of Conflicts. Journal of Information Sciences 109 pp. 65-78.
29. Pawlak, Z. and Skowron, A. (1993). A Rough Set Approach to Decision Rules Generation. Institute of Computer Science Reports, 23/93, Warsaw University of Technology.
30. Polkowski, L. and Skowron, A. (Eds.) (1998). Rough Sets in Knowledge Discovery 1: Methodology and Applications, Physica-Verlag, Heidelberg.
31. Polkowski, L. and Skowron, A. (Eds.) (1998). Rough Sets in Knowledge Discovery 2: Applications, Case Studies and Software Systems, Physica-Verlag, Heidelberg.
32. Puget, J-F. (1998). Constraint Programming: A great AI Success. 13th ECAI 98. Edited by Henri Prade. Published by John Wiley & Sons, Ltd.
33. Rosenheim, J.S. and Zlotkin, G. (1994). Designing Conventions for Automated Negotiation. AI Magazine 15(3) pp. 29-46. American Association for Artificial Intelligence.
34. Rosenheim, J.S. and Zlotkin, G. (1994). Rules of Encounter: Designing Conventions for Automated Negotiations among Computers. The MIT Press, Cambridge.
35. Sandholm, T. (1996). Negotiation among Self-Interested Computationally Limited Agents. Ph.D. Dissertation. University of Massachusetts at Amherst, Department of Computer Science. 297 pages.
36. Sandholm, T. (1992). Automatic Cooperation of Area-Distributed Dispatch Centers in Vehicle Routing. International Conference on Artificial Intelligence Applications in Transportation Engineering, San Buenaventura, California, pp. 449-467.
37. Sandholm, T. and Lesser, V. (1995). Equilibrium Analysis of the Possibilities of Unenforced Exchange in Multiagent Systems. Fourteenth International Joint Conference on Artificial Intelligence (IJCAI-95), Montreal, Canada, pp. 694-701.
38. Sandholm, T. and Lesser, V. (1995). Issues in Automated Negotiation and Electronic Commerce: Extending the Contract Net Framework. Proceedings of the International Conference on Multiagent Systems pp. 328-335. American Association for Artificial Intelligence.
39. Sandholm, T. and Lesser, V. (1997). Coalitions among Computationally Bounded Agents. Artificial Intelligence 94(1), 99-137, Special issue on Economic Principles of Multiagent Systems.
40. Schehory, O and Kraus, S. (1996). A Kernel-oriented model for Coalition-formation in General Environments: Implementation and Results, Proceedings of the National Conference on Artificial Intelligence, (AAAI-96), Portland.
41. Selman, B., Kautz H. and McAllester D. (1997). Ten Challenges in Propositional Reasoning and Search. Proceedings of the Fifteenth International Joint Conference on Artificial Intelligence (IJCAI- 97), Nagoya, Aichi, Japan.
42. Skowron, A. and Rauszer, C. (1991). The Discernibility Matrix and Functions in Information Systems. Institute of Computer Science Reports, 1/91, Warsaw University of Technology, and Fundamenta Informaticae.
43. Skowron, A. and Grzymala-Busse, J. (1991). From the Rough Set Theory to the Evidence Theory. Institute of Computer Science Reports, 8/91, Warsaw University of Technology.
44. Sycara, K. (1996). Coordination of Multiple Intelligent Softwareagents. International Journal of Cooperative Information Systems 5(2-3) pp. 181-211.

45. Tohme, F. and Sandholm, T. (1997). Coalition Formation Processes with Belief Revision among Bounded Rational Self-Interested Agents. Fifteenth International Joint Conference on Artificial Intelligence (IJCAI-97), Workshop on Social Interaction and Communityware, Nagoya, Japan, August 25.
46. Wellman, M. (1995). A Computational Market Model for Distributed Configuration Design. AI EDAM 9 pp. 125-133. Cambridge University Press.
47. Wiederhold, G. (1992). Mediators in the Architecture of Future Information Systems. IEEE Computer 25(3) pp. 38-49.
48. Zlotkin, G. and Rosenchein, J. (1993). The Extend of Cooperation in State-oriented Domains: Negotiations among Tidy Agents. Computers and Artificial Intelligence, 12(2) pp. 105-122.
49. Zlotkin, G. and Rosenchein, J. (1993). Negotiation with Incomplete Information about Worth: Strict versus Tolerant Mechanism. Proceedings of the First International Conference on Intelligent and Cooperative Information Systems, pp. 175-184, Rotterdam, The Netherlands.
50. Żakowski, W.(1991). On Conflicts and Rough Sets. Bulletin of the Polish Academy of Science, Technical Science, 39, 3/1991.
51. Żakowski, W.(1991). Conflicts, Configurations, Situations and Rough Sets. Bulletin of the Polish Academy of Science, Technical Science, 4/1991.

Chapter 9

Logical and Algebraic Techniques for Rough Set Data Analysis

Ivo Düntsch[1] *and Günther Gediga*[2]

[1] School of Information and Software Engineering,
University of Ulster, Newtownabbey,
BT 37 0QB, N. Ireland

[2] FB Psychologie / Methodenlehre, Universität Osnabrück,
49069 Osnabrück, Germany,
emails: I.Duentsch@ulst.ac.uk; Guenther@Gediga.de

Abstract. In this paper, we shall give an introduction to, and an overview of, the various relational, algebraic, and logical tools that are available to handle rule based reasoning.

1 Introduction

Operationalisation (or representation) of knowledge can be done in various ways. One of the simplest types stems from the observation that we often describe objects by listing a finite number of their properties. In other words, we are using an operationalisation of the form

OBJECT ↦ ATTRIBUTE VALUES.

Inherent in this type of representation is the assumption, that the world (i.e. the objects under consideration) is perceived up to the granularity given by their respective attribute vectors. This is the view of rough set data analysis (RSDA) [60] which leads to a rule based form of reasoning, exploiting only the algebraic and order structure of the given information. Numerical information outside the data at hand is not part of the model, even though, of course, statistical circumstances have to be taken into account when validating rules or ascertaining their significance. We cover these aspects of RSDA in our companion paper [35].

The simplest form of rules in such a setting are deterministic statements of the form

$$\text{If } q_0(a) \text{ and } q_1(a) \text{ and } \ldots \text{ and } q_n(a), \text{ then } p(a), \tag{1}$$

where the q_i and p are unary predicates representing attribute values. However, if there is an uncertainty about $p(a)$, given $q_0(a)$ and $q_1(a), \ldots q_n(a)$, rules of type 1 are not sufficient to express our knowledge – or the lack of it –, and we need indeterministic rules such as

$$\text{If } q_0(a) \text{ and } q_1(a) \text{ and } \ldots \text{ and } q_n(a), \text{ then } p_0(a) \text{ or } p_1(a) \text{ or } \ldots \text{ or } p_k(a). \tag{2}$$

On the other hand, we may have systems of rules of the form 1 with the same conclusion, which generates rules of the form

If $q_0(a)$ or $q_1(a)$ or ... or $q_n(a)$, then $p(a)$. (3)

Rules of this type lead to data compression, since the properties $q_0, \ldots q_n$ can be collected into one global property.

There is a long standing tradition to develop algebraic semantics for various logics, the most famous example being Boolean algebras as semantic structures for classical propositional logic, Heyting algebras for intuitionistic logic [62], and Boolean algebras with operators for modal logics [41]; thus, we shall put special emphasis on the algebraic structures arising from various rule systems, whenever this is appropriate.

Deterministic dependence relations along with various fields of application are introduced in Section 3. In Section 4 we introduce approximation spaces, leading to the simplest form of rule–based reasoning under uncertainty, and introduce various knowledge operators and the algebras derived from them. Our framework of modal logics, their frame semantics, and their algebraic counterparts are presented in Section 5. Logics for information systems, along with frame and algebraic semantics are discussed in Section 6. Finally, we will give an outlook to further tasks in Section 7.

2 Information systems

The formal concept to capture the intuitive notion of the preceding Section as that of information systems [59]. A *single valued information system*, or just *information system* is a tuple

$$\mathcal{I} = \langle U, \Omega, \{V_x : x \in \Omega\}\rangle, \tag{4}$$

where

1. U is a nonempty set, and
2. Ω is a finite, nonempty set of mappings $x : U \to V_x$.

If $V_q = \{0, 1\}$ for some $x \in \Omega$, then x is called a *binary attribute*; we call $\mathcal{I}$ *binary*, if each attribute is binary. In what follows, we will use $\mathcal{I}$ as defined in 4 as our generic information system.

We interpret U as a set of objects (or situations) and Ω as a set of attribute mappings each of which assigns to an object a a value which a may take under the respective attribute. An alternative interpretation regards Ω as a set of agents up to whose knowledge the objects can be classified [50].

For each subset Q of Ω, we will denote an element of $\prod\{V_x : x \in Q\}$ by $\mathbf{x}^Q$. The attribute functions can be extended over subsets of Ω in the following way: If $\emptyset \neq Q \subseteq \Omega, a \in U$, we regard Q as a mapping $U \xrightarrow{Q} \prod\{V_x : x \in Q\}$ defined by

$$Q(a) = \langle x(a)\rangle_{x \in Q}. \tag{5}$$

Thus, $Q(a)$ is the *feature vector of a with respect to Q*.

Each set of attributes defines an equivalence relation on the object set up to the classes of which objects are discernible. We let θ_Q be the relation defined by

$$a \theta_Q b \iff Q(a) = Q(b). \tag{6}$$

Thus,

$$a \theta_Q b \iff x(a) = x(b) \text{ for all } x \in Q.$$

The classes of θ_Q determine the granularity up to which objects are discernible with the knowledge provided by the features in Q.

Information systems as defined above are single valued in the sense that an object is assigned exactly one value under each attribute. This can be generalised in the following way [57]: A *multivalued information system* is a structure

$$\mathcal{I} = \langle U, \Omega, \{V_x : x \in \Omega\} \rangle, \tag{7}$$

where

1. U is a nonempty set, and
2. Ω is a finite, nonempty set of mappings $x : U \to 2^{V_x}$.

Thus, the difference to single valued information systems is that each object is assigned a set of values under an attribute function. There is an ambiguity on how to interpret $x(a)$, namely as

"a possesses all properties in $x(a)$" or \hfill (8)

"a possesses some property in $x(a)$". \hfill (9)

These possibilities are, in a way, extremal, since other quantifiers are possible, e.g. "a possesses exactly one of the properties in $x(a)$" or "a possesses no more than five properties in $x(a)$".

The conjunctive interpretation 8 is logically equivalent to a single valued information system via binarisation: Suppose without loss of generality that for $x \in \Omega$, $\bigcup \{x(a) : a \in U\} = V_x$. For each $t \in V_x$ define a binary attribute x_t by

$$x_t(a) = \begin{cases} 1, & \text{if } t \in x(a), \\ 0, & \text{otherwise.} \end{cases}$$

In this way, information is shifted from the columns to extended rows, and the resulting binary system has the same information content as the conjunctively interpreted multi valued system. Therefore, we will always interpret multi valued information systems disjunctively in the sense of 9; this will enable us to capture rules of the form 3. This type of information system has also recently been used for data filtering and compression [20, 73]. A detailed investigation of the rule systems associated with multi-valued information systems is the forthcoming [22].

3 Deterministic reasoning

In this Section, we are concerned with deterministic rules of type 1. This has been studied, among others, in [46, 45, 67]. We will show that formalisms behind these occur in diverse situations [21].

We assume the reader is familiar with the basic definitions and properties of lattice theory and order theory – suitable references being [11] or [36] –, and we shall just recall a few concepts. A *sup–semilattice* is a partially ordered set $\langle L, \leq \rangle$, in which the supremum $a \vee b$ exists for all $a, b \in L$. A *closure system* is a family of subsets of Ω which is closed under intersection, and a *knowledge space* is a family of subsets of Ω which is closed under union, and contains $\emptyset$ and Ω. A *closure operator on* Ω is a mapping $c : 2^\Omega \to 2^\Omega$ for which

1. $A \subseteq c(A) = c(c(A))$,
2. $f(\emptyset) = z$.

Suppose that $\mathcal{L} = \langle L, \vee \rangle$ is a sup – semilattice. A *congruence* on $\mathcal{L}$ is an equivalence relation θ on L which satisfies the substitution property

$a\theta b$ and $c\theta d$ imply $(a \vee c)\theta(b \vee d)$.

It is well known that each class K of a congruence θ is a subsemilattice of $\mathcal{L}$, and thus it has a maximum, written as $\max K$, if $\mathcal{L}$ is finite.

Let Ω be a set, and $T \subseteq 2^\Omega \times \Omega$ be a relation between subsets of Ω and elements of Ω. Consider the following properties of T, where $Q, P \subseteq \Omega, x, y \in \Omega$:

If $x \in Q$, then $\langle Q, x \rangle \in T$ \hfill (Reflexivity) (10)

If $\langle Q, x \rangle \in T$, then $\langle Q \cup P, x \rangle \in T$ \hfill (Monotony) (11)

If $\langle Q, y \rangle \in T$ for all $y \in P$ and $\langle Q \cup P, x \rangle \in T$, then $\langle Q, x \rangle \in T$ (Cut). (12)

Let Fml be the set of formulas of a logic $\mathcal{L}$. A *consequence relation* $\vdash$ is a subset of $2^{Fml} \times Fml$ which satisfies 3, 3, and 3 [63]; intuitively, $Q \vdash x$, if Q syntactically implies x. More generally, we call a relation $T \subseteq 2^\Omega \times \Omega$ which satisfies these conditions a *consequence relation* on Ω. With each consequence relation T on Ω one can associate an operator f on 2^Ω by

$$f(Q) = \{p \in \Omega : \langle Q, p \rangle \in T\}.$$

The conditions on T ensure that f is a closure operator. Tarski and Scott have shown that each closure operator on a set Ω is induced by the consequence relation of a monotone logic [32].

Now, each rule of the form 1 determines a pair $\langle Q, p \rangle \subseteq 2^\Omega \times \Omega$ by

If $Q(a) = \mathbf{x}^Q$, then $p(a) = x^p$ \hfill (13)

for all $a \in U$; in this case, we write $Q \to p$, and say that *p is dependent on Q*. If $Q \to p$, then the value of each a under p is uniquely determined by the value of a under (the attributes in) Q. In other words,

$$Q \to p \text{ iff } \theta_Q \subseteq \theta_p.$$

It is easy to see that $\to$ is a consequence relation on Ω. We extend $\to$ over $2^\Omega \times 2^\Omega$ by setting

$$Q \to P \text{ if and only if } Q \to p \text{ for all } p \in P.$$

The (extended) relation $\to$ has the following properties:

If $P \subseteq Q$, then $Q \to P$ (14)

$\to$ is transitive, (15)

If $Q \to P_i$ for all $i \in I$, then $Q \to \bigcup_{i \in I} P_i$. (16)

We call a binary relation R on 2^Ω which satisfies 14, 15, and 16 a *dependence relation* on Ω Such R determines a congruence $K(R)$ on the semilattice $\langle 2^\Omega, \cup \rangle$ via the assignment

$$K(R) = R \cap R^{-1},$$

cf [46]. This assignment is one – one as the following shows:

Proposition 1. *Let* $R, S \subseteq 2^\Omega \times 2^\Omega$ *satisfy 14 – 16 and suppose that* $K(R) = K(S)$; *then,* $R = S$.

Proof. By symmetry it suffices to show $R \subseteq S$; thus, let $\langle X, Y \rangle \in R$. Then, $\langle X, X \cup Y \rangle \in R$ follows from $\langle X, X \rangle \in R$ by 14, and 16. Again from 14 we have $\langle X \cup Y, X \rangle \in R$, and thus $\langle X, X \cup Y \rangle \in R \cap R^{-1} = S \cap S^{-1}$. We use 14 together with 15 once more to conclude $\langle X, Y \rangle \in S$.

Conversely, it is shown in [46] that every congruence ψ on $\langle 2^\Omega, \cup \rangle$ determines a relation R_ψ on 2^Ω which satisfies 14 – 16 via the assignment

$$\psi \mapsto \{\langle X, Y \rangle : \max(\psi Y) \subseteq \max(\psi X)\},$$

where for $Z \subseteq \Omega$, ψZ is the equivalence class of Z with respect to ψ, and the maximum is taken with respect to $\subseteq$.

We remark in passing that [46] has shown that the congruences on $\langle 2^\Omega, \cup \rangle$ are in a bijective correspondence to the closure operators on $\langle 2^\Omega, \subseteq \rangle$, which in turn are in a bijective correspondence to the closure systems on Ω [cf 6].

In the other direction, a relation with the properties 14 – 16 is determined by its restriction to $2^\Omega) \times \{\{u\} : u \in \Omega\}$: If R is a binary relation on 2^Ω, we let

$$\text{sng}(R) = \{\langle X, y \rangle : \langle X, \{y\} \rangle \in R\} \subseteq 2^\Omega \times \Omega.$$

Conversely, if $T \subseteq 2^\Omega \times \Omega$, we set

$$\text{ext}(T) = \{\langle X, Y \rangle : \langle X, y \rangle \in T \text{ for all } y \in Y\}.$$

The following was noted in [44], 4.2.(iii):

Proposition 2. Let $T \subseteq 2^\Omega \times \Omega$. Then, $ext(T) = R$ for some relation R on 2^Ω which satisfies 14 – 16, if and only if for all $X, Y \subseteq \Omega$, $x, y \in \Omega$,

$$\langle X, x \rangle \in T, \text{ if } x \in X, \tag{17}$$

$$\text{If } \langle X, y \rangle \in T \text{ for all } y \in Y, \text{ and } \langle Y, x \rangle \in T, \text{ then } \langle X, x \rangle \in T. \tag{18}$$

Conversely, if R is a relation which satisfies 14 – 16, then $\text{sng}(R)$ satisfies 17 and 18, and $ext(\text{sng}(R)) = R$. □

Apart from information systems and logical consequence, dependence relations arise in the context of querying experts to obtain knowledge structures, where they are called *entail relations*, see for example, [18, 21, 44]. In the theory of knowledge spaces [30], entail relations are used to compute an unknown knowledge space by querying an expert.

Let Ω be a set of problems, and $\mathfrak{K}$ be a family of subsets of Ω, containing $\emptyset$ and Ω. The pair $\langle \Omega, \mathfrak{K} \rangle$ is called a *knowledge structure*, and the elements of $\mathfrak{K}$ are interpreted as the knowledge states of individuals. Define a relation $R_\mathfrak{K}$ on 2^Ω by setting

$$\langle A, B \rangle \in R_\mathfrak{K} \iff B \cap K \neq \emptyset \text{ implies } A \cap K \neq \emptyset \text{ for all } K \in \mathfrak{K}. \tag{19}$$

The interpretation of $\langle A, B \rangle \in R_\mathfrak{K}$ is that if subjects master some problem in B they also master at least one problem from A; the relation $R_\mathfrak{K}$ satisfies 14 – 16, and thus, it is a dependence relation.

It may be worthy to mention that the assignment $\mathfrak{K} \mapsto \mathfrak{R}_\mathfrak{K}$ is just one possibility to obtain a dependence (or entail) relation from a family of subsets of Ω. In [18] we describe all Galois connections between knowledge structures and binary relations on 2^Ω in which the Galois closed relations are exactly the entail relations, and hence, the dependency relations of information systems.

The algebraic form of consequence relations arising from dependencies are equations in the inf - semilattice of equivalence relations on U; these are explored in [19].

4 Approximation spaces and their algebras

The main characteristic of the rough set approach is that it is able to handle not only deterministic rules but also imperfect knowledge in the form of indeterministic rules of the type 2. The simplest situation, which we will consider first, is the case that knowledge is given by the classes of an equivalence relation R on a set U, and that we want to determine membership of some $a \in U$ in a subset X of U. We will call a pair $\langle U, R \rangle$ an *approximation space*, and also set

$$R(x) = \{y \in U : xRy\}, \tag{20}$$

$$\overset{\smile}{R}(x) = \{y \in U : yRx\}. \tag{21}$$

Within the paradigm of information systems, we can think of R as some θ_Q, and X as a class belonging to some attribute p, where

$$p(a) = \begin{cases} 1, & \text{if } a \in X, \\ 0, & \text{otherwise.} \end{cases}$$

There are three types of rules:

If $Q(a) = \mathbf{x}^Q$, then $a \in X$, (22)

If $Q(a) = \mathbf{x}^Q$, then $a \notin X$, (23)

If $Q(a) = \mathbf{x}^Q$, then $a \in X$ or $a \notin X$,. (24)

The first case is true, when the R – class containing a is totally contained in X, the second is true, when it does not intersect X, and the third case occurs, when $R(a) \cap X \neq \emptyset$ and $R(a) \cap -X \neq \emptyset$; in this case, uncertainty arises, and we cannot determine membership in X. The situation is pictured in Figure 1.

Fig. 1. Rough set approximation

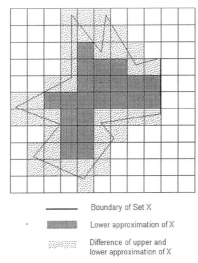

— Boundary of Set X
▓ Lower approximation of X
░ Difference of upper and lower approximation of X

Generalising this to all subsets of U leads to the following definitions: The *lower approximation with respect to R* of $X \subseteq U$ is defined by the operator

$$i(X) = \{x \in U : R(x) \subseteq X\}, \tag{25}$$

i.e.

$$i(X) = \{x \in U : xRy \text{ implies } y \in X\}. \tag{26}$$

i(X) is the set of all $x \in U$ which, with the granularity given by R, can be classified as being certainly in X.

The *upper approximation of $X \subseteq U$ with respect to R* is

$$h(X) = \{x \in U : R(x) \cap X \neq \emptyset\}, \tag{27}$$

i.e.

$$h(X) = \{x \in U : (\exists y \in X)xRy\} = \bigcup_{y \in X} R\check{\ }(y). \tag{28}$$

$h(X)$ is the set of all elements of U which are possibly in X.

The algebraic connection between the operators i and h is

$$i(X) = -h(-X), \tag{29}$$
$$h(X) = -i(-X). \tag{30}$$

where $-$ is the Boolean complement on 2^U. Two operators which satisfy 29 (or, equivalently, 30, are said to be *dual* to each other.

The collection of *rough sets on U* is the set

$$\{\langle i(X), h(X)\rangle : X \subseteq U\}. \tag{31}$$

Each subset X of U, which is either empty or a union of R – classes satisfies

$$i(X) = h(X) = X.$$

Such a set will be called *definable* or *crisp*. It is obvious that the collection B of crisp rough sets forms a Boolean algebra, namely, the subalgebra of 2^U whose atoms are the equivalence classes of R; furthermore, the collection 2^U_R of rough sets on U is a sublattice of $2^U \times 2^U$. With some abuse of notation, we assign to each rough set Z its lower bound $i(Z)$ and its upper bound $h(Z)$.

One can regard the collection of rough sets as an abstract algebra in the following way [27]: An *approximation algebra* (AA) is an algebra $\langle L, +, \cdot, 0, 1, i, h\rangle$ of type $\langle 2, 2, 0, 0, 1, 1\rangle$ such that for all $x, y \in L$,

1. $\langle L, +, \cdot, 0, 1\rangle$ is a bounded distributive lattice.
2. i is a co-normal multiplicative interior operator on L, i.e.

$$i(1) = 1,$$
$$x \leq y \Rightarrow i(x) \leq i(y),$$
$$i(x) \leq x,$$
$$i(i(x)) = i(x),$$
$$i(x \cdot y) = i(x) \cdot i(y).$$

3. h is a normal additive closure operator on L, i.e.

$$h(0) = 0,$$
$$x \leq y \Rightarrow h(x) \leq h(y),$$
$$x \leq h(x),$$
$$h(h(x)) = h(x),$$
$$h(x + y) = h(x) + h(y).$$

4. L satisfies

$$i(h(x)) = h(x), \ h(i(x)) = i(x),$$
$$i(x) = i(y) \text{ and } h(x) = h(y) \text{ imply } x = y.$$

5. Each closed element has a complement.

There are the following representation theorems:

Proposition 3. *1. [27] Each algebra of rough sets is an AA.*
2. [15] Each AA is isomorphic to a subalgebra of some 2_R^U.

Another class of algebras, appropriate for approximation spaces, is the following:
A *regular double Stone algebra* (RDSA) $\mathbf{L} = \langle L, *, {}^+, \mathbf{0}, \mathbf{1} \rangle$ is an algebra of type $\langle 2, 2, 1, 1, 0, 0 \rangle$ such that

1. L is a bounded distributive lattice,
2. x^* is the pseudocomplement of x, i.e. $y \leq x^* \iff y \cdot x = 0$,
3. x^+ is the dual pseudocomplement of x, i.e. $y \geq x^+ \iff y + x = 1$,
4. $x^* + x^{**} = 1$, $x^+ \cdot x^{++} = 0$.
5. $x \cdot x^+ \leq y + y^*$.

The last condition is equivalent to

$$x^+ = y^+ \text{ and } x^* = y^* \text{ imply } x = y. \tag{32}$$

[72] has shown that RDSA is an equational class which is generated by the three element chain $\mathbf{3} = \{\mathbf{0}, \mathbf{a}, \mathbf{1}\}$. The connection between RDSAs and rough sets is given by

Proposition 4. *1. [39, 61, 10] Suppose that $\langle U, R \rangle$ is an approximation space. Then, 2_R^U is a regular double Stone algebra with the operations*

$$\langle [R](X), \langle R \rangle (X) \rangle + \langle [R](Y), \langle R \rangle (Y) \rangle = \langle [R](X \cup Y), \langle R \rangle (X \cup Y) \rangle$$
$$\langle [R](X), \langle R \rangle (X) \rangle \cdot \langle [R](Y), \langle R \rangle (Y) \rangle = \langle [R](X \cap Y), \langle R \rangle (X \cap Y) \rangle$$
$$\langle [R](X), \langle R \rangle (X) \rangle^* = \langle -\langle R \rangle (X), -\langle R \rangle (X) \rangle$$
$$\langle [R](X), \langle R \rangle (X) \rangle^+ = \langle -[R](X), -[R](X) \rangle.$$

2. [9, 10] Let L be a regular double Stone algebra. Then, there is an approximation space $\langle U, R \rangle$ such that L is isomorphic to a subalgebra of 2_R^U. □

The following now does not come as a surprise:

Proposition 5. *[28] Let AA and RDSA be as above, regarded as categories with the appropriate homomorphisms. Define mappings γ and δ in the following way:*

Let $\langle L, +, \cdot, i, h, 0, 1 \rangle \in$ AA, and let $*$ and $^+$ be defined by

$$x^+ = -i(x),$$
$$x^* = -h(x).$$

Set $\gamma(L) = \langle L, +, \cdot, *, +, 0, 1\rangle$, and $\gamma(f) = f$ for any AA morphism. Conversely, let $\langle L, +, \cdot, *, +, 0, 1\rangle \in$ RDSA, and set $i(x) = x^{++}$ and $h(x) = x^{**}$. Define $\delta(L) = \langle L, +, \cdot, i, h, 0, 1\rangle$, and $\delta(f) = f$ for any RDSA morphism. Then, γ : AA $\to$ RDSA and δ : RDSA $\to$ AA are covariant functors, and $\gamma \circ \delta = \delta \circ \gamma = id$.
□

The third connection we would like to mention, is the one to the class 3Ł of three valued Łukasiewicz algebras see e.g. [7, 1, 4]. Here we can show

Proposition 6. *The class RDSA is term-equivalent to $3Ł\%$* [3].

Proof. It is enough to show that the 3Ł operations $\to$ and $\neg$ can be defined with the RDSA operators, and that the RDSA operators $*$ and $+$ can be defined by the 3Ł operators. We only need to do this in 3 since both classes are varieties generated by 3; the action of the operations on 3 is given in Table 1.

Table 1. 3Ł and RDSA operations

$\to$	0	a	1	$\neg$	$*$	$+$
0	1	1	1	1	1	1
a	a	1	1	a	0	1
1	0	a	1	0	0	0

1. From 3Ł to RDSA: Set $s(x) = \neg(\neg x \to x)$, and $t(x) = x \to \neg x$. Then, $s(x) = x^*$, and $t(x) = x^+$, as is easily checked.
2. From RDSA to 3Ł: Set $s(x) = x^* + x \cdot x^+$; then, $s(x) = \neg x$. Next, let

$$t(x, y) = (x^+ \cdot y^{++} + x^{++} \cdot y^+ + x^* \cdot y^{**} + x^{**} \cdot y^*)^*. \tag{33}$$

Claim:

$$t(x, y) = \begin{cases} 1, & \text{if } x = y, \\ 0, & \text{otherwise.} \end{cases}$$

Proof.

$$x = y \iff x^{++} = y^{++} \text{ and } x^{**} = y^{**}$$
$$\iff x^+ \cdot y^{++} + x^{++} \cdot y^+ = 0 \text{ and } (x^* \cdot y^{**} + x^{**} \cdot y^*)^* = 0,$$
$$\iff x^+ \cdot y^{++} + x^{++} \cdot y^+ + x^* \cdot y^{**} + x^{**} \cdot y^* = 0$$
$$\iff (x^+ \cdot y^{++} + x^{++} \cdot y^+ + x^* \cdot y^{**} + x^{**} \cdot y^*)^* = 1.$$

If $x \neq y$, then $t(x, y) = 0$, since in 3, $z^* \in \{0, 1\}$ for all z. This proves the claim.

Now, set

$$s(x, y) = x^* + y + x \cdot x^+ + t(x, y).$$

It is straightforward to verify that $s(x, y) = x \to y$.

---

[3] We have not been able to find a proof of this in the literature, but the result must have been known to [71].

A relational proof system for the class AA is given in [28], and the class RSDA is the semantics of the rough set logic of [17]. Regular double Stone algebras are equipollent to many other algebraic structures, for example, semi-simple Nelson algebras. Thus, the classical logic with strong negation developed in [64] is also appropriate for approximation spaces. A comprehensive treatment of these connections is [58], and related material can be found in [38].

5 Logics, algebras, and approximation spaces

A *frame* is a pair $\langle U, \mathfrak{R} \rangle$, where U is a set, and $\mathfrak{R}$ a family of binary relations on W. If $\mathfrak{R} = \{\mathfrak{R}\}$, we will just write $\langle U, R \rangle$. We observe that approximation spaces are frames in this sense. Frames can serve as semantics for modal logics. We shall briefly recall the theory for just one modality; this can be easily generalised to the multi-modal logics we will need to develop logics of information systems.

The alphabet of a propositional (uni –) modal logic consists of a set P of propositional variables, and the logical operators $\{\vee, \bot, \wedge, \top, \neg, \langle\,\rangle\}$, where $\langle\,\rangle$ is the possibility operator. As usual, we define $p \to q$ as $\neg p \vee q$, and $p \longleftrightarrow q$ is $(p \to q) \wedge (q \to p)$.

The necessity operator is $[\,] = \neg \langle\,\rangle \neg$, and the set Fml of formulas is defined recursively in the usual way.

A *modal logic* is a set Γ of formulas with the following properties:

1. All tautologies of the classical propositional calculus belong to Γ.
2. Γ is closed under substitution of variables.
3. $\langle \varphi \vee \psi \rangle \in \Gamma$ iff $\langle \varphi \rangle \vee \langle \psi \rangle \in \Gamma$.
4. Modus ponens, i.e. φ and $\varphi \to \psi \in \Gamma$ imply $\psi \in \Gamma$.
5. Necessitation, i.e. $\varphi \in \Gamma$ implies $[\varphi] \in \Gamma$.

If Γ is a modal logic, and $\Sigma \subseteq Fml$, $\varphi \in Fml$, we write $\Sigma \vdash_\Gamma \varphi$ if φ is derivable from $\Sigma \cup \Gamma$ via modus ponens and necessitation.

A model $\mathfrak{M}$ of Fml is a triple $\langle W, R, v \rangle$, where $\underline{W} = \langle W, R \rangle$ is a frame and $v : P \to 2^W$ a mapping, called a *valuation*; we can think of $v(p)$ as the set of states at which p is true.

We define *local truth* of a formula with respect to v via a forcing relation $\Vdash_v$ as follows: Let $w \in W$.

1. If $p \in P$, then $w \Vdash_v p \Longleftrightarrow w \in v(p)$.
2. If $\varphi, \psi \in Fml$, then

$$w \Vdash_v \neg \varphi \Longleftrightarrow w \not\Vdash_v \varphi, \tag{34}$$

$$w \Vdash_v \varphi \wedge \psi \Longleftrightarrow w \Vdash_v \varphi \text{ and } w \Vdash \psi, \tag{35}$$

$$w \Vdash_v \varphi \vee \psi \Longleftrightarrow w \Vdash_v \varphi \text{ or } w \Vdash \psi, \tag{36}$$

$$w \Vdash_v \langle \varphi \rangle \Longleftrightarrow (\exists u \in W)[wRu \text{ and } u \Vdash_v \varphi] \tag{37}$$

For $\mathfrak{M} = \langle W, R, v \rangle \in \mathsf{Mod}$ we now set

$$\mathsf{mng}(\varphi, \mathfrak{M}) = \{\mathfrak{w} \in \mathfrak{W} : \mathfrak{w} \Vdash_\mathfrak{v} \varphi\},$$
$$\mathfrak{M} \models \varphi \iff \mathsf{mng}_\mathfrak{v}(\varphi, \mathfrak{M}) = \mathfrak{W}.$$

If $\mathfrak{M} = \langle \mathfrak{W}, \mathfrak{R}, \mathfrak{v} \rangle$ is understood, we just write $\mathsf{mng}(\varphi)$.

Note that

$$w \Vdash \langle \varphi \rangle \iff R(w) \cap \mathsf{mng}(\varphi) \neq \emptyset, \tag{38}$$
$$w \Vdash [\varphi] \iff R(w) \subseteq \mathsf{mng}(\varphi). \tag{39}$$

We invite the reader to consult, for example, [8] or [31] for an introduction to basic modal logic.

The algebraic version of these semantics is as follows: With each frame $\langle U, R \rangle$ we associate a mapping $\langle R \rangle : 2^U \to 2^U$ by

$$\langle R \rangle(X) = \{u \in U : (\exists w \in X) u R w\}. \tag{40}$$

It is easy to see that

$$\langle R \rangle(X) = \bigcup \{R\,\breve{}(w) : w \in X\}, \tag{41}$$

and that furthermore,

$$\langle R \rangle(\emptyset) = \emptyset, \qquad \text{(Normal)} \tag{42}$$
$$\langle R \rangle(X \cup Y) = \langle R \rangle(X) \cup \langle R \rangle(Y). \qquad \text{(Additive)} \tag{43}$$

The mapping $\langle R \rangle$ is called a *modal* or *possibility operator*, and its dual $[R]$,

$$[R](X) = -\langle R \rangle(-X), \tag{44}$$

is called a *necessity operator*. More generally, we call an operator on a Boolean algebra B which satisfies 42 and 43 a modal operator on B. If f is a modal operator on B, then $\langle B, f \rangle$ is a Boolean algebra with operators in the sense of [40], and we call $\langle B, f \rangle$ a *modal algebra*. It follows from the definitions of lower approximation 25 and upper approximation 27, that i is the necessity operator $[R]$, and its dual h the modal operator $\langle R \rangle$. Since R is an equivalence relation, the following hold:

$$[R](X) \subseteq X \qquad \text{Reflexive} \tag{45}$$
$$\langle R \rangle([R](X)) \subseteq X \qquad \text{Symmetric} \tag{46}$$
$$\langle R \rangle(\langle R \rangle(X)) \subseteq \langle R \rangle. \qquad \text{Transitive} \tag{47}$$

Therefore, the appropriate modal logic for approximation spaces is a (normal) S5 logic.

Even though many relational properties can be expressed by equations of modal algebras such as 45 – 47, this is not true for all first order properties. It is well known that irreflexivity cannot be so expressed [70]. To remedy this situation, modal – like operations other than $\langle R \rangle$ and $[R]$ have been considered [37, 34]. We call operators $f(R), g(R)$ *orthogonal*[4] to each other, if for all X

$$f(R)(X) = g(-R)(-X) \tag{48}$$

holds. This leads to pairs of orthogonal operators

$$\langle\langle R \rangle\rangle(X) = \langle -R \rangle(-X), \tag{49}$$
$$[[R]](X) = [-R](-X), \tag{50}$$

and

$$\langle R \rangle(X) = \langle\langle -R \rangle\rangle(-X), \tag{51}$$
$$[R](X) = [[-R]](-X). \tag{52}$$

The operator $[[R]](X)$ of 50 is called a *sufficiency operator* [34, 23, 24].

Fig. 2. Duality and orthogonality

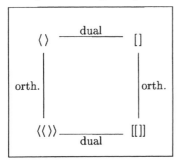

It is easy to see that

$$\langle\langle R \rangle\rangle(X) = \{x \in U : (\exists y \notin X) x(-R) y\}, \tag{53}$$
$$[[R]](X) = \{x \in U : y \in X \text{ implies } xRy\}. \tag{54}$$

More generally, a mapping $g : B \to B$ on a Boolean algebra B is called *sufficiency operator*, if it has the properties

$$g(0) = 1, \tag{55}$$
$$g(a+b) = g(a) \cdot g(b), \tag{56}$$

A *sufficiency algebra* (SUA) is a Boolean algebra B with additional sufficiency operators.

---

[4] What we call "orthogonal" is called "conjugate" in [56]. We choose the former since the latter means something different in the context of Boolean algebras with operators.

Table 2. Relations of indistinguishability

Strong indiscernibility	Weak indiscernibility
$\langle a,b \rangle \in ind(Q)$ iff $x(a) = x(b)$ for all $x \in Q$.	$\langle a,b \rangle \in wind(Q)$ iff $x(a) = x(b)$ for some $x \in Q$.
Strong similarity	**Weak similarity**
$\langle a,b \rangle \in sim(Q)$ iff $x(a) \cap x(b) \neq \emptyset$ for all $x \in Q$.	$\langle a,b \rangle \in wsim(Q)$ iff $x(a) \cap x(b) \neq \emptyset$ for some $x \in Q$.
Strong incomplementarity	**Weak incomplementarity**
$\langle a,b \rangle \in icom(Q)$ iff $x(a) \neq -x(b)$ for all $x \in Q$.	$\langle a,b \rangle \in wicom(Q)$ iff $x(a) \neq -x(b)$ for some $x \in Q$.

6 Logics for information systems

6.1 Information relations

The first step in building formal systems for reasoning about information needs to be to determine which ground relations we want to consider. The basic notion of RSDA is indiscernability, but there are other relations which naturally arise in reasoning about information. For example, it has often been asked whether the transitivity condition of indiscernability is not too strong an assumption, since one might easily fall into the "Sorites Paradox" trap: Sorites (meaning "heap" in Greek) is a paradox which has its root in old Greek philosophy, and could be stated thus: Suppose you have a heap of stones, and you remove one stone and ask yourself: Is it still a heap? The answer is: Yes it is. You remove another and ask yourself the same question, and the answer is still yes. But eventually it will cease to be a heap, and become just a few stones. (Russell used an example with a man from which you repeatedly pluck a hair, and asked whether the man was bald. etc.)[5]. A similar observation can be made when considering the precision of numerical measurements; in dynamical systems, for example, errors which are beyond the capability of measurement tend to come forward when the process is iterated, and everything will go into chaos.

At times, one may not be interested in indiscernability, but in the properties of those relations which distinguish objects; thus, it is also important to study and formalise relations of diversity.

A multivalued information system is called *deterministic*, if $|x(a)| = 1$ for all $a \in U$, $x \in \Omega$; these are the information systems in the previous sense. In what follows we will use "information system" for "multivalued information system".

Suppose that $\mathcal{I} = \langle U, \Omega, V_x \rangle_{x \in \Omega}$ is an information system. For each set Q of attributes we define relations of indistinguishability in Table 2 and of distinguishability in Table 3, each in a "strong" and "weak" version; we call relations

---

[5] We should like to thank Jenny Ottosson of the Department of Philosophy, University of Lund, Sweden, for explaining the paradox to us.

Table 3. Relations of distinguishability

Strong diversity	Weak diversity
$\langle a,b \rangle \in div(Q)$ iff $x(a) \neq x(b)$ for all $x \in Q$.	$\langle a,b \rangle \in wdiv(Q)$ iff $x(a) \neq x(b)$ for some $x \in Q$.
Strong orthogonality	**Weak orthogonality**
$\langle a,b \rangle \in ort(Q)$ iff $x(a) \cap x(b) = \emptyset$ for all $x \in Q$.	$\langle a,b \rangle \in wort(Q)$ iff $x(a) \cap x(b) = \emptyset$ for some $x \in Q$.
Strong complementarity	**Weak complementarity**
$\langle a,b \rangle \in com(Q)$ iff $x(a) = -x(b)$ for all $x \in Q$.	$\langle a,b \rangle \in wcom(Q)$ iff $x(a) = -x(b)$ for some $x \in Q$.

of any of these types *information relations*; other interesting information relations can be found in [56]. Note that the strong version of a relation $R(Q)$ always is a subset of the weak version. A list of some other equational properties (for want of a better word) of these relations can be found in Table 4.

Table 4. Equational properties

$$\begin{aligned}
\text{ind}(Q) &= \text{-wdiv}(Q) & \text{wind}(Q) &= \text{-div}(Q) \\
\text{sim}(Q) &= \text{-wort}(Q) & \text{wsim}(Q) &= \text{-ort}(Q) \\
\text{icom}(Q) &= \text{-wcom}(Q) & \text{wicom}(Q) &= \text{-com}(Q) \\
\text{ind}(Q) &\subseteq \text{sim}(Q) & \text{wind}(Q) &\subseteq \text{wsim}(Q) \\
\text{ort}(Q) &\subseteq \text{div}(Q) & \text{wort}(Q) &\subseteq \text{wdiv}(Q).
\end{aligned}$$

In the sequel, we shall call each of $ind, wind, sim, \ldots$ a *type* of information relation, using the generic name R. We can think of R as a mapping from 2^Ω into $2^{U \times U}$, which assigns to each set of attributes a binary relation on U according to the given constraints. Each of these types can be described by relational properties, as given in Table 5 [56]. There, we call a relation S

- a *tolerance*, if S is reflexive and symmetric,
- *weakly reflexive*, if aSb implies aSa,
- a *weak tolerance*, if S is weakly reflexive and symmetric,
- *3-transitive*, if $aSbScSd$ implies aSd,
- *intransitive*, if $aSbSc$ implies $a(-S)c$.

We have not only to consider relations among objects, but also dependencies among attributes which are reflected in the relational properties, and thus have an effect on the objects as well; recall, for example, the considerations in Section 3.

Table 5. Relational properties

ind	:	equivalence
wind	:	weak tolerance, $ind(\{x\})$ transitive
sim	:	tolerance
wsim	:	weak tolerance
icom	:	symmetric, $-icom(\{x\})$ irreflexive, 3–transitive
wicom	:	symmetric, -icom(Q) irreflexive, 3–transitive
div	:	$-div(Q)$ a tolerance, $-div(\{x\})$ transitive
wdiv	:	$-wdiv(Q)$ an equivalence
ort	:	$-ort(Q)$ a weak tolerance
wort	:	$-ort(Q)$ a tolerance
com	:	symmetric, irreflexive, 3–transitive
wcom	:	symmetric, $wcom(\{x\})$ irreflexive, 3–transitive

Given an information system, each of the types of information relations defined in Tables 2 and 3 determines a multi-frame by $\langle U, R(Q)\rangle_{Q\subseteq\Omega}$; we call this type of frame an *information frame*. The accessibility relations are sometimes called *relative*, namely, to the set of attributes involved [2, 3, 48], and a general structure $\langle U, R(Q)\rangle_{Q\subseteq\Omega}$ where $R(Q)$ is a binary relation on U will be called a *relative frame*.

Suppose that R is a type of information relation, and $P, Q \subseteq \Omega$. Several conditions can arise [48]:

$$R(P \cup Q) = R(P) \cap R(Q), \quad \text{(strong)} \tag{57}$$
$$R(\emptyset) = U \times U, \quad \text{(co-normal)} \tag{58}$$
$$R(P \cup Q) = R(P) \cup R(Q), \quad \text{(weak)} \tag{59}$$
$$R(\emptyset) = \emptyset. \quad \text{(normal)} \tag{60}$$

An information frame $\langle U, R(Q)\rangle_{Q\subseteq\Omega}$ is called *strong*, resp. *weak*, if the relations satisfy 57 and 58, resp. 59 and 60. It is not hard to see that each type of strong (weak) information relation of Tables 2, 3 induces a strong (weak) frame.

Each relative frame $\mathcal{K} = \langle \mathcal{U}, \mathcal{R}(\mathcal{Q})\rangle_{Q\subseteq\Omega}$ gives rise to an associated multi–modal, respectively, multi–sufficiency algebra

$$\mathfrak{B}_{\langle \mathcal{K}\rangle} = \langle 2^U, \cap, \cup, -, \emptyset, U, \langle R(Q)\rangle\rangle_{Q\subseteq\Omega}, \tag{61}$$
$$\mathfrak{B}_{\langle\langle \mathcal{K}\rangle\rangle} = \langle 2^U, \cap, \cup, -, \emptyset, U, \langle\langle R(Q)\rangle\rangle\rangle_{Q\subseteq\Omega} \tag{62}$$

in the sense of Section 5. If R is one of *ind, sim, icom, div, ort, com*, then we will write $\mathfrak{B}_{\langle\mathfrak{R}\rangle}$, resp. $\mathfrak{B}_{\langle\langle\mathfrak{R}\rangle\rangle}$, and $\mathfrak{B}_{\langle\mathfrak{w}\mathfrak{R}\rangle}$, resp. $\mathfrak{B}_{\langle\langle\mathfrak{w}\mathfrak{R}\rangle\rangle}$ for the weak version.

Extending the duality between frames and Boolean algebras with operators and motivated by the two concepts described before, E. Orłowska [53] defines four major types of information algebras according to the conditions 57 – 60: Let $\mathfrak{B} = \langle \mathfrak{B}, +, \cdot, -\text{o}, 1\mathfrak{f}(\mathfrak{P})\rangle_{\mathfrak{P} \subseteq \Omega}$ be a Boolean algebra with a unary operator $f(P): B \to B$ for each $P \subseteq \Omega$. $\mathfrak{B}$ is called an

1. *Information algebra with strong normal operators* (SN), if for all $P, Q \subseteq \Omega$, $a, a_i, \in B$ and atoms c of B,

$$f(P)(0) = 0, \tag{63}$$
$$a = \sum_i a_i \text{ implies } f(P)(a) = \sum_i f(P)(a_i), \tag{64}$$
$$f(\emptyset)(a) = 1, \text{ if } a \gneq 0, \tag{65}$$
$$f(P \cup Q)(c) = f(P)(c) \cdot f(Q)(c). \tag{66}$$

2. *Information algebra with weak normal operators* (WN), if for all $P, Q \subseteq \Omega$, $a, a_i \in B$,

$$f(P)(0) = 0, \tag{67}$$
$$a = \sum_i a_i \text{ implies } f(P)(a) = \sum_i f(P)(a_i), \tag{68}$$
$$f(\emptyset)(a) = 0, \tag{69}$$
$$f(P \cup Q)(a) = f(P)(a) + f(Q)(a). \tag{70}$$
$$\tag{71}$$

Observe that SN and WN algebras are modal algebras, and thus Boolean algebras with operators in the sense of [40].

3. *Information algebra with strong conormal operators* (SCN), if for all $P, Q \subseteq \Omega$, $a, a_i, \in B$,

$$f(P)(1) = 0, \tag{72}$$
$$a = \prod_i a_i \text{ implies } f(P)(a) = \sum_i f(P)(a_i), \tag{73}$$
$$f(\emptyset)(a) = 1, \text{ if } a \neq 1, \tag{74}$$
$$f(P \cup Q)(a) = f(P)(a) + f(Q)(a). \tag{75}$$

4. *Information algebra with weak conormal operators* (WCN), if for all $P, Q \subseteq \Omega$, $a, a_i, \in B$ and co-atoms c of B,

$$f(P)(1) = 0, \tag{76}$$
$$a = \prod_i a_i \text{ implies } f(P)(a) = \sum_i f(P)(a_i), \tag{77}$$
$$f(\emptyset)(a) = 1, \text{ if } a \neq 1, \tag{78}$$
$$f(P \cup Q)(c) = f(P)(c) \cdot f(Q)(c). \tag{79}$$

Note that SCN and WCN algebras are sufficiency algebras. We now have

Proposition 7. *[53]* Let $\mathcal{I} = \langle U, \Omega \rangle$ be an information system. Then,

$$\mathfrak{B}_{\langle\text{ind}\rangle}, \mathfrak{B}_{\langle\text{sim}\rangle} \in \mathfrak{SN}, \tag{80}$$

$$\mathfrak{B}_{\langle\text{wind}\rangle}, \mathfrak{B}_{\langle\text{wsim}\rangle} \in \mathfrak{WN}, \tag{81}$$

$$\mathfrak{B}_{\langle\langle\text{ort}\rangle\rangle}, \mathfrak{B}_{\langle\langle\text{div}\rangle\rangle} \in \mathfrak{SCN}, \tag{82}$$

$$\mathfrak{B}_{\langle\langle\text{wort}\rangle\rangle}, \mathfrak{B}_{\langle\langle\text{wdiv}\rangle\rangle} \in \mathfrak{WCN}. \tag{83}$$

There are, in fact, finer classifications of information algebras, and the algebras for each type can be described more fully. For our introductory purposes, however, the types above shall suffice, and the interested reader is invited to consult [53] for the details.

An important issue in the study of algebras is the question of representability, i.e. whether every abstract algebra of a certain class arises in some way from what one considers to be a "standard model". If the class is generated by the models of a logic, then, in a way, representability results correspond to completeness results of the logic. In our case, the standard models of the classes of algebras above are those which arise from information systems with the prescribed type of relations. Results in this direction can be found in [14] and the references given therein.

6.2 Information logics

In analogy to the single relation frames considered in Section 5, we can build a duality theory with multi-modal algebras and multi-modal logics on the basis of the information algebras of the preceding Section.

An *normal information logic* Γ is a modal logic as defined in Section 5, with an extra set PAR of parameters, and a possibility (necessity) operator $\langle\ \rangle_P$ ($[\ ]$ for each $P \subseteq PAR$ such that $\varphi \in \Gamma$ implies $[\varphi]_P \in \Gamma$ for all $P \subseteq PAR$. Models of Γ are tuples $\langle U, R(Q), v\rangle_{Q \subseteq \Omega}$, such that v maps PAR bijectively to Ω, and atomic formulas of Γ to subsets of U; without loss of generality, we will assume that $PAR = \Omega$. Local truth of possibility 37 has to be redefined in such a way that for every $P \subseteq \Omega$,

$$w \Vdash_v \langle \varphi \rangle_P \iff (\exists u \in W)[wR(P)u \text{ and } u \Vdash_v \varphi] \tag{84}$$

This is the relative version of the modal logic K. Conditions on the strongness, weakness or otherwise of the frames (and/or the corresponding algebras) now lead to other logics. For example, if $\mathcal{G}$ is the class of all frames for which

$$R(P \cap Q) = R(P) \cap R(Q), \tag{85}$$

then the most general logic which is sound and complete with respect to $\mathcal{G}$, is the logic which has the axiom scheme

$$[\varphi]_P \vee [\varphi]_Q \to [\varphi]_{P \cap Q}.$$

If $\mathcal{H}$ is the class of all strong frames, i.e. where

$$R(P \cup Q) = R(P) \cap R(Q),$$

then the most general logic which is sound and complete with respect to $\mathcal{H}$ is the logic DK with the axiom scheme

$$[\varphi]_P \vee [\varphi]_Q \to [\varphi]_{P \cup Q}.$$

Extensions of DK lead to information logics for strong indiscernability relations, e.g. the logic KR of [51] which has as axiom schemes

$$[\varphi]_P \to \varphi, \tag{86}$$
$$\varphi \to [\neg[\neg\varphi]_P]_P, \tag{87}$$
$$[\varphi]_P \to [[\varphi]_P]_P. \tag{88}$$

A moment's reflection tells us that these schemes correspond to the conditions on an equivalence relation. Leaving out axiom scheme 88 leads to strong similarity frames [42, 43]. Ontology and logic of similarity relations have been studied in great detail in [68].

Other normal information logics have been presented, for example, in [47], [49], [50], [65], [66]. Recent completeness results can be found in [2], [3], [12], and [13].

6.3 Relational proof systems

A valuable tool in the formalisation and implementation of information logics are the relational proof systems introduced by [52], which are sound and complete for standard modal logics. Such systems are in the style of [62], and consist of decomposition rules, specific rules and (sequences of) axiomatic expressions. The application of a decomposition rule syntactically simplifies a formula, while the specific rules are the counterparts of the properties satisfied by the accessibility relations. Axiomatic sequences play the role of axioms and are universally true. One advantage of relational proof systems is their modularity: The decomposition rules can be re-used for various systems. Furthermore, with relational proof systems it is possible to capture properties that are not expressible in standard modal or sufficiency logic. As an example, consider a frame $\langle W, R \rangle$, where R is the complementarity relation of Table 3. Since we consider only one relation, the concepts of "strong" and "weak" do not apply. It is shown in [24] that the defining properties of R are not expressible in either modal or sufficiency logic alone, but need a "mixed" kind of logic. On the other hand, a sound and complete relational proof system for complementarity frames was given in [25]. As an example for such a system, we present the decomposition rules in Table 6. For more details, we invite the reader to consult [52, 54, 25, 26].

7 Outlook

Even though the landscape of logics and their algebras of strong frames, e.g. indiscernability or similarity, seems to be well understood, research on logics and algebraic duality of the distinguishability relations still needs to be done.

Table 6. Decomposition rules for binary relations

$(\cup_2)$	$\dfrac{\Gamma, x(R \cup S)y, \Delta}{\Gamma, xRy, xSy, \Delta}$	$(-\cup_2)$	$\dfrac{\Gamma, x-(R \cup S)y, \Delta}{\Gamma, x(-R)y, \Delta \mid \Gamma, x(-S)y, \Delta}$
$(\cap_2)$	$\dfrac{\Gamma, x(R \cap S)y, \Delta}{\Gamma, xRy, \Delta \mid \Gamma, xSy, \Delta}$	$(-\cap_2)$	$\dfrac{\Gamma, x-(R \cap S)y, \Delta}{\Gamma, x(-R)y, x(-S)y, \Delta}$
$(\check{\ })$	$\dfrac{\Gamma, xR\check{\ }y, \Delta}{\Gamma, yRx, \Delta}$	$(-\check{\ })$	$\dfrac{\Gamma, x(-R\check{\ })y, \Delta}{\Gamma, y(-R)x, \Delta}$
$(--2)$	$\dfrac{\Gamma, x(--R)y, \Delta}{\Gamma, xRy, \Delta}$		
$(;)$	$\dfrac{\Gamma, x(R;S)y, \Delta}{\Gamma, xRz, \Delta, x(R;S)y \mid \Gamma, zSy, \Delta, x(R;S)y}$		where z is any variable
$(-;)$	$\dfrac{\Gamma, x-(R;S)y, \Delta}{\Gamma, x(-R)z, z(-S)y, \Delta}$		where z is a restricted variable

More application fields for these logics and their algebras still need to be identified, in particular, for the distinguishability relations. The rule systems based on these relations and their statistical underpinnings have not yet been fully exploited. A promising area of application is qualitative spatial reasoning, e.g. in geographical information systems, which requires logical tools for handling vagueness of regions. A four valued logic based on lower and upper approximation of regions, much in the spirit of the three valued logic of [17] has been proposed independently in [74]; a modal logic arising from vague spatial data can be found in [5]. Approximation algebras can be used for reasoning with imperfect information about regions [28, 27]. There are also interesting connections to algebras of binary relations [29], which could lead to a more developed theory of rough relation algebras and their logics in the spirit of [9], [16], and, recently, [69].

References

1. Balbes, R. and Dwinger, P. (1974). *Distributive Lattices*. University of Missouri Press, Columbia.
2. Balbiani, P. (1997). Axiomatisation of logics based on Kripke models with relative accessibility relations. In [55].
3. Balbiani, P. and Orłowska, E. (1999). A hierarchy of modal logics with relative accessibility relations. *Journal of Applied Non-Classical Logics*. To appear.
4. Becchio, D. (1978). Logique trivalente de Łukasiewicz. *Ann. Sci. Univ. Clermont-Ferrand*, 16:38–89.
5. Bennett, B. (1998). Modal semantics for knowledge bases dealing with vague concepts. Submitted for publication.
6. Birkhoff, G. (1948). *Lattice Theory*, volume 25 of *Am. Math. Soc. Colloquium Publications*. AMS, Providence, 2 edition.
7. Boicescu, V., Filipoiu, A., Georgescu, G., and Rudeanu, S. (1991). *Łukasiewicz–Moisil Algebras*, volume 49 of *Annals of Discrete Mathematics*. North Holland, Amsterdam.
8. Bull, R. and Segerberg, K. (1984). Basic modal logic. In [33], pages 1–88.

9. Comer, S. (1991). An algebraic approach to the approximation of information. *Fundamenta Informaticae*, 14:492–502.
10. Comer, S. (1993). On connections between information systems, rough sets, and algebraic logic. In Rauszer, C., editor, *Algebraic Methods in Logic and Computer Science*, volume 28 of *Banach Center Publications*, pages 117–124. Polish Academy of Science, Warszawa.
11. Davey, B. A. and Priestley, H. A. (1990). *Introduction to Lattices and Order*. Cambridge University Press.
12. Demri, S. (1996). A class of information logics with a decidable validity problem. In *21st International Symposium on Mathematical Foundations of Computer Science*, volume 1113 of *LNCS*, pages 291–302. Springer–Verlag.
13. Demri, S. (1997). A completeness proof for a logic with an alternative necessity operator. *Studia Logica*, 58:99–112.
14. Demri, S. and Orłowska, E. (1998). Informational representability of models for information logic. In Orłowska, E., editor, *Logic at Work*. Physica – Verlag, Heidelberg. To appear.
15. Düntsch, I. (1983). On free or projective Stone algebras. *Houston J. Math.*, 9:455–463.
16. Düntsch, I. (1994). Rough relation algebras. *Fundamenta Informaticae*, 21:321–331.
17. Düntsch, I. (1997). A logic for rough sets. *Theoretical Computer Science*, 179(1-2):427–436.
18. Düntsch, I. and Gediga, G. (1996). On query procedures to build knowledge structures. *J. Math. Psych.*, 40(2):160–168.
19. Düntsch, I. and Gediga, G. (1997). Algebraic aspects of attribute dependencies in information systems. *Fundamenta Informaticae*, 29:119–133.
20. Düntsch, I. and Gediga, G. (1998). Simple data filtering in rough set systems. *International Journal of Approximate Reasoning*, 18(1–2):93–106.
21. Düntsch, I. and Gediga, G. (1999). A note on the correspondences among entail relations, rough set dependencies, and logical consequence. Submitted for publication.
22. Düntsch, I., Gediga, G., and Orłowska, E. (1999a). Relational attribute systems. Submitted for publication.
23. Düntsch, I. and Orłowska, E. (1999). Mixing modal and sufficiency operators. *Bulletin of the Section of Logic, Polish Academy of Sciences*, 28(2):99–106.
24. Düntsch, I. and Orłowska, E. (2000a). Beyond modalities: Sufficiency and mixed algebras. *Algebra Universalis*. Submitted for publication.
25. Düntsch, I. and Orłowska, E. (2000b). Logics of complementarity in information systems. *Mathematical Logic Quarterly*, 46(4). To appear.
26. Düntsch, I. and Orłowska, E. (2000c). A proof system for contact relation algebras. *Journal of Philosophical Logic*. To appear.
27. Düntsch, I., Orłowska, E., and Wang, H. (2000a). An algebraic and logical approach to the approximation of regions. In *Proc of the 5th Seminar of Relational Methods in Computer Science, Banff, Jan 2000*. To appear.
28. Düntsch, I., Orłowska, E., and Wang, H. (2000b). Approximating regions. In preparation.
29. Düntsch, I., Wang, H., and McCloskey, S. (1999b). Relation algebras in qualitative spatial reasoning. *Fundamenta Informaticae*, 39:229–248.

30. Falmagne, J.-C., Koppen, M., Villano, M., Doignon, J.-P., and Johannesen, J. (1990). Introduction to knowledge spaces: How to build, test and search them. *Psychological Review*, 97:202–234.
31. Fitting, M. (1993). Basic modal logic. In Gabbay, D. M., Hogger, C. J., and Robinson, J. A., editors, *Logical foundations*, volume 1 of *Handbook of Logic in Artificial Intelligence and Logic Programming*, pages 368–448. Clarendon Press, Oxford.
32. Gabbay, D. M. (1985). Theoretical foundations for non–monotonic reasoning in expert systems. In Apt, K. R., editor, *Logics and Models of Concurrent Systems*, volume F13 of *NATO Advanced Studies Institute*, pages 439–457. Springer, Berlin.
33. Gabbay, D. M. and Guenthner, F., editors (1984). *Extensions of classical logic*, volume 2 of *Handbook of Philosophical Logic*. Reidel, Dordrecht.
34. Gargov, G., Passy, S., and Tinchev, T. (1987). Modal environment for Boolean speculations. In Skordev, D., editor, *Mathematical Logic and Applications*, pages 253–263, New York. Plenum Press.
35. Gediga, G. and Düntsch, I. (2000). Statistical techniques for rough set data analysis. In Polkowski, L., editor, *Rough sets: New developments*. Physica Verlag, Heidelberg. To appear.
36. Grätzer, G. (1978). *General Lattice Theory*. Birkhäuser, Basel.
37. Humberstone, I. L. (1983). Inaccessible worlds. *Notre Dame Journal of Formal Logic*, 24:346–352.
38. Iturrioz, L. and Orłowska, E. (1996). A Kripke–style and relational semantics for logics based on Łukasiewicz algebras. Conference in honour of J. Łukasiewicz, Dublin.
39. Iwinski, T. B. (1987). Algebraic approach to rough sets. *Bull. Polish Acad. Sci. Math.*, 35:673–683.
40. Jónsson, B. and Tarski, A. (1951). Boolean algebras with operators I. *Amer. J. Math.*, 73:891–939.
41. Jónsson, B. and Tarski, A. (1952). Boolean algebras with operators II. *Amer. J. Math.*, 74:127–162.
42. Konikowska, B. (1987). A formal language for reasoning about indiscernibility. *Bulletin of the Polish Academy of Sciences, Mathematics*, 35:239–249.
43. Konikowska, B. (1997). A logic for reasoning about similarity. *Studia Logica*, 58:185–226.
44. Koppen, M. and Doignon, J.-P. (1990). How to build a knowledge space by querying an expert. *J. Math. Psych.*, 34:311–331.
45. Novotný, M. (1997a). Applications of dependence spaces. In [55], pages 247–289.
46. Novotný, M. (1997b). Dependence spaces of information systems. In [55], pages 193–246.
47. Orłowska, E. (1984). Modal logics in the theory of information systems. *Zeitschr. f. Math. Logik und Grundlagen der Math.*, 30:213–222.
48. Orłowska, E. (1988a). Kripke models with relative accessibility relations and their applications to inference with incomplete information. In Mirkowska, G. and Rasiowa, H., editors, *Mathematical Problems in Computation Theory*, volume 21 of *Banach Center Publications*, pages 327–337. PWN.
49. Orłowska, E. (1988b). Logical aspects of learning concepts. *Journal of Approximate Reasoning*, 2:349–364.
50. Orłowska, E. (1989). Logic for reasoning about knowledge. *Zeitschr. f. Math. Logik und Grundlagen der Math.*, 35:559–572.

51. Orłowska, E. (1990). Kripke semantics for knowledge representation logics. *Studia Logica*, 49:255–272.
52. Orłowska, E. (1991). Relational interpretation of modal logics. In Andréka, H., Monk, J. D., and Németi, I., editors, *Algebraic Logic*, volume 54 of *Colloquia Mathematica Societatis János Bolyai*, pages 443–471. North Holland, Amsterdam.
53. Orłowska, E. (1995). Information algebras. In *Proceedings of AMAST 95*, volume 639 of *Lecture Notes in Computer Science*. Springer–Verlag.
54. Orłowska, E. (1996). Relational proof systems for modal logics. In Wansing, H., editor, *Proof theory of modal logic*, pages 55–78. Kluwer, Dordrecht.
55. Orłowska, E., editor (1997a). *Incomplete Information – Rough Set Analysis*. Physica – Verlag, Heidelberg.
56. Orłowska, E. (1997b). Introduction: What you always wanted to know about rough sets. In [55], pages 1–20.
57. Orłowska, E. and Pawlak, Z. (1987). Representation of nondeterministic information. *Theoretical Computer Science*, 29:27–39.
58. Pagliani, P. (1997). Rough sets theory and logic-algebraic structures. In [55], pages 109–190.
59. Pawlak, Z. (1973). Mathematical foundations of information retrieval. ICS Research Report 101, Polish Academy of Sciences.
60. Pawlak, Z. (1991). *Rough sets: Theoretical aspects of reasoning about data*, volume 9 of *System Theory, Knowledge Engineering and Problem Solving*. Kluwer, Dordrecht.
61. Pomykala, J. and Pomykala, J. A. (1988). The Stone algebra of rough sets. *Bull. Polish Acad. Sci. Math.*, 36:495–508.
62. Rasiowa, H. and Sikorski, R. (1963). *The Mathematics of Metamathematics*, volume 41 of *Polska Akademia Nauk. Monografie matematyczne*. PWN, Warsaw.
63. Tarski, A. (1930). Fundamentale Begriffe der Methodologie der deduktiven Wissenschaften. *Monatsh. Math. Phys.*, 37:361–404.
64. Vakarelov, D. (1977). Notes on N–lattices and constructive logic with strong negation. *Studia Logica*, 36:109–125.
65. Vakarelov, D. (1989). Modal logics for knowledge representation systems. In Meyer, A. R. and Zalessky, M., editors, *Symposium on Logic Foundations of Computer Science, Pereslavl-Zalessky*, volume 363 of *LNCS*, pages 257–277. Springer–Verlag.
66. Vakarelov, D. (1991). Modal logics for knowledge representation systems. *Theoretical Computer Science*, 90:433–456.
67. Vakarelov, D. (1992). Consequence relations and information systems. In Słowiński, R., editor, *Intelligent decision support: Handbook of applications and advances of rough set theory*, volume 11 of *System Theory, Knowledge Engineering and Problem Solving*, pages 391–399. Kluwer, Dordrecht.
68. Vakarelov, D. (1997a). Information systems, similarity relations, and modal logics. In [55], pages 492–550.
69. Vakarelov, D. (1997b). Modal logics of arrows. In de Rijke, M., editor, *Advances in intensional logic*, pages 137–171. Kluwer.
70. van Benthem, J. (1984). Correspondence theory. In [33], pages 167–247.
71. Varlet, J. C. (1968). Algèbres des Łukasiewicz trivalentes. *Bull. Soc. Roy. Sci. Liège*, 36:399–408.
72. Varlet, J. C. (1972). A regular variety of type $\langle 2,2,1,1,0,0 \rangle$. *Algebra Universalis*, 2:218–223.

73. Wang, H., Düntsch, I., and Gediga, G. (2000). Classificatory filtering in decision systems. *International Journal of Approximate Reasoning*. To appear.
74. Worboys, M. (1998). Imprecision in finite resolution spatial data. *Geoinformatica*, 2(3):257–280.

Chapter 10

Statistical Techniques for Rough Set Data Analysis

Günther Gediga[1] *and Ivo Düntsch*[2]

[1] FB Psychologie / Methodenlehre, Universität Osnabrück,
49069 Osnabrück, Germany,

[2] School of Information and Software Engineering,
University of Ulster, Newtownabbey, BT 37 0QB,
N. Ireland
emails: Guenther@Gediga.de; I.Duentsch@ulst.ac.uk

1 Introduction

Concept forming and classification in the absence of complete or certain information has been a major concern of artificial intelligence for some time. Traditional "hard" data analysis based on statistical models or are in many cases not equipped to deal with uncertainty, relativity, or non–monotonic processes. Even the recently popular "soft" computing approach with its principal components "... fuzzy logic, neural network theory, and probabilistic reasoning" [16] uses quite hard parameters outside the observed phenomena, e.g. representation and distribution assumptions, prior probabilities, beliefs, or membership degrees, the origin of which is not always clear; one should not forget that the results of these methods are only valid up to the – stated or unstated – model assumptions. The question arises, whether there is a step in the modelling process which is informative for the researcher and, at the same time, does not require additional assumptions about the data. To make this clearer, we follow [9] in assuming that a data model consists of

1. A domain $\mathcal{D}$ of interest.
2. An empirical system $\mathcal{E}$, which consists of a body of data and relations among the data, and a mapping $e : \mathcal{D} \to \mathcal{E}$, called *operationalisation*.
3. A (structural or numerical) model $\mathcal{M}$, and a mapping $m : \mathcal{E} \to \mathcal{M}$, called *representation*.
4. The agent (literally: the acting subject),

see Figure 1. The agent with her/his objectives is the central part of the modelling process. Agents choose operationalisation and representation according to their objectives and their view of the world. The numerical models are normally a reduction of the empirical models, and thus of the domain of interest which results in further decontextualisation. We observe that even the soft computing methods reside on the level of the numerical models.

Fig. 1. The modelling process

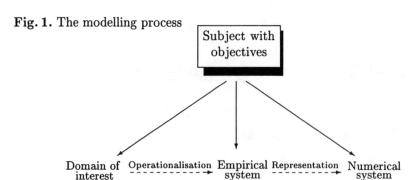

Rough set data analysis (RSDA) which has been developed by Z. Pawlak [10] and his co–workers since the early 1970s is a structural method which stays on the level of the empirical model; more formally, the representation mapping is the identity, and thus, there is a one–one relationship between the elements of the empirical model and the representation. In this way, we avoid further reduction, stay closer to the data, and keep the model assumptions to a minimum.

Although designed as a structural – in particular, a non statistical – approach to data analysis, application of RSDA only makes sense, if some basic statistical assumptions are observed. We will show that the application of these assumptions leads quite naturally to

- Statistical testing schemes for the significance of inference rules,
- Entropy measures for model selection, and
- A probabilistic version of RSDA.

2 Operationalisation in RSDA

Operationalisation of domain data in RSDA is done via a tabularised OBJECT $\mapsto$ ATTRIBUTE relationship: An *information system* is a tuple $\mathcal{I} = \langle U, \Omega, V_x \rangle_{x \in \Omega}$, where

1. $U = \{a_1, \ldots, a_N\}$ is a finite set.
2. $\Omega = \{x_1, \ldots, x_T\}$ is a finite set of mappings $x : U \to V_x$.

We interpret U as a set of objects and Ω as a set of attributes or features each of which assigns to an object a its value under the respective attribute. For each nonempty $Q \subseteq \Omega$ we define

$$V_Q = \prod_{a \in Q} V_a. \tag{1}$$

For $a \in U$, we also let

$$Q(a) = \langle x(a) \rangle_{a \in Q}, \tag{2}$$

written as $\mathbf{x}^Q(a)$ or just $\mathbf{x}^Q$ if a is understood or not relevant in the context. Each $Q(a)$ is called a *Q-granule*; the collection of all Q-granules is denoted by G_Q. A Q-granule can be understood as a piece of information about objects in U given by the features in Q. The equivalence relation on U induced by Q is denoted by ψ_Q, i.e. for $a_i, a_j \in U$,

$$a_i \equiv_{\psi_Q} a_j \iff Q(a_i) = Q(a_j). \tag{3}$$

Objects which in this sense belong to the same granule cannot be distinguished with the knowledge of Q. We denote the set of classes of ψ_Q by $\mathcal{P}(Q)$.

Suppose that $\emptyset \neq Q, P \subseteq \Omega$. Our aim is to describe the world according to P with our knowledge according to Q. If, for example, a class M of ψ_Q is contained totally within a class of ψ_P, then $Q(a)$ determines $P(b)$ for all $a, b \in M$. Such an M is called a *P-deterministic class of Q*, and

$$\text{If } Q(a) = \mathbf{x}^Q, \text{ then } P(a) = \mathbf{x}^P \tag{4}$$

is called a *deterministic Q,P – rule*. Otherwise, M intersects exactly the classes $L_1, \ldots, L_k$ of $\mathcal{P}(P)$ with associated $\mathbf{x_1}^P, \ldots, \mathbf{x_k}^P \in G_P$, and we call

$$\text{If } Q(a) = \mathbf{x}^Q, \text{ then } P(a) = \mathbf{x_1}^P \text{ or } \ldots \text{ or } P(a) = \mathbf{x_k}^P \tag{5}$$

an *indeterministic Q,P – rule*. The collection of all Q, P – rules is denoted by $Q \to P$, and with some abuse of language, will be sometimes called a rule (of the information system). In writing rules, we will usually identify singleton sets with the element they contain, e.g. we write $Q \to d$ instead of $Q \to \{d\}$.

Note that all constructions above use only the information given by the observed system, and no additional outside parameters.

Throughout this paper, we use $\mathcal{I}$ as above with the given parameters as a generic information system. For further information on RSDA we refer the reader to [11] or [6].

3 Basic statistics

Even though rough set analysis is a structural method, it makes basic statistical assumptions which we briefly want to describe in this section. Suppose that ψ is an equivalence relation on U which may be of the form ψ_Q; the only numerical information we have are the cardinality T of U, and the cardinalities of the classes of ψ.

For $X \subseteq U$, we call

$$\underline{X}_\psi \stackrel{def}{=} \bigcup \{\psi x : \psi x \subseteq X\} \tag{6}$$

the *lower approximation* or *positive region of X*. These are those elements of U which can be classified with certainty as being in X. The *upper approximation* or *possible region* of X with respect to ψ is defined as

$$\overline{X}^\psi \stackrel{def}{=} U \setminus (\underline{U \setminus X}_\psi). \tag{7}$$

The lower approximation function leads to the statistic

$$\mu_*^\psi(X) \stackrel{def}{=} \frac{|\underline{X}|}{|U|}. \tag{8}$$

We now define $\mu^{\psi*}(X) = 1 - \mu_*^\psi(-X)$, and it is easy to see that

$$\mu^{\psi*}(X) \stackrel{def}{=} \frac{|\overline{X}|}{|U|}. \tag{9}$$

We say that a probability measure p on 2^U is *compatible with* ψ, if

$$\mu_*^\psi(X) \leq p(X) \leq \mu^{\psi*}(X),$$

for all $X \in B_\psi$. Compatibility of p expresses the fact that $p(X)$ is within the bounds of uncertainty given by ψ. It is easy to see that the only probability measure on 2^U which is compatible to all functions μ_*^ψ is given by

$$p(X) = \frac{|X|}{|U|}, \tag{10}$$

so that $p(x) = \frac{1}{|U|}$ for all $x \in U$. In other words, RSDA assumes the *principle of indifference*, where in the absence of further knowledge all basic events are assumed to be equally likely. Unlike statistical models, RSDA does not model the dependency structure of attributes, but assumes that the principle of indifference is the only valid basis for an estimation of probability. If we assume marked dependencies among attributes, there may be better statistics than μ_*^ψ for the computation of $p(X)$, but even in this situation μ_*^ψ will remain a reasonable choice for $p(X)$.

The statistics derived from μ_*^ψ which is normally used in RSDA, the *approximation quality*, is defined as

$$\gamma_\psi(X) = \mu_*(X) + \mu_*(-X). \tag{11}$$

Clearly,

$$\gamma_\psi(X) \stackrel{def}{=} \frac{|\underline{X}_\psi| + |\underline{-X}_\psi|}{|U|}, \tag{12}$$

so that $\gamma_\psi(X)$ is the relative frequency of all elements of U which are correctly classified under the granulation of information by ψ with respect to being an element of X or not.

Generalising 12 to partitions induced by attribute sets, we define the *quality of an approximation* of a an attribute set Q with respect to an attribute set P by

$$\gamma(Q \to P) = \frac{|\bigcup \{X \in \mathcal{P}(Q) : X \text{ is } P\text{-deterministic}\}|}{|U|}. \tag{13}$$

The approximation is *perfect*, if $\gamma(Q \to P) = 1$; in this case, $\psi_Q \subseteq \psi_P$, and all Q, P – rules are deterministic.

If P is fixed, an attribute set which is $\subseteq$ – minimal with respect to $\gamma(Q \to P) = 1$ is called a *reduct* of P.

4 Significance testing

4.1 Rule significance

If we use RSDA for supervised learning, then its results must be controlled by statistical testing procedures; otherwise, we may read more into the results than what is actually in them. For example, if each Q, P – rule is based on a singleton class of ψ_Q – for example, a running number –, then the prediction of P (in fact, of any attribute set) will be perfect, but the rule will usually be rather useless for a different data sample. The underlying assumption on which prediction is based is that the information system $\mathcal{I}$ is a representative sample of the situation.

The assumption of representativeness is a problem of any analysis in most real life data bases. The reason for this is the huge state complexity of the space of possible rules, even when there are only a few number of features (Tab. 1).

Table 1. State complexity

# of attr. values	# of attributes		
	10	20	30
	$\log_{10}$ (states)		
2	3.01	6.02	9.03
3	4.77	9.54	14.31
4	6.02	12.04	18.06
5	6.99	13.98	20.97

In [4] we have developed two procedures, both based on randomisation techniques, to compute the conditional probability of a rule $Q \to P$, assuming that the null hypothesis

H_0: "Objects are randomly assigned to rules"

is true. Randomisation procedures are particularly suitable to RSDA since they do not require outside information; in particular, it is not assumed that the information system under discussion is a representative sample.

Suppose that $\emptyset \neq Q, P \subseteq \Omega$, and that we want to evaluate the statistical significance of the rule $Q \to P$. Let Σ be the set of all permutations of U, and $\sigma \in \Sigma$. We define new attribute functions x^σ by

$$x^\sigma(a) \stackrel{def}{=} \begin{cases} x(\sigma(a)), & \text{if } x \in Q, \\ x(a), & \text{otherwise.} \end{cases}$$

The resulting information system $\mathcal{I}_\sigma$ permutes the Q–columns according to σ, while leaving the P–columns constant; we let Q^σ be the result of the permutation in the Q–columns, and $\gamma(Q^\sigma \to P)$ be the approximation quality of the prediction of P by Q^σ in $\mathcal{I}_\sigma$.

The value

$$p(\gamma(Q \to P)|H_0) := \frac{|\{\gamma(Q^\sigma \to P) \geq \gamma(Q \to P) : \sigma \in \Sigma\}|}{|U|!} \tag{14}$$

now measures the significance of the observed approximation quality. If $p(\gamma(Q \to P)|H_0)$ is low, traditionally below 5%, then the rule $Q \to P$ is deemed significant, and the (statistical) hypothesis "$Q \to P$ is due to chance" can be rejected. Otherwise, if $p(\gamma(Q \to P)|H_0) \geq 0.05$, we call $Q \to P$ a *casual rule*.

As an example, consider the following information system [4]:

U	x_1	x_2	d
1	0	0	0
2	0	1	1
3	1	0	2

The rule $\{x_1, x_2\} \to d$ is perfect, since $\gamma(\{x_1, x_2\} \to d) = 1$. Furthermore, $p(\gamma(\{x_1, x_2\} \to d)|H_0) = 1$, because every instance is based on a single observation, and thus, the rule is casual.

Now suppose that we have collected three additional observations:

U	x_1	x_2	d	U	x_1	x_2	d
1	0	0	0	1'	0	0	0
2	0	1	1	2'	0	1	1
3	1	0	2	3'	1	0	2

To decide whether the given rule is casual under the statistical assumption, we have to consider all 720 possible rules as given in 14 and their approximation qualities. The distribution of the approximation qualities of these 720 rules is given in Table 2, with $\alpha = p(\gamma(\{x_1, x_2\} \to d)|H_0)$.

Table 2. Results of randomisation analysis; 6 observ.

γ	No of cases	α	Example of σ
1.00	48	0.067	1, 1', 2, 2', 3, 3'
0.33	288	0.467	1, 1', 2, 3, 2', 3'
0.00	384	1.000	1, 2, 2', 3, 1', 3'

Given the 6-observation example, the probability of obtaining a perfect approximation of d by $\{x_1, x_2\}$ under the assumption of random matching, is 0.067 which is by far smaller than in the 3-observation example, but, using conventional $\alpha = 0.05$, not convincing enough to decide that the rule is sufficiently significant to be not casual.

A small scale simulation study done in [4] indicates that the randomisation procedure has a reasonable power if the rule structure of the attributes is known.

We have applied the procedures to three well known data sets, and have found that not all claimed results, based on γ alone, can be called significant, and that other significant results were overlooked. Details and more examples can be found in [4].

4.2 Conditional casual attributes

In pure RSDA, the decline of the approximation quality when omitting one attribute is usually used to determine whether an attribute within a perfect rule $Q \to P$ is of high value for the prediction. This interpretation does not take into account that the decline of approximation quality may be due to chance.

As in the preceding section, our approach is to compare the actual $\gamma(Q \to P)$ with the results of a random system; here we randomise the value of a single attribute $t \in Q$ as follows: For each permutation σ of U we obtain a new attribute function $x^{\sigma,t}$ by setting

$$x^{\sigma,t}(a) \stackrel{def}{=} \begin{cases} x(\sigma(a)) & \text{if } x = t, \\ x(a), & \text{otherwise.} \end{cases}$$

Here, only the values in the t–column are permuted, and we denote the set of the resulting Q–granules by $Q^{\sigma,t}$. Now,

$$p_t(\gamma(Q \to P)|H_0) := \frac{|\{\gamma(Q^{\sigma,t} \to P) \geq \gamma(Q \to P) : \sigma \in \Sigma\}|}{|U|!} \qquad (15)$$

measures the significance of attribute t within Q for the prediction of P. If $\alpha = p_t(\gamma(Q \to P)|H_0) \leq 0.05$, the assumption of (random) conditional casualness can be rejected; otherwise we shall call the attribute t *conditional casual within* Q.

The example given in Table 3 shows that, depending on the nature of an attribute, statistical evaluation leads to different expectations of the increase of approximation quality which is not visible under ordinary RSDA methods.

Table 3.

U	x	r_1	r_2	r_3	d	U	x	r_1	r_2	r_3	d
1	0	1	1	1	a	5	1	5	5	3	c
2	0	2	1	1	a	6	1	6	4	3	c
3	0	3	3	3	b	7	2	7	7	3	d
4	0	4	3	3	b	8	2	8	7	3	d

The prediction rule $x \to d$ has the approximation quality $\gamma(x \to d) = 0.5$. Assume that an additional attribute r is conceptualised in three different ways:

- A fine grained measure r_1 using 8 categories,
- A medium grained description r_2 using 4 categories, and
- A coarse description r_3 using 2 categories.

For $1 \leq i \leq 3$ we have $\gamma(\{x, r_i\}) \to d) = 1$, so that each of these approximations is perfect, and the drop of the approximation quality is 0.5 when r_i is left out. Therefore, we have a situation in which standard RSDA does not distinguish between the different properties of the additional attribute $r_i, 1 \leq i \leq 3$.

If we consider the expectation $E[\gamma(\{x, r_i\}^{\sigma, r_i} \to p)]$, we observe that

$E[\gamma(\{x, r_1\}^{\sigma, r_1} \to p)] = 1.000,$
$E[\gamma(\{x, r_2\}^{\sigma, r_2} \to p)] = 0.880,$
$E[\gamma(\{x, r_3\}^{\sigma, r_3} \to p)] = 0.624.$

Whereas the statistical evaluation of the additional predictive power differs for each of the three realizations of the new attribute r, the analysis of the decline of the approximation quality tells us nothing about these differences. Therefore, rather than using the decline of approximation quality as a global measure of influence, it is more appropriate to compare the influence of an attribute using the proposed statistical testing procedure.

4.3 Sequential significance testing

One can see that randomisation is a computationally expensive procedure, and it might be said that this fact limits its usefulness in practical applications. We have argued in [4] that, if randomisation is too costly for a data set, RSDA itself will not be applicable in this case, and have suggested several simple criteria to speed up the computations.

Another, fairly simple, tool to shorten the processing time of the randomisation test is the adaptation of a sequential testing scheme to the given situation. Because this sequential testing scheme can be used as a general tool in randomisation analysis, we present the approach in a more general way.

Suppose that θ is a a statistic with realizations θ_i, and a fixed realization θ_c. We can think of θ_c as $\gamma(Q \to P)$ and θ_i as $\gamma(Q^\sigma \to P)$. Recall that the statistic θ is called α − *significant*, if the true value $p(\theta \geq \theta_c|H_0)$ is smaller than α. Traditionally, $\alpha = 0.05$, and in this case, one speaks just of *significance*.

An evaluation of the hypothesis $\theta \geq \theta_c$ given the hypothesis H_0 can be done by using a sample of size n from the θ distribution, and counting the number k of θ_i for which $\theta_i \geq \theta_c$. The evaluation of $p(\theta \geq \theta_c|H_0)$ can now be done by the estimator $\hat{p}_n(\theta \geq \theta_c|H_0) = \frac{k}{n}$, and the comparison $\hat{p}_n(\theta \geq \theta_c|H_0) < \alpha$ will be performed to test the significance of the statistic. For this to work we have to assume that the simulation is asymptotically correct, i.e. that

$$lim_{n \to \infty} \hat{p}_n(\theta \geq \theta_c|H_0) = p(\theta \geq \theta_c|H_0). \tag{16}$$

In order to find a quicker evaluation scheme of the significance, it should be noted that the results of the simulation k out of n can be described by a binomial distribution with parameter $p(\theta \geq \theta_c|H_0)$. The fit of the approximation of $\hat{p}_n(\theta \geq \theta_c|H_0)$ can be determined by the confidence interval of the binomial distribution.

In order to control the fit of the approximation more explicitly, we introduce another procedure within our significance testing scheme. Let

$$H_b : p(\theta \geq \theta_c|H_0)) \in [0, \alpha) \tag{17}$$
$$H_a : p(\theta \geq \theta_c|H_0)) \in [\alpha, 1] \tag{18}$$

be another pair of statistical hypotheses, which are strongly connected to the original ones: If H_b holds, we can conclude that the test is α-significant, if H_a holds, we conclude that it is not.

Because we want to do a finite approximation of the test procedure, we need to control the precision of the approximation; to this end, we define two additional error components:

1. $r =$ probability that H_a is true, but H_b is the outcome of the approximative test.
2. $s =$ probability that H_b is true, but H_a is the outcome of the approximative test.

The pair (r, s) is called the *precision* of the approximative test. To result in a good approximation, the values r, s should be small (e.g. $r = s = 0.05$); at any rate, we assume that $r + s \lessapprox 1$, so that $\frac{s}{1-r} \lessapprox \frac{1-s}{r}$, which will be needed below.

Using the Wald-procedure [15], we define the likelihood ratio

$$LQ(n) = \frac{\sup_{p \in [0,\alpha)} p^k (1-p)^{n-k}}{\sup_{p \in [\alpha,1]} p^k (1-p)^{n-k}}, \qquad (19)$$

and we obtain the following approximative sequential testing scheme:

1. If
$$LQ(n) \lessapprox \frac{s}{1-r},$$
then H_a is true with probability at most s.
2. If
$$LQ(n) \gtrapprox \frac{1-s}{r},$$
then H_b is true with probability at most r.
3. Otherwise
$$\frac{s}{1-r} \leq LQ(n) \leq \frac{1-s}{r},$$
and no decision with precision (r, s) is possible. Hence, the simulation must continue.

With this procedure, which is implemented in our rough set engine GROBIAN[3] [3], the computational effort for the significance test in most cases breaks down dramatically, and a majority of the tests need less than 100 simulations.

---
[3] e-mail:http:/www.infj.ulst.ac.uk/ cccz23/grobian/grobian.html

5 Model selection

In conventional RSDA, the approximation quality $\gamma(Q \to d)$ is used as a conditional measure to describe prediction success of a dependent decision attribute d from a set Q of independent attributes. However, approximation qualities cannot be compared, if we use different attribute sets Q and R for the prediction of d.

To define an unconditional measure of prediction success, one can use the *minimum description length principle* [13] by combining

- Program complexity (i.e. to find a deterministic rule in RSDA) and
- Statistical uncertainty (i.e. a measure of uncertainty when applying an indeterministic rule)

to a global measure of prediction success. In this way, dependent and independent attributes are treated similarly.

In [5], we combine the principle of indifference with the maximum entropy principle (where worst possible cases are compared) to arrive at an objective method which combines feature selection with data prediction.

Suppose that R is any nonempty set of attributes and $\mathcal{P}(R) = \{X_i : 1 \leq i \leq k\}$. We define the *entropy of R* by

$$H(R) \stackrel{def}{=} \sum_{i=1}^{k} \frac{r_i}{n} \cdot \log_2(\frac{n}{r_i}),$$

where $r_i \stackrel{def}{=} \frac{|X_i|}{n}$.

Now, suppose that the classes of θ_Q are $X_0, \ldots, X_m$, and that the probability distribution of the classes is given by $\hat{\pi}_i = \frac{|X_i|}{n}$; let $X_0, \ldots X_c$ be the deterministic classes with respect to d, and W be their union.

Since our data is the partition obtained from Q, and we know the world only up to the equivalence classes of θ_Q, an indeterministic observation y is a result of a random process whose characteristics are totally unknown. Given this assumption, no information within our data set will help us to classify the element y, and we conclude that each such y requires a rule (or class) of its own. This results in a new partition of the object set associated with the equivalence relation θ_Q^+ defined by

$$x \equiv_{\theta_Q^+} y \stackrel{def}{\iff} x = y \text{ or there is some } i \leq c \text{ such that } x, y \in X_i.$$

Its associated probability distribution is given by $\{\hat{\psi}_i : i \leq c + |U \setminus W|\}$ with

$$\hat{\psi}_i \stackrel{def}{=} \begin{cases} \hat{\pi}_i, & \text{if } i \leq c, \\ \frac{1}{n}, & \text{otherwise.} \end{cases} \qquad (20)$$

We now define the *entropy of rough prediction* (with respect to $Q \to d$) as

$$H_{\text{rough}}(Q \to d) \stackrel{def}{=} H(\theta_Q^+) = \sum_i \hat{\psi}_i \cdot \log_2(\frac{1}{\hat{\psi}_i}).$$

To obtain an objective measurement we define the *normalised rough entropy* (NRE) by

$$\text{NRE}(Q \to d) \stackrel{def}{=} 1 - \frac{H_{\text{rough}}(Q \to d) - H(d)}{\log_2(|U|) - H(d)}. \tag{21}$$

If the NRE has a value near 1, the entropy is low, and the chosen attribute combination is favourable, whereas a value near 0 indicates statistical casualness in the sense of Section 4. The normalisation does not use moving standards as long as we do not change the decision attribute d. Therefore, any comparison of NRE values between different predicting attribute sets given a fixed decision attribute makes sense.

The implemented procedure searches for attribute sets with a high NRE; since finding the NRE of each feature set is computationally expensive, we use a genetic – like algorithm to determine sets with a high NRE.

We have named the method SORES[4], an acronym for Searching Optimal Rough Entropy Sets; SORES is implemented in GROBIAN [3].

In order to test the procedure, we have used 14 datasets available from the UCI repository[5] from which the appropriate references of origin can be obtained.

The validation by the training set – testing set method was performed by splitting the full data set randomly into two equal sizes 100 times, assuming a balanced distribution of training and testing data; the mean error value is our measure of prediction success.

In Table 4 we compare the SORES results with the C4.5 performance given in [12]. Column 2 indicates how many attributes were used for prediction out of the total; for example, in the Annealing database, SORES has used 11 out of the 38 attributes.

The results indicate that SORES, even in its present unoptimised version, can be viewed as an effective machine learning procedure, because its performance compares well with that of the well established C4.5 method: The odds are 7:7 (given the 14 problems) that C4.5 produces better results. However, since the standard deviation of the error percentages of SORES is higher than that of C4.5, we conclude that C4.5 has a slightly better performance than the current SORES.

Details can be found in [5].

6 Probabilistic RSDA

RSDA concentrates on finding deterministic rules for the description of dependencies among attributes. Once a (deterministic) rule is found, it is assumed to

---

[4] All material relating to SORES, e.g. datasets, validation data, and a description of the algorithm can be obtained from the SORES website e-mail:http:/www.psycho.uni-osnabrueck.de/sores/

[5] e-mail:http:/www.ics.uci.edu/ mlearn/MLRepository.html

Table 4. Datasets and SORES validation

Dataset	SORES		C4.5 (Rel. 8)
Name	# attr.	Error	Error
Annealing	11/38	6.26	7.67
Auto	2/25	11.28	17.70
Breast-W	2/9	5.74	5.26
Colic	4/22	21.55	15.00
Credit–A	5/15	18.10	14.70
Credit–G	6/20	32.92	28.40
Diabetes	3/8	31.86	25.40
Glass	3/9	21.79	32.50
Heart–C	2/23	22.51	23.00
Heart–H	5/23	19.43	21.50
Hepatitis	3/19	17.21	20.40
Iris	3/4	4.33	4.80
Sonar	3/60	25.94	25.60
Vehicle	2/18	35.84	27.10
Std. Deviation		10.33	8.77

hold without any error. If a measurement error is assumed to be an immeasurable part of the data – as e.g. statistical procedures do – the pure RSDA approach will not produce acceptable results, because "real" measurement errors cannot be explained by any rule.

In order to formulate a probabilistic version of RSDA, which is able to handle measurement errors as well, we enhance some of the concepts defined before. A *replicated decision system* $\mathcal{D}$ is a structure $\langle U, \Omega, Y, V_x \rangle_{x \in \Omega}$, where

- $\langle U, \Omega, V_x \rangle_{x \in \Omega}$ is an information system,
- $Y = \{y_1, \ldots, y_S\}$ is a set of replicated decision attributes, explained more fully below.

We shall use the parameters of the information system $\mathcal{I}$ of p. 2; the set of Ω-granules is $G = \{\mathbf{x_1}, \ldots, \mathbf{x_M}\}$. In the sequel, we shall omit reference to Ω, if no confusion can arise; in particular, we will just speak of *granules* instead of Ω - *granules*.

Although not common in RSDA, the introduction of replicated decision variables offers the opportunity to control the effect of a measurement error: The smaller the agreement among multiple replications of the decision attribute, the more measurement error has to be assumed. This concept of replicated measurements is a way to estimate the reliability of, for example, psychometric tests, using the retest-reliability estimation, which in turn uses a linear model to estimate reliability and error of measurement as well.

With some abuse of notation, we assume that the decision attributes $y_1, \ldots, y_S$ are realizations of an unknown underlying distribution Y considered as a mapping

$$Y : U \times \{r_1, \ldots, r_Y\} \to [0, 1].$$

Y assign to each element a of U and each value r_j of the decision attribute the probability that $Y(a) = r_j$.

We suppose that (each replica of) the decision attribute takes the values $V_Y = \{r_1, r_2, \ldots, r_Y\}$. The classes of ψ_{y_t} are denoted by $M_{t,1}, \ldots, M_{t,r_Y}$; for each granule $\mathbf{x}_i$, we let $\xi(i, t, j)$ be the number of objects described by $\mathbf{x}_i$ which are in class $M_{t,j}$. In other words,

$$\xi(i, t, j) = |\{a \in U : \Omega(a) = \mathbf{x}_i \text{ and } a \in M_{t,j}\}| \tag{22}$$

We also let

$$\nu(\mathbf{x}_i) = |\{a \in U : \Omega(a) = \mathbf{x}_i\}|. \tag{23}$$

Clearly, $\sum_j \xi(i, t, j) = \nu(\mathbf{x}_i)$ for fixed i, and $\sum_{i,j} \xi(i, t, j) = |U|$. Each set $\{\xi(i, t, j) : 1 \leq t \leq s\}$ can be assigned an unknown value $\pi(i, j)$, which is the probability that an element $a \in U$ is assigned to a class r_j where $1 \leq j \leq r_Y$ and $\Omega(a) = \mathbf{x}_i$.

An example of the parameters of a decision system with one replica of the decision attribute is shown in Table 5.

Table 5. A decision system

$\mathbf{x}_i$	Ω $x_1\ x_2$	$Y = r_1$ $\xi(i, 1, 1)$	$Y = r_2$ $\xi(i, 1, 2)$	$\nu(\mathbf{x}_i)$
$\mathbf{x}_1$	0 1	5	1	6
$\mathbf{x}_2$	1 0	2	8	10
	Σ	7	9	16

The example given in Table 5 shows that indeterministic rules alone do not use the full information given in the database. There is no deterministic rule to predict a value of the decision attribute y_1, given a value of the independent attributes $\langle x_1, x_2 \rangle$: Both indeterministic rules will predict both possible outcomes in the decision variable. The pure rough set approach now concludes that no discernible assignment is possible. But if we inspect the table, we see that the error of assigning $\langle 0, 1 \rangle$ to 1 is small (1 observation) and that $\langle 1, 0 \rangle \mapsto 2$ is true up to 2 observations.

There are several possibilities to reduce the precision of prediction to cope with measurement error. One possibility is the so called *variable precision rough set model* [17], which assumes that rules are only valid within a certain part of the population. The advantage of this approach is that it uses only two parameters (the precision parameter and γ) to describe the quality of a rule system; the disadvantages are that precision and γ are partially exchangeable, and that there is no theoretical background to judge which combination is best suited to the data.

Another approach – based upon standard statistical techniques – is the idea of predicting random variables instead of fixed values. Conceptually, each realization of the distribution Y can be described by a mixture

$$Y = \sum_{1 \leq r \leq R} \omega_r Y_r, \tag{24}$$

with $\sum_r \omega_r = 1$, based on an index R of unknown size, and unknown basic distributions Y_r with unknown weights ω_r.

If we use the granules $\mathbf{x}_j$ to predict Y, the maximal number R of basic distributions is bounded by the number M of granules; equality occurs just when each granule $\mathbf{x}_j$ determines its own Y_j. In general, this need not to be the case, and it may happen that the same Y_j can be used to predict more than one granule; this can be indicated by an onto function

$$g : \{1, ..., M\} \twoheadrightarrow \{1, ..., R\},$$

mapping the (indices of) the granules to a smaller set of mixture components of Y.

Probabilistic prediction rules are of the form

$$\mathbf{x}_j \to Y_{g(j)}, \ 1 \leq j \leq M,$$

where each $Y_{g(j)} : V_Y \to [0,1]$ is a random variable. If the probabilities are understood, we shall often just write $\mathbf{x} \to Y$, with Y possibly indexed, for the rule system $\langle \mathbf{x}_j \to Y_{g(j)} \rangle_{1 \leq j \leq M}$.

In the example of Table 5 there are two possibilities for R, and we use maximum likelihood to optimise the binomial distribution, the application of which is straightforward, if we additionally assume that the observations stem from a simple sampling scheme. In case $R = 1$, both granules use the same distribution Y_1. In this case, the likelihood function $L_1 = L(Y_1|\langle 0,1\rangle, \langle 1,0\rangle)$ is given by

$$L_1 = \binom{16}{9} \pi^9 (1-\pi)^7 \tag{25}$$

which, as expected, has a maximum at $\hat{\pi} = \frac{9}{16}$. This leads to the rule system

$$\langle 0,1\rangle \text{ or } \langle 1,0\rangle \to \{\langle 1, \frac{9}{16}\rangle, \langle 2, \frac{7}{16}\rangle\}. \tag{26}$$

If $R = 2$, the samples belonging to $\langle 0,1\rangle$ and $\langle 1,0\rangle$ are assumed to be different in terms of the structure of the decision attribute, and the likelihood of the sample has to be built from the product of the likelihoods of both subsamples. If we have the rules $\langle 0,1\rangle \to Y_1$ and $\langle 1,0\rangle \to Y_2$, then

$$L_2 = \binom{6}{1} \pi_1^1 (1-\pi_1)^5 \binom{10}{8} \pi_2^8 (1-\pi_2)^2. \tag{27}$$

Using standard calculus, the maximum of L_2 is ($\hat{\pi}_1 = \frac{1}{6}$, $\hat{\pi}_2 = \frac{8}{10}$), which gives us the rule system

$$\begin{cases} \langle 0,1 \rangle \to \{\langle 1, \frac{5}{6} \rangle, \langle 2, \frac{1}{6} \rangle\}, \\ \langle 1,0 \rangle \to \{\langle 1, \frac{2}{10} \rangle, \langle 2, \frac{8}{10} \rangle\}. \end{cases} \quad (28)$$

In going from from L_1 to L_2 we change the sampling structure – the estimation of L_2 needs 2 samples, whereas L_1 needs only one sample – and we increase the number of probability parameters π_i by one.

Changing the sampling structure is somewhat problematic, because comparison of likelihoods can only be done within the same sample. A simple solution is to compare the likelihoods based on elements, thus omitting the binomial factors. Because the binomial factors are unnecessary for parameter estimation (and problematic for model comparison) they will be skipped in the sequel. Letting

$$L_1(\max) = \hat{\pi}^9(1-\hat{\pi})^7 \qquad = 0.0000173, \quad (29)$$
$$L_2(\max) = \hat{\pi}_1^1(1-\hat{\pi}_1)^6 \hat{\pi}_2^8(1-\hat{\pi}_2)^2 = 0.0003746, \quad (30)$$

we have to decide which rule offers a better description of the data. Although $L_2(\max)$ is larger than $L_1(\max)$, it is not obvious to conclude that the two rules are really 'essentially' different, because the estimation of L_2 depends on more free parameters than L_1.

There are – at least – two standard procedures for model selection, which are based on the likelihood and the number of parameters: The Akaike Information Criterion (AIC) [1] and the Schwarz Information Criterion (SIC) [14]. If $L(\max)$ is the maximum likelihood of the data, P the number of parameters, and K the number of observations, these are defined by

$$AIC = 2(P - \ln(L(\max))) \quad (31)$$
$$SIC = 2\left(\frac{\ln(K)}{2} \cdot P - \ln(L(\max))\right). \quad (32)$$

The lower AIC (and SIC respectively), the better the model. AIC and SIC are rather similar, but the penalty for parameters is higher in SIC then in AIC.

In the example, we have used one parameter π to estimate L_1. Therefore,

$$AIC(L_1(\max)) = 2(1 - \ln(0.0000173)) = 23.930, \quad (33)$$
$$SIC(L_1(\max)) = 2(\frac{\ln(16)}{2} - \ln(0.0000173)) = 24.702. \quad (34)$$

There are three free parameters to estimate L_2: First, the probabilities π_1, π_2; furthermore, one additional parameter is used, because we need to distinguish between the two granules. Therefore,

$$AIC(L_2(\max)) = 2(3 - \ln(0.0003746)) = 21.779 \quad (35)$$
$$SIC(L_2(\max)) = 2(3\ln(16) - \ln(0.0003746)) = 24.090. \quad (36)$$

and we can conclude that the rule – system 28 is better suited to the data then the simple 1-rule – system 26.

6.1 Finding probabilistic rules

The algorithm of finding probabilistic rules starts by searching for the optimal granule mapping based on a set Ω of (mutually) predicting attributes and a set Y of replicated decision attributes.

R=0;
$\Delta(AIC) = +\infty$;
While R < M and $\Delta(AIC) > 0$ do
    R=R+1;
    Compute the best mapping $g : \{1,\ldots,M\} \to \{1,\ldots,R\}$ in terms of the product of the maximum likelihood of the Y replicas;
    Compute number of parameters;
    Compute AIC_R;
    if $(R > 1)$ then $\Delta(AIC) = AIC_{R-1} - AIC_R$;
    else $\Delta(AIC) = AIC_1$;

Finding the best mapping g is a combinatorial optimisation problem, which can be approximated by hill-climbing methods, whereas the computation of the maximum likelihood estimators, given a fixed mapping g, is straightforward: One computes the multinomial parameters $\hat{\pi}_t(i_k)$ of the samples i defined by g for every replication y_t of Y and every value $r_k \in \{r_1,\ldots,r_Y\}$, and computes the mean value

$$\hat{\pi}(i_k) = \frac{\sum_{t=1}^{s} \hat{\pi}_t(i_k)}{s}, \tag{37}$$

from which the likelihood can be found. The number of parameters (np) depends on R and r_Y because

$$np = R \times r_Y - 1;$$

the computation of the AIC is now possible.

The result of algorithm offers the most parsimonious description of a probabilistic rule system (in terms of AIC). In order to reduce the number of independent attributes within the rules, a classical RSDA reduct analysis of these attributes can be applied, using the results of the mapping g as a decision attribute.

6.2 Application: Unsupervised learning and nonparametric distribution estimates

The most interesting feature of the probabilistic granule approach is that the analysis can be used for clustering, i.e. unsupervised learning. In this case the predicting attribute is the identity and any granule consists of one element. If we use more than one replication of the decision attribute, it will be possible to estimate the number of mixture components of Y and the distribution of the mixtures.

The Figures 2 and 3 show the result of the mixture analysis based on granules using the mixture

$$Y = \frac{1}{2}N(-2.0, 1.0) + \frac{1}{2}N(0.0, 1.0). \tag{38}$$

$N(\mu, \sigma)$ is the normal distribution with parameters μ and σ. 1024 observations per replication were simulated; one simulation was done with 2 replications (Figure 2), and another with 5 replications (Figure 3). The simulated data were grouped into 32 intervals with approximately the same frequencies in the replications, and the searching algorithm outlined in section 6.1 was applied.

Fig. 2. Nonparametric estimates of a $\frac{(N(-2,1)+N(0,1))}{2}$ mixture distribution (2 replications; lines denotes theoretical distributions)

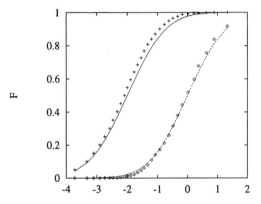

Fig. 3. Nonparametric estimates of a (N(-2,1)+N(0,1))/2 mixture distribution (5 replications; lines denotes theoretical distributions)

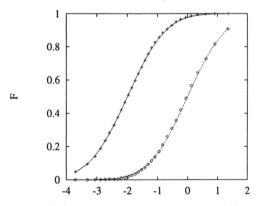

The result shows that the underlying distributions can be approximated quite successfully, although

- No parametric distributional assumption was used,
- Y has a inimical shape,
- Only a few replications were considered.

The next numerical experiment was performed with the famous Iris data [7]. These were used by Fisher to demonstrate his discriminant analysis; it consists of 50 specimen of each of the Iris species *Setosa, Versicolor,* and *Virginica,* which were measured by Sepal length, Petal length, Sepal width, Petal width. It is well known (e.g. [2]) that Sepal width attribute is not very informative; therefore we shall skip it for the subsequent analysis.

Table 6. Iris: Classification results

Setosa	50	0	0
Versicolor	7	41	2
Virginica	0	14	36

If we assume that the three remaining attributes measure the same variable up to some scaling constants, we can use the z–transformed attributes as a basis for the analysis. The unsupervised AIC search algorithm clearly votes for three classes in the unknown joint dependent attribute. If we use the estimated distribution functions (Figures 4, 5, 6) for the classification of the elements, we find a classification quality of about 85%, which is not too bad for an unsupervised learning procedure.

Fig. 4. Setosa distributions of 3 attributes and its estimation

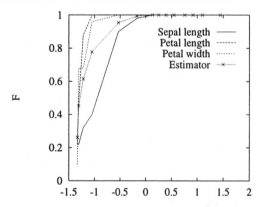

The procedure does not offer only classification results, but also estimators of the distributions of dependent attributes within the groups without having a prior knowledge about the group structure. The Figures 4, 5, 6 compare three estimated distributions with the respective the distributions of three (normalised)

variables within the groups. The results show that the "Sepal length" attribute does not fit very well and that the estimated distributions summarise this aspect of both "Petal" measures.

Fig. 5. Versicolor distributions of 3 attributes and its estimation

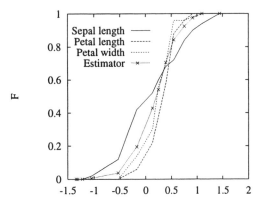

Fig. 6. Virginica distributions of 3 attributes and its estimation

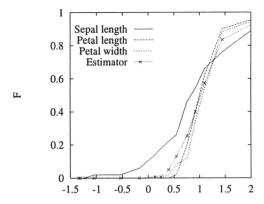

7 Summary and outlook

The paper has introduced a framework for applying statistical tools in concept forming and classification using rule based data analysis. Three different aspects were discussed:

1. Significance testing of rules, rule systems and parts of rules systems can be performed by randomisation procedures, which can be sped up by sequential testing plans.

2. The application of tailored rule based definitions of entropy, which are compatible to the non-numerical philosophy of RSDA, achieves machine learning procedures with excellent reclassification behaviour.
3. The concept of probabilistic granule analysis enables the researcher to use a noise reducing algorithm in advance to standard rule based data analysis. Probabilistic granule analysis itself turns out to be a valuable tool for unsupervised learning and non-parametric distribution estimation.

Whereas significance testing uses the same theoretical assumptions as the classical RSDA, the rough entropy based method SORES and the probabilistic granule analysis (as well as other approaches such as the variable precision model [17]) allow some error within prediction rules. Although all of these are RSDA based, from a strictly modelling point of view, these methods are partially incompatible competitors; their particular strengths and weaknesses need to be determined by further investigation.

References

1. Akaike, H. (1973). Information theory and an extension of the maximum likelihood principle. In Petrov, B. N. and Cáski, F., editors, *Second International Symposium on Information Theory*, pages 267–281, Budapest. Akademiai Kaidó. Reprinted in *Breakthroughs in Statistics*, eds Kotz, S. & Johnson, N. L. (1992), volume I, pp. 599–624. New York: Springer.
2. Browne, C., Düntsch, I., and Gediga, G. (1998). IRIS revisited: A comparison of discriminant and enhanced rough set data analysis. In Polkowski, L. and Skowron, A., editors, *Rough sets in knowledge discovery, Vol. 2*, pages 345–368, Heidelberg. Physica–Verlag.
3. Düntsch, I. and Gediga, G. (1997a). The rough set engine GROBIAN. In Sydow, A., editor, *Proc. 15th IMACS World Congress, Berlin*, volume 4, pages 613–618, Berlin. Wissenschaft und Technik Verlag.
4. Düntsch, I. and Gediga, G. (1997b). Statistical evaluation of rough set dependency analysis. *International Journal of Human–Computer Studies*, 46:589–604.
5. Düntsch, I. and Gediga, G. (1998). Uncertainty measures of rough set prediction. *Artificial Intelligence*, 106(1):77–107.
6. Düntsch, I. and Gediga, G. (2000). Rough set data analysis. In *Encyclopedia of Computer Science and Technology*. Marcel Dekker. To appear (Tech. report version avaliable at e-mail http:/www.infj.ulst.ac.uk/ cccz23/papers/rsda.html).
7. Fisher, R. A. (1936). The use of multiple measurements in taxonomic problems. *Ann. Eugen.*, 7:179–188.
8. Gediga, G. and Düntsch, I. (1999). Probabilistic granule analysis. Draft paper.
9. Gigerenzer, G. (1981). *Messung und Modellbildung in der Psychologie*. Birkhäuser, Basel.
10. Pawlak, Z. (1982). Rough sets. *Internat. J. Comput. Inform. Sci.*, 11:341–356.
11. Pawlak, Z. (1991). *Rough sets: Theoretical aspects of reasoning about data*, volume 9 of *System Theory, Knowledge Engineering and Problem Solving*. Kluwer, Dordrecht.
12. Quinlan, R. (1996). Improved use of continuous attributes in C4.5. *Journal of Artificial Intelligence Research*, 4:77–90.

13. Rissanen, J. (1978). Modeling by the shortest data description. *Automatica*, 14:465–471.
14. Schwarz, G. (1978). Estimating the dimension of a model. *Annals of Statistics*, 6:461–464.
15. Wald, A. (1947). *Sequential Analysis*. Wiley, New York.
16. Zadeh, L. A. (1994). What is BISC? e-mail: http: /http.cs.berkeley. edu/projects/Bisc/bisc.memo.html, University of California.
17. Ziarko, W. (1993). Variable precision rough set model. *Journal of Computer and System Sciences*, 46.

Chapter 11

Data Mining in Incomplete Information Systems from Rough Set Perspective

Marzena Kryszkiewicz and *Henryk Rybiński*

Institute of Computer Science
Warsaw University of Technology
Nowowiejska 15/19 00-665 Warsaw Poland
email: {mkr, hrb}@ii.pw.edu.pl

Abstract. Mining rules is of a particular interest in Rough Sets applications. Inconsistency and incompleteness issues in the information system are considered. Algorithms, which mine in very large incomplete information systems for certain, possible and generalized decision rules are presented. The algorithms are based on efficient data mining techniques devised for association rules generation from large data bases. The algorithms are capable to generate rules both supported by the system directly and *hypothetical*. The rules generated from incomplete system are not contradictory with any plausible extension of the system.
Keywords: Knowledge Discovery, Incomplete Information Systems, Rough Sets

1 Introduction

Very often, when trying to discover knowledge from information systems, one may face the problem of missing values. Missing data may result from difficulties in measuring some values, measurement failures (very often in medical files and scientific data), changes in the database schema (often in corporate information resources) etc. Several approaches to the problem have been proposed. The simplest among them consists in removing examples with unknown values or replacing unknown values with the most common values. More complex approaches were presented in [5, 15]. A Bayesian formalism was used in [5] to determine the probability distribution of the unknown value over the possible values from the domain. This method could either choose the most likely value or divide the example into fractional examples, each with one possible value weighted according to the determined probabilities. In [15] a method of prediction of an attribute value, based on the values of other attributes and the class information was proposed.

In [10,13] a framework for treating missing values in data mining has been provided. In particular, formulae for pessimistic and optimistic support and confidence of an association rule have been derived.

The Rough Sets Theory (RST) approach [14] has been proved to be successful in dealing with inexact, uncertain or vague knowledge. Although classical RST deals only with information systems in which values of all attributes for all objects are known, a number of approaches to missing values were proposed and discussed also in this area (e.g. [2, 4, 5, 6, 7, 9, 12, 17]).

Generally, one can distinguish two approaches to the rule computing: (1) the one based on Boolean reasoning [16], and (2) the other one based on data mining techniques [1], which have been devised for large databases. In this Chapter, we will concentrate on the latter one. In [8,11] we have proposed data mining methods for computing decision rules from complete information systems. This approach was extended over incomplete information systems in [12]. Algorithms generating optimal certain and possible rules which we present here are extended versions of algorithms we offered in [12]. The algorithm for generalized decision rule generation is new.

The construction of the Chapter is the following: in the next section we introduce basic notions referring to the incomplete information system. We discuss and illustrate problems resulting from incompleteness. Section 3 is devoted to formal properties of decision rules that can be inferred from an incomplete system. These properties form a basis for algorithms described in Section 4.

2 Information Systems

An *information system* (IS) is a pair $S = (O, AT)$, where O is a non-empty finite set of *objects*, and AT is a non-empty finite set of *attributes*, such that $a: O \to V_a$ for any $a \in AT$, where V_a is called the *value set* of a. In classical RST approach, only *complete* information systems are considered, i.e. systems with known values of all attributes for all objects. A system containing unknown values will be called an *incomplete information system* (IIS).

Each subset of attributes $A \subseteq AT$ determines a binary *indiscernibility relation IND(A)* as follows: $IND(A) = \{(x,y) \in O \times O: \forall a \in A.\ a(x)=a(y)\}$. $I_A(x)$ denotes the set of objects indiscernible to x wrt. A (i.e. $I_A(x) = \{y \in O: (x,y) \in IND(A)\}$).

In the case when the system contains unknown values, the notion of indiscernibility fails. For example, if we compare two objects being identical on a common set of attributes with known values, and the values of other attributes of one of the objects are unknown, the real values can be different, hence we cannot classify the objects as indiscernible. On the other hand there is a chance that the real values of the given attributes are the same for both objects. Therefore, for the systems with unknown values (denoted by *) we have introduced a notion of similarity [6].

A *similarity relation* wrt. the set A of attributes is denoted by $SIM(A)$, and it is defined as follows: $SIM(A) = \{(x,y) \in O \times O: \forall a \in A.\ a(x)=a(y) \lor a(x)=* \lor a(y)=*\}$.

$S_A(x)$ will denote the set of objects that are *similar to* x wrt. A (i.e. $S_A(x) = \{y \in O: (x,y) \in SIM(A)\}$). The meaning of *similar* objects is that they are likely to be indiscernible.

Let $X \subseteq O$ and $A \subseteq AT$. $\underline{A}X = \{x \in O: S_A(x) \subseteq X\}$ is the *lower approximation* of X. $\overline{A}X = \{x \in O: S_A(x) \cap X \neq \emptyset\}$ is the *upper approximation* of X. $\underline{A}X$ is the set of objects that belong to X with certainty, while $\overline{A}X$ is the set of objects that possibly belong to X. Clearly, $\underline{A}X \subseteq X \subseteq \overline{A}X$ and for every $A \subseteq B \subseteq AT$ we have $\underline{A}X \subseteq \underline{B}X$ and $\overline{B}X \subseteq \overline{A}X$.

A *decision table* (DT) is an information system: $(O, AT \cup \{d\})$, where d such that $d \notin AT$, $* \notin V_d$, is a distinguished attribute called *decision;* the elements of AT are called *conditions*.

Let us define a function $\partial_A: O \rightarrow P(V_d)$, $A \subseteq AT$, as follows: $\partial_A(x) = \{d(y) | y \in S_A(x)\}$. ∂_A will be called *generalized decision* in DT. If $card(\partial_{AT}(x)) = 1$ for all $x \in O$, we say that DT is *consistent*, otherwise it is *inconsistent*.

In the sequel, we consider *decision rules* which have the form: $t \Rightarrow s$, where $t = \wedge(c,v)$, $c \in A \subseteq AT$, $v \in V_c$, and $s = \vee(d,w)$, $w \in V_d$. The formula t is called the *antecedent (condition part)* of the rule $t \Rightarrow s$, whereas s is called the *consequent (decision part)* of the rule. If the consequent contains only one decision value, the rule is called *definite*; otherwise, it is called *indefinite*. The set of objects described by a conjunction t of attribute-value pairs will be denoted by $\|t\|$. By $\|t\|^*$ we denote the set of objects possibly satisfying the property t, i.e. $\|t\|^* = \{x \in O: \forall \ a \in A. \ (a,a(x))$ is in t or $a(x) = *\}$, where A is the set of attributes in t.

A decision rule $t \Rightarrow s$ is called *certain* in DT if it is definite, $\|t\|^* \subseteq \|s\|$, and $\|t\|^* \neq \emptyset$ in DT. If the rule $t \Rightarrow s$ is certain, then for every object x in $\|t\|^*$, $s = (d, d(x))$ and $\partial_{AT}(x) = \{d(x)\}$. The decision rule $t \Rightarrow s$ is called *possible* if it is definite, $\|t\|^* \subseteq \overline{AT}\|s\|$, and $\|t\|^* \neq \emptyset$ in DT. A decision rule $t \Rightarrow s$ is an *optimal certain (optimal possible)* rule if there is no other certain (possible) rule $t' \Rightarrow s$, where t' consists of a proper subset of conditions in t. The decision rule $t \Rightarrow s$ is called a *generalized decision rule* in DT if s is a subset of a generalized decision of an object from $\|t\|^*$ and $\|t\|^* \subseteq \|s\|$ and $\|t\|^* \neq \emptyset$ in DT. Thus, certain rules are a special case of generalized decision rules. A decision rule $t \Rightarrow s$ is an *optimal generalized decision* rule if there is no other generalized decision rule $t' \Rightarrow s'$, where t' consists of a proper subset of conditions in t and s' consists of a proper subset of decisions in s.

Let us consider an incomplete system $S = (O, AT)$. For such a system there is a number of systems, each of which is an extension of S to a complete system, such that it differs from S only in the positions where S contains null values [7]. An important feature we would like to have is that the rules generated from an incomplete system are not contradictory with any complete extension of S, and at least one complete extension of S would support each rule. In addition, by acquiring missing knowledge (leading to a complete system S') we would like to receive from S' the rules that resolve some indefiniteness of generalized decision rules obtainable from S. In the presented approach we guarantee that the postulates above are satisfied. It results from the fact that consequents of generalized decision rules are restricted by generalized decisions on objects. As shown in [7], for an incomplete system S and its complete extension S' the following holds: $\partial'_{AT} \subseteq \partial_{AT}$. Hence, consequents of rules generated from a complete system are more restricted. To illustrate this feature let us consider an example.

Example 1

Let us consider Table 1 which describes an incomplete decision table S containing information about cars. Attribute domains of S are as follows:

$V_{Price}=\{high, low\}$, $V_{Mileage}=\{high, low\}$, $V_{Size}=\{full, compact\}$, $V_{Max\text{-}Speed}=\{high, low\}$, and $V_d=\{poor, good, excel\}$.

Table 1 An incomplete car decision table

Car	Price	Mileage	Size	Max-Speed	d	∂_{AT}
1	high	high	full	low	good	{good}
2	low	*	full	low	good	{good}
3	*	*	comp.	high	poor	{poor}
4	high	*	full	high	good	{good,excel}
5	*	*	full	high	excel.	{good,excel}
6	low	high	full	*	good	{good,excel}

In this system, the generalized decision function indicates potential inconsistency, which may result from missing information. The following set of rules can be derived directly from the above incomplete system

1. $(Max\text{-}Speed, low) \Rightarrow (d, good)$.
2. $(Size, compact) \Rightarrow (d, poor)$.
3. $(Size, full) \Rightarrow (d, excel) \vee (d, good)$.

The indefiniteness of the rule (3) could perhaps be avoided, if we knew real values of unknown attributes.

Provided that we are able to get all missing information, we can obtain a complete extension of the system S. The information system S' presented in Table 2 is such a complete extension of S.

Table 2 A complete extension of the car decision table

Car	Price	Mileage	Size	Max-Speed	d	∂'_{AT}
1	High	high	full	low	good	{good}
2	Low	low	full	low	good	{good}
3	High	high	comp.	high	poor	{poor}
4	High	high	full	high	good	{good}
5	Low	low	full	high	excel	{excel}
6	Low	high	full	low	good	{good}

Let us note that the generalized decision function ∂'_{AT} becomes definite in S', and adding new knowledge to the system has made it consistent.

By adding missing information to S, we can directly derive from S' the rule

4. (Price, low)∧(Mileage, low)∧(Size, full)∧(Max-Speed, high) ⇒ (d, excel).

It is worth noting that this rule is also an optimal generalized decision rule in S, although not supported by any object in S. The rules which can be derived from an incomplete system, though not supported in the system, play roles of hypotheses. The more missing information we get, the better the hypothetical rule is supported.

The example above shows also another problem. As we can see, rule (4) cannot be applied directly to the object 5 because the information on *Price* and *Mileage* is missing (actually it cannot be applied directly to any object in S). However, the rule may turn out to be useful for classifying new objects. In general, if a rule uses attributes which are often unknown (e.g. difficult to measure) even if it is definite, its value is more cognitive than pragmatic. On the other extreme, an indefinite rule may turn out to be more pragmatic if it does not refer to attributes with unknown values. For example, referring to the system S from Table 1, provided that *Mileage* is difficult to judge, rule (4) although definite may become difficult to use, whereas the indefinite rule (3) may be applied easily. As in general case it is difficult to say which rules may be more useful, we claim that the tools provided to the user should generate hypothetical generalized decision rules on demand.

3 Decision Rules

Natural measures of usefulness of decision rules in a complete information system are *support* and *confidence*. Let t be a conjunction of attribute-value pairs. The cardinality of the object set $\|t\|$ will be called *support* of t and it will be denoted by $sup(t)$. We say that an object x, $x \in O$, supports a rule $t \Rightarrow s$ if x has both property t and s (i.e. $x \in \|t \cap s\|$). The support of a rule $t \Rightarrow s$ is defined as the number of objects satisfying both t and s (i.e. $sup(t \Rightarrow s) = sup(t) \cap sup(s)$).

The *confidence* of a rule $t \Rightarrow s$ is denoted by $conf(t \Rightarrow s)$ and is defined as follows:
$$conf(t \Rightarrow s) = sup(t \wedge s) / sup(t).$$

Further on we define analogous notions for the case of an incomplete information system. In this case one may not be able to compute the support of a property t precisely. Nevertheless, one can estimate the actual support by the pair of two numbers meaning the lowest possible support (pessimistic case) and the greatest possible support (optimistic case). We will denote these two numbers by $pSup(t)$ and $oSup(t)$, respectively. Clearly, $pSup(t) = sup(t)$ and $oSup(t) = card(\|t\|^*)$. The difference between $oSup(t)$ and $pSup(t)$ will be denoted by $dSup(t)$ (i.e. $dSup(t) = oSup(t) - pSup(t)$), that is $dSup(t)$ is equal to the number of objects that are likely to possess the property t, but our knowledge is not sufficient to confirm it.

A similar problem arises when one tries to compute the confidence of a rule in an IIS. Let $pConf(t \Rightarrow s)$ and $oConf(t \Rightarrow s)$ denote the lowest possible confidence and the greatest possible confidence of $t \Rightarrow s$, respectively. The property below shows how to compute these values (see [12] for proof).

Property 3.1 [12] Let $t \Rightarrow s$ be a rule in an incomplete DT.
a) $pConf(t \Rightarrow s) = sup(t \wedge s) / [sup(t) + dSup(t \wedge \neg s)]$;
b) $oConf(t \Rightarrow s) = [sup(t \wedge s) + dSup(t \wedge s)] / [sup(t) + dSup(t \wedge s)]$;
c) $pConf(t \Rightarrow s) = sup(t \wedge s) / [sup(t \wedge s) + oSup(t \wedge \neg s)]$;
d) $oConf(t \Rightarrow s) = oSup(t \wedge s) / [sup(t) + oSup(t \wedge s) - sup(t \wedge s)]$.

Property 3.2 [12] Let $\|t \Rightarrow s\|^* \neq \emptyset$.
$t \Rightarrow s$ is a generalized decision rule if ad only if $oSup(t)=oSup(t \wedge s)$ if and only if $oSup(t \wedge \neg s) = 0$ if and only if $pConf(t \Rightarrow s) = 1$.

We now show how approximations of decision classes can be expressed in terms of the generalized decision function and we discuss how certainty and possibility of a rule can be tested by evaluating this function.

Property 3.3 [12] Let $DT = (O, AT \cup \{d\})$ and $A \subseteq AT$.
a) $S_A(x) \subseteq \|(d,i)\|$ if and only if $\partial_A(x) = \{i\}$;
b) $S_A(x) \cap \|(d,i)\| \neq \emptyset$ if and only if $i \in \partial_A(x)$;
c) $\underline{A}\|(d,i)\| = \{x \in O: \partial_A(x) = \{i\}\}$;
d) $\overline{A}\|(d,i)\| = \{x \in O: i \in \partial_A(x)\}$.

Proof. Properties (c-d) are simple consequences of definitions of approximations and properties (a-b). Therefore, we will prove only first two properties.

Ad. (a) $S_A(x) \subseteq \|(d,i)\|$ iff $\forall y \in S_A(x). y \in \|(d,i)\|$ iff $\forall y \in S_A(x). d(y) = i$ iff $\partial_A(x) = \{i\}$.

Ad. (b) The proof is similar.

Property 3.4. Let DT = $(O, AT \cup \{d\})$ be a decision table and $\|t\|^* \neq \emptyset$ in DT.
a) A decision rule $t \Rightarrow (d,i)$ is certain if and only if $\forall x \in \|t\|^*. \partial_{AT}(x) = \{i\}$;
b) A decision rule $t \Rightarrow (d,i)$ is possible if and only if $\forall x \in \|t\|^*. i \in \partial_{AT}(x)$;
c) A rule $t \Rightarrow (d,i_1) \vee ... \vee (d,i_n)$ is a generalized decision rule if and only if
d) $\forall x \in \|t\|^*. d(x) \in \{i_1, ..., i_n\}$ and $\exists y \in \|t\|^*. \{i_1, ..., i_n\} \subseteq \partial_{AT}(y)$.

Proof.

Ad. (a) $t \Rightarrow (d,i)$ is certain if $\|t\|^* \subseteq \|(d,i)\|$ if $\|t\|^* \subseteq \underline{A}\|(d,i)\|$ where $A \subseteq AT$ is the set of attributes in t if $\|t\|^* \subseteq \underline{AT}\|(d,i)\|$ if $\forall x \in \|t\|^*. \partial_{AT}(x) = \{i\}$ (by Property 3.3.c); the proof of the converse is similar.

Ad. (b) $t \Rightarrow (d,i)$ is possible if $\|t\|^* \subseteq \overline{AT}\|(d,i)\|$ if $\forall x \in \|t\|^*. i \in \partial_{AT}(x)$ (by Property 3.3.d).

Ad. (c) $t \Rightarrow (d,i_1) \vee ... \vee (d,i_n)$ is a generalized decision rule if $\|t\|^* \subseteq \|(d,i_1) \cup ... \cup (d,i_n)\|$ and $\exists y \in \|t\|^*. \{i_1, ..., i_n\} \subseteq \partial_{AT}(x)$ if $\forall x \in \|t\|^*. d(x) = i_1 \vee ... \vee d(x) = i_n$ and $\exists y \in \|t\|^*. \{i_1, ..., i_n\} \subseteq \partial_{AT}(x)$ if $\forall x \in \|t\|^*. d(x) \in \{i_1, ..., i_n\}$ and $\exists y \in \|t\|^*. \{i_1, ..., i_n\} \subseteq \partial_{AT}(x)$.

Properties 3.3-3.4 are extensions of analogous properties for complete information systems [8,11].

4. Algorithms

Usually the user is interested in the rules that satisfy some conditions referring to the support and confidence. In the sequel, the rules that have their support and confidence above a user-specified threshold will be called *strong*. In this section we provide three algorithms, which generate certain, possible and generalized decision rules respectively. The algorithms allow us to specify the thresholds for support and confidence. In the algorithms the following user thresholds are provided

1. *minPSup* - the minimum pessimistic rule support in DT.
2. *minOSup* - the minimum optimistic rule support in DT.
3. *minPConf* - the minimum pessimistic rule confidence in DT.
4. *minOConf* - the minimum optimistic rule confidence in DT.

In order to guarantee a positive support in a complete extension of the incomplete system DT we assume that $minOSup \geq 1$, whereas all the other parameters are ≥ 0. If the pessimistic support threshold is set to 0, the algorithms below will generate all rules (respectively certain, possible and generalized), including the hypothetical ones; otherwise the hypothetical rules are not generated.

4.1 Certain Rules

The presented *IAprioriCertain* algorithm (see Fig. 1a) can be used for generating optimal certain rules from an incomplete system DT = $(O, AT \cup \{d\})$. The algorithm guarantees that the pessimistic and optimistic supports of the found rules are above the respective thresholds and their pessimistic and optimistic confidence is 1.

IAprioriCertain is a modification of the very efficient data mining algorithm *Apriori* [1], which is applied in mining the association rules in very large databases. The algorithm exploits the following property: if the pessimistic (optimistic) support of $t \Rightarrow s$ is below the threshold, then the pessimistic (optimistic) support of $t \wedge p \Rightarrow s$ is below the threshold, too. It is a slightly extended version of the algorithm proposed in [12]. Contrary to the previous version, the new algorithm takes into account optimistic support. In the sequel, a rule with k attribute-value pairs in the antecedent will be called *kant*-rule. For the algorithm, the attribute-value pair sets are assumed to be ordered lexicographically with regard to attributes and values. The following notation is used in the algorithm: $Inf(x) = \{(a, a(x)): a \in AT \cup \{d\}\}$; $Inf^*(x) = \{(a, a(x)): a \in AT \cup \{d\}, a(x) \neq *\} \cup \{(a,v): a \in AT, a(x) = *, v \in V_a\}$; C_k - set of candidate *kant*-rules; R_k - set of optimal certain *kant*-rules.

Each element of C_k and R_k is characterized by the following fields

1. *ant* (antecedent), which is a set of k pairs (*condition_attribute, value*).
2. *cons* (consequent), which is a pair (*decision_attribute, value*).
3. *ruleCount*, which holds the pessimistic estimation of the support of the rule.
4. *ruleOCount*, which holds the optimistic estimation of the support of the rule.
5. *certain* - a Boolean field indicating whether the rule is certain or not.

Let us describe briefly the *IAprioriCertain* algorithm. In the beginning, the function *InitRules* initializes the set C_1 of candidate rules with the rules such that for each object in DT the antecedent of the rule is taken from a pair (*condition_attribute, value*), whereas the consequent is equal to the pair (*decision_attribute, value*) of that object. If the condition attribute is unknown, a number of candidate rules are created in such a way that the antecedent takes any value from the attribute domain. Then, an iteration loop is performed, where each k-th iteration consists of the following operations:

1. In one pass over DT, pessimistic and optimistic supports (*ruleCount* and *ruleOCount* respectively) for the candidate rules in C_k are determined. At the same time the candidate rules that are not certain are identified.
2. Each candidate rule in C_k whose pessimistic or optimistic support does not exceed the minimum supports (pessimistic or optimistic, respectively) is discarded, or if a given candidate is a certain rule, it is moved from C_k to the set of optimal certain rules R_k. The remaining rules in C_k are not certain.
3. The function *AprioriRuleGen* is called to generate the candidate $(k+1)ant$-rules C_{k+1} from the remaining rules in C_k. First, the function constructs candidate $(k+1)ant$-rules by merging pairs of rules in C_k, which have the same consequent. Then, each $t \Rightarrow s$ in C_{k+1} for which a *kant*-rule $t' \Rightarrow s$, $t' \subset t$, is not present in C_k, is pruned.

The iteration is stopped when the set C_k for a given k is empty. For a detailed description of *AprioriRuleGen* see [7].

Algorithm *IAprioriCertain*
$C_1 = InitRules(DT)$;
for ($k = 1$; $C_k \neq \varnothing$; k++) **do begin**
  forall candidates $c \in C_k$ **do begin**
    $c.ruleCount = c.ruleOCount = 0$;
    $c.certain =$ **true**;
  endfor;
  forall objects x in DT **do**
    forall candidates $c \in C_k$ **do**
      if $c.ant \subset Inf^*(x)$ **then**
        if $c.cons \subset Inf(x)$ **then**
          $c.ruleOCount$++;
          if $c.ant \subset Inf(x)$ **then**
$c.ruleCount$++;
        else
          $c.certain =$ **false**;
        endif;
      endif;
    endfor;
  endfor;
  $R_k = \varnothing$;
  forall candidates $c \in C_k$ **do**
    if $c.ruleCount < minPSup$ **or**
      $c.ruleOCount < minOSup$ **then**
      delete c from C_k;
    elseif $c.certain =$ **true then**
      move c from C_k to R_k;
    endif;
  endfor;
  $C_{k+1} = AprioriRuleGen(C_k)$;
endfor;
return $\cup_k R_k$;

function *InitRules*(decision table DT);
$C_1 = \varnothing$;
forall objects x in DT **do**
  $C_1 = C_1 \cup \{c \notin C_1 | \ c.cons=(d,d(x))$ **and** $a \in AT$,
    $c.ant=(a,a(x))$ if $a(x) \neq *$ **or**
    $c.ant=(a,v), v \in V_a$ otherwise$\}$;
endfor;
return C_1;

function *AprioriRuleGen*(C_k);
$C_{k+1} = \{(ant, cons)| f,c \cup C_k, cons = f.cons$
$= c.cons$,
  $ant = f.ant[1] \cup f.ant[2] \cup ...$

Algorithm *IAprioriPossible*
  forall objects $x \in$ DT **do**
    compute generalized decision value $\partial_{AT}(x)$;
  endfor;
  $C_1 = InitRules(DT)$;
  for ($k = 1$; $C_k \neq \varnothing$; k++) **do begin**
    forall candidates $c \in C_k$ **do**

$c.ruleCount = c.ruleOCount = c.antCount =$
      $c.negRuleOCount = 0$;
      $c.possible =$ **true**;
    endfor;
    forall objects x in DT **do**
      forall candidates $c \in C_k$ **do**
        if $c.ant \subset Inf^*(x)$ **then**
          if $c.cons \in Inf(x)$ **then**
            $c.ruleOCount$++;
          else
            $c.negRuleOCount$++;
          /* $x \notin \overline{AT} \ ||(d, (c.cons).value|| \ ?$ */
          if $(c.cons).value \notin \partial_{AT}(x)$ **then**
            $c.possible =$ **false**;
          endif;
          if $c.ant \subset Inf(x)$ **then**
            $c.antCount$++;
            if $c.cons \in Inf(x)$ **then**
$c.ruleCount$++;
          endif;
        endif;
      endfor;
    endfor;
    $R_k = \varnothing$;
    forall candidates $c \in C_k$ **do**
      if $c.ruleCount < minPSup$ **or**
        $c.ruleOCount < minOSup$ **then**
        delete c from C_k;
      else
        $c.pConf = c.ruleCount \ /$
          $(c.ruleCount +$
$c.negRuleOCount)$;
        $c.oConf = c.ruleOCount \ /$
          $(c.antCount+c.ruleOCount-$ $c.ruleCount)$;
        if $c.possible$ **and**
          $c.pConf \geq minPConf$ **and**
          $c.oConf \geq minOConf$ **then**

```
f.ant[k]∪c.ant[k],                          move c from C_k to R_k;
  f.ant[1]=c.ant[1] and...and f.ant[k-        endif;
1]=c.ant[k-1]                               endfor;
  and (f.ant[k]).attribute <                C_{k+1} = AprioriRuleGen(C_k);
(c.ant[k]).attribute};                      endfor;
 forall c ∈C_{k+1} do                       return ∪_k R_k;
  if  |{f∈C_k  :  f.cons=c.cons  and
f.ant⊂c.ant}| < k+1
   then delete c from C_{k+1};
 endfor;
return C_{k+1};
```

Fig. 1. Rule generation algorithms: (a) *IAprioriCertain*; (b) *IAprioriPossible*

4.2 Possible Rules

Now we present the algorithm *IAprioriPossible* (see Fig. 1b). The algorithm extends the algorithm proposed in [12] in that it allows specifying conditions imposed not only on pessimistic support and confidence of possible rules to be found but also on their optimistic support and confidence. It generates optimal possible rules. The notation in *IAprioriPossible* is similar to the one used in *IAprioriCertain*. All the rule fields used in *IAprioriCertain*, except for *certain*, are also used in this algorithm. In addition new fields are introduced:

6. *possible* - a Boolean field indicating whether the rule is possible or not.
7. *antCount*, which holds the pessimistic estimation of the support of the rule's antecedent.
8. *negRuleOCount*, which holds the number of objects that possibly violate the candidate rule.
9. *pConf* - pessimistic confidence of the rule.
10. *oConf* - optimistic confidence of the rule.

The values of *pConf* and *oConf* are derived according to Prop. 3.1 (c-d). The Boolean field *possible* indicates whether a respective upper approximation is preserved. General structure of the algorithm *IAprioriPossible* is the same as for *IAprioriCertain*. The main difference consists in additional checking (according to Property 3.4 (b) if a candidate rule is possible (i.e. whether it preserves an appropriate upper approximation) and if its pessimistic and optimistic supports are sufficiently high.

4.3 Generalized Decision Rules

We propose here the algorithm *IAprioriGeneralized* (see Fig. 2), which generates optimal generalized decision rules. The notation in *IAprioriGeneralized* is similar to the one used in *IAprioriCertain*. The rule fields used in *IAprioriCertain*, except for *certain*, are also used in this algorithm. Also the meaning of the fields in this algorithm is the same as before, except for the field *cons*; in this algorithm *cons* is devoted to store a set of decision values of objects that are likely to satisfy the antecedent *ant* of the rule. In addition new fields will be used:

11. *expCons* - expected consequent of the rule, which is a set of decision values.

12. *optExpCons* - a Boolean field indicating that the candidate rule with *cons* equal to *expCons* is an optimal generalized decision rule.
13. ∂_{AT} - the field holds an original generalized decision of an object in DT supporting the candidate rule.

Now, let us describe the *IAprioriGeneralized* algorithm. At the beginning the function *InitIndefRules* initializes the set C_1 of candidate rules with the rules, so that for each object in DT (1) the antecedent of the rule is taken from a pair (*condition_attribute, value*); (2) the anticipated consequent *expCons* and the field $\square_{AT}$ are set to the generalized decision value of that object; (3) *optExpCons* is set to *true* meaning that any candidate rule in C_1 with the actual consequent *cons* not greater than *expCons* is optimal generalized. Then, an iteration loop is performed, where each k-th iteration consists of the following operations

1. Pessimistic and optimistic supports (*ruleCount* and *ruleOCount* respectively), as well as, the actual consequent (*cons*) for the candidate rules in C_k are determined in one pass over DT. As a result, *cons* contains decision values of all objects that possibly satisfy the antecedent of the rule.

2. All candidate rules in C_k whose pessimistic or optimistic supports do not exceed the minimum supports (pessimistic or optimistic, respectively) are discarded.

3. The remaining candidate rules with single decision in the current consequent of the rule are moved from C_k to the set of optimal generalized decision rules R_k (more precisely, if there is a number of single decision rules with the same *ant* and *cons*, then only one of them is moved to R_k and the others are discarded).

4. The remaining rules in C_k whose actual consequent *cons* is less than *expCons* are optimal generalized. Also rules with actual consequent equal to *expCons* for which *optExpCons* is *true* are optimal generalized. Optimal generalized decision rules are copied to the set R_k (actually, only one rule is copied for each group of rules with the same *ant* and *cons*, the one having the least ∂_{AT} in a group). Optimal generalized decision rules with non-single consequents remain in C_k - they will be used in the construction of longer candidates, as there is still a chance to obtain optimal generalized decision rules with shorter consequent.

5. Now the function *AprioriIndefRuleGen* is called in order to generate the candidate $(k+1)$ant-rules C_{k+1} from C_k. The function constructs candidate $(k+1)$ant-rules by merging pairs of rules in C_k, with the same value of ∂_{AT}. Then, for each rule r in C_{k+1} the set of sub-rules F is determined as the subset of those rules in C_k that have the antecedent contained in r and the same ∂_{AT}. If cardinality of F is smaller than the number of all *k*ant-sub-rules for r (i.e. $|F| < k+1$) is pruned. For each remaining r in C_{k+1} the fields *expCons* and *optExpCons* are determined according to the following observation: anticipated consequent of any optimal generalized decision rule cannot be greater than ∂_{AT} and cannot be greater than any actual consequent of any sub-rule in F. If there is a sub-rule in F whose actual consequent is equal to the anticipated consequent of a given candidate rule r, then such a rule is a generalized decision rule. If it turns out that the consequent of r is equal to its anticipated consequent, then r is not optimal, since there is a generalized decision sub-rule of r with the same actual consequent. In this case the field *optExpCons* has to be set to *false*. If there is no sub-rule in F whose actual consequent is equal to anticipated consequent of a given candidate rule r, then *optExpCons* is set to *true*.

Algorithm *IAprioriGeneralized*
  forall objects $x \in$ DT **do**
    compute generalized decision value $\partial_{AT}(x)$;
  endfor;
  C_1 = *InitIndefRules*(DT);
  for ($k = 1$; $C_k \neq \emptyset$; k++) **do**
    forall candidates $c \in C_k$ **do**
      $c.ruleCount = c.ruleOCount = 0$;
      $c.cons = \emptyset$;
    endfor;
    forall objects x in DT **do**
      forall candidates $c \in C_k$ **do**
        if $c.ant \subset Inf^*(x)$ **then**
          $c.cons = c.cons \cup \{d(x)\}$;
          if $d(x) \in c.expCons$ **then**
            $c.RuleOCount$++;
            if $c.ant \subset Inf(x)$ **then**
$c.ruleCount$++;
          endif;
        endif;
      endfor;
    endfor;
    $R_k = \emptyset$;
    forall candidates $c \in C_k$ **do**
      if $c.ruleCount < minPSup$ **or**
      $c.ruleOCount < minOSup$ **then**
        delete c from C_k;
    endfor;
    forall candidates $c \in C_k$ **do**
      if $c.cons$ is a singleton set **then**
        move c from C_k to R_k;
        delete all $c' \in C_k$ such that $c'.ant=c.ant$
        and $c'.cons = c.cons$;
      endif;
    endfor;
    forall candidates $c \in C_k$ **do**
      if {($c.cons \subset c.expCons$) **or**
      ($c.optExpCons$ = **true** and
      $c.cons = c.expCons$)} and
      ($\neg \exists c' \in C_k$, $c'.ant=c.ant$ **and** $c'.\partial_{AT} \subset$
$c.\partial_{AT}$)   **then**
        copy c from C_k to R_k;
    endfor;
    C_{k+1} = *AprioriIndefRuleGen*(C_k);
  endfor;
  return $\cup_k R_k$;

function *InitIndefRules*(decision table DT);
  $C_1 = \emptyset$;
  forall objects x in DT **do**
    $C_1 = C_1 \cup \{c \notin C_1 |\ c.expCons = c.\partial_{AT} = \partial_{AT}(x)$
      and $a \in AT$, $c.ant=(a,a(x))$ if $a(x) \neq *$ **or**
        $c.ant=(a,v)$, $v \in V_a$ otherwise$\}$;
  endfor;
  forall candidates $c \in C_k$ **do**
    $c.optExpCons =$ **true**;
  endfor;
  return C_1;

function *AprioriIndefRuleGen*(C_k);
  $C_{k+1} = \{(ant, \partial_{AT}): f, c \cup C_k, \partial_{AT} = f.\partial_{AT} = c.\partial_{AT}$,
    $ant = f.ant[1] \cup f.ant[2] \cup ... \cup$
$f.ant[k] \cup c.ant[k]$,
    $f.ant[1]=c.ant[1]$ **and**...**and** $f.ant[k-1]=c.ant[k-1]$
    and $(f.ant[k]).attribute <$
$(c.ant[k]).attribute\}$;
  /* Rules pruning */
  forall $c \in C_{k+1}$ **do**
    $F = \{f \in C_k: f.\partial_{AT} = c.\partial_{AT}$ and $f.ant \subset c.ant\}$;
    if $|F| < k+1$ **then**
      delete c from C_{k+1};
    else
      $c.expCons = c.\partial_{AT} \cap (\cap_{f \in F} f.cons)$;
      if $\exists f \in F$, $c.expCons = f.cons$ **then**
        $c.optExpCons =$ **false**
      else
        $c.optExpCons =$ **true**;
      endif;
    endif;
  endfor;
  return C_{k+1};

Fig. 2 The *IAprioriGeneralized* algorithm

5 Conclusions

In the presented approach, rules generated from an incomplete system are not contradictory with any complete extension of the system, and for each rule generated there is at least one complete extension that supports the rule. The presented algorithms are efficient in that they generate such rules directly from the incomplete system without referring to its extensions.

The algorithms allow specifying the pessimistic and optimistic support and confidence thresholds. Setting the minimum pessimistic support to 0 the user allows for hypothetical rules generation. Otherwise only rules directly supported in the system are generated. The algorithms do not generate hypothetical rules for the complete system regardless of the value of minimum pessimistic support.

We have shown that hypothetical rules in an incomplete system resolve or at least decrease indefiniteness of a rule which is directly derivable from the system and has a positive support. On the other hand such a hypothetical rule has a larger antecedent part in comparison with the corresponding rule supported by the system. At least one among additional conditions in the antecedent of the hypothetical rule refers to an attribute with unknown values in the original system. By acquiring missing knowledge some hypothetical rules may get support.

References

[1] Agraval R., Mannila H., Srikant R., Toivonen H., and Verkamo A.I., Fast Discovery of Association Rules, in: Fayyad, U.M., Piatetsky-Shapiro, G. , Smyth, P. , and Uthurusamy, R. (eds.), *Advances in Knowledge Discovery and Data Mining*, AAAI Press, 1996, pp. 307-328.

[2] Chmielewski M.R., Grzymala-Busse J.W., Peterson N.W., and Than S., The Rule Induction System LERS - A Version for Personal Computers, *Foundations of Computing and Decision Sciences* 18(3-4), 1993, pp. 181-212.

[3] Deogun J.S., Raghavan V.V., Sarkar A., and Sever H., Data Mining: Trends in Research and Development, in: Lin, T.Y. and Cercone, N. (eds.), *Rough Sets and Data Mining*, Kluwer Academic Publishers, 1997, pp. 9-45.

[4] Greco S., Matarazzo B., Słowiński R., and Zanakis S., Rough Set Analysis of Data Tables with Missing Values, in: *Proceedings: the 5th Intern. Conference of the Decision Sciences Institute*, DSI '99, Athens, Greece, Vol. 2 , 1999, pp. 1359-1362.

[5] Kononenko, I., Bratko, I., and Roskar, E., Experiments in Automatic Learning of Medical Diagnostic Rules, *Technical Report*, Jozef Stefan Institute, Ljubljana, 1984.

[6] Kryszkiewicz, M., Rough Set Approach to Incomplete Information Systems, *Journal of Information Sciences* 112, 1998, pp. 39-49.

[7] Kryszkiewicz, M., Properties of Incomplete Information Systems in the Framework of Rough Sets, in: Polkowski, L. and Skowron, A. (eds.), *Rough Sets in Knowledge Discovery 1*, Physica-Verlag, Heidelberg, 1998, pp. 442-450.

[8] Kryszkiewicz, M., Strong Rules in Large Databases, in: *Proceedings*: IPMU '98, Paris, France, Vol. 2, 1998, pp. 1520-1527.

[9] Kryszkiewicz, M., Rules in Incomplete Information Systems, *Journal of Information Sciences* 113, 1999, pp. 271-292.

[10] Kryszkiewicz, M., Association Rules in Incomplete Databases, in: *Proceedings: the 3rd Pacific-Asia Conference*, PAKDD '99, Beijing, China, LNAI 1574, *Methodologies for Knowledge Discovery and Data Mining*, Springer –Verlag, Berlin, 1999, pp. 84-93.

[11] Kryszkiewicz M. and Rybiński H., Strong Rules in Large Databases, in: *Proceedings*: EUFIT '98, Aachen, Germany, Vol. 1. 1998, pp. 85-89.

[12] Kryszkiewicz M and Rybiński H., Incompleteness Aspect in Rough Set Approach, in: *Proceedings: the Intern. Joint Conference of Information Sciences*, Raleigh NC, Vol. 2, 1998, pp. 371-374.

[13] Kryszkiewicz M. and Rybiński H., Incomplete Database Issues for Representative Association Rules, in: *Proceedings: the 11th International Symposium*, ISMIS '99, Warsaw, Poland, LNAI 1609, *Foundations of Intelligent Systems*, 1999, pp. 583-591.

[14] Pawlak Z., *Rough Sets: Theoretical Aspects of Reasoning about Data*, Kluwer Academic Publishers, 1991.

[15] Quinlan J.R., Induction of Decision Trees, in: Shavlik J. W. and Dietterich T. G. (eds.), *Readings in Machine Learning*, Morgan Kaufmann Publishers, 1990, 57-69.

[16] Skowron A. and Rauszer C., The Discernibility Matrices and Functions in Information Systems, in: Słowiński R. (ed.), *Intelligent Decision Support: Handbook of Applications and Advances of Rough Sets Theory*, Kluwer Academic Publishers, 1992, pp. 331-362.

[17] Słowiński R. and Stefanowski J., Rough-Set Reasoning about Uncertain Data, *Fundamenta Informaticae* 27 (2-3), 1996, pp. 229-244.

PART 5:

AFTERWORD

Chapter 12

Rough Sets and Rough Logic: A KDD Perspective

Zdzisław Pawlak[1], *Lech Polkowski*[2,] *and Andrzej Skowron*[3]

[1] University of Information Technology and Management
Newelska 6, 01–447 Warsaw, Poland

[2] Polish-Japanese Institute of Information Technology
Koszykowa 86, 02–008 Warszawa, Poland
and
Department of Mathematics and Information Sciences
Warsaw University of Technology
Pl. Politechniki 1, 00–665 Warszawa, Poland

[3] Institute of Mathematics Warsaw University
Banacha 2, 02–097 Warszawa, Poland
emails: zpw@ii.pw.edu.pl; polkow@pjwstk.waw.pl; skowron@mimuw.edu.pl

Abstract. Basic ideas of rough set theory were proposed by Zdzisław Pawlak [90, 91] in the early 1980's. In the ensuing years, we have witnessed a systematic, world-wide growth of interest in rough sets and their applications. There are numerous areas of successful applications of rough set software systems [101]. Many interesting case studies are reported (for references see e.g., [100, 101], [87] and the bibliography in these books, in particular [19], [46], [57], [132], [146]).

The main goal of rough set analysis is induction of approximations of concepts. This main goal is motivated by the basic fact, constituting also the main problem of KDD, that languages we may choose for knowledge description are incomplete with respect to expressibility. A fortiori, we have to describe concepts of interest (features, properties, relations etc.) known not completely but by means of their reflections (i.e., approximations) in the chosen language. The most important issues in this induction process are:

- construction of relevant primitive concepts from which approximations of more complex concepts are assembled,
- measures of inclusion and similarity (closeness) on concepts,
- construction of operations producing complex concepts from the primitive ones.

Basic tools of rough set approach are related to concept approximations. They are defined by approximation spaces. For many applications, in particular for KDD problems, it is necessary to search for relevant approximation spaces in the large space of parameterized approximation spaces. Strategies for tuning parameters of approximation spaces are crucial for inducing concept approximations of high quality.

Methods proposed in rough set approach are kin to general methods used to solve Knowledge Discovery and Data Mining (KDD) problems like feature selection, feature extraction (e.g., discretization or grouping of symbolic value), data reduction, decision rule generation, pattern extraction (templates, association rules), or decomposition of large data tables. In this Chapter we examine rough set contributions to Knowledge Discovery from the perspective of KDD as a whole.

This Chapter shows how several aspects of the above problems are solved by the classical rough set approach and how they are approached by some recent extensions to the classical theory of rough sets. We point out the role of Boolean reasoning in solving discussed problems. Rough sets induce via its methods a specific logic, which we call rough logic. We also discuss rough logic and related logics from a wider perspective of logical approach in KDD. We show some relationships between these logics and potential directions for further research on rough logic.

Keywords: concept, concept approximation, rough sets, rough mereology, knowledge discovery and data mining, reducts, patterns, decision and association rules, feature selection and extraction, classical deductive systems, logics for reasoning under uncertainty, rough logic.

1 Rough logic: a perspective on logic in KDD

Logic understood as a study of mechanisms of inference, involving inference about knowledge from data, has evolved into many deductive schemes differing by understanding of semantics of inference $p \vdash q$. In this Chapter, basic schemes are outlined: classical calculi, many–valued logics, modal logics along with deductive mechanisms: axiomatized schemes, natural deduction, resolution, sequent calculus. Also outlined are various approaches to reasoning in inconsistent situations: belief revision, non–monotonic logics. In complex tasks of AI and KDD like pattern recognition or machine learning, inductive reasoning is more frequent in applications aimed at defining relevant concepts and dependencies among them. Examples of fuzzy logic, rough logic, mereological logic are presented. However, all these logical systems have one drawback. They allow to reason about properties of given constructs, like concept approximations, but they do not offer efficient strategies for synthesis of such constructs which is crucial for KDD applications. A step towards this goal is offered by rough mereology.

A general view of the logical process for KDD can be outlined as follows. First, relevant relational structures and a suitable language allowing to express relevant properties (features) of these structures should be discovered from data. These in turn are used to construct solutions in the form of concept approximations. The term *relevant* is relative to a given specified task, e.g., classification. This process is in particular related to problems of feature extraction and selection and in general to discovering of relevant data models. The aim is to discover data models with minimal description length [109] still relevant for solving a given task. Certainly, such models should preserve the properties of data important from the point of view of the considered task and should contain as little as possible information not relevant with respect to that task. The great effort has been made in such areas like Machine Learning, Pattern Recognition, or KDD to develop methods discovering relevant data models. However, much more should be done to solve complex tasks, e.g., related to spatial reasoning [110]. In particular, logic of perception should be developed.

Perception [40], description and analysis of real life phenomena has been a dominant feature in intellectual activity of a human being; a fortiori, this activity has been assigned to machine systems exploiting various computing paradigms. Perception refers to the construction of knowledge from sensory data for subsequent usage in reasoning [55]. Observation of a real life phenomenon may be passive or active: by the former we mean perceiving the phenomenon and rendering its impression possible while by the latter we mean the process in which we create tools for a quantitative description of the phenomenon in terms of measurements, recordings, expert's knowledge etc. This latter process leads to the record of the phenomenon in the form of data. Data may therefore be of many various types: numerical data, symbolic data, pattern data including time series data, etc. These types of data may be further made into more complex types, e.g., arrays of numerical data or audio or video series (e.g., documentary films). The choice of a way of data collecting as well as a type of data depends on a particular problem which we are going to solve about the given phenomenon; this data elicitation process may have a great complexity and it is thoroughly studied by many authors [42]. Data elicited from a phenomenon should undergo a representation process in which they are modeled by a certain structure. This structure allows for efficient knowledge representation and reasoning about it in order to solve some queries or problems. The relationship of data and knowledge, in particular how knowledge can be acquired from data, has attracted attention of many philosophers and logicians cf. e.g., [112].

From logical point of view, data represent a model of a phenomenon, i.e., a set of entities arranged in a certain space – time structure. Clearly, a real phenomenon may have associated with it many distinct data structures.

Usually, the nature of the phenomenon suggests us certain primitive concepts, i.e., sets of entities in data structure in terms of which we build more complex concepts and we carry reasoning about the phenomenon. The properties of data structures, concepts and their relationships or actions were found to be abstracted best by means of logics. A logic involves a set of formulas (i.e. well–formed expressions in a symbolic language) along with a family of relational structures (models) in which formulas are interpreted as concepts (i.e., sets of entities) of various relational complexities as well as a mechanism allowing us to reason about properties of models. The choice of the language of formulas may be critical: on one hand, this language should be expressive enough to render all essential concepts in data structures. On the other hand, too expressive a language may cause too high complexity of inference process (the phenomenon of language bias in Machine Learning, Pattern Recognition, KDD [42], [73], [78],

[29]). The primitive data structures constructed according to a selected way(s) of recording a phenomenon present itself as possible models for various logics. Strategies for discovery of a particular logic, relevant for problems to be solved, present a challenge for KDD.

In fitting a logic to a data structure, some important intermediate steps are to be taken:

- in the data structure, certain sets of entities (concepts) and relationships among them are selected giving admissible relational structures in data;
- mechanisms of inference about properties of these admissible structures are selected (e.g., some deductive systems (see below)).

Inference mechanisms may provide us with descriptions of complex concepts hidden in data structures; for various reasons, formulas of logic may not describe concepts in data structures exactly but approximately only: one of reasons is that we may not know exactly the concept in question (we usually know some positive or negative examples of this concept – this is quite often a case for Machine Learning, Data Mining and Knowledge Discovery). This fact makes it necessary to invoke in addition to deductive systems also inductive ones allowing us to build inference models from sets of examples. This leads to new logical systems for approximate reasoning allowing us to carry out reasoning about properties of data structures on the basis of uncertain, incomplete or insufficient data [*logics for reasoning under uncertainty*]. In these logics we encounter various phenomena not experienced by classical logics like non–monotonicity, necessity of belief revision etc. (see below). When a logic is selected which approximately fits a data structure, this logic becomes an inference engine for using knowledge about a given phenomenon.

2 Concepts

In general, the idea of a *concept* is associated with a set of entities; given a set U of entities, we call a concept any subset of (the universe set) U. For instance, a concept may be any subset of the domain D of objects listed in data table. This notion of a concept is what may be called a *crisp* (*theoretical*) *concept*: the subset is understood here in the classical sense, i.e., for each element of U we can decide whether it is in X or not. However, concepts in data structures are often *non-crisp* (*vague*): there are elements about which we cannot determine their membership in X with certainty. A typical example is provided by concepts expressed in natural language, e.g., *high*: it may be a matter of dispute whether a man of height 175 cm is high or not. Also concepts known by examples only (observational concepts) are such. To cope with such concepts various theories have been proposed, e.g., fuzzy set theory [138], rough set theory [91], multi-valued logics [108] etc.

Concept description may be two–fold: *syntactical* as well as *semantic*. In classical case, syntactical description of a concept is provided by a formula of a logic. Semantic description relative to a model (the concept meaning) is provided by the meaning of this formula, i.e., a set of entities in a model which satisfy this formula. This becomes more complex in non–classical cases, e.g., for fuzzy or rough concepts where models learned from training data have to be tested against new cases.

A concept may be also characterized with respect to a set of formulas: given a crisp concept X and a set F of formulas, the *intension* of X with respect to F is the subset F' of F consisting of those formulas which are satisfied by each of elements of X; assigning to a subset F' of F the family of all elements which satisfy each formula in F', we obtain a set (concept) called the *extension* of F' (in particular, we may start with X and find the extension X' of the intension F' of X; here we work in the frame of Galois connections [136]). This idea becomes more complicated in case of non–crisp concepts where a formula is satisfied by an object in a degree usually less than 1.

In many applications there is the need for analyzing dynamic structures involving changes of situations (concepts) by actions; from our point of view, actions are binary relations on concepts, i.e., sets of pairs (pre–condition, post–condition) [113].

Reasoning about crisp concepts may be carried out by means of *classical deductive systems*. In non–classical cases, where observational concepts are approximated by theoretical concepts, an important additional ingredient in reasoning is provided by some measures of similarity (distance) among the concept and its approximation as well as by some mechanisms for propagation of these closeness measures or uncertainty coefficients [139]. In many applications one uses common–sense non–deductive reasoning [69] and proper knowledge representation involves ingredients from logic as well as other tools like procedural representation schemes, semantic networks or frames for representing common–sense knowledge [112].

2.1 Rough sets: General view on concepts and their approximations

In the simplest case the data exploited by rough set analysis are represented in the form of attribute–value tables, see Chapter 1. Such tables are also used by Pattern Recognition, Machine Learning and KDD. These simple tables are combined when forming hierarchical or distributed data structures.

A data set can be represented by a table where each row represents, for instance, an object, a case, or an event. Every column represents an attribute, or an observation, or a property that can be measured for each object; it can also be supplied by a human expert or user. This table is called an *information system*.

The choice of attributes is subjective (they are often called *conditional attributes*) and reflects our intuition about factors that influence the classification of objects in question. The chosen attributes determine in turn primitive descriptors that provide *intensions* of primitive concepts.

In many cases the target of the classification, that is, the family of concepts to be approximated is represented by an additional attribute d called *decision*.

Information systems of this kind are called *decision systems* and they are written down as triples $\mathcal{A} = (U, A, d)$.

We try in this case to approximate concepts that are defined by the known decision. This is known in Machine Learning as *supervised learning*.

In some applications it may be necessary to work with conditional attributes which are compound in the sense that they depend on other simpler attributes which in turn depend on other attributes, etc. (e.g., this happens when the values of attributes are complex structural objects, like images or algorithms to be performed). In this case it is necessary to work with hierarchical data tables. A still more general case occurs when the data is distributed between a number of processing units (agents). We have to work then in a *multi-agent* environment. In such a case the process of induction of approximations to concepts is more complicated. This is due to the necessity of adding interface mechanisms that translate concepts and their similarity degrees from agent to agent. These problems are addressed by rough mereology (for references see [99]).

Indiscernibility

A decision system expresses all currently available knowledge about the objects in question. Such a table may be unnecessary large: some objects may be indiscernible or some attributes redundant.

Let $\mathcal{A} = (U, A)$ be an information system; then, with any $B \subseteq A$ there is associated an equivalence relation $IND_{\mathcal{A}}(B)$:

$$IND_{\mathcal{A}}(B) = \{(x, x') \in U^2 \mid \forall a \in B \; a(x) = a(x')\}$$

$IND_{\mathcal{A}}(B)$ is called the *B-indiscernibility relation*, its classes are denoted by $[x]_B$. By X/B we denote the partition of U defined by the indiscernibility relation $IND(B)$. See Chapter 1 for more details.

Lower and upper approximation of sets, boundary regions, positive regions

We have already mentioned that vague concepts may be only approximated by crisp concepts; these approximations are recalled now.

Let $\mathcal{A} = (U, A)$ be an information system, $B \subseteq A$, and $X \subseteq U$. We can approximate X using only the information contained in B by constructing the *B-lower* and *B-upper approximations of* X, denoted $\underline{B}X$ and $\overline{B}X$ respectively, where $\underline{B}X = \{x \mid [x]_B \subseteq X\}$ and $\overline{B}X = \{x \mid [x]_B \cap X \neq \emptyset\}$.

The lower approximation induces *certain rules* while the upper approximation induces *possible rules* (i.e., rules with confidence greater than 0). The set $BN_B(X) = \overline{B}X - \underline{B}X$ is called the *B-boundary region of* X thus consisting of those objects that on the basis of the attribute set B cannot be unambiguously classified into X. The set $U - \overline{B}X$ is called the *B-outside region of* X and consists of those objects which can be with certainty classified as not belonging to X on the basis of the attribute set B. A set (concept) X is said to be *rough* (respectively, *crisp*) if the boundary region of X is non–empty (respectively, empty).

The following properties of approximations can easily be verified:

(1) $\underline{B}(X) \subseteq X \subseteq \overline{B}(X)$,
(2) $\underline{B}(\emptyset) = \overline{B}(\emptyset) = \emptyset, \underline{B}(U) = \overline{B}(U) = U$,
(3) $\overline{B}(X \cup Y) = \overline{B}(X) \cup \overline{B}(Y)$,
(4) $\underline{B}(X \cap Y) = \underline{B}(X) \cap \underline{B}(Y)$,
(5) $X \subseteq Y$ implies $\underline{B}(X) \subseteq \underline{B}(Y)$ and $\overline{B}(X) \subseteq \overline{B}(Y)$,
(6) $\underline{B}(X \cup Y) \supseteq \underline{B}(X) \cup \underline{B}(Y)$,
(7) $\overline{B}(X \cap Y) \subseteq \overline{B}(X) \cap \overline{B}(Y)$,
(8) $\underline{B}(-X) = -\overline{B}(X)$,
(9) $\overline{B}(-X) = -\underline{B}(X)$,
(10) $\underline{B}(\underline{B}(X)) = \overline{B}(\underline{B}(X)) = \underline{B}(X)$,
(11) $\overline{B}(\overline{B}(X)) = \underline{B}(\overline{B}(X)) = \overline{B}(X)$,

where $-X$ denotes $U - X$.

One can single out the following four basic classes of rough sets:

a) X is *roughly B-definable* iff $\underline{B}(X) \neq \emptyset$ and $\overline{B}(X) \neq U$,
b) X is *internally B-undefinable* iff $\underline{B}(X) = \emptyset$ and $\overline{B}(X) \neq U$,
c) X is *externally B-definable* iff $\underline{B}(X) \neq \emptyset$ and $\overline{B}(X) = U$,
d) X is *totally B-undefinable* iff $\underline{B}(X) = \emptyset$ and $\overline{B}(X) = U$.

These categories of vagueness have a clear intuitive meaning.

It is important to note that for many problems, in particular in KDD, in synthesis of concept approximation one should take into account a fact that usually information about the concept is incomplete. In reality, for complexity or cost of information gathering reasons, we are given an information system (U, A) which presents to us a part of the potential knowledge about the problem which may be encoded as an information system of the form (U^∞, A^∞) where U^∞ (the universe of all objects pertaining to the problem as well as A^∞ (the set of all available attributes) may be countably infinite. The approximation of a concept $X \subseteq U^\infty$ is constructed then on the basis of a sample $X \cap U$ of objects from $U \subseteq U^\infty$. Hence, one should induce approximation of $X \cap U$ in such a way that its extension in U^∞ will approximate X with sufficiently high quality. We will discuss this problem later.

Measures of quality of concept approximation and measures of inclusion and closeness of concepts

Let us consider an example of quality measure of approximations.

Accuracy of approximation. A rough set X can be characterized numerically by the following coefficient

$$\alpha_B(X) = \frac{|\underline{B}(X)|}{|\overline{B}(X)|},$$

called the *accuracy of approximation*, where $|X|$ denotes the cardinality of $X \neq \emptyset$. Obviously $0 \leq \alpha_B(X) \leq 1$. If $\alpha_B(X) = 1$, X is *crisp* with respect to B (X is exact with respect to B), and otherwise, if $\alpha_B(X) < 1$, X is *rough* with respect to B (X is *vague* with respect to B).

We now present some examples of inclusion and closeness measures. These are *rough membership functions, measure of positive region*, and *dependencies in a degree*. These notions are instrumental in evaluating the strength of rules and closeness of concepts as well as being applicable in determining plausible reasoning schemes [98], [103]. Important role is also played by entropy measures (see, e.g., [25] and Chapter by ŚLĘZAK).

Rough membership functions. In classical set theory either an element belongs to a set or it does not. The corresponding membership function is the characteristic function of the set, i.e., the function which takes values 1 and 0, respectively, depending on whether its current argument is in the set or not. In the case of rough sets the notion of membership is different. The *rough membership function* quantifies the degree of relative overlap between the set X and the equivalence class to which the current argument belongs. It is defined as follows:

$$\mu_X^B(x) : U \longrightarrow [0,1] \text{ and } \mu_X^B(x) = \frac{|[x]_B \cap X|}{|[x]_B|}$$

The rough membership function can be interpreted as a frequency–based estimate of $\Pr(y \in X \mid u)$, the conditional probability that object y belongs to set X, given knowledge of the information signature $u = Inf_B(x)$ of x with respect to attributes B. The value $\mu_X^B(x)$ measures degree of inclusion of $\{y \in U : Inf_B(x) = Inf_B(y)\}$ in X.

Positive region and its measure. Assuming that $X_1, \ldots, X_{r(d)}$ are decision classes of a decision system $\mathcal{A}$, the set $\underline{B}X_1 \cup \ldots \cup \underline{B}X_{r(d)}$ is called the *B–positive region of* $\mathcal{A}$ and is denoted by $POS_B(d)$. The number $|POS_B(d)|/|U|$ measures a degree of inclusion of the partition defined by attributes from B into the partition defined by the decision.

Dependencies in a degree. Another important issue in data analysis is discovering dependencies among attributes. Intuitively, a set of attributes D depends totally on a set of attributes C, denoted $C \Rightarrow D$, if all values of attributes from D are uniquely determined by values of attributes from C. In other words, D depends totally on C, if there exists a functional dependency between values of D and C. Dependency can be formally defined in the following way cf. Chapter 1.

Let D and C be subsets of $\mathcal{A}$. We will say that D *depends on* C *to a degree* k ($0 \leq k \leq 1$), denoted $C \Rightarrow_k D$, if

$$k = \gamma(C, D) = \frac{|POS_C(D)|}{|U|},$$

where $POS_C(D) = POS_C(d_D)$.

Obviously

$$\gamma(C,D) = \sum_{X \in U/D} \frac{|\underline{C}(X)|}{|U|}.$$

If $k=1$ we say that D *depends totally* on C, and if $k<1$, we say that D *depends partially* (to a *degree* k) on C. $\gamma(C,D)$ describes the closeness of the partition U/D and its approximation with respect to conditions in C.

The coefficient k expresses the ratio of all elements in the universe which can be properly classified to blocks of the partition U/D by employing attributes C. It will be called the *degree of the dependency*.

For two non–empty concepts $X, Y \subseteq U$, one can define their closeness by $min\left(|X \cap Y|/|X|, |X \cap Y|/|Y|\right)$.

In [119] the reader can find examples of how to construct inclusion and closeness measures for complex information granules. They are used to build calculi on information granules for solving complex problems, in particular for spatial reasoning [110].

All the measures mentioned above are constructed on the basis of the available attributes. Two important problems are the extraction of relevant parameterized closeness measures and methods of their tuning in the process of concept approximation.

Extensions of rough sets

We discuss in this section shortly some extensions of rough sets. Many other generalizations have been investigated and some of them have been used for real–life data analysis. Among them are: abstract approximation spaces (see e.g., [15]); nondeterministic information systems (see e.g., [89]); extensions of rough set approach to deal with preferential ordering on attributes (criteria) in multi–criteria decision making (see e.g., [45]); extensions of rough set methods for incomplete information systems (see e.g [60] cf. Chapter by KRYSZKIEWICZ and RYBINSKI); formal language approximations (see e.g., [88]); neighborhood systems (see e.g., [64]); distributed systems (see e.g., [107]). For discussion of other possible extensions see [99].

The variable precision rough set model. The formulae for the lower and upper set approximations can be generalized to some arbitrary level of precision $\pi \in (\frac{1}{2}, 1]$ by means of the *variable precision rough membership* function [144] (see below).

Note that the lower and upper approximations as originally formulated are obtained as a special case with $\pi = 1.0$.

$$\underline{B}_\pi X = \{x \mid \mu_X^B(x) \geq \pi\}$$

$$\overline{B}_\pi X = \{x \mid \mu_X^B(x) > 1 - \pi\}$$

Sets of patients, events, outcomes, etc. can be approximated by variable precision rough sets with a varied precision that depends on the parameter π.

Assuming that data are influenced by noise, one can tune the threshold π to find the "best" concept approximation. One can, e.g., start from π "close" to 1 and incrementally decrease the value of π. In each step, e.g., lower approximations of decision classes are calculated and corresponding decision rules are induced. As lower approximations of decision classes are becoming larger when the parameter π is decreasing, induced decision rules are becoming stronger, e.g., being supported by more objects. Decrease in value of the parameter π should be stopped when the quality of new object classification by induced rules starts to decrease.

The lower and upper approximations are just one example of possible approximations. In the terminology of Machine Learning they are approximations of subsets of objects known from the training sample. It is also desirable to approximate subsets of all objects (including also new unseen objects). The well known technique for such applications is the so-called *boundary region thinning*. It is related to the variable precision rough set approach [144]. Another technique may be used for tuning of decision rules. For instance, better quality of new objects classification may be achieved by introducing some degree of inconsistency of the rules on the training objects. This technique is an analogue of the well-known techniques for decision tree pruning. These approaches can be characterized in the following way: parameterized approximations of sets are defined and better approximations of sets (or decision rules) are obtained by tuning the parameters.

Tolerance based rough set model. Tolerance relations provide an attractive and general tool for studying indiscernibility phenomena. The importance of investigations of tolerance relations had been noticed by Poincaré and Carnap.

Any tolerance relation defines a covering of the universe of objects (by neighborhoods defined by so called *tolerance classes*). Tolerance relations for objects can be defined by similarity relation on feature (attribute) value vectors of objects cf. Chapter 1 by STEPANIUK.

A relation $\tau \subseteq X \times U$ is called a *tolerance relation* on U if (i) τ is *reflexive*: $x\tau x$ for any $x \in U$ and (ii) τ is *symmetric*: $x\tau y$ implies $y\tau x$ for any pair x, y of elements of U. The pair (U, τ) is called a *tolerance space*. It leads to a metric space with the distance function

$$d_\tau(x,y) = \min\{k : \exists_{x_0, x_1, \ldots, x_k} x_0 = x \land x_k = y \land (x_i \tau x_{i+1} \text{ for } i = 0, 1, \ldots, k-1)\}$$

Sets of the form $\tau(x) = \{y \in U : x\tau y\}$ are called *tolerance sets*. These sets as well as metrics above can be used to define more general approximations and their clusters. This is done by substituting tolerance classes for indiscernibility classes.

Definitions of the lower and upper approximations of sets can be easily generalized. [15] defines approximations of sets which are in some sense closer to X than the classical ones. They are defined as follows: $\tau_* X = \{x \in U : \exists y(x\tau y \& \tau(y) \subseteq X)\}$ and $\tau^* X = \{x \in U : \forall y(x\tau y \Rightarrow \tau(y) \cap X \neq \emptyset)\}$. It is easy to check that $\underline{\tau}X \subseteq \tau_* X \subseteq X \subseteq \tau^* X \subseteq \overline{\tau}X$.

It follows that in the process of learning a proper concept approximation there are more possibilities when using tolerance relations but at a greater computational cost that is due to a larger search space.

Tolerance relations can be defined for information systems or decision systems: by a *tolerance information system* we understand a triple $\mathcal{A}' = (U, A, \tau)$ where $\mathcal{A}' = (U, A)$ is an information system and τ is a tolerance relation on *information signatures* $Inf_B(x) = \{(a, a(x)) : a \in B\}$ where $x \in U$, $B \subseteq A$.

Tolerance reducts and tolerance-based decision rules can be generated by standard methods adapted to the tolerance case.

Some efficient techniques for discovery of relevant tolerance relations from data have been developed cf. Chapter by HOA SINH NGUYEN. For references see the bibliography in [100, 101].

We also recall general definition of approximation space (see [118] and Chapter 1 by STEPANIUK).

Parameterized approximation space is a system $AS_{\#,\$} = (U, I_\#, \nu_\$)$, where

- U is a non-empty set of objects,
- $I_\# : U \to P(U)$, where $P(U)$ denotes the powerset of U, is an uncertainty function,
- $\nu_\$: P(U) \times P(U) \to [0, 1]$ is a rough inclusion function.

The uncertainty function defines for every object x a set of similarly described objects. A constructive definition of uncertainty function can be based on the assumption that some metrics (distances) are given on attribute values. For example, if for some attribute $a \in A$ a metric $\delta_a : V_a \times V_a \longrightarrow [0, \infty)$ is given, where V_a is the set of all values of attribute a, then one can define the following uncertainty function:

$$y \in I_a^{f_a}(x) \text{ if and only if } \delta_a(a(x), a(y)) \leq f_a(a(x), a(y)),$$

where $f_a : V_a \times V_a \to [0, \infty)$ is a given threshold function.

A set $X \subseteq U$ is *definable in* $AS_{\#,\$}$, if it is a union of some values of the uncertainty function.

The rough inclusion function defines the degree of inclusion between two subsets of U [118], [98].

Thee lower and the upper approximations of subsets of U are defined as follows.

For a parameterized approximation space $AS_{\#,\$} = (U, I_\#, \nu_\$)$ and any subset $X \subseteq U$ the lower and the upper approximations are defined by

$LOW(AS_{\#,\$}, X) = \{x \in U : \nu_\$(I_\#(x), X) = 1\},$

$UPP(AS_{\#,\$}, X) = \{x \in U : \nu_\$(I_\#(x), X) > 0\}$, respectively.

Approximations of concepts (sets) are constructed on the basis of background knowledge. Obviously, concepts are also related to new (unseen) objects. Hence it is very useful to define parameterized approximations with parameters tuned in the searching process for approximations of concepts. This idea is crucial for construction of concept approximations using rough set methods. In our notation #, $ denote vectors of parameters which can be tuned in the process of concept approximation.

Rough mereology. The approach based on inclusion in a degree has been generalized to the *rough mereological approach* (see e.g., [103], [98]) cf. Chapter 3 by POLKOWSKI and SKOWRON. The inclusion relation $x\mu_r y$ with the intended meaning *x is a part of y in a degree r* has been taken as the basic notion of the rough mereology being a generalization of the Leśniewski mereology. Rough mereology offers a methodology for synthesis and analysis of objects in distributed environment of intelligent agents, in particular, for synthesis of objects satisfying a given specification in satisfactory degree, i.e., objects sufficiently close to standard objects (prototypes) satisfying the specification. Moreover, rough mereology has been recently used for developing foundations of the *information granule calculus*, an attempt towards formalization of the Computing with Words paradigm, recently formulated by Lotfi Zadeh.

Let us also note that one of the prospects for rough mereological applications is to look for algorithmic methods of extracting logical structures from data such as, for instance, finding relational structures corresponding to relevant feature extraction, synthesizing default rules (approximate decision rules), constructing connectives for uncertainty coefficients propagation and synthesizing schemes of approximate reasoning. A progress in this direction is crucial for further development of applications, in particular, we believe it is one of the central issues for KDD [29]. Rough set approach combined with rough mereology can be treated as an inference engine for computing with words and granular computing [140], [141], [99].

3 General view on logic

In spite of many views, often contradictory, on logic and its usefulness in Artificial Intelligence, in particular in Knowledge Discovery and Data Mining, it seems that every one should agree with the statement that logic gives us a mechanism for creating aggregates (collections) of statements (regardless for now of a language in which these statements are expressed) in which one statement (called the *conclusion*) is the consequence of all the remaining (called *premises*) in the sense that whenever we believe the premises we should also believe the conclusion (no matter now what we do understand by belief). The conclusion and the premises are then in the *consequence* relation. The process of passing from believed premises to the believed conclusion (called the *inference process*) is at heart of reasoning. Inference processes may be composed leading to chains of inferences and the overall inference mechanism may be very complex. Formal logic attempts at capturing essential features and properties of inferential mechanisms applied in many various contexts. Various logics differ with respect to the language in which they construct their statements, the way in which they construct their consequence relations, and the way in which they understand the notion of belief. For instance, in classical logics, the belief is understood as the (absolute) truth therefore consequence relations in these logics lead from true premises to the true conclusion. On the contrary, in non–classical logics, belief is understood, e.g., as the probability of a statement to be true, the possibility of a statement to be true, the degree of belief in a statement to be true etc. Therefore consequence relations express the degree of belief in the conclusion as a function of degrees of belief in premises.

Syntax, semantics

Any logic needs a language in which its statements are constructed and its inference mechanisms are represented. Hence we should define an alphabet of symbols over which well-formed expressions (formulas) of logic are to be constructed. Usually, the process of constructing formulas is of a generative character: one starts with simple (elementary, atomic) formulas and applies some generative rules for producing more complex formulas. Syntactic characteristics of a logic involves a specification of an alphabet, a class of atomic formulas as well as rules for generating formulas. This purely syntactic aspect of logic has its counterpart in the semantic aspect dealing with the meaning of formulas and the semantic aspect of consequence relations. Building semantics for a logic involves therefore a certain world (or worlds) external to the set of formulas of the logic in which we interpret formulas assigning to each of them its meaning usually being a relation(s) in this world (worlds); a good example is a relational database providing a semantic frame for logic. With respect to this world (worlds), we can define truth values (in general, belief degrees) of formulas. When this is done, we can study semantically acceptable consequence relations as these inference mechanisms which lead from true premises to true conclusions (respectively, from premises in which we believe in satisfactory degree to the conclusions in which we believe sufficiently) (in both cases with respect to a chosen set of worlds).

Although technically often difficult, problems of relationships between syntax and semantics are solved satisfactorily for majority of formal logics; on the other hand, in applicational problems we face the serious obstacle due to the fact that our models have to reflect a particular real phenomenon about which we reason hence we have to assure ourselves that our model conforms to reality. We face here the distinction between *closed worlds* (informally, worlds about which we know all what is important) and *open worlds*, i.e., worlds our knowledge about which is incomplete and gets augmented in the process of reasoning and exploration; in the former case our inference mechanisms are learned in ultimate form, in the latter, we learn them gradually and we need to correct and adapt them to the current knowledge.

With a logic we associate therefore two basic relations: the relation of syntactic consequence denoted $\vdash$, and the relation of semantic consequence (*entailment*) denoted $\models$.

Exemplary logical systems (calculi)

We review now some basic logical systems and we discuss their existing or potential connections with rough sets.

Propositional logic

Propositional reasoning turns out to be very useful in searching for rough set constructs like reducts, reduct approximations or decision rules. It seems [114], [116] that further research can give more results based on propositional reasoning, especially for solving problems in KDD. One of the promising methodologies is based on approximate Boolean reasoning, which we discuss in the end of this section.

In propositional logic we consider *propositions*, i.e., declarative statements like *London is the capital of Great Britain* or $2+2=3$ about which we can establish with certainty whether they are true or false. The calculus of propositions is effected by means of *propositional connectives* which allow for constructions of complex propositions from simpler ones. Formally, we begin with the *alphabet* consisting of a countably many *propositional symbols* $p_1, p_2, ..., p_k, ...$, *functor symbols* $\neg, \Rightarrow$ and *auxiliary symbols* (parentheses): ,(,],[. An *expression* is any word over this alphabet. The set of *formulas* of propositional logic is defined as the set X such that (i) X contains all propositional symbols (ii) X with all expressions u,v contains expressions $\neg u$ and $u \Rightarrow v$ (iii) if a set Y satisfies (i), (ii) then $X \subseteq Y$. To describe X, a generative approach is also used: one specifies a set $A \subset X$ of formulas called *axioms* along with *derivation rules* for generating formulas from axioms. Axioms may be chosen in many distinct ways; a simple axiomatics [71] consists of the following axiom schemes (meaning that in each of these expressions we may substitute for $p, q, r, ...$ any formula and we obtain an axiom formula):

(Ax1) $(p \Rightarrow (q \Rightarrow p))$;
(Ax2) $(p \Rightarrow (q \Rightarrow r)) \Rightarrow ((p \Rightarrow q) \Rightarrow (p \Rightarrow r))$;
(Ax3) $((\neg p \Rightarrow \neg q) \Rightarrow ((\neg p \Rightarrow q) \Rightarrow r))$.

The set of derivation rules consists of a single relation on expressions called *modus ponens* MP being the set of triples of the form $(p, p \Rightarrow q, q)$ meaning that: if $p, p \Rightarrow q$ are already derived from axioms, then q is regarded also as derived. The set of *theorems* is the set of formulas which can be obtained from instances of axioms by means of applying MP a finite number of times. The basic tool in investigating syntactic properties of propositional logic is the *Herbrand deduction theorem*: For any set of formulas Γ from $\Gamma, p \vdash q$ it follows that $\Gamma \vdash p \Rightarrow q$.

Now, we discuss semantics of propositional logic. We evaluate formulas with respect to their truth values: *truth* (denoted by T) and *falsity* (denoted by F) assuming that functors are truth-functional, i.e., they are functions on truth values and they do not depend on particular type of a formula. Under these assumptions one can characterize functors semantically by means of tables. We give the tables for $\neg, \Rightarrow$.

p	$\neg p$
0	1
1	0

p	q	$p \Rightarrow q$
0	0	1
0	1	1
1	0	0
1	1	1

Table 1. Negation functor $\neg$

Table 2. Implication functor $\Rightarrow$

Semantics of propositional logic is defined with respect to a *model* being the set of all Boolean (i.e., 0, 1 - valued) functions (called valuations) on the set of propositional symbols: given a formula $\alpha(p_{i_1}, p_{i_2}, ..., p_{i_k})$ (which means that the propositional symbols $p_{i_1}, p_{i_2}, ..., p_{i_k}$ are the only variable symbols in α) and a valuation v we define the value $v(\alpha)$ with respect to α. An admissible state in the model is any valuation v satisfying α, i.e., such that $v(\alpha) = 1$. A formula α is true when all states are admissible, i.e., $v(\alpha) = 1$ for every valuation v.

Two important properties of this deductive system are: *soundness* (meaning that every theorem is true) and *completeness* (meaning that every true formula is a theorem). It is straightforward to check the soundness of propositional logic by induction on the formula length. Less obvious is the completeness of propositional logic established first by Gödel [71]. Propositional logic is *decidable*: for each formula it is sufficient to check finitely many partial valuations restricted to the finite set of propositional symbols occurring in this formula to decide whether the formula is true. It is also *effectively axiomatizable*: for each formula one can decide in a finite number of steps whether the formula is an instance of an axiom scheme. Completeness implies *consistency*: for no formula α both α and $\neg \alpha$ can be theorems.

Let us add finally that in practical usage additional functors are introduced: the conjunction functor $\wedge$ defined by taking $\alpha \wedge \beta$ as a shortcut for $\neg(\alpha \Rightarrow \neg\beta)$, the disjunction functor $\vee$ defined by taking $\alpha \vee \beta$ as the shortcut for $\neg\alpha \Rightarrow \beta$ and the logical equivalence functor $\leftrightarrow$ defined by taking $\alpha \leftrightarrow \beta$ as the shortcut for $(\alpha \Rightarrow \beta) \wedge (\beta \Leftarrow \alpha)$. Truth tables for these functors follow immediately from these definitions and tables 1,2.

Model checking for propositional formulas (i.e., checking if a given propositional formula is satisfiable; in case of satisfiable formula a valuation satisfying it is expected) turned out to be successful for solving of many AI problems [114] despite of the high computational complexity of the satisfiability problem of propositional calculus (it is NP-complete problem [36]).

In the paper we discuss applications of Boolean reasoning [13], [14]. The idea of Boolean reasoning consists in constructing – for a given problem P – a corresponding Boolean function f_P with the following property: solutions for the problem P can be recovered from prime implicants of the Boolean function f_P. Let us recall that an *implicant* of a Boolean function f is any conjunction of *literals* (variables or their negations) such that if the values of these literals are true under an arbitrary valuation v of variables then the value of the function f under v is also true. A *prime implicant* is a minimal (with respect to the number of literals) implicant. In the examples presented in the paper we are interested in implicants of monotone Boolean functions only, i.e., functions constructed without negation. The main idea is based on observation that constraints which should be preserved while searching for solutions to problems can be expressed by Boolean functions. Moreover, (prime) implicants of these functions define the (approximate) solutions to problems. For real-life applications, it is necessary to deal with large Boolean functions. Hence, it was necessary to develop efficient heuristics searching for (prime) implicants of Boolean functions (problem of finding the shortest prime implicant of a given Boolean function in CNF form is NP-hard). In the paper we discuss applications of Boolean reasoning for constructing efficient heuristics solving various KDD problems.

One of the challenging problems is to develop methods for *approximate Boolean reasoning*. These methods should transform a given large Boolean function into a simpler one (i.e., of a feasible size) with (prime) implicants representing approximate solutions encoded by the original function. Hence, the approximate solutions can be found efficiently by computing (prime) implicants of Boolean functions of feasible size.

Predicate logic

Propositional logic renders us good service by formalizing the calculus of propositions; however, in many practical situations, KDD applications including, we are concerned with properties of objects expressed as concepts, i.e., sets of objects and with relations among these properties. In order to ensure the expressibility of relations, e.g., inclusions (like *every ripe tomato is red*) we need to quantify statements involving object descriptors over concepts. The predicate logic is an extension of propositional logic enabling us to manipulate concept descriptors. We will write $P(x)$ to denote that the object denoted x has the property denoted P; the symbol P is a (unary) predicate symbol. With the expression $P(x)$ we associate two expressions: $\forall x P(x)$ and $\exists x P(x)$; the symbol $\forall x$ is called the universal quantifier and $\forall x P(x)$ is read *for each object x the property P holds* and the symbol $\exists x$ is called the existential quantifier and $\exists x P(x)$ is read *there exists an object x such that P holds for x*. Predicate calculus does formalize such utterances. It will be useful to keep generality of our discussion, hence we will give a formal analysis of deductive systems known as first order theories of which predicate calculus is a specialization. To give a formal description of first order logical calculi on lines of deductive systems, we begin with an alphabet which in general case consists of few types of symbols:

(i)   *individual variables* $x_1, x_2, ..., x_k, ...$;
(ii) *individual constants* $c_1, c_2, ..., c_k, ...$;
(iii) *predicate symbols* $P_1^1, P_2^1, ..., P_{i_k}^k, ...$ where the upper index gives the arity of the predicate denoted thus;
(iv) *functional symbols* $f_1^1, f_2^1, ..., f_{i_k}^k, ...$;
(v)   *symbols* $\neg, \Rightarrow, \forall x$ (where x is a variable) ,) , (.

First we define the set of *terms* by requiring it to be the set X with the properties that (i) each individual variable or constant is in X (ii) if $t_1, .., t_k$ are elements of X and $f_{i_k}^k$ is a functional symbol of arity k then $f_{i_k}^k(t_1, t_2, ..., t_k)$ is in X (iii) if Y satisfies (i), (ii) then $X \subseteq Y$.

Next, the set of *formulas* is defined as the set Z with the properties (i) for each predicate symbol $P_{i_k}^k$ and any set $\{t_1, t_2, .., t_k\}$ of terms, the expression $P_{i_k}^k(t_1, t_2, ..., t_k)$ is in Z (ii) for each pair α, β of elements of Z, the expressions $\neg \alpha, \alpha \Rightarrow \beta, \forall x \alpha$ are in Z for each variable x (iii) if Y satisfies (i), (ii) then $Z \subseteq Y$.

The existential quantifier is defined by duality clear on intuitive basis: $\exists x P(x)$ is the shortcut for $\neg \forall x \neg P(x)$. A standard distinction on occurrences of a variable x in a formula α is between free and bound occurrences which informally means that an occurrence is bound when this occurrence happens in a part of the formula (sub–formula) preceded by the quantifier sign; otherwise the occurrence is free. It is intuitively clear that a formula in which all occurrences are bound is a proposition, i.e., either true or false in a given model.

To define the syntax on generative lines, one should specify the axioms of logic. Axioms of T can be divided into two groups: the first group consists of general logical axioms, the second group consists of specific axioms; when the second group is present, we speak of a *first order theory*. The axiom schemes of the first group may be chosen as follows:

- (Ax1), (Ax2), (Ax3) are axiom schemes for propositional logic.
- (Ax4) $\forall x \alpha(x) \Rightarrow \alpha(t)$ where x is a variable, t is a term and t contains no variable such that x occurs in a sub–formula quantified with respect to that variable;
- (Ax5) $\forall x(\alpha \Rightarrow \beta) \Rightarrow (\alpha \Rightarrow \forall x \beta)$ where the variable x is not free in α.

Specific axioms depend on T; for instance the theory of equivalence relation may be expressed by means of a binary predicate symbol P_1^2 and axioms
(1) $\forall x P_1^2(x, x)$;
(2) $\forall x \forall y (P_1^2(x, y) \Rightarrow P_1^2(y, x))$;
(3) $\forall x \forall y \forall z (P_1^2(x, y) \Rightarrow (P_1^2(y, z) \Rightarrow P_1^2(x, z)))$.

The set of *derivation rules* consists of two rules: modus ponens (MP) known from propositional logic and *quantification* (generalization) rule Q which is the binary relation on expressions consisting of pairs of the form $(\alpha, \forall x \alpha)$ where x is any variable. A *theorem* of predicate logic is any formula which may be obtained from an instance of an axiom by applying a derivation rule a finite number of times. Semantics of a first order theory T is defined according to Tarski [129] as follows.

A *model* M for the theory T is a pair (D, f) where D is a set and f is an interpretation of T in D, i.e., f assigns to each individual constant c an element $f(c) \in D$, to each predicate symbol P_i^k a relation $f(P_i^k)$ on D of arity k and to each functional symbol f_i^k a function $f(f_i^k)$ from D^k to D. Truth of a formula, relative to M, is defined inductively on complexity of the formula. To this end, we consider the states of the model M as sequences $\sigma = (a_i)_i$ of elements of D. Given a formula α and a state σ, we need to declare when the formula α is satisfied by σ, in symbols, $\sigma \models \alpha$. We define a map F_σ which assigns an element of D to each term of T. Individual variables are interpreted via a given state σ:

$F_\sigma(x_i) = a_i$. The mapping F_σ is equal to f on individual constants. The inductive condition is as follows: if F_σ is already defined on terms $t_1, t_2, ..., t_k$ and f_i^k is a functional symbol then $F_\sigma(f_i^k(t_1, t_2, ..., t_k)) = f(f_i^k)(F_\sigma(t_1), F_\sigma(t_2), .., F_\sigma(t_k))$.

The satisfiability $\models$ is defined inductively as follows:
(i) $\sigma \models P_i^k(t_1, t_2, ..., t_k)$ if and only if $f(P_i^k)(F_\sigma(t_1), .., F_\sigma(t_k))$ holds;
(ii) $\sigma \models \neg\alpha$ if and only if it is not true that $\sigma \models \alpha$;
(iii) $\sigma \models (\alpha \Rightarrow \beta)$ if and only if $\sigma \models \alpha$ implies that $\sigma \models \beta$;
(iv) $\sigma \models \forall x \alpha(x)$ if and only if $\sigma^x \models \alpha$ for each state σ^x where (letting x to be the variable x_i) σ^x is like σ except that the i-th member of σ^x need not be equal to a_i.

These conditions allow to check for each formula whether it is *satisfied* by a given state. A formula is *true in the model* M if and only if it is satisfied by every state σ. A formula is *true (tautology)* if and only if it is true in every model M. Observe that a formula $\alpha(x_1, ..., x_n)$ is true if and only if its closure $\forall x_1...\forall x_n \alpha(x_1, ..., x_n)$ is true.

The first order theory PC without specific axioms is called the *predicate calculus*. It is obvious that properties of first order theories depend on specific axioms so we here recapitulate the facts about the predicate calculus. The soundness of predicate calculus can be easily established by structural induction: each theorem of PC is true as all instances of axiom schemes (Ax1)-(Ax3) are true and truth is preserved by derivation rules MP and Q. The important Gödel completeness theorem [43], [71] states that predicate calculus is complete: each true formula is a theorem. Decidability problems for first order theories involve questions about the formalizations of the intuitive notion of a *finite procedure* and can be best discussed in the frame of the fundamentally important first order theory of Arithmetic (cf. [71]): a predicate calculus without functional symbols and individual constants is called pure predicate calculus PP while predicate calculus with infinite sets of functional symbols and individual constants is called functional calculus PF. The classical theorem of Church [20], [71] states that both PP, PF are recursively undecidable (algorithmically unsolvable). On the other hand, many problems are recursively decidable (algorithmically solvable) however their time–or space–complexity makes them not feasible, e.g., satisfiability problem for propositional calculus is NP-complete [36].

Predicate logic turns out to be very successful in many AI applications related, e.g., to knowledge representation or representation of structural objects.

Deductive systems (DS)

Here we sum up the features of deductive systems like propositional logic or predicate calculus. By a *deductive system*, we understand a tuple $(Ax, Gen, \vdash)$ where Ax is a set of *axioms* (meaning by an axiom a formula which is assumed to be well-formed and desirably true), Gen is a set of *inference (derivation) rules*, each rule R being a relation on the set of formulas and $\vdash$ is a relation on formulas such that whenever $\Gamma \vdash \alpha$ holds this means that there exists a *formal proof* of α from Γ (α is derivable from Γ), i.e., there exists a finite sequence (the formal proof) $\alpha_1, ..., \alpha_k$ such that (i) α_1 is either an axiom or an element of Γ

(ii) α_k is α (iii) each α_i ($i = 2, ..., k$) is either an axiom or is in Γ or satisfies $R(\alpha_{j_1}, ..., \alpha_{j_m}, \alpha_i)$ for some $R \in Gen$ and a subset $\{\alpha_{j_1}, ..., \alpha_{j_m}\}$ of $\{\alpha_1, ..., \alpha_{i-1}\}$. Any formula α such that $\vdash \alpha$ (meaning $\emptyset \vdash \alpha$) is said to be a *theorem of the deductive system*. From these definitions, the properties of $\vdash$ follow: (a) $\Gamma \subseteq Cn(\Gamma)$ where $Cn(\Gamma) = \{\alpha : \Gamma \vdash \alpha\}$; (b) $Cn(\Gamma) = Cn(Cn(\Gamma))$; (c) $Cn(\Gamma) \subseteq Cn(\Gamma')$ whenever $\Gamma \subseteq \Gamma'$ (the Tarski axioms for syntactic consequence [128]). Semantics of a deductive system is defined with respect to a class of specified structures called models: there exists a mechanism which for each model and each formula assigns to this formula a subset of the model domain (called the *interpretation* of the formula in the model). A formula is true with respect to a given model in case its interpretation in the model equals the model domain. A formula is *true* (is a *tautology*) in case it is true with respect to all models (in the assumed class of models).

The semantic consequence $\models$ (entailment) is defined on sets of formulas as follows: $\Gamma \models \Gamma'$ if for any model M the truth of each formula in Γ in M implies the truth of each formula from Γ' in M.

Properties of DS: soundness, consistency, completeness, decidability, expressiveness, complexity

Among properties of deductive systems there are some whose importance deserves them to be mentioned separately. The first of them is *soundness* (of axiomatization) which means that all theorems of the system are true. The dual property of *completeness* means that each true formula has a formal proof in the system. As a rule verification of soundness is straightforward while the completeness proofs are usually non–trivial. Another important property often intervening in completeness proofs is *consistency*: a set Γ of formulas is *consistent* if there is no formula α such that both α and its negation are derivable from Γ. Another important question is whether there exists an algorithm which for each formula can *decide* if this formula is a theorem; if yes, we say that the deductive system is *decidable*. In this case we may ask about the time– , or space– complexity of the decidability problem. We may study complexity of other problems like *satisfiability* (i.e. whether the interpretation of the formula is non–empty). From the point of view of knowledge representation, it is important to decide what properties can be expressed by means of formulas of the deductive system. A useful meta–rule is: the greater expressibility, the greater complexity of problems about the deductive system.

Let us observe that for building effective reasoning systems for complex tasks (e.g., in spatial reasoning [110]) there is a need to develop methods for approximate reasoning not only because of incomplete information but also because of huge search space for proofs – one would like to obtain *proof – construction* returning object satisfying a given specification in *satisfactory degree* (possibly, not exactly) but in an acceptable time. This is one of the main task of granular computing [142], new emerging paradigm of computing based on calculi on information granules (see e.g., [141], [142], [103]) using ideas of soft computing approaches like fuzzy sets and rough sets, among others.

Semantic tableaux, natural deduction, sequent calculus

Semantic tableaux method provides a method of determining validity in propositional or predicate calculus. A tableaux proof of $\Gamma \models \alpha$ begins with $\Gamma \cup \{\neg\alpha\}$. A tableaux proof of $\Gamma \models \alpha$ is a binary tree labeled by formulae and constructed from $\Gamma \cup \{\neg\alpha\}$ by using rules for each logical connective specifying how the tree branches. A branch of the tree closes if it contains some sentence and its negation; the tableau closes if all branches close. If the tableau closes then $\Gamma \models \alpha$ is valid. If tableau does not close and none of the rules can be applied to extend it, then $\Gamma \cup \{\neg\alpha\}$ is satisfiable and $\Gamma \models \alpha$ does not hold. For the propositional calculus, semantic tableau gives a decision procedure. In case of predicate calculus if the set $\Gamma \cup \{\neg\alpha\}$ is satisfiable, i.e., $\Gamma \models \alpha$ is not true the method may never terminate (the rules for the universal quantifier can be applied repeatedly). It terminates if and only if the set $\Gamma \cup \{\neg\alpha\}$ is unsatisfiable, i.e., $\Gamma \models \alpha$ holds. We say that predicate calculus is *semi-decidable*.

Let us mention that deduction may be formalized as so called *natural deduction* [39] and its form known as the *sequent calculus*. The sequent calculus is a set of rules for transforming sequents, i.e., expressions of the form $\Gamma \vdash \Delta$ where Γ, Δ are sets of formulas. Gentzen proposed [39] a set of sequent rules for classical predicate calculus.

The tableaux method can be treated as another way of writing sequent calculus derivations.

Resolution and logic programming

It is desirable from computing point of view to have systems for automated deduction; the widely accepted technique for this is *resolution* due to J.A.Robinson [27]. It requires *clausal* form of formulas, i.e., a conjunction of disjunctions of literals (a literal is a variable or its negation). A symbol like $\{p,q\}$ means disjunction of literals p, q and a symbol like $\{.\}; \{..\}; ...; \{...\}$ means a conjunction of disjunctions, i.e., a *clause*. Resolution uses *refutational proof technique*: instead of checking validity of α it checks unsatisfiability of $\neg\alpha$; to this end $\neg\alpha$ is represented in clausal form and the resolution rule:

from clauses $a \cup p$ and $b \cup \neg p$ the clause $a \cup b$ is derived

is applied a finite number of times. Final appearance of the empty clause $\square$ witnesses unsatisfiability of $\neg\alpha$ hence validity of α. The resolution calculus is sound and complete with respect to entailment $\models$. Resolution in predicate calculus involves *unification*, i.e., the process of finding substitutions making two terms containing free variables identical. For extensions and refinements see [27].

Particularly important from computational point of view is the *Horn clausal logic* [50] based on Horn clauses of which *Horn facts* are of the form
$$\forall x_1...\forall x_m P_{i_k}^k(\tau_1, ..., \tau_k)$$
and *Horn rules* are of the form
$$\forall x_1...\forall x_k \alpha_1(x) \wedge .... \wedge \alpha_n(x) \Rightarrow \beta(x).$$
A set of Horn clauses is a *Horn clausal theory* T. Inferences for T are based on inference rules of the form: $\alpha_1(c)\wedge....\wedge\alpha_n(c)/\beta(c)$ where c is a *ground term*, i.e., a term without variables corresponding to Horn rules in T. A proof $T \vdash \gamma$ is a finite

sequence of inferences starting from an inference based on a fact and ending with γ. This calculus is sound and complete. The Horn clausal logic can be considered as a generative device for incremental buildup of a set from Horn facts (alphabet) and Horn rules (generating rules). It has been applied in implementations of PROLOG and DATALOG in particular in logic programming [66].

The idea behind logic programming is that the logic program is a specification written as a formula in a logical language and the inference engine for the construction solution consists of a deduction system for this language. The system of deduction in logic programming is resolution. The logic programs can be of the form known as definite clause programs (a definite clause is a universally quantified disjunction of one or more literals, only one of which is negated). They are executed by adding a goal being a clause in a special form.

Semantics for logic programs can be defined by Herbrand interpretations. A *Herbrand interpretation* is based on the *Herbrand universe*, i.e., the set of all ground atoms constructed from constants, function and predicate symbols in the program. The least Herbrand model of a logic program can be defined as the least fixed point of a certain function from the Herbrand universe into itself. Any predicate calculus sentence can be transformed into a set of clauses and then resolution, like in the tableau method, can be used to test, by refutation, the validity of entailment in predicate calculus.

An example of logic programming system is PROLOG (Colmerauer, 1972). More details can be found in [66]. These systems may be regarded as engines for constructing *knowledge bases* from data.

One of the main tasks of inference is to obtain a description of a target object satisfying (exactly or in satisfactory degree) a given specification (formulated in some logical language). In the *constraint programming logic* [131] the construction schemes of such objects can be extracted from logical proofs.

Recently some research results have been reported on combining rough set approach with logic programming (see e.g., [65]).

Theorem provers

Automated theorem proving was initiated in 1950's and by 1960 various computer programs for theorem proving were implemented (Newell, Davis and Putnam, Gilmore, Prawitz, Hao Wang) and able to prove very simple theorems. Resolution technique (J.A. Robinson, 1965) proved to be much more powerful and by now most theorem provers use resolution. In spite of the progress, still much remains to be done in this area; in the first place, in discovering proof strategies. This will need in particular to introduce some similarity measures on proofs. Moreover, KDD stimulates research towards revision of exact formal proofs by introducing instead of them schemes for approximate reasoning extracted from data [103].

Exemplary logics for reasoning under uncertainty

Modal logic

In many applications in KDD, when our knowledge is incomplete or uncertain, e.g., in mining association rules in databases with inconsistent decision [3], we cannot have exact logical statements, but we only may express certain modalities like *property P is possible*. There are strong relationships of rough logic and modal logic (see section on rough logic; cf. also a Chapter by DÜNTSCH and GEDIGA). It seems also that for developing a *perception logic* modal formulas can be very useful as a tool for expressing properties of structural objects (see e.g., [9]).

Modal propositional logics deal with formalizations of phrases like *it is possible that ..., it is necessary that ....* These modalities are formally rendered as generalized quantifiers: $[\alpha]$ is read as *it is necessary that* α, $\langle \alpha \rangle$ is read as *it is possible that* α. These operators are related by duality: $\langle \alpha \rangle$ is the shortcut for $\neg[\neg \alpha]$. The syntax of modal calculus is defined over an alphabet much like that of propositional logic: the only addition is the introduction of modal operator symbols $[.]$ and $\langle . \rangle$. The set of formulas of modal logic is defined as the smallest set X such that (i) X contains all propositional variables (ii) with each pair α, β, X contains $\neg \alpha$, $\alpha \Rightarrow \beta$ and $[\alpha]$.

Way of axiomatization of a modal logic depends essentially on properties of necessity which we intuitively deem as desirable; their rendering in axioms leads to various systems of modal calculi. We will briefly review the most important ones.

The simplest modal system K is obtained by adding to the axiom schemes (Ax1)-(Ax3) of propositional logic the axiom scheme (K) of the following form:

(K) $[\alpha \Rightarrow \beta] \Rightarrow ([\alpha] \Rightarrow [\beta])$.

(K) expresses our basic intuition about necessity: if both an implication and its precedent are necessarily true then the consequent should be necessarily true also. The derivation rules of modal propositional logic are: *modus ponens MP* and the *necessity rule N* which is the relation on formulas consisting of pairs of the form $(\alpha, [\alpha])$. This calculus is consistent [52]: to see it, it suffices to collapse formulas of K onto formulas of propositional calculus by omitting all modal operator symbols. Then theorems of modal logic are in one-to-one correspondence with theorems of propositional calculus.

Among syntactic properties of necessity valid in K we may mention the following expressed by theorems of K: (i) $(\alpha \Rightarrow \beta) \Rightarrow ([\alpha] \Rightarrow [\beta])$; (ii) $[\alpha \vee \beta] \Rightarrow ([\alpha] \vee [\beta])$; (iii) $[\alpha \wedge \beta] \Leftrightarrow ([\alpha] \wedge [\beta])$.

Semantics of modal logic is defined as the *possible worlds (Kripke) semantics* [30]. A model for a modal logic system is a triple $M = (W, R, v)$ where W is a collection of *states (worlds)* and R is a binary relation on W (called *accessibility relation*); the symbol v denotes a state of the model (a valuation), i.e., the Boolean function on the set of all pairs of the form (w, p) where $w \in W$ is a world and p is a propositional variable. The notion of satisfiability $M, v, w \models \alpha$ is defined by structural induction: (i) $M, v, w \models p$ if and only if $v(w, p) = 1$; (ii) $M, v, w \models \neg \alpha$ if and only if it is not true that $M, v, w \models \alpha$; (iii) $M, v, w \models \alpha \Rightarrow \beta$ if and only if either it is not true that $M, v, w \models \alpha$ or it is true that $M, v, w \models \beta$; (iv) $M, v, w \models [\alpha]$ if and only if $M, v, w_1 \models \alpha$ for each world w_1 such that $R(w, w_1)$.

A formula α is *true in the model M* if and only if $M, v, w \models \alpha$ for each world

$w \in W$ and every state v; a formula is *true* if and only if it is true in each model.

It is straightforward to check that the system K is sound: all instances of axioms are true and derivation rules MP, N preserve truth of formulas. Completeness of K can be proved by, e.g., the Lemmon-Scott extension of Henkin's technique of canonical models [30], [52]. Decidability of K follows from its completeness and the collapse property mentioned above along with decidability property of propositional calculus.

Properties of necessity axiomatized in K are by no means the only desirable; one may ask, e.g., whether the property $[\alpha] \Rightarrow \alpha$ holds in K. It is easy to see that by completeness, truth of this formula requires the relation R be reflexive hence this formula is not true in general system K.

Adding to the axiom schemes of K the axiom scheme

(T) $[p] \Rightarrow p$

we obtain a new modal system T. The completeness of T can be now expressed as follows: a formula of T is a theorem of T if and only if this formula is true in all models where the accessibility relation is reflexive.

Another property of necessity is the following :

(S4) $[p] \Rightarrow [[p]]$.

Adding to axiom schemes of T the axiom scheme (S4), we obtain a new system called S4. Theorems of S4 are those formulas of K which are true in all models with the accessibility relation R reflexive and transitive.

Finally, we may consider the formula

(S5) $\langle[p]\rangle \Rightarrow [p]$.

The formula (S5) is true in all models with the accessibility relation R being an equivalence relation. As (S5) implies syntactically (S4), the system S5 obtained from T by adding the axiom scheme (S5) contains the system S4. Completeness of S5 is expressed as follows: theorems of S5 are those formulas which are true in all models with the accessibility relation R being an equivalence.

We have therefore a strictly increasing hierarchy $K, T, S4, S5$ of modal logic systems; these systems do not exhaust all possibilities [52]. Recently modal logics play an important role in many theoretical branches of Computer Science and Artificial Intelligence, e.g., in formalization of reasoning by groups of intelligent agents [28]; in applicational domain, we may mention hand–written digit recognition [9] where modal formulas are used to express properties discerning between structural objects.

Temporal and dynamic logics

There are some particular contexts in which formulas of modal logic may be specialized and tailored to a specific usage. Let us mention two such cases, i.e., temporal as well as dynamic logics. These logics are useful to express knowledge in a changing environment, e.g., in geographic information systems (GIS') [26]. Rough sets offer a possibility to make inference on approximate concepts. Hence by combining rough set methods with temporal reasoning one can develop tools for efficient temporal reasoning.

In temporal logics, the set W of possible worlds is interpreted as the set of time instants and the accessibility relation R is the precedence in time relation, i.e., wRw_1 means that w precedes w_1 in time (in particular, it may happen that $w = w_1$). Clearly, R is reflexive and transitive hence this logic is of S4 type [11].

Dynamic logic is applied in the context of properties of programs [48]. In this case the set W is the set of states of an abstract computing machine. Given a program P, the modality $[.]_P$ acts on formulas describing states of the machine and its semantics is defined by the accessibility relation R_P which holds on a pair (w, w_1) if and only if an execution of P starting at w terminates at w_1; specifically, a state w satisfies the formula $[\alpha]_P$ if and only if each state w_1 such that $R_P(w, w_1)$ satisfies α. This means that the state in which P terminates necessarily satisfies α.

Epistemic and doxastic logics

These are *logics of knowledge and belief*. Logics for reasoning about knowledge, belief, obligations, norms etc. have to deal with statements which are not merely true or false but which are known or believed etc. to be true at a moment; an additional complication is of pragmatic character: knowledge, belief etc. may be relativized to particular intelligent reasoners (agents) hence we may need also to express statements about group or common knowledge, belief etc. Modal logics have proved suitable as a general vehicle for carrying out the task of constructing such logics. There has been made a great effort to use rough sets for reasoning about knowledge (see e.g., [86], [100], [101] and the bibliography in these books). Logics of knowledge and belief are useful in KDD in, e.g., tasks of describing problems of distributed/many-agent nature, in building networks of reasoning agents (e.g., belief networks) etc. Epistemic logics for reasoning about knowledge [49], [137], [47] are built as modal logics with a family $K_i : i = 1, 2, ..., n$ of necessity operators, K_i interpreted as the modal operator *the agent i knows that*.... Syntax of such logic is like that of modal propositional logic except for the above family K_i instead of a single $[.]$ modal operator symbol. Formulas are defined as usual, in particular given a formula α, the expression $K_i\alpha$ is a formula, each i. We have therefore formulas like $K_i K_j \alpha$ read as *the agent i knows that the agent j knows that α* etc. Semantics is the usual Kripke semantics of possible worlds except that to accommodate all $K_i's$, a model is now a tuple $M = (W, v, R_1, ..., R_n)$ where R_i is an accessibility relation of K_i, i.e., $v, w \models K_i\alpha$ if and only if $v, w' \models \alpha$ for every w' with $R_i(w, w')$. One may want to express also in this logic statements like *every agent in a group G knows that...* or *it is common knowledge among agents in G that....* This may be done by introducing additional symbols E_G, C_G for each subset $G \subseteq \{1, 2, ..., n\}$, requiring that for each formula α and each G, expressions $E_G\alpha, C_G\alpha$ be formulas and defining semantics of these formulas as follows: $v, w \models E_G\alpha$ if and only if $v, w \models K_i\alpha$ for each $i \in G$ and $v, w \models C_G\alpha$ if and only if $v, w \models K_{i_1} K_{i_2}, ..., K_{i_j}\alpha$ for each sequence $i_1 i_2...i_j$ over G. In other words, the accessibility relation for E_G is $\cap\{R_i : i \in G\}$ and the accessibility relation for C_G is the transitive closure of $\{R_i : i \in G\}$. These logics are axiomatized soundly and completely exactly as modal logics of respective types; additional axiom schemes for logics endowed with operators E_G, C_G may be chosen as follows [47]: $E_G p \leftrightarrow \wedge K_i p$; $C_G p \leftrightarrow E_G(p \wedge C_G p)$ along with the additional derivation rule $(p \Rightarrow E_G(\alpha \wedge \beta)) \Rightarrow (\alpha \Rightarrow C_G\beta)$. These logics are decidable (to check validity it is sufficient to examine at most 2^n worlds where n is the formula length).

Belief logics (*doxastic logics*) [37] may model belief states either as consistent sets of formulas or as sets of possible worlds. To introduce the former approach, assume that a propositional logic L is given along with its consequence relation Cn about which one usually requires *monotonicity* ($\Gamma \subseteq \Delta$ implies $Cn(\Gamma) \subseteq Cn(\Delta)$, *compactness* ($\alpha \in Cn(\Gamma)$ implies $\alpha \in Cn(\Delta)$ for a finite $\Delta \subseteq \Gamma$), *cut property* ($\alpha \in Cn(\Gamma \cup \{\beta\})$ and $\beta \in Cn(\Gamma)$ imply $\alpha \in Cn(\Gamma)$), *deduction property* ($\alpha \in Cn(\Gamma \cup \{\beta\}$ implies $\beta \Rightarrow \alpha \in Cn(\Gamma)$). Belief states are represented by sets of formulas of L; a principal assumption may be that believing in a formula should imply believing in all consequences of this formula hence representing sets should be closed under consequence Cn. In this approach calculus of beliefs is reduced to calculus on closed sets of formulas of L. The latter approach in which belief states are represented as sets of possible worlds in which a formula is believed to be true may be shown to be equivalent to the former.

Deontic logics, i.e., logics of *normative concepts* like obligations, prohibitions, permissions, commitments [8] are also built as modal logics. A deontic logic has an alphabet of propositional logic endowed with modal operators O, P (obligation, permission) modeled as necessity, resp. possibility. Axiomatics depends on type of modal logic one wants to obtain and it is sound and complete. These logics are decidable for reasons as above.

Para–consistent and relevant logics

Investigations on *para–consistent logics* [5], [6] have been initiated in order to challenge the logical principle saying that for any two formulas α, β it follows from $\{\alpha, \neg\alpha\}$ (syntactically or semantically) that β. Para–consistent logics are motivated by philosophical considerations as well as by computer science applications. It is quite often necessary for information systems to deal with inconsistent information because of multiple sources of information or noisy data. Another area of applications is related to *belief revision*. Mechanisms of belief revision should work on inconsistent sets of beliefs. Other applications of para–consistent logics concern theories of mathematical significance. Among systems of para–consistent logics are non–adjunctive systems (initiated by Jaśkowski's discussive or discursive logic); non–truth functional systems (initiated by da Costa); para–consistent logic generated by multi-valued logic (introduced by Asenjo).

Relevant (*relevance*) *logics* are developed as systems to avoid the so-called paradoxes of material and strict implications. In addition relevance logic is trying to avoid to infer conclusions having to do nothing with the premise. These logics were pioneered by Anderson and Belnap. Relevant logics have been used in computer science and in philosophy, e.g., the linear logic (discovered by Girard) is a weak relevant logic with the addition of two operators.

Conditional logics

Conditional logics [85] investigate logical properties of declarative conditional phrases of natural language: *if.... then....* The problem studied in these logics is whether *material (propositional) implication* $\Rightarrow$ represents adequately such phrases especially in case of counter–factual conditionals where the premise as well as the consequent are false. Negative answers to this question lead to various systems of conditional logic. Denoting by $\rangle$ the conditional *if... then...*, one may exploit modal logic and interpret $\alpha\rangle\beta$ as true when β is true in a world closest to world(s) in which α is true (with respect to some closeness measure) where worlds may be also constructed as closed sets of propositional logic. Accordingly, a conditional logic may be built over propositional logic by adding a new symbol $\rangle$, requiring that $\alpha\rangle\beta$ be a formula in case α, β are formulas and adding a new derivation rule: (R_1) from $\alpha \Rightarrow \beta, \gamma\rangle\alpha$ derive $\gamma\rangle\beta$. Axiom schemes of this logic are (Ax1)-(Ax3) of propositional logic plus (Ax4) $\alpha\rangle\alpha$; (Ax5) $\alpha\rangle\beta \Rightarrow (\alpha \Rightarrow \beta)$; (Ax6) $\neg\alpha\rangle\alpha \Rightarrow (\beta\rangle\alpha)$; (Ax7) $\alpha\rangle\beta \wedge \beta\rangle\alpha \Rightarrow (\alpha\rangle\gamma \Leftrightarrow \beta\rangle\gamma)$; (Ax8) $\alpha\rangle\beta \wedge \neg(\alpha\rangle\neg\gamma) \Rightarrow (\alpha \wedge \gamma\rangle\beta)$; (Ax9) $\alpha\rangle\beta \vee \neg\alpha\rangle\beta$. This is the syntax of conditional logic C2 of Stalnaker [126]. Models of C2 are of the form of a quadruple $M = (W, R, S, E)$ where W is a set of possible worlds, R is an accessibility relation, S is a selector assigning to any formula α and any world w a world $S(\alpha, w)$ such that $S(\alpha, w) \models \alpha$ and $S(\alpha, w)$ is closest to w and E assigns to any formula its extension i.e the set of worlds in which the formula is true. There may be additional postulates on S, e.g., $R(w, S(\alpha, w))$ etc. One obtains a sound and complete axiomatization of $C2$. Replacing (Ax9) by (Ax9'): $\alpha \wedge \beta \Rightarrow \alpha\rangle\beta$ one obtains conditional logic VC of Lewis [85].

One can expect interesting relationships between conditional logic and granular computing based on closeness and inclusion of information granules [119], [103].

Logics for dealing with inconsistent situations

When we apply logic to reason in real situations we often face the problem of dealing with inconsistencies: statements we encounter at a given moment may be inconsistent with our current knowledge. A way out of this problem has been proposed as belief revision: we have to modify our knowledge in order to remove inconsistency. A guiding principle may be the economy principle of minimal changes: when modifying our knowledge we should make changes as little as possible. Doxastic logics offer a convenient playground for treating inconsistent statements by means of logics for belief revision [23]. Main operations on belief sets (i.e., closed under Cn sets of formulas) are: -*expansion*: adding α to a set Γ results in $Cn(\Gamma \cup \{\alpha\})$; -*revision*: when an inconsistent α is added, a maximal consistent subset $\Delta \subseteq \Gamma \cup \{\alpha\}$ is selected resulting in $Cn(\Delta)$; -*contraction*: removing a set $\Delta \subseteq \Gamma$ results in a closed set $\Psi \subseteq \Gamma - \Delta$. Some postulates have been proposed [37]: denoting by $\Gamma r \alpha$ resp. $\Gamma c \alpha$ the result of revision resp. contraction of Γ relative to α, we may require that: (r1) $\Gamma r \alpha$ be closed; (r2) $\alpha \in \Gamma r \alpha$; (r3) $\Gamma r \alpha \subseteq Cn(\Gamma \cup \alpha)$; (r4) if $not\ \neg\alpha \in \Gamma$ then $Cn(\Gamma \cup \alpha) \subseteq \Gamma r \alpha$; (r5) $\Gamma r \alpha$ is the set of all formulas if and only if $\vdash \neg\alpha$; (r6) $\alpha \leftrightarrow \beta \Rightarrow \Gamma r \alpha = \Gamma r \beta$. The Levi and Harper identities: $\Gamma r_c \alpha = Cn(\Gamma c(\neg\alpha) \cup \alpha)$ resp. $\Gamma c_r \alpha = \Gamma \cap \Gamma r \alpha$ allow for defining one of these operations from the other and establish a duality:

$\Gamma c_{r_c}\alpha = \Gamma c\alpha$, $\Gamma r_{c_r}\alpha = \Gamma r\alpha$.

Various heuristics have been proposed for these operations, e.g., -*choice contraction function*: given Γ, α, define $Max(\Gamma, \alpha)$ as the set of maximal closed sets M such that not $\alpha \in \Gamma$ and next select a member of $Max(\Gamma, \alpha)$ as $\Gamma c\alpha$; -*meet contraction function*: take $\cup Max(\Gamma, \alpha)$ as $\Gamma c\alpha$; - *epistemic entrenchment*: assign various degrees of importance to formulas in Γ and initiate contraction with formulas of lowest degrees.

Belief revision logics may be perceived in a wider perspective as tools for reasoning in situations when the standard consequence fails: in presence of inconsistencies non–monotonicity may arise, i.e., a greater set of premises may lead to a smaller set of consequents because of need for revision of our knowledge. Attempts at formal rendering of this phenomenon has led to non–monotonic logics [69]. Non–monotonic reasoning is central for intelligent systems dealing with common–sense reasoning being non–monotonic.

Non–monotonic logics deal with non–monotonic consequence $\vdash_{nm}$; a general idea for rendering $\vdash_{nm}$ may be as follows: try to define $\alpha \vdash_{nm} \beta$ as holding when there exists a belief set Γ of formulas, $\alpha \vdash \beta$ and $\alpha \vdash \gamma$ for sufficiently many $\gamma \in \Gamma$.

This idea is realized in various ways: in *default logic* [104], *probabilistic logic* [1], *circumscription* [56], *auto–epistemic logic* [62]. Reiter's default logic is built over predicate calculus L by enriching it with inference rules (called *defaults*) of the form $\alpha(x); \beta(x)/\gamma(x)$ where $\alpha(x), \beta(x), \gamma(x)$ are formulas called resp. the precondition, the test condition and the consequent of the default. For a constant a, the default permits to derive $\gamma(a)$ from $\alpha(a)$ under the condition that *not* $\vdash \neg\beta(a)$; we denote this consequence by $\vdash_d$. Formally, a default theory T may be represented as a pair (K, E) where K, a background context, contains rules and E, the evidence set, contains facts. Rules in K are of two kinds: rules of L and defaults D.

Let us consider one example of a characterization of non–monotonic inference $\vdash_{K,E}$ from given (K, E). It may be done by a set of postulates of which we mention: (*Defaults*) $p \vdash_d q \in D$ implies $p \vdash_{K,E} q$; (*Deduction*) $\vdash p$ implies $\vdash_{K,E} p$; (*Augmentation*) $\vdash_{K,E} p$ and $\vdash_{K,E} q$ imply $\vdash_{K,E\cup\{p\}} q$; (*Reduction*) $\vdash_{K,E} p$ and $\vdash_{K,E\cup\{p\}} q$ imply $\vdash_{K,E} q$; (*Disjunction*) $\vdash_{K,E\cup\{p\}} r$ and $\vdash_{K,E\cup\{q\}} r$ imply $\vdash_{K,E\cup\{p\vee q\}} r$. As shown in [38], these rules are sound and complete under a probabilistic interpretation of ϵ – entailment [1]. In this interpretation, probability distributions P_K ϵ-consistent with K in the sense of ϵ-entailment, i.e., such that $P_K(\alpha) = 1$ for each $\alpha \in L$, $P_K(\beta|\alpha) \geq 1 - \epsilon$ and $P_K(\alpha) \geq 0$ for each rule $\alpha \vdash_d \beta$ in D (where ϵ is a fixed parameter) are considered. A proposition p is ϵ-entailed by T when for each ϵ there exists a δ such that $P_K(p|E) \geq 1 - \epsilon$ for each δ-consistent probability distribution P_K.

One of main problems for non–monotonic logics is to define for a given set A of sentences a family $E(A)$ of all its *extensions*, i.e., sets of sentences acceptable by an intelligent agent as a description of the world. Different formal attempts to solve this problem are known. Some of them are trying to implement the principle called *Closed World Assumption* (facts which are not known are false). Having the set of extensions $E(A)$ one can define the skeptical consequence relation by taking the intersection of all possible extensions of A.

Let us observe that, e.g., classification problems can be treated as problems of constraints satisfaction with constraints specified by discernibility conditions and some optimality measures. Two consistent sets describing situations (objects) are satisfying the discernibility relation if their union creates an inconsistent (or inconsistent in some degree) set.

Rough set approach can be used to build a bridge between research on non–monotonic reasoning and practical applications. For example, one can develop methods for extracting rules for non–monotonic reasoning from data, e.g., to extract from data default rules as a special type of association rules [4], [76], [83].

Another area of research is to develop strategies for adaptive computing of rough set constructs. This is on one side related, e.g., to belief revision. However, using rough set approach one can expect rather to learn these strategies from data than to search for a very general mechanism behind belief revision strategies.

Many valued logics

Yet another treatment of inference was proposed by Jan Łukasiewicz [68] by assigning to propositions other – than *truth* and *falsity* – logical values. In the first 3–valued logic Ł$_3$ propositions were assigned additionally the value $1/2$ (*possible*); the meaning of implication $p \Rightarrow q$ was determined as $min(1, 1-v(p)+v(q))$ where $v(p)$ is the logical value of the proposition p; similarly negation was determined by the condition $v(\neg p) = 1 - v(p)$.

The same formulas were used to define semantics of the $n--valued$ Łukasiewicz logic Ł$_n$ and the infinite–valued logic Ł$_\omega$ (where logical values are rational numbers from the interval $[0,1]$). A complete axiomatization for these logics was proposed by Wajsberg [111]. Other systems for many-valued logic were proposed by Post, Kleene and others [111], [134].

In general, one may consider, e.g., Łukasiewicz implication, negation etc. as interpreted as functions of suitable arity on the interval $[0,1]$ a fortiori truth values may be regarded as real numbers from the interval $[0,1]$. In real–valued logics, truth-functional definitions of propositional functors rely on real functions on $[0,1]$ in particular on so–called $t-norms$ and $t-co-norms$ which resp. define semantics of conjunction and disjunction. Implications are usually defined by Łukasiewicz or Kleene implications or so-called *residuated implications* and *negations* are interpreted as decreasing idempotent functions on $[0,1]$ [59]. The interest in many–valued logics grew rapidly after introduction of fuzzy logic by Lotfi A. Zadeh (see below).

For many interesting relationships of rough logic and many–valued logic the reader is referred to section on rough logic and the bibliography there.

Inductive reasoning

Inductive reasoning (inference) can be described as an art of hypothesizing a set of premises P for a given set C of consequences with respect to a given background knowledge BK, i.e., $P \cup BK \models C$ [72].

In the above scheme, the unknown element is a set P of premises but also the semantic inference $\models$ has to be specified. Contrary to logical deductive semantic consequence, the inductive inference $\models$ is not concerned with absolute truth–preserving but deals with approximations of concepts and along with mechanisms for concept approximations it should also possess mechanisms for generating degrees of closeness between any concept approximated and its approximation. These degrees may be expressed as numerical values or logical expressions. It is hardly expected that the inductive inference $\models$ may be defined abstracting from the specific background knowledge BK, i.e., from the applicational context; one should rather expect a variety of inference mechanisms dependent on the context and extracted from data by using appropriate algorithmic tools. This seems to be a challenge for further development of logic.

A general scheme for inductive reasoning may thus consist of a mechanism for primitive concept formation and a mechanism for construction of complex concepts from primitives. All concepts are of approximative character and mechanisms for their construction must rely on some measures of closeness among concepts. It is important to realize that concepts may be defined in various languages and their comparison is effected by imposing measures of closeness on their extensions.

Particular areas of importance for inductive reasoning are Machine Learning, Pattern Recognition and Knowledge Discovery in Data. Various specific approaches have been proposed for inductive (approximative) inference in these areas. In addition some universal paradigms for inductive reasoning like inductive logic programming [77] have been developed, e.g., fuzzy inference, rough inference, probabilistic and statistical reasoning. In what follows we will outline the basic ideas of these approaches.

General view: experimental concepts and approximate definability

Background knowledge is often expressed by means of data tables (e.g., training and test examples in Machine Learning, Pattern Recognition, Inductive Logic Programming). These data tables contain positive as well as negative examples for concepts to be learned or recognized; most often, the given examples form a relatively small part of the concept extension so we may not learn these concepts exactly but approximately only. In constructing approximations to concepts, the choice of a language for concept description (e.g., a language of logical (Boolean) formulas) involving the choice of primitive formulas as well as inference mechanisms is very important. Finding a suitable language is itself a challenging problem in scientific discovery. Approximate definability is effected by means of some measure μ of closeness (similarity) on concept extensions; one of widely used measures is the Lukasiewicz measure μ_L [67] based on frequency count, $\mu_L(A, B) = card(A \cap B)/card(A)$ where A, B are concept extensions, rediscovered by Machine Learning and KDD communities recently.

Some natural constraints can be put on concept approximations (i) the extension of concept approximation should be consistent with the concept extension or, at least almost consistent, i.e., consistent on training examples and having *small* error on test examples; (ii) some minimality (economy) conditions, e.g., minimal length of concept description, universality of description or best adaptability. These conditions are applied in Machine Learning, Pattern Recognition and KDD. Satisfying (i) as well as (ii) may be computationally hard (finding a minimal consistent description is NP-hard for propositional logic [7], [75]) hence there are strategies for suboptimal approximations. Hence as a solution to concept approximation problem we obtain as a rule a family of concept descriptions parameterized by languages chosen or strategies applied for particular solutions. Choice of a particular solution may be motivated by ease of adaptation, computational complexity of procedures of parameter tuning etc. Let us observe that the choice of primitive concepts is actually a choice of an initial model (relational structure) which itself is to be discovered; an essential criterion is its expresiveness for concept approximations (see also Section 1).

Relationships with machine learning, pattern recognition, inductive logic programming

To illustrate the general scheme, we borrow an example from KDD. A *decision rule* may be expressed in the form: $a_1 = v_1 \wedge a_2 = v_2 \wedge \wedge a_k = v_k \Rightarrow d = v$ where $a_1, a_2, ..., a_k$ are features (predicates) used to build formulas expressing approximating concepts and d is a (vector of) feature(s) used in formulas discerning concepts (decisions) approximated. An *association rule* [3], [29] is a decision rule in which the concept defined by the premise of the rule approximates the concept defined by the consequent of the rule in high degree (high confidence), i.e., sufficient fraction of examples satisfying the premise satisfy the consequent as well and there is a sufficient (defined by a set threshold) number of examples supporting both the premise and the consequent. Similar ideas are exploited in Machine Learning and Pattern Recognition for estimating a strength of a rule.

The problem of selecting relevant features [78], [75] involves some searching procedures like discretization, grouping of symbolic values, clusterization, morphological filtering. These preliminary procedures define primitive concepts (features) for a given problem of concept approximation. This process leads from primitive features (variables) (e.g., real–valued features, pixel–valued features) to new intrinsic features (variables), the gain being a more compact and a more general description of concepts. Another alternative in search for features is to search for hidden features (variables) – possibly better suited for concept approximation – in terms of which one may define (possibly near–to–functionally) the existing features (variables).

Inductive reasoning about knowledge

Inductive reasoning about knowledge presented above aims at evaluating the degree of assurance in validity of inferences $p \vdash q$. In this process, the semantics of inference is established relative to a parameterized family of connectives. The next step consists in finding approximations to connectives making inferences conforming with reality. Tasks of finding those approximations are often local. From these approximations, elementary inference schemes are built and mechanisms for composing these schemes are found. Inference schemes are often desirable to be distributed (autonomous or hierarchical) due to high complexity of inference problems. On these schemes auto–epistemic modalities may be superposed permitting the system to evaluate its knowledge and express meta–statements about it. We will discuss an example on constructing such schemes in the following section.

Reasoning about knowledge

Approximate reasoning about knowledge

Logics for approximate reasoning about knowledge attempt to use symbolic calculi in order to express concepts by their approximations being extensions of formulas of these logics.

Fuzzy logic

Fuzzy logic [139] is a logic of vague concepts whose descriptors (very often in natural language) are familiar to everyone but whose extensions are subjective due to distinct understanding; Lotfi Zadeh proposed to interpret such concepts as *fuzzy sets*. A subset $X \subseteq U$ is a *fuzzy set* in U when the membership in X is not crisp (binary) but it is subject to gradation; formally, this is expressed by requiring the characteristic function μ_X of X to be a function from U into the interval $[0,1]$ (not into $\{0,1\}$ as in classical set theory); usually, a fuzzy set X is identified with its fuzzy membership function μ_X. A model for a concept X is a domain D_X of this concept along with a finite set $A_1, A_2, ..., A_k$ of its *values (features)* interpreted as fuzzy subsets μ_{A_i} of D_X. Fuzzy logic is built over an alphabet consisting of propositional logic symbols along with symbols for concepts and their features and a symbol ι. Elementary formulas are of the form $X\iota A$ (read *X is A*) where X is a concept symbol and A is a feature symbol of that concept. Formulas are formed from elementary formulas by means of propositional functors. Models for fuzzy logics consist of a family of domains for concepts, fuzzy membership functions of concept features and certain operations for building complex domains from the simple ones. To illustrate the workings of this mechanism, consider, e.g., a formula $\alpha : X\iota A \vee Y\iota B$. Interpretations of elementary formulas $X\iota A$, $Y\iota B$ are resp. fuzzy membership functions μ_A on D_X and μ_B on D_Y. These functions have first to be lifted to the Cartesian product $D_X \times D_Y$ by cylindrical extensions. The resulting functions μ'_A, μ'_B may be compared and for a chosen t-co-norm T, the meaning of α is $T(\mu'_A(x,y), \mu'_B(x,y))$. Inferences in fuzzy logic are of the form of implications: **if** $X\iota A$ **then** $Y\iota B$ interpreted as functions of the form $I((\mu'_A(x,y), \mu'_B(x,y))$ where I is a many-valued logic implication. Fuzzy logic is the underlying logic of input–output signals in fuzzy controllers [59].

There have been reported many applications based on rough–fuzzy hybridization (see e.g., [87]) and results showing the relationships between rough and fuzzy approaches. Recently developed rough mereology (see section on mereological logic) can be treated as a generalization of rough and fuzzy approaches. It seems that hybridization of rough and fuzzy reasoning can be especially useful for further development of granular computing.

Rough logic

Logic of ambiguous concepts whose extensions are defined only approximately due to the incompleteness of knowledge was proposed by Zdzisław Pawlak [91] as *rough logic*. This logic is built over data structures known as information systems or data tables formalized as pairs (U, A) where U is a universe of objects and A is a finite set of attributes (features) where any attribute a is modeled as a mapping on U with values in a value set V_a. Rough logic is built from elementary formulas (descriptors) of the form (a, v) where $v \in V_a$ and propositional connectives. The model for this logic is the universe U and the meaning of an elementary formula (a, v) is $[(a, v)] = \{x \in U : a(x) = v\}$. Letting $[\alpha \vee \beta] = [\alpha] \cup [\beta]$, $[\alpha \wedge \beta] = [\alpha] \cap [\beta]$ and $[\neg \alpha] = U - [\alpha]$ extends the meaning to all formulas. Inference rules in rough logic are of the form : **if** α **then** β (decision rules, dependencies). True inference rules are those whose meaning is the universe U; in general, one may assign to an inference rule its truth degree defined, e.g., as the Łukasiewicz measure of closeness of the extension of the premise and the extension of the consequent. *Definable (exact, crisp)* concepts are extensions of formulas of rough logic; other concepts are *inexact (rough)*. Their description is approximate: for each such concept X there exist two formulas: α_{-X}, α_{+X} such that $[\alpha_{-X}] \subseteq X \subseteq [\alpha_{+X}]$ and $[\alpha_{-X}], [\alpha_{+X}]$ are resp. the maximal and the minimal exact concepts with this property; these approximating concepts are called resp. the lower and the upper approximation of X. Reasoning in terms of exact concepts is carried out on lines of classical logic while reasoning with and about general concepts is approximate but degrees of approximation can be found from data. Several attempts has been made to built formal deductive systems for rough logic [86].

For any information system $\mathcal{A} = (U, A)$ one can define a family $RS(\mathcal{A})$ of *rough set representations*, i.e., pairs $(\underline{A}X, \overline{A}X)$, where $X \subseteq U$. Two questions arise immediately: (i) How to characterize the set of all rough set representations in a given information system? and (ii) What are "natural" algebraic operations on rough set representations?

A pair $(\underline{A}X, U - \overline{A}X)$ can be assigned to any rough set $(\underline{A}X, \overline{A}X)$ in $\mathcal{A}$. The following "natural" operations on those pairs of sets are: $(X_1, X_2) \wedge (Y_1, Y_2) = (X_1 \cap Y_1, X_2 \cup Y_2)$, $(X_1, X_2) \vee (Y_1, Y_2) = (X_1 \cup Y_1, X_2 \cap Y_2)$, $\sim (X_1, X_2) = (X_2, X_1)$ or $\neg(X_1, X_2) = (U - X_1, X_1)$, or $\div(X_1, X_2) = (X_2, U - X_2)$. The defined operations are not accidental: we are now very close (still the implication operation should be defined properly!) to basic varieties of abstract algebras, such as *Nelson* or *Heyting* algebras, extensively studied in connection with different logical systems. The reader can find formal analysis of relationships of rough sets with Nelson, Heyting, Łukasiewicz, Post or double Stone algebras, in particular the representation theorems for rough sets in different classes of algebras in [100]. Let us also note that the properties of the negation operations defined above show that they correspond to the well-known negations studied in logic: strong (constructive) negation or weak (intuitionistic) negation.

Rough algebras can be derived from rough equality. Some relationships of rough algebras with many-valued logics have been shown such as, for example, soundness and completeness of Łukasiewicz's 3-valued logic with respect to rough semantics have been proven. The rough semantics defined by rough algebras is a special kind of a topological quasi-Boolean algebra; relationships of rough sets with 4-valued logic have been found. The relationships of rough logic with Kleene 3-valued logic has been established [70].

There is a number of results on logics, both propositional and predicate, that touch upon various rough set aspects. They have some new connectives (usually modal ones) reflecting different aspects of the approximations. On the semantic level they allow to express, among other possibilities, how the indiscernibility classes (or tolerance classes) are "matching" the interpretations of formulae in a given model M. For example, in the case of necessity connective the meaning $(\Box \alpha)_M$ of the formula α in the model M is the lower approximation of α_M, in case of possibility connective $(\Diamond \alpha)_M$ it is the upper approximation of α_M. Many other connectives have been introduced and logical systems with these connectives have been characterized. For example, rough quantifiers can be defined for predicate logic. Results related to the completeness of axiomatization, decidability as well as expressibility of these logical systems are typical. The reader can find more information on rough logic in [86], [70] and in the bibliography in [101].

It is finally worth to mention a research direction related to the rough mereological approach to approximate synthesis of objects satisfying a given specification in a satisfactory degree. Let us note here that an interesting prospect for applied logic is to look for algorithmic methods of extracting logical structures from data. This goal is related to aims of rough mereology, several aspects in the KDD research and to the calculi on information granules and to computing with words [140] and [141]. For further references see [86], [100] and [101].

Probabilistic logic

A number of logics for approximate reasoning employ probabilistic semantics [1], [95]. It is based on evaluating evidence based probabilities of inferences $p \Rightarrow q$. In these evaluations often one applies *Bayesian reasoning*: in many applications, e.g., in medicine, it is relatively easy to establish probabilities $P(q \Rightarrow p)$, $P(p), P(q)$; from those, $P(p \Rightarrow q)$ is found via the Bayes formula as $P(q \Rightarrow p)P(p)$. Complex inferences may be carried out in semantic networks known as *Bayesian belief networks* which are graphical representations of causal relations in a domain. Having joint probability table $P(U)$ where $U = \{A_1, ..., A_n\}$, one can calculate $P(A_i)$ or $P(A_i \mid E)$ where E is an evidence. However the size of $P(U)$ grows exponentially with the number of variables. Therefore we look for compact representation of $P(U)$ from which $P(U)$ can be calculated if needed. A Bayesian network over U is such a representation. If conditional independences in the Bayesian network hold for U then $P(U)$ can be calculated from the conditions specified in the network.

Relationships of decision rules to the Bayes theorem are shown in [92]. Recently [121], rough set methods have been developed for extraction of Bayesian belief networks from data.

Localization (mereological logics)

Lukasiewicz measure μ_L of concept closeness may be generalized to a notion of rough inclusion [98], i.e., a predicate $\mu(X, Y)$ which for concepts X, Y returns the value of degree in which X is *a part of Y*. This predicate allows for approximate reasoning based on the notion of a partial containment: the inference $p \Rightarrow q$ is true in the degree $\mu([p], [q])$ where $[p]$ is the extension of the statement p. It seems that the understanding of μ should be local, i.e., each intelligent agent should have its own collection of specific forms of rough inclusion and apply them locally again, i.e., in distinct data substructures different rough inclusions may be valid. This idea involves necessity of a calculus for fusion of local rough inclusions at any agent as well as for propagation of rough inclusions among agents.

In search of a proper framework we turn to mereological theory of St. Leś - niewski [61]. This set theory has as the primitive notion the notion of a (*proper*) *part predicate*: $XpartY$ is required to be irreflexive and transitive; the improper part predicate *ipart* is defined as follows: $XipartY$ if $XpartY$ or $X = Y$. The notion of a set is relativized to non–void properties: for any such property P, an object X is a set of objects with the property P ($XsetP$) if and only if given any Y with $YipartX$ there exist objects Q, R with $QipartY, QipartR, RipartX, R$ having P. A universal set for P is a class of P; one requires the uniqueness of class for each P. In this theory X is said to be an element of Y if for some property P, X has P and Y is class of P. The notion of a subset is extensional: X is a subset of Y if for each Z: Z element of X implies Z element of Y. It turns out that to be an element, to be a subset, and to be an improper part are all equivalent.

One may want to define a *rough inclusion* μ in such a way that a hierarchy of partial relations created by it contains a hierarchy of objects according to a part relation in the sense of Leśniewski. A set of postulates to this end may be as follows [98]: (1) $\mu(x,x) = \omega$; (2) $\mu(x,y) = \omega \Rightarrow (\mu(z,y) \geq \mu(z,x))$; (3) $(\mu(x,y) = \omega \land \mu(y,x) = \omega) \Rightarrow (\mu(x,z) \geq \mu(y,z))$; (4) $(\mu(z,x) = \omega) \Rightarrow \exists w : (\mu(w,z) = \omega \land \mu(w,y) = \omega) \Rightarrow (\mu(x,y) = \omega)$. An additional set of axiom schemes may guarantee class uniqueness. A model is a set M along with a 2-ary function F from M into a lattice L (let us set $L = [0,1]$) and the constant ω interpreted as $maxL$ (in our case, 1); in this interpretation $\mu(x,y)$ becomes $F(v(x), v(y))$, i.e., degree in which the element $v(x)$ of D is a part of $v(y)$. Letting in D: $v(x)partv(y)$ if $F(v(x), v(y)) = 1$ gives a relation of part in the sense of Leśniewski. Hence, a model for μ is also a model for fuzzy inference.

Reasoning on basis of μ requires a calculus for fusion of local inclusions at any agent reasoning about the world as well as a calculus for propagation of uncertainty among agents in order to ensure the correctness of inference over a scheme of agents. Logical calculus for an agent ag may be defined over a predicate calculus $L(ag)$: elementary formulas are of the form $\langle ag, \Phi(ag), \epsilon(ag) \rangle$ where $\Phi(ag)$ is a predicate of $L(ag)$ and $\epsilon(ag) \in [0,1]$. Fusion of local calculi is achieved by selecting a subset $St(ag) \subseteq D(ag)$ of *standards*: to each standard st a local rough inclusion μ_{st} is attached. Then we say that an object $x \in D(ag)$

satisfies $\langle ag, \Phi(ag), \epsilon(ag) \rangle$ if and only if there exists a standard st such that : (i) st satisfies $\Phi(ag)$ (ii) $\mu_{st}(x, st) \geq 1 - \epsilon(ag)$. Propagation of uncertainty is achieved by means of mereological connectives f extracted from data; their role is to state that when children of ag submit objects satisfying their formulas in degrees say $\epsilon_1, ..., \epsilon_n$ then the object composed them satisfies a formula at ag in degree at least $f(\epsilon_1, ..., \epsilon_n)$ [98]. Schemes of agents may be composed to create inference engines for approximate proofs of complex formulas from atomic ones.

As with fuzzy logic and bayesian reasoning, application of this logic is analytical: first, the necessary ingredients have to be learned on training data and then tested on test data.

4 KDD and rough logic

4.1 KDD as a logical process

Data mining can be described as searching in data for relevant structures (semantic models of logics) and their primitive properties, i.e., *patterns*. These relevant constructs are used to discover knowledge. The process of knowledge discovery can be treated as a kind of inference process (classical, common-sense etc.) based on the constructs found in data mining stage leading to efficient solutions of tasks like classification, prediction, etc. The solutions provide us with approximate descriptions of concepts of interest having satisfactory quality. The inference process has its own logic: in some cases it may be based on classical logic but in many cases, due to uncertainty, the logic of inference should be extracted from data as schemes for approximate reasoning [98], [139]. In this latter case a very important issue is knowledge granulation and reasoning with information granules [140, 141], [103] making feasible the reasoning process in case of complex problems like spatial reasoning [125] where the perception mechanisms play important role. Finally, let us mention that the inference process should be dynamically adapted to changing data. This calls for a development of adaptive reasoning strategies tuning parameters of logical models (structures) and formulas to induce the optimal (sub-optimal) concept approximations.

4.2 Basic philosophy of rough set methods in KDD

Rough set tools for KDD are based on parameterized approximation spaces and strategies for parameter tuning leading to approximations of (whole or parts of) sets of objects called *decision classes* (or *concepts*).

Approximation spaces (cf. Chapter by STEPANIUK) consist of two main constructs: sets of objects called *neighborhoods* and a measure of inclusion on sets. Neighborhoods are described by some parameterized formulas. The inclusion measure is indexed by some parameters describing a degree of set inclusion. By parameter tuning the relevant neighborhoods as well as the relevant degree of inclusion are chosen with respect to a given problem. In this section we explain the main steps in inducing concept (or its part) description on the basis of rough set methods.

We assume a finite universe U of objects, represented in an information system $\mathcal{A} = (U, A)$ (see Chapter 1), being a subset of the whole universe U^∞ of objects. By a *concept (decision class)* in U^∞ we understand any subset of U^∞.

One of the main goals of such areas as Machine Learning, Pattern Recognition as well as Knowledge Discovery and Data Mining is to develop methods for approximate descriptions of concepts (or their interesting parts) on the basis of partial information about them represented by information about objects from the universe U. More formally, let $X \subseteq U^\infty$ be a concept and let $\mathcal{A} = (U, A)$ be a given information system. (The information system together with the characteristic function of $X \cap U$ are represented by so called *decision systems*.) The problem is to induce an approximate description of $U \cap X$ using attributes from A so that its extension to U^∞ will approximate X with sufficiently high quality. One could formulate this by saying that the induced concept description should estimate the real concept well.

Let us mention that there are some other factors due to which a concept approximation can be induced rather than its exact description. Among them are well known problems with missing values or noise in data.

Rough set approach allows to precisely define the notion of a concept approximation. It is based [91] on the *indiscernibility* (or, a weaker form viz. a similarity) relation between objects defining a *partition* (or, covering) of the universe U of objects. There are some issues to be discussed here.

The first one concerns the way in which inclusion of sets, in particular, of neighborhoods is measured.

From the point of view of KDD, it is important to consider two notions related to formulas, namely *inclusion* and *closeness*. Instead of classical crisp inclusion or equivalence it is more appropriate to consider inclusion in a degree and closeness in a degree. An example of such inclusion is used in case of association rules [4]. Assuming a given set of formulas and a given finite universe of objects the task is to search for pairs of formulas α, β such that α is sufficiently included in β. The degree of inclusion for association rules is expressed by two thresholds consisting of coefficients known as *support* and *confidence*.

The second issue concerns the definition of quality measure used to estimate quality of concept description by neighborhoods from a given partition (or covering) in U and in U^∞.

To solve the first problem, i.e., approximation in U, rough sets are offering the approximation operations on sets of objects (see Chapter 1). The classical rough set approach [91] is using crisp inclusion and by taking the union of all neighborhoods included in a given set it yields the *lower approximation* of this set. One can also define the *upper approximation* of the given set of objects by taking union of all objects having non-empty intersection with a given set. In the complement of the upper approximation of a given set there are all objects which can be with certainty classified as not belonging to the set (using available neighborhoods). The difference between the upper and lower approximations is called the *boundary region* of the approximated set and it is created by objects which cannot be classified with certainty neither to the approximated set nor to its complement (using given neighborhoods defining partition or covering).

Consequently, each rough set exhibits *boundary-line cases*, i.e., objects which cannot be with certainty classified neither as members of the set nor of its complement.

As a consequence vague concepts, in contrast to precise concepts, cannot be characterized in terms of information about their elements. Therefore, in the proposed approach, we assume that any vague concept is replaced by a pair of precise concepts – the lower and the upper approximation of this vague concept. The difference between the upper and the lower approximation constitutes the boundary region of the vague concept.

For applications in KDD, a more relaxed approach has been developed: instead of crisp inclusion it seems more appropriate to use inclusion in a degree and to define set approximations on the basis of such inclusion [144], [118], [98], [94], [119]. One can search for proper degrees in the sense that induced by their means concept descriptions are showing better classification quality on new objects or, more generally, they give better quality of concept description in U^∞.

There are some other aspects of concept description which should be taken into consideration when one would like to estimate its quality in U^∞. One can follow, e.g., the *minimal description length principle* [109] (cf. Chapter by ŚLĘZAK) suggesting that more coarser partitions of U allow for better concept approximation assuming they describe the set $X \cap U$ still with a sufficient quality (measured, e.g., by the size of boundary region or other measures like entropy). Hence, in particular, it follows that preprocessing methods, leading to partitions more coarser yet still relevant for concept approximation, are important for rough set approach. Moreover, it is necessary to develop methods for achieving an appropriate tradeoff between coarseness of partitions and quality of concept approximation determined by the chosen partition (or covering). There are several approaches, known in Machine Learning or Pattern Recognition realizing this tradeoff. Among them one can distinguish the following ones:

1. **Feature selection** is realized by searching for relevant indiscernibility relations in the family of all indiscernibility relations $IND(B)$ where $B \subseteq A$ (see Chapter 1).
2. **Feature extraction** is realized by searching for relevant indiscernibility relations in the family of all indiscernibility relations $IND(B)$ where $B \subseteq C$ and C is a given set of attributes in A (see Chapter 1).
3. **Discovery of a relevant subspace C of attributes definable by A.** Next, the relevant attributes are extracted within a given subspace C of attributes definable by A. In this way the searching process for relevant attributes can be made more efficient because the computational complexity of the searching problem for relevant attributes in the whole space of attributes definable by A makes the process of relevant feature extraction infeasible.

It is worth to mention that very little is known about how to discover relevant subspaces C of attributes definable by A and this is still a challenge for knowledge discovery and data mining.

The (described above) process of searching for appropriate (coarser yet relevant) partitions (defined by C) is related in rough set theory to searching for constructs called *reducts* and their approximations. Let us observe that in practice the generated models of partitions (coverings) are usually sub-optimal with respect to the minimal description length principle. This is because of the computational complexity of searching problems for the optimal models. Moreover, it is necessary to tune the extracted models to receive satisfactory quality, e.g., of classification or description of induced models.

The combination of rough set approach with Boolean Reasoning [14] has created a powerful methodology allowing to formulate and efficiently solve searching problems for different kinds of reducts and approximations to them.

It is necessary to deal with Boolean functions of large size to solve real-life problems. A successful methodology based on the discernibility of objects and Boolean reasoning (see Chapter 1) has been developed for computing many important for applications constructs like reducts and their approximations, decision rules, association rules, discretization of real value attributes, symbolic value grouping, searching for new features defined by oblique hyper-planes or higher order surfaces, pattern extraction from data as well as conflict resolution or negotiation. Reducts are also basic tools for extracting functional dependencies or functional dependencies in a degree from data (for references see the papers and bibliography in [116], [87], [100], [101]).

A great majority of problems related to generation of the above mentioned constructs are of high computational complexity (i.e., they are NP-complete or NP-hard). This is also showing that most problems related to, e.g., feature selection or pattern extraction from data have intrinsic high computational complexity. However, using the methodology based on discernibility and Boolean reasoning it has been possible to discover efficient heuristics returning suboptimal solutions to these problems. The reported results of experiments on many data sets are very promising. They show very good quality of solutions (expressed by the classification quality of unseen objects and time necessary for solution construction) generated by proposed heuristics in comparison to other methods reported in literature. Moreover, for large data sets the decomposition methods based on patterns called templates have been developed (see e.g., [84], [81]) as well as a method to deal with large relational databases (see e.g., [80]).

It is important to note that this methodology allows for constructing heuristics having a very important *approximation property* which can be formulated as follows: expressions generated by means of heuristics (i.e., implicants) and sufficiently *close* to prime implicants define approximate solutions to the problem [114].

Parameterized approximation spaces and strategies for parameter tuning are basic tools for rough set approach in searching for data models under a given partition of objects (see e.g., [84], [81],[116], [123], [124], [100], [101]). Parameters to be tuned are thresholds used to select elements of partitions (coverings), to measure degree of inclusion (or closeness) of sets, or parameters measuring quality of approximation.

Rough sets offer methods for exploratory data analysis, i.e., methods for hypothesis generation rather than hypothesis testing [41]. Data mining without proper consideration of statistical nature of the inference problem is indeed to be avoided. We now shortly discuss how statistical methods are used in combination with rough set methods.

There is an important issue about the statistical validity and significance of constructs generated using rough set methods. Rough set data analysis becomes the most popular non-invasive method [25], [24] today. Non-invasive methods of data analysis aim to derive conclusions by taking into account only the given data without stringent additional model assumption. Using fewer model assumption results in more generality, wider applicability, and reduced costs and time. Non-invasive methods of data analysis use only few parameters which require only simple statistical estimation procedures. However, the generated constructs, like reducts or decision rules, should be controlled using statistical testing procedures.

Hence any non-invasive method needs to be complemented by methods for statistical significance, statistical validation, estimation of errors and model selection using only parameters supplied by the data at hand. This can be related to the ideas of statisticians looking for methods which could weaken the assumptions necessary for good estimators [41], [51].

The reader can find in literature (e.g., in [24],[25], [10], [2]), [54], [100], [101]) such methods, together with discussion of results of their applications. Among them are the following ones: data filtering based on classification error [24],[25], methods for significance testing using randomization techniques (e.g., dynamic reducts and rules [10]), model selection (e.g., by rough set entropy [24],[25]; by rough modeling allowing for generation of compact and accurate models based on, e.g., receiver operating characteristic analysis and rough modeling [2],[57] by decision rule approximation [54] - with the quality described by entropy [54], [24] and quality of classification of new objects, by reduct approximation [54] - with the quality expressed by positive region [54] and quality of classification of new objects, by boundary region thinning [54] using variable precision rough set model [144]). For more details the reader is referred to the bibliography on rough sets (see e.g., the bibliography in [100],[101]). Rough set approach backed up by non-invasive methods becomes a fully fledged instrument for data mining [25].

There are some other interesting current research topics which we can only mention. Let us observe that when defining approximation of concepts, in most cases we deal with concepts creating a partition of U^∞. Classifying new objects one should resolve *conflicts* between votes, coming from different approximations, for or against particular concepts. Recently a hybrid rough-neuro method has been applied for learning the strategy of conflict resolution for a given data [127]. Other current research topic deals with more complex information granules than discussed so far, defined by simple formulas, and methods for fusion of information granules to induce complex information granule approximations. The interested reader is referred to papers related to *rough mereological* approach (see e.g., [98], [103]) and papers related to spatial data mining (see e.g., [97], [119]).

In what follows we discuss in more details rough set approach vs. a general task of KDD and we refer to basic notions and examples of Chapter 1. For more information the reader is referred to [87], [100], [101] and to bibliographies in these books.

4.3 Reducts

In the previous section we investigated one dimension of reducing data which aimed at creating equivalence classes. The gain is apparent: only one element of the equivalence class is needed to represent the entire class. The other dimension in reduction is to store only those attributes that suffice to preserve the chosen indiscernibility relation and, consequently, the concept approximations. The remaining attributes are redundant since their removal does not worsen the classification.

Given an information system $\mathcal{A} = (U, A)$ a *reduct* is a minimal set of attributes $B \subseteq A$ such that $IND_\mathcal{A}(B) = IND_\mathcal{A}(A)$, cf. Chapter 1. In other words, a reduct is a minimal set of attributes from A that preserves the original classification defined by the set A of attributes. Finding a minimal reduct is NP-hard [117]. One can also show that the number of reducts of an information system with m attributes can be equal to

$$\binom{m}{\lfloor m/2 \rfloor}$$

There exist fortunately good heuristics that compute sufficiently many reducts in an often acceptable time. Boolean reasoning can be successfully applied in the task of reduct finding, cf. Chapter 1. We recall this algorithmic procedure here.

For $\mathcal{A}$ with n objects, the *discernibility matrix* of $\mathcal{A}$ is a symmetric $n \times n$ matrix with entries c_{ij} as given below. Each entry consists of the set of attributes upon which objects x_i and x_j differ.

$$c_{ij} = \{a \in A \mid a(x_i) \neq a(x_j)\} \quad i, j = 1, ..., n$$

A *discernibility function* $f_\mathcal{A}$ for an information system $\mathcal{A}$ is a Boolean function of m Boolean variables $a_1^*, ..., a_m^*$ (corresponding to the attributes $a_1, ..., a_m$) defined below, where $c_{ij}^* = \{a^* \mid a \in c_{ij}\}$.

$$f_\mathcal{A}(a_1^*, ..., a_m^*) = \bigwedge \left\{ \bigvee c_{ij}^* \mid 1 \leq j \leq i \leq n, c_{ij} \neq \emptyset \right\}$$

The set of all prime implicants of $f_\mathcal{A}$ determines the set of all reducts of $\mathcal{A}$. In the sequel we will write a_i instead of a_i^*.

The intersection of all reducts is the so-called *core* (which may be empty).

In general, the decision is not constant on the indiscernibility classes. Let $\mathcal{A} = (U, A \cup \{d\})$ be a decision system. The *generalized decision in* $\mathcal{A}$ is the function $\partial_\mathcal{A} : U \longrightarrow \mathcal{P}(V_d)$ defined by $\partial_\mathcal{A}(x) = \{i \mid \exists x' \in U \; x' \; IND(A) \; x \text{ and } d(x') = i\}$. A decision system $\mathcal{A}$ is called *consistent (deterministic)*, if $|\partial_\mathcal{A}(x)| = 1$ for any $x \in U$, otherwise $\mathcal{A}$ is *inconsistent (non-deterministic)*. Any set consisting of all objects with the same generalized decision value is called a *generalized decision class*.

It is easy to see that a decision system $\mathcal{A}$ is consistent if, and only if, $POS_A(d) = U$. Moreover, if $\partial_B = \partial_{B'}$, then $POS_B(d) = POS_{B'}(d)$ for any pair of non-empty sets $B, B' \subseteq A$. Hence the definition of a decision-relative reduct: a subset $B \subseteq A$ is a *relative reduct* if it is a minimal set such that $POS_A(d) = POS_B(d)$. Decision-relative reducts may be found from a discernibility matrix: $M^d(\mathcal{A}) = (c_{ij}^d)$ assuming $c_{ij}^d = \emptyset$ if $d(x_i) = d(x_j)$ and $c_{ij}^d = c_{ij} - \{d\}$, otherwise. Matrix $M^d(\mathcal{A})$ is called *the decision-relative discernibility matrix of $\mathcal{A}$*. Construction of *the decision-relative discernibility function* from this matrix follows the construction of the discernibility function from the discernibility matrix. It has been shown [117] that the set of *prime implicants* of $f_M^d(\mathcal{A})$ defines the set of all *decision-relative reducts* of $\mathcal{A}$.

In some applications, instead of reducts we prefer to use their approximations called α-*reducts*, where $\alpha \in [0,1]$ is a real parameter. For a given information system $\mathcal{A} = (\mathcal{U}, \mathcal{A})$, the set of attributes $B \subset A$ is called an α-reduct in case B has a non–empty intersection with at least $\alpha \cdot 100\%$ of non–empty entries $c_{i,j}$ in the discernibility matrix of $\mathcal{A}$.

Here, it will be important to make some remarks because the most methods discussed later are based on generation of some kinds of reducts.

The discernibility matrix creates a kind of universal *board game* used to develop efficient heuristics (see e.g., [116]). High computational complexity of analyzed problems (like NP-hardness of minimal reduct computation problem [117]) is not due to their formulation using Boolean reasoning framework but it is the intrinsic property of these problems. One cannot expect that by using other formalization the computational complexity of these problems can be decreased.

One should take into account the fact that discernibility matrices are of large size for large data sets. Nevertheless, it was possible to develop efficient and high quality heuristics for quite large data sets (see e.g., papers and bibliography in [100], [101]). This was possible due to the fact that in general it is not necessary to store the whole discernibility matrix and analyze all of its entries. This follows from reasons like: (i) only some short reducts should be computed; (ii) for some kinds of reducts, like reducts relative to objects only one column of the matrix is important; (iii) in discretization of real value attributes, some additional knowledge about the data can be used in searching for relevant (for computing reducts) Boolean variables. Let us also note that our approach is strongly related to propositional reasoning [114] and further progress in propositional reasoning will bring further progress in discussed methods.

For data sets too large to be analyzed by developed heuristics, several approaches have been developed. The first one is based on decomposition of large data into regular sub–domains of size feasible for developed methods. We will shortly discuss this method later. The second one, a statistical approach, is based on different sampling strategies. Samples are analyzed using the developed strategies and stable constructs for sufficiently large number of samples are considered as relevant for the whole table. This approach has been successfully used for generating different kinds of so called *dynamic reducts* (see e.g., [10]). It yields so called *dynamic decision rules*. Experiments with different data sets have proved these methods to be promising in case of large data sets. Another interesting method (see e.g., [80]) has shown that Boolean reasoning methodology can be extended to large relational data bases. The main idea is based on observation that relevant Boolean variables for very large formula (corresponding to analyzed relational data base) can be discovered by analyzing some statistical information. This statistical information can be efficiently extracted from large data bases.

4.4 Rough sets and Boolean reasoning: examples of applications

Decision rules

Reducts serve the purpose of inducing *minimal* decision rules. Any such rule contains the minimal number of descriptors in the conditional part so that their conjunction defines the largest subset of a generalized decision class (decision class, if the decision table is deterministic). Hence, information included in conditional part of any minimal rule is sufficient for prediction of the generalized decision value for all objects satisfying this part. Conditional parts of minimal rules define neighborhoods relevant for generalized decision classes approximation. It turns out that conditional parts of minimal rules can be computed (by means of Boolean reasoning) as so called *reducts relative to objects* (see e.g., [115], [10]). Once these reducts have been computed, conditional parts of rules are easily constructed by laying reducts over the original decision system and reading off the values. In the discussed case the generalized decision value is preserved during the reduction. One can consider stronger constraints which should be preserved. For example, in [120] the constraints are described by probability distributions corresponding to information signatures of objects cf. Chapter by ŚLĘZAK. Again, the same methodology can be used to compute *generalized reducts* corresponding to these constraints.

We recall, cf. Chapter 1, that rules are defined as follows.

Let $\mathcal{A} = (U, A, d)$ be a decision system. Atomic formulae over $B \subseteq A \cup \{d\}$ and V are expressions of the form $a = v$; they are called *descriptors* over B and V, where $a \in B$ and $v \in V_a$. The set $\mathcal{F}(B, V)$ of formulae over B and V is the least set containing all atomic formulae over B and V and closed under propositional connectives $\wedge$ (conjunction), $\vee$ (disjunction) and $\neg$ (negation).

The semantics (meaning) of the formulae is also defined recursively. For $\varphi \in \mathcal{F}(B, V)$, the meaning of φ in the decision system $\mathcal{A}$ denoted $[\varphi]_\mathcal{A}$ is the set of all objects in U with the property φ:

1. if φ is of the form $a = v$ then $[\varphi]_\mathcal{A} = \{x \in U \mid a(x) = v\}$
2. $[\varphi \wedge \varphi']_\mathcal{A} = [\varphi]_\mathcal{A} \cap [\varphi']_\mathcal{A}$; $[\varphi \vee \varphi']_\mathcal{A} = [\varphi]_\mathcal{A} \cup [\varphi']_\mathcal{A}$; $[\neg \varphi]_\mathcal{A} = U - [\varphi]_\mathcal{A}$

The set $\mathcal{F}(B, V)$ is called the set of *conditional formulae of* $\mathcal{A}$ and is denoted $\mathcal{C}(B, V)$.

A *decision rule* for $\mathcal{A}$ is any expression of the form $\varphi \Rightarrow d = v$, where $\varphi \in \mathcal{C}(B, V)$, $v \in V_d$ and $[\varphi]_\mathcal{A} \neq \emptyset$. Formulae φ and $d = v$ are referred to as the *predecessor* and the *successor* of the decision rule $\varphi \Rightarrow d = v$.

A decision rule $\varphi \Rightarrow d = v$ is *true* in $\mathcal{A}$ if and only if $[\varphi]_\mathcal{A} \subseteq [d = v]_\mathcal{A}$.

For a systematic overview of rule induction see the bibliography in [101].

Several numerical factors can be associated with a synthesized rule. For example, the support of a decision rule is the number of objects that match the predecessor of the rule. Various frequency-related numerical quantities may be computed from such counts.

The main challenge in inducing rules from decision systems lies in determining which attributes should be included in the conditional part of the rule. First, one can compute minimal rules. Their conditional parts describe largest object sets (definable by conjunctions of descriptors) with the same generalized decision value in a given decision system. Hence, they create largest neighborhoods still relevant for defining the decision classes (or sets of decision classes when the decision system is inconsistent). Although such minimal decision rules can be computed, this approach may result in a set of rules of not satisfactory classification quality. Such detailed rules might be over–fit and they will poorly classify new cases. Instead, shorter rules should rather be synthesized. Although they will not be perfect on the known cases (as they may be influenced by noise) there is a good chance that they will be of high quality in classifying new cases. They can be constructed by computing approximations in the above mentioned sense to reducts. The quality of approximation is characterized by a degree α of approximation. This degree can be tuned to obtain relevant neighborhoods. Using reduct approximations in place of reducts, we can obtain larger neighborhoods still relevant for decision classes description in the universe U^∞. Approximations to reducts received by dropping some descriptors from conditional parts of minimal rules define more general neighborhoods, not purely included in decision classes but included in them in a satisfactory degree. It means that when the received neighborhood descriptions are considered in U^∞ they can be more relevant for decision class (concept) approximation than neighborhoods described by exact reducts, e.g., because all (or, almost all) objects from the neighborhood not included in the approximated decision class are those listed in U. Hence, one can expect that when by dropping a descriptor from the conditional part we receive the description of the neighborhood almost included in the approximated decision class than this descriptor is a good candidate for dropping.

For estimation of the quality of decision classes approximation global measures based on the positive region [115] or entropy [25] are used. Methods of boundary region thinning can be based, e.g., on the idea of variable precision rough set model [144]. The idea is based on an observation that neighborhoods included in decision classes in satisfactory degree can be treated as parts of the lower approximations of decision classes. Hence, lower approximations of decision classes are enlarged and decision rules generated from them are usually stronger (e.g., they are supported by more examples). The degree of inclusion is tuned experimentally to achieve, e.g., high classification quality on new cases.

Another way of approaching reduct approximations is by computing reducts for random subsets of the universe of a given decision system and selecting the most stable reducts, i.e., reducts that occur in "most" subsystems. These reducts, called *dynamic reducts*, are usually inconsistent for the original table, but rules synthesized from them are more tolerant to noise and other abnormalities; rules synthesized from such reducts perform better on unseen cases since they cover most general patterns in the data (for references see the bibliography in [101] and [100]).

When a set of rules has been induced from a decision system containing a set of training examples, they can be inspected to see if they reveal any novel relationships between attributes that are worth further research. Furthermore, the rules can be applied to a set of unseen cases in order to estimate their classificatory power.

Several application schemes can be envisioned but a simple one that has proved useful in practice is the following:

1. When a *rough set classifier* (i.e., a set of decision rules together with a method for conflict resolving when they classify new cases) is confronted with a new case, then the rule set is scanned to find applicable rules, i.e., rules whose predecessors match the case.
2. If no rule is found (i.e., no rule "fires"), the most frequent outcome in the training data is chosen.
3. If more than one rule fires, these may in turn indicate more than one possible outcome.
4. A voting process is then performed among the rules that fire in order to resolve conflicts and to rank the predicted outcomes. All votes in favor of the rule outcome are summed up and stored as its support count. Votes from all the rules are then accumulated and divided by the total number of votes in order to arrive at a numerical measure of certainty for each outcome. This measure of certainty is not really a probability but may be interpreted as an approximation to such if the model is well calibrated.

The above described strategy to resolve conflicts is the simplest one. For a systematic overview of rule application, see the bibliography in [100] and [101]. Rough set methods can be used to learn from data the strategy for conflict resolving between decision rules when they are classifying new objects.

Several methods based on the rough set approach have been developed to deal with large data tables, e.g., to generate strong decision rules from them. We will discuss one of these methods based on decomposition of tables by using patterns, called *templates*, see Chapter 1, describing regular sub–domains of the universe (e.g., they describe a large number of customers having a large number of common features); cf. also Chapter by NGUYEN SINH HOA .

Let $\mathcal{A} = (\mathcal{U}, \mathcal{A})$ be an information system. The notion of a descriptor can be generalized by using terms of the form $(a \in S)$, where $S \subseteq V_a$ is a set of values. By a *template* we mean the conjunction of descriptors, i.e., $\mathbf{T} = D_1 \wedge D_2 \wedge ... \wedge D_m$, where $D_1, ... D_m$ are either simple or generalized descriptors. We denote by $length(\mathbf{T})$ the number of descriptors in $\mathbf{T}$.

An object $u \in U$ is satisfying the template $\mathbf{T} = (a_{i_1} = v_1) \wedge ... \wedge (a_{i_m} = v_m)$ if and only if $\forall_j a_{i_j}(u) = v_j$. Hence the template $\mathbf{T}$ describes the set of objects having the common property: *"the values of attributes $a_{j_1}, ..., a_{j_m}$ on these objects are equal to $v_1, ..., v_m$, respectively"*.

The *support* of **T** is defined by $support(\mathbf{T}) = |\{u \in U : u \text{ satisfies } \mathbf{T}\}|$. Long templates with a large support are preferred in many Data Mining tasks. We consider several quality functions which can be used to compare templates. The first function is defined by $quality_1(\mathbf{T}) = support(\mathbf{T}) + length(\mathbf{T})$. The second can be defined by $quality_2(\mathbf{T}) = support(\mathbf{T}) \times length(\mathbf{T})$.

Let us consider the following problems [81]:

1. **Optimal Template Support (OTS) Problem:**
   Instance: Information system $\mathcal{A} = (\mathcal{A}, \mathcal{U})$, and a positive integer L.
   Question: Find a template **T** with the length L and the maximal support.
2. **Optimal Template Quality (OTQ) Problem:**
   Instance: An information system $\mathcal{A} = (\mathcal{U}, \mathcal{A})$,
   Question: Find a template for $\mathcal{A}$ with optimal quality.

In [81] it has been proved that the optimal support problem (**OPT**) is NP-hard. The second problem is NP-hard with respect to $quality_1(\mathbf{T})$ and it is not known if this problem is NP-hard in case of $quality_2(\mathbf{T})$.

Large templates can be found quite efficiently by *Apriori* algorithms and its modifications (see [4, 143]). Some other methods for large template generation have been proposed (see e.g., [81]).

Templates extracted from data are used to decompose large data tables. In consequence the decision tree is built with internal nodes labeled by (extracted from data) templates, and edges outgoing from them labeled by 0 (false) and 1 (true). Any leaf is labeled by a sub–table (sub–domain) consisting of all objects from the original table matching all templates or their complements appearing on the path from the root of the tree to the leaf. The process of decomposition is continued until sub–tables attached to leaves can be efficiently analyzed by existing algorithms (e.g., decision rules for them can be generated efficiently) based on existing rough set methods. The reported experiments are showing that such decomposition returns many interesting regular sub–domains (patterns) of the large data table in which the decision classes (concepts) can be approximated with high quality (for references see [81], [84], [100] and [101]).

It is also possible to search for patterns that are almost included in the decision classes defining default rules [76]. For a presentation of default rules see the bibliography in [100] and [101].

Feature extraction: discretization and symbolic attribute value grouping

The rough set community have been committed to constructing efficient algorithms for (new) feature extraction. Rough set methods combined with Boolean reasoning [14] lead to several successful approaches to feature extraction. The most successful methods are:

- discretization techniques,
- methods of partitioning of nominal attribute value sets and
- combinations of the above methods.

Searching for new features expressed by multi-modal formulae can be mentioned here. Structural objects can be interpreted as models (so called Kripke models) of such formulas and the problem of searching for relevant features reduces to construction of multi-modal formulas expressing properties of the structural objects discerning objects or sets of objects [86].

For more details the reader is referred to the bibliography in [101].

Non-categorical attributes must be discretized in a pre–processing step. The discretization step determines how coarsely we want to view the world. Discretization, cf. Chapter 1, is a step that is not specific to the rough set approach. A majority of rule or tree induction algorithms require it in order to perform well. The search for appropriate cutoff points can be reduced to search for prime implicants of an appropriately constructed Boolean function.

There are two reasons that we include the discussion on discretization here. First of all it is related to the general methodology of rough sets discussed at the beginning of this section. Discretization can be treated as a searching for more coarser partitions of the universe still relevant for inducing concept description of high quality. We will also show that this basic problem can be reduced to computing reducts in some appropriately defined systems. It follows that we can estimate the computational complexity of the discretization problem. Moreover, heuristics for computing reducts and prime implicants can be used here. The general heuristics can be modified to more optimal ones using a knowledge about the problem, e.g., the natural order on the set of reals, etc. Discretization is only an illustrative example of many other problems with the same property.

Reported results show that discretization problems and symbolic value partition problems are of high computational complexity (i.e., NP-complete or NP-hard) which clearly justifies the importance of designing efficient heuristics. The idea of discretization is illustrated with a simple example.

Example 1. Let us consider a (consistent) decision system (see Tab. ??(a)) with two conditional attributes a and b and seven objects $u_1, ..., u_7$. Values of attributes of these objects and values of the decision d are presented in Tab. ??.

(a)

$\mathcal{A}$	a	b	d
u_1	0.8	2	1
u_2	1	0.5	0
u_3	1.3	3	0
u_4	1.4	1	1
u_5	1.4	2	0
u_6	1.6	3	1
u_7	1.3	1	1

$\Longrightarrow$

(b)

$\mathcal{A}^P$	a^P	b^P	d
u_1	0	2	1
u_2	1	0	0
u_3	1	2	0
u_4	1	1	1
u_5	1	2	0
u_6	2	2	1
u_7	1	1	1

Table 3. The discretization process: (a) The original decision system $\mathcal{A}$. (b) The P-discretization of $\mathcal{A}$, where $\mathbf{P} = \{(a, 0.9), (a, 1.5), (b, 0.75), (b, 1.5)\}$

Sets of possible values of a and b are defined by:

$$V_a = [0,2)\,;\, V_b = [0,4).$$

Sets of values of a and b for objects from U are respectively given by:

$$a(U) = \{0.8, 1, 1.3, 1.4, 1.6\} \text{ and}$$
$$b(U) = \{0.5, 1, 2, 3\}$$

Discretization process produces partitions of value sets of conditional attributes into intervals in such a way that a new consistent decision system is obtained from a given consistent decision system by replacing original value of an attribute on an object in $\mathcal{A}$ with the (unique) name of the interval(s) in which this value is contained. In this way, the size of value sets of attributes may be reduced. In case a given decision system is not consistent, one can transform it into a consistent one by taking the generalized decision instead of the original decision. Discretization will then return cuts with the following property: regions bounded by them consist of objects with the same generalized decision. One can also consider *soft* (impure) cuts and induce the relevant cuts on their basis (see the bibliography in [100]).

The following intervals are obtained in our example system:

$[0.8, 1)$; $[1, 1.3)$; $[1.3, 1.4)$; $[1.4, 1.6)$ for a);
$[0.5, 1)$; $[1, 2)$; $[2, 3)$ for b).

The idea of cuts can be introduced now. Cuts are pairs (a, c) where $c \in V_a$. Our considerations are restricted to cuts defined by the middle points of the above intervals. In our example the following cuts are obtained:

$(a, 0.9)$; $(a, 1.15)$; $(a, 1.35)$; $(a, 1.5)$;
$(b, 0.75)$; $(b, 1.5)$; $(b, 2.5)$.

Any cut defines a new conditional attribute with binary values. For example, the attribute corresponding to the cut $(a, 1.2)$ is equal to 0 if $a(x) < 1.2$; otherwise it is equal to 1.

By the same token, any set P of cuts defines a new conditional attribute a_P for any a. Given a partition of the value set of a by cuts from P put the unique names for the elements of these partition.

Example 2. Let $P = \{(a, 0.9), (a, 1.5), (b, 0.75), (b, 1.5)\}$ be the set of cuts. These cuts glue together the values of a smaller then 0.9, all the values in interval $[0.9, 1.5)$ and all the values in interval $[1.5, 4)$. A similar construction can be repeated for b. The values of the new attributes a_P and b_P are shown in Tab. ?? (b).

The next natural step is to construct a set of cuts with a minimal number of elements. This may be done using Boolean reasoning.

Let $\mathcal{A} = (U, A \cup \{d\})$ be a decision system where $U = \{x_1, x_2, \ldots, x_n\}$, $A = \{a_1, \ldots, a_k\}$ and $d : U \longrightarrow \{1, \ldots, r\}$. We assume $V_a = [l_a, r_a) \subset \Re$ to be a real interval for any $a \in A$ and $\mathcal{A}$ to be a consistent decision system. Any pair (a, c) where $a \in A$ and $c \in \Re$ will be called a *cut on* V_a. Let $\mathbf{P}_a =$

$\{[c_0^a, c_1^a), [c_1^a, c_2^a), \ldots, [c_{k_a}^a, c_{k_a+1}^a)\}$ be a partition of V_a (for $a \in A$) into sub–intervals for some integer k_a, where $l_a = c_0^a < c_1^a < c_2^a < \ldots < c_{k_a}^a < c_{k_a+1}^a = r_a$ and $V_a = [c_0^a, c_1^a) \cup [c_1^a, c_2^a) \cup \ldots \cup [c_{k_a}^a, c_{k_a+1}^a)$. It follows that any partition $\mathbf{P}_a$ is uniquely defined and is often identified with the set of cuts

$$\{(a, c_1^a), (a, c_2^a), \ldots, (a, c_{k_a}^a)\} \subset A \times \Re$$

Given $\mathcal{A} = (U, A \cup \{d\})$ any set of cuts $\mathbf{P} = \bigcup_{a \in A} \mathbf{P}_a$ defines a new decision system $\mathcal{A}^{\mathbf{P}} = (U, A^{\mathbf{P}} \cup \{d\})$ called $\mathbf{P}$-*discretization of* $\mathcal{A}$, where $A^{\mathbf{P}} = \{a^{\mathbf{P}} : a \in A\}$ and $a^{\mathbf{P}}(x) = i \Leftrightarrow a(x) \in [c_i^a, c_{i+1}^a)$ for $x \in U$ and $i \in \{0, .., k_a\}$.

Two sets of cuts $\mathbf{P}'$ and $\mathbf{P}$ are *equivalent*, written $\mathbf{P}' \equiv_A \mathbf{P}$, iff $\mathcal{A}^{\mathbf{P}} = \mathcal{A}^{\mathbf{P}'}$. The equivalence relation $\equiv_A$ has a finite number of equivalence classes. Equivalent families of partitions will be not discerned in the sequel.

The set of cuts $\mathbf{P}$ is called $\mathcal{A}$-*consistent* if $\partial_{\mathcal{A}} = \partial_{\mathcal{A}^{\mathbf{P}}}$, where $\partial_{\mathcal{A}}$ and $\partial_{\mathcal{A}^{\mathbf{P}}}$ are generalized decisions of $\mathcal{A}$ and $\mathcal{A}^{\mathbf{P}}$, respectively. An $\mathcal{A}$-consistent set of cuts $\mathbf{P}^{irr}$ is $\mathcal{A}$-*irreducible* if $\mathbf{P}$ is not $\mathcal{A}$-consistent for any $\mathbf{P} \subset \mathbf{P}^{irr}$. The $\mathcal{A}$-consistent set of cuts $\mathbf{P}^{opt}$ is $\mathcal{A}$-*optimal* if $card(\mathbf{P}^{opt}) \leq card(\mathbf{P})$ for any $\mathcal{A}$-consistent set of cuts $\mathbf{P}$.

It can be shown that the decision problem of checking if for a given decision system $\mathcal{A}$ and an integer k there exists an irreducible set of cuts $\mathbf{P}$ in $\mathcal{A}$ such that $card(\mathbf{P}) < k$ (k–**minimal partition problem**) is NP-complete. The problem of searching for an optimal set of cuts $\mathbf{P}$ in a given decision system $\mathcal{A}$ (**optimal partition problem**) is NP-hard, see Chapters by HOA SINH NGUYEN as well as by BAZAN et AL.

Despite these complexity bounds, it is possible to devise efficient heuristics that return semi–minimal sets of cuts. Heuristics based on Johnson's strategy look for a cut discerning a maximal number of object pairs and eliminate all already discerned object pairs. This procedure is repeated until all object pairs to be discerned are discerned. It is interesting to note that this can be realized by computing the minimal relative reduct of the corresponding decision system. The *"MD heuristic"* searches for a cut with a maximal number of object pairs discerned by this cut. The idea is analogous to Johnson's approximation algorithm. It may be formulated as follows:

ALGORITHM: MD-heuristics (A semi–optimal family of partitions)

Step 1. *Construct table* $\mathcal{A}^* = (U^*, A^* \cup \{d\})$ *from* $\mathcal{A} = (U, A \cup \{d\})$ *where U^* is the set of pairs (x, y) of objects to be discerned by d and A^* consists of attribute c^* for any cut c and c^* is defined by $c^*(x, y) = 1$ if and only if c discerns x and y (i.e., x, y are in different half-spaces defined by c); set $\mathcal{B} = \mathcal{A}^*$;*

Step 2. *Choose a column from $\mathcal{B}$ with the maximal number of occurrences of 1's;*

Step 3. *Delete from $\mathcal{B}$ the column chosen in Step 2 and all rows marked with 1 in this column;*

Step 4. *If $\mathcal{B}$ is non-empty then go to Step 2 else Stop.*

This algorithm searches for a cut which discerns the largest number of pairs of objects (MD-heuristics). Then the cut c is moved from A^* to the resulting set of cuts $\mathbf{P}$; and all pairs of objects discerned by c are removed from U^*. The algorithm continues until U^* becomes empty.

Let n be the number of objects and let k be the number of attributes of decision system $\mathcal{A}$. The following inequalities hold: $card\,(A^*) \leq (n-1)\,k$ and $card\,(U^*) \leq \frac{n(n-1)}{2}$. It is easy to observe that for any cut $c \in A^*$ $O\left(n^2\right)$ steps are required in order to find the number of all pairs of objects discerned by c. A straightforward realization of this algorithm therefore requires $O\left(kn^2\right)$ of memory space and $O(kn^3)$ steps in order to determine one *cut*. This approach is clearly impractical. However, it is possible to observe that in the process of searching for the set of pairs of objects discerned by currently analyzed cut from an increasing sequence of cuts one can use information about such set of pairs of objects computed for the previously considered cut. The MD-heuristic using this observation [79] determines the best cut in $O\,(kn)$ steps using $O\,(kn)$ space only. This heuristic is reported to be very efficient with respect to the time necessary for decision rules generation as well as with respect to the quality of unseen object classification.

We report some results of experiments on data sets using MD-like heuristics. We would like to comment for example on the result of classification received by an application to Shuttle data (Table 5). The result concerning classification quality is the same as the best result reported in [74] but the time is of order better than for the best result from [74]. In this table we present also the results of experiments with heuristic searching for features defined by oblique hyper-planes. This has been developed using genetic algorithm allowing to tune the position of the hyperplane to get an optimal one [79]. In this way one can implement propositional reasoning using some background knowledge about the problem.

For experiments, several data tables with real value attributes were chosen from the U.C. Irvine repository. For some tables, taking into account the small number of their objects, the approach based on the five-fold cross-validation $(CV-5)$ was adopted. The obtained results (Table 5) can be compared with those reported in [22],[74] (Table 4). For predicting decisions on new cases we apply only decision rules generated either by the decision tree (using hyperplanes) or by rules generated in parallel with discretization.

For some tables the classification quality of this algorithm is better than that of the C4.5 or Naive–Bayes induction algorithms [106] even when used with different discretization methods [22, 74, 16].

Comparing this method with the other methods reported in [74], we can conclude that algorithms based on the presented approach have the shortest runtime and a good overall classification quality (in many cases our results were the best in comparison to many other methods reported in literature).

We would like to stress that inducing the minimal number of the relevant cuts is equivalent to computing the minimal reduct in a decision system constructed from the discussed above system $\mathcal{A}^*$ [79]. This in turn, as we have shown, is equivalent to the problem of computing a minimal prime implicant of Boolean

Names	Nr of class.	Train. table	Test. table	Best results
Australian	2	690×14	CV5	85.65%
Glass	7	214×9	CV5	69.62%
Heart	2	270×13	CV5	82.59%
Iris	3	150×4	CV5	96.00%
Vehicle	4	846×19	CV5	69.86%
Diabetes	2	768×8	CV5	76.04%
SatImage	6	4436×36	2000	90.06%
Shuttle	6	43500×7	14500	99.99%

Table 4. Data tables stored in the UC Irvine Repository

Data tables	Diagonal cuts		Hyperplanes	
	#cuts	quality	#cuts	quality
Australian	18	79.71%	16	82.46%
Glass	14±1	67.89%	12	70.06%
Heart	11±1	79.25%	11±1	80.37%
Iris	7±2	92.70%	6±2	96.7%
Vehicle	25	59.70%	20±2	64.42%
Diabetes	20	74.24%	19	76.08%
SatImage	47	81.73%	43	82.90%
Shuttle	15	99.99%	15	99.99%

Table 5. Results of experiments on machine learning data.

function. This is only illustration of a wide class of basic problems of Machine Learning, Pattern Recognition and KDD which can be reduced to problems of relevant reduct computation.

The presented approach may be extended to the case of symbolic (nominal, qualitative) attributes as well as to the case of mixed nominal and numeric attributes.

In case of symbolic value attribute (i.e., without pre-assumed order on values of given attributes) the problem of searching for new features of the form $a \in V$ is, in a sense, from practical point of view more complicated than the for real value attributes. However, it is possible to develop efficient heuristics for this case using Boolean reasoning.

Any function $P_a : V_a \to \{1, \ldots, m_a\}$ (where $m_a \leq card(V_a)$) is called a *partition on* V_{a_i}. The *rank of* P_{a_i} is the value $rank(P_i) = card(P_{a_i}(V_{a_i}))$. The family of partitions $\{P_a\}_{a \in B}$ is *consistent with* B ($B - consistent$) if the condition $[(u, u') \notin IND(B/\{d\})$ implies $\exists_{a \in B}[P_a(a(u)) \neq P_a(a(u'))]]$ holds for any $(u, u') \in U$. It means that if two objects u, u' are discerned by B and d, then they must be discerned by partition attributes defined by $\{P_a\}_{a \in B}$. We consider the following optimization problem:

PARTITION PROBLEM: SYMBOLIC VALUE PARTITION PROBLEM:

Given a decision table $\mathcal{A}$ and a set of attributes $B \subseteq A$, search for the minimal $B-consistent$ family of partitions (i.e. ,such $B-consistent$ family $\{P_a\}_{a \in B}$ that $\sum_{a \in B} rank(P_a)$ is minimal).

To discern between pairs of objects, we will use new binary features $a_v^{v'}$ (for $v \neq v'$) defined by $a_v^{v'}(x,y) = 1$ if $a(x) = v \neq v' = a(y)$. One can apply Johnson's heuristics for the new decision table with these attributes to search for a minimal set of new attributes that discerns all pairs of objects from different decision classes. After extracting these sets, for each attribute a_i we construct a graph $\Gamma_{a_i} = \langle V_{a_i}, E_{a_i} \rangle$ where E_{a_i} is defined as the set of all new attributes (propositional variables) found for the attribute a_i. Any vertex coloring of Γ_{a_i} defines a partition of V_{a_i}. The colorability problem is solvable in polynomial time for $k = 2$, but remains NP-complete for all $k \geq 3$. But, similarly to discretization, one can apply some efficient heuristics searching for an optimal partition.

Let us consider an example of a decision table presented in Table 1 and (a reduced form) of its discernibility matrix (Table 1).

Fig. 1. The decision table and the discernibility matrix

$\mathcal{A}$	a	b	d
u_1	a_1	b_1	0
u_2	a_1	b_2	0
u_3	a_2	b_3	0
u_4	a_3	b_1	0
u_5	a_1	b_4	1
u_6	a_2	b_2	1
u_7	a_2	b_1	1
u_8	a_4	b_2	1
u_9	a_3	b_4	1
u_{10}	a_2	b_5	1

$\mathcal{M}(\mathcal{A})$	u_1	u_2	u_3	u_4
u_5	$b_{b_4}^{b_1}$	$b_{b_4}^{b_2}$	$a_{a_2}^{a_1}, b_{b_4}^{b_3}$	$a_{a_3}^{a_1}, b_{b_4}^{b_1}$
u_6	$a_{a_2}^{a_1}, b_{b_2}^{b_1}$	$a_{a_2}^{a_1}$	$b_{b_3}^{b_2}$	$a_{a_3}^{a_2}, b_{b_2}^{b_1}$
u_7	$a_{a_2}^{a_1}$	$a_{a_2}^{a_1}, b_{b_2}^{b_1}$	$b_{b_3}^{b_1}$	$a_{a_3}^{a_2}$
u_8	$a_{a_4}^{a_1}, b_{b_2}^{b_1}$	$a_{a_4}^{a_1}$	$a_{a_4}^{a_2}, b_{b_3}^{b_2}$	$a_{a_4}^{a_3}, b_{b_2}^{b_1}$
u_9	$a_{a_3}^{a_1}, b_{b_4}^{b_1}$	$a_{a_3}^{a_1}, b_{b_4}^{b_2}$	$a_{a_3}^{a_2}, b_{b_3}^{b_3}$	$b_{b_4}^{b_1}$
u_{10}	$a_{a_2}^{a_1}, b_{b_5}^{b_1}$	$a_{a_2}^{a_1}, b_{b_5}^{b_2}$	$b_{b_5}^{b_3}$	$a_{a_3}^{a_2}, b_{b_5}^{b_1}$

Fig. 2. Coloring of attribute value graphs and the reduced table.

From the Boolean function $f_\mathcal{A}$ with Boolean variables of the form $a_{v_1}^{v_2}$ one can find the shortest prime implicant: $\mathbf{a}_{a_2}^{a_1} \wedge \mathbf{a}_{a_3}^{a_2} \wedge \mathbf{a}_{a_4}^{a_1} \wedge \mathbf{a}_{a_4}^{a_3} \wedge \mathbf{b}_{b_4}^{b_1} \wedge \mathbf{b}_{b_4}^{b_2} \wedge \mathbf{b}_{b_3}^{b_2} \wedge \mathbf{b}_{b_3}^{b_1} \wedge \mathbf{b}_{b_5}^{b_3}$ which can be treated as graphs presented in the Figure 2.

We can color vertices of those graphs as shown in Figure 2. Colors are corresponding to partitions:

$$P_a(a_1) = P_a(a_3) = 1; \quad P_a(a_2) = P_a(a_4) = 2$$
$$P_b(b_1) = P_b(b_2) = P_b(b_5) = 1; \quad P_b(b_3) = P_b(b_4) = 2.$$

At the same time one can construct the new decision table (Table 2).

One can extend the presented approach (see e.g., [82]) to the case when in a given decision system nominal and numeric attribute appear. The received heuristics are of very good quality.

Experiments for classification methods (see [82]) have been carried over decision systems using two techniques called *train–and–test* and *n-fold-cross-validation*. In Table 6 some results of experiments obtained by testing the proposed methods MD (using only discretization based on MD-heuristics using the Johnson approximation strategy) and MD-G (using discretization and symbolic value grouping) for classification quality on well known data tables from the "UC Irvine repository" are shown. The results reported in [31] are summarized in columns labeled by S-ID3 and C4.5 in Table 6). It is interesting to compare those results with regard both to the classification quality. Let us note that the heuristics MD and MD-G are also very efficient with respect to the time complexity.

Names of	Classification accuracies			
Tables	S-ID3	C4.5	MD	MD-G
Australian	78.26	85.36	83.69	84.49
Breast (L)	62.07	71.00	69.95	69.95
Diabetes	66.23	70.84	71.09	76.17
Glass	62.79	65.89	66.41	69.79
Heart	77.78	77.04	77.04	81.11
Iris	96.67	94.67	95.33	96.67
Lympho	73.33	77.01	71.93	82.02
Monk-1	81.25	75.70	100	93.05
Monk-2	69.91	65.00	99.07	99.07
Monk-3	90.28	97.20	93.51	94.00
Soybean	100	95.56	100	100
TicTacToe	84.38	84.02	97.7	97.70
Average	78.58	79.94	85.48	87.00

Table 6. Quality comparison of various decision tree methods. Abbreviations: MD: MD-heuristics; MD-G: MD-heuristics with symbolic value partition

In the case of real value attributes one can search for features in the feature set that contains the characteristic functions of half–spaces determined by hyper–planes or parts of spaces defined by more complex surfaces in multi–dimensional spaces.

Genetic algorithms have been applied in searching for semi–optimal hyper–planes. The reported results are showing substantial increase in the quality of classification of unseen objects but at the cost of increased time for searching for the semi-optimal hyperplane.

Feature selection

Selection of relevant features is an important problem and has been extensively studied in Machine Learning and Pattern Recognition (see e.g., [75]). It is also a very active research area in the rough set community.

One of the first ideas [91] was to consider the *core* of the reduct set of the information system $\mathcal{A}$ as the source of relevant features. One can observe that relevant feature sets, in a sense used by the machine learning community, can be interpreted in most cases as the decision–relative reducts of decision systems obtained by adding appropriately constructed decisions to a given information system.

Another approach is related to dynamic reducts (for references see e.g., [100]) cf. Chapter by BAZAN et AL. Attributes are considered relevant if they belong to dynamic reducts with a sufficiently high stability coefficient, i.e., they appear with sufficiently high frequency in random samples of a given information system. Several experiments (see [100]) show that the set of decision rules based on such attributes is much smaller than the set of all decision rules. At the same time the quality of classification of new objects increases or does not change if one only considers rules constructed over such relevant features.

Another possibility is to consider as relevant the features that come from approximate reducts of sufficiently high quality.

The idea of attribute reduction can be generalized by introducing a concept of *significance of attributes* which enables to evaluate attributes not only in the two–valued scale *dispensable – indispensable* but also in the multi–value case by assigning to an attribute a real number from the interval $[0,1]$ that expresses the importance of an attribute in the information table.

Significance of an attribute can be evaluated by measuring the effect of removing the attribute from an information table.

Let C and D be sets of condition and decision attributes, respectively, and let $a \in C$ be a condition attribute. It was shown previously that the number $\gamma(C, D)$ expresses the degree of dependency between attributes C and D, or the accuracy of the approximation of U/D by C. It may be now checked how the coefficient $\gamma(C, D)$ changes when attribute a is removed, i.e., what is the difference between $\gamma(C, D)$ and $\gamma((C - \{a\}, D)$. The difference is normalized and the significance of attribute a is defined as

$$\sigma_{(C,D)}(a) = \frac{(\gamma(C,D) - \gamma(C - \{a\}, D))}{\gamma(C,D)} = 1 - \frac{\gamma(C - \{a\}, D)}{\gamma(C,D)},$$

Coefficient $\sigma_{C,D}(a)$ can be understood as a classification error which occurs when attribute a is dropped. The significance coefficient can be extended to sets of attributes as follows:

$$\sigma_{(C,D)}(B) = \frac{(\gamma(C,D) - \gamma(C - B, D))}{\gamma(C,D)} = 1 - \frac{\gamma(C - B, D)}{\gamma(C,D)}.$$

Any subset B of C is called an *approximate reduct* of C and the number

$$\varepsilon_{(C,D)}(B) = \frac{(\gamma(C,D) - \gamma(B,D))}{\gamma(C,D)} = 1 - \frac{\gamma(B,D)}{\gamma(C,D)},$$

is called an *error of reduct approximation*. It expresses how exactly the set of attributes B approximates the set of condition attributes C with respect to determining D.

The following equations are obvious: $\varepsilon(B) = 1 - \sigma(B)$ and $\varepsilon(B) = 1 - \varepsilon(C - B)$. For any subset B of C, we have $\varepsilon(B) \leq \varepsilon(C)$. If B is a reduct of C, then $\varepsilon(B) = 0$.

The concept of an approximate reduct (with respect to the positive region) is a generalization of the concept of a reduct that was considered previously. A minimal subset B of condition attributes C, such that $\gamma(C,D) = \gamma(B,D)$, or $\varepsilon_{(C,D)}(B) = 0$ is a reduct in the previous sense. The idea of an approximate reduct can be useful in these cases where a smaller number of condition attributes is preferred over the accuracy of classification.

Several other methods of reduct approximation based on measures different from positive region have been developed. All experiments confirm the hypothesis that by tuning the level of approximation the quality of the classification of new objects may be increased in most cases. It is important to note that it is once again possible to use Boolean reasoning to compute different types of reducts and to extract from them relevant approximations.

α-reducts and association rules

In this section we discuss the relationship between association rules [4] and approximations of reducts [115], [116], [83].

Association rules can be defined in many ways (see [4]). Here, according to our notation, association rules can be defined as implications of the form $(\mathbf{P} \Rightarrow \mathbf{Q})$, where $\mathbf{P}$ and $\mathbf{Q}$ are different simple templates, i.e., formulas of the form

$$(a_{i_1} = v_{i_1}) \wedge \ldots \wedge (a_{i_k} = v_{i_k}) \Rightarrow (a_{j_1} = v_{j_1}) \wedge \ldots \wedge (a_{j_l} = v_{j_l}) \qquad (1)$$

These implications can be called *generalized association rules*, because association rules were originally defined by formulas of the form $\mathbf{P} \Rightarrow \mathbf{Q}$ where $\mathbf{P}$ and $\mathbf{Q}$ were sets of items (e.g., goods or articles in stock market), e.g., $\{A, B\} \Rightarrow \{C, D, E\}$ (see [4]). One can see that this form can be obtained from 1 by replacing values of descriptors by 1, i.e.,: $(A = 1) \wedge (B = 1) \Rightarrow (C = 1) \wedge (D = 1) \wedge (E = 1)$.

Usually, for a given information table $\mathcal{A}$, the quality of the association rule $\mathcal{R} = \mathbf{P} \Rightarrow \mathbf{Q}$ can be evaluated by two measures called *support* and *confidence* with respect to $\mathcal{A}$. Support of the rule $\mathcal{R}$ is defined by the number of objects from $\mathcal{A}$ satisfying the condition $(\mathbf{P} \wedge \mathbf{Q})$, i.e.,

$$support(\mathcal{R}) = support(\mathbf{P} \wedge \mathbf{Q})$$

The second measure – confidence of $\mathcal{R}$ – is the ratio between the support of $(\mathbf{P} \wedge \mathbf{Q})$ and the support of $\mathbf{P}$, i.e.,

$$confidence(\mathcal{R}) = \frac{support(\mathbf{P} \wedge \mathbf{Q})}{support(\mathbf{P})}$$

The following problem has been investigated by many authors (see e.g., [4, 143])

> FOR A GIVEN INFORMATION TABLE $\mathcal{A}$, AN INTEGER s, AND A REAL NUMBER $c \in [0, 1]$, FIND AS MANY AS POSSIBLE ASSOCIATION RULES $\mathcal{R} = (\mathbf{P} \Rightarrow \mathbf{Q})$ SUCH THAT $support(\mathcal{R}) \geq s$ AND $confidence(\mathcal{R}) \geq c$

All existing association rule generation methods consist of two main steps:

1. Generate as many as possible templates $\mathbf{T} = D_1 \wedge D_2 ... \wedge D_k$ such that $support(\mathbf{T}) \geq s$ and $support(\mathbf{T} \wedge D) < s$ for any descriptor D (i.e., maximal templates among those which are supported by more than s objects).
2. For any template $\mathbf{T}$, search for a partition $\mathbf{T} = \mathbf{P} \wedge \mathbf{Q}$ such that:
   (a) $support(\mathbf{P}) < \frac{support(\mathbf{T})}{c}$
   (b) $\mathbf{P}$ is the smallest template satisfying the previous condition

We show that the second step can be solved using rough set methods and Boolean reasoning approach.

Let us assume that a template $\mathbf{T} = D_1 \wedge D_2 \wedge \ldots \wedge D_m$ supported by at least s objects, has been found. For the given confidence threshold $c \in (0; 1)$, the decomposition $\mathbf{T} = \mathbf{P} \wedge \mathbf{Q}$ is called c-irreducible if $confidence(\mathbf{P} \Rightarrow \mathbf{Q}) \geq c$ and for any decomposition $\mathbf{T} = \mathbf{P}' \wedge \mathbf{Q}'$ such that $\mathbf{P}'$ is a sub–template of $\mathbf{P}$, $confidence(\mathbf{P}' \Rightarrow \mathbf{Q}') < c$.

We show that the problem of searching for c-irreducible association rules from the given template is equivalent to the problem of searching for local α-reducts (for some α) from a decision table.

Let us define a new decision table $\mathcal{A}|_\mathbf{T} = (U, A|_\mathbf{T} \cup d)$ from the original information table $\mathcal{A}$ and the template $\mathbf{T}$ by

1. $A|_\mathbf{T} = \{a_{D_1}, a_{D_2}, ..., a_{D_m}\}$ is a set of attributes corresponding to the descriptors of $\mathbf{T}$ such that $a_{D_i}(u) = \begin{cases} 1 & \text{if the object } u \text{ satisfies } D_i, \\ 0 & \text{otherwise.} \end{cases}$
2. the decision attribute d determines if the object satisfies template $\mathbf{T}$, i.e., $d(u) = \begin{cases} 1 & \text{if the object } u \text{ satisfies } \mathbf{T}, \\ 0 & \text{otherwise.} \end{cases}$

The following facts [116], [83] describe the relationship between association rules and approximations of reducts.

For a given information table $\mathcal{A} = (\mathcal{U}, \mathcal{A})$, a template $\mathbf{T}$, and a set of descriptors $\mathbf{P}$, an implication $\left(\bigwedge_{D_i \in \mathbf{P}} D_i \Rightarrow \bigwedge_{D_j \notin \mathbf{P}} D_j \right)$ is

1. an 100%-*irreducible association rule* from $\mathbf{T}$ if and only if $\mathbf{P}$ is a reduct in $\mathcal{A}_\mathbf{T}$.

2. a *c-irreducible association rule* from **T** if and only if **P** is an α-reduct in $\mathcal{A}_\mathbf{T}$, where $\alpha = 1 - (\frac{1}{c} - 1)/(\frac{n}{s} - 1)$, n is the total number of objects from U, and $s = support(\mathbf{T})$.

Searching for minimal α-reducts is a well known problem in rough set theory. One can show, that the problem of searching for the shortest α-reduct is NP-hard.

The following example illustrates the main idea of our method. Let us consider the following information table $\mathcal{A}$ (see Table 5) with 18 objects and 9 attributes.

Table 7. The example of information table $\mathcal{A}$ and template **T** support by 10 objects and the new decision table $\mathcal{A}|_\mathbf{T}$ constructed from $\mathcal{A}$ and template **T**

$\mathcal{A}$	a_1	a_2	a_3	a_4	a_5	a_6	a_7	a_8	a_9
u_1	0	1	1	1	80	2	2	2	3
u_2	0	1	2	1	81	0	aa	1	aa
u_3	0	2	2	1	82	0	aa	1	aa
u_4	0	1	2	1	80	0	aa	1	aa
u_5	1	1	2	2	81	1	aa	1	aa
u_6	0	2	1	2	81	1	aa	1	aa
u_7	1	2	1	2	83	1	aa	1	aa
u_8	0	2	2	1	81	0	aa	1	aa
u_9	0	1	2	1	82	0	aa	1	aa
u_{10}	0	3	2	1	84	0	aa	1	aa
u_{11}	0	1	3	1	80	0	aa	2	aa
u_{12}	0	2	2	2	82	0	aa	2	aa
u_{13}	0	2	2	1	81	0	aa	1	aa
u_{14}	0	3	2	2	81	2	aa	2	aa
u_{15}	0	4	2	1	82	0	aa	1	aa
u_{16}	0	3	2	1	83	0	aa	1	aa
u_{17}	0	1	2	1	84	0	aa	1	aa
u_{18}	1	2	2	1	82	0	aa	2	aa

$\mathcal{A}|_\mathbf{T}$	D_1 $a_1 = 0$	D_2 $a_3 = 2$	D_3 $a_4 = 1$	D_4 $a_6 = 0$	D_5 $a_8 = 1$	d
u_1	1	0	1	0	0	
u_2	1	1	1	1	1	1
u_3	1	1	1	1	1	1
u_4	1	1	1	1	1	1
u_5	0	1	0	0	1	
u_6	1	0	0	0	1	
u_7	0	0	0	0	1	
u_8	1	1	1	1	1	1
u_9	1	1	1	1	1	1
u_{10}	1	1	1	1	1	1
u_{11}	1	0	1	1	0	
u_{12}	1	0	0	1	0	
u_{13}	1	1	1	1	1	1
u_{14}	1	1	0	0	0	
u_{15}	1	1	1	1	1	1
u_{16}	1	1	1	1	1	1
u_{17}	1	1	1	1	1	1
u_{18}	0	1	1	1	0	

Assume that the template

$$\mathbf{T} = (a_1 = 0) \land (a_3 = 2) \land (a_4 = 1) \land (a_6 = 0) \land (a_8 = 1)$$

has been extracted from the information table $\mathcal{A}$. One can see that $support(\mathbf{T}) = 10$ and $length(\mathbf{T}) = 5$. The newly constructed decision table $\mathcal{A}|_\mathbf{T}$ is presented in Table 5. The discernibility function for $\mathcal{A}|_\mathbf{T}$ can be explained as follows

$$f(D_1, D_2, D_3, D_4, D_5) = (D_2 \lor D_4 \lor D_5) \land (D_1 \lor D_3 \lor D_4) \land (D_2 \lor D_3 \lor D_4)$$
$$\land (D_1 \lor D_2 \lor D_3 \lor D_4) \land (D_1 \lor D_3 \lor D_5)$$
$$\land (D_2 \lor D_3 \lor D_5) \land (D_3 \lor D_4 \lor D_5) \land (D_1 \lor D_5)$$

After simplification the condition presented in Table 8 we obtain six reducts: $f(D_1, D_2, D_3, D_4, D_5) = (D_3 \land D_5) \lor (D_4 \land D_5) \lor (D_1 \land D_2 \land D_3) \lor (D_1 \land D_2 \land D_4) \lor (D_1 \land D_2 \land D_5) \lor (D_1 \land D_3 \land D_4)$ for the decision table $\mathcal{A}|_\mathbf{T}$. Thus, we have found from **T** six association rules with (100%)-confidence.

Table 8. The simplified version of discernibility matrix $\mathcal{M}(\mathcal{A}|\mathbf{T})$ and association rules

$\mathcal{M}(\mathcal{A}|\mathbf{T})$	
	u_2, u_3, u_4, u_8, u_9
	$u_{10}, u_{13}, u_{15}, u_{16}, u_{17}$
u_1	$D_2 \vee D_4 \vee D_5$
u_5	$D_1 \vee D_3 \vee D_4$
u_6	$D_2 \vee D_3 \vee D_4$
u_7	$D_1 \vee D_2 \vee D_3 \vee D_4$
u_{11}	$D_1 \vee D_3 \vee D_5$
u_{12}	$D_2 \vee D_3 \vee D_5$
u_{14}	$D_3 \vee D_4 \vee D_5$
u_{18}	$D_1 \vee D_5$

= 100% ⟹

$D_3 \wedge D_5 \Rightarrow D_1 \wedge D_2 \wedge D_4$
$D_4 \wedge D_5 \Rightarrow D_1 \wedge D_2 \wedge D_3$
$D_1 \wedge D_2 \wedge D_3 \Rightarrow D_4 \wedge D_5$
$D_1 \wedge D_2 \wedge D_4 \Rightarrow D_3 \wedge D_5$
$D_1 \wedge D_2 \wedge D_5 \Rightarrow D_3 \wedge D_4$
$D_1 \wedge D_3 \wedge D_4 \Rightarrow D_2 \wedge D_5$

= 90% ⟹

$D_1 \wedge D_2 \Rightarrow D_3 \wedge D_4 \wedge D_5$
$D_1 \wedge D_3 \Rightarrow D_2 \wedge D_4 \wedge D_5$
$D_1 \wedge D_4 \Rightarrow D_2 \wedge D_3 \wedge D_5$
$D_1 \wedge D_5 \Rightarrow D_2 \wedge D_3 \wedge D_4$
$D_2 \wedge D_3 \Rightarrow D_1 \wedge D_4 \wedge D_5$
$D_2 \wedge D_5 \Rightarrow D_1 \wedge D_3 \wedge D_4$
$D_3 \wedge D_4 \Rightarrow D_1 \wedge D_2 \wedge D_5$

For $c = 90\%$, we would like to find α-reducts for the decision table $\mathcal{A}_\mathbf{T}$, where $\alpha = 1 - \frac{\frac{1}{c}-1}{\frac{n}{s}-1} = 0.86$. Hence we would like to search for a set of descriptors that covers at least $\lceil (n-s)(\alpha) \rceil = \lceil 8 \cdot 0.86 \rceil = 7$ elements of discernibility matrix $\mathcal{M}(\mathcal{A}|\mathbf{T})$. One can see that the following sets of descriptors: $\{D_1, D_2\}$, $\{D_1, D_3\}$, $\{D_1, D_4\}$, $\{D_1, D_5\}$, $\{D_2, D_3\}$, $\{D_2, D_5\}$, $\{D_3, D_4\}$ have nonempty intersection with exactly 7 entries of the discernibility matrix $\mathcal{M}(\mathcal{A}|\mathbf{T})$. In Table 8 we present all association rules induced from those sets. Heuristics searching for α-reducts are discussed, e.g., in [83].

4.5 Conclusions

We have discussed rough set approach to KDD in the logical framework. In particular we have shown exemplary applications of Boolean reasoning combined with rough set methods for solving various KDD problems. In the paper we have presented some relationships of rough set approach with different logical systems.

Rough set theory constitutes a sound basis for Data Mining and Knowledge Discovery applications. The theory offers mathematical tools to discover hidden patterns in data. It identifies partial or total dependencies (i.e., cause–effect relations) in data bases, eliminates redundant data, proposes an approach to deal with feature selection and extraction problems, missing data, dynamic data and others. Also methods of data mining in very large data bases using rough sets recently have been proposed and investigated.

A substantial progress has been done in developing rough set methods for Data Mining and Knowledge Discovery (see methods and cases reported in, e.g., in [17], [19], [21], [46], [53], [63], [76], [79], [87], [100], [101], [102], [146]).

New methods for extracting patterns from data (see e.g., [58], [84], [76]), [57], [96]), decomposition of decision systems (see e.g., [84]) as well as a new methodology for data mining in distributed and multi–agent systems (see e.g., [99]) have been developed.

We have also stressed that logical systems build for solving KDD problems should be based on logical constructs extracted from data. In particular, approximate reasoning should be based on approximate schemes of reasoning extracted from data rather than on proofs generated by classical deductive systems. KDD faces currently problems related to cognitive and information aspects of perception and reasoning [110]. This certainly stimulates investigations on foundations of logic towards the revision and redefining of its traditional notions [12]. Here again, the notion of a proof seems to evolve towards the notion of an approximate scheme of reasoning (extracted from data) due to uncertainty or complexity of search in possible proof space. The above problems will certainly influence the further development of logic.

Acknowledgement. This work has been supported by the grant from the ESPRIT-CRIT 2 project #20288. Lech Polkowski has also been partially supported by the grant No. 8T11C02417 from the State Committee for Scientific Research (KBN) of the Republic of Poland. Andrzej Skowron has also been partially supported by a grant of the Wallenberg Foundation and by a grant from the State Committee for Scientific Research (KBN) of the Republic of Poland.

References

1. Adams, E.W.: The Logic of Conditionals, An Application of Probability to Deductive Logic. D. Reidel Publishing Company, Dordrecht, 1975.
2. Agotnes, T., Komorowski, J., Loken, T.: Taming large rule models in rough set approaches. Proceedings of the 3rd European Conference of Principles and Practice of Knowledge Discovery in Databases, September 15-18, 1999, Prague, Czech Republic, Lecture Notes in Artificial Intelligence **1704**, Springer-Verlag, Berlin, 1999, pp. 193-203.
3. Agrawal, R., Imieliński, T., Swami, A.: Mining association rules between sets of items in large databases. In: Proceedings ACM SIGMOD Conference on Management of Data, Washington, 1993, pp. 207-213.
4. Agrawal, R., Mannila, H., Srikant, R., Toivonen, H., Verkano, A.: Fast discovery of association rules. Fayyad U.M., Piatetsky-Shapiro G., Smyth P., Uthurusamy R. (Eds.): Advances in Knowledge Discovery and Data Mining, The AAAI Press/The MIT Press 1996, pp. 307-328.
5. Anderson, A.R., Belnap. N.D.: Entailment: The Logic of Relevance and Necessity. Princeton University Press, Vol.1, 1975.
6. Anderson, A.R., Belnap. N.D., Dunn, J.M: Entailment: Vol.2. Oxford University Press, 1992.
7. Anthony, M., Biggs, N.: Computational Learning Theory. Cambridge University Press, Cambridge, 1992.
8. Aquist, L.: Deontic logic. In: [32], pp. 605-714.
9. Bazan, J.G., Nguyen,Son Hung, Nguyen, Tuan Trung, Skowron, A., Stepaniuk, J.: Decision rules synthesis for object classification. In: E. Orlowska (ed.), Incomplete Information: Rough Set Analysis, Physica - Verlag, Heidelberg, 1998, pp. 23–57.
10. Bazan, J.G.: A comparison of dynamic and non-dynamic rough set methods for extracting laws from decision system. In: [100], 1998, pp. 321–365.
11. van Bentham, J.: Temporal logic. In: [35],1995, pp. 241-350.

12. van Bentham, J.: Logic after the Golden Age. IILC Magazine, December 1999, Institute for Logic, Language and Computation, Amsterdam University, p. 12.
13. Boole, G.: An Investigation of the Laws of Thought. Walton and Maberley, London, 1854.
14. Brown, E.M.: Boolean Reasoning. Kluwer Academic Publishers, Dordrecht, 1990.
15. Cattaneo, G.: Abstract approximation spaces for rough theories. In: Polkowski and Skowron [100], 1998, pp. 59–98.
16. Chmielewski, M.R., Grzymala-Busse, J.W.: Global discretization of attributes as preprocessing for machine learning. In: Proceedings of the Third International Workshop on Rough Sets and Soft Computing (RSSC'94),San Jose State University, San Jose, California, USA, November 10–12, 1994, pp. 294–301.
17. Cios, J., Pedrycz, W., Swiniarski, R.W.: Data Mining in Knowledge Discovery. Kluwer Academic Publishers, Dordrecht, 1998.
18. Codd, E.F.: A relational model for large shared data banks. Comm. ACM. Vol.13, No 6, 1970, pp. 377-382.
19. Czyżewski, A.: Soft processing of audio signals. In: Polkowski and Skowron [101], 1998, pp. 147–165.
20. Davis, M.: Computability and Unsolvability. Mc Graw-Hill, New York, 1958.
21. Deogun, J., Raghavan, V., Sarkar, A., Sever, H.: Data mining: Trends in research and development. In: Lin and Cercone [63], 1997, pp. 9–45.
22. Dougherty J., Kohavi R., and Sahami M.: Supervised and unsupervised discretization of continuous features. In: Proceedings of the Twelfth International Conference on Machine Learning, Morgan Kaufmann, San Francisco, CA, 1995..
23. Dubois, D., Prade, H.: Belief Change. Vol. 3 in: Gabbay, D.M., Smets Ph. (eds.), Handbook of Defeasible Reasoning and Uncertainty Management Systems. Kluwer Academic Publishers, Dordrecht, 1998.
24. Duentsch, I., Gediga, G.: Statistical evaluation of rough set dependency analysis. International Journal of Human-Computer Studies **46**, 1997, pp. 589-604.
25. Duentsch, I., Gediga, G.: Rough set data analysis. In: Encyclopedia of Computer Science and Technology, Marcel Dekker (to appear).
26. Egenhofer, M.J., Golledge, R.G. (eds.): Spatial and Temporal Reasoning in Geographic Information Systems. Oxford University Press, Oxford, 1997.
27. Eisinger, M., Ohlbach, H. J.: Deduction systems based on resolution. In: [33], pp. 183-271.
28. Fagin, R., Halpern. J.Y., Moses, Y., and Vardi, M.Y.: Reasoning about Knowledge. MIT Press, Cambridge MA, 1995.
29. Fayyad, U., Piatetsky-Shapiro, G. (eds.): Advances in Knowledge Discovery and Data Mining. MIT and AAAI Press, Cambridge MA, 1996.
30. Fitting, M.: Basic modal logic. In: [34], pp. 368-438.
31. Friedman, J., Kohavi, R., Yun, Y.: Lazy decision trees. In: Proc. of AAAI-96, 1996, pp. 717–724.
32. Gabbay, D., Guenthner, F. (eds.): Handbook of Philosophical Logic Vol.2. Kluwer Academic Publishers, Dordrecht, 1994.
33. Gabbay, D.M., Hogger, C.J., and Robinson, J.A. (eds.): Handbook of Logic in Artificial Intelligence and Logic Programming Vol.1. Oxford University Press, New York, 1993.
34. Gabbay, D.M., Hogger, C.J., and Robinson, J.A. (eds.): Handbook of Logic in Artificial Intelligence and Logic Programming vol.3. Oxford University Press, New York, 1993.

35. Gabbay, D.M., Hogger, C.J., and Robinson, J.A. (eds.): Handbook of Logic in Artificial Intelligence and Logic Programming Vol.4. Oxford University Press, New York, 1995.
36. Garey, M.R., Johnson, D.S.: Computers and Intractability. A Guide to the Theory of NP-completeness. W.H. Freeman and Company, New York, 1979.
37. Gärdenfors, P., Rott, H.: Belief revision. In: [35], 1995, pp. 35-132.
38. Geffner, H.: Default Reasoning: Causal and Conditional theories. MIT Press, Cambridge, MA, 1992.
39. Gentzen, G.: Untersuchungen über das logische Schliessen, Math. Zeitschrift **39**, 1934, pp. 176-210, 405-431.
40. Gibson, J.I.: The Ecological Approach to Visual Perception, Hillsdale: Lawrence Erlbaum, 2nd edition, 1986.
41. Glymour, C., Madigan, D., Pregibon, D., Smyth, P.: Statistical themes and lessons for data mining. Data Mining and Knowledge Discovery **1**, 1996, pp. 25-42.
42. Gonzalez. A.J., Dankel, D.D.: The Engineering of Knowledge Based Systems: Theory and Practice. Prentice Hall, Englewood Cliffs, NJ, 1993.
43. Gödel, K.: die Vollständigkeit der Axiome des Logischen Funktionenkalküls, Monatshefte für Mathematik und Physik **37**, 1930, pp. 349-360.
44. Gödel, K.: Über formal unetscheidbare Sätze der Principia Mathematica und Verwandtersysteme I, Monatshefte für Mathematik und Physik **38**, 1931, pp. 173-198.
45. Greco, S., Matarazzo, B., Słowinski, R.: Rough Approximation of a Preference Relation in a Pairwise Comparison Table. In: Polkowski and Skowron [101], 1998, pp. 13–36.
46. Grzymała–Busse, J.W.: Applications of the rule induction system LERS. In: Polkowski and Skowron [100], 1998, pp. 366–375.
47. Halpern, J.Y.: Reasoning about knowledge: a survey. In: [35], 1995, pp. 1–34.
48. Harel D.: Dynamic logic. In: [32], 1994, pp. 497–604.
49. Hintikka, J.: Knowledge and Belief. Cornell University Press, Ithaca, N.Y., 1962.
50. Hodges, W.: Logical features of Horn clauses. In: [33], 449-503.
51. Huber, P.J.: Robust statistics. Wiley, New York, 1981.
52. Hughes, G.E., Creswell, M.J.: An Introduction to Modal Logic. Methuen, London, 1968.
53. Komorowski, J., Żytkow, J. (Eds.): The First European Symposium on Principles of Data Mining and Knowledge Discovery (PKDD'97). June 25–27, Trondheim, Norway. Lecture Notes in Artificial Intelligence **1263**, Springer-Verlag, Berlin, 1997, pp. 1–396.
54. Komorowski, J., Pawlak, Z., Polkowski, L., Skowron, A.: Rough sets: A tutorial. In: S.K. Pal and A. Skowron (eds.), Rough–Fuzzy Hybridization: A New Trend in Decision–Making. Springer-Verlag, Singapore, 1997, pp. 3-98.
55. Konar, A.: Artificial Intelligence and Soft Computing, Behavioral and Cognitive Modeling of the Human Brain, CRC Press, Boca Raton, 1999, pp. 49.
56. Konolige, K.: Autoepistemic logic. In: [34], 1994, 217-295.
57. Kowalczyk, W.: Rough data modelling, A new technique for analyzing data. In: Polkowski and Skowron [100], 1998, pp. 400–421.
58. Krawiec, K., Słowiński, R., Vanderpooten, D.: Learning decision rules from similarity based rough approximations. In: Polkowski and Skowron [101], 1998, pp. 37–54.
59. Kruse, R., Gebhardt, J., and Klawonn, F.: Foundations of Fuzzy Systems. J. Wiley, New York, 1994.

60. Kryszkiewicz, M.: Generation of rules from incomplete information systems. In: Komorowski and Żytkow [53], 1997, pp. 156–166.
61. Leśniewski, S.: On the foundations of mathematics. In: Surma, S., Srzednicki,J.T., Barnett, D.I., Rickey, F.V. (eds), Stanislaw Leśniewski. Collected Works. Kluwer Academic Publishers, Dordrecht (1992), pp. 174-382.
62. Lifschitz, V.: Circumscription. In: [34], 297-352.
63. Lin, T.Y., Cercone, N. (Eds.): Rough Sets and Data Mining. Analysis of Imprecise Data. Kluwer Academic Publishers, Boston, 1997.
64. Lin, T.Y.: Granular computing on binary relations I, II. In: Polkowski and Skowron [100], 1998, pp. 107–140.
65. Liu. C., Zhong, N.: Rough problem settings for inductive logic programming, In: N. Zhong, A. Skowron, S. Ohsuga (eds.): New Directions in Rough Sets, Data Mining, and Granular – Soft Computing, Proceedings of the 7th International Workshop RSFDGSC'99, Yamaguchi, Japan, November 1999, Lecture Notes in Artificial Intelligence 1711, Springer-Verlag, Berlin, 1999, pp. 168–177.
66. Lloyd, J. W.: Foundations of Logic Programming. Springer- Verlag, Berlin, 1984.
67. Łukasiewicz, J.: Die logischen Grundlagen der Wahrscheinchkeitsrechnung. Kraków, 1913.
68. Łukasiewicz, J.: Philosophische Bemerkungen zu mehrwertigen Systemen des Aussagenkalküls. Comptes rendus de la Société des Sciences et des Lettres de Varsovie **23**, 1930, pp. 57-77.
69. Makinson, D.: General patterns in non-monotonic reasoning. In: [34], 35-110.
70. Marek, V. W., Truszczyński, M.: Contributions to the theory of rough sets, Fundamenta Informaticae **39(4)**, 1999, pp. 389-409.
71. Mendelson, E.: Introduction to Mathematical Logic. Van Nostrand, New York, 1960.
72. Michalski, R.: Inferential theory of learning as a conceptual basis for multistrategy. Machine Learning **11**, 1993, pp. 111-151.
73. Michalski R., Tecuci G.: Machine Learning. A Multistrategy Approach Vol.4. Morgan Kaufmann, San Francisco, 1994.
74. Michie, D., Spiegelhalter, D.J., Taylor, C.C. (Eds.): Machine Learning, Neural and Statistical Classification. Ellis Horwood, New York, 1994.
75. Mitchell, T.M.: Machine Learning. Mc Graw-Hill, Portland, 1997.
76. Mollestad, T., Komorowski, J.: A Rough Set Framework for Propositional Default Rules Data Mining. In: S.K. Pal and A. Skowron (Eds.): Rough–Fuzzy Hybridization: A New Trend in Decision Making. Springer-Verlag, Singapore, 1999.
77. Muggleton, S.: Foundations of Inductive Logic Programming. Prentice Hall, Englewood Cliffs, 1995.
78. Nadler M., Smith E.P.: Pattern Recognition Engineering. Wiley, New York, 1993.
79. Nguyen, H.S.: Discretization of Real Value Attributes, Boolean Reasoning Approach. Ph.D. Dissertation, Warsaw University, 1997, pp. 1–90.
80. Nguyen, H.S.: Efficient SQL-learning method for data mining in large data bases. In: Proceedings of the Sixteenth International Joint Conference on Artificial Intelligence (IJCAI'99), 1999, pp. 806–811.
81. Nguyen, S.H.: Data Regularity Analysis and Applications in Data Mining. Ph.D. Dissertation, Warsaw University, 1999.
82. Nguyen, H.S., Nguyen, S.H.: Pattern extraction from data. Fundamenta Informaticae **34**, 1998, pp. 129–144.
83. Nguyen, H.S., Nguyen, S.H.: Rough sets and association rule generation. Fundamenta Informaticae **40/4** (in print).

84. Nguyen, S.H., Skowron, A., Synak, P.: Discovery of data patterns with applications to decomposition and classification problems. In: Polkowski and Skowron [101], 1998, pp. 55–97.
85. Nute, D.: Conditional logic. In: [32], 387-440.
86. Orłowska, E. (ed.): Incomplete Information: Rough Set Analysis. Physica–Verlag, Heidelberg, 1998.
87. Pal, S.K., Skowron, A.: Rough–fuzzy Hybridization: A New Trend in Decision Making. Springer–Verlag, Singapore, 1999.
88. Paun, G., Polkowski, L., Skowron, A.: Parallel communicating grammar systems with negotiations. Fundamenta Informaticae **28/3-4**, 1996, pp. 315–330.
89. Pawlak, Z.: Information systems–theoretical foundations. Information Systems **6**, 1981, pp. 205–218.
90. Pawlak, Z.: Rough sets. International Journal of Computer and Information Sciences **11**, 1982, pp. 341–356.
91. Pawlak, Z.: Rough Sets: Theoretical Aspects of Reasoning about Data. Kluwer Academic Publishers, Dordrecht, 1992.
92. Pawlak, Z.: Decision Rules, Bayes' Rule and Rough Sets, In: N. Zhong, A. Skowron, S. Ohsuga (eds.): New Directions in Rough Sets, Data Mining, and Granular – Soft Computing, Proceedings of the 7th International Workshop RSFDGSC'99, Yamaguchi, Japan, November 1999, Lecture Notes in Artificial Intelligence 1711, Springer-Verlag, Berlin, 1999. 1–9.
93. Pawlak, Z., Ras, Z. (Eds.): Proceedings: Ninth International Symposium on Methodologies for Intelligent Systems (ISMIS'96),Springer–Verlag, Berlin, 1996.
94. Pawlak, Z., Skowron, A.: Rough set rudiments. Bulletin of the International Rough Set Society **3/4**, 1999, pp. 181-185.
95. Pearl, J.: Probabilistic Reasoning in Intelligent Systems: Networks of Plausible Inference. Morgan Kaufmann, San Francisco, 1988.
96. Piasta, Z., Lenarcik, A.: Rule induction with probabilistic rough classifiers. Machine Learning (to appear).
97. Polkowski, L.: On synthesis of constructs for spatial reasoning via rough mereology. Fundamenta Informaticae (to appear).
98. Polkowski, L., Skowron, A.: Rough mereology: A new paradigm for approximate reasoning. International Journal of Approximate Reasoning **15/4**,1994, pp. 333–365.
99. Polkowski, L., Skowron, A.: Rough sets: A perspective. In: Polkowski and Skowron [100], 1998, pp. 31–58.
100. Polkowski, L., Skowron, A. (Eds.): Rough Sets in Knowledge Discovery 1: Methodology and Applications. Physica-Verlag, Heidelberg, 1998.
101. Polkowski, L., Skowron, A. (Eds.: Rough Sets in Knowledge Discovery 2: Applications, Case Studies and Software Systems. Physica-Verlag, Heidelberg, 1998.
102. Polkowski, L., Skowron, A. (Eds.): Proc. First International Conference on Rough Sets and Soft Computing, RSCTC'98, Warszawa, Poland, LNAI **1424**, Springer-Verlag, Berlin, 1998.
103. Polkowski, L., Skowron, A.: Towards adaptive calculus of granules. In: [142], **1**, 1999, pp. 201–227.
104. Poole, D.: Default logic. In: [34], 189-216.
105. Prawitz, D.: Natural deduction, a proof theoretic study. Stockholm Studies in Philosophy Vol.3, Almquist & Wiksell, Stockholm, 1965.
106. Quinlan, J.R.: C4.5. Programs for machine learning. Morgan Kaufmann, San Mateo, CA, 1993.

107. Ras, Z.W.: Cooperative knowledge-based systems. Journal of the Intelligent Automation Soft Computing **2/2** (special issue edited by T.Y. Lin), 1996, pp. 193–202.
108. Rescher, N.: Many-valued Logics. Mc Graw Hill, New York, 1969.
109. Rissanen, J.J.: Modeling by Shortest Data Description. Automatica **14**, 1978, pp. 465-471.
110. Roddick J.F., Spiliopoulou M.: A Bibliography of Temporal, Spatial, and Temporal Data Mining Research. Newsletter of the Special Interest Group (SIG) on Knowledge Discovery & Data Mining **1/1**, 1999, pp. 34-38.
111. Rosser, J.B., Turquette, A.R.: Many-valued Logics. North-Holland, Amsterdam, 1952.
112. Russel, S.J., Norvig, P.: Artificial Intelligence. A Modern Approach. Prentice Hall, Englewood Cliffs, 1995.
113. Sandewall, E., Shoham, Y.: Non-monotonic temporal reasoning. In: [35], 439-498.
114. Selman, B., Kautz, H., McAllester, D.: Ten challenges in propositional reasoning and search. Proc. IJCAI'97, Japan.
115. Skowron, A.: Synthesis of adaptive decision systems from experimental data. In: A. Aamodt, J. Komorowski (eds): Proc. of the Fifth Scandinavian Conference on Artificial Intelligence (SCAI'95), May 1995, Trondheim, Norway. IOS Press, Amsterdam, 1995, pp. 220–238.
116. Skowron, A., Nguyen, H.S.: Boolean reasoning scheme with some applications in data mining. Proceedings of the 3-rd European Conference on Principles and Practice of Knowledge Discovery in Databases, September 1999, Prague, Czech Republic. Lecture Notes in Computer Science **1704**, 1999, pp. 107–115.
117. Skowron, A., Rauszer, C.: The Discernibility Matrices and Functions in Information Systems. In: Słowiński [122], 1992, pp. 331–362.
118. Skowron, A., Stepaniuk, J.: Tolerance Approximation Spaces. Fundamenta Informaticae **27**, 1996, pp. 245–253.
119. Skowron, A., Stepaniuk, J., Tsumoto, S.: Information Granules for Spatial Reasoning. Bulletin of the International Rough Set Society **3/4**, 1999, pp. 147-154.
120. Ślęzak, D.: Approximate reducts in decision tables. In: Proceedings of the Sixth International Conference, Information Processing and Management of Uncertainty in Knowledge-Based Systems (IPMU'96) vol. 3, July 1–5, Granada, Spain, 1996, pp. 1159–1164.
121. D. Ślęzak: Decomposition and synthesis of approximate functional dependencies, PhD Thesis, Warsaw University (forthcoming).
122. Słowiński, R. (Ed.): Intelligent Decision Support – Handbook of Applications and Advances of the Rough Sets Theory. Dordrecht, Kluwer Academic Publishers, 1992.
123. Słowiński, R., Vanderpooten, D.: Similarity relation as a basis for rough approximations. In: P. Wang (Ed.): Advances in Machine Intelligence & Soft Computing. Bookwrights, Raleigh NC, 1997, pp. 17–33.
124. Słowiński, R., Vanderpooten, D.: A generalized definition of rough approximations based on similarity. IEEE Trans. on Data and Knowledge Engineering (to appear).
125. http:agora.leeds.ac.ukspacenet.html
126. Stalnaker, R.: A theory of conditionals. In: N. Rescher (ed.): Studies in Logical Theory. Blackwell, Oxford, 1968.
127. Szczuka, M.: Symbolic and neural network methods for classifiers construction. Ph.D. Dissertation, Warsaw University, 1999.

128. Tarski, A.: On the concept of logical consequence. In: Logic, Semantics, Metamathematics. Oxford University Press, Oxford, 1956.
129. Tarski, A.: Der Wahrheitsbegriff in den Formalisierten Sprachen, Studia Philosophica 1, 1936, pp. 261-405.
130. Troelstra, A. S.: Aspects of constructive mathematics. In: Barwise, J. (ed.): Handbook of Mathematical Logic. North Holland, Amsterdam, 1977, pp. 973-1052.
131. Tsang, E.: Foundations of Constraint Satisfaction. Academic Press, London 1993.
132. Tsumoto, S.: Modelling diagnostic rules based on rough sets. In: Polkowski and Skowron [102], 1998, pp. 475–482.
133. Ullman, J.D., Widom, J.: A First Course in Database Systems. Prentice–Hall, Inc., Englewood Cliffs, 1997.
134. Urquhart, A.: Many-valued logic. In: [34], 71-116.
135. Wasilewska, A., Vigneron, L.: Rough algebras and automated deduction. In: Polkowski and Skowron [100], 1998, pp. 261–275.
136. Wille, R.: Formale Begriffsanalyse: Mathematische Grundlagen. Springer-Verlag, Berlin, 1996.
137. Von Wright, G.H.: An Essay in Modal Logic. North Holland, Amsterdam, 1951.
138. Zadeh, L.A.: Fuzzy sets. Information and Control 8, 1965, pp. 333-353.
139. Zadeh, L.A.: A theory of approximate reasoning. In: Hayes, J.E., Michie, D., Mikulich, L.C. (eds.): Machine Intelligence Vol.9. J. Wiley, New York, 1979, pp. 149-194.
140. Zadeh, L.A.: Fuzzy logic = computing with words. IEEE Trans. on Fuzzy Systems 4, 1996, pp. 103-111.
141. Zadeh, L.A: Toward a theory of fuzzy information granulation and its certainty in human reasoning and fuzzy logic. Fuzzy Sets and Systems 90, 1997, pp. 111-127.
142. Zadeh, L.A., Kacprzyk, J. (eds.): Computing with Words in Information/ Intelligent Systems Vol. 1-2. Physica-Verlag, Heidelberg, 1999.
143. Zaki, M.J., Parthasarathy, S., Ogihara, M., Li, W.: New parallel algorithms for fast discovery of association rules. In: Data Mining and Knowledge Discovery : An International Journal (special issue on Scalable High-Performance Computing for KDD) 1/4, 1997, pp. 343–373.
144. Ziarko, W.: Variable Precision Rough Set Model. J. of Computer and System Sciences 46, 1993, pp. 39–59.
145. Ziarko, W. (ed.): Rough Sets, Fuzzy Sets and Knowledge Discovery (RSKD'93). Workshops in Computing, Springer–Verlag & British Computer Society, London, Berlin, 1994.
146. Ziarko, W.: Rough sets as a methodology for data mining. In: Polkowski and Skowron [101], 1998, pp. 554–576.

APPENDIX:

SELECTED BIBLIOGRAPHY ON ROUGH SETS

Bibliography

The bibliography begins with a list [A] of books and conference proceedings dedicated to rough set theory and its applications. Then selected papers on rough sets are listed in [B].

This bibliography is a sequel to the bibliography of 1077 papers appendiced in: L. Polkowski and A. Skowron (eds.), *Rough Sets in Knowledge Discovery 2. Applications, Case Studies and Software Systems*, this Series, vol. 19

[A] Books and conference proceedings

1. S. K. Pal and A. Skowron (eds.), *Rough Fuzzy Hybridization: A New Trend in Decision-Making*, Springer–Verlag, Singapore, 1999.
2. L. Polkowski and A. Skowron (eds.), *Rough Sets in Knowledge Discovery 1. Methodology and Applications*, this Series, vol. 18, Physica–Verlag, Heidelberg, 1998.
3. L. Polkowski and A. Skowron (eds.), *Rough Sets in Knowledge Discovery 2. Applications, Case Studies and Software Systems*, this Series, vol. 19, Physica–Verlag, Heidelberg, 1998.
4. T. Y. Lin and N. Cercone (eds.), *Rough Sets and Data Mining. Analysis of Imprecise Data*, Kluwer Academic Publishers, Dordrecht, 1997.
5. N. Zhong, A. Skowron, and S. Ohsuga (eds.), *New Directions in Rough Sets, Data Mining, and Granular-Soft Computing*, Proceedings: the 7th International Workshop (RSFDGrC'99), Ube–Yamaguchi, Japan, November 1999, LNAI 1711, Springer–Verlag, Berlin, 1999.
6. L. Polkowski and A. Skowron (eds.), *Rough Sets and Current Trends in Computing*, Proceedings: the First International Conference on Rough Sets and Current Trends in Computing (RSCTC'98), Warsaw, Poland, June 1998, LNAI 1424, Springer–Verlag, Berlin, 1998.

[B] Journal and monograph articles, and conference papers

1. P. Apostoli and A. Kanda, Parts of the Continuum: Towards a modern ontology of science, to appear in: Poznań Series on the Philosophy of Science and the Humanities.

2. G. Arora, F. Petry, and T. Beaubouef, New information measures for fuzzy sets, in: *Proceeings: the 7th IFSA World Congress*, Prague, the Czech Republic, June 1997.
3. G. Arora, F. Petry, and T. Beaubouef, Information measure of type β under similarity relations, in: *Proceedings: the 6th IEEE International Conference on Fuzzy Systems* (FUZZ-IEEE'97), Barcelona, Spain, July 1997.
4. C. Baizán, E. Menasalvas, and S. Millán, Multi-valued dependencies as inference rules on a deductive process under a relational data model set theory approach, *Computers and Applications*, Slovak Acad. Sci., to appear.
5. C. Baizán, E. Menasalvas, J. Peña, A new approach to efficient calculation of reducts in large databases, in: *Proceedings: the Fifth International Workshop on Rough Sets and Soft Computing* (RSSC'97) at *Proceedings: the 3rd Joint Conference on Information Sciences* (JCIS'97), Research Triangle Park NC, March 1997, pp. 340–344.
6. C. Baizán, E. Menasalvas, J. Peña, Using rough sets to mine socio–economic data, in: *Proceedings: SMC'97*, 1997, pp. 567–571.
7. C. Baizán, E. Menasalvas, J. Peña, S. Millán, and E. Mesa, Calculating approximative reducts by means of step–wise linear regression, *Intern. Journal Computer Science*, in print.
8. C. Baizán, E. Menasalvas, J. Peña, S. Millán, and E. Mesa, Rough dependencies as a particular case of correlation: Application to the calculation of approximate reducts, in: *Proceedings: Principles of Data Mining and Knowledge Discovery* (PKDD'99), Prague, Czech Republic, September 1999, LNAI 1704, Springer-Verlag, Berlin, 1999, pp. 335–341.
9. C. Baizán, E. Menasalvas, J. Peña, and J. Pastrana, Integrating KDD algorithms and RDBMS code, in: *Proceedings: the First International Conference on Rough Sets and Current Trends in Computing* (RSCTC'98), Warsaw, Poland, June 1998, LNAI 1424, Springer-Verlag, Berlin, 1998, pp. 210–214.
10. C. Baizán, E. Menasalvas, J. Peña, and J. Pastrana, RSDM system, *Bull. Intern. Rough Set Society*, 2, 1998, pp. 21–24.
11. C. Baizán, E. Menasalvas, J. Peña, C. P. Peréz and E. Santos, The lattices of generalizations in a KDD process, in: *Proceedings: Cybernetics and Systems'98*, 1998, pp. 181–184.
12. C. Baizán, E. Menasalvas, and A. Wasilewska, A model for RSDM Implementation, in: *Proceedings: the First international Conference on Rough Sets and Current Trends in Computing* (RSCTC'98), Warsaw, Poland, June 1998, LNAI 1424, Springer–Verlag, Berlin, 1998, pp. 186–196.
13. C. Baizán, E. Menasalvas, A. Wasilewska, and J. Peña, The lattice structure of the KDD process: Mathematical expression of the KDD process, *Intern. Journal Computer Science*, in print.
14. M. Banerjee, S. Mitra, and S.K. Pal, Rough Fuzzy MLP: Knowledge encoding and classification, *IEEE Trans. Neural Networks* 9(6), 1998, pp. 1203–1216.
15. M. Banerjee and S.K. Pal, Roughness of a fuzzy set, *Information Science* 93(3/4), 1996, pp. 235–246.
16. W. Bartol, X. Caicedo, and F. Rosselló, Syntactical content of finite approximations of partial algebras, in: *Proceedings: the First International*

Conference on Rough Sets and Current Trends in Computing (RSCTC'98), Warsaw, Poland, June 1998, LNAI 1424, Springer-Verlag, Berlin, 1998, pp. 408–415.
17. J.G. Bazan, Approximate reasoning in decision rule synthesis, in: *Proceedings of the Workshop on Robotics, Intelligent Control and Decision Support Systems*, Polish–Japanese Institute of Information Technology, Warsaw, Poland, February 1999, pp. 10–15.
18. J.G. Bazan, Discovery of decision rules by matching new objects against data tables, in: *Proceedings: the First International Conference on Rough Sets and Current Trends in Computing* (RSCTC-98), Warsaw, Poland, June 1998, LNAI 1424, Springer–Verlag, Berlin, 1998, pp. 521–528.
19. J.G. Bazan, A comparison of dynamic and non–dynamic rough set methods for extracting laws from decision table, in: L. Polkowski, A. Skowron (eds.), *Rough Sets in Knowledge Discovery 1. Methodology and Applications*, Physica–Verlag, Heidelberg, 1998, pp. 321–365.
20. J. G. Bazan, Approximate reasoning methods for synthesis of decision algorithms (in Polish), Ph.D. Dissertation, supervisor A. Skowron, Warsaw University, 1998, pp. 1–179.
21. J.G. Bazan, Nguyen Hung Son, Nguyen Tuan Trung, A. Skowron, and J. Stepaniuk, Decision rules synthesis for object classification, in: E. Orłowska (ed.), *Incomplete Information: Rough Set Analysis*, Physica–Verlag, Heidelberg, 1998, pp. 23–57.
22. T. Beaubouef and R. Lang, Rough set techniques for uncertainty management in automated story generation, in: *the 36th Annual ACM Southeast Conference*, Marietta GA, April 1998.
23. T. Beaubouef, F. Petry, and G. Arora, Information measures for rough and fuzzy sets and application to uncertainty in relational databases, in: S. Pal and A. Skowron (eds.), *Rough-Fuzzy Hybridization: A New Trend in Decision-Making*, Springer-Verlag, Singapore, 1998, pp. 200–214.
24. T. Beaubouef, F. Petry, and G. Arora, Information–theoretic measures of uncertainty for rough sets and rough relational databases, *Information Sciences* 109(1–4),1998, pp. 185–195.
25. T. Beaubouef, F. Petry, and G. Arora, Information–theoretic measures of uncertainty for rough sets and rough relational databases, in: *Proceedings: the 5th International Workshop on Rough Sets and Soft Computing* (RSSC'97), Research Triangle Park NC, March 1997.
26. T. Beaubouef, F. Petry, and J. Breckenridge, Rough set based uncertainty management for spatial databases and Geographical Information Systems, in: *Proceedings: Fourth On–line World Conference on Soft Computing in Industrial Applications* (WSC4), September 1999, pp. 21–30.
27. Chien-Chung Chan and J. W. Grzymala–Busse, On the lower boundaries in learning rules from examples, in: E. Orlowska (ed.), *Incomplete Information: Rough Set Analysis*, Physica–Verlag, Heidelberg, 1998, pp. 58–74.
28. I. V. Chikalov, On average time complexity of decision trees and branching programs, *Fundamenta Informaticae* 39(4), 1999, pp. 337–357.

29. I. V. Chikalov, On decision trees with minimal average depth, in: *Proceedings: the First International Conference on Rough Sets and Current Trends in Computing* (RSCTC'98), Warsaw, Poland, LNAI 1424, Springer–Verlag, Berlin, 1998, pp. 506–512.
30. I. V. Chikalov, Bounds on average weighted depth of decision trees depending only on entropy, in: *Proceedings: the 7th International Conference of Information Processing and Management of Uncertainty in Knowledge-Based Systems*, La Sorbonne, Paris, France, July 1998, pp. 1190–1194.
31. B. Chlebus and Nguyen Sinh Hoa, On finding optimal discretization on two attributes, in: *Proceedings: the First International Conference on Rough Sets and Current Trends in Computing* (RSCTC'98), Warsaw, Poland, June 1998, LNAI 1424, Springer–Verlag, Berlin, 1998, pp. 537–544.
32. A. Czyżewski, Soft Processing of Audio Signals, in: L. Polkowski and A. Skowron (eds.), *Rough Sets in Knowledge Discovery 2. Applications, Case Studies and Software Systems*, Physica–Verlag,Heidelberg, 1998, pp. 147–165.
33. A. Czyżewski, Speaker – independent recognition of isolated words using rough sets, *J. Information Sciences* 104, 1998, pp. 3–14.
34. A. Czyżewski, Learning algorithms for audio signal enhancement. Part 2: Implementation of the rough set method for the removal of hiss, *J. Audio Eng. Soc.* 45(11), 1997, pp. 931-943.
35. A. Czyżewski, Speaker–independent recognition of digits – experiments with neural networks, fuzzy logic and rough sets, *J. Intelligent Automation and Soft Computing* 2(2), 1996, pp. 133-146.
36. A. Czyżewski and B. Kostek, Tuning the perceptual noise reduction algorithm using rough sets, in: *Proceedings: the First international Conference on Rough Sets and Current Trends in Computing* (RSCTC'98), Warsaw, Poland, June 1998, LNAI 1424, Springer–Verlag, Berlin, 1998, pp. 467–474.
37. A. Czyżewski and B. Kostek, Rough set–based filtration of sound applicable to hearing prostheses, In: *Proceedings: the 4th Intern. Workshop on Rough Sets, Fuzzy Sets, and Machine Discovery* (RSFD'96), Tokyo, Japan, November 1996, pp. 168–175.
38. A. Czyżewski and B. Kostek, Restoration of old records employing Artificial Intelligence methods, In: *Proceedings: the IASTED Internat. Conference – Artificial Intelligence, Expert Systems and Neural Networks*, Honolulu, Hawaii, 1996, pp.372–375.
39. A. Czyżewski, B. Kostek , H. Skarżyński, and R. Królikowski, Evaluation of some properties of the human auditory system using rough sets, In: *Proceedings: the 6th European Congress on Intelligent Techniques and Soft Computing* (EUFIT'98), Aachen, Germany, September 1998, Verlag Mainz, Aachen, 1998, pp. 965-969.
40. A. Czyżewski and R. Królikowski, Noise reduction in audio signals based on the perceptual coding approach, in: *Proceedings: the IEEE Workshop on Applications of Signal Processing to Audio and Acoustics*, New Paltz NY, October 1999, pp. 147–150.
41. A. Czyżewski and R. Królikowski, Noise reduction algorithms employing an

intelligent inference engine for multimedia applications, in: *Proceedings: the IEEE 2nd Workshop on Multimedia Signal Processing*, Redondo Beach CA, December 1998, pp.125–130.
42. A. Czyżewski, R. Królikowski, S. K. Zieliński, and B. Kostek, Echo and noise reduction methods for multimedia communication systems, in: *Proceedings: the IEEE Signal Processing Society 1999 Workshop on Multimedia Signal Processing*, Copenhagen, Denmark, September 1999, pp. 239–244.
43. A. Czyżewski, R. Królikowski, S. K. Zieliński, and B. Kostek, Intelligent echo and noise reduction, in: *Proceedings: the 3rd World Multi-conference on Systemics, Cybernetics and Informatics* (SCI'99) and *the 5th International Conference on Information System Analysis and Synthesis* (ISAS'99), Orlando Fla., August 1999, pp.234–238.
44. A. Czyżewski, H. Skarżyński, B. Kostek, and R. Królikowski, Rough set analysis of electro-stimulation test database for the prediction of post-operative profits in cochlear implanted patients, in: *Proceedings: the 7th Intern. Workshop on Rough Sets, Fuzzy Sets, Data Mining, and Granular-Soft Computing* (RSFDGrC'99), Ube–Yamaguchi, Japan, November 1999, LNAI 1711, Springer–Verlag, Berlin, 1999, pp. 109–117.
45. A. Czyżewski, H. Skarżyński, and B. Kostek, Multimedia databases in hearing and speech pathology, in: *Proceedings: the World Automation Congress* (WAC'98), Anchorage, Alaska, May 1998, pp. IFMIP-052.1–052.6.
46. R. Deja, Conflict analysis, in: *Proceedings: the 7th European Congress on Intelligent Techniques and Soft Computing* (EUFIT'99), Aachen, Germany, September 1999.
47. S. Demri, A class of decidable information logics, *Theoretical Computer Science*, 195(1), 1998, pp. 33–60.
48. S. Demri and B. Konikowska, Relative similarity logics are decidable: reduction to FO^2 with equality, in: *Proceedings: JELIA '98*, LNAI 1489, Springer–Verlag, Berlin, 1998, pp. 279–293.
49. S. Demri and E. Orłowska, Informational representability: Abstract models versus concrete models, in: D. Dubois and H. Prade (eds.), *Fuzzy sets, Logics and Reasoning about Knowledge*, Kluwer Academic Publishers, Dordrecht, 1999, pp. 301–314.
50. S. Demri and E. Orłowska, Informational representability of models for information logics, in: E. Orłowska (ed.), *Logic at Work. Essays Dedicated to the Memory of Helena Rasiowa*, Physica–Verlag, Heidelberg, 1998, pp. 383–409.
51. S. Demri, E. Orłowska, and D. Vakarelov, Indiscernibility and complementarity relations in Pawlak's information systems, in: *Liber Amicorum for Johan van Benthem's 50th Birthday*, 1999.
52. A.I. Dimitras, R. Słowiński, R. Susmaga, and C. Zopounidis: Business failure prediction using rough sets, *European Journal of Operational Research* 114, 1999, pp. 49–66.
53. Ju. V. Dudina and A. N. Knyazev, On complexity of language word recognition generated by context-free grammars with one non-terminal symbol (in Russian), *Bulletin of Nizhny Novgorod State University. Mathematical Simulation and Optimal Control* 19, 1998, pp. 214–223.

54. A. E. Eiben, T. J. Euverman, W. Kowalczyk, and F. Slisser, Modeling customer retention with statistical techniques, rough data models, and genetic programming, in: S. K. Pal and A. Skowron (eds.), *Rough Fuzzy Hybridization: A New Trend in Decision–Making*, Springer–Verlag, Singapore, 1999, pp. 330–348.
55. P. Ejdys and G. Góra, The More We Learn the Less We Know? - On Inductive Learning from Examples, *Proceedings: the 11th International Symposium on Methodologies for Intelligent Systems, Foundations of Intelligent Systems* (ISMIS'99), Warsaw, Poland, June 1999, LNAI, Springer–Verlag, Berlin, in print.
56. L. Goodwin, J. Prather, K. Schlitz, M. A. Iannacchione, M. Hage, W. E. Hammond Sr., and J. W. Grzymala-Busse, Data mining issues for improved birth outcomes, *Biomedical Sciences Instrumentation* 34, 1997, pp. 291–296.
57. S. Greco, B. Matarazzo, and R. Słowiński, Dominance–based rough set approach to rating analysis, *Gestion 2000 Magazine*, to appear.
58. S. Greco, B. Matarazzo, and R. Słowiński, Decision rules, in: *Encyclopedia of Management*, 4th edition, 2000, to appear.
59. S. Greco, B. Matarazzo, and R. Słowiński, Rough set processing of vague information using fuzzy similarity relation, in: C. Calude and G. Paun (eds.), *Finite versus Infinite – Contributions to an Eternal Dilemma*, Springer–Verlag, Berlin, to appear.
60. S. Greco, B. Matarazzo, and R. Słowiński, Rough approximation of a preference relation by dominance relations, *European Journal of Operational Research* 117, 1999, pp. 63–83.
61. S. Greco, B. Matarazzo, and R. Słowiński, The use of rough sets and fuzzy sets in MCDM, in: T. Gal, T. Stewart, and T. Hanne (eds.), *Advances in Multiple Criteria Decision Making*, Kluwer Academic Publishers, Boston, 1999, Chapter 14: pp. 14.1–14.59.
62. S. Greco, B. Matarazzo, R. Słowiński, Fuzzy dominance as basis for rough approximations, in: *Proceedings: the 4th Meeting of the EURO WG on Fuzzy Sets and 2nd Internat. Conf. on Soft and Intelligent Computing*, (EUROFUSE-SIC'99), Budapest, Hungary, May 1999, pp. 273-278.
63. S. Greco, B. Matarazzo, and R. Słowiński, Handling missing values in rough set analysis of multi–attribute and multi–criteria decision problems, in: *Proceedings: New Directions in Rough Sets, Data Mining and Granular-Soft Computing* (RSFSGrC'99), Ube – Yamaguchi, Japan, November 1999, LNAI 1711, Springer–Verlag, Berlin, 1999, pp. 146–157.
64. S. Greco, B. Matarazzo, and R. Słowiński, Fuzzy dominance as a basis for rough approximations, in: *Proceedings: Workshop Italiano sulla Logica Fuzzy* (Wilf'99), Genova, Italy, June 1999, pp.14–16.
65. S. Greco, B. Matarazzo, and R. Słowiński, Misurazione congiunta e incoerenze nelle preferenze, in: *Atti del Ventitreesimo Convegno A.M.A.S.E.S.*, Rende-Cosenza, Italy, September 1999, pp. 255–269.
66. S. Greco, B. Matarazzo, and R. Słowiński, L'approcio dei rough sets all'analisi del rating finanziario, in: *Atti del Ventitreesimo Convegno A. M. A. S. E. S.*, Rende-Cosenza, Italy, September 1999, pp. 271–286.

67. S. Greco, B. Matarazzo, and R. Słowiński, On joint use of indiscernibility, similarity and dominance in rough approximation of decision classes, in: *Proceedings: the 5th International Conference of the Decision Sciences Institute*, Athens, Greece, July 1999, pp. 1380–1382; also in: *Research Report RA-012/98*, Inst. Comp. Sci., Poznań Univ. Technology, 1998.
68. S. Greco, B. Matarazzo, and R.Słowiński, A new rough set approach to evaluation of bankruptcy risk, in: C. Zopounidis (ed.), *Operational Tools in the Management of Financial Risks*, Kluwer Academic Publishers, Dordrecht, 1998, pp. 121–136.
69. S. Greco, B. Matarazzo, and R. Słowiński, Fuzzy similarity relation as a basis for rough approximations, in: *Proceedings: Rough Sets and Current Trends in Computing* (RSCTC'98), Warsaw, Poland, June 1998, LNAI 1424, Springer–Verlag, Berlin, 1998, pp. 283–289.
70. S. Greco, B. Matarazzo, and R. Słowiński, A new rough set approach to multi–criteria and multi–attribute classification, in: *Proceedings: Rough Sets and Current Trends in Computing* (RSCTC'98), Warsaw, Poland, June 1998, LNAI 1424, Springer–Verlag, Berlin, 1998, pp. 60–67.
71. S. Greco, B. Matarazzo, and R. Słowiński, Rough approximation of a preference relation in a pair–wise comparison table, in: L. Polkowski and A. Skowron (eds.), *Rough Sets in Knowledge Discovery 2. Applications, Case Studies and Software Systems*, Physica-Verlag, Heidelberg, 1998, pp. 13–36.
72. S. Greco, B. Matarazzo, and R. Słowiński, Rough set theory approach to decision analysis, in: *Proceedings: the 3rd European Workshop on Fuzzy Decision Analysis and Neural Networks for Management, Planning and Optimization* (EFDAN'98), Dortmund, Germany, June 1998, pp. 1–28.
73. S. Greco, B. Matarazzo, and R. Słowiński: Conjoint measurement, preference inconsistencies and decision rule model, in: *Proceedings: the 2nd International Workshop on Preferences and Decisions*, Trento, Italy, July 1998, pp. 49–53.
74. S. Greco, B. Matarazzo, and R. Słowiński, New developments in the rough set approach to multi–attribute decision analysis, in: *Tutorials and Research Reviews: 16th European Conference on Operational Research* (EURO XVI), Brussels, Belgium, July 1998, 37 pp.
75. S. Greco, B. Matarazzo, and R. Słowiński, Rough set handling of ambiguity, in: *Proceedings: the 6th European Congress on Intelligent Techniques and Soft Computing* (EUFIT'98), Aachen, Germany, September 1998, Verlag Mainz, Aachen, 1998, pp. 3–14.
76. S. Greco, B. Matarazzo, and R. Słowiński, Fuzzy measures as a technique for rough set analysis, in: *Proceedings: the 6th European Congress on Intelligent Techniques and Soft Computing* (EUFIT'98), Aachen, Germany, September 1998, Verlag Mainz, Aachen, 1998, pp. 99–103.
77. S. Greco, B. Matarazzo, and R. Słowiński, Un nuovo approccio dei rough sets alla classificazione multiattributo e multicriteriale, in: *Atti del Ventiduesimo Convegno A. M. A. S. E. S.*, Genova, Italy, September 1998, Bozzi Editore, Genova, 1998, pp. 249–260.
78. S. Greco, B. Matarazzo, and R. Słowiński, Modellizzazione delle preferenze

per mezzo di regole di decisione, in: *Atti del Ventiduesimo Convegno A. M. A. S. E. S.*, Genova, Italy, September 1998, Bozzi Editore, Genova, 1998, pp. 233–247.

79. S. Greco, B. Matarazzo, and R. Słowiński, The rough set approach to decision support, in: *Proceedings: the Annual Conference of the Operational Research Society of Italy* (AIRO), Treviso, Italy, September 1998, pp. 561–564.

80. S. Greco, B. Matarazzo, R. Słowiński, and A. Tsoukias, Exploitation of a rough approximation of the outranking relation in multi–criteria choice and ranking, in: T.J. Stewart and R.C. van den Honert (eds.), *Trends in Multi– criteria Decision Making*, LNEMS 465, Springer–Verlag, Berlin, 1998, pp. 45–60.

81. S. Greco, B. Matarazzo, R. Słowiński, and S. Zanakis, Rough set analysis of information tables with missing values, in: *Proceedings: the 5th International Conference of the Decision Sciences Institute*, Athens, Greece, July 1999, pp. 1359–1362.

82. J. W. Grzymala–Busse, Applications of the rule induction system LERS, in: L. Polkowski and A. Skowron (eds.), *Rough Sets in Knowledge Discovery 1. Methodology and Applications*, Physica–Verlag, Heidelberg, 1998, pp. 366–375.

83. J. W. Grzymala–Busse, LERS : A knowledge discovery system, in: L. Polkowski and A. Skowron (eds.),*Rough Sets in Knowledge Discovery 2. Applications, Case Studies and Software Systems*, Physica–Verlag, Heidelberg, 1998, pp. 562–565.

84. J. W. Grzymala–Busse, Rule induction system LERS, *Bull. of Intern. Rough Set Society* 2, 1998, pp. 18–20.

85. J. W. Grzymala–Busse, Classification of unseen examples under uncertainty, *Fundamenta Informaticae* 30, 1997, pp. 255–267.

86. J. W. Grzymala–Busse, A new version of the rule induction system LERS, *Fundamenta Informaticae* 31, 1997, pp. 27–39.

87. J. W. Grzymala–Busse and L. K. Goodwin, Predicting pre–term birth risk using machine learning from data with missing values, *Bull. of Intern. Rough Set Society* 1, 1997, pp. 17–21.

88. J. W. Grzymala–Busse, L. K. Goodwin, and Xiaohui Zhang, Pre–term birth risk assessed by a new method of classification using selective partial matching, in: *Proceedings: the 11th International Symposium on Methodologies for Intelligent Systems* (ISMIS'99), Warsaw, Poland, June 1999, LNAI 1609, Springer Verlag, Berlin, 1999, pp. 612–620.

89. J. W. Grzymala–Busse, L. K. Goodwin, and Xiaohui Zhang, Increasing sensitivity of pre–term birth by changing rule strengths, in: *Proceedings: the 8th Workshop on Intelligent Information Systems* (IIS'99), Ustroń, Poland, June 1999, pp. 127–136.

90. J. W. Grzymala–Busse, W. J. Grzymala–Busse, and L. K. Goodwin, A closest fit approach to missing attribute values in pre–term birth data, in: *Proceedings: the 7th International Workshop on Rough Sets, Fuzzy Sets, Data Mining and Granular–Soft Computing* (RSFDGrC'99), Ube-Yamaguchi, Japan, November 1999, LNAI 1711, Springer–Verlag, Berlin, 1999, pp.

405–413.
91. J. W. Grzymala–Busse and L. J. Old, A machine learning experiment to determine part of speech from word-endings, in: *Proceedings: the 10th Intern. Symposium on Methodologies for Intelligent Systems* (ISMIS'97), Charlotte NC, October 1997, LNAI 1325, Springer-Verlag, Berlin, 1997, pp. 497–506.
92. J. W. Grzymala–Busse, S. Y. Sedelow, and W. A. Sedelow Jr., Machine learning and knowledge acquisition, rough sets, and the English semantic code, in: T. Y. Lin and N. Cercone (eds.), *Rough Sets and Data Mining. Analysis of Imprecise Data*, Kluwer Academic Publishers, Dordrecht, 1997, pp. 91–107.
93. J. W. Grzymala–Busse and Soe Than, Inducing simpler rules from reduced data, in: *Proceedings: the Seventh Workshop on Intelligent Information Systems* (IIS'98), Malbork, Poland, June 1998, pp. 371–378.
94. J. W. Grzymala–Busse and J. Stefanowski, Two approaches to numerical attribute discretization for rule induction, in: *Proceedings: the 5th International Conference of the Decision Sciences Institute*, Athens, Greece, July 1999, pp. 1377–1379.
95. J. W. Grzymala–Busse and J. Stefanowski, Discretization of numerical attributes by direct use of the rule induction algorithm LEM2 with interval extension, in: *Proceedings: the Sixth Symposium on Intelligent Information Systems* (IIS'97), Zakopane, Poland, June 1997, pp. 149–158.
96. J. W. Grzymala–Busse and Ta-Yuan Hsiao, Dropping conditions in rules induced by ID3, in: *Proceedings: the 6th International Workshop on Rough Sets, Data Mining and Granular Computing* (RSDMGrC'98) at *the 4th Joint Conference on Information Sciences* (JCIS'98), Research Triangle Park NC, October 1998, pp. 351–354.
97. J. W. Grzymala–Busse and A. Y. Wang, Modified algorithms LEM1 and LEM2 for rule induction from data with missing attribute values, in: *Proceedings: the 5th Intern. Workshop on Rough Sets* (RSSC'97) at *the 3rd Joint Conference on Information Sciences* (JCIS'97), Research Triangle Park NC, March 1997, pp. 69–72.
98. J. W. Grzymala–Busse and P. Werbrouck, On the best search method in the LEM1 and LEM2 algorithms, in :E. Orłowska (ed.), *Incomplete Information: Rough Set Analysis*, Physica–Verlag, Heidelberg, 1998, pp. 75–91.
99. J. W. Grzymala–Busse and Xihong Zou, Classification strategies using certain and possible rules, in: *Proceedings: the First International Conference on Rough Sets and Current Trends in Computing*(RSCTC'98), Warsaw, Poland, June 1998, LNAI 1424, Springer Verlag, Berlin, 1998, pp. 37–44.
100. Hoang Kiem and Do Phuc, A combined multi-dimensional Genetic Algorithm and Kohonen Neural Network for cluster discovery in Data Mining, in: *Proceedings: the 3rd International Conference on Data Mining* (PAKDD'99), Beijing, China, 1999.
101. Hoang Kiem and Do Phuc, A Rough Genetic Kohonen Neural Network for conceptual cluster discovery, in: *Proceedings: the 7th WorkShop on Rough Set, Fuzzy Set, Granular Computing and Data Mining* (RSFDGrC'99), Ube–Yamaguchi, Japan, November 1999, LNAI 1711, Springer Verlag, Berlin,

1999, pp. 448–452.
102. Hoang Kiem and Do Phuc, On the association rules based extension of the dependency of attributes in rough set theory for classification problem, *Magazine of Science and Technology* 1, 1999, Vietnam National University.
103. Hoang Kiem and Do Phuc, Discovering the binary and fuzzy association rules from database, *Magazine of Science and Technology* 4, 1999, Vietnam National University.
104. V. Jog, W. Michałowski, R. Słowiński, and R. Susmaga, The rough set analysis and the neural networks classifier – a hybrid approach to predicting stocks' performance, in: *Proceedings: the 5th International Conference of the Decision Sciences Institute*, Athens, Greece, July 1999, pp. 1386–1388.
105. R. E. Kent, Soft concept analysis, in: S. K. Pal and A. Skowron (eds.), *Rough Fuzzy Hybridization: A New Trend in Decision–Making*, Springer–Verlag, Singapore, 1999, pp. 215–232.
106. A. N. Knyazev, On word recognition in language generated by 1-context-free grammar (in Russian), in: *Proceedings: 12th International Conference on Problems of Theoretical Cybernetics*, Nizhny Novgorod, Russia, 1999, Moscow State University Publishers, Moscow, Part 1 (1999), pp. 96.
107. A. N. Knyazev, On recognition of words from languages generated by linear grammars with one non–terminal symbol, in: *Proceedings: the First International Conference on Rough Sets and Current Trends in Computing* (RSCTC'98), Warsaw, Poland, 1998, LNAI 1424, Springer–Verlag, Berlin, 1998, pp. 111–114.
108. J. Komorowski, L. Polkowski, and A. Skowron, Rough Sets: A Tutorial, in: *Lecture Notes for ESSLLI'99: the 11th European Summer School in Language, Logic and Information*, Utrecht, Holland, August 1999, 111 pp.
109. J. Komorowski, Z. Pawlak, L. Polkowski, and A. Skowron, Rough sets: A tutorial, in: S. K. Pal and A. Skowron (eds.), *Rough Fuzzy Hybridization: A New Trend in Decision Making*, Springer Verlag, Singapore, 1999, pp. 3–98.
110. J. Komorowski, L. Polkowski, and A. Skowron, Towards a rough mereology–based logic for approximate solution synthesis, *Studia Logica* 58(1), 1997, pp. 143–184.
111. J. Komorowski, L. Polkowski, and A. Skowron, Rough sets for Data Mining and Knowledge Discovery (Tutorial–abstract), in: *Proceedings: the First European Symposium on Principles of Data Mining and Knowledge Discovery*, Trondheim, Norway, June 1997, LNAI 1263, Springer–Verlag, Berlin, pp. 395–395.
112. B. Kostek, *Soft Computing in Acoustics, Applications of Neural Networks, Fuzzy Logic and Rough Sets to Musical Acoustics* in the Series: *Studies in Fuzziness and Soft Computing* (J. Kacprzyk (ed.)), vol. 31, Physica–Verlag, Heilderberg, 1999.
113. B. Kostek, Assessment of concert hall acoustics using rough set and fuzzy set approach, in: S. K. Pal and A. Skowron (eds.), : *Rough–Fuzzy Hybridization: A New Trend in Decision Making*, Springer-Verlag, Singapore, 1999, pp. 381–396.

114. B. Kostek, Rough–fuzzy method of subjective test result processing, in: *Proceedings: the 8th International Symposium on Sound Engineering and Mastering* (ISSEM'99), Gdańsk, Poland, September 1999, pp. 11–18.
115. B. Kostek, Soft computing–based recognition of musical sounds, in: L. Polkowski and A. Skowron (eds.), *Rough Sets in Knowledge Discovery 2. Applications, Case Studies and Software Systems*, Physica–Verlag, Heidelberg, 1998, pp. 193–213.
116. B. Kostek, Computer–based recognition of musical phrases using the rough set approach, *J. Information Sciences* 104, 1998, pp. 15–30.
117. B. Kostek, Soft set approach to the subjective assessment of sound quality, in: *Proceedings: the Conference FUZZ-IEEE'98 at the World Congress on Computational Intelligence*(WCCI'98), Anchorage, Alaska, May 1998, pp. 669–674.
118. B. Kostek, Automatic recognition of sounds of musical instruments: An expert media application, in: *Proceedings: the World Automation Congress* (WAC'98), pp. IFMIP–053.
119. B. Kostek, Sound quality assessment based on the rough set classifier, in: *Proceedings: the 5th European Congress on Intelligent Techniques and Soft Computing* (EUFIT'97), Aachen, Germany, September 1997, Verlag Mainz, Aachen, 1997, pp. 193–195.
120. B. Kostek, Soft set approach to the subjective assessment of sound quality, in: *Proceedings: the 9th Intern. Conference on Systems Research Informatics and Cybernetics* (InterSymp'97), Baden–Baden, Germany, 1997.
121. B. Kostek, Rough set and fuzzy set methods applied to acoustical analyses, *J. Intelligent Automation and Soft Computing*, 2(2), 1996, pp. 147–160.
122. K. Krawiec, R. Słowiński, and D. Vanderpooten, Learning of decision rules from similarity based rough approximations, in: L. Polkowski, A. Skowron (eds.), *Rough Sets in Knowledge Discovery 2. Applications, Case Studies and Software Systems*, Physica–Verlag, Heidelberg, 1998, pp. 37–54.
123. R. Królikowski and A. Czyżewski, Noise reduction in telecommunication channels using rough sets and neural networks, in: *Proceedings: the 7th Intern. Workshop on Rough Sets, Fuzzy Sets, Data Mining, and Granular-Soft Computing* (RSFDGrC'99), Ube–Yamaguchi, Japan, November 1999, LNAI 1711, Springer–Verlag, Berlin, 1999, pp. 109–117.
124. R. Królikowski and A. Czyżewski, Applications of rough sets and neural nets to noisy audio enhancement, in: *proceedings: the 7th European Congress on Intelligent Techniques and Soft Computing* (EUFIT'99), Aachen, Germany, September 1999.
125. M. Lifantsev and A. Wasilewska, A decision procedure for rough sets equalities, in: *Proceedings: the 18th International Conference of the North American Fuzzy Information Processing Society* (NAFIPS'99), New York NY, June 1999, pp.786–791.
126. T. Y. Lin, Data Mining and machine oriented modeling: A granular computing approach, *Journal of Applied Intelligence*, in print.
127. T. Y. Lin, Theoretical sampling for Data Mining, in: *Proceedings: the 14th Annual International Symposium Aerospace/Defense Sensing, Simulation,*

and *Controls* (SPIE) 4057, Orlando Fla., April 2000, to appear.
128. T. Y. Lin, Attribute transformations on numerical databases: Applications to stock market data, in: *Methodologies for Knowledge Discovery and Data Mining*, LNAI, Springer–Verlag, Berlin, 2000, to appear.
129. T. Y. Lin, Belief functions and probability of fuzzy sets, in:*Proceedings: the 8th International Fuzzy Systems Association World Congress* (IFSA'99), Taipei, Taiwan, August 1999, pp. 219–223.
130. T. Y. Lin, Discovering patterns in numerical sequences using rough set theory, in: *Proceedings: the 3rd World Multi–Conference on Systemics, Cybernetics and Informatics*, Orlando Fla., July 1999, 5, pp. 568–572.
131. T. Y. Lin, Measure theory on granular fuzzy sets, in : *Proceedings: the 18th International Conference of North America Fuzzy Information Processing Society*, June 1999, pp. 809–813.
132. T. Y. Lin, Data Mining: Granular computing approach, in: *Methodologies for Knowledge Discovery and Data Mining, the 3rd Pacific–Asia Conference*, Beijing, China, April 1999, LNAI 1574, Springer–Verlag, Berlin, 1999, pp. 24–33.
133. T. Y. Lin, Granular computing: Fuzzy logic and rough sets, in: L.A. Zadeh and J. Kacprzyk (eds), *Computing with Words in Information/Intelligent Systems 1*, Physica–Verlag, Heidelberg, 1999, pp. 183–200.
134. T. Y. Lin, Granular computing on binary relations II: Rough set representations and belief functions, in: L. Polkowski and A. Skowron (eds), *Rough Sets In Knowledge Discovery 1.Methodology and Applications*, Physica–Verlag, Heidelberg, 1998, pp. 121–140.
135. T. Y. Lin, Granular computing on binary relations I: Data Mining and neighborhood systems, in: L. Polkowski and A. Skowron (eds.), *Rough Sets In Knowledge Discovery 1. Methodology and Applications*, Physica–Verlag, Heidelberg, 1998, pp. 107–121.
136. T. Y. Lin, Context free fuzzy sets and information tables, in: *Proceedings: the Sixth European Congress on Intelligent Techniques and Soft Computing* (EUFIT'98), Aachen, Germany, September 1998, Verlag Mainz, Aachen, pp. 76–80.
137. T. Y. Lin, Granular fuzzy sets: Crisp representation of fuzzy sets, in: *Proceedings: the Sixth European Congress on Intelligent Techniques and Soft Computing* (EUFIT'98), Aachen, Germany, September 1998, Verlag Mainz, Aachen, pp. 94–98.
138. T. Y. Lin, Fuzzy partitions : Rough set theory, in: *Proceedings: the Conference on Information Processing and Management of Uncertainty in Knowledge–Based Systems* (IPMU'98), La Sorbonne, Paris, France, July 1998, pp. 1167–1174.
139. T. Y. Lin, Sets with partial memberships: A Rough sets view of fuzzy sets, in: *Proceedings: the FUZZ-IEEE International Conference, 1998 IEEE World Congress on Computational Intelligence* (WCCI'98), Anchorage, Alaska, May 1998.
140. T. Y. Lin and Q. Liu, First–order rough logic revisited, in: *Proceedings: the 7th International Workshop on rough Sets, Fuzzy Sets, Data Mining and*

Granular–Soft computing (RSFSGrC'99), Ube–Yamaguchi, Japan, November 1999, LNAI 1711, Springer–Verlag, Berlin, pp. 276–284.

141. T. Y. Lin and E. Louie, A Data Mining approach using machine oriented modeling: finding association rules using canonical names, in: *Proceedings: the 14th Annual International Symposium Aerospace/Defense Sensing, Simulation, and Controls* (SPIE) 4057, Orlando Fla., April 2000, to appear.

142. T. Y. Lin, Ning Zhong, J. J. Dong, and S. Ohsuga, An incremental, probabilistic rough set approach to rule discovery, in: *Proceedings: the FUZZ-IEEE International Conference, 1998 IEEE World Congress on Computational Intelligence* (WCCI'98), Anchorage, Alaska, May 1998.

143. T. Y. Lin, Ning Zhong, J. J. Dong, and S. Ohsuga, Frameworks for mining binary relations in data, in : *Proceedings: the First International Conference on Rough Sets and Current Trends in Computing* (RSCTC'98), Warsaw, Poland, June 1998, LNAI 1424, Springer–Verlag, Berlin, 1998, pp. 387–393.

144. T. Y. Lin and S. Tsumoto, Context-free fuzzy sets in Data Mining context, in: *Proceedings: the 7th International Workshop on Rough Sets, Fuzzy Sets, Data Mining and Granular–Soft computing* (RSFSGrC'99), Ube–Yamaguchi, Japan, November 1999, LNAI 1711, Springer–Verlag, Berlin, pp. 212–220.

145. B. Marszał–Paszek, Linking α–approximation with evidence theory, in: *Proceedings: the 6th International Conference Information Processing and Management of Uncertainty in Knowledge-Base System* (IPMU'96), Granada, Spain, 1996, pp. 1153–1158.

146. B. Marszał–Paszek and P. Paszek, Searching for attributes which well determinate decision in the decision table, in: *Proceedings: Intelligent Information Systems VIII*, Ustroń, Poland, 1999, pp. 146–148.

147. B. Marszał–Paszek and P. Paszek, Extracting strong relationships between data from decision table, in: *Proceedings: Intelligent Information Systems VII*, Malbork, Poland, 1998, pp. 396–399.

148. V. W. Marek and M. Truszczyński, Contributions to the theory of rough sets, *Fundamenta Informaticae* 39(4), 1999, pp. 389–409.

149. E. Martienne and M. Quafafou, Learning fuzzy relational descriptions = using the logical framework and rough set theory, in: *Proceedings: the 7th IEEE International Conference on Fuzzy Systems* (FUZZ-IEEE'98), IEEE Neural Networks Council, 1998.

150. E. Martienne and M. Quafafou, Learning logical descriptions for document understanding: a rough sets based approach, in: *Proceedings: the First International Conference on Rough Sets and Current Trends in Computing* (RSCTC'98), Warsaw, Poland, June 1998, LNAI 1424, Springer–Verlag, Berlin, 1998.

151. E. Martienne and M. Quafafou, Vagueness and data reduction in learning of logical descriptions, in: *Proceedings: the 13th European Conference on Artificial Intelligence* (ECAI'98), Brighton, UK, August 1998, John Wiley and Sons, Chichester, 1998.

152. P. Mitra, S. Mitra, and S.K. Pal, Staging of cervical cancer with Soft Computing, *IEEE Trans. Bio-Medical Engineering*, in print.

153. S. Mitra, M. Banerjee, and S.K. Pal, Rough Knowledge-based networks, fuzziness and classification, *Neural Computing and Applications* 7, 1998, pp. 17–25.
154. S. Mitra, P. Mitra, and S. K. Pal, Evolutionary design of modular Rough Fuzzy MLP, *Neurocomputing*, communicated.
155. S. Miyamoto and Kyung Soo Kim, Images of fuzzy multi–sets by one-variable functions and their applications (in Japanese), *Journal of Japan Society for Fuzzy Theory and Systems*10(1), 1998, pp. 150–157.
156. S. Miyamoto, Application of rough sets to information retrieval, *Journal of the American Society for Information Science* 47(3), 1998, pp. 195–205.
157. S. Miyamoto, Indexed rough approximations and generalized possibility theory, in: *Proceedings: FUZZ-IEEE'98*, May 4-9, 1998, Anchorage, Alaska, pp. 791-795.
158. S. Miyamoto, Fuzzy multi–sets and a rough approximation by multi–set–valued function, in: L. Polkowski and A. Skowron (eds.), *Rough Sets in Knowledge Discovery 1. Methodology and Applications*, Physica–Verlag, Heidelberg,1998, pp. 141–159.
159. H. Moradi, J. W. Grzymala–Busse, and J. A. Roberts, Entropy of English text: Experiments with humans and a machine learning system based on rough sets, *Information Sciences. An International Journal* 104, 1998, pp. 31–47.
160. M. Ju. Moshkov, Time complexity of decision trees (in Russian), in: *Proceedings: the 9th Workshop on Synthesis and Complexity of Control Systems*, Nizhny Novgorod, Russia, 1998, Moscow State University Publishers, Moscow, 1999, pp. 52–62.
161. M. Ju. Moshkov, Local approach to construction of decision trees, in: S.K.Pal and A. Skowron (eds.), *Rough Fuzzy Hybridization. A New Trend In Decision–Making*, Springer-Verlag, Singapore, 1999, pp. 163–176.
162. M. Ju. Moshkov, On complexity of deterministic and nondeterministic decision trees (in Russian), in: *Proceedings: the 12th International Conference on Problems of Theoretical Cybernetics*, Nizhny Novgorod, Russia, 1999, Moscow State University Publishers, Moscow, 1999, p. 164.
163. M. Ju. Moshkov, On the depth of decision trees (in Russian), *Doklady RAN*, 358(1), 1998, p. 26.
164. M. Ju. Moshkov, Some relationships between decision trees and decision rule systems, in: *Proceedings: the First International Conference on Rough Sets and Current Trends in Computing* (RSCTC'98), Warsaw, Poland, June 1998, LNAI 1424, Springer–Verlag, Berlin, 1998, pp. 499–505.
165. M. Ju. Moshkov, Three ways for construction and complexity estimation of decision trees, in: *Program: the 16th European Conference on Operational Research*(EURO XVI), Brussels, Belgium, July 1998, pp. 66-67.
166. M. Ju. Moshkov, Rough analysis of tree–program time complexity, in: *Proceedings: the 7th International Conference of Information Processing and Management of Uncertainty in Knowledge-Based Systems*, La Sorbonne, Paris, France, July 1998, pp. 1376–1380.

167. M. Ju. Moshkov, On time complexity of decision trees, in: L. Polkowski and A. Skowron (eds.), *Rough Sets in Knowledge Discovery 1. Methodology and Applications*, Physica–Verlag, Heidelberg, 1998, pp. 160–191.
168. M. Ju. Moshkov, On time complexity of decision trees (in Russian), in: *Proceedings: International Siberian Conference on Operational Research*, Novosibirsk, Russia, 1998, pp. 28–31.
169. M. Ju. Moshkov, On complexity of decision trees over infinite information systems, in: *Proceedings: the Third Joint Conference on Information Sciences* (JCIS'97), Duke University, USA, 1997, pp. 353–354.
170. M. Ju. Moshkov, Algorithms for constructing of decision trees, in: *Proceedings: the First European Symposium on Principles of Data Mining and Knowledge Discovery* (PKDD'97), Trondheim, Norway, 1997, LNAI 1263, Springer–Verlag, Berlin, 1997, pp. 335–342.
171. M. Ju. Moshkov, Unimprovable upper bounds on time complexity of decision trees, *Fundamenta Informaticae* 31(2), 1997, pp. 157–184.
172. M. Ju. Moshkov, Rough analysis of tree-programs, in: *Proceedings: the 5th European Congress on Intelligent Techniques and Soft Computing* (EUFIT' 97), Aachen, Germany, September 1997, Verlag Mainz, Aachen, pp. 231–235.
173. M. Ju. Moshkov, Complexity of deterministic and nondeterministic decision trees for regular language word recognition, in: *Proceedings : the 3rd International Conference on Developments in Language Theory*, Thessaloniki, Greece, 1997, pp. 343–349.
174. M. Ju. Moshkov, Comparative analysis of time complexity of deterministic and nondeterministic tree–programs (in Russian), in: *Actual Problems of Modern Mathematics* 3, Novosibirsk University Publishers, Novosibirsk, 1997, pp. 117–124.
175. M. Ju. Moshkov and I. V. Chikalov, On effective algorithms for conditional test construction (in Russian), in: *Proceedings: the 12th International Conference on Problems of Theoretical Cybernetics*, Nizhny Novgorod, Russia, 1999, Moscow State University Publishers, Moscow, 1999, pp. 165.
176. M. Ju. Moshkov and I. V. Chikalov, Bounds on average depth of decision trees, in: *Proceedings: the Fifth European Congress on Intelligent Techniques and Soft Computing* (EUFIT'97), Aachen, Germany, September 1997, Verlag Mainz, Aachen, pp. 226–230.
177. M. Ju. Moshkov and I. V. Chikalov, Bounds on average weighted depth of decision trees, *Fundamenta Informaticae* 31(2), 1997, pp. 145–156.
178. M. Ju. Moshkov and A. Moshkova, Optimal bases for some closed classes of Boolean functions, in: *Proceedings: the 5th European Congress on Intelligent Techniques and Soft Computing* (EUFIT 97), Aachen, Germany, September 1997, Verlag Mainz, Aachen, pp. 1643–1647.
179. A. M. Moshkova, On complexity of "retaining" fault diagnosis in circuits (in Russian), in: *Proceedings: the 12th International Conference on Problems of Theoretical Cybernetics*, Nizhny Novgorod, Russia, 1999, Moscow State University Publishers, Moscow, 1999, p. 166.
180. A. M. Moshkova, On diagnosis of retaining faults in circuits, in: *Proceedings: the First International Conference on Rough Sets and Current Trends in*

Computing (RSCTC'98), Warsaw, Poland, June 1998, LNAI 1424, Springer-Verlag, Berlin, 1998, pp. 513–516.
181. A. M. Moshkova, Diagnosis of "retaining" faults in circuits (in Russian), *Bulletin of Nizhny Novgorod State University. Mathematical Simulation and Optimal Control* 19, 1998, pp. 204–213.
182. A. Nakamura, Conflict logic with degrees, in: S. K. Pal and A. Skowron (eds.), *Rough Fuzzy Hybridization: A New Trend in Decision–Making*, Springer–Verlag, Singapore, 1999, pp. 136–150.
183. Nguyen Hung Son, From optimal hyperplanes to optimal decision trees, *Fundamenta Informaticae* 34(1-2), 1998, pp. 145–174.
184. Nguyen Hung Son, Discretization problems for rough set methods, in: *Proceedings: the First International Conference on Rough Sets and Current Trend in Computing* (RSCTC'98), Warsaw, Poland, June 1998, LNAI 1424, Springer–Verlag, Berlin, 1998, pp. 545–552.
185. Nguyen Hung Son, Discretization of real value attributes. Boolean reasoning approach, Ph.D. Dissertation, supervisor A. Skowron, Warsaw University, Warsaw, 1997, pp. 1–90.
186. Nguyen Hung Son, Rule induction from continuous data, in: *Proceedings: the 5th International Workshop on Rough Sets and Soft Computing* (RSSC'97) at *the 3rd Annual Joint Conference on Information Sciences* (JCIS'97), Durham NC, March 1997, pp. 81–84.
187. Nguyen Hung Son and Nguyen Sinh Hoa, An application of discretization methods in control, in: *Proceedings: the Workshop on Robotics, Intelligent Control and Decision Support Systems*, Polish-Japanese Institute of Information Technology, Warsaw, Poland, February 1999, pp. 47–52.
188. Nguyen Hung Son and Nguyen Sinh Hoa, Discretization methods in Data Mining, in: L. Polkowski and A. Skowron (eds.): *Rough Sets in Knowledge Discovery 1. Methodology and Applications*, Physica–Verlag, Heidelberg, 1998, pp. 451–482.
189. Nguyen Hung Son and Nguyen Sinh Hoa, Discretization methods with back-tracking, in: *Proceedings: the 5th European Congress on Intelligent Techniques and Soft Computing* (EUFIT'97), Aachen, Germany, September 1997, Verlag Mainz, Aachen, 1997, pp. 201–205.
190. Nguyen Hung Son, Nguyen Sinh Hoa, and A. Skowron, Decomposition of task specifications, in: *Proceedings: the 11th International Symposium on Methodologies for Intelligent Systems, Foundt-ons of Intelligent Systems* (IS-MIS'99), Warsaw, Poland, June 8–11, LNAI 1609, Springer–Verlag, Berlin, pp. 310–318.
191. Nguyen Hung Son and A. Skowron, Boolean reasoning scheme with some applications in Data Mining, in: *Proceedings: Principles of Data Mining and Knowledge Discovery* (PKDD'99), Prague, Czech Republic, September 1999, LNAI 1704, Springer–Verlag, Berlin, 1999, pp. 107–115.
192. Nguyen Hung Son and A. Skowron, Task decomposition problem in multi-agent system, in: *Proceedings: the Workshop on Concurrency, Specification and Programming*, Berlin, Germany, September 1998, Informatik Bericht 110, Humboldt–Universität zu Berlin, pp. 221–235.

193. Nguyen Hung Son and A. Skowron, Boolean reasoning for feature extraction problems, in: *Proceedings: the 10th International Symposium on Methodologies for Intelligent Systems, Foundations of Intelligent Systems* (ISMIS'97), Charlotte NC, October 1997, LNAI 1325, Springer–Verlag, Berlin, 1997, pp. 117–126.
194. Nguyen Hung Son and A. Skowron, Quantization of real value attributes: Rough set and boolean reasoning approach, *Bulletin of International Rough Set Society* 1(1), 1997, pp. 5–16.
195. Nguyen Hung Son, M. Szczuka, and D. Ślęzak, Neural network design: Rough set approach to continuous data, in: *Proceedings: the First European Symposium on Principles of Data Mining and Knowledge Discovery* (PKDD'97), Trondheim, Norway,June 1997, LNAI 1263, Springer–Verlag, Berlin, 1997, pp. 359–366.
196. Nguyen Hung Son and D. Ślęzak, Approximate reducts and association rules: correspondence and complexity results, in: *Proceedings: the 7th International Workshop on New Directions in Rough Sets, Data Mining, and Granular-Soft Computing* (RSFDGrC'99), Ube–Yamaguchi, Japan, November 1999, LNAI 1711, Springer–Verlag, Berlin, 1999, pp. 137–145.
197. Nguyen Sinh Hoa, Discovery of generalized patterns, in: *Proceedings: the 11th International Symposium on Methodologies for Intelligent Systems, Foundations of Intelligent Systems* (ISMIS'99), Warsaw, Poland, June 1999, LNAI 1609, Springer–Verlag, Berlin, in print.
198. Nguyen Sinh Hoa, Data regularity analysis and applications in data mining, Ph.D. Dissertation, supervisor B. Chlebus, Warsaw University, Warsaw, Poland, 1999.
199. Nguyen Sinh Hoa and Nguyen Hung Son, Pattern extraction from data, in: *Proceedings: the Conference of Information Processing and Management of Uncertainty in Knowledge-Based Systems* (IPMU'98), La Sorbonne, Paris, France, July 1998, pp. 1346–1353.
200. Nguyen Sinh Hoa and Nguyen Hung Son, Pattern extraction from data, *Fundamenta Informaticae* 34(1-2), 1998, pp. 129–144.
201. Nguyen Sinh Hoa, Nguyen Tuan Trung, L. Polkowski, A. Skowron, P. Synak, and J. Wróblewski, Decision rules for large data tables, in: *Proceedings: CESA'96 IMACS Multiconference: Computational Engineering in Systems Applications* (CESA'96), Lille, France, July 1996, pp. 942–947.
202. Nguyen Sinh Hoa and A. Skowron, Searching for relational patterns in data, in: *Proceedings: the First European Symposium on Principles of Data Mining and Knowledge Discovery* (PKDD'97), Trondheim, Norway, June 1997, LNAI 1263, Springer–Verlag, Berlin, 1997, pp. 265–276.
203. Nguyen Sinh Hoa, A. Skowron, and P. Synak, Discovery of data patterns with applications to decomposition and classification problems, in: L. Polkowski and A. Skowron (eds.), *Rough Sets in Knowledge Discovery 2. Applications, Case Studies and Software Systems*, Physica–Verlag, Heidelberg, 1998, pp. 55–97.
204. Nguyen Sinh Hoa, A. Skowron, P. Synak, and J. Wróblewski, Knowledge discovery in data bases: Rough set approach, in: *Proceedings: the 7th Inter-*

national Fuzzy Systems Association World Congress (IFSA'97), Prague, the Czech Republic, June 1997, Academia, Prague, 1997, pp. 204–209.
205. A. Øhrn, J. Komorowski, A. Skowron, and P. Synak, The design and implementation of a knowledge discovery toolkit based on rough sets– The ROSETTA system, in: L. Polkowski and A. Skowron (eds.), *Rough Sets in Knowledge Discovery 1. Methodology and Applications*, Physica–Verlag, Heidelberg, 1998, pp. 376–399.
206. A. Øhrn, J. Komorowski, A. Skowron, and P. Synak, The ROSETTA software system, In: L. Polkowski and A. Skowron (eds.), *Rough Sets in Knowledge Discovery 2. Applications, Case Studies and Software Systems*, Physica–Verlag, Heidelberg, 1998, pp. 572–576.
207. A. Øhrn, J. Komorowski, A. Skowron, and P. Synak, A software system for rough data analysis, *Bulletin of the International Rough Set Society* 1(2), 1997, pp. 58–59.
208. P. Paszek and A. Wakulicz–Deja, Optimalization diagnose in progressive encephalopathy applying the rough set theory, in: *Intelligent Information Systems V*, Dęblin, Poland, 1996, pp. 142–151.
209. G. Paun, L. Polkowski, and A. Skowron, Rough set approximations of languages, *Fundamenta Informaticae* 32(2), 1997, pp. 149–162.
210. G. Paun, L. Polkowski, and A. Skowron, Parallel communicating grammar systems with negotiations, *Fundamenta Informaticae* 28(3-4), 1996, pp. 315–330.
211. G. Paun, L. Polkowski, and A. Skowron, Rough–set–like approximations of context–free and regular languages, in: *Proceedings: Information Processing and Management of Uncertainty in Knowledge Based Systems* (IPMU-96), Granada, Spain, July 1996, pp. 891–895.
212. Z. Pawlak, Granularity of knowledge, indiscernibility, and rough sets, in: *Proceedings: IEEE Conference on Evolutionary Computation*, Anchorage, Alaska, May 5-9, 1998, pp. 106–110; also in: *IEEE Transactions on Automatic Control* 20, 1999, pp. 100–103.
213. Z. Pawlak, Rough set theory for intelligent industrial applications, in: *Proceedings: the 2nd International Conference on Intelligent Processing and Manufacturing of Materials*, Honolulu, Hawaii, 1999, pp. 37–44.
214. Z. Pawlak, Data Mining - a rough set perspective, in: *Proceedings: Methodologies for Knowledge Discovery and Data Mining. The 3rd Pacific–Asia Conference*, Beijing, China, Springer– Verlag, Berlin, 1999, pp. 3–11.
215. Z. Pawlak, Rough sets, rough functions and rough calculus, in: S.K. Pal, A. Skowron (eds.), *Rough Fuzzy Hybridization: A New Trend in Decision Making*, Springer-Verlag, Singapore, 1999, pp. 99–109.
216. Z. Pawlak, Logic, Probability and Rough Sets, in: J. Karhumaki, H. Maurer, G. Paun, and G. Rozenberg (eds.), *Jewels are Forever. Contributions to Theoretical Computer Science in Honor of Arto Salomaa*, Springer–Verlag, Berlin, 1999, pp. 364–373.
217. Z. Pawlak, Decision rules, Bayes' rule, and rough sets, in: *Proceedings: the 7th International Workshop on rough Sets, Fuzzy Sets, Data Mining and*

Granular–Soft computing (RSFSGrC'99), Ube–Yamaguchi, Japan, November 1999, LNAI 1711, Springer–Verlag, Berlin, pp. 1–9.
218. Z. Pawlak, An inquiry into anatomy of conflicts, *Journal of Information Sciences* 109, 1998, pp. 65–78.
219. Z. Pawlak, Sets, fuzzy sets, and rough sets, in: *Proceedings: Fuzzy–Neuro Systems – Computational Intelligence*, Muenchen, Germany, March 18-20, 1998, pp. 1–9.
220. Z. Pawlak, Reasoning about data–a rough set perspective, in: *Proceedings: the First International Conference on Rough Sets and Current Trends in Computing*(RSCTC'98), LNAI 1424, Springer–Verlag, Berlin, 1998, pp. 25–34.
221. Z. Pawlak, Rough sets theory and its applications to data analysis, *Cybernetics and Systems* 29, 1998, pp. 661–688.
222. Z. Pawlak, Rough set elements, in: L. Polkowski and A. Skowron (eds.), *Rough Sets in Knowledge Discovery 1. Methods and Applications*, Physica–Verlag, Heidelberg, 1998, pp. 10–30.
223. Z. Pawlak, Rough Modus Ponens, in: *Proceedings: the 7th Conference on Information Processing and Management of Uncertainty in Knowledge Based Systems* (IPMU'98), La Sorbonne, Paris, France, July 1998, pp. 1162–1165.
224. Z. Pawlak, Rough set approach to knowledge-based decision support, *European Journal of Operational Research* 29(3), 1997, pp. 1–10.
225. Z. Pawlak, Rough sets and Data Mining, in: *Proceedings: the International Conference on Intelligent Processing and Manufacturing Materials*, Gold Coast, Australia, 1997, pp. 1–5.
226. Z. Pawlak, Rough sets, in: T.Y. Lin, N. Cercone (eds.), *Rough Sets and Data Mining. Analysis of Imprecise Data*, Kluwer Academic Publishers, Dordrecht, 1997, pp. 3–8.
227. Z. Pawlak, Rough real functions and rough controllers, in: T.Y. Lin, N. Cercone (eds.), *Rough Sets and Data Mining. Analysis of Imprecise Data*, Kluwer Academic Publishers, Dordrecht, 1997, pp. 139–147.
228. Z. Pawlak, Conflict analysis, in: *Proceedings: the 5th European Congress on Intelligent Techniques and Soft Computing* (EUFIT'97), Aachen, Germany, September 9-11, Verlag Mainz, Aachen, 1997, pp. 1589–1591.
229. Z. Pawlak, Rough sets and their applications, *Proceedings: Fuzzy Sets 97*, Dortmund, Germany, 1997.
230. Z. Pawlak, Vagueness–a rough set view, in: *Structures in Logic and Computer Science*, LNCS 1261, Springer–Verlag, Berlin, 1997, pp. 106–117.
231. Z. Pawlak, Data analysis with rough sets, in: *Proceedings: CODATA'96*, Tsukuba, Japan, October 1996.
232. Z. Pawlak, Rough sets, rough relations and rough functions, in: *Fundamenta Informaticae* 27(2-3), 1996, pp. 103–108.
233. Z. Pawlak, Data versus Logic: A rough set view, in: *Proceedings: the 4th International Workshop on Rough Sets, Fuzzy Sets, and Machine Discovery* (RSFD'96), Tokyo, November 1996, pp. 1–8.
234. Z. Pawlak, Rough sets: Present state and perspectives, in: *Proceedings: the Sixth International Conference, Information Processing and Management of*

Uncertainty in Knowledge-Based Systems (IPMU'96), Granada, Spain, July 1996, pp. 1137–1145.
235. Z. Pawlak, Some remarks on explanation of data and specification of processes, *Bulletin of International Rough Set Society* 1(1), 1996, pp. 1–4.
236. Z. Pawlak, Why rough sets?, in: *Proceedings: the 5th IEEE International Conference on Fuzzy Systems* (FUZZ-IEEE'96), New Orleans, Louisiana, September 1996, pp. 738–743.
237. Z. Pawlak, Rough Sets and Data Analysis, in: *Proceedings: the 1996 Asian Fuzzy Systems Symposium - Soft Computing in Intelligent Systems and Information Processing*, Kenting, Taiwan ROC, December 1996, pp. 1–6.
238. Z. Pawlak, On some Issues Connected with Indiscernibility, in: G. Paun (ed.), *Mathematical Linguistics and Related Topics*, Editura Academiei Romane, Bucureşti, 1995, pp. 279–283.
239. Z. Pawlak, Rough sets, in: *Proceedings of ACM : Computer Science Conference*, Nashville TN, February 28–March 2, 1995, pp. 262–264.
240. Z. Pawlak, Rough real functions and rough controllers, in: *Proceedings: the Workshop on Rough Sets and Data Mining at 23rd Annual Computer Science Conference*, Nashville TN, March 1995, pp. 58–64.
241. Z. Pawlak, Vagueness and uncertainty: A Rough set perspective, *Computational Intelligence: An International Journal* 11(2), 1995 (a special issue: W. Ziarko (ed.)), pp. 227–232.
242. Z. Pawlak, Rough set approach to knowledge-based decision support, in: *Towards Intelligent Decision Support. Semi–Plenary Papers: the 14th European Conference of Operations Research - 20th Anniversary of EURO*, Jerusalem, Israel, July 1995.
243. Z. Pawlak, Rough set theory, in: *Proceedings: the 2nd Annual Joint Conference on Information Sciences* (JCIS'95), Wrightsville Beach NC, September 28–October 1, 1995, pp. 312–314.
244. Z. Pawlak, Rough sets: Present state and further prospects, in: T. Y. Lin and A. M. Wildberger (eds.), *Soft Computing: Rough Sets, Fuzzy Logic, Neural Networks, Uncertainty Management, Knowledge Discovery*, Simulation Councils Inc., San Diego CA, 1995, pp. 78–85.
245. Z. Pawlak, Hard and soft sets, in: W. Ziarko (ed.), *Rough Sets, Fuzzy Sets and Knowledge Discovery* (RSKD'93), Workshops in Computing, Springer–Verlag and British Computer Society, Berlin and London, 1994, pp. 130–135.
246. Z. Pawlak, Knowledge and uncertainty - A rough sets approach, in: *Proceedings: Incompleteness and Uncertainty in Information Systems; SOFTEKS Workshop on Incompleteness and Uncertainty in Information Systems*, Concordia Univ., Montreal, Canada,1993; also in: W. Ziarko (ed.), *Rough Sets, Fuzzy Sets and Knowledge Discovery* (RSKD'93), Workshops in Computing, Springer–Verlag and British Computer Society, Berlin and London, 1994, pp. 34–42.
247. Z. Pawlak, An inquiry into vagueness and uncertainty, in: *Proceedings: the 3rd International Workshop on Intelligent Information Systems*, Wigry, Poland, June 1994, Institute of Computer Science, Polish Academy of Sciences, Warsaw, 1994, pp. 338–359.

248. Z. Pawlak, Rough sets: Present state and further prospects, in: *Proceedings: the 3rd International Workshop on Rough Sets and Soft Computing* (RSSC94), San Jose, California, November 10-12, pp. 3–5.
249. Z. Pawlak, Rough sets and their applications, *Microcomputer Applications* 13(2), 1994, pp. 71–75.
250. Z. Pawlak, Anatomy of conflict, *Bull. of the European Association for Theoretical Computer Science* 50, 1993, pp. 234–247.
251. Z. Pawlak, Rough sets. Present state and the future, in: *Proceedings: the First International Workshop on Rough Sets: State of the Art and Perspectives*, Kiekrz – Poznań, Poland, September 1992, pp. 51–53.
252. Z. Pawlak, E. Czogała, and A. Mrózek, Application of a rough fuzzy controller to the stabilization of an inverted pendulum, in: *Proceedings: the 2nd European Congress on Intelligent Techniques and Soft Computing* (EUFIT'94), Aachen, Germany, September 1994, Verlag Mainz, Aachen, pp. 1403–1406.
253. Z. Pawlak, E. Czogała and A. Mrózek, The idea of a rough fuzzy controller and its applications to the stabilization of a pendulum-car system, *Fuzzy Sets and Systems* 72, 1995, pp. 61–73.
254. Z. Pawlak, J.W. Grzymala–Busse, W. Ziarko, and R. Słowiński, Rough sets, *Communications of the ACM* 38/11, 1995, pp. 88–95.
255. Z. Pawlak, A.G. Jackson, and S.R. LeClair, Rough sets and the discovery of new materials, *Journal of Alloys and Compounds*, 1997, pp. 1-28.
256. Z. Pawlak and T. Munakata, Rough control: Application of rough set theory to control, in: *Proceedings: the 4th European Congress on Intelligent Techniques and Soft Computing* (EUFIT'96), Aachen, Germany, September 1996, Verlag Mainz, Aachen, 1996, pp. 209–218.
257. Z. Pawlak and A. Skowron, Helena Rasiowa and Cecylia Rauszer's research on logical foundations of Computer Science, in: A. Skowron (ed.), *Logic, Algebra and Computer Science. Helena Rasiowa and Cecylia Rauszer in Memoriam, Bulletin of the Section of Logic* 25(3-4), 1996 (a special issue), pp. 174–184.
258. Z. Pawlak and A. Skowron, Rough membership functions, in: R.R. Yaeger, M. Fedrizzi, and J. Kacprzyk (eds.), *Advances in the Dempster–Shafer Theory of Evidence*, John Wiley and Sons Inc., New York, 1994, pp. 251–271.
259. Z. Pawlak and A. Skowron, Rough membership functions: A tool for reasoning with uncertainty, in: C. Rauszer (ed.), *Algebraic Methods in Logic and Computer Science*, Banach Center Publications 28, Polish Academy of Sciences, Warsaw, 1993, pp. 135–150.
260. Z. Pawlak and A. Skowron, A rough set approach for decision rules generation, in: *Proceedings: the Workshop W12: The Management of Uncertainty in AI* at the 13th IJCAI, Chambery Savoie, France, August 30, 1993.
261. Z. Pawlak and R. Słowiński, Decision analysis using rough sets, *International Transactions in Operational Research* 1(1), 1994, pp. 107–114.
262. Z. Pawlak and R. Słowiński, Rough set approach to multi–attribute decision analysis, *European Journal of Operational Research* 72, 1994, pp. 443–45.
263. W. Pedrycz, Shadowed sets : bridging fuzzy and rough sets, in: S. K. Pal and

A. Skowron (eds.), *Rough Fuzzy Hybridization: A New Trend in Decision-Making*, Springer-Verlag, Singapore, 1999, pp. 179–199.
264. J. E. Peters, W. Pedrycz, S. Ramanna, A. Skowron, and Z. Suraj, Approximate real – time decision making: Concepts and rough Petri net models, *Intern. Journal Intelligent Systems*, 14(8), 1999, pp. 805–840.
265. J. E. Peters and S. Ramanna, A rough set approach to assessing software quality: concepts and rough Petri net models, in: S. K. Pal and A. Skowron (eds.), *Rough Fuzzy Hybridization: A New Trend in Decision-Making*, Springer-Verlag, Singapore, 1999, pp. 349–380.
266. J. E. Peters, S. Ramanna, A. Skowron, and Z. Suraj, Graded transitions in rough Petri nets, in: *Proceedings: the 7th European Congress on Intelligent Techniques and Soft Computing* (EUFIT'99), Aachen, Germany, September 1999.
267. J. E. Peters, A. Skowron, and Z. Suraj, An application of rough set methods in control design, in : *Proceedings: the Workshop on Concurrency, Specification and Programming* (CS&P'99), Warsaw, Poland, September 1999, pp. 214–235.
268. J. E. Peters, A. Skowron, Z. Suraj, S. Ramanna, and A. Paryzek, Modeling real–time decision–making systems with rough fuzzy Petri nets, in: *Proceedings: the Sixth European Congress on Intelligent Techniques and Soft Computing* (EUFIT'98) , Aachen, Germany, September 1998, Verlag Mainz, Aachen, pp. 985–989.
269. L. Polkowski, On synthesis of constructs for spatial reasoning via rough mereology, *Fundamenta Informaticae*, in print.
270. L. Polkowski, Approximate mathematical morphology. Rough set approach, in: S. K. Pal and A. Skowron (eds.), *Rough Fuzzy Hybridization: A New Trend in Decision-Making*, Springer-Verlag, Singapore, 1999, pp. 151–162.
271. L. Polkowski, Rough set approach to mathematical morphology: Approximate compression of data, in: *Proceedings: the 7th International Conference on Information Processing and Management of Uncertainty in Knowledge – Based Systems* (IPMU'98), La Sorbonne, Paris, France, July, pp. 1183–1189.
272. L. Polkowski and M. Semeniuk–Polkowska, Towards usage of natural language in approximate computation: A granular semantics employing formal languages over mereological granules of knowledge, *Scheda Informaticae (Fasc. Jagiellonian University)*, in print.
273. L. Polkowski and M. Semeniuk–Polkowska, Concerning the Zadeh idea of computing with words: Towards a formalization, in: *Proceedings: Workshop on Robotics, Intelligent Control and Decision Support Systems*, Polish–Japanese Institute of Information Technology, Warsaw, Poland, February 1999, pp. 62–67.
274. L. Polkowski and A. Skowron, Towards adaptive calculus of granules, in: L.A. Zadeh and J. Kacprzyk (eds.), *Computing with Words in Information/Intelligent Systems*, Physica–Verlag, Heidelberg, 1999, pp. 201–228.
275. L. Polkowski and A. Skowron, Grammar systems for distributed synthesis of approximate solutions extracted from experience, in: Gh. Paun, A.Salomaa

(eds.), *Grammar Systems for Multi-agent Systems*, Gordon and Breach Science Publishers, Amsterdam, 1999, pp. 316–333.

276. L. Polkowski and A. Skowron, Rough mereology and analytical morphology, in: E. Orlowska (ed.), *Incomplete Information: Rough Set Analysis*, Physica–Verlag, Heidelberg, 1998, pp. 399–437.

277. L. Polkowski and A. Skowron, Rough sets: A perspective, in: L. Polkowski and A. Skowron (eds.), *Rough Sets in Knowledge Discovery 1. Methodology and Applications*, Physica–Verlag, Heidelberg, 1998, pp. 31–56.

278. L. Polkowski and A. Skowron, Introducing the book, in: L. Polkowski and A. Skowron (eds.), *Rough Sets in Knowledge Discovery 1. Methodology and Applications* , Physica–Verlag, Heidelberg, 1998, pp. 3–9.

279. L. Polkowski and A. Skowron, Introducing the book, in: L. Polkowski and A. Skowron (eds.), *Rough Sets in Knowledge Discovery 2. Applications, Case Studies and Software Systems*, Physica–Verlag, Heidelberg, 1998, pp. 1–9.

280. L. Polkowski and A. Skowron, Rough mereological foundations for design, analysis, synthesis, and control in distributed systems, *Information Sciences. An International Journal* 104(1-2), Elsevier Science, New York, 1998, pp. 129–156.

281. L. Polkowski and A. Skowron, Rough mereological approach - A survey, *Bulletin of International Rough Set Society* 2(1), 1998, pp. 1–13.

282. L. Polkowski and A. Skowron, Rough mereological formalization, in: W. Pedrycz and J. F. Peters III (eds.), *Computational Intelligence and Software Engineering*, World Scientific, Singapore, 1998, pp. 237–267.

283. L. Polkowski and A. Skowron, Towards adaptive calculus of granules, in: *Proceedings: the FUZZ-IEEE International Conference, 1998 IEEE World Congress on Computational Intelligence* (WCCI'98), Anchorage, Alaska, May 1998, pp. 111–116.

284. L. Polkowski and A. Skowron, Synthesis of complex objects: Rough mereological approach, in: *Proceedings: Workshop W8 on Synthesis of Intelligent Agents from Experimental Data* (at ECAI'98), Brighton, UK, August 1998, pp. 1–10 .

285. L. Polkowski and A. Skowron, Calculi of granules for adaptive distributed synthesis of intelligent agents founded on rough mereology, in: *Proceedings: the 6th European Congress on Intelligent Techniques and Soft Computing* (EUFIT'98), Aachen, Germany, September 1998, Verlag Mainz, Aachen, 1998, pp. 90–93.

286. L. Polkowski and A. Skowron, Towards information granule calculus, in: *Proceedings: the Workshop on Concurrency, Specification and Programming* (CS&P'98), Berlin, Germany, September 1998, Humboldt University Berlin, Informatik Berichte 110, pp. 176–194.

287. L. Polkowski and A. Skowron, Synthesis of decision systems from data tables, in: T.Y. Lin, N. Cercone (eds.), *Rough sets and data mining: Analysis of imprecise data*, Kluwer Academic Publishers, Dordrecht, 1997, pp. 259–299.

288. L. Polkowski and A. Skowron, Decision algorithms: A survey of rough set theoretic methods, *Fundamenta Informaticae* 30(3-4), 1997, pp. 345–358.

289. L. Polkowski and A. Skowron, Approximate reasoning in distributed systems, in: *Proceedings of the Fifth European Congress on Intelligent Techniques and Soft Computing* (EUFIT'97), Aachen, Germany, September 1997, Verlag Mainz, Aachen, 1997, pp. 1630–1633.
290. L. Polkowski and A. Skowron, Mereological foundations for approximate reasoning in distributed systems (plenary lecture), in: *Proceedings of the Second Polish Conference on Evolutionary Algorithms and Global Optimization*, Rytro, September 1997, Warsaw University of Technology Press, 1997, pp. 229–236.
291. L. Polkowski and A. Skowron, Adaptive decision–making by systems of cooperative intelligent agents organized on rough mereological principles, *Intelligent Automation and Soft Computing, An International Journal* 2(2), 1996, pp. 121–132.
292. L. Polkowski and A. Skowron, Rough mereology: A new paradigm for approximate reasoning, *Intern. Journal Approx. Reasoning* 15(4), 1996, pp. 333–365.
293. L. Polkowski and A. Skowron, Analytical morphology: Mathematical morphology of decision tables, *Fundamenta Informaticae* 27(2-3), 1996, pp. 255–271.
294. L. Polkowski and A. Skowron, Rough mereological controller, in: *Proceedings of The Fourth European Congress on Intelligent Techniques and Soft Computing* (EUFIT'96), Aachen, Germany, September 1996, Verlag Mainz, Aachen, 1996, pp. 223–227.
295. L. Polkowski and A. Skowron, Learning synthesis scheme in intelligent systems, in: *Proceedings: the 3rd International Workshop on Multi–strategy Learning* (MSL-96), Harpers Ferry, West Virginia, May 1996, George Mason University and AAAI Press 1996, pp. 57–68.
296. L. Polkowski and A. Skowron, Implementing fuzzy containment via rough rough inclusions: rough mereological approach to distributed problem solving, in: *Proceedings: the 4th IEEE International Conference on Fuzzy Systems* (FUZZ-IEEE'96), New Orlean LA, September 1996, pp. 1147–1153.
297. L. Polkowski, A. Skowron, and J. Komorowski, Approximate case-based reasoning: A rough mereological approach, in: *Proceedings: the 4th German Workshop on Case-Based Reasoning. System Development and Evaluation*, Berlin, Germany, April 1996, Informatik Berichte 55, Humboldt University, Berlin, pp. 144–151.
298. B. Prędki, R. Słowiński, J. Stefanowski, R. Susmaga, and S. Wilk, ROSE – software implementation of the rough set theory, in: *Proceedings: the First International Conference on Rough Sets and Current Trends in Computing* (RSCTC'98),Warsaw , Poland, June 1998, LNAI 1424, Springer-Verlag, Berlin, 1998, pp. 605–608.
299. M. Quafafou, α–RST : A generalization of Rough Set Theory, *Information Systems*, 1999, to appear.
300. M. Quafafou, Learning flexible concepts from uncertain data, in: *Proceedings: the 10th International Symposium on Methodologies for Intelligent Systems* (ISMIS'97), Charlotte NC, 1997.

301. M. Quafafou and M. Boussouf, Generalized rough sets based feature selection, *Intelligent Data Analysis Journal* 4(1), 1999.
302. M. Quafafou and M. Boussouf, Induction of strong feature subsets, in: *Proceedings: the First European Symposium on Principles of Data Mining and Knowledge Discovery*, Trondheim, Norway, June 1997, LNAI 1263, Springer-Verlag, Berlin, 1997.
303. S. Radev, Argumentation systems, *Fundamenta Informaticae* 28(3-4), 1996, pp. 331–346.
304. Z. W. Ras, Discovering rules in information trees, in: *Proceedings: Principles of Data Mining and Knowledge Discovery* (PKDD'99), Prague, Czech Republic, September 1999, LNAI 1704, Springer–Verlag, Berlin, 1999, pp. 518–523.
305. Z. W. Ras, Intelligent query answering in DAKS, in: O. Pons, M. A. Vila, and J. Kacprzyk (eds.), *Knowledge Management in Fuzzy Databases*, Physica-Verlag, Heidelberg, 1999, pp. 159–170.
306. Z. W. Ras, Answering non-standard queries in distributed knowledge–based systems, in: L. Polkowski and A. Skowron (eds.), *Rough Sets in Knowledge Discovery 2. Applications, Case Studies and Software Systems*, Physica Verlag, Heidelberg, 1998, pp. 98–108.
307. Z. W. Ras, Handling queries in incomplete CKBS through knowledge discovery, in: *Proceedings: the First International Conference on Rough Sets and Current Trends in Computing*(RSCTC'98), Warsaw, Poland, June 1998, LNAI 1424, Springer–Verlag, Berlin, 1998, pp. 194–201.
308. Z. W. Ras, Knowledge discovery objects and queries in distributed knowledge systems, in: *Proceedings: Artificial Intelligence and Symbolic Computation*(AISC'98), LNAI 1476, Springer–Verlag, Berlin, 1998, pp. 259–269.
309. Z. W. Ras and A. Bergmann, Maintaining soundness of rules in distributed knowledge systems, in: *Proceedings: the Workshop on Intelligent Information Systems* (IIS'98), Malbork, Poland, June 1998, Polish Academy of Sciences, Warsaw, 1998, pp. 29–38.
310. Z. W. Ras and J. M. Żytkow, Mining for attribute definitions in a distributed two-layered DB system, *Journal of Intelligent Information Systems* 14(2-3), 2000, in print.
311. Z. W. Ras and J. M. Żytkow, Mining distributed databases for attribute definitions, in: *Proceedings: SPIE. Data Mining and Knowledge Discovery: Theory, Tools, and Technology*, Orlando, Florida, April 1999, pp. 171–178.
312. Z. W. Ras and J. M. Żytkow, Discovery of equations and the shared operational semantics in distributed autonomous databases, in: *Proceedings: Methodologies for Knowledge Discovery and Data Mining*(PAKKD'99), Beijing, China, 1999, LNAI 1574, Springer-Verlag, Berlin, 1999, pp. 453–463.
313. L. Rossi, R. Słowiński, and R. Susmaga, Rough set approach to evaluation of storm water pollution, *International Journal of Environment and Pollution*, to appear.
314. L. Rossi, R. Słowiński, and R. Susmaga, Application of the rough set approach to evaluate storm water pollution, in: *Proceedings: the 8th International*

Conference on Urban Storm Drainage (8th ICUSD), Sydney, Australia, 1999, vol. 3, pp. 1192–1200.

315. H.Sakai, Some issues on non–deterministic knowledge bases with incomplete and selective information, in: *Proceedings: the First International Conference on Rough Sets and Current Trends in Computing* (RSCTC'98), LNAI 1424, Springer–Verlag, Berlin, 1998, pp. 424–431.

316. H. Sakai, Another fuzzy Prolog, in: *Proceedings: The Fourth International Workshop on Rough Sets, Fuzzy Sets, and Machine Discovery* (RSFD'96), Tokyo, November 1996, pp. 261–268.

317. H. Sakai and A. Okuma, An algorithm for finding equivalence relations from tables with non–deterministic information, in: *Proceedings: the 7th International Conference on Rough Sets, Fuzzy Sets, Data Mining and Granular-Soft Computing* (RSFDGrC99), LNAI 1711, Springer–Verlag, Berlin, 1999, pp. 64–72.

318. M. Semeniuk–Polkowska, On Applications of Rough Set Theory in Humane Sciences (in Polish), Warsaw University Press, Warsaw, 2000.

319. M. Semeniuk–Polkowska, On Rough Set Theory in Library Sciences (in Polish), Warsaw University Press, Warsaw, 1996.

320. V. I. Shevtchenko, On the depth of decision trees for diagnosis of non–elementary faults in circuits (in Russian), in: *Proceedings: the 9th Workshop on Synthesis and Complexity of Control Systems*, Nizhny Novgorod, Russia, 1998, Moscow State University Publishers, Moscow, 1999, pp. 94–98.

321. V. I. Shevtchenko, On complexity of non–elementary fault detection in circuits (in Russian), in: *Proceedings: the 12th International Conference on Problems of Theoretical Cybernetics*, Nizhny Novgorod, Russia, 1999, Moscow State University Publishers, Moscow, 1999, p. 254.

322. V. I. Shevtchenko, On complexity of confused connection diagnosis in circuits (in Russian), in: *Proceedings: the 9th All–Russian Conference on Mathematical Methods of Pattern Recognition*, Moscow, Russia, 1999, pp. 129–131.

323. V. I. Shevtchenko, On the depth of decision trees for diagnosing of nonelementary faults in circuits, in: *Proceedings: the First International Conference on Rough Sets and Current Trends in Computing* (RSCTC'98), Warsaw, Poland, June 1998, LNAI 1424, Springer–Verlag, Berlin, 1998, pp. 517–520.

324. V. I. Shevtchenko, On complexity of "OR" ("AND")–closing detection in circuits (in Russian), in: *Proceedings: the International Conference on Discrete Models in Theory of Control Systems*, Moscow, Russia, 1997 pp. 61–62.

325. V. I. Shevtchenko, On complexity of "OR" ("AND")–closing diagnosis in circuits (in Russian), in: *Proceedings: the 8th All-Russian Conference on Mathematical Methods of Pattern Recognition*, Moscow, Russia, 1997, pp. 125–126.

326. A. Skowron and J. Stepaniuk, Information granules: Towards foundations for spatial and temporal reasoning, *Journal of Indian Science Academy*, in print.

327. A. Skowron and J. Stepaniuk, Information granules in distributed systems, in: *Proceedings: the 7th International Workshop on Rough Sets, Fuzzy Sets, Data Mining and Granular-Soft Computing* (RSFDGrC'99), Ube – Yama-

guchi, Japan, November 1999, Lecture Notes in Artificial Intelligence 1711, Springer–Verlag, Berlin, 1999, pp. 357–365.
328. A. Skowron and J. Stepaniuk, Towards discovery of information granules, in: *Proceedings: Principles of Data Mining and Knowledge Discovery* (PKDD-'99), Prague, the Czech Republic, September 1999, LNAI 1704, Springer–Verlag, Berlin, 1999, pp. 542–547.
329. A. Skowron and J. Stepaniuk, Information granules and approximation spaces, in: *Proceedings: the 7th International Conference on Information Processing and Management of Uncertainty in Knowledge – Based Systems* (IPMU'98), La Sorbonne, Paris, France, July 1998, pp. 1354–1361.
330. A. Skowron and J. Stepaniuk, Information Reduction Based on Constructive Neighborhood Systems, in: *Proceedings: the 5th International Workshop on Rough Sets Soft Computing* (RSSC'97) at the 3rd Annual Joint Conference on Information Sciences (JCIS'97), Durham NC, October 1997, pp. 158–160.
331. A. Skowron and J. Stepaniuk, Constructive information granules, in: *Proceedings: the 15th IMACS World Congress on Scientific Computation, Modeling and Applied Mathematics*, Berlin, Germany, August 1997; also in: *Artificial Intelligence and Computer Science* 4, pp. 625–630.
332. A. Skowron and J. Stepaniuk, Tolerance approximation spaces, *Fundamenta Informaticae* 27(2-3), 1996, pp. 245–253.
333. A. Skowron, J. Stepaniuk, and S. Tsumoto, Information granules for spatial reasoning, *Bulletin Intern. Rough Set Society* 3(4), 1999, pp. 147–154.
334. A. Skowron and Z. Suraj, A parallel algorithm for real-time decision making: A rough set approach, *Journal of Intelligent Information Systems* 7, 1996, pp. 5–28.
335. K. Słowiński and J. Stefanowski, Medical information systems – problems with analysis and ways of solutions, in: S. K. Pal and A. Skowron (eds.), *Rough Fuzzy Hybridization: A New Trend in Decision–Making*, Springer–Verlag, Singapore, 1999, pp. 301.
336. R. Słowiński, Rough set data analysis – a new way of solving some decision problems in transportation, *Proceedings: Modeling and Management in Transportation* (MMT'99), Kraków – Poznań, October 1999, pp. 63–66.
337. R. Słowiński, Multi–criterial decision support based on rules induced by rough sets (in Polish), in: T.Trzaskalik (ed.),*Metody i Zastosowania Badań Operacyjnych*, Part 2, Wydawnictwo Akademii Ekonomicznej w Katowicach, Katowice, 1998, pp. 19–39.
338. R. Słowiński and J. Stefanowski, Handling inconsistency of information with rough sets and decision rules, in: *Proceedings: Intern. Conference on Intelligent Techniques in Robotics, Control and Decision Making*, Polish–Japanese Institute of Information Technology, Warsaw, February 1999, pp. 74–81.
339. R. Słowiński and J. Stefanowski, Rough family – software implementation of the rough set theory, in: L. Polkowski, A. Skowron (eds.), *Rough Sets in Knowledge Discovery 2. Applications, Case Studies and Software Systems*, Physica-Verlag, Heidelberg, 1998, pp. 581–586.
340. R. Słowiński, J. Stefanowski, S. Greco, and B. Matarazzo, Rough sets processing of inconsistent information in decision analysis, *Control and Cybernetics*,

to appear.
341. R. Słowiński and D. Vanderpooten, A generalized definition of rough approximations based on similarity, *IEEE Transactions on Data and Knowledge Engineering*, to appear.
342. R. Słowiński, C. Zopounidis, A. I. Dimitras, and R. Susmaga, Rough set predictor of business failure, in: R. A. Ribeiro, H.-J. Zimmermann, R. R. Yager, and J. Kacprzyk (eds.), *Soft Computing in Financial Engineering*, Physica–Verlag, Heidelberg, 1999, pp. 402–424.
343. J. Stepaniuk, Rough set based data mining in diabetes mellitus data table, in: *Proceedings: the 6th European Congress on Intelligent Techniques and Soft Computing* (EUFIT'98), Aachen, Germany, September 1998, pp. 980–984; for extended version see: : *Proceedings: the 11th International Symposium on Methodologies for Intelligent Systems, Foundations of Intelligent Systems* (ISMIS'99), Warsaw, Poland, June 1999, LNAI 1609, Springer–Verlag, Berlin, 1999.
344. J. Stepaniuk, Optimizations of rough set model, *Fundamenta Informaticae* 36(2-3), 1998, pp. 265–283.
345. J. Stepaniuk, Rough relations and logics, in: L. Polkowski, A. Skowron (eds.), *Rough Sets in Knowledge Discovery 1. Methodology and Applications*, Physica–Verlag, Heidelberg, 1998, pp. 248–260.
346. J. Stepaniuk, Approximation spaces, reducts and representatives, in: L. Polkowski, A. Skowron (eds.), *Rough Sets in Knowledge Discovery 2. Applications, Case Studies and Software Systems*, Physica–Verlag, Heidelberg, 1998, pp. 109–126.
347. J. Stepaniuk, Approximation spaces in extensions of rough set theory, in: *Proceedings: the First International Conference on Rough Sets and Current Trends in Computing* (RSCTC'98), Warsaw, Poland, June 1998, LNAI 1424, Springer–Verlag, Berlin, 1998, pp. 290–297.
348. J. Stepaniuk, Rough sets similarity based learning, in: *Proceedings: the 5th European Congress on Intelligent Techniques and Soft Computing* (EUFIT' 97), Aachen, Germany, September 1997, Verlag Mainz, Aachen, 1997, pp. 1634–1639.
349. J. Stepaniuk, Similarity relations and rough set model, in: *Proceedings: the International Conference MENDEL97*, Brno, the Czech Republic, June 1997.
350. J. Stepaniuk, Attribute discovery and rough sets, in: *Proceedings:the First European Symposium on Principles of Data Mining and Knowledge Discovery* (PKDD'97), Trondheim, Norway, June 1997, LNAI 1263, Springer–Verlag, Berlin, 1997, pp. 145–155.
351. J. Stepaniuk, Searching for optimal approximation spaces, in: *Proceedings: the 6th International Workshop on Intelligent Information Systems* (ISMIS' 97), Zakopane, Poland, June 1997, Publ. Institute of Computer Science, Polish Academy of Sciences, pp. 86–95.
352. Z. Suraj, An application of rough sets and Petri nets to controller design, in: *Workshop on Robotics, Intelligent Control and Decision Support Systems*, Polish-Japanese Institute of Information Technology, Warsaw, Poland, February 1999, pp. 86–96.

353. Z. Suraj, The synthesis problem of concurrent systems specified by dynamic information systems, in: L. Polkowski and A. Skowron (eds.), *Rough Sets in Knowledge Discovery 2. Applications, Case Studies and Software Systems*, Physica–Verlag, Heidelberg, 1998, pp. 418–448.

354. Z. Suraj, Reconstruction of cooperative information systems under cost constraints: A rough set approach, *Journal of Information Sciences* 111, 1998, pp. 273–291.

355. Z. Suraj, Reconstruction of cooperative information systems under Cost constraints: A rough set approach, in: *Proceedings: the First International Workshop on Rough Sets and Soft Computing* (RSSC'97), Durham NC, March 1997, pp. 364–371.

356. Z. Suraj, Discovery of concurrent data models from experimental tables, *Fundamenta Informaticae* 28(3-4), 1996, pp. 353–376.

357. R. Susmaga, R. Słowiński, S. Greco, and B. Matarazzo, Computation of reducts for multi–attribute and multi–criteria classification, in: *Proceedings: the 7th Workshop on Intelligent Information Systems* (IIS'99), Ustroń, Poland, June 1999, pp. 154–163.

358. M. Szczuka, Refining decision classes with neural networks, in: *Proceedings: the 7th International Conference on Information Processing and Management of Uncertainty in Knowledge–Based Systems* (IPMU'98), La Sorbonne, Paris, France, July 1998, pp. 1370–1375.

359. M. Szczuka, Rough Sets and Artificial Neural Networks, in: L. Polkowski and A. Skowron (eds.), *Rough Sets in Knowledge Discovery 2. Applications, Case Studies and Software Systems*, Physica–Verlag, Heidelberg, 1998, pp. 449–470.

360. M. Szczuka, Rough set methods for constructing neural networks, in: *Proceedings: the 3rd Biennial Joint Conference On Engineering Systems Design Analysis, Session on Expert Systems*, Montpellier, France, 1996, pp. 9–14.

361. M. Szczuka, D. Ślęzak, and S. Tsumoto, An application of reduct networks to medicine – chaining decision rules, in: *Proceedings: the 5th International Workshop on Rough Sets and Soft Computing* (RSSC'97) at *the 3rd Annual Joint Conference on Information Sciences* (JCIS'97), Duke University, Durham NC, USA, 1997, pp. 395–398.

362. D. Ślęzak, Foundations of entropy based bayesian networks: Theoretical results & rough set based extraction from data, in : *Proceedings: the 8th International Conference on Information Processing and Management of Uncertainty in Knowledge–Based Systems* (IPMU'00), Madrid, Spain, July 2000, in print.

363. D. Ślęzak, Normalized decision functions and measures for inconsistent decision tables analysis, *Fundamenta Informaticae*, in print.

364. D. Ślęzak, Decomposition and synthesis of decision tables with respect to generalized decision functions, in: S. Pal and A. Skowron (eds.), *Rough Fuzzy Hybridization: A New Trend in Decision–Making*, Springer–Verlag, Singapore, 1999, pp. 110–135.

365. D. Ślęzak, Decision information functions for inconsistent decision tables analysis, in: *Proceedings: the 7th European Congress on Intelligent Techni-*

ques & Soft Computing (EUFIT'99), Aachen, Germany, September 1999, p. 127.

366. D. Ślęzak, Searching for Dynamic Reducts in Inconsistent Decision Tables, in: *Proceedings: the 7th International Conference on Information Processing and Management of Uncertainty in Knowledge–Based Systems* (IPMU'98), La Sorbonne, Paris, France, July 1998, pp. 1362–1369.

367. D. Ślęzak, Searching for frequential reducts in decision tables with uncertain objects, in: *Proceedings: the First International Conference on Rough Sets and Current Trends in Computing* (RSCTC'98), Warsaw, Poland, June 1998, LNAI 1424, Springer-Verlag, Berlin, 1998, pp. 52–59.

368. D. Ślęzak, Rough set reduct networks, in: *Proceedings: the 5th International Workshop on Rough Sets Soft Computing* (RSSC'97) at *the 3rd Annual Joint Conference on Information Sciences* (JCIS'97), Durham NC, 1997, pp. 77–81.

369. D. Ślęzak, Attribute set decomposition of decision tables, in: *Proceedings: the 5th European Congress on Intelligent Techniques and Soft Computing* (EUFIT'97), Aachen, Germany, September 1997, Verlag Mainz, Aachen, 1997, pp. 236–240.

370. D. Ślęzak, Decision value oriented decomposition of data tables, in: *Proceedings: the 10th International Symposium on Methodologies for Intelligent Systems, Foundations of Intelligent Systems* (ISMIS'97), Charlotte NC, October 1997, LNAI 1325, Springer–Verlag, Berlin, 1997, pp. 487–496.

371. D. Ślęzak, Approximate reducts in decision tables, in: *Proceedings: the 6th International Conference, Information Processing and Management of Uncertainty in Knowledge–Based Systems* (IPMU'96), Granada, Spain, July 1996, pp. 1159–1164.

372. D. Ślęzak, Tolerance dependency model for decision rules generation, in: *Proceedings: the 4th International Workshop on Rough Sets, Fuzzy Sets, and Machine Discovery* (RSFD'96), Tokyo, Japan, November 1996, pp. 131–138.

373. D. Ślęzak and M. Szczuka, Hyperplane–based neural networks for real–valued decision tables, in: *Proceedings: the 5th International Workshop on Rough Sets Soft Computing* (RSSC'97) at the 3rd Annual Joint Conference on Information Sciences (JCIS'97), Durham NC, 1997, pp. 265–268.

374. D. Ślęzak and J. Wróblewski, Classification algorithms based on linear combinations of features, in: *Proceedings: Principles of Data Mining and Knowledge Discovery* (PKDD'99), Prague, Czech Republic, September 1999, LNAI 1704, Springer-Verlag, Berlin, 1999, pp. 548–553.

375. I. Tentush, On minimal absorbents and closure properties of rough inclusions: new results in rough set theory, Ph.D. Dissertation, supervisor L. Polkowski, Institute of Fundamentals of Computer Science, Polish Academy of Sciences, Warsaw, Poland, 1997.

376. S. Tsumoto, Induction of expert decision rules using rough sets and set–inclusion, in: S.K. Pal and A. Skowron (eds.), *Rough Fuzzy Hybridization: A New Trend in Decision–Making*, Springer–Verlag, Singapore, 1999, pp. 316–329.

377. S. Tsumoto, Discovery of rules about complications, in: *Proceedings: the*

7th International Workshop on Rough Sets, Fuzzy Sets, Data Mining and Granular-Soft Computing (RSFDGrC'99), Ube–Yamaguchi, Japan, November 1999, Lecture Notes in Artificial Intelligence 1711, Springer–Verlag, Berlin, 1999, pp. 29–37.

378. S. Tsumoto, Extraction of expert's decision rules from clinical databases using rough set model, *J. Intelligent data Analysis* 2(3), 1998.

379. S. Tsumoto, Automated induction of medical expert system tules from clinical databases based on rough set theory, *Information Sciences* 112, 1998, pp. 67–84.

380. A. Wakulicz–Deja, M. Boryczka, and P. Paszek, Discretization of continuous attributes on decision system in mitochondrial encephalomyopathies, in: *Proceedings: the First International Conference on Rough Sets and Current Trends in Computing* (RSCTC'98), Warsaw, Poland, June 1998, LNAI 1424, 1998, pp. 483–490.

381. A. Wakulicz–Deja, B. Marszał–Paszek, P. Paszek, and E. Emich–Widera, Applying rough sets to diagnose in children's neurology, in: *Proceedings: the 6th International Conference Information Processing and Management of Uncertainty in Knowledge-Base Systems* (IPMU'96), Granada, Spain, 1996, pp. 1463–1468.

382. A. Wakulicz–Deja and P. Paszek, Optimalization of decision problems on medical knowledge bases, in: *Proceedings: Intelligent Information Systems VI*, Zakopane, Poland, 1997, pp. 204–210.

383. A. Wakulicz–Deja and P. Paszek, Optimalization of decision problems on medical knowledge bases, in: *Proceedings: the 5th European Congress on Intelligent Techniques and Soft Computing* (EUFIT'97), Aachen, Germany, September 1997, Verlag Mainz, Aachen, 1997, pp. 1607–1610.

384. A. Wakulicz–Deja and P. Paszek, Diagnose progressive encephalopathy applying the rough set theory, *International Journal of Medical Informatics* 46, 1997, pp. 119–127.

385. A. Wakulicz–Deja and P. Paszek, Optimalization of diagnose in progressive encephalopathy applying the rough set theory, in: *Proceedings: the 4th European Congress on Intelligent Techniques and Soft Computing* (EUFIT'96), Aachen, Germany, September 1996, Verlag Mainz, Aachen, 1996, pp. 192–196.

386. A. Wakulicz–Deja, P. Paszek, and B. Marszał–Paszek, Optymalizacja procesu podejmowania decyzji (diagnozy) w medycznych bazach wiedzy (in Polish), in: *Proceedings: II Krajowa Konferencja Techniki Informatyczne w Medycynie*, Jaszowiec, Poland, 1997, pp. 279–286.

387. H. Wang and Nguyen Hung Son, Text classification using Lattice Machine, in: *Proceedings: the 11th International Symposium on Methodologies for Intelligent Systems, Foundations of Intelligent Systems* (ISMIS'99), Warsaw, Poland, June 1999 , LNAI 1609, Springer–Verlag, Berlin, 1999.

388. A. Wasilewska, Topological rough algebras, in: T. Y. Lin and N. Cercone (eds.), *Rough Sets and Data Mining. Analysis of Imprecise Data*, Kluwer Academic Publishers, Dordrecht, 1997, pp. 411–425.

389. A. Wasilewska, E. Menasalvas, and M. Hadjimichael, A generalization model

for implementing a Data Mining system, in: *Proceedings: IFSA'99*, Taipei, Taiwan, August 1999, pp. 245–251.

390. A. Wasilewska and L. Vigneron, Rough algebras and automated deduction, in: L. Polkowski and A. Skowron (eds.), *Rough Sets in Knowledge Discovery 1. Methodology and Applications*, Physica–Verlag, Heidelberg, 1998, pp. 261–275.

391. A. Wasilewska and L. Vigneron, On Generalized rough sets, in: *Proceedings: the 5th Workshop on Rough Sets and Soft Computing* (RSSC'97) at the 3rd Joint Conference on Information Sciences (JCIS'97), Research Triangle Park NC, March 1997.

392. P. Wojdyłło, Wavelets, rough sets and artificial neural networks in EEG analysis, in: *Proceedings: the First International Conference on Rough Sets and Current Trends in Computing* (RSCTC'98), Warsaw, Poland, June 1998, LNAI 1424, Springer-Verlag, Berlin, pp. 444–449.

393. J. Wróblewski, Genetic algorithms in decomposition and classification problem, in: L. Polkowski and A. Skowron (eds.), *Rough Sets in Knowledge Discovery 2. Applications, Case Studies and Software Systems*, Physica–Verlag, Heidelberg, 1998, pp. 471–487.

394. J. Wróblewski, Covering with reducts – a fast algorithm for rule generation, in: *Proceedings: the First International Conference on Rough Sets and Current Trends in Computing* (RSCTC'98), Warsaw, Poland, June 1998, LNAI 1424, Springer-Verlag, Berlin, 1998, pp. 402–407.

395. J. Wróblewski, A Parallel Algorithm for Knowledge Discovery System, in: *Proceedings: the International Conference on Parallel Computing in Electrical Engineering* (PARELEC'98), Białystok, Poland, September 1998, The Press Syndicate of the Technical University of Białystok, 1998, pp. 228–230.

396. J. Wróblewski, Theoretical Foundations of Order-Based Genetic Algorithms, *Fundamenta Informaticae* 28(3-4), 1996, pp. 423–430.

397. L. Vigneron, Automated deduction techniques for studying rough algebras, *Fundamenta Informaticae* 33(1), 1998, pp. 85–103.

398. L. Vigneron and A. Wasilewska, Rough sets based proofs visualisation, in: *Proceedings: the 18th International Conference of the North American Fuzzy Information Processing Society* (NAFIPS'99)(invited session on Granular Computing and Rough Sets), New York NY, 1999, pp. 805–808.

399. L. Vigneron and A. Wasilewska, Rough sets congruences and diagrams, in: *Proceedings: the 16th European Conference on Operational Research* (EURO XVI), Brussels, Belgium, July 1998.

400. L. Vigneron and A. Wasilewska, Rough diagrams, in: *Proceedings: the sixth Workshop on Rough Sets, Data Mining and Granular Computing* (RSDM-GrC'98) at the 4th Joint Conference on Information Sciences (JCIS'98), Research Triangle Park NC, October 1998.

401. L. Vigneron and A. Wasilewska, Rough and modal algebras, in: *Proceedings: the International Multi–conference (Computational Engineering in Systems Applications), Symposium on Modelling, Analysis and Simulation* (IMACS/-IEEE CESA'96), Lille, France, July 1996, pp. 1107–1112.

402. Zhang Qi and Han Zhenxiang, Rough sets : theory and applications, *Control Theory and Applications* 16(2), 1999, pp. 153–157, S. China Univ. Technology Press, Guangzhou, China.
403. Zhang Qi and Han Zhenxiang, A new method for alarm processing in power systems using rough set theory, *Electric Power* 31(4), 1998, pp. 32–38, China Electric Power Press, Beijing.
404. Zhang Qi, Han Zhenxiang, and Wen Fushuan, Analysis of Rogers ratio table for transformer fault diagnosis using rough set theory, in: *Proceedings: CUS–EPSA'98*, Harbin, China, 1998, pp. 386–391.
405. Zhang Qi, Han Zhenxiang, and Wen Fushuan, A new approach for fault diagnosis in power systems based on rough set theory, in: *Proceedings: APS-COM'97*, Hong Kong, 1997, pp. 597–602.
406. W. Ziarko, Decision making with probabilistic decision tables, in: *: Proceedings: the 7th International Workshop on Rough Sets, Fuzzy Sets, Data Mining and Granular-Soft Computing* (RSFDGrC'99), Ube–Yamaguchi, Japan, November 1999, Lecture Notes in Artificial Intelligence 1711, Springer–Verlag, Berlin, 1999, pp. 463–471.
407. W. Ziarko, Rough sets as a methodology for data mining, in: L. Polkowski and A. Skowron (eds.), *Rough Sets in Knowledge Discovery 1. Methodology and Applications*, Physica–Verlag, Heidelberg, 1998, pp. 554–576.
408. W. Ziarko, KDD–R: rough sets based data mining system, in: L. Polkowski and A. Skowron (eds.), *Rough Sets in Knowledge Discovery 2. Applications, Case Studies and Software Systems*, Physica–Verlag, Heidelberg, 1998, pp. 598–601.
409. W. Ziarko, Approximation region–based decision tables, in : *Proceedings: the First International Conference on Rough Sets and Current Trends in Computing* (RSCTC'98), Warsaw, June 1998, Lecture Notes in Artificial Intelligence 1424, Springer–Verlag, Berlin, 1998, pp. 178–185.
410. C. Zopounidis, R. Słowiński, M. Doumpos, A.I. Dimitras, and R. Susmaga, Business failure prediction using rough sets – a comparison with multivariate analysis techniques, *Fuzzy Economic Review* 4, 1999(1), pp. 3–33.

[C] **Books of related interest**

1. K. Cios, W. Pedrycz, and R. Świniarski, *Data Mining Methods for Knowledge Discovery*, Kluwer Academic Publishers, Boston, 1998.
2. T. Munakata, *Fundamentals of the New Artificial Intelligence. Beyond Traditional Paradigms*, Springer–Verlag, New York, 1998.
3. E. Orłowska, *Incomplete Information: Rough Set Analysis*, this Series, vol. 13, Physica–Verlag, Heidelberg, 1998.

Studies in Fuzziness and Soft Computing

Editor-in-chief
Prof. Janusz Kacprzyk
Systems Research Institute
Polish Academy of Sciences
ul. Newelska 6
01-447 Warsaw, Poland
E-mail: kacprzyk@ibspan.waw.pl
http://www.springer.de/cgi-bin/search_book.pl?series=2941

Vol. 3. A. Geyer-Schulz
*Fuzzy Rule-Based Expert Systems
and Genetic Machine Learning, 2nd ed. 1996*
ISBN 3-7908-0964-0

Vol. 4. T. Onisawa and J. Kacprzyk (Eds.)
Reliability and Safety Analyses under Fuzziness, 1995
ISBN 3-7908-0837-7

Vol. 5. P. Bosc and J. Kacprzyk (Eds.)
Fuzziness in Database Management Systems, 1995
ISBN 3-7908-0858-X

Vol. 6. E. S. Lee and Q. Zhu
Fuzzy and Evidence Reasoning, 1995
ISBN 3-7908-0880-6

Vol. 7. B. A. Juliano and W. Bandler
Tracing Chains-of-Thought, 1996
ISBN 3-7908-0922-5

Vol. 8. F. Herrera and J. L. Verdegay (Eds.)
Genetic Algorithms and Soft Computing, 1996
ISBN 3-7908-0956-X

Vol. 9. M. Sato et al.
Fuzzy Clustering Models and Applications, 1997
ISBN 3-7908-1026-6

Vol. 10. L. C. Jain (Ed.)
*Soft Computing Techniques in Knowledge-based
Intelligent Engineering Systems, 1997*
ISBN 3-7908-1035-5

Vol. 11. W. Mielczarski (Ed.)
Fuzzy Logic Techniques in Power Systems, 1998,
ISBN 3-7908-1044-4

Vol. 12. B. Bouchon-Meunier (Ed.)
*Aggregation and Fusion of Imperfect Information,
1998*
ISBN 3-7908-1048-7

Vol. 13. E. Orłowska (Ed.)
Incomplete Information: Rough Set Analysis, 1998
ISBN 3-7908-1049-5

Vol. 14. E. Hisdal
*Logical Structures for Representation of Knowledge
and Uncertainty, 1998*
ISBN 3-7908-1056-8

Vol. 15. G. J. Klir and M. J. Wierman
Uncertainty-Based Information, 2nd ed., 1999
ISBN 3-7908-1242-0

Vol. 16. D. Driankov and R. Palm (Eds.)
Advances in Fuzzy Control, 1998
ISBN 3-7908-1090-8

Vol. 17. L. Reznik, V. Dimitrov and J. Kacprzyk (Eds.)
Fuzzy Systems Design, 1998
ISBN 3-7908-1118-1

Vol. 18. L. Polkowski and A. Skowron (Eds.)
Rough Sets in Knowledge Discovery 1, 1998
ISBN 3-7908-1119-X

Vol. 19. L. Polkowski and A. Skowron (Eds.)
Rough Sets in Knowledge Discovery 2, 1998
ISBN 3-7908-1120-3

Vol. 20. J. N. Mordeson and P. S. Nair
Fuzzy Mathematics, 1998
ISBN 3-7908-1121-1

Vol. 21. L. C. Jain and T. Fukuda (Eds.)
Soft Computing for Intelligent Robotic Systems, 1998
ISBN 3-7908-1147-5

Vol. 22. J. Cardoso and H. Camargo (Eds.)
Fuzziness in Petri Nets, 1999
ISBN 3-7908-1158-0

Vol. 23. P. S. Szczepaniak (Ed.)
Computational Intelligence and Applications, 1999
ISBN 3-7908-1161-0

Vol. 24. E. Orłowska (Ed.)
Logic at Work, 1999
ISBN 3-7908-1164-5

Vol. 25. J. Buckley and Th. Feuring
Fuzzy and Neural: Interactions and Applications, 1999
ISBN 3-7908-1170-X

Vol. 26. A. Yazici and R. George
Fuzzy Database Modeling, 1999
ISBN 3-7908-1171-8

Vol. 27. M. Zaus
*Crisp and Soft Computing with Hypercubical
Calculus, 1999*
ISBN 3-7908-1172-6

Studies in Fuzziness and Soft Computing

Vol. 28. R. A. Ribeiro, H.-J. Zimmermann,
R. R. Yager and J. Kacprzyk (Eds.)
Soft Computing in Financial Engineering, 1999
ISBN 3-7908-1173-4

Vol. 29. H. Tanaka and P. Guo
Possibilistic Data Analysis for Operations Research, 1999
ISBN 3-7908-1183-1

Vol. 30. N. Kasabov and R. Kozma (Eds.)
Neuro-Fuzzy Techniques for Intelligent Information Systems, 1999
ISBN 3-7908-1187-4

Vol. 31. B. Kostek
Soft Computing in Acoustics, 1999
ISBN 3-7908-1190-4

Vol. 32. K. Hirota and T. Fukuda
Soft Computing in Mechatronics, 1999
ISBN 3-7908-1212-9

Vol. 33. L. A. Zadeh and J. Kacprzyk (Eds.)
Computing with Words in Information/ Intelligent Systems 1, 1999
ISBN 3-7908-1217-X

Vol. 34. L. A. Zadeh and J. Kacprzyk (Eds.)
Computing with Words in Information/ Intelligent Systems 2, 1999
ISBN 3-7908-1218-8

Vol. 35. K. T. Atanassov
Intuitionistic Fuzzy Sets, 1999
ISBN 3-7908-1228-5

Vol. 36. L. C. Jain (Ed.)
Innovative Teaching and Learning, 2000
ISBN 3-7908-1246-3

Vol. 37. R. Słowiński and M. Hapke (Eds.)
Scheduling Under Fuzziness, 2000
ISBN 3-7908-1249-8

Vol. 38. D. Ruan (Ed.)
Fuzzy Systems and Soft Computing in Nuclear Engineering, 2000
ISBN 3-7908-1251-X

Vol. 39. O. Pons, M. A. Vila and J. Kacprzyk (Eds.)
Knowledge Management in Fuzzy Databases, 2000
ISBN 3-7908-1255-2

Vol. 40. M. Grabisch, T. Murofushi and M. Sugeno (Eds.)
Fuzzy Measures and Integrals, 2000
ISBN 3-7908-1255-2

Vol. 41. P. Szczepaniak, P. Lisboa and J. Kacprzyk (Eds.)
Fuzzy Systems in Medicine, 2000
ISBN 3-7908-1263-4

Vol. 42. S. Pal, G. Ashish and M. Kundu (Eds.)
Soft Computing for Image Processing, 2000
ISBN 3-7908-1217-X

Vol. 43. L. C. Jain, B. Lazzerini and U. Halici (Eds.)
Innovations in ART Neural Networks, 2000
ISBN 3-7908-1270-6

Vol. 44. J. Aracil and F. Gordillo (Eds.)
Stability Issues in Fuzzy Control, 2000
ISBN 3-7908-1277-3

Vol. 45. N. Kasabov (Ed.)
Future Directions for Intelligent Information Systems on Information Sciences, 2000
ISBN 3-7908-1276-5

Vol. 46. J. N. Mordeson and P. S. Nair
Fuzzy Graphs and Fuzzy Hypergraphs, 2000
ISBN 3-7908-1286-2

Vol. 47. E. Czogała† and J. Łęski
Fuzzy and Neuro-Fuzzy Intelligent Systems, 2000
ISBN 3-7908-1289-7

Vol. 48. M. Sakawa
Large Scale Interactive Fuzzy Multiobjective Programming, 2000
ISBN 3-7908-1293-5

Vol. 49. L. I. Kuncheva
Fuzzy Classifier Design, 2000
ISBN 3-7908-1298-6

Vol. 50. F. Crestani and G. Pasi (Eds.)
Soft Computing in Information Retrieval, 2000
ISBN 3-7908-1299-4

Vol. 51. J. Fodor, B. De Baets and P. Perny (Eds.)
Preferences and Decisions under Incomplete Knowledge, 2000
ISBN 3-7908-1303-6

Vol. 52. E. E. Kerre and M. Nachtegael (Eds.)
Fuzzy Techniques in Image Processing, 2000
ISBN 3-7908-1304-4

Vol. 53. G. Bordogna and G. Pasi (Eds.)
Recent Issues on Fuzzy Databases, 2001
ISBN 3-7908-1319-2

Vol. 54. P. Sinčák and J. Vaščák (Eds.)
Quo Vadis Computational Intelligence?, 2000
ISBN 3-7908-1324-9

Vol. 55. J. N. Mordeson, D. S. Malik and S.-C. Cheng
Fuzzy Mathematics in Medicine, 2000
ISBN 3-7908-1325-7

Druck: Strauss Offsetdruck, Mörlenbach
Verarbeitung: Schäffer, Grünstadt